9th International Conference on Computational Information Technologies for Environmental Sciences (CITES 2019)

IOP Conference Series: Earth and Environmental Science
Volume 386

Moscow, Russia
3 – 6 June 2019

Editors:

Evgeny Gordov
Vasily Lykosov

ISBN: 978-1-7138-0254-9
ISSN: 1755-1307

Printed from e-media with permission by:

Curran Associates, Inc.
57 Morehouse Lane
Red Hook, NY 12571

Some format issues inherent in the e-media version may also appear in this print version.

Additional copies of this publication are available from:

Curran Associates, Inc.
57 Morehouse Lane
Red Hook, NY 12571 USA
Phone: 845-758-0400
Fax: 845-758-2633
Email: curran@proceedings.com
Web: www.proceedings.com

TABLE OF CONTENTS

IOP Conf. Series: Earth and Environmental Science **386** (2019) 011001 doi:10.1088/1755-1315/386/1/011001

Preface

These Conference Proceedings contain the selected papers of the 9th International Conference on Computational Information Technologies for Environmental Sciences (CITES) held on June 3-6, 2019 in Moscow, Russia. The conference was preceded by Young Scientists School which was held on May 27 - 31, 2019.

The main theme of the event was subseasonal to decadal weather and climate predictions, covering aspects from the modeling and data assimilation to forecast information delivery and relevant practical applications. At the school 12 lectures were delivered and 14 hours of practical classes were held by experts in the field of climate variability and forecasting, members of Working Group of the World Climate Research Program on seasonal and multi-year forecasts, members of World Meteorological Organization's Panel of Experts on Seasonal to Long-Range Forecasts (ET-OPSLS) and WMO CCl / CBS Intergovernmental Panel on Regional Climate Activities.

The conference sessions were devoted to different aspects of Earth system modeling, long range forecasts and applications. The sixteenth session of the CIS Climate Forum on Seasonal Forecasts (NEACOF-16) was also held on the first day of the conference in the form of a parallel session.

During the conference, 4 lectures and 90 reports were presented and discussed in a series of successive presentations. All materials are presented in the conference proceedings and posted on the conference web-site http://www.scert.ru/en/conferences/cites2019/.

All contributions, selected for publication in this volume, have been peer reviewed according to rigorous international standards. We are grateful to the reviewers for maintaining the standards and quality of the manuscripts throughout the reviewing process. However, the authors take full responsibility for the content of their papers.

The event was organized by Institute of Monitoring of Climatic and Ecological Systems SB RAS and G.I. Marchuk Institute of Computational Mathematics RAS. It was supported by International Center for Theoretical Physics, the World Climate Research Program, Ministry of Science and Higher Education of Russian Federation and Russian Foundation for Basic Research. We gratefully acknowledge all our sponsors.

Editors

Professor Evgeny Gordov

RAS Corresponding Member Vasily Lykosov

IOP Conf. Series: Earth and Environmental Science **386** (2019) 011002 doi:10.1088/1755-1315/386/1/011002

Peer review statement

All papers published in this volume of *IOP Conference Series: Earth and Environmental Science* have been peer reviewed through processes administered by the proceedings Editors. Reviews were conducted by expert referees to the professional and scientific standards expected of a proceedings journal published by IOP Publishing.

October snow cover and winter atmospheric conditions in Siberia

Yu V Martynova

Institute of Monitoring of Climatic and Ecological Systems SB RAS, Tomsk, 634055, Russia

E-mail: FoxyJ13@gmail.com

Abstract. In this study, the relationship between autumn Siberian snow cover and atmospheric conditions of the following winter is evaluated in time and space. NOAA observational data, NCEP2, and ERA-Interim reanalysis data, and results of a climate model INMCM4 are used. The study is carried out for a territory of Western Siberia of 55° – 74° N and 60° – 90° E. The results obtained show that the relationship does not have a constant manifestation in time. It is hypothesized that the abnormal amount of snow in autumn in Siberia and the anomalies of meteorological parameters in the following winter are initiated by some third process which occurred earlier, possibly at the beginning of autumn or at the end of summer.

1. Introduction

The influence of autumn snow cover anomalies in the middle and high latitudes of the Northern Hemisphere on the following winter atmospheric conditions has been of considerable interest to the scientific community for many years. Various aspects of this influence are being studied. The influence of autumn snow cover anomalies on the tropospheric-stratospheric interaction [1-3] in general, and on the Arctic Oscillation (AO) [4, 5] and the air temperature at the surface in the subsequent winter [6] in particular, were considered. The stationarity of the influence and its possible relationship with various atmospheric processes, for example, with quasi-biennial fluctuations [7, 8] were analyzed. The relationship between the state of snow cover and regional atmospheric blockings was also studied [9-11]. The polar vortex weakening contributes to a stratospheric temperature increase and weakening and a southward shift of the jet stream. As a result, this secondary circulation often takes the form of blocking. Thus, variation of the autumn snow cover can be considered as a predictor both for the abnormal following winter atmospheric conditions at the surface and for sudden stratospheric warming.

Currently, there are works showing that the identification of the manifestation of this mechanism may strongly depend on the time period chosen for the study [12-14]. Generally, the mechanism of the influence of autumn snow cover anomalies on the atmospheric conditions during the next winter in the Northern Hemisphere has not been thoroughly studied yet. However, researchers agree that it is very complex and requires a comprehensive investigation [5, 7]. There is also a hypothesis that the effect of snow variation on the AO is not subject to any physical mechanism but is stochastic [8]. Our studies suggest that the mechanism proposed by Cohen J. in 2007 [1] is not a governing mechanism in itself; it is likely to be launched and phased in some successful combination of factors [12, 13].

Despite a rather long history of the studies of individual aspects of the interseasonal interaction of snow cover and atmosphere, a conclusion on it has not yet been formulated in detail [15]. Although this knowledge, in addition to their weight in themselves, could be helpful for seasonal and subseasonal forecasts [16, 17], especially in the context of the current and possible future climate change.

In this work, manifestations in time and height of the relationship between the autumn Siberian snow cover and the following winter atmospheric conditions were evaluated using various datasets.

2. Data and method

The study was performed using several datasets. NOAA satellite data on snow cover from the Rutgers University Global Snow Lab were used (GSL) [18]. These data have a weekly time resolution and contain information for the entire globe. The data on the values of the Arctic Oscillation Index from The National Weather Service Climate Prediction Center (NWS CPC) were also used [19]. Although the AO index values were obtained as a result of the EOF decomposition of the geopotential height at 1000 hPa from the NCEP2 reanalysis [20], they are, de facto, taken as the observed index value. Geopotential height fields from the NCEP2 reanalysis were also involved in the study. These data have a horizontal resolution of 2.5°x2.5° and 17 vertical isobaric levels with an upper boundary at 10 hPa. An analysis using these data was carried out for 1979-2016.

The study also used data on the snow cover and the geopotential height from the ERA-Interim reanalysis [21] and INMCM4 global climate model output [22]. The fields from ERA-Interim have a horizontal resolution of 0.75°x0.75° and 37 vertical isobaric levels with an upper atmospheric boundary at 1 hPa. The period of 1979-2015 was considered. The INMCM4 data were taken from the CMIP5 databank for 1976-2005 from the Historical experiment [23]. These data have a horizontal resolution of 2.0°x1.5°, 17 vertical isobaric levels with an upper boundary at 10 hPa. The AO index considered in this work was calculated using the technique applied by NOAA [24].

In this study, attention was focused on the territory of Western Siberia, 55°–74° N and 60°– 90°E.

In Siberia during the fall in general, and in October in particular, the snow cover is steadily increasing (Figure 1). Consequently, as an indicator of the snow cover area formed by the end of October, the maximum value of the area for this month was considered (S_{snow}).

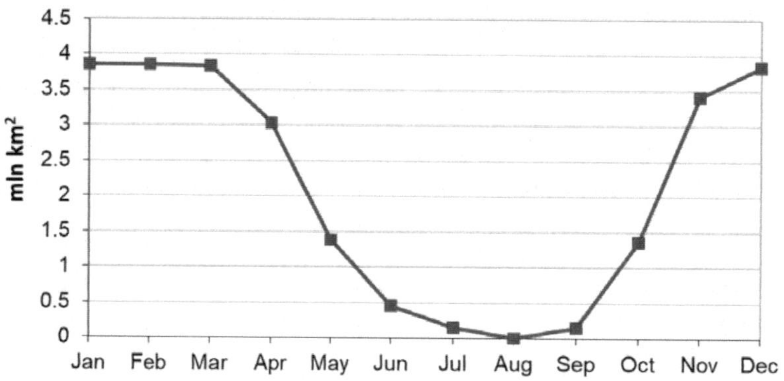

Figure 1. Climatological S_{snow} over Western Siberia, GSL, 1979-2016.

The geopotential height anomalies (Z_{anom}) were also used for the analysis. Z_{anom} at each considered moment of time is the deviation of the daily geopotential height from the mean period value. This parameter was used as an indicator of wave propagation anomalies in the atmosphere.

The manifestation of the linear relationship between the S_{snow} and AO index was examined using the procedure of checking all nested periods (PCANP) based on the calculation of Pearson correlation coefficients [13]. Both original (ODS) and detrended data series (DDS) were considered.

3. Results

3.1. October S_{snow} and winter AO index

The PCANP allowed identifying all subperiods for all considered datasets with a statistically significant linear relationship between the S_{snow} and AO index. A set of map-schemes was used for analysis and demonstration of the obtained results (Figure 2). Here the abscissa axis shows the start year of the period estimated, and the ordinate axis shows the length of the period in years.

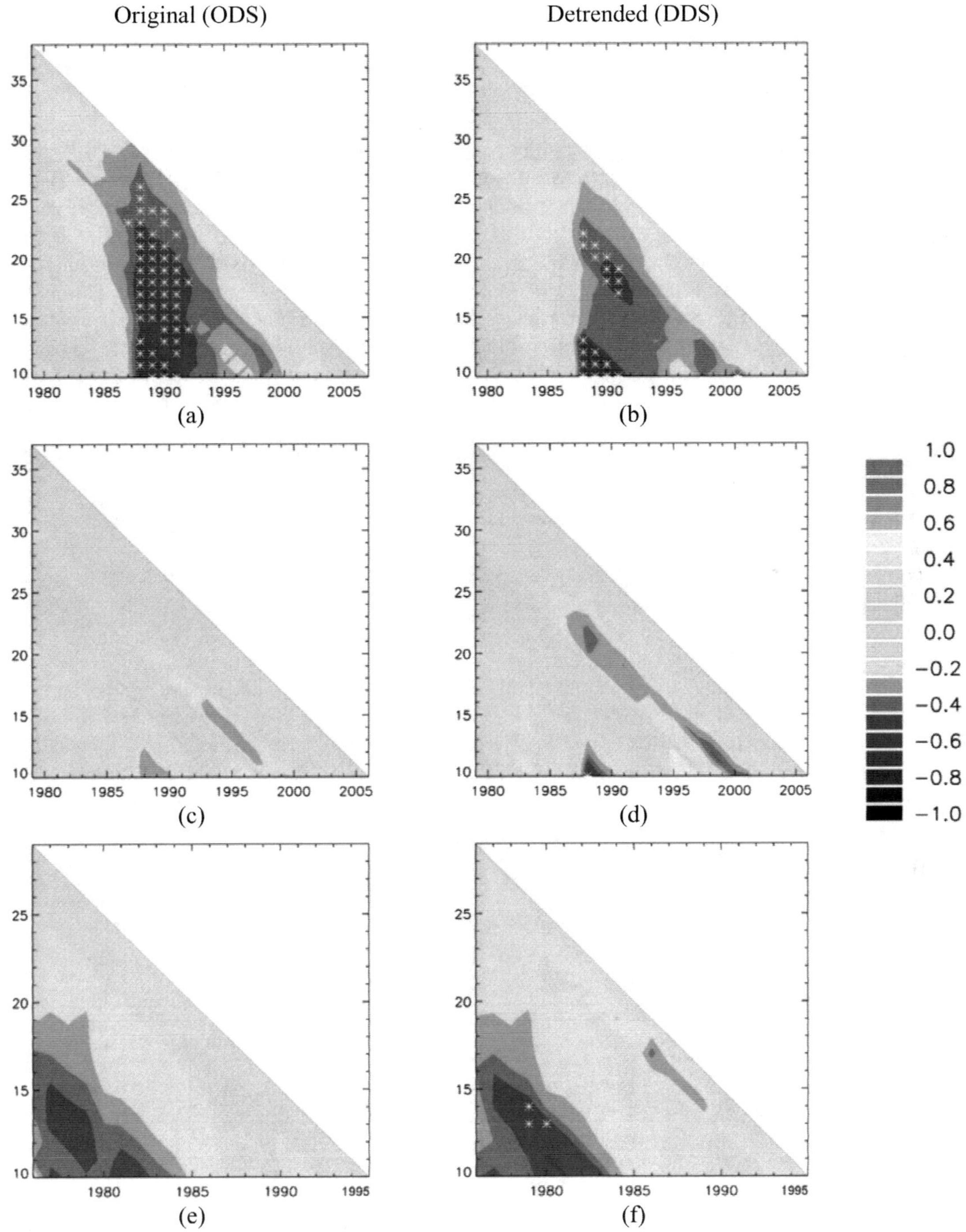

Figure 2. Correlation between S_{snow} and AO index from observations (GSL, NOAA) (a, b), ERA-Interim reanalysis (c, d) and climate model INMCM4 (e, f) for original (a, c, e) and detrended (b, d, f) datasets. Axis X is the start of the subperiod, and axis Y is its length; white symbols denote periods with a significant linear relationship for $\alpha = 0.1$.

The results obtained for different datasets and for the ODS and DDS vary significantly. In particular, when analyzing the ODS for the winter season as a whole, a linear relationship was obtained only for the observational data: S_{snow} from the GSL and the AO index from NOAA (Figure 2). The other datasets did not demonstrate a significant linear relationship for the ODS. The situation with the DDS is moderately different. All three datasets examined demonstrated the presence of a significant linear relationship. However, the number of such periods varies significantly. ERA-Interim showed a significant linear relationship for only one short subperiod (1988-1997). The results obtained using the INMCM4 data contain three subperiods with a significant linear relationship between the S_{snow} and AO index of 13 and 14 years (1979-1991, 1979-1992, and 1980-1992). At the same time, an analysis of the DDS of the observational data revealed more than 10 periods with a manifestation of a statistically significant linear relationship.

A similar picture in the context of differences between the ODS and DDS, as well as between different data sets, was obtained by considering the AO index for each winter month separately.

To analyze the results obtained using the PCANP, for each data set for each month, for the ODS and DDS, the whole variety of cases of manifestation of the relationship between the S_{snow} and AO index was divided into the following categories:

- the relationship manifested in individual subperiods of the period under consideration (1-2 subperiods);
- the relationship manifested in a small number of neighboring and/or nested periods (3-10 subperiods);
- the relationship manifested in a relatively large number of neighboring and/or nested periods (more than 10 sub-periods).

The obtained sign of the considered relationship was also taken into account.

Based on the information obtained, a generalizing scheme was constructed (Figure 3). For almost all subperiods for which a statistically significant linear relationship has been established, the sign of the correlation coefficient is negative. The only exception is the result obtained by the ERA-Interim reanalysis. In the analysis of the ODS for February, a positive sign of the linear relationship under consideration was obtained. For the DDS for January, a relationship with both a positive and a negative sign was taken. Analysis of observational data (GSL and NOAA) showed the presence of a statistically significant linear relationship for December, February, and the entire winter season. The nature of this manifestation is similar for the ODS and DDS.

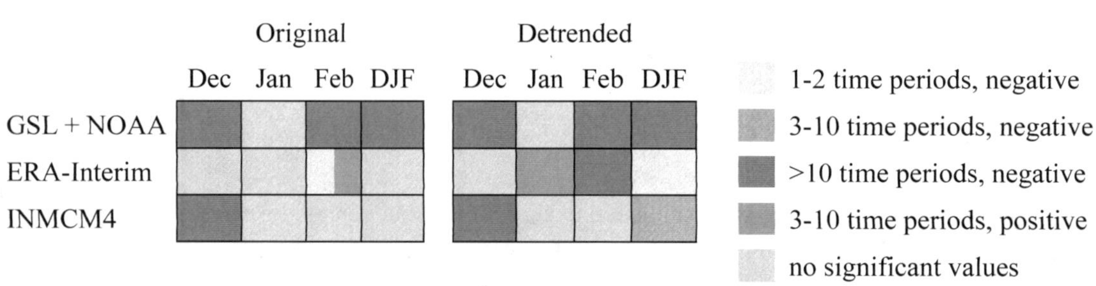

Figure 3. A generalizing scheme for manifestation of a statistically significant linear relationship between S_{snow} and AO index. Blue color indicates the negative sign of the relationship; red color is the positive sign, $\alpha = 0.1$.

The INMCM4 climate model data revealed a manifestation of a statistically significant relationship for December for the ODS and DDS, and for the entire winter season only for the DDS and only in three short subperiods. The nature of the manifestation of the relationship in the ERA-Interim reanalysis data is moderately different from the other datasets being considered. Here the relationship

IOP Conf. Series: Earth and Environmental Science **386** (2019) 012001 doi:10.1088/1755-1315/386/1/012001

appears in February in the analysis of the ODS, and there are periods with both positive and negative signs. In the analysis of the DDS, the relationship appears in February in more than ten subperiods, in the entire winter season in one subperiod with a negative sign, and also in January, but only with a positive sign of the relationship.

Moreover, a significant linear relationship between the S_{snow} and AO index was obtained only for subperiods with length of no more than 30 years. Besides, a sharp transition from statistically significant to insignificant values of the considered correlation coefficient was obtained. A shift of the beginning and/or the end of the considered subperiod by at least one year can change the value of the correlation coefficient from significant to insignificant (and vice versa) with an enormous difference.

3.2. October S_{snow} and winter Z_{anom}

According to the mechanism proposed by Cohen J., when snow cover of an anomalously large area is established, strong cooling should occur, and a wave propagating from the surface through the troposphere to the stratosphere should appear [Cohen, 2007]. The wave propagation can be detected by the propagation of geopotential height anomalies through the atmosphere. Analyzing the direct atmospheric response to the S_{snow} variation, the correlation coefficients between S_{snow} and the Siberian mean Z_{anom} were considered for each isobaric level and each day of winter using the ODS.

In this part of the work, attention was focused on situations where a period shift of one year on the time scale dramatically changes the value of the correlation coefficient between the S_{snow} and AO index from statistically significant to insignificant. Such a manifestation was considered on the example of the subperiods 1988-1997 and 1987-1996 according to the data of GSL, NCEP2, and ERA-Interim (Figure 4). When analyzing the DDS for the winter season, the correlation coefficient between the S_{snow} and AO index for 1988-1997 is statistically significant, and it is -0.639 for GSL and NOAA (NCEP2) observations and -0.580 for ERA-Interim. When the period shifts one year to the left on the time scale, the correlation coefficient becomes insignificant, and it is 0.040 for GSL and NOAA data (NCEP2) and 0.086 for ERA-Interim. In this case, for a series of such length the absolute values of correlation coefficients exceeding 0.54 are significant.

The distribution of the correlation coefficients with height over time obtained using snow cover observations (GSL), for periods with a significant linear relationship, 1988-1997, differs significantly from the periods without it, 1987-1996 (Figure 4 a, b). For 1988-1997 a significant positive relationship is shown between S_{snow} and Z_{anom} in the stratosphere and at significantly lower altitudes in the troposphere than for 1987-1996. Moreover, for 1987-1996 a negative significant linear relationship between S_{snow} and Z_{anom} was obtained at the surface for individual days of the second half of winter.

The ERA-Interim reanalysis data did not show such significant differences (Figure 4 c, d).

4. Discussion and concluding remarks

This study examined a calendar period of snow cover formation and a possible subsequent development of interaction between the thus formed surface anomaly and the atmosphere. Based on a general climatic assessment of snow cover duration in Siberia (Figure 1), its extent was calculated strictly for October. However, situations with late, closer to November, formation of snow cover are quite typical for this territory. The rate of increase in snow cover extent can also vary significantly from year to year. These features could affect the results. In the future, it makes sense to take into account the actual start and end times of snow cover for the territory under consideration.

The differences between the results obtained with the observations, reanalysis, and climate modeling are probably due to differences in the quality of reproduction by the models of the direct relationships and feedbacks between the processes on the surface, in the atmospheric boundary layer, troposphere and stratosphere, and at their interfaces. For this kind of research, it is essential to use data that well represent the real state of the atmosphere at all altitude levels.

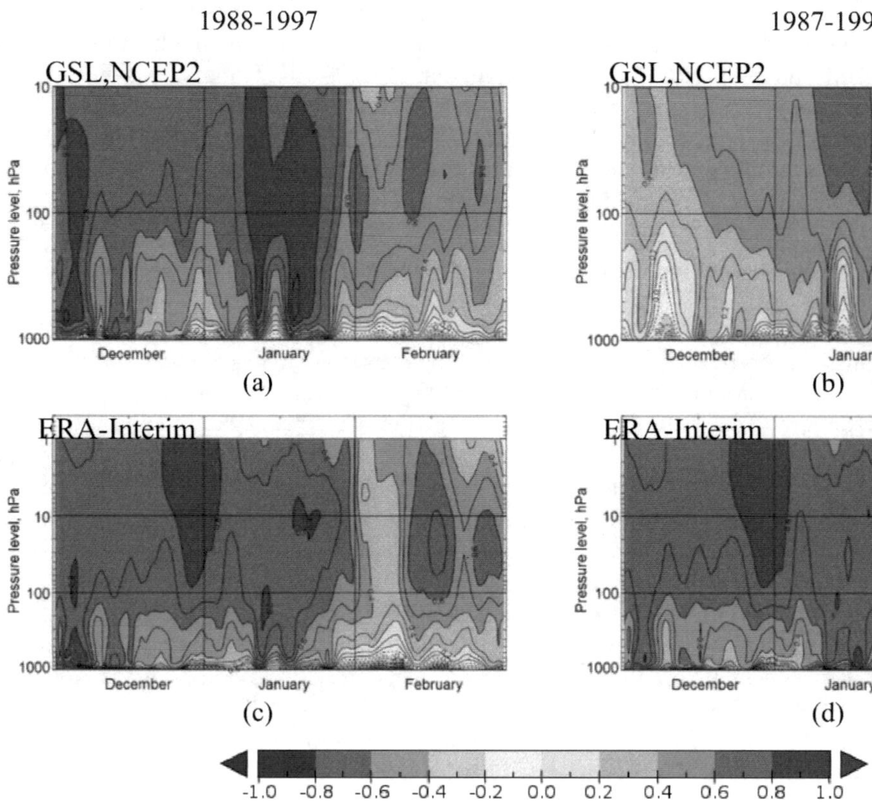

Figure 4. Correlation between S_{snow} and Z_{anom} from GSL and NCEP2 (a, b) and ERA-Interim (c, d) for winter of periods with (a, c) and without (b, d) significant relationship.

The results obtained above show that the relationship between the autumn period of snow cover extent in Siberia and the meteorological conditions in the following winter does not have a constant manifestation in time. Perhaps it would manifest itself in a subsequent combination of some circumstances. It is hypothesized that the abnormal amount of snow in autumn in Siberia and the anomalies of meteorological parameters are initiated by some third process which occurred earlier, possibly at the very beginning of autumn or even at the end of summer.

Acknowledgments. This study was supported by the RAS under Program «Climate Change: Causes, Risks, Consequences, Problems of Adaptation and Regulation». The NCEP Reanalysis 2 data provided by the NOAA/OAR/ESRL PSD, Boulder, Colorado, USA, were taken from their Web site at https://www.esrl.noaa.gov/psd/.

References

[1] Cohen J, Barlow M, Kushner P J and Saito K 2007 Stratosphere-troposphere coupling and links with Eurasian land-surface variability *J. Climate.* **20** 5335–43
[2] Martynova Yu V and Krupchatnikov V N 2010 A Study of the sensitivity of the surface temperature in Eurasia in winter to snow cover anomalies: the role of the stratosphere *Izv. - Atm. Ocean. Phys.* **46** (6) 757–69
[3] Furtado J C *et al.* 2015 Eurasian snow cover variability and links to winter climate in the CMIP5 models *Climate Dyn.* **45** 2591–605
[4] Allen R J and Zender C S D 2010 Effects of continental-scale snow albedo anomalies on the wintertime Arctic oscillation *J. Geophys. Res.* **115** D23105

[5] Sae-Rim Yeo, WonMoo Kim and Kwang-Yul Kim 2017. Eurasian snow cover variability in relation to warming trend and Arctic Oscillation *Climate Dynamics* **48** 499–511
[6] Foster J, Owe M and Rango A 1983 Snow cover and temperature relationships in North America and Eurasia *J. Clim. appl. Met.* **22** 460–9
[7] Peings Y, Brun E, Mauvais V and Douville H 2013 How stationary is the relationship between Siberian snow and Arctic Oscillation over the 20 th century? *Geophys. Res. Lett.* **40** 183–8
[8] Douville H, Peings Y and Saint-Martin D 2017 Snow-(N)AO relationship revisited over the whole twentieth century *Geophys. Res. Lett.* **44** 569–77 DOI:10.1002/2016GL071584
[9] García-Herrera R and Barriopedro D 2006 Northern Hemisphere snow cover and atmospheric blocking variability *J. Geophys. Res.* **111** D21104
[10] Choi Y W and Ahn J B 2017. The Effect of Boreal Late Autumn Snow Cover over Western and Central China on the Northern Hemisphere Wintertime Blocking Frequency *J. of Climate* **30** (22) 9027–39
[11] Antokhina O Yu, Antokhin P N, Devyatova E V and Martynova Yu V 2018 Atmospheric blockings in Western Siberia. Part 2. Long-term variations in blocking frequency and their relation with climatic variability over Asia *Russian Meteorology and Hydrology* **43** (3) 143–51 DOI:10.3103/S1068373918030020
[12] Martynova Yu V, Kharyutkina E V, Krupchatnikov V N and Antokhina O Yu 2018 Relationship between variations in autumn snow cover and temperature and circulation regimes of the following winter in Western Siberia *Fund. and Appl. Clim.* **1** 71–83 DOI:10.21513/2410-8758-2018-1-71-83
[13] Martynova Yu 2018 Relationship between Siberian autumn snow cover and winter Arctic Oscillation: features of its time manifestation *IOP Conf. Series: Earth and Environmental Science* **211** 012007 DOI: 10.1088/1755-1315/211/1/012007
[14] Popova V V, Shiryaeva A V and Morozova P A 2014 Snow cover setting-up dates in the north of Eurasia: relations and feedback to the macro-scale atmospheric circulation *Ice and Snow* **54** (3) 39–49
[15] Henderson G R, Peings Y, Furtado J C and Kushner P J 2018 Snow–atmosphere coupling in the Northern Hemisphere *Nature Climate Change* **8** 954–64
[16] Cohen J 2011 Eurasian snow cover variability and links with stratosphere-troposphere coupling and their potential use in seasonal to decadal climate predictions. *In Climate Test Bed Joint Seminar Series. NCEP, Camp Springs, Maryland*
[17] Orsolini Y J, Senan R, Vitart F, Balsamo G, Weisheimer A and Doblas-Reyes F J 2016 Influence of the Eurasian snow on the negative North Atlantic Oscillation in subseasonal forecasts of the cold winter 2009/2010 *Climate Dyn.* **47** (3–4) 1325–34
[18] Robinson David A, Estilow Thomas W and NOAA CDR Program 2012 NOAA Climate Data Record (CDR) of Northern Hemisphere (NH) Snow Cover Extent (SCE), Version 1. [NH SCE] *NOAA National Centers for Environmental Information* DOI: 10.7289/V5N014G9 [25 Jul 2017] (https://climate.rutgers.edu/snowcover/)
[19] http://www.cpc.ncep.noaa.gov/products/precip/CWlink/daily_ao_index/ao.shtml
[20] Kanamitsu M, Ebisuzaki W, Woollen J, Yang S-K, Hnilo J J, Fiorino M, and Potter G L 2002 NCEP-DOE AMIP-II Reanalysis (R-2) *Bulletin of the American Meteorological Society* 1631–43
[21] Dee D P, Uppala S M, Simmons A J, Berrisford P, Poli P, Kobayashi S, Andrae U et al. 2011 The ERA-Interim reanalysis: Configuration and performance of the data assimilation system *Quarterly Journal of the royal meteorological society* **137** (656) 553–97
[22] Volodin E M, Dianskii N A and Gusev A V 2010 Simulating present-day climate with the INMCM4.0 coupled model of the atmospheric and oceanic general circulations *Izv. - Atm. Ocean. Phys.* **46** (4) 414–31
[23] http://cmip-pcmdi.llnl.gov/cmip5/
[24] https://www.cpc.ncep.noaa.gov/products/precip/CWlink/daily_ao_index/history/method.shtml

Influence of the stratospheric polar vortex on the Barents Sea ice extent in early 2012

V V Zuev and E S Savelieva

Institute of Monitoring of Climatic and Ecological Systems of the Siberian Branch of the Russian Academy of Sciences, 10/3, Academichesky Ave., Tomsk, Russia

vzuev@list.ru, ekat.savelieva@gmail.com

Abstract. A significant decrease in the sea ice extent in the Barents Sea was observed in February 2012 due to a large increase in Arctic surface temperature. In early January 2012 a sudden stratospheric warming occurred simultaneously with a stratospheric polar vortex displacement. Based on the ERA-Interim reanalysis data, we show that the displacement of the stratospheric polar vortex led to a perturbation and splitting of the tropospheric polar vortex. An inflow of warm air masses into the polar region, mainly over the Barents Sea, occurred during the splitting of the tropospheric vortex, which resulted in an increase in Arctic surface temperature and a decrease in the sea ice extent.

1. Introduction

Stratospheric polar vortices usually form in autumn, reach peak intensity in mid-winter, and decay from late winter to spring [1]. The dynamics of the polar vortices plays an important role in the stratospheric ozone distribution, the movement of air masses in the polar stratosphere, and the change in the stratospheric temperature over the polar region [2]. The area and extent of the stratospheric ozone depletion over the polar region depend on the polar vortex strength and persistence in spring [3]. The polar vortices act as containment vessels and prevent the inflow of warm and ozone-rich air masses from the midlatitude stratosphere into the polar region [4, 5]. At the same time, polar stratospheric clouds (PSCs) form at extremely low temperatures (lower than –78 °C) inside the polar vortex. Heterogeneous reactions, releasing photochemically active molecular chlorine, occur on and in PSC particles [6]. The stratospheric ozone depletion over the polar region is a result of catalytic cycles involving chlorine species in the presence of weak solar radiation from late winter to spring [7].

One of the most prominent features of the Arctic polar vortex is its large interannual variability associated with the frequency of occurrence of sudden stratospheric warmings (SSWs) [8]. The SSWs are characterized by a sudden increase in the latitudinal mean temperature poleward from 60° latitude at the 10 hPa level (~ 30 km) or below. Such temperature increase occurs during the breaking and dissipation of westward propagating planetary waves, which weaken or reverse the eastward zonal flow (i.e. the stratospheric polar vortex) and induce heat by adiabatic processes [8]. The SSWs are divided into major and minor ones [9]. A strong displacement of the polar vortex occurs during minor SSWs, whereas major SSWs are often accompanied by the polar vortex splitting or even its breakdown.

Displacement, splitting, and breakdown of the polar vortex usually develop from the middle stratosphere to the lower stratosphere [10]. Most studies on this subject focus on the stratospheric effects of the polar vortex dynamics, such as changes in the gas and aerosol composition of the polar stratosphere [9, 11, 12] and its temperature fluctuations [13, 14]. The aim of this study is to investigate the influence of the stratospheric polar vortex displacement which occurred during a SSW on January 12, 2012 on the tropospheric polar vortex dynamics and its effects on the Arctic surface temperature and sea ice extent.

2. Data and methods

The data on the zonal wind and temperature for 30°–90° N on the 50 and 500 hPa pressure surfaces and the temperature data for 60°–90° N on the 1000 hPa pressure surface were retrieved from the European Centre for Medium-Range Weather Forecasts (ECMWF) ERA-Interim reanalysis (http://apps.ecmwf.int/datasets/data/interim-full-daily/levtype=pl/) [15]. The monthly mean values of the sea ice extent over the Barents Sea were taken from the National Snow and Ice Data Center (NSIDC, http://nsidc.org) [16].

To explore the dynamics of the Arctic polar vortices we obtained 38-year monthly climatological means of the zonal wind and temperature for 30°–90° N on the 50 and 500 hPa pressure surfaces over 1979–2016. To analyze variations of the Arctic surface temperature from January to March 2012 we obtained a 38-year climatological mean seasonal cycle of the air temperature for 60°–90° N on the 1000 hPa pressure surface over 1979–2016 with standard deviations (σ). The climatological mean and standard deviations were smoothed by a 15-point fast Fourier transform (FFT) filter.

3. Results

3.1. Stratospheric and tropospheric polar vortices

The polar vortex is a large-scale cyclone located over the polar region. The strong polar vortex leads to a temperature decrease inside the vortex and prevents the inflow of air masses into the polar region. The stratospheric polar vortex exists from autumn to spring and extends from the tropopause to the stratopause [2]. The tropospheric polar vortex extends from the ground to the upper troposphere and exists all year. Figure 1 shows 38-year monthly climatological means (1979–2016) of the zonal wind and temperature at the 50 and 500 hPa pressure levels (~ 20 and 5 km) over the Arctic from October to March obtained using the ERA-Interim reanalysis. As seen in Figure 1, the tropospheric polar vortex is larger than the stratospheric one, has lower velocities of the zonal wind and a more curved trajectory of the zonal flow (which forms the vortex edge). The tropospheric polar vortex usually strengthens from autumn to spring during the existence of the stratospheric vortex [10]. The splitting of the stratospheric polar vortex occurring under the influence of vertically propagating planetary Rossby waves can appear in the dynamics of the tropospheric vortex [17].

3.2. Stratospheric polar vortex displacement and tropospheric polar vortex splitting in early 2012

The SSW which occurred on January 12, 2012 [11] was accompanied by the stratospheric polar vortex displacement, preceding the weakening and breakdown of the vortex. Figure 2 shows the monthly means of the zonal wind and temperature at the 50 and 500 hPa pressure levels over the Arctic from October 2011 to March 2012 retrieved from the ERA-Interim reanalysis. In December 2011 the stratospheric polar vortex was strong and centered over the pole. The displacement and subsequent breakdown of the stratospheric polar vortex were observed in January 2012: it is almost not manifested in the zonal wind distributions for February and March (Figure 2). The changes in the dynamics of the stratospheric vortex appeared in the perturbation of the tropospheric polar vortex in January and its short-term splitting, in February (Figure 2), then the tropospheric vortex recovered in March.

IOP Conf. Series: Earth and Environmental Science **386** (2019) 012002 doi:10.1088/1755-1315/386/1/012002

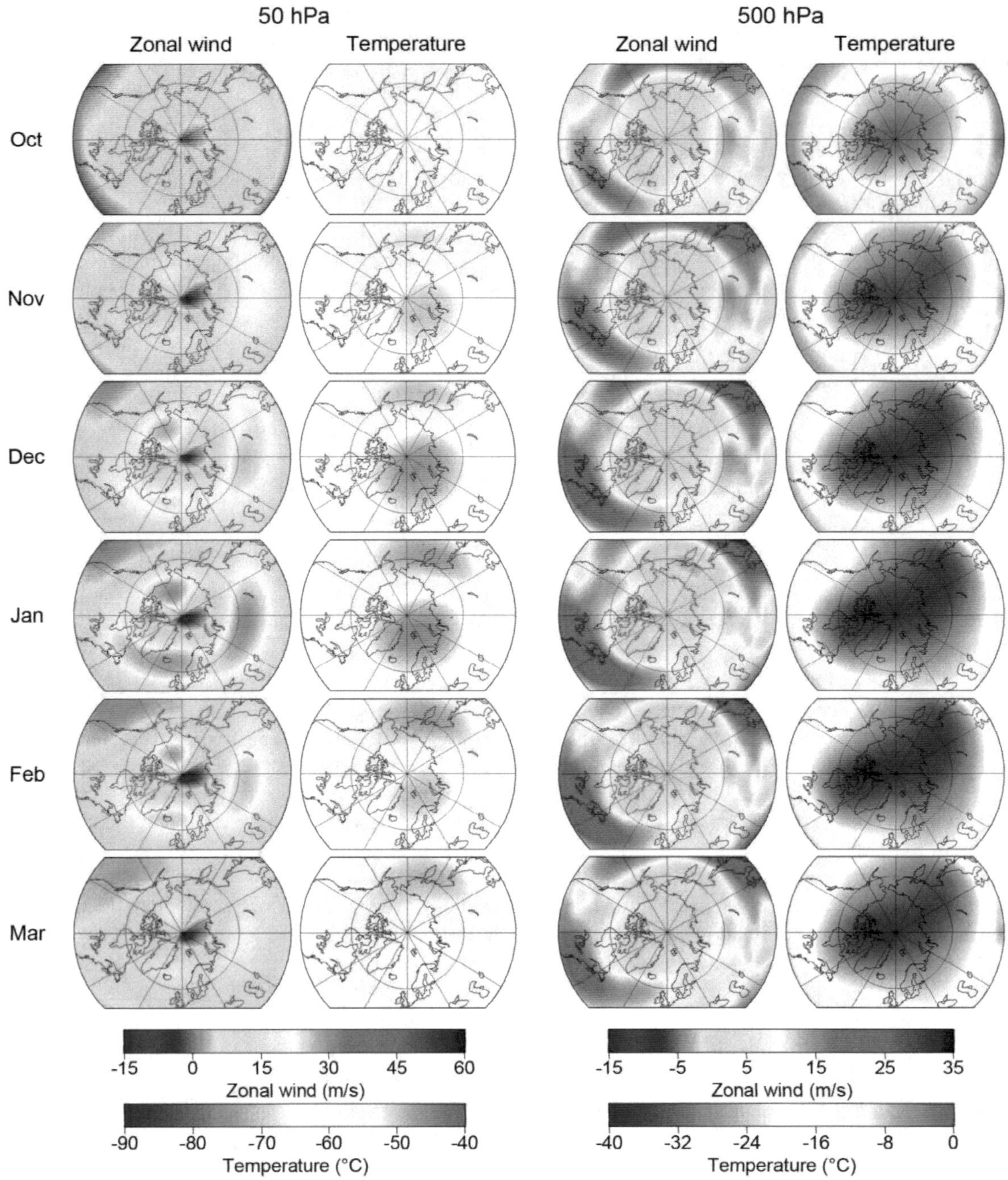

Figure 1. Monthly climatological means of zonal wind and temperature at 50 and 500 hPa pressure levels over the Arctic from October to March over 1979–2016.

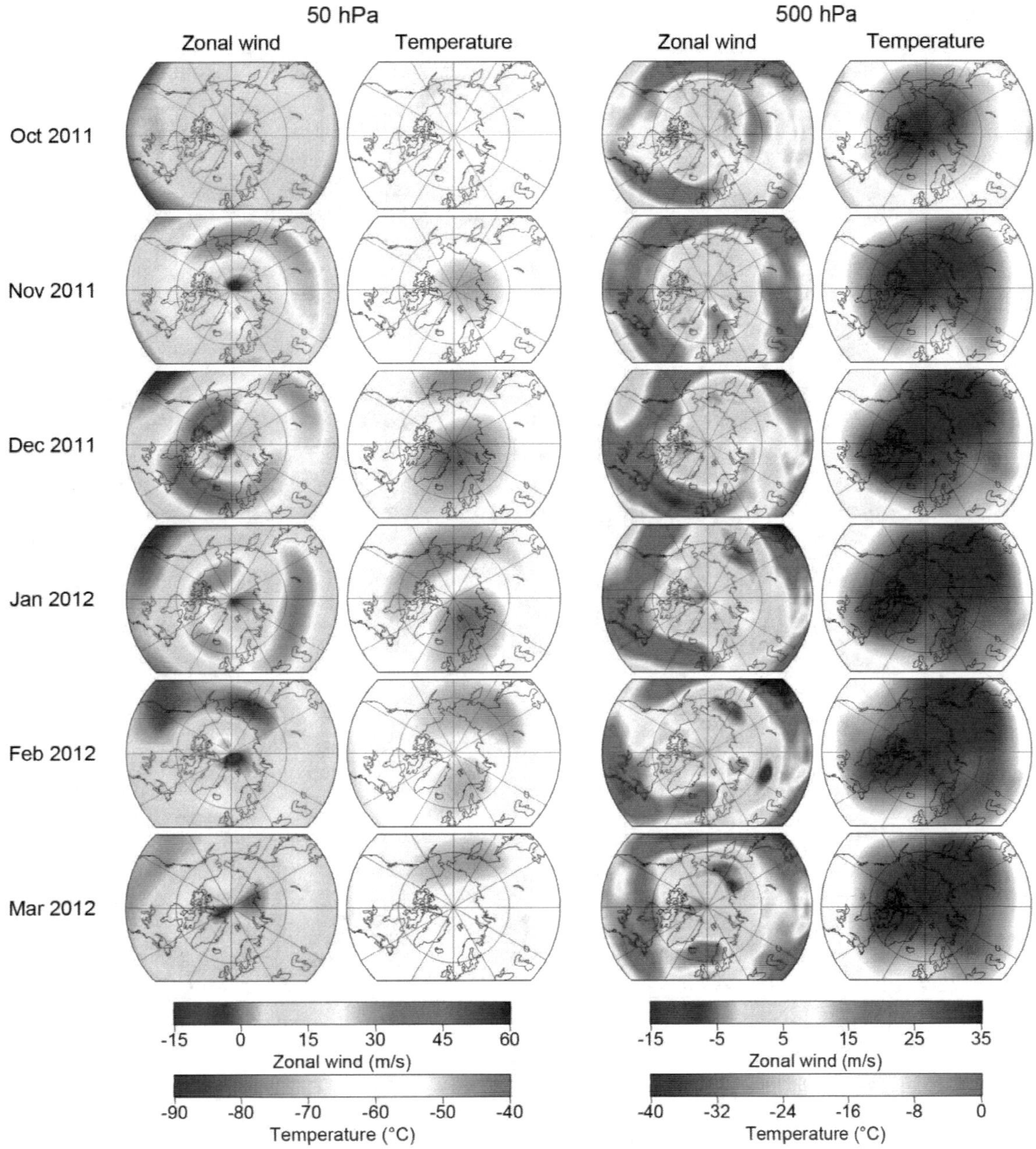

Figure 2. Monthly means of zonal wind and temperature at 50 and 500 hPa pressure levels over the Arctic from October 2011 to March 2012.

Figure 3 shows the zonal wind and temperature distributions at the 50 and 500 hPa pressure levels over the Arctic from January to February 2012 retrieved from the ERA-Interim reanalysis. As seen in Figure 3, both the stratospheric and tropospheric polar vortices appear clearly in the zonal wind and temperature distributions in early January. The displacement and elongation of the stratospheric polar vortex were observed on January 12, resulting in the perturbation of the tropospheric polar vortex. The tropospheric vortex splitting occurred in early February. The Beaufort and Barents Seas appeared to be located outside the tropospheric vortex (Figure 3). The Beaufort Sea was covered with cold air masses of the polar vortex in a few days. The Barents Sea was located outside the tropospheric vortex during February 2012. The inflow of warm air masses in the troposphere over the Barents Sea led to an increase in the surface temperature and a decrease in the sea ice extent over the Barents Sea.

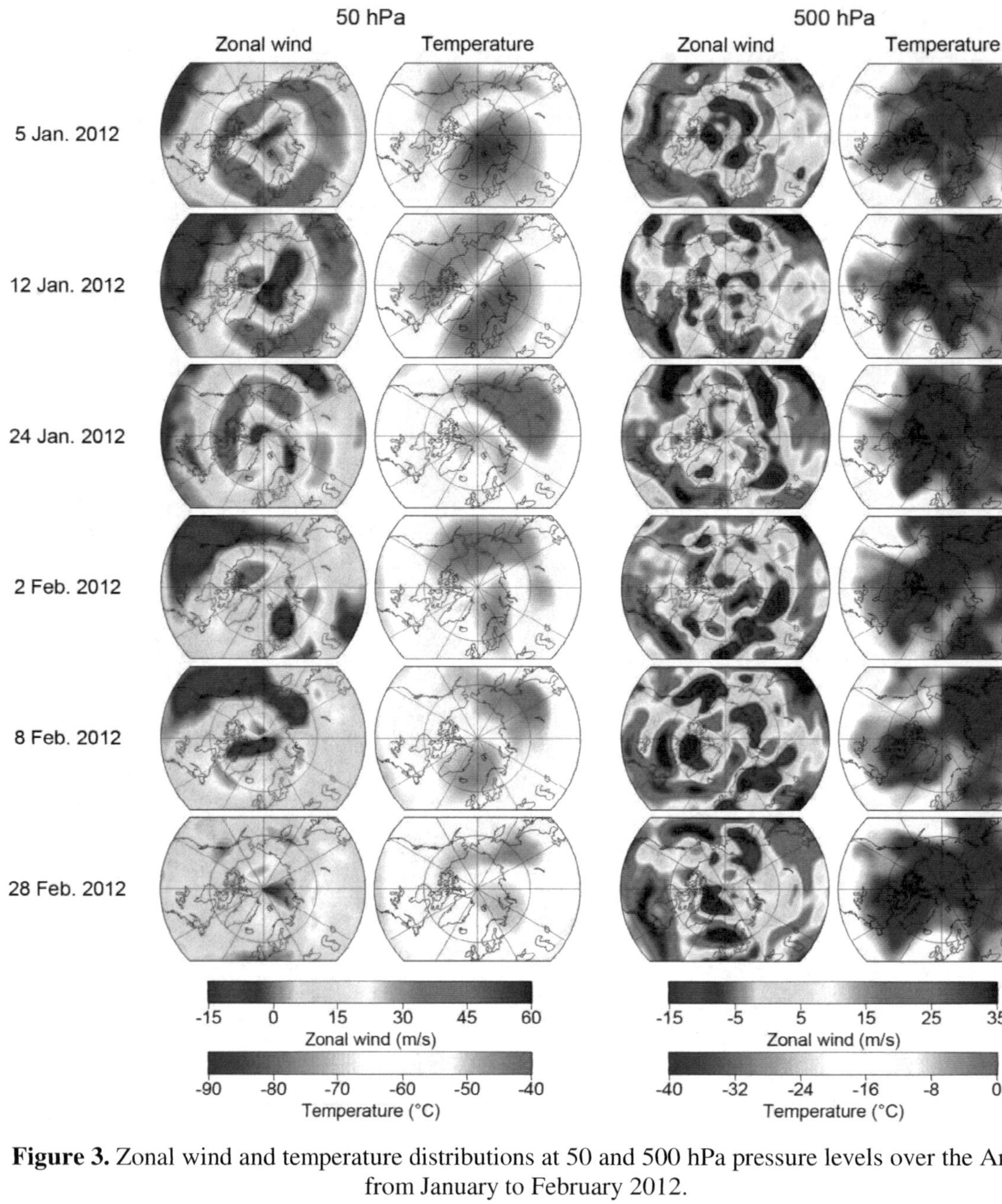

Figure 3. Zonal wind and temperature distributions at 50 and 500 hPa pressure levels over the Arctic from January to February 2012.

3.3. Arctic surface temperature and Barents Sea ice extent in early 2012

Figure 4 shows the time series of the air temperature for 60°–90° N at the 1000 hPa pressure level from January to March 2012 compared to its 38-year climatological mean seasonal cycle over 1979–2016 with ±1 standard deviations. The time series and seasonal cycle are retrieved from the ERA-Interim reanalysis. As seen in Figure 4, an increase in the Arctic surface temperatures was observed in the second half of January and then from 6 to 26 February. Thus, it occurred a few days after the stratospheric polar vortex displacement in January, during the tropospheric vortex perturbation, and then in February during the splitting of the tropospheric vortex. In early 2012 the monthly mean surface temperature T_s reached -17.1 °C and -16.7 °C in January and February, respectively, in comparison with the 38-year monthly mean values for January T_s = -20.0±2.0 °C and February T_s = -19.8±2.3 °C.

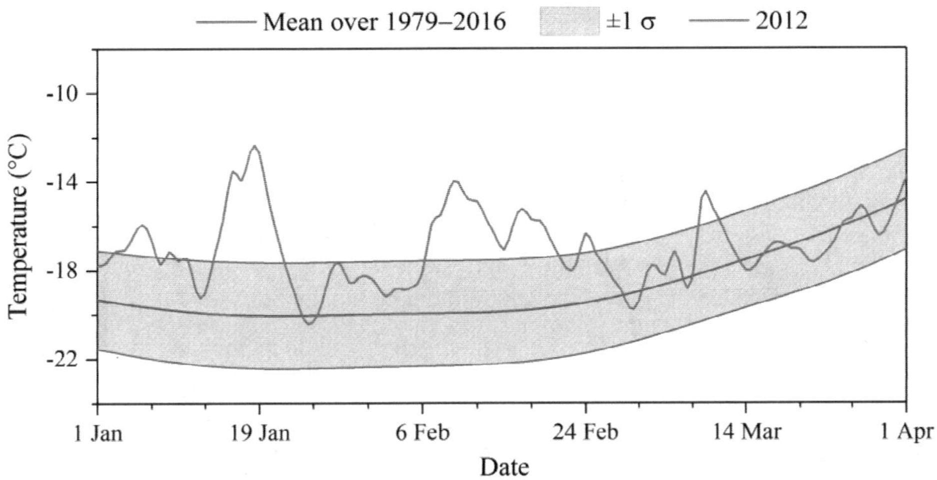

Figure 4. Time series of temperature for 60°–90° N at 1000 hPa pressure level from January to March 2012 (red line) and 1979–2016 climatological means (green line) with ±1 standard deviations (green areas).

The main inflow of warm air masses into the polar troposphere was observed over the Barents Sea during the perturbation and splitting of the tropospheric polar vortex in January and February 2012 (Figure 3). Figure 5 shows the time series of the monthly mean sea ice extent over the Barents Sea from February 1979 to 2018 retrieved from the NSIDC dataset. As seen in Figure 5, one of the minimum values of the sea ice extent over the Barents Sea for 1979–2018 was observed in February 2012 and reached $3.9 \cdot 10^5$ km^2.

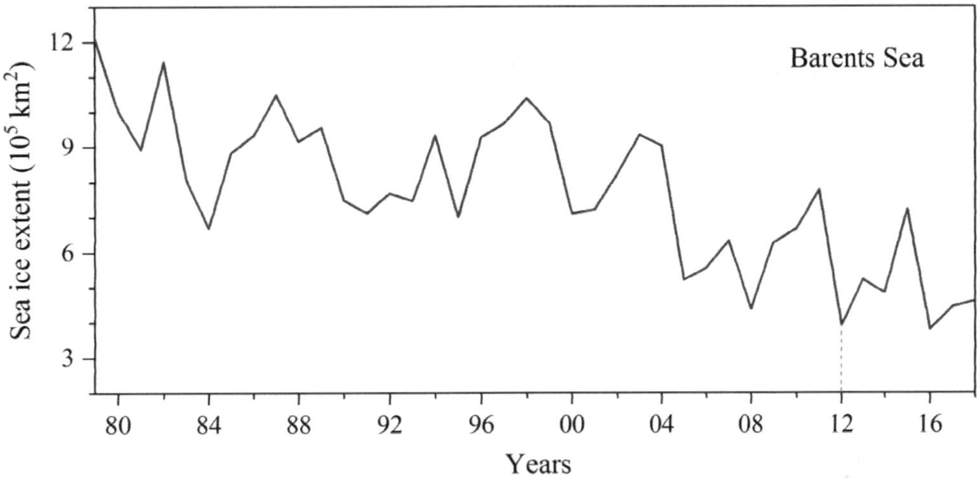

Figure 5. Time series of February mean sea ice extent over the Barents Sea from 1979 to 2018.

The dynamics of the tropospheric polar vortex plays an important role in the movement of air masses in the polar troposphere and the temperature variations over the polar region [10]. The incursions of cold polar air into the middle latitudes (cold air outbreaks) resulting in extreme negative anomalies of the surface air temperature occur due to a change in the polar vortex edge [18–21]. The Arctic sea ice extent is also influenced by the tropospheric polar vortex, especially in winter when the surface temperature significantly decreases under the conditions of vortex strengthening [22, 23]. Some papers [24, 25] investigated the effect of regional cyclone activity on the Arctic sea ice extent in winter. Overland and Wang in [22] showed an increase in the Arctic surface temperature in the winter of 2015/2016 as a result of splitting of the tropospheric polar vortex.

4. Conclusions

In this study we showed that a stratospheric polar vortex displacement occurring under the influence of vertically propagating planetary Rossby waves can lead to a splitting of the tropospheric polar vortex resulting in an increase in Arctic surface temperature. The displacement and subsequent breakdown of the stratospheric polar vortex were observed during a SSW which occurred in January 2012 and appeared, firstly, in a perturbation and splitting of the tropospheric polar vortex and, secondly, in a large increase in Arctic surface temperature in January and February 2012. As a result, a significant decrease in the sea ice extent in the Barents Sea was observed in February 2012.

References

[1] Waugh D W and Randel W J 1999 Climatology of Arctic and Antarctic polar vortices using elliptical diagnostics *J. Atmos. Sci.* **56** 1594–613

[2] Waugh D W and Polvani L M 2010 Stratospheric polar vortices *The Stratosphere: Dynamics, Transport, and Chemistry (Geophys. Monogr. Ser.)* vol 190 (Washington: AGU) pp 43–57

[3] Newman P A 2010 Chemistry and dynamics of the Antarctic ozone hole *The Stratosphere: Dynamics, Transport, and Chemistry (Geophys. Monogr. Ser.)* vol 190 (Washington: AGU) pp 157–171

[4] Manney G L and Zurek R W 1994 On the motion of air through the stratospheric polar vortex *J. Atmos. Sci.* **51** 2973–94

[5] Sobel A H, Plumb R A and Waugh D W 1997 Methods of calculating transport across the polar vortex edge *J. Atmos. Sci.* **54** 2241–60

[6] Finlayson-Pitts B J and Pitts J N 2000 *Chemistry of the Upper and Lower Atmosphere: Theory, Experiments, and Applications* (California: Academic Press) pp 657–726

[7] Solomon S, Garcia R R, Rowland F S and Wuebbles D J 1986 On the depletion of Antarctic ozone *Nature* **321** 755–8

[8] Kuttippurath J and Nikulin G 2012 A comparative study of the major sudden stratospheric warmings in the Arctic winters 2003/2004–2009/2010 *Atmos. Chem. Phys.* **12** 8115–29

[9] Flury T, Hocke K, Haefele A, Kämpfer N and Lehmann R 2009 Ozone depletion, water vapor increase, and PSC generation at midlatitudes by the 2008 major stratospheric warming *J. Geophys. Res.* D **114** D18302

[10] Waugh D W, Sobel A H and Polvani L M 2017 What is the polar vortex and how does it influence weather? *B. Am. Meteorol. Soc.* **98** 37–44

[11] Ageyeva V Yu, Gruzdev A N, Elokhov A S, Mokhov I I and Zueva N E 2017 Sudden stratospheric warmings: statistical characteristics and influence on NO_2 and O_3 total contents *Izv. Atmos. Oceanic Phys.* **53** 477–86

[12] Sofieva V F, Kalakoski N, Verronen P T, Päivärinta S-M, Kyrölä E, Backman L and Tamminen J 2012 Polar-night O_3, NO_2 and NO_3 distributions during sudden stratospheric warmings in 2003–2008 as seen by GOMOS/Envisat *Atmos. Chem. Phys.* **12** 1051–66

[13] Limpasuvan V, Thompson D W J and Hartmann D L 2004 The life cycle of the Northern Hemisphere sudden stratospheric warmings *J. Climate* **17** 2584–96

[14] Manney G L, Schwartz M J, Krüger K, Santee M L, Pawson S, Lee J N, Daffer W H, Fuller R A and Livesey N J 2009 Aura Microwave Limb Sounder observations of dynamics and transport during the record-breaking 2009 Arctic stratospheric major warming *Geophys. Res. Lett.* **36** L12815

[15] Dee D P *et al* 2011 The ERA-Interim reanalysis: configuration and performance of the data assimilation system *Q. J. Roy. Meteor. Soc.* **137** 553–97

[16] Fetterer F, Knowles K, Meier W, Savoie M and Windnagel A K 2017, updated daily. Sea Ice Index, Version 3. Boulder, Colorado USA. NSIDC: National Snow and Ice Data Center. https://doi.org/10.7265/N5K072F8

[17] Black R X 2002 Stratospheric forcing of surface climate in the Arctic Oscillation *J. Climate* **15** 268–77

[18] Vera C S and Vigliarolo P K 2000 A diagnostic study of cold-air outbreaks over South America *Mon. Weather Rev.* **128** 3–24

[19] Cellitti M P, Walsh J E, Rauber R M and Portis D H 2006 Extreme cold air outbreaks over the United States, the polar vortex, and the large-scale circulation *J. Geophys. Res.* **111** D02114

[20] Vavrus S, Walsh J E, Chapman W L and Portis D 2006 The behavior of extreme cold air outbreaks under greenhouse warming *Int. J. Climatol.* **26** 1133–47

[21] Li Zh, Manson A H, Li Y and Meek C 2017 Circulation characteristics of persistent cold spells in central–eastern North America *J. Meteorol. Res.* **31** 250–60

[22] Overland J E and Wang M 2016 Recent extreme Arctic temperatures are due to a split polar vortex *J. Climate.* **29** 5609–16

[23] Li F, Wang H and Gao Y 2015 Extratropical ocean warming and winter Arctic sea ice cover since the 1990s *J. Climate.* **28** 5510–22

[24] Sorteberg A and Kvingedal B 2006 Atmospheric forcing on the Barents Sea winter ice extent *J. Climate.* **19** 4772–84

[25] Cullather R I, Lim Y-K, Boisvert L N, Brucker L, Lee J N and Nowicki S M J 2016 Analysis of the warmest Arctic winter, 2015–2016 *Geophys. Res. Lett.* **43** 10808–16

Current Climatic Changes in the Troposphere, Stratosphere, and Mesosphere, and Inter-Layer Interactions

Y P Perevedentsev[1], K M Shantalinskii[1], B G Sherstyukov[2] and V V Guryanov[1]

[1] Department of Meteorology, Kazan (Volga Region) Federal University, Kazan, Russia
[2] Russian Research Institute for Hydrometeorological Information - World Data Centre, Obninsk, Russia

Abstract. Using new data about near-surface air temperature in the Northern Hemisphere (from 1850 to 2018) and ERA5 reanalysis of temperature distributions up to s height of 80 km (from 1979 to 2018), the authors calculate and analyse mean and root-mean square deviations, linear trend slope coefficients, and low-frequency components of temperature at various levels and latitude zones. A 60-year quasi-cyclicity of the temperature near the Earth's surface has been discovered. The authors evaluate the rates of tropospheric warming and stratospheric cooling observed over the last decades. Much attention is given to the search for correlations between the atmospheric layers, thermal regime characteristics, and the Arctic Oscillation. A predictive valuation of temperature oscillations is obtained for Kazan until 2051, taking into account distant relationships revealed between the temperature status of the ocean surface and the region considered.

1. Introduction

Over recent years interest in dynamic and physico-chemical processes taking place in the thick atmospheric mass (up to an altitude of 80 km) has been growing thanks to intensive development of computers and information technologies, and the invention of "Big Data". At the same time, the ocean surface temperature (OST) exerts certain influence on the circulation processes observed in the lower stratosphere. Thus, strong correlations have been revealed in paper [3] between the dynamics of the stratospheric vortex above the Arctic and the variability of OST anomalies in the northern part of the Atlantic and Pacific Oceans. Correspondingly, explosive winter stratospheric warmings are taken into consideration when studying the interaction between the troposphere and stratosphere. As before, the wave mechanism is used to explain the correlations between the atmospheric layers. Simulation exercises with general atmospheric circulation (GAC) models helped to identify new details in the physical mechanism of interaction between neighbouring layers and construct a general Earth system model [1,2].

The present research is a follow-up of the previous papers of the authors [6-8, 10-14], which are devoted to peculiar features of the thermodynamic processes both in the troposphere of the Northern Hemisphere in the 19th – 21st centuries, and in the higher atmospheric layers (up to 100 km). The objectives of this paper are: to study the distribution of the air temperature (AT) characteristics up to a level of 80 km; evaluate the impact on the thermal state of the troposphere and lower stratosphere by the circulation factor and the OST status; construct a predicative physical-statistical model for long-term regional forecasting of the air temperature near the Earth's surface up to 2051.

The air temperature time series formed for 1850–2018 at regular latitude-longitude geographic grid points by the Climatic Research Unit of the University of East Anglia and Hadley Centre [5,9] (CRU data) were used as raw data. Monthly air temperature (AT) means for the Northern Hemisphere (NH) were used to characterise the processes taking place in the troposphere, stratosphere, and mesosphere (up to 80 km). They were drawn on 51 isobaric surfaces at the geographic grid points 1°x1° (latitude x longitude) for 1979-2018, as presented in ERA5 reanalysis. Correspondingly, long-term AT series for Moscow, Kazan, and Tomsk were used to establish correlations with the OST. The ocean surface temperature was taken at the grid points with the step 2x2°, which permitted identifying synchronous and asynchronous correlations between the ocean temperature and the near-surface AT, and to set up prognostic regression equations.

Near-surface temperatures kept in CRU and ERA5 data pools were compared. To make this comparison, the annual near-surface air temperature means, as well as the winter and summer AT means in the Northern Hemisphere were calculated for 1979-2018. Then the authors took their differences, built linear trends, and found the determination coefficients. High consistency was proven for all the above-listed parameters.

The mean values (norms), the mean square deviations (MSD) near the Earth's surface and on 51 surfaces of the NH were calculated. The mean vertical profiles and distribution maps were plotted for the AT. The aforementioned manipulations permitted analyzing the processes observed in the troposphere, stratosphere, and mesosphere. In order to identify the spatial differences, the AT means were calculated for the polar (90-65° N), temperate (65-30° N), and tropical (30-0° N) zones of the Northern Hemisphere. Means in 3 sectors of the temperate zone (Atlantic-European, Asia-Pacific and American) were calculated. The monthly AT means were subjected to linear trend analysis and low-frequency filtration to pick out oscillations with a period exceeding 10 years. The linear trend slope coefficients (LTSCs) and the determination coefficients for the linear trend and low-frequency component (LFC) with a period exceeding 10 years were calculated as well. The vertical correlation coefficients were calculated to evaluate the correlations between the levels.

The Hayashi spectrum components were calculated to evaluate the wave activity and the correlations with the climatic and geophysical indices.

2. Results

Let us study key features of the climate pattern behaviour in the atmospheric mass from the Earth's surface up to an altitude of 80 km. The multiyear curve of the annual air temperature mean (AATM) constructed according to CRU data for 1850-2018 tends to rise in the Northern Hemisphere (Figure 1). In Figure 1 the LFC depicts global long-period oscillations caused by natural forces. The positive AATM anomaly, intensively growing in the NH from 0.1 to 0.8°C from the late seventies, is explained by the greenhouse effect. Figure 1 vividly presents a quasi-60-year oscillation of the AT rise with the amplitude build-up, which is determined by taking the year-on-year first-order differences of the LFCs. Only at the very end of the period the temperature variation slightly slacks up. The global climate warming has resonance for certain Russian regions, in particular, for the Middle Volga Region.

In order to identify fundamental patterns of climatic changes in the Volga Federal Region (VFR), the authors of the paper have studied the curve of the annual near-surface air temperature mean averaged for the territory of the VFR for the period between 1955 and 2018 and two subperiods: 1955–1999 and 2000–2018 (Table 1).

One can see from Table 1 and Figure 2a that a dramatic jump in the temperature mean (by 1.2°C) was registered in the early 21st century. At the same time, interannual temperature variability halved, while the minimum value of the annual air temperature mean (AATM) sharply rose from 0.55°C to 3.58°C. All these facts evidence a significant change in the thermal regime of the region at the turn of the century. As this takes place, the contribution of the global factor to the regional temperature variation is 37% in winter and 23% in summer.

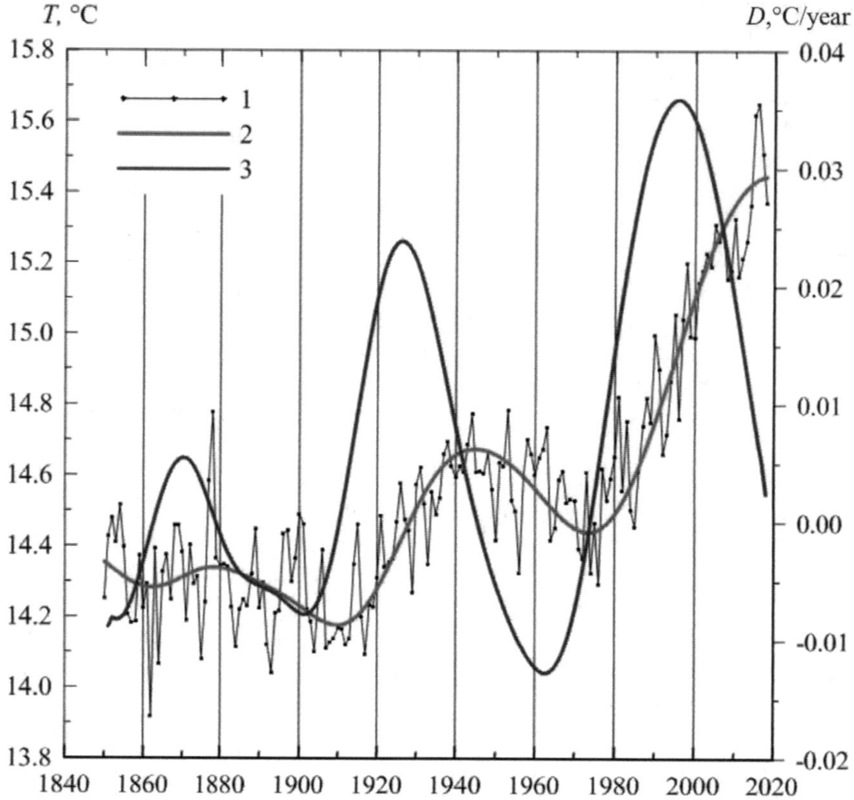

Figure 1. Multiyear curve of annual air temperature mean in the Northern Hemisphere, 1850–2018.
1 – basic series, 2 – LFC with a period exceeding 30 years, 3 – first-order LFC differences.

Table 1. Characteristics of annual air temperature mean in the Volga Federal Region.

Period, years	Mean value, °C	MSD, °C	Maximum, °C	Minimum, °C
1955–2018	3.49	1.04	5.49 1995	0.55 1969
1955-1999	3.14	1.00	5.49 1995	0.55 1969
2000–2018	4.34	0.47	5.33 2008	3.58 2011

Figure 2b illustrates the transition of the thermal regime from one state to another and linear trends of the AATM in the period being studied.

Thus, for the whole period the linear trend slope coefficient (LTSC) is equal to 0.32°C/10 years (the adjusted coefficient of determination R2=31%, the trend determination reliability is over 0.99). In the earlier period, from 1955 to 1999, the LTSC=0.23°C/10 years (R2= 5%, the reliability is 0.96). In the latest period, from 2000 to 2018, the LTSC=0.08°C/10 years (R2= 0, the reliability is 0.33). It means that no sizeable rise in the annual air temperature mean has been observed over the 21st century.

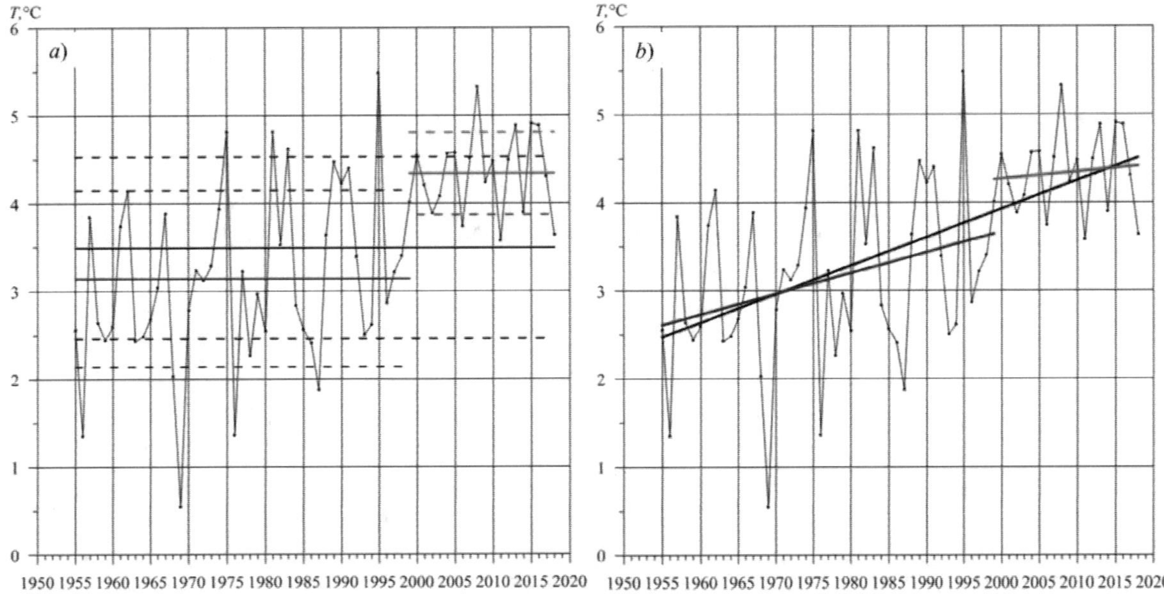

Figure 2. Multiyear curve of the basic statistical characteristics of the annual near-surface air temperature mean averaged for the territory of VFR (1955 – 2018), (a – full lines – mean values, dashed lines – mean-square deviation from the mean, b – linear trends). Black lines – characteristics of the whole period, blue lines – 1955 – 1999, red lines – 2000 – 2018.

Changes in the regional climate had a substantial impact on the time variation of intensive AT anomalies. Thus, in the 21st century (2000–2018) the number of negative AT anomalies drastically decreased, while the number of positive anomalies grew in all months of the year in comparison with the period between 1955 and 1999. Winters became warmer and snowier.

Let us study the thermodynamic processes in the higher atmospheric layers. Table 2 presents a vertical distribution of long-term air temperature means Av (°C) and linear trend slopes A (°C/year) on 51 isobaric surfaces. The data averaged for the NH both for January and July show Av temperature fall with altitude in the troposphere, its rise in the stratosphere, and a new fall in the mesosphere.

Table 2. Long-term temperature means Av(°C) and linear trend slope coefficients A(°C/year) presented on 51 isobaric surfaces for the period between 1979 and 2018.

P, (hPa)	H, km	Year (I-XII)				Summer (VI-VIII)				Winter (XII-II)			
		[a] Av	[b] Rms	[c] A	[d] R^2L	[a] Av	[b] Rms	[c] A	[d] R^2L	[a] Av	[b] Rms	[c] A	[d] R^2L
1000	0.1	16.17	0.32	0.0255	82	21.69	0.28	0.0212	77	10.52	0.36	0.0258	66
975	0.3	14.65	0.32	0.0248	81	20.19	0.27	0.0201	74	9.04	0.36	0.0256	66
950	0.5	13.38	0.30	0.0232	79	18.97	0.26	0.0185	68	7.83	0.35	0.0244	65
925	0.7	12.31	0.29	0.0222	77	17.92	0.26	0.0185	66	6.83	0.33	0.0233	64
900	1.0	11.27	0.28	0.0216	76	16.87	0.26	0.0188	66	5.89	0.32	0.0225	64
875	1.2	10.25	0.28	0.0213	75	15.78	0.27	0.0189	65	5.00	0.31	0.0221	64
850	1.5	9.22	0.28	0.0210	74	14.62	0.27	0.0186	64	4.13	0.31	0.0220	64
825	1.7	8.12	0.28	0.0206	72	13.37	0.26	0.0181	60	3.23	0.32	0.0220	62
800	1.9	6.93	0.27	0.0201	70	12.04	0.26	0.0178	58	2.24	0.32	0.0218	60
775	2.2	5.65	0.27	0.0198	68	10.63	0.27	0.0177	56	1.12	0.32	0.0213	57
750	2.5	4.29	0.27	0.0194	65	9.16	0.27	0.0176	54	-0.10	0.32	0.0208	54

700	3.0	1.30	0.27	0.0185	60	5.97	0.28	0.0172	49	-2.84	0.32	0.0195	47
650	3.6	-2.07	0.27	0.0177	54	2.43	0.28	0.0168	44	-6.01	0.31	0.0182	42
600	4.2	-5.76	0.26	0.0167	50	-1.38	0.28	0.0160	41	-9.58	0.30	0.0171	39
550	4.9	-9.70	0.26	0.0158	47	-5.35	0.28	0.0152	37	-13.52	0.29	0.0163	38
500	5.6	-14.14	0.27	0.0162	46	-9.74	0.29	0.0157	37	-18.00	0.29	0.0165	39
450	6.4	-19.22	0.28	0.0171	47	-14.74	0.31	0.0170	38	-23.09	0.31	0.0175	40
400	7.2	-25.09	0.30	0.0185	48	-20.63	0.33	0.0189	40	-28.88	0.33	0.0192	43
350	8.1	-31.82	0.31	0.0194	48	-27.56	0.35	0.0197	39	-35.36	0.34	0.0202	44
300	9.2	-39.29	0.30	0.0179	44	-35.53	0.34	0.0182	35	-42.36	0.34	0.0190	39
250	10.4	-46.94	0.28	0.0149	35	-44.07	0.31	0.0152	28	-49.27	0.34	0.0166	28
225	11.1	-50.55	0.26	0.0128	28	-48.17	0.31	0.0138	23	-52.42	0.36	0.0143	17
200	11.8	-53.96	0.25	0.0094	15	-52.08	0.30	0.0106	12	-55.34	0.39	0.0109	6
175	12.7	-57.41	0.24	0.0070	6	-56.10	0.31	0.0068	2	-58.28	0.40	0.0099	3
150	13.6	-61.18	0.26	0.0066	4	-60.43	0.32	0.0047	-2	-61.65	0.40	0.0136	11
125	14.8	-65.05	0.29	0.0007	-5	-64.51	0.37	-0.0034	-4	-65.46	0.38	0.0128	10
100	16.2	-67.90	0.35	-0.0153	22	-66.62	0.46	-0.0232	30	-69.06	0.31	-0.0012	-5
70	18.5	-65.96	0.64	-0.0412	54	-63.11	0.76	-0.0433	41	-68.66	0.65	-0.0379	43
50	20.6	-61.38	0.72	-0.0482	58	-58.33	0.84	-0.0502	45	-64.19	0.76	-0.0441	42
30	23.9	-55.83	0.75	-0.0531	65	-52.37	0.86	-0.0525	47	-59.01	0.78	-0.0478	48
20	26.6	-51.48	0.75	-0.0547	70	-47.53	0.81	-0.0498	47	-55.14	0.80	-0.0537	58
10	31.2	-44.21	0.47	-0.0200	20	-39.82	0.69	-0.0145	1	-48.33	0.63	-0.0176	6
7	33.6	-39.57	0.59	0.0035	-5	-35.16	0.90	0.0240	5	-43.62	0.81	-0.0087	-4
5	36.0	-32.96	0.92	-0.0494	35	-28.38	0.99	-0.0214	1	-36.98	1.26	-0.0804	52
3	39.7	-21.13	2.19	-0.1680	78	-15.64	2.33	-0.1741	73	-25.79	2.56	-0.1953	76
2	42.7	-14.03	2.04	-0.1534	74	-8.38	2.30	-0.1744	75	-19.22	2.23	-0.1622	69
1	48.2	-11.75	1.62	-0.0869	35	-7.55	1.73	-0.1119	53	-16.79	1.51	-0.0474	9
0.97	48.4	-11.88	1.62	-0.0858	34	-7.75	1.73	-0.1105	52	-16.88	1.52	-0.0449	7
0.78	50.2	-14.27	1.64	-0.0822	30	-10.82	1.70	-0.1017	45	-18.88	1.55	-0.0353	2
0.62	52.0	-17.93	1.69	-0.0757	23	-15.27	1.70	-0.0854	30	-21.98	1.66	-0.0295	-1
0.49	53.9	-22.60	1.77	-0.0647	14	-20.85	1.71	-0.0604	12	-25.80	1.88	-0.0245	-3
0.38	55.9	-28.02	1.89	-0.0430	2	-27.39	1.80	-0.0247	-3	-30.07	2.19	-0.0123	-5
0.29	57.9	-34.02	2.08	-0.0156	-4	-34.72	2.03	0.0146	-5	-34.56	2.43	0.0031	-5
0.22	60.1	-40.31	2.34	0.0193	-4	-42.62	2.43	0.0592	3	-39.08	2.59	0.0238	-4
0.16	62.3	-46.81	2.72	0.0609	2	-50.85	3.00	0.1107	14	-43.71	2.81	0.0495	-1
0.12	64.5	-53.23	3.22	0.1064	10	-59.07	3.76	0.1670	22	-48.36	3.19	0.0790	3
0.08	66.9	-59.27	3.71	0.1449	16	-66.91	4.56	0.2145	26	-52.84	3.53	0.1036	7
0.06	69.4	-64.73	3.98	0.1619	18	-74.00	5.05	0.2357	25	-57.05	3.68	0.1118	8
0.04	71.9	-69.74	4.07	0.1703	19	-80.30	5.23	0.2505	27	-61.30	3.70	0.1160	9
0.03	74.6	-74.68	3.98	0.1734	21	-85.67	5.29	0.2707	31	-66.24	3.59	0.1184	10
0.01	80.3	-82.59	3.43	0.1494	21	-91.82	4.48	0.2339	33	-75.21	3.23	0.1121	12

[a] Av – mean value, °C.
[b] Rms – mean square deviation, °C.
[c] A – linear trend slope coefficient, °C/year.
[d] R^2L – linear trend determination coefficient (%).

As one can see from Table 2, a marked annual cycle is registered at all levels. The largest amplitude of annual oscillations is observed in the troposphere and mesosphere. In the troposphere, on the whole, the temperature rises both in winter and summer (A>0). Thus, at the level of 500 hPa A=0.0165°C/year in winter and A=0.0157°C/year in summer. In the stratosphere, cooling is observed, as at the level of 10 hPa A= -0.0176°C/year in winter and A= -0.0145°C/year in summer. In the mesosphere, where the lowest air temperatures are registered (at the level of 0.01 hPa (80 km) the AT drops to -91.82°C in summer), the LTSCs become positive. At the level of 0.01 hPa A=0.1121°C/year in winter and A=0.2339°C/year in summer. Thus, we have obtained atmosphere stratification by behaviour of the thermal regime trend. Taken together, the LTSC distribution corresponds to the global warming theory, however, there are perceptible differences conditioned upon the radiation and dynamic factors between the year seasons and latitudinal zones.

The cooling tendency is more pronounced in the summer stratosphere than in the winter stratosphere, since during this period the vertical wave action is weak and no sudden stratospheric warmings (SSWs) typical for winters and resulting in the air warming in the polar zone occur. Furthermore, in summer at high altitudes the behaviour of the processes is more homogeneous, and owing to the ozone screen interaction with ultraviolet radiation the air masses warm up in the middle stratosphere, and the temperature trends vertically change their sign. In the mesosphere, the temperature rise is more noticeable in summer than in winter. The LTSCs in summer are larger than in winter. Seasonal heterogeneity is observed in the air temperature trend.

In the troposphere, the MSD has low values both in winter and summer (less than 0.4°C). In the stratosphere, the interannual oscillations intensify and reach their maximum in the mesosphere, where the MSD attains 5°C in summer, which is an evidence of its unstable behaviour.

To identify the geographical and seasonal peculiarities in long-term variations of the air temperature, the authors drew vertical sections for the multiyear curve of the first-order differences of LFCs with a period exceeding 10 years AT (°C/year) for the polar, temperate, and tropical zones (1979–2018). An analysis of the results demonstrates that the most sophisticated picture is observed in the January polar zone, where starting from a level of 10 hPa and higher significant interannual AT variations occur, and cooling and warming centres alternate, while in July the situation is less disturbed.

At the same time, the AT rise tendency prevails both in the winter and summer troposphere. In the temperate and tropical zones the section structures are similar: the temperature increment prevails in the troposphere in the seasons concerned, and starting from an altitude of 20 km and higher strongly pronounced quasi-cyclic AT oscillations prevail. The above-described phenomenon is especially noticeable starting from a level of 40 km, where considerable temperature drops registered between 1979 and 2000 give place to the temperature rise. In the latest period (2000–2018) in the upper atmospheric layers (at 60 km and higher) cooling is registered everywhere except for the winter polar zone.

The Asia-Pacific region stands out from the other regions of the temperate zone, since in winter in a layer of 10–40 km the air temperature there is higher by 6.5°C than in the neighbouring Atlantic-European and American sectors.

Two approaches were used to estimate the vertical correlations between the AT at different levels. In the first one, the correlation coefficients (r) were calculated between the AT of a concrete level and all upper levels. In the second approach, the calculations were made between the neighbouring surfaces, which allowed one to identify the geographical peculiarities of the correlation nature. The calculations performed have revealed that in the lower troposphere the AT correlations between the neighbouring levels throughout the Northern Hemisphere are close both in January and July (r>0.9), in the upper troposphere between levels of 300 and 250 hPa in the polar and temperate zones they considerably weaken in January (r~0.3), while in July the correlations are weak in the polar zone due to the influence of the tropopause. In the middle stratosphere (between the levels of 20 and 10 hPa), the correlations significantly weaken in January at the subtropical latitudes. However, in the polar zone, where in winter stratospheric warmings are registered and wave activity is high, the correlation is strong: r=0.9÷1.0, while in July, on the contrary, r has minimum values (~0.3) in the polar zone. Thus, spatial heterogeneity is manifested. At the higher stratospheric levels, the correlations between the surfaces strengthen all over the hemisphere (r=0.9). In such a way, the heterogeneous nature of the interlayer interaction has been established: in the troposphere the correlations are tight, in the tropopause region and in the layer of 25-45 km they weaken, and then strengthen again in the upper stratospheric layers and lower and the middle mesosphere (r>0) (Figure 3).

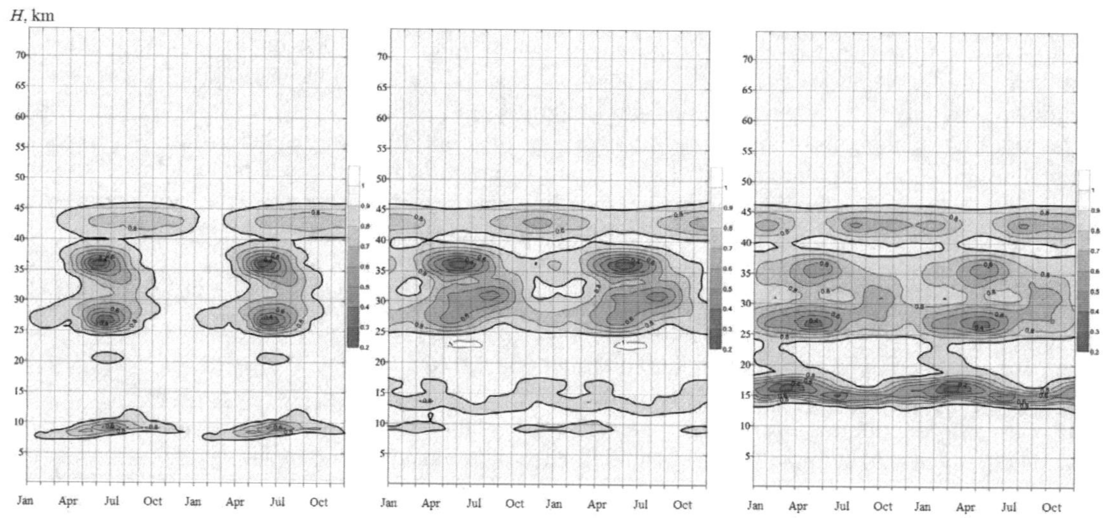

Figure 3. Temperature correlation coefficients (1979–2018) at neighbouring levels in a layer of 0.1 – 80 km in three latitudinal zones of the NH: polar zone (left), temperate zone (centre), and tropical zone (right).

It is a well-known fact that circulation processes play an important role in variations of the temperature conditions. For this reason, the correlation coefficients for 27 levels were calculated for winter, summer, and the whole year to reveal the correlation between the air temperature oscillations and the Arctic Oscillation (AO). It has been found that the correlation coefficients are nonsignificant ($r < 0.32$) if we correlate the temperature averaged throughout the NH with the AO index. The strongest correlation was identified for the polar zone in winter. At an altitude of 10 km r = -0.60, in summer a strong correlation with the AO index (r = 0.32) is detected at the levels of 11-12 km and 30 km. In the temperate zone r = 0.56 in the upper troposphere in winter. It is well-known that when planetary Rossby waves propagate from the troposphere to the stratosphere, winter stratospheric warmings occur, resulting in the circumpolar cyclone breakdown. As the AO manifests itself mainly in winters, the correlation coefficients for summer turned out to be nonsignificant.

Long waves are referred to the most important components of the atmosphere dynamics [8]. There are waves propagating to the east (E), west (W), and also stationary waves (S). Using the data about the AT at the grid points 1×1° in a layer of 1000–1 hPa, the authors calculated the Hayashi wave spectra characteristics for the zonal waves numbered 1 through 10 for November – March, 1979–2016. Plotted thereafter were the height-latitude sections of the coefficients of correlation between the E, W, S wave intensity and the AO index. It has been revealed that height-latitude distributions of the correlations between the components and the AO index significantly differ for W and E waves. The strongest positive correlations were found between E waves and the AO index in the troposphere of the temperate latitudes and the major part of the stratosphere at the extra-tropical latitudes, with maximum values (0.7–0.8) reached at 55–60°N in the lower stratosphere. The correlation distributions for W waves are of the opposite nature, to a certain extent: the correlations are negative (-0.7-0.8) in the troposphere to the north of 60°N and positive (0.7–0.8) in the troposphere of the low latitudes.

The estimates obtained are consistent with the fact that during the positive AO phase E waves develop, the zonal circulation strengthens, and the stratosphere becomes colder. During the negative AO phase the zonal circulation weakens, and the developed W waves promote the wave energy transmission to the stratosphere and foster the stratospheric warmings.

Since the radiation and circulation factors, along with the underlying surface state, refer to the key climate forcing factors, to assess the impact of the underlying surface (land, ocean) on the thermal regime of the atmosphere the authors calculated the air temperature characteristics over the land and ocean. These calculations demonstrate that in winter warming is registered in the troposphere and over the land and ocean, while in the stratosphere cooling is observed. In the lower troposphere it is warmer over the land than over the ocean in summer. In summer the linear trend slope coefficient over the land significantly exceeds that over the ocean. Thus, in the near-surface layer over the land the LTSC is equal to 0.035°C/year, while over the ocean it is 0.011°C/year, which means that the warming is more noticeable over the land than over the ocean. In the stratosphere the nature of the underlying surface has no impact.

Taking into account the significant role played by the ocean in the climate system, as well as its impact on the weather and climate, in this paper the authors analysed asynchronous correlations of long-period oscillations of the annual air temperature mean (AATM) in Moscow (1879-2016), Kazan (1854-2016), and Tomsk (1875-2016) with oscillations of the annual ocean surface temperature mean at points of the World Ocean geographic grid 2x2° (latitude x longitude) plotted for 1854-2016.

Distant relationships were evaluated by the best (maximum) asynchronous correlation coefficients between the OST at each individual geographical grid point and the AT at the stations with an AT delay of 0 to 36 years. At each studied point of the series of asynchronous correlation coefficients, the maximum coefficient was picked out, and the OST oscillation advance was fixed at the point relative to the subsequent AT oscillations at each of the stations. The ultimate correlation shift was set considering the fact that according to [15] the propagation time for the temperature variations in the ocean within the hemisphere is 30 years.

The study has shown that long-period AT oscillations in Eurasia are, to a great extent, the consequence of global changes in the climate system, whose sources can be found in a special region of the atmosphere interaction with the Southern Ocean in the area of the Antarctic circumpolar current of westerly winds. This assumes that the long-range transport of the temperature oscillations is performed by the World Ocean's water cycling system, known under the name of Broecker conveyor [4].

The delay in the air temperature oscillations at the stations is 30-35 years in Moscow and Kazan, and 35-45 years in Tomsk. The intensity of atmospheric interaction with the Southern ocean is strengthened by natural physico-geographical conditions of the circumpolar ocean current.

The detected delay in the fluctuation transmission has made it possible to derive asynchronous regression equations, allowing one to calculate long-period prognostic temperature oscillations at stations for the coming decades. Figure 4 presents results of a long-term AATM forecast, according to which in 2019–2025 the temperature in Kazan will be lower than the mean value (5,1°C) by 0.3–0.5°C, in 2028–2037 it will rise above this norm by 0.4–0.5°C, and by 2045 the AT will become 0.3 – 0.4°C lower than the norm again.

IOP Conf. Series: Earth and Environmental Science **386** (2019) 012003 doi:10.1088/1755-1315/386/1/012003

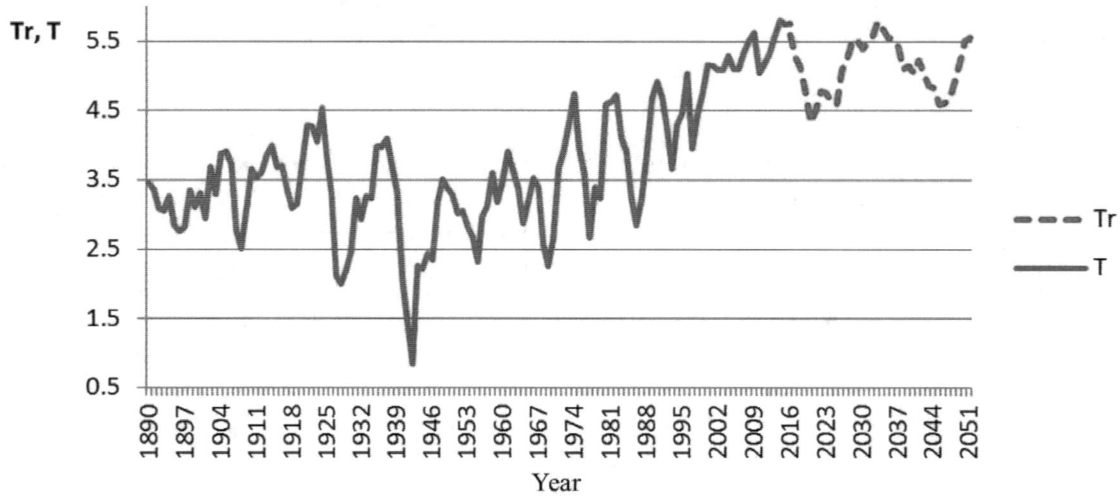

Figure 4. Annual temperature mean in Kazan, *T*, according to observation data for 1890-2016 (triennially smoothed) and temperature in Kazan calculated by the regression equation for the period until 2051 (Tr).

Conclusions

A trend towards a rising surface average annual air temperature in the Northern Hemisphere has been revealed. In the most active phase (from the 1970s to 2018) it increased by 0.7°C.

A quasi-60-year variation was detected in the long-term Northern Hemisphere's average annual air temperatures, whose statistical significance cannot be estimated due to the short observation range (169 years).

In the Volga Federal Region at the beginning of the 21st century, the average annual temperature increased by 1.2°C, with a minimum value rising sharply from 0.55°C to 3.58°C. The contribution of global factors to the regional temperature change in winter was 37%, and in summer 23%.

According to the above linear trends in the average annual temperature, its growth slowed greatly from 2000 to 2018 compared to the period from 1955 to 1999 (the LTSC decreased from 0.23°C/10 years to 0.08°C /10 years).

On average, in the Northern Hemisphere the linear trend slope coefficient is positive in the troposphere (1000-200 hPa) both in winter and summer; starting from a surface of 100 hPa and up to 0.5 hPa the stratosphere cools (LTSC < 0). According to the ERA5 data, in the mesosphere from a level of 0.1 hPa a warming is observed again (with a significant temperature rise from 1995 to 2015).

The strongest air cooling in the stratosphere is observed in the layer of 35 to 40 km in winter and in a layer of 35 to 45 km in summer.

The influence of the underlying surface type (land, ocean) is powerful up to a surface of 200 hPa. At the same time, in the troposphere the air temperature increase over the land significantly surpasses its increase over the ocean (in July the LTSC is equal to 0.035°C/year and 0.011 °C/year, respectively). From year on year these differences persist (0.037 and 0.017 °C/year).

According to the above analysis of the first-order differences of the LFCs with a period exceeding 10 years, in the upper stratosphere and lower mesosphere the temperature rise or drop centres are registered in winter with a cyclicity of about 10 years.

A correlation analysis revealed a heterogeneous nature of the statistical correlations in the temperature field: the correlations between the layers sharply weaken in the region of tropopause both in winter and summer. In the lower stratosphere, the vertical correlations in the temperature field strengthen, and in the layer from 25 to 45 km the thermal processes are weakly correlated. According to the ERA5 data, a strengthening of the correlation is registered in the mesosphere.

In the summer period, negative correlations set in between the troposphere and lower stratosphere, which is an evidence of antiphase nature of the temperature variation.

According to the correlation analysis, in winter the Arctic Oscillation has the most powerful influence on the near-surface tropospheric layer (with a temperature increase) and the layer from 7 to 3 hPa in the stratosphere, where the temperature decreases.

During the positive AO phase, E waves develop, zonal circulation strengthens, and the stratosphere becomes colder.

During the negative AO phase, the zonal circulation weakens, and the developed W and S waves promote wave energy transmission to the stratosphere and foster stratospheric warmings.

A wave activity intensification has been discovered in the tropical troposphere and stratosphere (E, W, S waves), and the upper stratosphere (E, S waves) of the NH.

The distant relationships (with a delay of 30 to 45 years) identified between the regional ocean surface temperature variations in the area of Antarctic circumpolar current of westerly winds (a region near the Drake Strait) permitted running the prognostic AATM model until 2051.

Acknowledgements
We thank the Copernicus Climate Change Service (C3S) (2017): ERA5: Fifth generation of ECMWF atmospheric reanalyses of the global climate. Copernicus Climate Change Service Climate Data Store (CDS), https://cds.climate.copernicus.eu/cdsapp#!/home
This research was supported by the Russian Foundation for Basic Research (grants 18-05-00721 and 18-45-160006).

References

[1] Vargin P N and Volodin E M 2016 Analysing the simulation of dynamic processes in the stratosphere with a climatic model built in Marchuk Institute of Numerical Mathematics, Russian Academy of Science *Izvestiya RAN. Atmospheric and Ocean Physics* **152** 3-18

[2] Dymnikov V P, Lykosov V N and Volodin E M Earth system modelling 2015 *Proceedings of the VII All-Russian Meteorological Congress* (Saint-Petersburg) 43-65

[3] Zhadin E A, Zyulyaeva Y A and Volodin E M 2008 Correlations between year-to-year variations of stratospheric warmings, tropospheric circulation and ocean surface temperature in the Northern Hemisphere *Izvestiya RAN. Atmosphere and Ocean Physics* **44** 641-53

[4] Broecker W S 1991 The great ocean conveyor *Oceanography* **4** 79-89

[5] Brohan P *et al* 2006 Uncertainty estimates in regional and global observed temperature changes: A new dataset from 1850 *J. Geophys. Res.* **111** 1-35

[6] Fahrutdinova A N, Perevedentsev Y P, Guryanov V V and Kulikov V V 2001 Dynamical processes and correlations at midlatitudes in the lower and middle atmosphere *Advances in Space Research* **27** 1667-72

[7] Fahrutdinova A N, Guryanov V V and Korotyshkin D V 2003 Structure of rotational spectra of vectorial wave fields in the middle atmosphere *Proceedings of SPIE* **5027** 248-57

[8] Guryanov V V, Eliseev A V, Mokhov I I and Perevedentsev Y P 2018 Wave Activity and its Changes in the Troposphere and Stratosphere of the Northern Hemisphere in Winters of 1979-2016 *Izvestiya RAN. Atmosphere and Ocean Physics* **54** 114-26

[9] Jones P D *et al* Surface air temperature and its variations over the last 150 years 1999 *Rev. Geophys.* **37** 173-99

[10] Perevedentsev Y P, Shantalinskii K M and Vazhnova N A 2014 Spatiotemporal variations of major parameters of temperature and humidity regime in the Volga Federal District *Russian Meteorology and Hydrology* **39** 228-39

IOP Conf. Series: Earth and Environmental Science **386** (2019) 012003 doi:10.1088/1755-1315/386/1/012003

[11] Perevedentsev Y P and Shantalinskii K M 2014 Estimation of contemporary observed variations of air temperature and wind speed in the troposphere of the Northern Hemisphere *Russian Meteorology and Hydrology* **39** 650-59
[12] Perevedentsev Y P, Vasilev A A, Shanntalinskii K M and Guryanov V V 2017 Long-term variations in surface air pressure and surface air temperature in the Northern Hemisphere mid-latitudes *Russian Meteorology and Hydrology* **42** 461-70
[13] Ermolaev O P, Kotlyakov V M, Perevedentsev Y P and Selivanovskaya S Y 2009 Environment and sustainable development of the regions new methods and technologists of studies (review of the conference. Kazan university. May 2009) *Izvestiya Akademii Nauk. Seriya Geograficheskaya* **6** 119-22
[14] Vereshyagin M A, Perevedentsev Y P, Shantalinsky K M and Tudriy V D 2004 Factor analyses of perennial dynamic of the global thermal regime of the near-surface air layer *Izvestiya Akademii Nauk. Seriya Geograficheskaya* **5** 34-41
[15] Wyatt M G, Kravtsov S and Tsonis A A 2012 Atlantic multidecadal oscillation and Northern hemisphere's climate variability *Climate Dynamics* **38** 929-49

A study of feedbacks and the formation of climate trends in the Arctic climate system

G Platov[1,2,4], V Krupchatnikov[1,2,3] and I Borovko[1]

[1] Institute of Computational Mathematics and Mathematical Geophysics, SB RAS, Novosibirsk, prospect Lavrentieva 6, Russia
[2] Novosibirsk State University, 1, Pirogova str., Novosibirsk, 630090, Russia

[3] Siberian Regional Hydrometerological Research Institute, 30, Sovetskaya str., Novosibirsk, 630099, Russia

[4] Platov.G@gmail.com

Abstract. The role of the ocean in the response of the climate system to an increase in the atmospheric CO_2 concentration is investigated by using a system of numerical models, ICMMG-PlaSim. The results of this study are summarized as follows: a) the ocean, to some extent (up to 20%), contributes to the increase in the annual mean state and to the decrease in the amplitude of seasonal oscillations (by 2-3%), which ultimately leads to insignificant changes in the summer period and to a significant mitigation of winter, b) the ocean stabilizes the annual mean state of the Arctic oscillation, making it practically unchanged with increasing CO_2 concentration but, at the same time, contributes to the significant increase in the amplitude of the seasonal cycle of this oscillation, c) the ocean enhances the temperature (or thermal) component of the seasonal variation associated with the appearance of additional areas freed from ice cover, with an additional average increase in the temperature of the atmosphere at the ice edge. Besides, the ocean enhances the seasonal oscillations of this component, so that the summer manifestations become much stronger, d) our tests have revealed that the role of the Arctic dipole under global warming is insignificant. These conclusions, though, may undergo significant changes under a more detailed consideration of carbon cycles in the atmosphere, ocean, and land.

1. Introduction

The problem of future climate change in the Arctic is associated with a number of scientific problems, as well as socioeconomic and environmental ones. The projected effects of climate change on the natural environment of the Arctic are large and can have a significant reverse impact on the global climate, which means the global importance of climate change in the Arctic [1]. Most of the supposed feedbacks that are formed due to climate change are concentrated in the Arctic and are still poorly understood. The most significant consequences are brought about by changes in the cryosphere, in particular, a reduction in the area and volume of floating ice. The specificity of the Arctic region: the atmospheric polar vortex, the three-phase boundary layer, radiation-cloud connections, the formation of the thermohaline structure of the Arctic Ocean and North Atlantic, and a number of other features make the Arctic an extremely complex and interesting object of study.

Much attention of scientists is given to improving forecasts of climate change on a time scale from decades to centuries based on scenarios of carbon emissions, sulfur emissions, changes in land use and other anthropogenic (technogenic) factors affecting the climate. Closely related to this are studies devoted to explaining the observed climate change by various types of impacts, both man-made and natural.

The most significant climate changes occur when the global energy balance between the incoming energy from the Sun and the outgoing heat from the Earth is disturbed. There are a number of natural mechanisms that can upset this balance, for example, fluctuations in the Earth's orbit, changes in the circulation of the ocean, and changes in the composition of the Earth's atmosphere. The latter, as is obvious, is a consequence not only of natural processes, but also of anthropogenic pollution caused by emissions of greenhouse gases and aerosols. Each year, the absorption of CO_2 by the ocean is approximately equal to a quarter of what is released annually into the atmosphere as a result of human activity. This is primarily controlled by the difference between the CO_2 concentrations in the atmosphere and the ocean with slight fluctuations from year to year and due to changes in the circulation of the ocean and in the biology of the Earth. Due to the fact that anthropogenic emissions of CO_2 and other greenhouse gases continue to grow, and also because they can remain in the atmosphere for decades or centuries (depending on the gas), we will have a warmer climate in the future. The IPCC [2] predicts an increase in global temperature by 2-6°C by 2100, while the temperature in the polar regions may rise even faster.

Melting due to global warming is expected to reduce the size and extent of polar ice caps, even after allowance for the accumulation of more snow and ice on ice sheets due to increased precipitation. The melting of polar ice and ground glaciers is expected to contribute to an increase in the sea level of approximately 0.2–1.5 m predicted by the IPCC in the 21st century [2], and a sudden collapse of the West Antarctic ice sheet would raise the sea levels by 5–6 m due to further warming.

Recently, a number of assessments of climate impact on the polar regions have been carried out, including a summary report on snow, water, ice, and permafrost in the Arctic [3, 4], on the climate of the Antarctic [5, 6]; the results of the International Polar Year are presented in [7]. According to these studies, the surface temperatures in the Arctic have risen substantially since the mid-20th century, and the warming rate exceeds the global one. In the recent decades, the sea-ice extent of the summer minimum has decreased significantly, and the Arctic Ocean is predicted to become almost ice-free in summer during this century. The duration of the snow cover and its depth are decreasing in North America and increasing in Eurasia. Since the late 1970s the permafrost temperature has increased by 0.5 to 2°C. Since the 1950s the oceans absorbed 93% of the excess heat in the Earth's system, which accumulated as a result of an increase in the concentration of greenhouse gases in the atmosphere. A significant increase in the heat content was observed in the upper 2000 m of the ocean since the 1960s, when the surface layers of the ocean became warmer by about 0.7±0.1°C from 1900 to 2016.

A reduction in the sea ice affects the energy balance due to changes in the surface albedo and as a result of a reduction in the insulating layer between the ocean and the atmosphere [8]. These manifestations mean that the loss of sea ice initiates a feedback, which directly contributes to the Arctic amplification of near-surface warming [9]. In addition, the loss of Arctic sea ice was associated with climatic changes in the lower latitudes through changes in the oceanic and atmospheric circulation [10]. For example, large-scale atmospheric circulation in winter due to a reduction in the sea ice in September is characterized by an increase in the meridional structure of the flow resembling the negative phase of Arctic Oscillations (AOs), which supports cold winters in northern Eurasia. Rossby waves with anomalously large amplitudes penetrate the stratosphere in February and weaken the stratospheric polar vortex, which subsequently gives rise to negative AO anomalies [11]. However, the remote influence is still difficult to interpret, and its mechanisms are still under investigation.

In this paper, we are going to find out the role of the ocean in shaping climate trends associated with an increase in greenhouse gas concentrations, using CO_2 as an example, and whether the ocean is a deterrent or, conversely, it strengthens the atmospheric trends.

In our previous work [12], which studied atmospheric forcing trends and their role in reducing the Arctic ice based on EOF decomposition, we found that among the first non-degenerate EOF decomposition modes (Figures 1, 3, 5, and 7) the third mode is a temperature mode, that is, it is absent

IOP Conf. Series: Earth and Environmental Science **386** (2019) 012004 doi:10.1088/1755-1315/386/1/012004

during decomposition using only surface pressure and appears only when a temperature field of 10 m is included in the state vector. An analysis of the spatial pattern of this EOF mode shows that its region of action coincides with zones of additional opening from the ice cover in the recent decades in summer and, therefore, it possibly arises as a result of the climatic trend. It would be interesting to find out the further role of this mode and of all the others in the formation of atmospheric and ocean dynamics in the 21st century.

2. Estimation of radiation time scales

If we assume that radiative transfer is the only process occurring in the Earth's atmosphere, then the surface temperature will be determined solely by the total radiation in the upper part of the atmosphere. There are two fundamental time scales that determine how the atmosphere transfers heat. The first one is associated with atmosphere dynamics. The other one is the radiation time scale τ_{rad}. It is the time taken by the atmosphere to reduce its temperature perturbation e-fold from the original value using radiation.

2.1. Atmosphere and terrain

To get started, we get a simple measure of the planetary radiation time scale. Consider an atmospheric column with surface pressure p_s. If there is no solar heating, the atmosphere will be cooled by thermal radiation. Assume that the entire column is initially in radiation equilibrium. Then at the terrain surface we have

$$(1-\alpha)\,S_0/4 + B = U, \tag{1}$$

where the first term is the mean incoming solar radiation (S_0 is the solar constant, α is the surface albedo), the second term is downward atmosphere thermal radiation, $U = \sigma T_e^4$ is the surface thermal radiation (T_e is the surface temperature). For a column of atmospheric air, under the condition of hydrostatics, one can write the equation of the energy balance depending on time (see, for example, [13])

$$\frac{c_p p_s}{g}\frac{dT_a}{dt} = \varepsilon U - 2B, \tag{2}$$

where ε is the optical transmissivity of the atmosphere. Suppose that the temperature of the atmosphere can be represented as $T_a = T_{a0} + \Delta T$, assuming that the temperature deviation from the temperature of radiation equilibrium is small. Substituting (1) into (2) to express U and using the Taylor form in a first approximation, we obtain the following equation:

$$\frac{c_p p_s}{g}\frac{d\Delta T}{dt} = \varepsilon\sigma T_e^4 - (\varepsilon-2)\varepsilon\sigma(T_{a0})^4 - 4(\varepsilon-2)\varepsilon\sigma(T_{a0})^3\Delta T\,. \tag{3}$$

The first two terms in the right-hand side of (3) represent the equilibrium state and, therefore, give a total of zero. Finally, we have

$$\frac{d\Delta T}{dt} = -\frac{4(\varepsilon-2)\varepsilon\sigma g(T_{a0})^3\Delta T}{c_p p_s} \tag{4}$$

giving exponentially vanishing temperature perturbation with the time scale of the e-fold decrease,

$$\tau_{erad} = \frac{c_p p_s}{4(\varepsilon-2)\varepsilon\sigma g(T_{a0})^3} \tag{5}$$

Assuming that T_{a0} = 300K, ε = 0.6, and c_p = 1000 J·(K kg)$^{-1}$, the scale of radiation equilibrium in (5) will be about 18 days.

2.2. *Upper ocean*

A simple energy balance of the system, taking into account the heat capacity of the upper ocean layer, can be written as

$$C_s \frac{d\Delta T_s}{dt} = \Delta Q + \lambda_f \Delta T_s, \quad (6)$$

where ΔT_s is the sea surface temperature perturbation, ΔQ is the radiation forcing, C_s is the heat capacity of the upper ocean layer (for the 200-m layer it is approximately equal to 8.36 10^8 J·K^{-1}·m^{-2}), and λ_f is the feedback parameter (approximately equal to -1.0 W·K^{-1}·m^{-2}). Thus, an estimate of the time scale for the transition to an almost equilibrium state $\tau_{srad} = -C_s \cdot \lambda_f{}^{-1}$ leads to a value of approximately 22 years.

3. Description of the numerical model and testing experiments

The purpose of the present research is to elucidate the role of the oceanic and ice response to climate changes in the atmosphere and on land in the further development of these changes. According to the IPCC scenarios [2] we will associate climate change with an increase in the atmospheric CO_2. A similar approach is used quite often when trying to reproduce the features of climate change, in particular, to assess the sensitivity of climate models to an increase in the CO_2 concentration in the atmosphere (see, for example, [14] and the references therein).

Generally speaking, of the atmospheric gases water vapor is the dominant greenhouse gas. If H_2O were the only greenhouse gas present, the greenhouse effect of a clear-sky atmosphere in the mid-latitudes, measured by the difference between the infrared flux on the surface and in the upper atmosphere, would be about 60-70% of this value taking into account all gases. If only CO_2 were present, the corresponding value would be about 25%. Although, due to the overlap between the absorption bands of different gases, the calculation of percentages is not strictly additive.

The additional heat fluxes in 2025 relative to the pre-industrial value calculated according to the IPCC scenario A ("business-as-usual case"), as well as their percentage contribution for various greenhouse gases, are as follows: CO_2 - 2.9 Wm^{-2} (63 %); CH_4 - 0.7 Wm^{-2} (15%); N_2O - 0.2 Wm^{-2} (4%); CFCs and HCFCs - 0.5 Wm^{-2} (11%); stratospheric H_2O - 0.2Wm^{-2} (5%). That is, the total additional heat flux is approximately 4.6 Wm^{-2}, which corresponds to the effective amount of CO_2 approximately twice the pre-industrial value.

As a research method we will use numerical modeling, and analyze the results of numerical experiments.

3.1. *ICMMG-PlaSim climate system model*

The PlaSim-ICMMG [15] system model was developed as a modular structure that allows one to create a range of models of the Earth system of intermediate complexity by choosing various options to describe the various components of the climate and carbon cycle. The model is able to integrate at different time scales. The modularity of its structure facilitates the connection of complex components with an increase in the computing power.

The system is based on the original version of the PlaSim model [16], in which the original modules for calculating the characteristics of the ocean and sea ice were replaced with more detailed models of the World Ocean (model SibCIOM [17, 18]) and ice dynamics (model CICE-3 [19, 20]). As part of the multi-modularity, an additional opportunity was realized, in comparison with PlaSim, to use an individual domain grid for each of the modules. An atmospheric module using the spectral approach is more convenient to describe in a spherical coordinate system (longitude, latitude) on isobaric surfaces, while it is undesirable for an oceanic module to have a singularity in the region of the north pole associated with these coordinates and, therefore, a three-pole coordinate system is used

[21] with poles shifted to the continents. In the ice module, it is undesirable to use a grid covering the entire oceans, it is sufficient to confine the model domain to the subpolar and polar regions, discarding latitudes where there is almost no ice, and instead improve the resolution in those regions where ice-snow cover is really observed.

The interaction of the system modules is carried out by a separate process, a coupler, in the framework of a multitasking parallel system (MPI), which provides the gathering of necessary information about the state of each of the modules at each time step and the scattering of the forcing fields for all components of the system on a spatial grid that is convenient for each module. The coupler allows the replacement of one module with another. In particular, we will take advantage of the fact that the real numerical model can be replaced by the so-called data model, that is, a module that is not a model of any component. Such a module emulates the operation of this component, and it is a program that sends the appropriate data set to the coupler at the right time, for example, read from files generated from observational data or from the results of other numerical experiments.

The atmospheric module of the system sets the atmospheric average CO_2 concentration as a parameter. This parameter plays an important role in calculating the optical properties of the atmosphere and in the formation of radiation fluxes. We will use this opportunity in numerical experiments simulating an increase in atmospheric CO_2 concentration similar to IPCC scenarios.

3.2. Numerical tests

In order to identify the potential role of the ocean and sea ice in the formation of climate trends resulting from atmospheric effects, the following series of numerical experiments with the PlaSim-ICMMG climate model is proposed:

test A Integration of the model at a fixed level of CO_2 concentration, approximately equal to the value in the mid-90s, 360 ppmv, with the full interaction of all modules before reaching the quasi-stationary mode, when most of the variability is reduced to seasonal changes and when every next year approximately repeats the previous one, it was about a hundred years, and in the future we will refer to this period of time as 1900-2000.

test B Restart of the complete system run from the quasi-stationary state of test A, with an abrupt increase in the CO_2 concentration value to the level of 450 ppmv. It seems to be a very close perspective of future changes in the atmosphere conditions since in May 2019 a CO_2 concentration of 415.26 ppmv was recorded at the Mauna Loa Observatory in Hawaii by researchers from the Scripps Institution of Oceanography. The value of 450 ppmv roughly corresponds to the mildest of the IPCC scenarios, RCP 2.6. This test is to show the reaction of the simulated climate system to a sudden change in the CO_2 concentration in the atmosphere.

test C Subsequent integration of the system, starting from the state of quasi-stationary state of test A, with an abrupt increase in the CO_2 concentration to 450 ppmv as in test B, but the ocean-ice system continues to reproduce the quasi-stationary state, which was recorded during the last year cycle of test A. This test is to show the reaction of the incomplete climate system to a sudden change in the CO_2 concentration in the atmosphere, the ocean and ice responses to this change are excluded from this test.

The time in tests B and C will be conditionally counted from 2000, the time of completion of the test A.

Thus, comparing the results of the three experiments, we can evaluate what processes arise from an increase in CO_2 in the coupled system, and from a comparison of the last two we can assess whether the ocean with sea ice (hereafter referred to as the ocean) enhances or weakens emerging trends, and is also possibly responsible for the formation of some of them. This study considers only the fast mode of atmosphere response, which is about 20 days' time scale, but we can also trace the first tendencies of the slower mode associated with the upper ocean response. It should also be borne in mind that in this study we do not consider changes in the absorption of carbon dioxide by the ocean, believing that in a first approximation this process can be neglected.

4. Results

In [12], using the EOF analysis of atmospheric forcing, it was found that in the case where the state vector includes the surface pressure, temperature, and horizontal velocity components at a level of 10 m and in the absence of seasonal filtering, four non-degenerate modes can be distinguished: the first is associated with seasonal variations, the second with the AO phase [22], the third is the temperature mode associated with the introduction of the surface temperature field into the state vector, and the fourth representing the so-called Arctic dipole [23].

Figure 1 shows the structural function of the first mode in the form of a distribution of the surface pressure field and temperature at a height of 10 m. It can be seen that with the onset of the positive phase the temperature over the continents increases, and in the case of the negative phase, on the contrary. Changes over the oceans are minimal, but it can be seen that the temperature in the surface layer of the Arctic Ocean behaves more similarly to the continental part of the region. A similar picture can be seen in the surface pressure field; however, in contrast to the temperature above the ocean, significant variations are found over the oceans (except for the Arctic Ocean) acting in antiphase to variations over the continents.

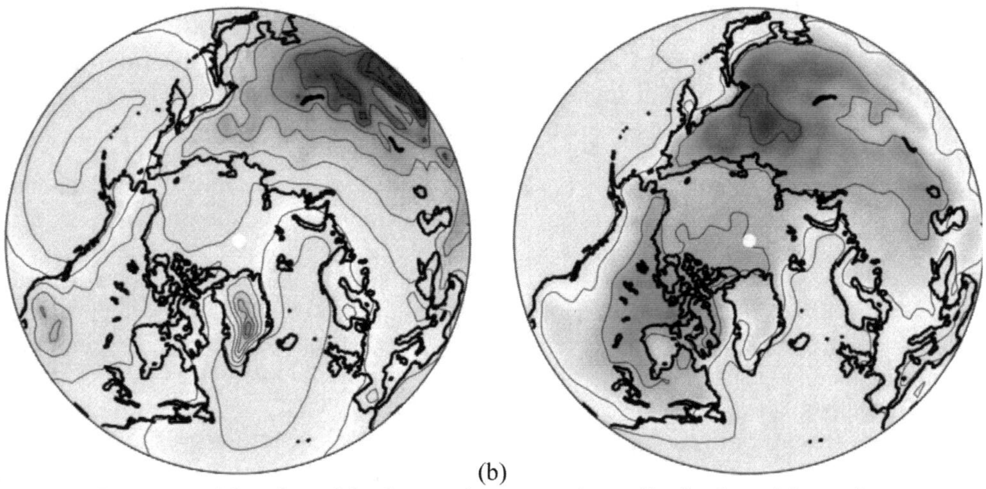

(a) (b)

Figure 1. The structural function of the first mode presented as a distribution of the surface pressure field (a) and temperature at a height of 10 m (b).

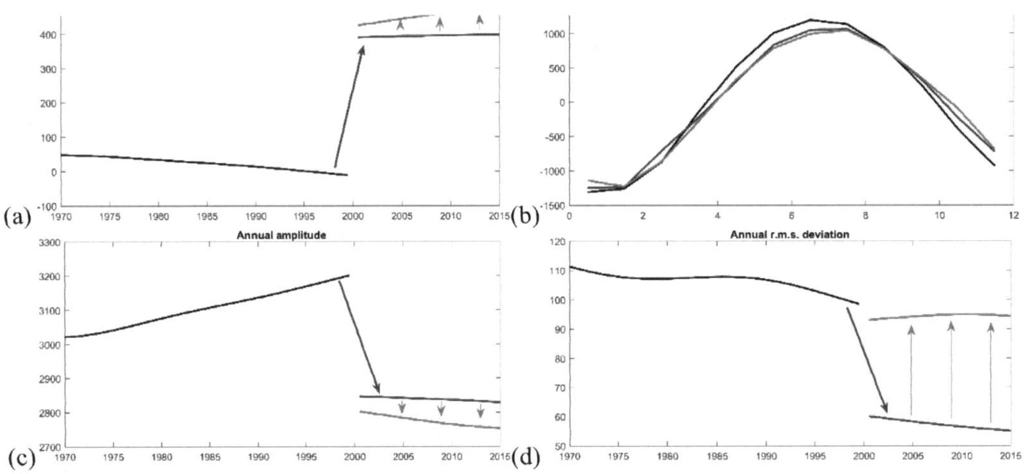

Figure 2. PC time series of the first EOF mode of decomposition of atmospheric forcing: (a) annual average values, (b) seasonal cycle averaged over years, (c) amplitude of seasonal cycle, (d) standard deviation from the seasonal variation. The black line is the results of test A, the blue one is test C, and the red one is test B. The blue arrow shows a jump in the transition to the mode of increased CO_2 concentration, and the red ones emphasize the correction introduced by the presence of active ocean.

When analyzing the trends of the first mode from 1980 to 2010, [12] noted an increase in the average value of the PC (principal component) time coefficient and a simultaneous decrease in the amplitude of the seasonal cycle, which in terms of temperature means the trend of warmer winters, with a relatively stable summer period. The results of tests A, B, and C are presented in Figure 2 as annual average values of the PC time series for the first EOF mode of the state vector. It can be seen that the trends associated with an increase in atmospheric CO_2 concentration coincide with those noted in [12]. The average value of the time-depending coefficient of EOF at the end of test A turned out to be near zero, and with increasing concentration it increased noticeably in test C. In the case of inclusion of an active response from the ocean (test B in comparison with test C), this trend even increases somewhat (up to 20%). According to the results of [12], the seasonal variability of the first mode (Figure 2b) describes about 99% of the total variability of this mode. The amplitude of the seasonal cycle changes in the direction of its decrease (Figure 2c), and the jump is about 10% of its value, with the ocean adding another 2-3%. In terms of the values of the decomposition coefficient, a decrease from the level of 3200 is by 300-400 units, while the increase in the average annual value is also in this range. Thus, as before, we state that the summer state has changed insignificantly, and the winter one has noticeably approached the summer one. The standard deviation from the seasonal cycle of Figure 2b with an amplitude of Figure 2c is shown in Figure 2d, from which it can be seen that in the case of a fixed ocean the deviations from the seasonal course decrease by almost 2 times with an increase in CO_2; however, in the case of an actively interacting ocean they remain approximately at the same level. Thus, the ocean supports intra-seasonal variability.

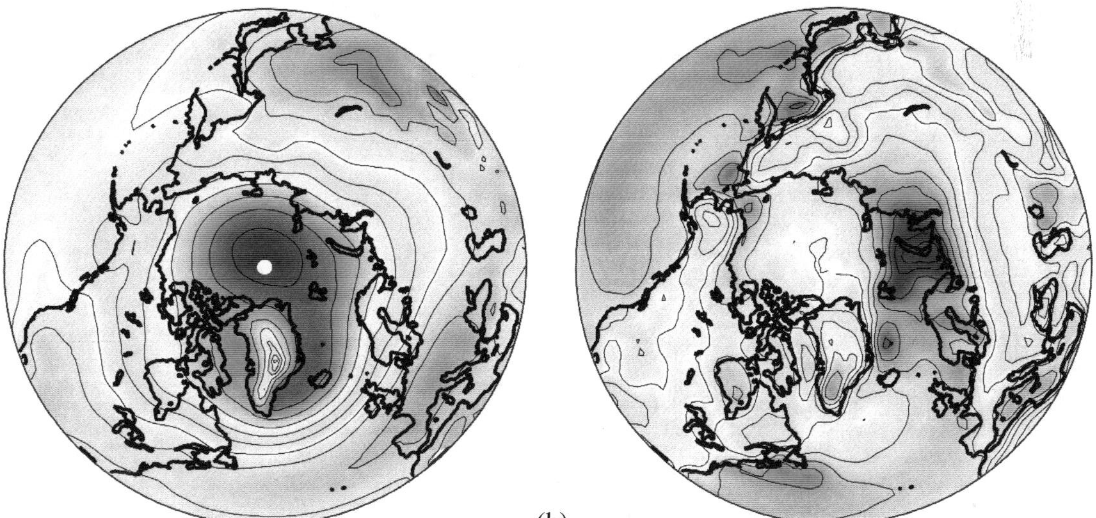
(a) (b)

Figure 3. The structural function of the second mode presented as a distribution of the surface pressure field (a) and temperature at a height of 10 m (b).

For the second mode (Figure 3), the previously noted trends of the period 1980–2010 consisted in a certain decrease in the average annual value of the time-dependent coefficient with a small increase in the amplitude of seasonal fluctuations, while the seasonal variability itself describes only the 20% variability of the second mode. The first means the strengthening of the positive phase of the AO, which was actually observed in the 90s [24]. However, Figure 4 shows that the atmosphere under the condition of a fixed ocean demonstrates the opposite tendency of an increase in the average annual value and a weakening of the AO, while the ocean strengthens the positive phase of this oscillation.

In the classical view, the AO is the first mode of the EOF decomposition of the pressure field averaged during the period of January-March (or December-February). If atmospheric forcing used in [12] is used to construct the eigenfunction of the AO, then having obtained it and comparing it with our modes, one can notice that the second mode is 70% correlated with the eigenfunction of the AO, but at the same time explains only 6% of the variability of atmospheric forcing, while the first mode correlates with the AO only by 5%, but at the same time explains 76% of the variability. Thus, the insignificant tendency of the average state of the second mode toward the amplification of the AO is excessively compensated by the reverse action from the trends of the first mode. As a result, the first two modes give a negative AO trend, which was also noted earlier in [11].

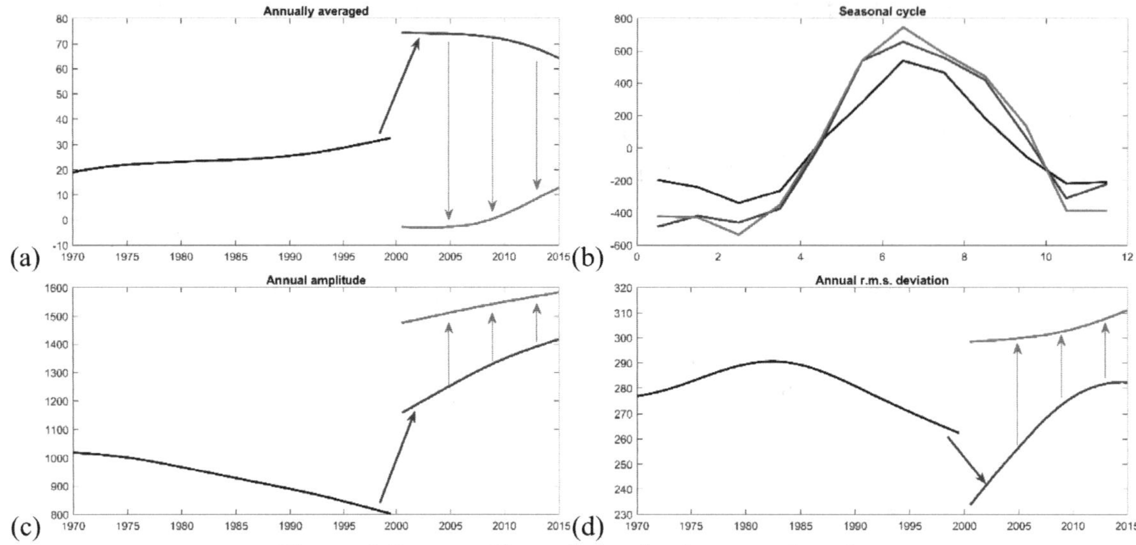

Figure 4. Same as Figure 2, but for the second mode.

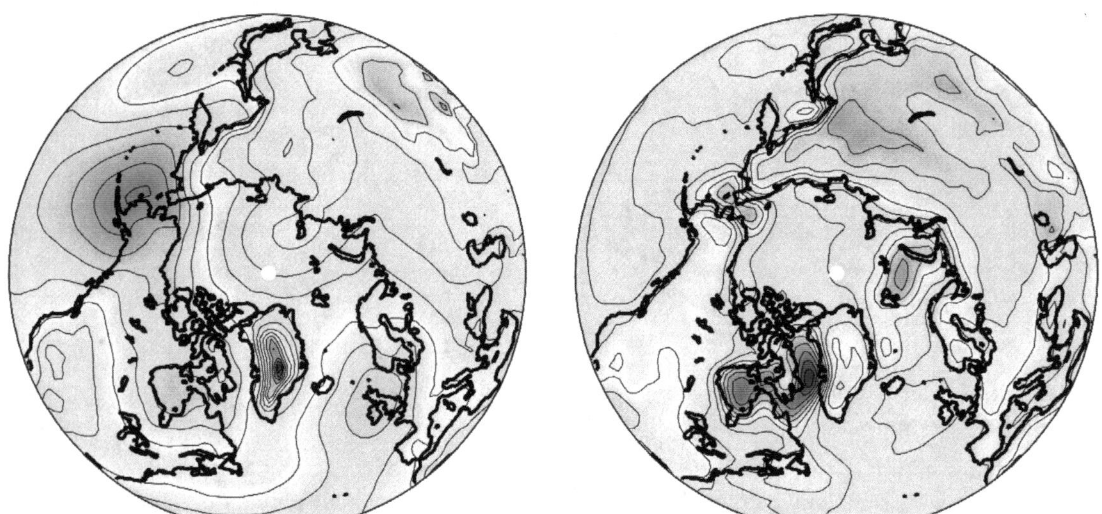

(a) (b)

Figure 5. The structural function of the third mode presented as a distribution of the surface pressure field (a) and temperature at a height of 10 m (b).

In Figure 4c one can also see that the amplitude of the seasonal cycle increased by almost 1.5 times, making the difference between the winter minimum (AO gain) and the summer maximum (AO attenuation) more significant. It is important that the ocean contributes to an increase in the amplitude by an order of magnitude greater than a decrease in the average state of this mode. The interseasonal variability, presented as the standard deviation from the seasonal variation in Figure 4d, as in the case of the first mode, is somewhat weakened in test C, but due to the ocean reaction in test B it turns out even stronger than in test A.

A feature of the third mode of the EOF decomposition is that its structure in the temperature field identifies zones additionally released from ice in summer due to climatic changes: these are the area of the Barents and Kara Seas, the vicinity of the Bering Strait and the straits of the Canadian Archipelago (Figure 5). For the rest, you can see that the oceans and continents behave in different directions, but unlike the first fashion, the Arctic Ocean behaves like an ocean, not land. This is also a consequence of climate change, as a result of which a significant part of the Arctic Ocean in summer opens up for active interaction with the atmosphere.

Along with an increase in the CO_2 concentration, there is a sharp increase in the average annual value of the time-dependent coefficient of this mode, up to 30% compared to the amplitude of the annual cycle (Figure 6a), and the role of the ocean in this increase is positive but insignificant. The amplitude of the seasonal cycle increases in test C by about 15%, and in test B by 25% (Figure 6c). The seasonal cycle accounts for 45% of the total variability of this mode, which makes it second after the first mode in sensitivity to the season. Thus, the ocean plays an important role in the fact that in summer there is a significant decrease in the area of ice, and in winter it is more actively restored.

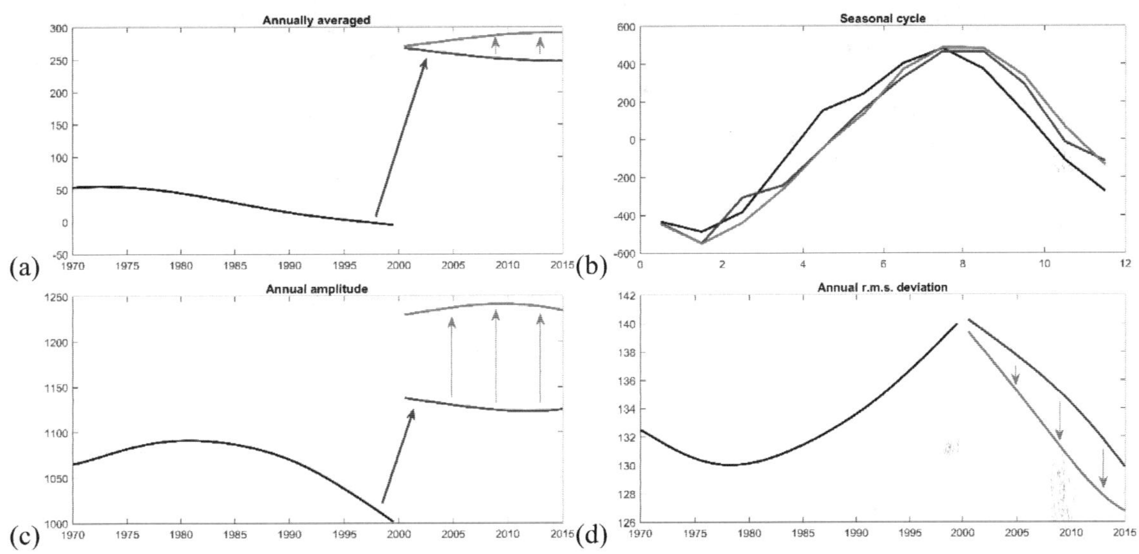

Figure 6. Same as Figure 2, but for the third mode.

The Arctic dipole, represented by the fourth mode of the EOF decomposition (Figure 7), is formed as a result of opposite pressure changes over Greenland on the one hand, and over the Kara Sea and the Taimyr Peninsula on the other. In the positive phase, when the maximum pressure is over the Taimyr Peninsula, the prevailing winds counteract the transfer of ice to the North Atlantic, and vice versa, intensify the transpolar drift in the negative phase. In the negative phase, one can also note the warming of the coastal regions of the Arctic Ocean, the Baffin Sea, and the Hudson Bay.

The trends revealed by changes in the carbon concentration in the atmosphere are shown in Figure 8. The change in the average annual values of the decomposition coefficient for this mode (Figure 8a) in test C does not exceed the value of fluctuations of this value in test A. However, the inclusion of the active ocean increases the deviation of this value to the positive side by about 2 times, and then relaxes downward. Seasonality (the annual cycle) (Figure 8b) is weakly expressed and accounts for only 13% of the total mode variability. An increase in the CO_2 concentration leads to an almost 2-fold increase in the amplitude of the seasonal cycle in test C. However, in test B the ocean reduces the amplitude of seasonal changes (Figure 8c) back to unperturbed values. Deviations from the seasonal cycle for this

IOP Conf. Series: Earth and Environmental Science **386** (2019) 012004 doi:10.1088/1755-1315/386/1/012004

mode (Figure 8d) vary over a wide range and, although an increase in CO_2 in the atmosphere leads to a slight decrease in this deviation, in general, this decrease is in the range of usual changes of this value. Thus, the conclusion on the increasing role of the Arctic dipole during global warming made in [23] was not confirmed in the course of our experiments, which may mean, for a number of different reasons, that its strengthening is not directly related to an increase in the concentration of greenhouse gases in the atmosphere.

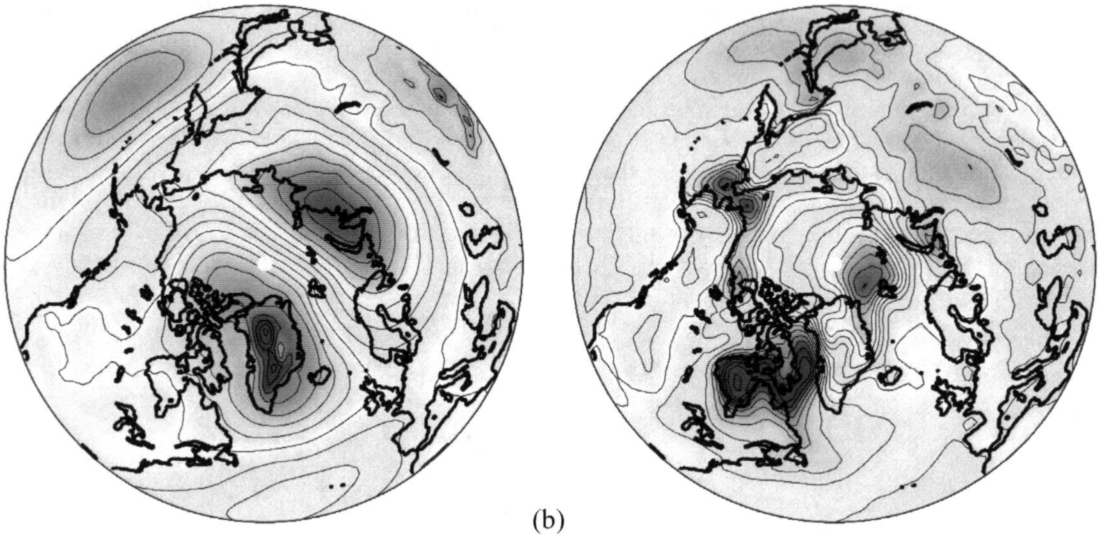

(a) (b)

Figure 7. The structural function of the fourth mode presented as a distribution of the surface pressure field (a) and temperature at a height of 10 m (b).

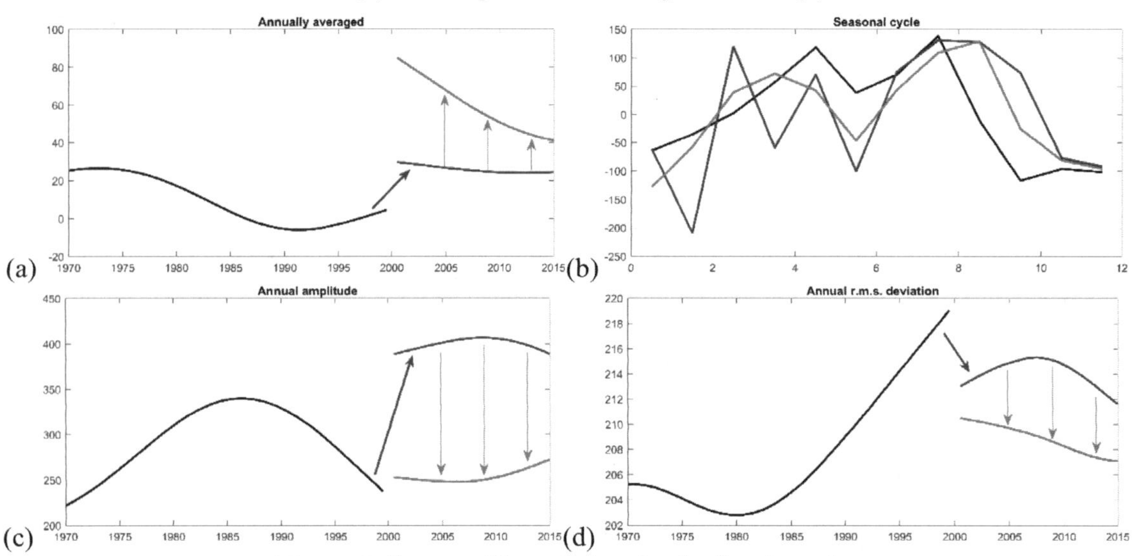

Figure 8. Same as Figure 2, but for the fourth mode.

5. Conclusions

Summarizing the results of the numerical experiments, we can state that the role of the ocean in the response of the climate system to an increase in the atmospheric CO_2 concentration, to a first approximation, is as follows:

- the ocean contributes to some extent (up to 20%) to the increase in the annual mean state of the seasonal cycle and to the decrease in the amplitude of seasonal oscillations (by 2-

3%), which ultimately leads to insignificant changes in the summer period and to a significant mitigation of the winter;

- the ocean reduces the changes in the annual mean state of the Arctic oscillation, making it practically stable with increasing CO_2 concentration but, at the same time, contributes to the significant increase in the amplitude of the seasonal cycle of this oscillation;
- the ocean enhances the temperature (or thermal) component of the seasonal variation associated with the appearance of additional areas freed from ice cover, with an additional average increase in the temperature of the atmosphere at the ice edge. Besides, the ocean enhances the seasonal oscillations of this component, so that the summer manifestations become much stronger;
- the role of the Arctic dipole under global warming is not significant, although the results of [23] show that this mode cannot be disregarded, since it can promote or restrict the Arctic ice export.

Thus, the direct response is that the ocean mainly enhances the trends resulting from an increase in CO_2 for the first three modes of the EOF decomposition. The third "temperature" mode actually arises due to the formation of a vast area free of ice.

However, these conclusions may undergo significant changes under a more detailed consideration of carbon cycles in the atmosphere, ocean, and land.

Acknowledgments
This work was supported by the Russian Science Foundation under grant no. 19-17-00154. The authors are grateful to the Siberian Supercomputer Center of the Siberian Branch of the Russian Academy of Sciences for providing their supercomputer facilities.

References

[1] Katsov V M, Porfiriev B N 2011 Climate Change in the Arctic: Environmental and Economic Impacts *Arctic: Ecology and Economy* **2**(6) 66-79 (in Russian)

[2] Le Treut H, Somerville R, Cubasch U, Ding Y, Mauritzen C, Mokssit A, Peterson T and Prather M 2007 Historical Overview of Climate Change *Climate Change 2007: The Physical Science Basis. Contribution of Working Group I to the Fourth Assessment Report of the Intergovernmental Panel on Climate Change* ed Solomon S, Qin D, Manning M, Chen Z, Marquis M, Averyt K B, Tignor M and Miller H L (Cambridge, United Kingdom and New York, NY, USA: Cambridge University Press)

[3] AMAP 2011 *Snow, Water, Ice and Permafrost in the Arctic (SWIPA): Climate Change and the Cryosphere. Arctic Monitoring and Assessment Programme (AMAP)* (Oslo, Norway) pp 538

[4] Larsen J N, Anisimov O A, Constable A, Hollowed A B, Maynard N, Prestrud P, Prowse T D, and Stone J M R 2014 Polar regions *Climate Change 2014: Impacts, Adaptation, and Vulnerability. Part B: Regional Aspects. Contribution of Working Group II to the Fifth Assessment Report of the Intergovernmental Panel on Climate Change* ed Barros V R *et al*, (Cambridge, United Kingdom and New York, NY, USA: Cambridge University Press) p 1567-612

[5] *Antarctic Climate Change and the Environment. Scientific Committee on Antarctic Research (SCAR)* 2009 ed Turner J,. Bindschadler R A, Convey P, Di Prisco G, Fahrbach E, Gutt J, Hodgson D A, Mayewski P A and Summerhayes C P (Cambridge, UK) p 526

[6] Turner J A *et al* 2013 Antarctic climate change and the environment: an update *Polar Record* **50** (3) doi:10.1017/S0032247413000296

[7] *Understanding Earth's Polar Challenges: International Polar Year 2007-2008 – Summary by the IPY Joint Committee. University of the Arctic Publications Series No. 4, International Council for Science (ICSU)/World Meteorological Organization (WMO) Joint Committee for International Polar Year 2007-2008* 2011ed Krupnik I, Allison I, Bell R, Cutler R, Hik D, López-Martínez J, Rachold V, Sarukhanian E and Summerhayes C (Edmonton, AB, Canada

and University of the Arctic, Rovaniemi, Finland: CCI Press) p 695
[8] Stroeve J C, Kattsov V M, Barrett A P, Serreze M C, Pavlova T, Holland M M and Meier W N, 2012 Trends in Arctic sea ice extent from CMIP5, CMIP3 and observations *Geophys. Res. Lett.* **39** L16502
[9] Serreze M C and Barry R G 2011 Processes and impacts of Arctic amplification: A research synthesis *Global Planet. Change* **77** 85–96
[10] Vihma T 2014 Effects of Arctic sea ice decline on weather and climate: A review *Surv. Geophys.* **35** 1175–214
[11] Screen J A, Simmonds I, Deser C, Tomas R 2013 The atmospheric response to three decades of observed Arctic sea ice loss *J Clim* **26** 1230–48
[12] Platov G, Golubeva E, Kraineva M, Malakhova V 2019 Modeling of climate tendencies in Arctic seas based on atmospheric forcing EOF decomposition *Ocean Dynamics* **69** (6) 747-67 DOI: 10.1007/s10236-019-01259-1
[13] Gordov E P, Lykosov V N, Krupchatnikov V N, Okladnikov I G, Titov A G and Shulgina T M 2013 *Computing and information technology for monitoring and modeling of climate change and its consequences* (Novosibirsk: Nauka) p 199 (in Russian)
[14] Dymnikov V P, Volodin E M, Galin V Ya, Glazunov A V, Gritsun A S, Dianskii N A and Lykosov V N 2004 Sensitivity of the Climate System to Small External Forcing *Meteorology and Hydrology* 4 77-92 (in Russian)
[15] Platov G, Krupchatnikov V, Martynova Y, Borovko I and Golubeva E 2017 A new earth's climate system model of intermediate complexity, PlaSim-ICMMG-1.0: description and performance *IOP Conf. Ser.: Earth Environ. Sci.* **96** 012005 DOI: 10.1088/1755-1315/96/1/012005
[16] Fraedrich K, Jansen H, Kirk E, Luksch U and Lunkeit F 2005 The Planet Simulator: Towards a user friendly model *Meteorologische Zeitschrift* **14** (3) 299 – 304 DOI: 10.1127/0941-2948/2005/0043
[17] Golubeva E N and Platov G A 2009 Numerical modeling of the Arctic Ocean ice system response to variations in the atmospheric circulation from 1948 to 2007 *Izvestiya. Atmospheric and Oceanic Physics* **45** (1) 137-51 DOI: 10.1134/S0001433809010095
[18] Kuzin V I, Platov G A, Golubeva E N and Malakhova V V 2012 Certain results of numerical simulation of processes in the Arctic Ocean *Izvestiya. Atmospheric and Oceanic Physics* **48** (1) 102-19 DOI: 10.1134/S0001433812010069
[19] Hunke E C and Dukowicz J K 1997 An elastic-viscous-plastic model for ice dynamics *J. Phys. Oceanogr.* **27** (9) 1849–67 DOI: 10.1175/1520-0485(1997)027<1849:AEVPMF>2.0.CO;2
[20] Lipscomb W H and Hunke E C 2004 Modeling Sea ice transport using incremental remapping *Monthly Weather Review* **132** (6) 1341–54 DOI: 10.1175/1520-0493(2004)132<1341:MSITUI>2.0.CO;2
[21] Murray R J 1996 Explicit generation of orthogonal grids for ocean models *J. Comput. Phys.* **126** (2) 251–73 DOI: 10.1006/jcph.1996.0136
[22] Thompson D W J and Wallace J M 1998 The Arctic oscillation signature in the wintertime geopotential height and temperature fields *Geophys. Res. Lett.* **25** (9) 1297-300 DOI: 10.1029/98GL00950
[23] Watanabe E, Wang J, Sumi A and Hasumi H 2006 Arctic dipole anomaly and its contribution to sea ice export from the Arctic Ocean in the 20th century *Geophys. Res. Lett.* **33** L23703 doi:10.1029/2006GL028112
[24] Lindsay R W and Zhang J 2005 The Thinning of Arctic Sea Ice, 1988–2003: Have We Passed a Tipping Point? *J. Climate* **18** 4879–94 DOI: 10.1175/JCLI3587.1

IOP Conf. Series: Earth and Environmental Science **386** (2019) 012005 doi:10.1088/1755-1315/386/1/012005

Differences in the formation of blocking processes between the Northern and Southern Hemispheres

S V Morozova[1], E A Polyanskaya[1], N K Kononova[2], K E Denisov[3] and A V Letuchiy[3]

[1] Saratov State University, Saratov, Russia
[2] Institute of Geography, Russian Academy of Sciences, Moscow Russia
[3] Saratov State Vavilov Agrarian University, Russia

E-mail: swetwl@yandex.ru

Abstract. Mechanisms of formation of blocking processes in the Southern Hemisphere are considered. They are compared with analogous mechanisms in the Northern Hemisphere by using daily synoptic maps of the Northern and Southern Hemispheres. It is found that in the circulation mechanisms of the Southern Hemisphere the process of merging of Antarctic and subtropical anticyclones takes place in a completely different way than in the Northern Hemisphere. The main difference is that in the Northern Hemisphere, as a rule, the ridge of an Arctic anticyclone extends to the low latitudes, but in the Southern Hemisphere the ridge of a subtropical anticyclone extends to the high latitudes, towards Antarctica. An assumption is made that such a peculiarity of circulation leads to the fact that the rear of coastal Antarctic cyclones does not receive cold air masses, as in the cyclones of the Northern Hemisphere, but gets warm in comparison with their frontal part.

1. Introduction

The study of blocking is of great scientific and practical interest. These are large-scale stable processes in the atmosphere. They cause weather and climate abnormalities. The study of these phenomena is important for monitoring the current climate, modeling atmospheric processes, medium-term and long-term weather forecasts.

The study of blocking is carried out by climatic methods, synoptic methods, and hydrodynamic methods. A climatic study of blockings has established areas of their increased and decreased frequency of occurrence. The seasonal distribution of blockings is described. The maximum frequency of blocking in the northern hemisphere is observed in the eastern Atlantic, Europe, and the eastern Pacific [1]. The maximum frequency of blocking in the East Atlantic and Europe is observed in winter. The maximum frequency of blocking over the east of the Pacific Ocean is observed in summer [2]. The main reason for the formation of blocking anticyclones is considered to be nonlinear instability of Rossby waves with intensive energy exchange both with planetary waves and with synoptic vortices (breaking) [3].

It is well-known that the nonlinear phase of instability development is characterized by an intense interaction of a growing wave with movements of other scales. Depending on the energy reserves that can be transmitted to the growing wave, its nonlinear development can lead to the establishment of one of the following nonlinear modes [4]:

1. Stationary mode. Ridges and hollows are formed in the stream, whose axes are oriented meridionally or at an angle to the meridian; they persist for a long time without changing the

amplitude, shape, and wavelength.

2. "Flicker" or pulsation mode: it is also a regular mode, but the ridges and hollows experience slow periodic changes: these are either changes in the amplitude of the waves, or changes in their orientation or, finally, a change in the wave number.

3. Irregular mode type of turbulence.

Blocking is a first or second mode for an unstable Rossby type wave.

The synoptic approach to the study of blocking anticyclones showed that for the anticyclone (blocking) to stay, it is necessary to combine the increase in the amplitude of the planetary tall frontal zone (PTFZ) with the anticyclone of arctic origin entering the region of maximum. The formation of a blocking anticyclone over Eastern Europe is classified as a form of E - circulation to Wangenheim – Girs. It is in this form of circulation that most abnormal weather conditions are most often established [5].

Blocking is formed when anticyclones from the Arctic move to the continent. The displacement to the east (on the continent) of anticyclones from tropical regions does not lead to blocking. This conclusion is confirmed by studies of synoptic processes in the Middle and Lower Volga regions. [6]. Subtropical anticyclones form blocks over the ocean. An example of this is the movement of the Azores anticyclone to the north over the Atlantic Ocean from the subtropics in January 2012 [7].

On the eastern and northeastern periphery of this high-altitude anticyclone, very cold air entered the European part of Russia. At the same time, the average daily temperatures decreased by 10 - 15 ° C compared to the perennial average.

In this work, using the synoptic approach, the task is to consider the mechanisms of blocking formation in the southern hemisphere and to identify the features of establishment of blocking regimes in the northern and southern hemispheres.

2. Materials and methods

To study blocking regimes in the Southern Hemisphere, synoptic conditions leading to the establishment of blockings were analyzed. The studies were performed using daily weather maps and altitude maps of standard isobaric surfaces. The temporary material for the research was 2000-2005. The blocking situation was determined in accordance with the approach described in [8],

$$G(50) = \frac{Z(\varphi_0)-Z(\varphi_S)}{\varphi_S-\varphi_S},$$

$$G(70) = \frac{Z(\varphi_n)-Z(\varphi_0)}{\varphi_n-\varphi_0},$$

where $\varphi_0 = 60° + \Delta$;
$\varphi_n = 80° + \Delta$;
$\varphi_s = 40° + \Delta$;
$\Delta = -4°; 0°; 4°$ of latitude.

On a specific longitude of 00 GMT, an implicit blocking was determined. There is blocking if, at least with one knowledge, the following conditions are met:

$$G(50) > 0,$$
$$G970) < -10 \text{ m/1 degree of latitude.}$$

The source material for studying the conjugation of processes in the Northern and Southern Hemispheres was the Calendar of Successive Changes in Elementary Circulation Mechanisms (ECMs) compiled by N. Kononova and posted on the site [9]. ECMs with one, two, three, and four blocking processes in the hemispheres were selected from the “Calendar ...”, and their development during the ECMs action was considered.

3. Results and discussion

In the Southern Hemisphere, 75 cases of blocking regimes were analyzed. The average duration of the blocking process in summer is five to six days, and in winter it is three to four days. The vast majority of blocking cases (82%) are characterized by the following: the blocking process begins by pulling the ridge of the subtropical anticyclone to the rear of the cyclone located off the coast of Antarctica. In this case the air flows are directed from north to south, that is, from the subtropics to the high latitudes. Moreover, sometimes there is no independent anticyclonic center over Antarctica.

As an example, we cite the process that took place from July 28 to August 1, 2005. On July 28, active cyclonic activity was around the coast of Antarctica. The cyclones were located directly off the coast, except for the Pacific one. There was an extensive anticyclone along the Pacific coast, elongated in the latitudinal direction. On July 29, only four of all cyclones remained, which became central and stationed over each of the three oceans and in the Antarctic Peninsula. In the next two days spurs from subtropical anticyclones began to spread to the rear of these stationary cyclones. This process culminated in the establishment of blocking regimes in four regions of the Southern Hemisphere near the coast of Antarctica. Three baric ridges are located above the Atlantic, Pacific, and Indian Oceans. The fourth ridge extended from South America through the Antarctic Peninsula all the way to the central part of the mainland (Figure 1). The extension of the baric ridge to the rear of the coastal Antarctic cyclones suggests that a warm air mass enters the rear of such cyclones but not the front.

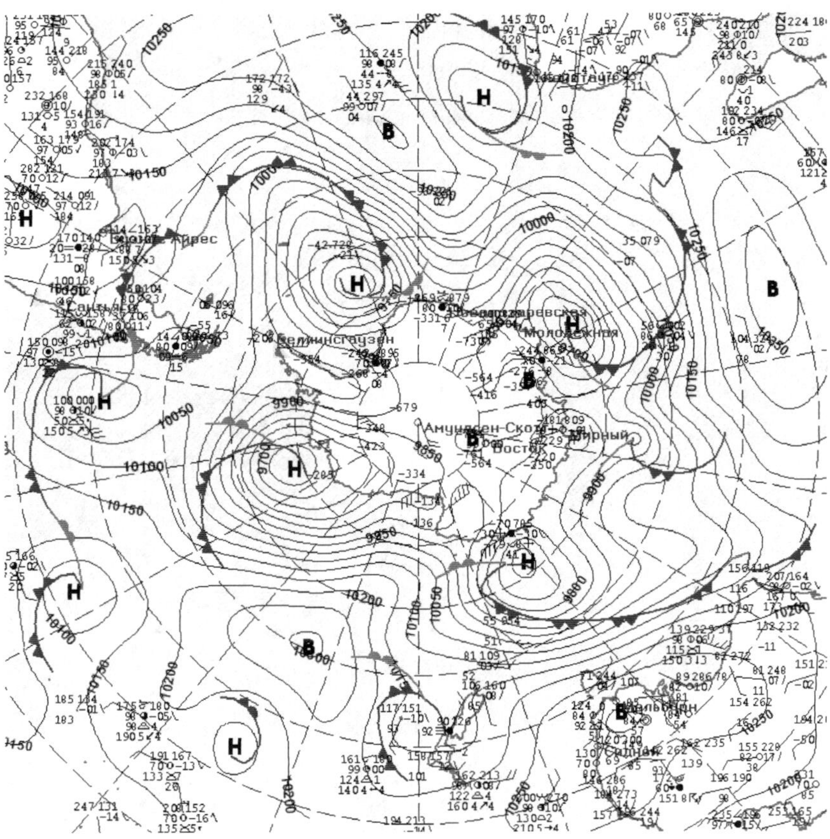

Figure 1. Weather map for August 1, 2005, ECM 12a, Southern Hemisphere.

IOP Conf. Series: Earth and Environmental Science **386** (2019) 012005 doi:10.1088/1755-1315/386/1/012005

It is interesting to compare the process of establishing blockages in the Northern and Southern Hemispheres.

For the Northern Hemisphere, the mechanism of occurrence, development, and establishment of blocking regimes is as follows.

In most cases, a combination of arctic and subtropical anticyclones occurs. In this case, the Arctic anticyclone moves with the southern (more often - southeastern) component and, as a result, the subtropical high-pressure band is replenished with fresh Arctic nuclei. This, in turn, maintains high pressure in the tropics [10].

In the 70s of the 20th century, when studying the blocking processes B. L. Dzerdzeevsky [11, 12] found that they develop simultaneously in the Northern and Southern Hemispheres at approximately the same longitudes. The authors of this study verified this statement on the material of 2000-2005. As a result of the analysis, it was found that in 69% of cases the processes in the Northern and Southern Hemispheres are coupled. That is, similar processes (blocking) develop simultaneously at the same longitudes.

In Figure 2 we give an example of a related process in the Northern Hemisphere considered earlier in the Southern Hemisphere. Figure 2 shows the ECM 12a weather map with four blocking processes: over the Atlantic, the Pacific Ocean, Eastern Siberia, and America. According to Figure 2, over the Atlantic an Arctic invasion occurs in the rear of a series of southern cyclones, over the Pacific Ocean it occurs in the rear of the Aleutian cyclone, in Alaska in the rear of the cyclone over America, and in Eastern Siberia in the rear of the Far Eastern cyclonic system. In all cases, in the rear of cyclones the air flows from north to south, that is, cold air masses spread to the rear of the cyclones.

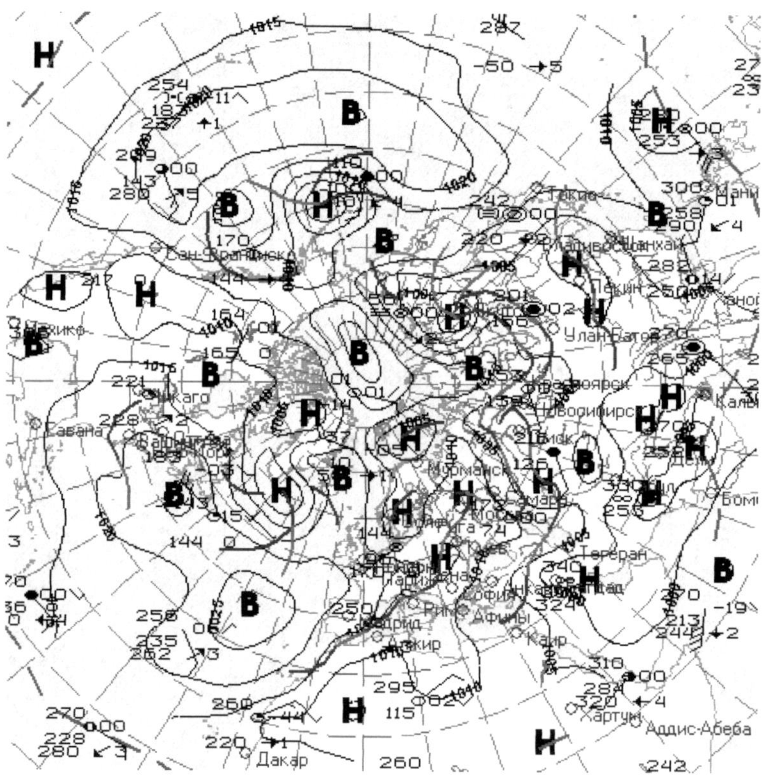

Figure 2. Weather map for August 1, 2005, ECM 12a, Northern Hemisphere.

Thus, the development of blocking processes in the Northern and Southern Hemispheres has its own characteristics. If in the Northern Hemisphere during an Arctic invasion cooling extends from north to south, in the Southern Hemisphere, with the extension of the ridge of the subtropical anticyclone, warm air also spreads from north to south with the anticyclone, but from the low latitudes to high ones.

It is possible that such a difference is due to the fact that the Antarctic anticyclone is much more intensive and larger in area than the Arctic one. It forces subtropical anticyclones to creep in the direction of the area occupied by it. The nature of changes in the air temperature in the Northern and Southern Hemispheres during the formation of blocking processes in the rear of the cyclone is also different. These features should be taken into account when analyzing and modeling the atmospheric circulation of the planet Earth system.

Conclusions

1. In the Northern Hemisphere, the formation of blocking processes usually begins with an arctic invasion which goes southward, combining with a subtropical anticyclone. In this case, a cold air mass flows into the rear of the cyclone.
2. In the Southern Hemisphere, as a rule, the development of a blocking process is carried out by pulling the ridge of a subtropical anticyclone to the pole and then connecting it with an Antarctic anticyclone. In this case, a warm air mass spreads to the rear of the cyclone but not to its front.
3. The correlation between the development of processes in the Northern and Southern Hemispheres has been confirmed with a more recent material (2000-2005).

Acknowledgements

This work was performed under Fundamental Research Program no. 0148-2019-0009, Institute of Geography RAS.

References

[1] Gruza G V and Korovkina L.V 1991 Seasonal features of the spatial distribution of blocking indices in the Northern Hemisphere *Meteorology and hydrology* **3** 108-10
[2] Gruza G V and Korovkina L V 1991 Climate monitoring of the processes of blocking the Western transfer in the Northern Hemisphere *Meteorology and hydrology* **8** 11-8
[3] Hide R and Mason P J 1975 Sloping convection in the rotating fluid *Adv. Physics* **24** (1) 47-100
[4] Pelly J.L and Hoskins B.J. 2003 A new perspective on blocking *J. Atmos.Sci.* **60** (3) pp 743-55
[5] Kononova N K and Cherenkova E A 2018 Reiteration of the Elementary Circulation Mechanisms in the Atmosphere of the Northern Hemisphere *Izvestiya Rossiiskoi Akademii Nauk. Seriya Geograficheskaya* **6** 17-25
[6] Polyanskaya E A 1986 Synoptic processes and weather phenomena in the Lower Volga region *Saratov State University Press* p 208
[7] Golubev A D, Kabak A M, Nikol'skaya N A, Butova G I and Berezhnaya T V 2013 Blocking of Westerlies over Eurasia in Winter 2012 and Associated Weather Anomalies *Proceedings of Hydrometcentre of Russia* **349** 188-213
[8] Tibaldi S and Molteni F 1990 On the operational predictability of blocking **42**A 343-65
[9] http: // www.atmospheric-circulation.ru
[10] Girs A A 1960 Basics of Long-Range Weather Forecasts Leningrad Gidrometpress p 560
[11] Dzerdzeyevskiy B L 1967 Comparison of the main patterns of atmospheric circulation over the Northern and Southern hemispheres *Newsletter of the Soviet Antarctic expedition* **65** 58-68
[12] Dzerdzeyevskiy B L 1975 Selected Works *Nauka - Press* 149-58

Global Atmospheric Oscillation in Geopotential Fields of the Free Atmosphere

I V Serykh, D M Sonechkin, V I Byshev, V G Neiman

Shirshov Institute of Oceanology, Russian Academy of Sciences,
36, Nahimovskiy prospekt, Moscow, Russia, 117997

E-mail: iserykh@ocean.ru

Abstract. The three-dimensional structure of geopotential anomalies arising in the troposphere and the lower stratosphere during a Global Atmospheric Oscillation (GAO) is studied. It was previously detected in temperature and pressure fields. It has been established that these anomalies have a high statistical significance. This serves as another formal evidence of the GAO authenticity. In the troposphere (850–100 hPa) and in the lower stratosphere (100–10 hPa) the structure of GAO substantially changes in comparison with the near-surface fields, gradually acquiring a zonal character. Apparently, this is due to the fact that the atmospheric polar tide, caused by the Chandler wobble of the Earth's poles and lunar-solar nutation, bends around the Earth almost unhindered, experiencing only a small topographical impact from the continents.

1. Introduction

In studies of the dynamics of the atmosphere and the ocean, the notion that the El Niño – Southern Oscillation (ENSO) phenomena that develop in the equatorial strip of the Pacific Ocean influence the climate almost everywhere on Earth is widely used. This, as a possible circumstance, was also mentioned by the pioneers of the ENSO research J. Walker, G. Berlage, A. Trope, and K. Wyrtki. Then works appeared in which it was established that at different phases ENSO develop their specific anomalies in areas of the Earth often very remote from the tropics of the Pacific Ocean. The first publications were probably [1, 2]. Now there are so many such works that it is impossible to list all of them in any journal article.

Although no direct evidence was presented that just ENSO was the cause, and these specific anomalies were the result, but the relationships found were interpreted precisely in the spirit of such causality. For example, K. Trenberth - one of the most well-known modern ENSO researchers - begins his article [3] with the phrase: "One of the most prominent sources of interannual variations in weather and climate around the world is the ENSO phenomenon".

This interpretation may be understood in a manner that says "tail" (ENSO) "sees the dog" (weather and climate everywhere on the Earth). Byshev et al. [4-6] have questioned this opinion. They carried out a thorough analysis of the differences in the global spatial structure of the near surface temperature (NST) and sea level pressure (SLP) fields at the extreme phases of ENSO processes, i.e. the occurrence of El Niño and La Niña, as well as in the period preceding these events. For this purpose, the most complete of the arrays of meteorological observations available at that time were used. As a result, the spatial structure of NST and SLP anomalies was found, called Global Atmospheric Oscillation (GAO). This structure is symmetric about the equator, despite the differences in the configuration of the continents within the Northern and Southern Hemispheres. It covers almost the entire Earth, including the anomalies attributed to the manifestation of long-range links with the ENSO. These anomalies are detected at the time of appearance far before either El Niño or La Niña event occurs. Therefore, the direct dependence of these anomalies on ENSO events may hardly be considered correct. From all this it was concluded that the GAO was primary and the ENSO was secondary, i.e. "The dog wags its tail, and not the tail wags the dog".

In a later work of Byshev et al. [7], the Student's t-test was applied to the differences in the spatial structures between the El Niño and La Niña events. The test showed that the differences in the values of SLP and NST, characteristic of the events of El Niño and La Niña, are statistically highly significant almost everywhere on the globe, and not only in the tropics of the Pacific Ocean (an area of the ENSO). Thus, an additional formal proof of the real existence of the GAO was obtained. Although the problem of the GAO and ENSO causality was not specifically considered in [7], based on the above its authors formulated and substantiated a working hypothesis that the ENSO should be considered as a structural element of the GAO in the Pacific region.

Careful computations of the ENSO and GAO power spectra were made in [8-10]. These computations reveal that a part of these spectra, corresponding to the periods from two years to one decade, is prominent by the existence of several narrow bands of increased spectral density centered at the subharmonics 2:1, 3:1, and 4:1 of the Chandler wobble in the Earth's pole motion (~ 1.2 years), the superharmonics 1:2, 1:3, and 1:4 of the luni-solar nutation of the Earth's rotation axis (~ 18.6 years), as well as the superharmonics 1:2, 1:3, and 1:4 of the 11-year cycle of the Sun spots. The existence of similar bands in the ENSO power spectra was recognized many years ago [11, 12]. However, it turns out that the above spectral bands have been also seen in spectra of the GAO characteristics outside of the tropics. Moreover, the respective climatic variations are globally synchronized. The synchronization takes place because the above-mentioned external periodicities must influence short-term climatic variations globally. It is very probable that the periods of the external periodicities indicated are incommensurable with each other. Therefore, if these periodicities actually influence short-term climatic variations, they would have to do it discordantly. As a result, no resonances can exist which could make the affected climatic variations chaotic. A linear dependence of the logarithms of serial numbers of the spectral bands on the logarithms of the band magnitudes, as well as a linear decrease in the accumulated sum of the squared autocorrelations of the respective atmospheric characteristics confirm that the dynamics of the interannual to decadal climatic variations are not chaotic.

In [13] it was established that an integrity named GAO coves both the ENSO and all its so-called extratropical teleconnections revealed as the North Atlantic Oscillation (NAO), the Arctic Oscillation (AO), the Northern Hemisphere annular (NHA) mode, as well as the Pacific – North American (PNA) pattern, and their analogs in the Southern Hemisphere in the interannual timescale. Investigations of the GAO indices revealed that the GAO extratropical components may be real irrespective of the ENSO, while the latter accompanies the GAO in all cases. Moreover, in view of a general eastward propagation of the GAO as a spatial structure, some of its extratropical components show their features before an El Niño (La Niña) begins to form.

The purpose of this work is as follows: without discussing the problem of causality, which requires special consideration due to its extreme complexity for such a nonlinear dynamic system as the

IOP Conf. Series: Earth and Environmental Science **386** (2019) 012006 doi:10.1088/1755-1315/386/1/012006

variability of the global climate (see paper on this subject [14]), to trace the manifestation of the GAO spatial structure throughout the entire troposphere and lower stratosphere.

2. Materials and methods of the study

The baseline data for this work were reanalyses of 20thC_ReanV2c, ERA-20C, and NCEP/NCAR global monthly fields of geopotential height and air temperature on standard isobaric levels of the atmosphere, as well as SLP. The NOAA CIRES 20th Century Global Reanalysis Version 2c (20thC_ReanV2c) data are specified at 2°x2° regular grid nodes on 24 main isobaric levels from 1000 to 10 hPa for the period 1851-2014 [15]. The ECMWF ERA-20C data are specified in the regular grid of 1°x1° on 37 main isobaric levels from 1000 to 1 hPa for 1900-2010 [16]. The NCEP/NCAR Reanalysis data are specified at regular grid points of 2.5°x2.5° on 17 basic isobaric levels from 1000 to 10 hPa for 1948-2018 [17].

For each time series of the listed data sets at each grid point, the average seasonal signal of the considered climatic characteristics was calculated, which was then subtracted from the initial data. According to the fields obtained in this way, the geopotential height and air temperature anomalies were calculated using the Extratropical GAO index (EGAO) [10]. It is calculated as the combination of the normalized (by standard deviations of their time series) anomalies of the average SLP in six geographical areas coinciding with the extratropical extremes (maxima and minima) in the GAO field: (55°-65°N, 95°-85°W) + (65°-55°S, 95°-85°W) – (45°-55°N, 175°-165°W) – (45°-55°N, 15°-5°W) – (55°-45°S, 15°-5°W) – (55°-45°S, 175°-165°W). To remove high frequency dynamics, the EGAO time series were smoothed by the three months running mean. Linear trends were removed from the obtained EGAO time series by the least squares method in order to suppress the interdecadal climate changes. The positive and negative GAO phases, corresponding just timely to the events of El Niño and La Niña, were determined by the centered and normalized EGAO index so that the value of this index either exceeded the sub-series +0.5 for 5 months or more, or it was less than -0.5, respectively.

The sequences of fields of the anomalies of the geopotential height and air temperature for the standard isobaric levels, corresponding in time to the positive and negative GAO phases, were selected, and their mean differences were calculated as described in [13]. In addition, the standard deviations of the global fields of the indicated characteristics for the time of each such specific event from the corresponding average field at the time of the positive and negative GAO phases were calculated. Then, for each node of the regular grid, the Student's t-test was calculated, estimating the statistical significance of the difference from zero of the average difference of the specified global fields at this node of the computational grid.

3. Results

A comparison of the global fields of geopotential height differences and air temperature anomalies between the opposite phases of the GAO, calculated in relation to the time evolution of the above EGAO index, constructed according to the indicated reanalyses, showed their close similarity. Based on this, these fields are illustrated in the paper only by the example of NCEP/NCAR Reanalysis, which contains the sample: 20 positive and 18 negative GAO phases for 1948-2018.

Figure 1 shows the average field of differences in anomalies of the air temperature at the level of 1000 hPa (a) and the corresponding field of values of probabilities according to the Student's t-test (b) between the two opposite phases of the GAO, which coincide with El Niño and La Niña. The region of the largest temperature differences in modulus (more than 1°C) is visible in the east and in the center of the equatorial zone of the Pacific Ocean, i.e. in the canonical area of the ENSO. The probability that these differences are non-zero, in the center of this area exceeds 99%. This is to be expected according to the definitions of the events of El Niño and La Niña as components of the GAO adopted in this work.

IOP Conf. Series: Earth and Environmental Science **386** (2019) 012006 doi:10.1088/1755-1315/386/1/012006

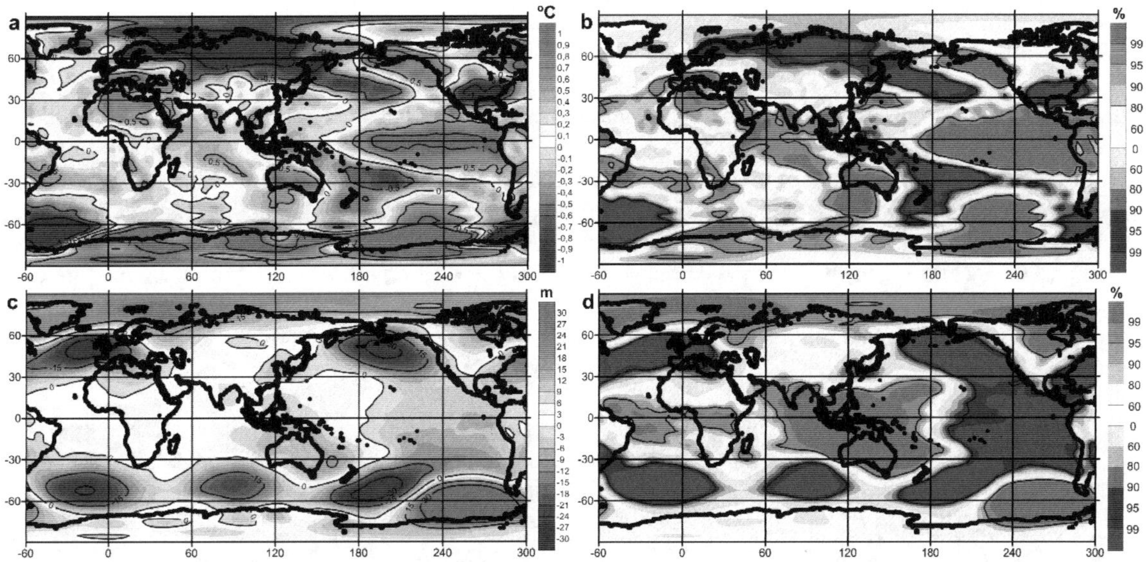

Figure 1. Fields of differences in air temperature (a) and geopotential (c) anomalies at the level of 1000 hPa between the opposite phases of the GAO and the corresponding probability fields according to the Student's t-test (b, d).

The probabilities of non-zero differences are quite large (95% and higher) not only in the canonical El Niño area. Thus, in the south of the Pacific Ocean, where the temperature difference between the events of El Niño and La Niña (the opposite phases of the GAO) is, on average, +0.5 °C, these probabilities exceed 99%. As one can see in Figure 1, the same significant positive temperature differences (+0.5 °C) between the opposite phases of the GAO occur in the near-coastal region of Alaska and western Canada symmetrically located relative to the equator. More than 95% of the probabilities of the positive temperature differences are more seen in the other three areas: 1) in the Indian Ocean, 2) in the tropics of the Atlantic Ocean and on the Atlantic coast of Brazil, 3) in the northern and southern Africa.

Statistically significant negative temperature differences (with probabilities greater than 95%) are observed in several geographic areas. Two of them are the middle latitudes of the western part of the Pacific Ocean. These two areas are located symmetrically with respect to the equator in relation to each other.

Turning to the manifestations of the GAO in the geopotential height fields (Figure 1 c,d), we note that at the level of 1000 hPa the GAO field is very similar to that characteristic of the field of mean SLP anomalies difference shown in figures in [4, 7, 13]. Negative values (up to 30 m by module) in the field of Figure 1c, whose probabilities reach 99% (Figure 1d), cover almost the entire water area of the Pacific Ocean to the east of the date line and the middle latitudes of the Atlantic Ocean, making up an X-shaped structure.

Positive differences (up to 10 m) almost completely cover the tropics of the Indian and Atlantic Oceans, as well as Africa, the archipelago of Indonesia, most of Australia, and the adjacent area of the Pacific Ocean within ±30° latitude. The probability values of the geopotential differences in these areas reach 99% due to the relatively small variability of the geopotential of the 1000 hPa level in the tropics. Outside the tropics, statistically significant (with probabilities of more than 95%) regions of the positive difference are in Canada and in the southeast Pacific. Although the average geopotential difference in the southeast Pacific region is very large (up to 30 m), it loses its significance from the point of view of statistical significance due to the enormous variability of the geopotential in this region, both real and caused by the errors of observations. Also statistically significant are the branches of the X-shaped structure, going from its center to the side to: Europe, South Atlantic, New Zealand, and the Chukotka Peninsula.

For the level of 850 hPa (Figure 2a), the spatial structure of the field of the average difference in the geopotential height anomalies remains, in general, similar to that of the 1000 hPa level. However, the center of the X-shaped structure of the negative difference at the level of 850 hPa is already beginning to collapse. The average geopotential difference directly in the canonical region of El Niño (the center and east of the near-equatorial area of the Pacific Ocean) is less than 5 m in absolute value, i.e. less than for the level of 1000 hPa, and it is already statistically insignificant (probability values less than 80%) (Figure 2b). On the other hand, the values of the positive geopotential difference and the corresponding probability values in the Indian Ocean, the tropics of the Atlantic Ocean, and in the region of Indonesia slightly increase. The regions of positive geopotential differences in the north of Canada and in the southeast of the Pacific Ocean are also somewhat intensified.

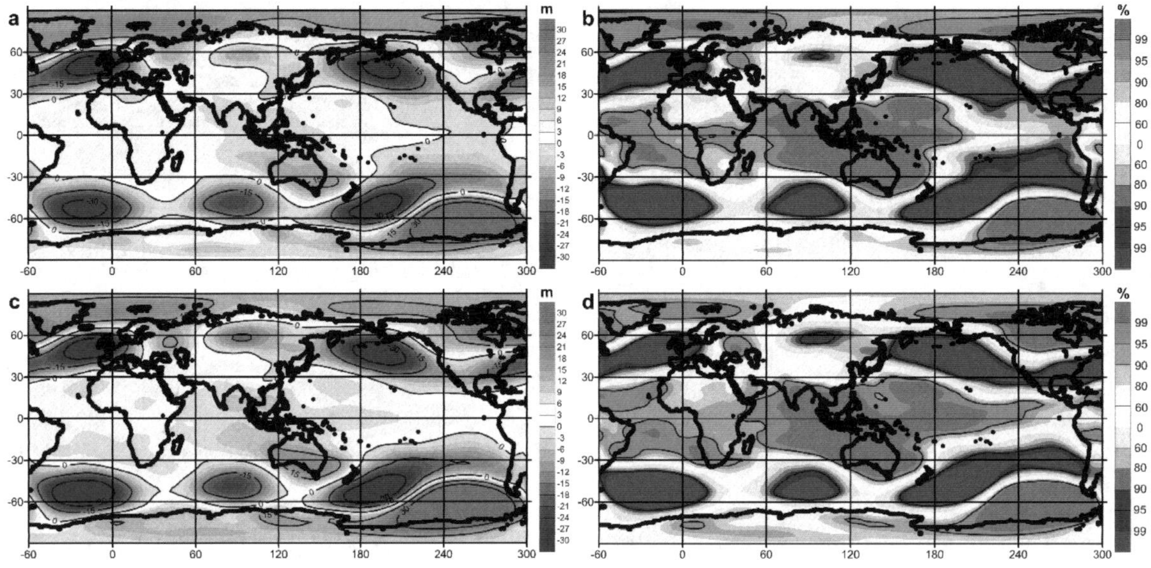

Figure 2. Fields of differences in geopotential anomalies at the levels of 850 hPa (a) and 700 hPa (c) between the opposite phases of the GAO, and the corresponding probability fields according to Student's t-test (b, d).

For the level of 700 hPa (Figure 2c), the spatial structure of the field of the average difference of the geopotential height anomalies changed to a more zonal one. The geopotential anomalies in the center and east of the equatorial Pacific Ocean are already positive. Therefore, the tropical belt at the level of 700 hPa has positive anomalies of the geopotential height differences between the opposite GAO phases.

For the level of 500 hPa, the global structure of the mean field of the geopotential difference is shown in Figure 3a. It is almost zonal, which corresponds to the results obtained in [18]. In the entire tropical belt, the average difference of the geopotential height exceeds 10 m, and the values of its probabilities exceed 95% (Figure 3b). Outside the tropical belt in the north and south of the Pacific Ocean, there are areas of very large negative differences of the geopotential (up to 30 m in absolute value) with a high level of the statistical significance (over 99%), as if preserved from the ends of the X-shaped structure described above for the geopotential field of the 1000 hPa level. The branches of the X-shaped structure stretching to the north and south of the Atlantic are also visible, and they are statistically significant. Two areas of positive mean geopotential differences corresponding to the anticyclones mentioned above in the south of the Pacific Ocean and Canada became even more powerful and statistically significant (the probabilities are over 99%). Thus, the reduction of the influence of the location of oceans and continents on the spatial structure of the GAO has an effect. Apparently, this is due to the fact that the atmospheric polar tide caused by the Chandler wobble of the Earth's poles and luni-solar nutation bends around the Earth, experiencing only a small topographic impact from the continents.

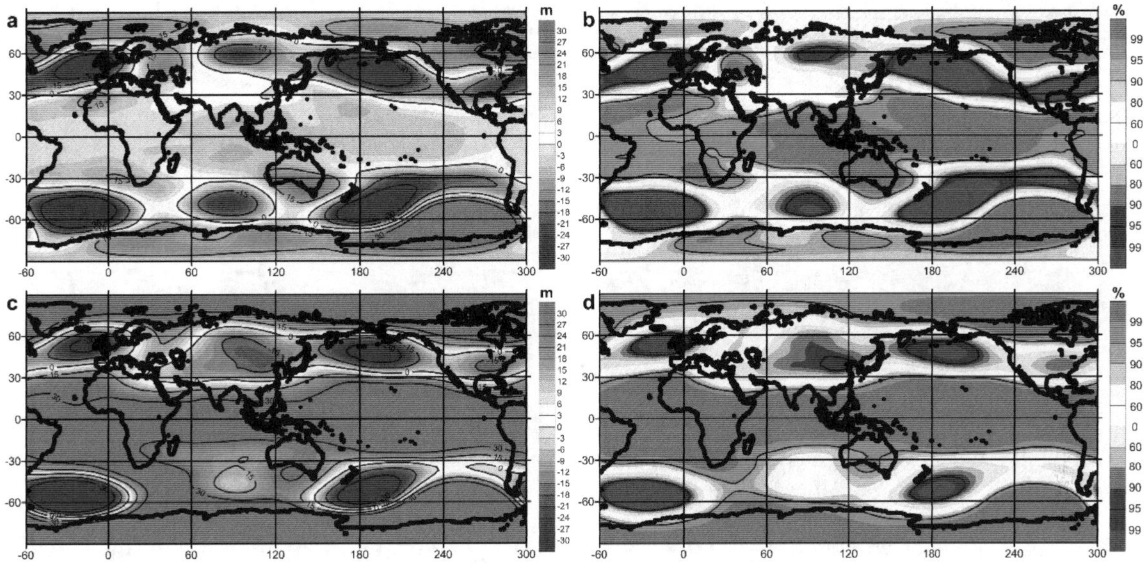

Figure 3. Fields of differences in geopotential anomalies at the levels of 500 hPa (a) and 100 hPa (c) between the opposite phases of the GAO, and the corresponding probability fields according to the Student's t-test (b, d).

Figure 3c shows the average difference field for the level of 100 hPa. The zonal structure of this field is even more pronounced than for the level of 500 hPa. In the tropical belt, positive geopotential differences exceed 30 m. All other areas of the significant differences also increased. However, due to much greater variability of the geopotential height field itself for the level of 100 hPa compared to the field for the level of 500 hPa, the values of the corresponding probabilities have somewhat decreased (Figure 3d), although the fairly high statistical significance of some areas of the largest values of the geopotential difference in the middle and high latitudes is preserved.

The field of average difference for the level of 50 hPa (Figure 4a) does not qualitatively differ from the field for the level of 100 hPa (Figure 3c). However, the statistical significance of regions in the 50 hPa level field is decreasing (Figure 4b). For the level of 10 hPa (Figure 4c), the formal statistical significance is lost almost everywhere (Figure 4d), but some elements of the GAO zonal structure are retained at this stratospheric level.

Thus, the GAO spatial structure gradually becomes zonal with height up to the level of 10 hPa, where it is destroying. The absence of such zonal structure in the GAO induced anomalies of surface fields is easily explained by the significant influence of the oceans and continents on it. In papers [10, 19] it is shown that a polar tide in the Pacific Ocean caused by the Chandler wobble of the Earth's poles and the luni-solar nutation of the Earth's axis of rotation transformed after its reflection from the western coasts of North and South America and turned into a wave running west along the equator. This wave gives rise to the El Niño's positive anomalies of surface ocean waters.

IOP Conf. Series: Earth and Environmental Science **386** (2019) 012006 doi:10.1088/1755-1315/386/1/012006

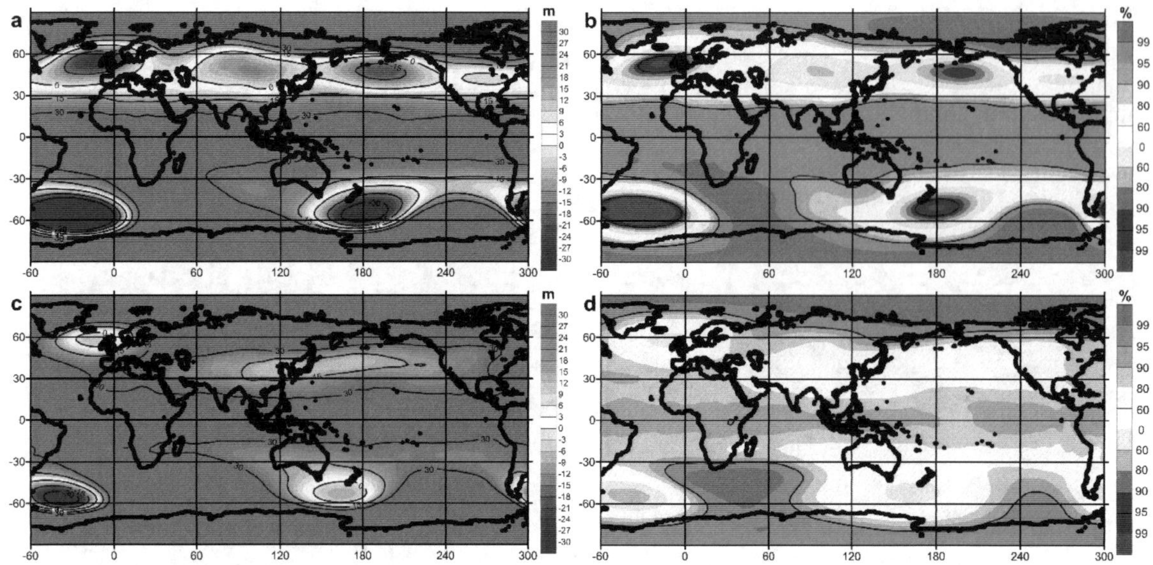

Figure 4. Fields of differences in geopotential anomalies at the levels of 50 hPa (a) and 10 hPa (c) between the opposite phases of the GAO, and the corresponding probability fields according to the Student's t-test (b, d).

Since the Chandler wobble in the Earth's pole motion is one of the external drivers of the climate system, it excites tides propagating eastward not only in the extratropical oceans, but also in the atmosphere. The phases of these tides are opposite in the Northern and Southern Hemispheres. Therefore, a general eastward zonal propagation must be a property inherent to the GAO-field. And the spatial structure of the GAO as a whole propagates eastward [13]. Continents affect the west-east distribution of the atmospheric tidal wave as topographic obstacles. However, this influence is much less significant than the reflection of the oceanic pole tide from the coasts of America.

4. Conclusions

It was found that the Global Atmospheric Oscillation, which was previously detected in sea level pressure and near surface temperature fields, extends to the entire troposphere and the lower stratosphere (up to 10 hPa), maintaining the symmetry of its spatial structure with respect to the equator.

The statistically significant differences between the opposite phases of the GAO in the air temperature field at the level of 1000 hPa are quite similar to those that were revealed earlier in the field of near surface temperature. The differences in the field of geopotential height at the level of 1000 hPa form an X-shaped structure similar to that previously found in the field of sea level atmospheric pressure. The branches of this X-shaped structure extend to the middle and high latitudes of the Pacific and Atlantic Oceans and lock themselves over Eurasia and the south of the Indian Ocean. However, the last statement has not been confirmed by Student's t-tests due to the large variability of the geopotential in this area.

The spatial structure of the GAO gradually becomes zonal with height. The absence of such zonal structure in the GAO induced anomalies of surface fields may be explained by the fact that the oceans and continents have a significant influence on it.

References

[1] Ropelewski C F and Halpert M S 1989 Precipitation Patterns Associated with the High Index Phase of the Southern Oscillation *J. Climate* **2** 268-84

[2] Glantz M H, Katz R W and Nicholls N 1991 *Teleconnections Linking Worldwide Climate Anomalies* (Cambridge Univ. Press, New York) p 535

[3] Trenberth K E, Caron J M 2000 The Southern Oscillation revisited: sea level pressures, surface temperatures, and precipitation *J. Climate* **13** 4358-65

[4] Byshev V I, Neiman V G, Romanov Yu A and Serykh I V 2012 El Niño as a consequence of the global oscillation in the dynamics of the Earth's climatic system *Doklady Earth Sciences* **446** 1 1089–94

[5] Byshev V I, Neiman V G, Romanov Y A and Serykh I V 2012 On El Niño's impact upon the climate characteristics of the Indian monsoon *Oceanology* **52** 2 147–56

[6] Byshev V I, Neiman V G, Ponomarev V I, Romanov Yu A, Serykh I V and Tsurikova T V 2014 The Influence of Global atmospheric oscillation on formation of climate anomalies in the Russian Far East *Doklady Earth Sciences* **458** 1 1116–20

[7] Byshev V I, Neiman V G, Romanov Yu A, Serykh I V and Sonechkin D M 2016 Statistical significance and climatic role of the Global Atmospheric Oscillation *Oceanology* **56** 2 165–71

[8] Serykh I V, Sonechkin D M 2017 Manifestations of motions of the Earth's pole in the El Niño – Southern Oscillation rhythms *Doklady Earth Sciences* **472** 2 256–59

[9] Serykh I V, Sonechkin D M 2017 Chaos and order in athmospheric dynamics: Part 2. Interannual rhythms of the El Niño – Southern oscillation *Izvestiya Vysshikh Uchebnykh Zavedeniy. Prikladnaya Nelineynaya Dinamika* **25** 5 5–25

[10] Serykh I V, Sonechkin D M 2019 Nonchaotic and globally synchronized short-term climatic variations and their origin *Theoretical and Applied Climatology* https://doi.org/10.1007/s00704-018-02761-0

[11] Trenberth K E 1976 Spatial and temporal variations of the Southern Oscillation *Quart. J. Roy. Meteor. Soc.* **102** 639–53

[12] Torrence C, Compo G P 1997 A practical guide to wavelet analysys *Bull. Amer. Meteorol. Soc.* **79** 1 61–78

[13] Serykh I V, Sonechkin D M., Byshev V I, Neiman V G and Romanov Yu A 2019 Global Atmospheric Oscillation: An Integrity of ENSO and Extratropical Teleconnections *Pure and Applied Geophysics* https://doi.org/10.1007/s00024-019-02182-8

[14] Sugihara G, May R, Ye H 2012 Detecting causality in complex systems *Science* **338** 496–500

[15] Compo G P, Whitaker J S, Sardeshmukh P D et al 2011 The Twentieth Century Reanalysis Project *Quarterly J. Roy. Meteorol. Soc.* **137** 1–28

[16] Stickler A, Brönnimann S, Valente M A et al 2014 ERA-CLIM: Historical Surface and Upper-Air Data for Future Reanalyses *Bull. Amer. Meteor. Soc.* **95** 9 1419–30

[17] Kalnay E, Kanamitsu M, Kistler R et al 1996 The NCEP / NCAR 40-year reanalysis project *Bull. Amer. Meteor. Soc.* **77** 437–71

[18] Brands S 2017 Which ENSO teleconnections are robust to internal atmospheric variability *Geophys. Res. Lett.* **44** 3 1483–93

[19] Serykh I V, Sonechkin D M 2016 Confirmation of the oceanic pole tide influence on El Niño *Sovremennye problemy distantsionnogo zondirovaniya Zemli iz kosmosa* **13** 2 44–52

IOP Conf. Series: Earth and Environmental Science **386** (2019) 012007 doi:10.1088/1755-1315/386/1/012007

Investigation of boreal storm tracks in historical simulations of INM CM5 and reanalysis data

P N Vargin[1], Y V Martynova[2], E M Volodin[3] and S V Kostrykin[3]

[1] Central Aerological Observatory, Dolgoprudny, Moscow region, Russia
[2] Institute of Monitoring of Climatic and Ecological Systems of the Russian Academy of Sciences, Tomsk, Russia
[3] Marchuk Institute of Numerical Mathematics of the Russian Academy of Sciences, Moscow, Russia

E-mail: p_vargin@mail.ru

Abstract. Ensemble simulations of the 5th version of the INM coupled climate model are employed to analyze the Northern Hemisphere storm track characteristics in the winter season. The results show similar features of the North Atlantic and North Pacific storm tracks that dominate in the Northern Hemisphere in the model simulations and reanalysis data. A composite analysis of the weakening /strengthening of Arctic stratospheric polar vortex events with their influence on the troposphere shows an equatorward/poleward shift in the North Atlantic storm track. As a response to the recent Arctic amplification, a poleward shift in the Pacific storm track and weakening of the North Atlantic storm track have been revealed when comparing two periods: 1998-2014 and 1980-1997.

1. Introduction

Storm tracks are marked by regions with the strongest meridional temperature gradient where extratropical cyclones are generated. These intense cyclones bring strong winds and heavy precipitation. The storm tracks transport large amounts of heat, momentum, and moisture poleward, and are observed in the mid-latitudes with a maximum at 40°N. The strongest activity of two dominant storm tracks in the Northern Hemisphere - over North Atlantic and North Pacific - is observed in the cold season [1, 2]. The North Atlantic storm track peaks in mid-winter, whereas the North Pacific storm track peaks in spring and fall and displays a mid-winter weakening (suppression) when the zonal wind speed exceeds a threshold of 44 m/s at 250 hPa, which, in turn, is accompanied by reduced baroclinic wave activity [3, 4]. The Europe and Mediterranean weather in winter is defined by cyclone propagation generated in the storm track area in North Atlantic. Shift or expansion of storm tracks leads to precipitation and weather changes. In the Southern Hemisphere storm tracks are observed between 60°S and 30°S over a more continuous band with a maximum over the Indian Ocean.

The interannual variability of storm tracks is influenced by ENSO / Southern Oscillation, quasi-biennial oscillation of stratospheric wind over equator (QBO), and Arctic stratospheric polar vortex changes. The North Pacific storm track area shifts equatorward and expands eastward in seasons of extreme El Niño. As a result, stronger cyclones bring heavy precipitation and mudflows to the western USA. The La Nino seasons are associated with a North Pacific storm track shift in the opposite direction [5]. During the easterly QBO, a poleward shift of the North Pacific storm track is observed. The North Atlantic storm track is characterized by downward shrinking in easterly QBO and upward expansion in westerly QBO winters [6]. Weakening of the Arctic stratosphere polar vortex is accompanied by an equatorward shift of the tropospheric jet stream related to storm tracks, whereas strengthening of the polar vortex leads to a poleward shift [7, 8, 9].

On the subseasonal time scale, the North Pacific storm track is influenced by the Madden–Julian oscillation (MJO) [6]. The eastward movement of anomalous tropical convection associated with the

MJO over the Indian Ocean and the western Pacific is accompanied by the extension of the North Pacific storm track northeastward.

The storm track time–mean position and intensity depend on the surrounding areas: the underlying surface which forms the lower boundary, the overlying stratosphere, and the tropical and high latitude troposphere on either side.

Considering significant impacts of storm tracks on the weather and regional climate and its response to climate change [10-13], our study is aimed at analyzing storm tracks realization in simulations of the 5th version of a coupled climate model developed at the Marchuk Institute of Numerical Mathematics of the Russian Academy of Sciences (INM CM5). The INM CM5 is the only Russian climate model involved in a Coupled Model Intercomparison Project (CMIP). It should be noted that notwithstanding the recent progress in the development of climate models mainly related to improvement of its horizontal resolution, increase of the vertical levels and higher upper boundary, there are still problems with realization, by such models, of the stratosphere-troposphere dynamic coupling and critical extratropical circulation processes, such as heat and cold waves and extreme precipitation events [14].

Analysis of 12 climate models has shown that biases in the winter European blocking frequency are related to the North Atlantic storm track tilt [15]. Therefore, improvement in the simulation of storm tracks in climate models is essential for better representation of other important extratropical dynamic processes.

2. Data and methods of analysis

To analyze the storm tracks we used data of five 50-year realizations of ensemble simulations with the INM CM5 with a longitude-latitude resolution of 2° × 1.5° in the atmosphere. These simulations differ from each other by slightly disturbing initial conditions. Also, two 50-year simulations of the coupled and the only atmospheric model version with a high resolution of 0.67° × 0.5° were analyzed. All these model data cover the final period (1965–2014) of the historical climate simulation for 1850–2014.

The INM CM5 includes atmosphere, ocean, sea ice, land, and aerosol modules [16]. The main novelties in comparison to the previous version of this model are improvement of the atmosphere model (increase in the vertical resolution for the upper stratosphere and lower mesosphere and improvement of the parameterization of large-scale condensation and cloudiness, and addition of the aerosol module [17]). Important model improvements in the description of the stratosphere are the ability to simulate the quasi-biennial oscillation of the equatorial zonal wind in the stratosphere and better statistics of Sudden Stratospheric Warming events. The land model includes modules responsible for the calculation of soil, underlying surface, and vegetation parameters.

Recommended for Phase 6 of the CMIP, the following external forcings on the climate system were specified for all model simulations: concentrations of CO_2, CH_4, and N_2O as average annual values; volcanic aerosol concentration as monthly mean fields depending on the latitude and height; ozone values as monthly mean fields depending on the longitude, latitude, and height and taking into account the ozone layer depletion since the early 1980s; the emission of SO_2, black and organic carbon as monthly mean fields depending on the longitude and latitude. The solar constant and the spectral distribution of radiation were specified in the form of monthly mean values.

The following storm track characteristics were calculated using a high-pass filter [18]:

the variance $vv = \overline{\left[v(t+24) - v(t)\right]^2}$

the eddy momentum flux: $uv = \overline{\left[u(t+24) - u(t)\right] \cdot \left[v(t+24) - v(t)\right]} \cos\varphi$

the eddy heat flux $vT = \overline{\left[v(t+24) - v(t)\right] \cdot \left[T(t+24) - T(t)\right]}$

the eddy moisture flux $vq = \overline{\left[v(t+24) - v(t)\right] \cdot \left[q(t+24) - q(t)\right]}$, where *u, v* are the zonal and meridional wind velocity, *T* is the temperature, *q* is the specific humidity, φ is the latitude, the overbar over the right-hand side denotes the average over a selected period (usually the monthly mean), and *t+24* means the next time step (or day).

If the storm track variance *vv* is a part of the eddy kinetic energy transferred from eddies to the zonal flow, the eddy momentum flux *uv* characterizes the interaction of baroclinic waves (that are components of the storm tracks) and a large-scale low-frequency circulation. The eddy heat flux *vT* defines the transfer of sensible heat, whereas the transfer of latent heat associated with evaporation and condensation is estimated by the moisture flux *vq*.

The storm track parameters calculated by the INM CM5 were compared with the ones derived using NCEP reanalysis data from 1968 to 2018 and ERA-Interim reanalysis data from 1979 to 2018. Since the storm track parameters in both reanalysis data are in agreement, only figures with NCEP reanalysis are shown.

3. Results

3.1. Realization of storm tracks in model simulations

The obtained results show that spatial structure and temporal variability of the storm tracks in model simulations are comparable with NCEP and ERA-Interim reanalysis data. However, the maximum variance of both storm tracks in the model simulation is 20-25% weaker (Figure 1a-b).

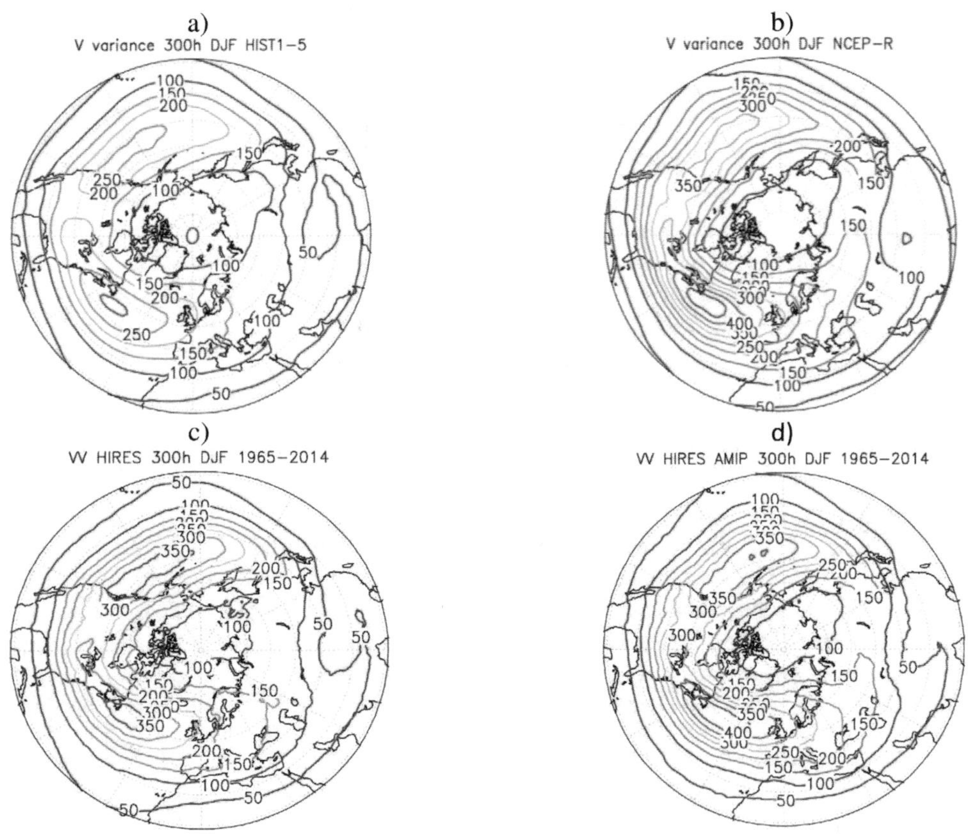

Figure 1. Storm track variance [m^2/s^2] at 300 hPa in December-February in model simulations (experiments HIST1-5) (a), with high resolution and interactive ocean HIRES-HIST (b), with high resolution and specified SST HIRES-AMIP (c) averaged over 1965 - 2014, the same but in NCEP reanalysis data averaged over 1968-2018 (d). Latitudes are from 20° N to 90° N.

Improvement of both storm-tracks variance realization in December-February averaged over 50 winters was found in the simulation with high-resolution INM CM5 in comparison with the low resolution model (Figure 1c). The realization of both storm tracks in the high-resolution atmospheric model simulation with specified sea surface temperature (SST) is better than with interactive ocean (Figure 1d). This is possibly due to biases in the simulated SST.

Analysis of the interseasonal variability of storm tracks in the upper troposphere reveals similar features in the model simulation and reanalysis data: the maximum variance of the North Pacific storm track is observed at 40-50° N in fall and spring and suppression, in the mid-winter (Figure 2a-b). The North Atlantic storm track displays the maximum variance at 50°N in December-January (Figure 2c-d).

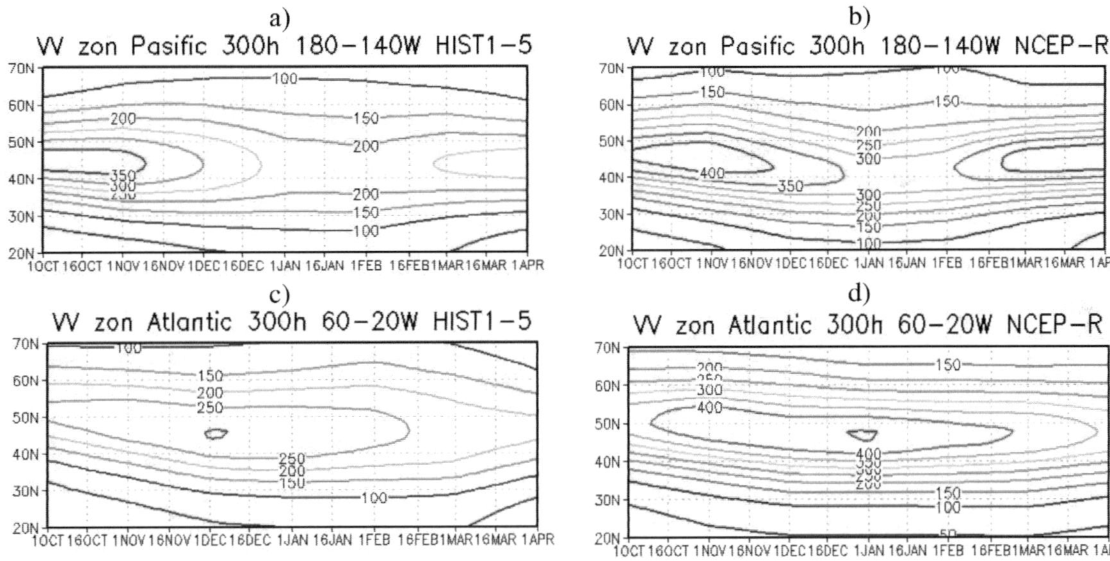

Figure 2. Latitude-time cross-section of North Pacific (180°-140° W) and North Atlantic (60°-20° W) storm tracks [m^2/s^2] at 300 hPa from October to March in the model simulation (a, c) and reanalysis data (b, d).

The eddy momentum and heat fluxes of storm tracks revealed in the model simulation and reanalysis data are also similar (not shown). Zonal mean momentum fluxes of storm tracks averaged over December-February peak between pressure levels of 300 hPa and 200 hPa and in the latitudinal band of 30-40°N. The zonal mean heat fluxes display two maximums: in the lower troposphere at 850 hPa and 40-50°N and the second one, in the upper troposphere with values that are about two times less than the first maximum. Note that the upper tropospheric maximum of the storm track heat flux in the model simulation is about two times weaker than in NCEP reanalysis.

Then the eddy moisture flux of the storm tracks responsible for the transfer of latent heat was analyzed. A maximum of the zonal mean eddy moisture fluxes of the storm tracks averaged over December-February is observed in the lower troposphere at 40°N, and their values in the model simulation and reanalysis data are comparable (Figure 3).

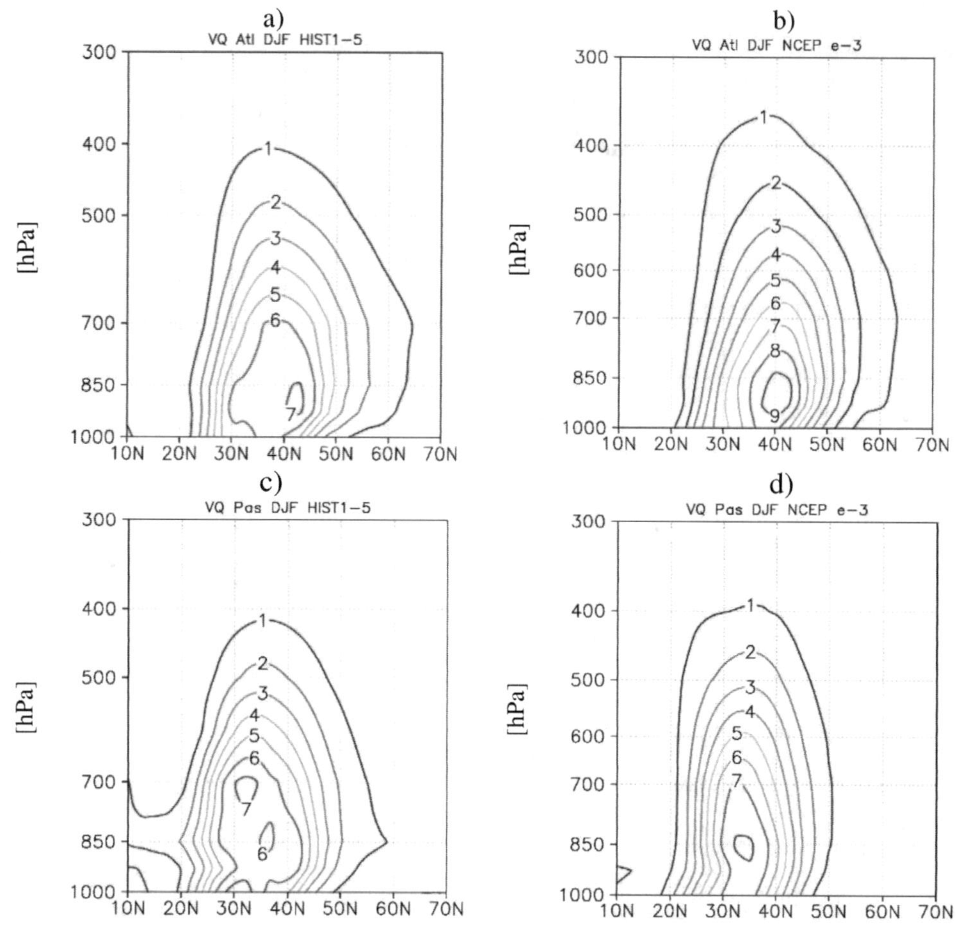

Figure 3. Altitude-latitude cross-section of the eddy moisture flux [10^{-3} m/s] of North-Atlantic storm track (100-30° W) and North-Pacific storm track (120-180° E) in December-February in the model simulation (left) and reanalysis data (right).

3.2. North Atlantic storm track response to weakening and strengthening of stratospheric polar vortex

Using composite analysis, the extension of the North Atlantic storm track caused by weakening and strengthening of the Arctic stratospheric polar vortex events with the detected influence on the troposphere was studied. These events in the model and NCEP reanalysis data were revealed early [19] and defined as follows: if the daily geopotential height anomalies from the climate mean averaged over 60-90°N and normalized with the standard deviation exceed a threshold value of $+/-1.5\ \sigma$ and propagate continuously downward from the middle stratosphere at a pressure level of 30 hPa to the upper troposphere at 300 hPa, such events were considered as troposphere-affecting, and vice versa.

The difference between the composites consisting of troposphere-affecting weakening and strengthening of the Arctic stratospheric polar vortex events shows that weakening of the polar vortex leads to an extension of the North Atlantic storm track towards lower latitudes (Figure 4a-b). The strengthening of the polar vortex, in turn, leads to an extension of this storm track towards the high latitudes. The strongest changes of the North Atlantic storm track occur at 300 hPa. Statistically significant responses of this storm track to strengthening/weakening of the stratospheric polar vortex were revealed in the model simulations and reanalysis data. Due to the mid-winter suppression of the North Pacific storm track, its response to variability of the Arctic polar vortex is less pronounced.

3.3. Storm tracks response to Arctic amplification

The warming trend in the Arctic in the recent two decades is almost twice as large as the global mean temperature trend. This result of the recent climate change is called Arctic amplification and has been accompanied by a rapid loss of sea ice, with a maximum in autumn. Observational and modeling studies suggest that the Arctic amplification may affect the middle latitudes weather [e.g., 20-24].

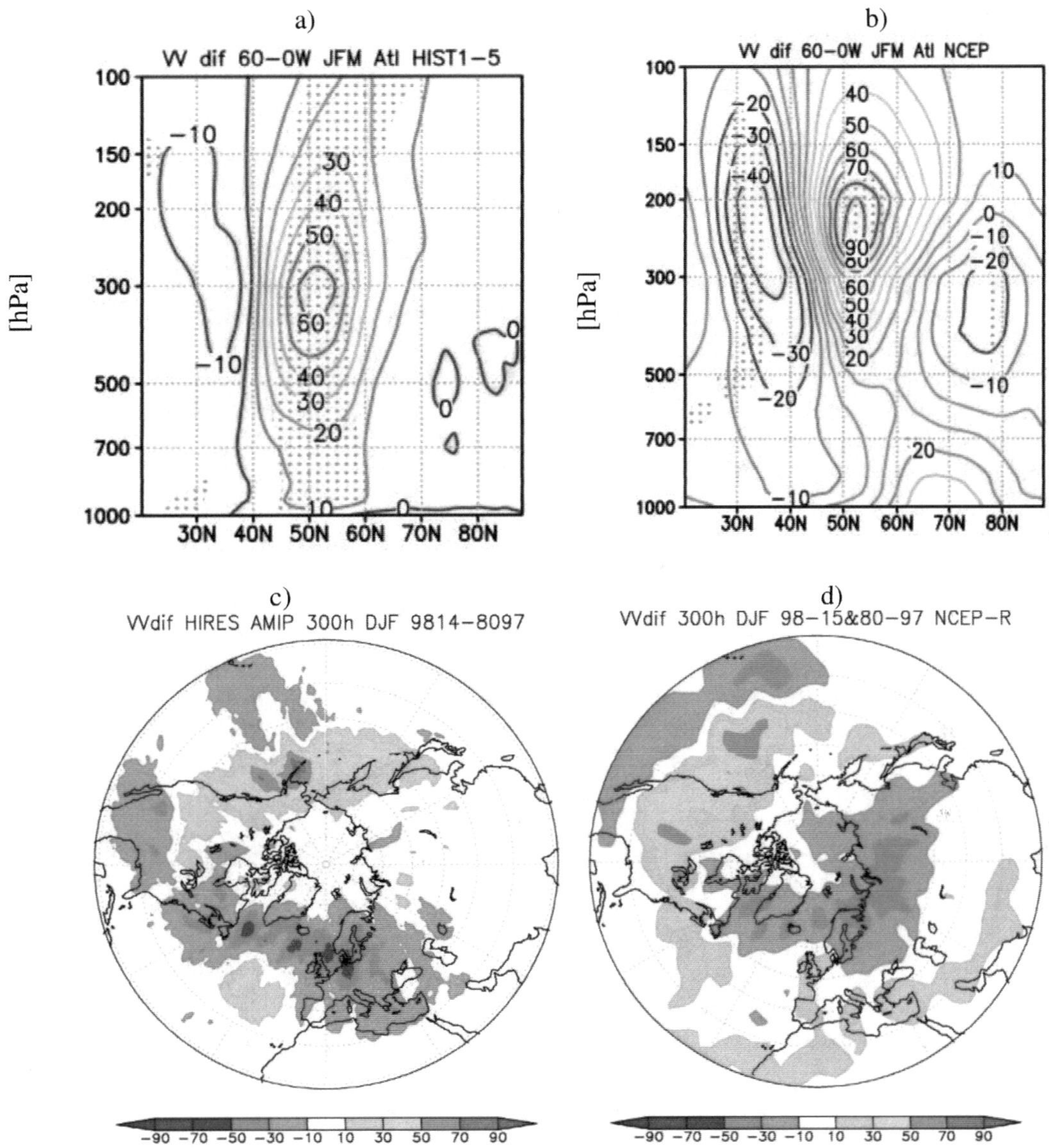

Figure 4. Altitude-latitude cross-section of the difference between the North Atlantic storm track variance [m^2/s^2] in the longitudinal band of 60° W - 0° in January-March between composites with strengthening and weakening of the Arctic stratospheric polar vortex with a tropospheric impact in the model simulation (a) and reanalysis data (b). Zones where the difference between the composites is statistically significant at a confidence level of 95% are shown by grey dots. The difference in the storm track variance [m^2/s^2] at 300 hPa in December-February between 1998 - 2014 and 1980 - 1997 in the model simulation (c) and reanalysis data (d).

The factors that contribute to the amplified warming in the Arctic include [20]:
- albedo-temperature feedback associated with a reduction of the sea ice;
- increased atmospheric humidity and the associated increase of the downwelling longwave radiation;
- increased poleward transports of heat and moisture by the ocean and atmosphere.

A number of mechanisms for middle latitude circulation response to the Arctic warming and sea ice reduction have been recently suggested, including the following [20]:
- the impact of the Arctic warming on the pressure fields and the subsequent changes of the blocking frequency;
- the Arctic and middle latitude connection via the terrestrial snow cover.

It was also shown that the Northern Hemisphere storm tracks are influenced by the Arctic amplification: a weakening of the North Atlantic storm track and a poleward shift of the North Pacific storm track were revealed using the Community Earth System Model Large Ensemble project [13].

Similar storm track changes were revealed in the model simulations with high resolution and specified SST: a poleward shift of the North Pacific storm track and a weakening of the North Atlantic storm track between two winter periods, 1998-2014 and 1980-1997 (Figure 4c-d). Similar changes of the storm tracks were revealed in reanalysis data. Both main areas of the variance change display a slight westward shift in comparison with reanalysis data.

The following explanation was suggested on the basis of analysis of the model simulations: the weakening of the North Atlantic storm track is caused by the decreased baroclinicity associated with the Arctic amplification, whereas the poleward shift of the North Pacific storm track is influenced by the La Niña–like change in the sea surface temperature [13].

4. Summary

The main characteristics of Northern Hemisphere storm tracks have been revealed and analyzed in five 50-year simulations of the INM-CM5 model version with a longitude-latitude resolution of 2.5°×2.5°, two 50-years simulations with a higher resolution (0.67°×0.5°), as well as NCEP and ERA-Interim reanalysis data. Although the spatial structure, variance, and other parameters of the simulated North Pacific and North Atlantic storm tracks are comparable with those revealed in reanalysis data, their maximum variance is weaker by about 20%. The high-resolution INM CM5 version reproduces the storm tracks better: their variance is comparable to that of reanalysis data.

The results of the composite analysis show that strengthening of the Arctic stratospheric polar vortex with its revealed tropospheric impact lead to an extension of the North Atlantic storm track toward high latitudes: its strengthening is observed at 50°N and weakening, at 30-40°N. Weakening of the Arctic polar vortex, in turn, leads to an extension of the North Atlantic storm track to lower latitudes.

A poleward shift of the Pacific storm track and weakening of the North Atlantic storm track have been revealed in INM CM5 simulations with high resolution and specified SST comparing two periods: 1998-2014 and 1980-1997. Similar changes in the storm track associated with the recent Arctic amplification have been revealed in reanalysis data. These are in agreement with the results of [13].

Finally, we conclude that the realization of Northern Hemisphere storm tracks in the INM CM5 historical simulations and their responses to the strengthening and weakening of the Arctic stratospheric polar vortex with the tropospheric impact and the recently observed Arctic amplification are comparable with the storm track parameters revealed in the reanalysis data.

Acknowledgments
This research was supported by the Russian Found of Basic Research under project (#19-05-00370) on "Investigation of Stratosphere-Troposphere Dynamical Coupling in Present and Future Climate".

References

[1] Chang E, Lee S and Swanson K 2002 Storm Track Dynamics *J. Climate* **15** 2163-82

[2] Chang E, Guo Y, Xia X and Zheng M 2013 Storm-Track Activity in IPCC AR4/CMIP3 Model Simulations *J. Climate* **26** 246-60
[3] Nakamura H 1992 Midwinter suppression of baroclinic wave activity in the Pacific *J. Atmos. Sci.* **49** 1629–42
[4] Christoph M, Ulbrich U and Speth P 1997 Midwinter suppression of Northern Hemisphere storm track activity in the atmosphere and in GCM experiments *J. Atmos. Sci.* **54** 1589–99
[5] Straus D and Shukla J 1997 Variations of midlatitude transient dynamics associated with ENSO. *J. Atmos. Sci.* **54** 777–90
[6] Wang J, Kim H-M and Chang E 2018 Interannual Modulation of Northern Hemisphere Winter Storm Tracks by the QBO *Geophys. Res. Lett.* **45** (6) 2786-94
[7] Walter K and Graf H-F 2005 The North Atlantic variability structure, storm tracks, and precipitation depending on the polar vortex strength. *Atmos. Chem. Phys.* **5** 239-48
[8] Baldwin M and Dunkerton T 1999 Propagation of the Arctic Oscillation from the stratosphere to the troposphere. *J. Geophys. Res.* **104** 30937–46
[9] Kidston J *et al* 2015 Stratospheric influence on tropospheric jet streams, storm tracks and surface weather *Nature Geoscience* **8** 433–40
[10] Bengtsson L, Hodges K and Roeckner E 2006 Storm Tracks and Climate Change *J. Climate* **19** 3518-42
[11] Martynova Yu V and Krupchatnikov V N 2015 Peculiarities of the Dynamics of the General Atmospheric Circulation in Conditions of the Global Climate Change *Izvestiya, Atmospheric and Oceanic Physics* **51** (3) 306–17
[12] Shaw T, Baldwin M, Barnes E *et al* 2016 Storm track processes and the opposing influences of climate change *Nature Geoscience* **8** 433–40
[13] Wang J, Kim H-M and Chang E 2017 Changes in Northern Hemisphere Winter Storm Tracks under the Background of Arctic Amplification *J. Climate* **30** 3705-24
[14] Shepherd T 2014 Atmospheric circulation as a source of uncertainty in climate change projections *Nature Geoscience* **7** (10) 703–08
[15] Zappa G, Masato G, Shaffrey L, Woollings T and Hodges K 2014 Linking Northern Hemisphere blocking and storm track biases in the CMIP5 climate models *Geophys. Res. Lett.* **41** 135–9
[16] Volodin E M, Mortikov E V, Kostrykin S V, Galin V Ya, Lykosov V N, Gritsun A S, Diansky N A, Gusev A V and Yakovlev N G 2017 Simulation of Modern Climate with the New Version of the INM RAS Climate Model *Izvestiya, Atmospheric and Oceanic Physics* **53** (2) 142–55
[17] Volodin E M and Kostrykin S V 2016 The Aerosol Module in the INM RAS Climate Model *Russian Meteorology and Hydrology* **41** (8) 519–28
[18] Wallace J, Lim G and Blackmon M 1988 Relationship between cyclone tracks, anticyclone tracks and baroclinic waveguides *J. Atmos. Sci.* **45** 439–62
[19] Vargin P, Kostrykin S and Volodin E 2018 Analysis of Simulation of Stratosphere-troposphere Dynamical Coupling with the INM-CM5 Climate Model *Russian Meteorology and Hydrology* **43** (11) 780-6
[20] Walsh J E 2014 Intensified warming of the Arctic: Causes and impacts on middle latitudes *Global and Planetary Change* **117** 52–63
[21] Cohen J *et al* 2014 Recent Arctic amplification and extreme mid-latitude weather *Nature Geoscience* **7** 627–37
[22] Overland J, Francis J, Hall R, Hanna E, Kim S-J and Vihma T 2015 The Melting Arctic and Midlatitude Weather Patterns: Are They Connected? J. Climate **28** 7917-32
[23] Semenov V A and Latif M 2015 Nonlinear winter atmospheric circulation response to Arctic sea ice concentration anomalies for different periods during 1966–2012 *Environ. Res. Lett.* **10**, 054020, doi:10.1088/1748-9326/10/5/054020
[24] Vihma T, Ruhe G, Chen L *et al* 2019 Effects of the tropospheric large-scale circulation on European winter temperatures during the period of amplified Arctic warming *Int. J. Climatology* DOI: 10.1002/joc.6225

Comparative analysis of hydrothermal conditions of Tomsk region by using different drought coefficients

A A Ryazanova[1] and N N Voropay[1,2]

[1]Institute of Monitoring of Climatic and Ecological Systems SB RAS, Russia, Tomsk, Akademicheskii 10/3
[2]V.B. Sochava Institute of Geography SB RAS, Russia, Ulan-Batorskaya 1

Abstract. The global and regional climatic changes that have been most intense in recent decades are characterized by an increase in the frequency of extreme natural events which include droughts and periods of extensive moistening. Air temperature and precipitation are the main characteristics of climate indicating the changes in the droughts/extensive moistening. According to observations of weather stations of Tomsk region, air temperature has been changing very much since the 70s of the 20th century (during 1967-1997 from -0.6 to + 0.56°C and during 1970-2009 from 0.4 to 0.6°C/10 years). Since the mid-1980s, an increase in precipitation has been observed. For a detailed analysis of drought indices (HTC, S, SPI, and SPEI) in the present paper, we use observation data from the Tomsk region weather stations from May to September, 1966-2017. When calculating the indices, warming of the atmosphere, expressed through air temperature, as well as atmospheric humidification, characterized by precipitation, must be taken into account. Also, for a correct comparison of the hydrothermal conditions of different landscapes it is necessary to use normalized indicators. A comparative analysis of the different hydrothermal indices has shown that the S and SPEI indices most closely correspond to these requirements. They also most synchronously indicate droughts.

1. Introduction

Global and regional climatic changes that have been intensively occurring in the recent decades [1-5] are characterized by an increase in the frequency of extremes, which include droughts and periods of extensive moistening. Air temperature and precipitation are the main characteristics of climate indicating changes in the droughts/extensive moistening. A large number of different complex indicators based on these characteristics have been developed to determine droughts. Some indicators require only precipitation for calculation, and other complex indicators require soil moisture in addition to air temperature and precipitation. Some indicators (SPI, SPEI, Palmer index, etc.) allow taking into account changes in meteorological characteristics not only during the summer months, but also during the previous cold period that affect the hydrothermal condition of the soil and atmosphere at the beginning of the growing season.

In Russia, the following indicators are most often used to determine droughts: the Selyaninov hydrothermal coefficient (HTC) [6] and the Ped drought index (S) [7]. In 2009 WMO [8], along with regional indicators, proposed a standardized precipitation index (SPI) [9] for worldwide use. Currently there are various modifications of this index. One of the most frequently used is the standardized precipitation and evapotranspiration index (SPEI) [10].

The main goal of this work is a comparative analysis of various indicators of droughts and the determination of the indicator or the groups of indicators that best describe the hydrothermal conditions in Siberia.

Many researchers use different GIS applications to perform calculations and visualize results. The main tool of our work is the previously developed web-GIS “CLIMATE” [11-13] aimed at climate-ecological studies in a selected region and realizing the necessary integration of the results of retrospective and predictive modeling and observations (http://climate.scert.ru/). The "CLIMATE" system, based on web and GIS technologies, is a part of the hardware and software complex for "cloud" analysis of climate data, including various sets of climate and meteorological data, as well as special interactive tools for their search, sampling, processing, and visualization. Using this system greatly facilitates and accelerates work with large volumes of geospatial climatic data, allowing a user who is not an expert in information technologies to remotely perform their statistical analysis using any modern desktop PC connected to the Internet.

2. Objects, data, and methods

In previous papers [14, 15], a comparative analysis of the most well-known Russian and international indicators of aridity, such as the Selyaninov hydrothermal coefficient (HTC), the Ped drought index (S), the standardized precipitation index (SPI), and the standardized precipitation and evapotranspiration index (SPEI), was carried out for the territory of Western Siberia. The study used the Era Interim reanalysis [16] data from 1979 to 2017. Since the obtained results did not give an understanding of which of the indicators is better, it was decided to narrow the study area and perform a more detailed analysis. Tomsk region was chosen as the study area, and we used air temperature and precipitation data from meteorological stations of Roshydromet (“RIHMI-WDC”) [17]. There are 10 stations on the territory of Tomsk region (Aleksandrovskoye, Sredniy Vasyugan, Napas, Vanzhil-Kynak, Pudino, Kolpashevo, Ust’-Ozernoe, Bakchar, Pervomayskoye, and Tomsk). It was also decided to take 3 (Severnoye, Bolotnoye, and Taiga) stations located close to the region (Figure 1).

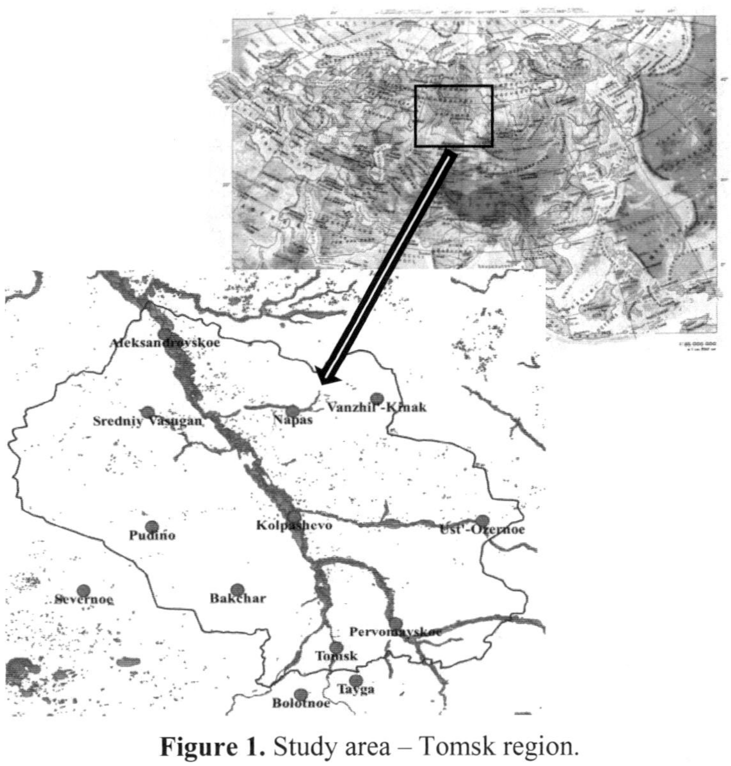

Figure 1. Study area – Tomsk region.

IOP Conf. Series: Earth and Environmental Science **386** (2019) 012008 doi:10.1088/1755-1315/386/1/012008

The study period was chosen from 1966 to 2017, according to the direction of Roshydromet [17]. The homogeneity of the time series of precipitation was disturbed at least three times since 1966. In the first one, in the 30s of the last century, there was a massive transfer of stations to an open space, representative of several elements, but not of precipitation. In the second one, at the beginning of the 50s, a massive change in the measuring device was made. A precipitation gauge with a Tretyakov shield was installed. In the third one, since January 1966, a correction to wetting began to be applied directly at the station. After 1966 there were no changes in the measurement and processing and, therefore, the precipitation series can be considered homogeneous.

Calculation of hydrothermal indices, as in previous works [4, 16, 18], was performed using the web-GIS "CLIMATE".

It is more correct to use normalized indicators when comparing the hydrothermal conditions in different landscapes and, therefore, the normalized anomaly of HTC (HTC_{norm}) (and not the HTC) was used in the work. When determining a drought, its intensity is taken into account according to the classification in Table 1.

Table 1. Classification of droughts based on HTC_{norm}, S, SPI, and SPEI.

Intensity of drought	HTC_{norm}	Ped index	SPI/SPEI
weak	$-1.25 < HTC_{norm} \leq -1$	$1 \leq S < 2$	$-2 < SPI/SPEI \leq -1$
medium	$-1.5 < HTC_{norm} \leq -1,25$	$2 \leq S < 3$	$-3 < SPI/SPEI \leq -2$
severe	$-1.75 < HTC_{norm} \leq -1,5$	$3 \leq S < 4$	$-4 < SPI/SPEI \leq -3$
extreme	$HTC_{norm} \leq -1,75$	$S \geq 4$	$SPI/SPEI \leq -4$

3. Results and discussion

3.1. Climatic characteristics of the territory

According to the observations of meteorological stations of Tomsk region, changes in the average annual air temperatures have been observed since the 70s of the 20th century. Their growth is especially pronounced in the subtaiga zone. According to the data of the Tomsk weather station, the average annual air temperature changed from -0.6 to +0.56 °C in 1967-1997. If we assume that the trend is linear, its parameters were + 0.67 ° C/10 years [19]. In the following decade, the warming process accelerated. This is confirmed by studies of other authors [20], who noted a rise in the air temperature from 1970 to 2009 according to meteorological stations throughout the region at a rate of 0.4-0.6 °C/10 years, and individual months of the cold period (February-March) trends reach 1.5-1.7 °C/10 years. Since the warming rate is greater than in the Northern Hemisphere, this may be due to the deviation from the long-term norms of the characteristics of the regional circulation [21]. Analysis of the obtained data showed that the greatest changes occurred in the cold part of the year and, especially, in the central phase of the winter season [22]. Since the mid-1980s, an increase in the precipitation amounts has been observed. The trends in the annual precipitation have a positive direction, which reaches a maximum due to the precipitation during the cold period (October-April), especially in Aleksandrovskoe, Tomsk, and Kolpashevo [20].

The average long-term spatial characteristics of the climate (air temperature and precipitation) and the current trends of their changes were estimated using the web-GIS "CLIMATE". Figure 2 shows the spatial distribution of the air temperature and precipitation amount for individual months according to the ERA-Interim reanalysis with a grid of 0.75 × 0.75° for 1979-2017 for the territory of Tomsk region.

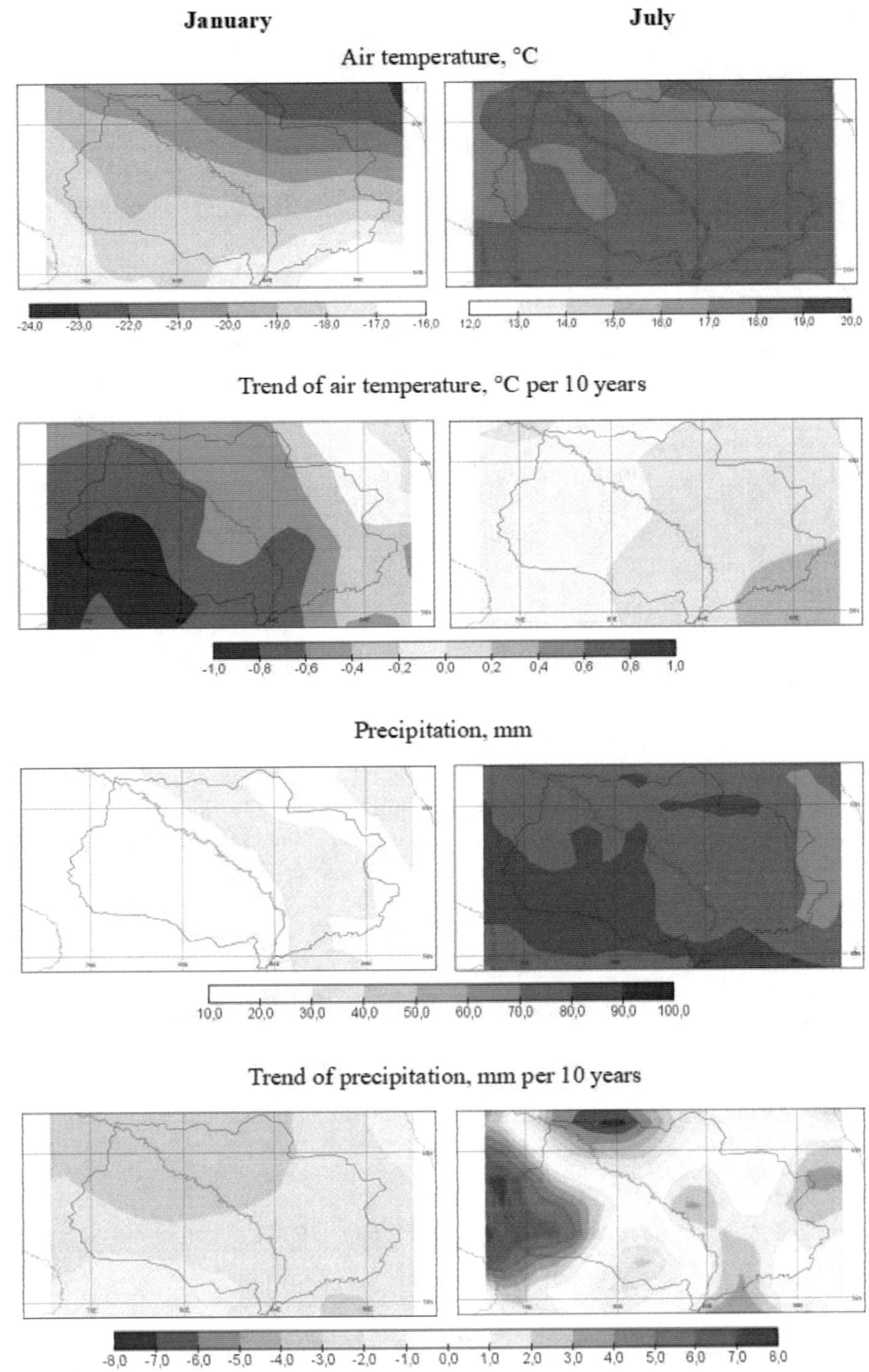

Figure 2. Average long-term values and trends of average monthly temperature and precipitation amounts for January, July (1979-2017). Field of the study - within box 55-61 N, 75-90 E.

In winter, there is a latitudinal distribution of the air temperature in the territory. The average long-term air temperatures in January vary from -23 ° C in the north of the region to -16 ° C in the south-east. In July, the temperature field is more homogeneous: the average monthly temperatures are from

17 to 19 °C. The values of air temperature trends according to reanalysis data for 1979-2017 differ significantly from the estimates obtained by other authors for 1967-1997 [19] and 1970–2009 [20]. The winter months are characterized by negative trends throughout the region (up to -1.0 °C/10 years in the south-west), and in summer, by positive ones (up to + 0.4 °C/10 years in the south-east). In the summer months in Tomsk region, the precipitation is about 2 times more than in winter. On average, over the period under consideration the precipitation reaches 60–90 mm in July and 20–40 mm in January. The long-term changes in individual months are mostly statistically insignificant (-1 ... -4 mm/10 years in January and -8 ... + 8 mm/10 years in July).

3.2. Analysis of the calculation of indices

Correlation analysis confirmed the synchronicity of changes in the considered indices of droughts. However, the identified extreme events in individual years (intensity, duration of droughts) do not always coincide. For a more detailed analysis of the indices, in this work we used the series of observations of meteorological stations of Tomsk region from May to September (the sum of precipitation and air temperature) for 1966-2017.

For the majority of meteorological stations in the region in May and September, the HTC was not calculated due to the absence of average daily air temperatures above 10 °C in the observation series (which is a necessary condition for the calculation). Consequently, the relationship between the HTC and other indices for May and September was not defined. In other cases, statistically significant correlation coefficients were obtained. The closest relations were observed at the beginning of the growing season (May, June).

Despite some differences in determining the intensity of droughts, the analysis for a long-term period made it possible to identify phenomena of severe and extreme intensity that correspond to the characteristics given in Table 1. The driest months during the study period were May (1999, 2004), June (1981, 1982, 1983, 2012), July (1998, 1999, 2012), August (2003, 2016), and September (1966, 2016). In this time most meteorological stations recorded a state of severe or extreme drought. In 2012 and 2016, the dry period lasted 2-4 months throughout Tomsk region. In most cases drought is observed in the first months of the growing season. However, in 1966, 2011, and 2016, according to the results for all calculated indices, droughts were observed in August-September at many meteorological stations, with medium humidification at the beginning of summer.

In recent years the frequency of hydroclimatic extremes has increased [2, 15]. In this connection, the hydrothermal conditions were analyzed in more detail since 2000. 100% coincidence (9 cases) in determining the state of drought was shown by a comparison of the S and SPEI indices (Figure 3). At the same time, according to the calculations of the Ped index, in most cases the drought was determined as severe, and in June 2012 and September 2016 as extreme. The SPEI values in all cases correspond to a weak drought. The SPI index behaves similarly; only a weak drought was observed in the territory (8 cases). The analysis of the HTC_{norm} showed coincidence with the other indices in 4 cases.

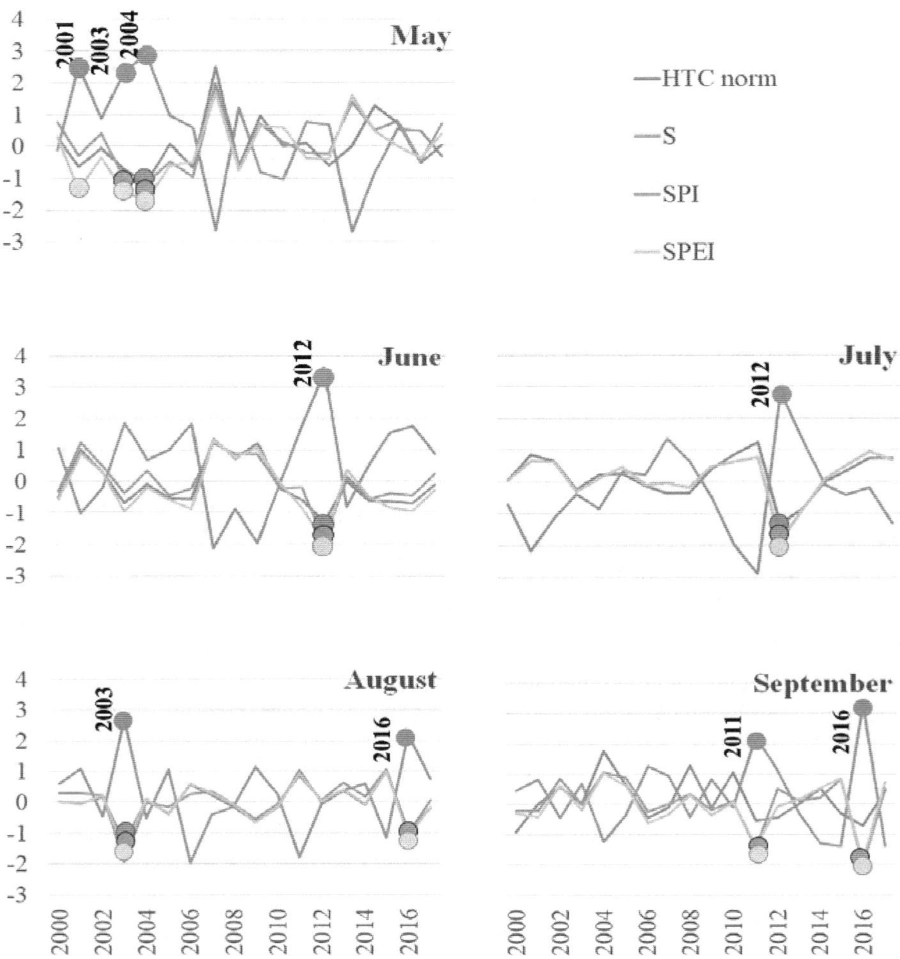

Figure 3. Changes in hydrothermal indices (circles indicate cases of drought) in Tomsk region average for all stations.

4. Conclusions

The study of atmospheric droughts is an important component of landscape monitoring. At the same time, the correct choice of indicators of droughts is important. A comparison of the most frequently used Russian and international drought indices yielded statistically significant correlation coefficients. The closest relations are observed for the beginning of the growing season (May, June). A cause of atmospheric droughts can be high air temperatures and/or lack of precipitation. The limiting role of these factors may vary during the growing season and depending on the landscape. Thus, when calculating the indices warming of the atmosphere, expressed through air temperature, as well as atmospheric humidification, characterized by the amounts of precipitation, must be taken into account. Also, for a correct comparison of the hydrothermal conditions of different landscapes it is necessary to use normalized indicators. The above comparative analysis of the different hydrothermal indices has shown that the S and SPEI indices most closely correspond to these requirements. They also most synchronously indicate droughts in individual months. The driest months during the study period were May (1999, 2004), June (1981, 1982, 1983, 2012), July (1998, 1999, 2012), August (2003, 2016), and September (1966, 2016). In these months, most of the meteorological stations recorded severe or extreme droughts.

Acknowledgments
This study was funded by the Russian Foundation for Basic Research and the Tomsk Region Administration under grant no. 18-45-700015 and the basic research program AAAA-A17-117013050037-0.

References
[1] Stocker T F, Qin D, Plattner G-K, Tignor M, Allen S K, Boschung J, Nauels A, Xia Y, Bex V and Midgley P M 2013 *Climate Change 2013: The Physical Science Basis. Contribution of Working Group I to the Fifth Assessment Report of the Intergovernmental Panel on Climate Change* (Cambridge: Cambridge University Press and NY, USA) p 1535
[2] Kattsov V M *et al* 2014 *Second Roshydromet assessment report on climate change and its consequences in Russian Federation* (Moscow: Federal service for hydrometeorology and environmental monitoring (Roshydromet)) 56
[3] Maksyutova E V *et al* 2012 Tendencies of hydroclimatic changes on the Baikal *Natural territory geography and natural resources* **4** 72-80
[4] Voropay N N, Maksyutova E V and Riazanova A A 2016 Hydrothermal conditions at the south of East Siberia during the ongoing warming *IOP Conference Series: Earth and Environmental Science* **48** (1) 012003
[5] Voropay N N, Kichigina N V 2018 Long-term changes in the hydroclimatic characteristics in the Baikal region *IOP Conference Series: Earth and Environmental Science* **107** 012042
[6] Selyanoinov G T 1928 About agricultural estimate of climate *Proc. of Agricultural Meteorology* **20** 165-77
[7] Ped D A 1975 About index of drought and excessive wetting *Proc. of the Hydrometcentre of USSR* **156** 19-38
[8] World Meteorological Organization 2012 *Standardized Precipitation Index User Guide* 1090 (World Meteorological Organization, Geneva, Switzerland)
[9] McKee T B, Doesken N J and Kleist J 1993 The relationship of drought frequency and duration of time scales *Eighth Conference on Applied Climatology American Meteorological Society* **8** 179-86
[10] Vicente-Serrano S M, Beguería S and López-Moreno J I 2010 A Multi-scalar drought index sensitive to global warming: The Standardized Precipitation Evapotranspiration Index – SPEI *Journal of Climate* **23** 1696-718
[11] Gordov E P *et al* 2016 Development of Distributed Research Center for analysis of regional climatic and environmental changes *IOP Conf. Series: Earth and Environmental Science* **48** 012033
[12] Gordov E P *et al* 2018 Development of Information-computational Infrastructure for Modern Climatology *Russian Meteorology and Hydrology* **43** (11) 722-728
[13] Riazanova A A *et al* 2016 Development of computational module of regional aridity for web-gis "CLIMATE" *IOP Conference Series: Earth and Environmental Science* **48** 012032
[14] Voropay N N, Ryazanova A A 2018 A comparative assessment of the aridity indices for analysis of the hydrothermal conditions *IOP Conf. Series: Earth and Environmental Science* **190** 012041
[15] Voropay N N, Ryazanova A A 2018 Atmospheric droughts in Southern Siberia in the late 20th and early 21st centuries *IOP Conf. Series: Earth and Environmental Science* **211** 012062
[16] Dee D P *et al* 2011 The ERA-Interim reanalysis: configuration and performance of the data assimilation system *Quarterly Journal of the Royal Meteorological Society* **137** 553–97
[17] Bulygina O N, Veselov V M, Razuvaev V N and Aleksandrova T M Description of the array of urgent data on the main meteorological parameters at Russian stations Certificate of State Registration of the Database 2014620549 (in Russian) Income accessed online on 28th July 2018 via http://meteo.ru/data/163-basic-parameters#description of the mass-data
[18] Ryazanova A A, Voropay N N 2017 Droughts and Excessive Moisture Events in Southern

Siberia in the Late XXth - Early XXIst Centuries *IOP Conference Series: Earth and Environmental Science* **96** 012015

[19] Romashova T V *et al* 2002 Hidden patterns of climate change in Tomsk in the period 1967-1997 *Proc. of the «ENVIROMIS 2002» (July 6-15, 2002, Tomsk, Russia)* 97

[20] Barashkova N K *et al* 2011 Evaluation of the contemporary climate component of Tomsk region economic complex during cold season *Tomsk state university journal* **351** 163-198

[21] Gorbatenko V P 2011 Circulating factors role in the climate warming of Siberia *Tomsk state university journal* **346** 174-180

[22] Romashova T V 2004 Seasonal climate rhythms and their influence on the progress of soil erosion (on the example of the south of the Tomsk region) *Ph.D. thesis* (Tomsk) p 239

Climate clusters of Eurasia against the background of global climate change

N N Cheredko, V A Tartakovsky and Y V Volkov

Institute of Monitoring of Climatic and Ecological Systems SB RAS
634055, Akademichesky Av., 10/3, Tomsk, Russia

E-mail: atnik3@rambler.ru, trtk@list.ru

Abstract In this paper we suggest a solution to the problem of identifying the patterns of climate system transformations against the background of global climate change by means of studying changes in the pattern of climate classifications. We present an implementation of the objective classification method using an algorithm proposed by us for the territory of Eurasia for two different time intervals corresponding to global temperature trends. The results of applying our approach to temperature series measured at 485 weather stations in Eurasia from 1955 to 2016 are presented. The classification algorithm is based on comparing changes in the phases of the temperature series. A comparative analysis is performed to show changes in the temperature synchronization regime against the background of climate warming and changes in the structure of climate classes as well. The Russian Arctic and Subarctic, South Western Siberia, as well as stations located in mountainous regions of Eurasia, have shown high sensitivity to the changing climate. We believe that our approach will allow the formation of a general view on the phenomenon under study. This approach can be useful for the analysis of observational data, analytical transformations, and climate modelling.

1. Introduction

Changes in the consistency structure of temperature fluctuations can be a reflection of a climate system transition from one quasi-stable state to another. Against the background of the global climate change, the problem of climatic classification is being extended, becoming both diagnostic and prognostic. An objective solution to such an extended problem can be provided by identifying restructuring patterns for fields of climate parameters [1]. It is important to study the influence of external factors on the formation and transformation of regional climate clusters in view of the increasing significance of improving predictability of medium-term and long-term climate forecasts for the purpose of optimizing systems of adaptation to changing environmental conditions.

The main factor of the current climate change is considered to be the increasing anthropogenic influence on the geosystem. However, a number of factors of the climate change do not depend on the human activity but are natural. In any case, one of the main mechanisms of climate regulation is the mode of large-scale atmospheric circulation. Its properties may change in the process of transition to a new state depending on the external influences. Changes in the structure of the Eurasian climate system can be consistent, in particular, with changes in the activity of the western transfer, which is reflected in the North Atlantic Oscillation index (NAO index) [2-5] whose influence can be traced throughout the greatest part of the Northern Hemisphere [5] and which is mainly expressed in the cold period of the year.

2. Data series and methods

This paper presents the results of assessing the consistency of the climatic component structure of the Eurasian geosystem with the long-term trends of the global temperature, which is recognized as one of the main integral indicators of the global climate change [6]. The material used for the study is the series of average monthly temperatures obtained from 485 meteorological stations of Eurasia from 1955 to 2011 [7].

Air temperature can be considered as a continuous integral characteristic reflecting the weather and climate conditions of an area, since it is a result of a combined action from a complex of various factors of environment-forming and climate-forming processes characterized by nonlinear interactions. Surface temperature data are fluctuations, which can be described in terms of amplitude (envelope) and phase. The fluctuation phase of the climate parameters can be considered a self-organization indicator for geosystems of different levels reflecting a certain equilibrium state of a system through the consistency structure of changes in its parameters. In the case of external influence, the response of most regional geosystems can be consistent with this impact in the form of a certain deviation of the values from the average data, with the subsequent return to the norm for a given state. With an increase in the duration of exposure, or in the case of exceeding the threshold level, either directly or by the accumulated effect, the dependence of any geosystem parameter on it cannot remain linear. The existing consistency in the dynamics of the parameters of a regional climate system is distorted. The differently directed phase shifts of fluctuations form a new structure of fields of the climate parameters.

We have considered different periods: 1976-2011, the period of the most pronounced trend of global warming, and 1955-1975, the period of accelerated climate warming. The modern warming which began in the mid-1970s and is most intensely manifested during the winter months [6] is associated with the increase in the zonal transfer during the positive phase of the NAO index in the winter period [3].

For each of the periods, considering each meteorological station, we determined the phase of the annual temperature variation. The classification was constructed using the numerical algorithm described in [1]. As a basis we used the assumption on the modulation of temperature fluctuations being geographically dependent on the external factors regulating the influx of solar radiation to the underlying surface and its energy exchange with the atmosphere. By inter-comparing all estimates of the phases of the annual temperature variation, we formed groups of stations with similar laws of phase changes within the period under consideration. The level of relationship closeness was determined by the correlation coefficient. This classification method corresponds to the principles of any regionalization: the differences within a single class are smaller than those between the classes, and each class is geographically clearly localized; additionally, the spatial arrangement of the classes corresponds to the well-known classical ideas about the climate types, which confirms the physical validity of the method.

3. Results and discussion

The classification results for the indicated periods allowed identifying areas which are most sensitive to changes in the global temperature (Figure 1).

The territories where the weather stations changed their structural affiliation within the period from 1976 to 2011 compared to 1950-1975 were considered to be sensitive to the global climate trends. The hemisphere regions where no changes were detected in the field structure against the background of the global temperature change were considered to be resistant to changes in the current climate.

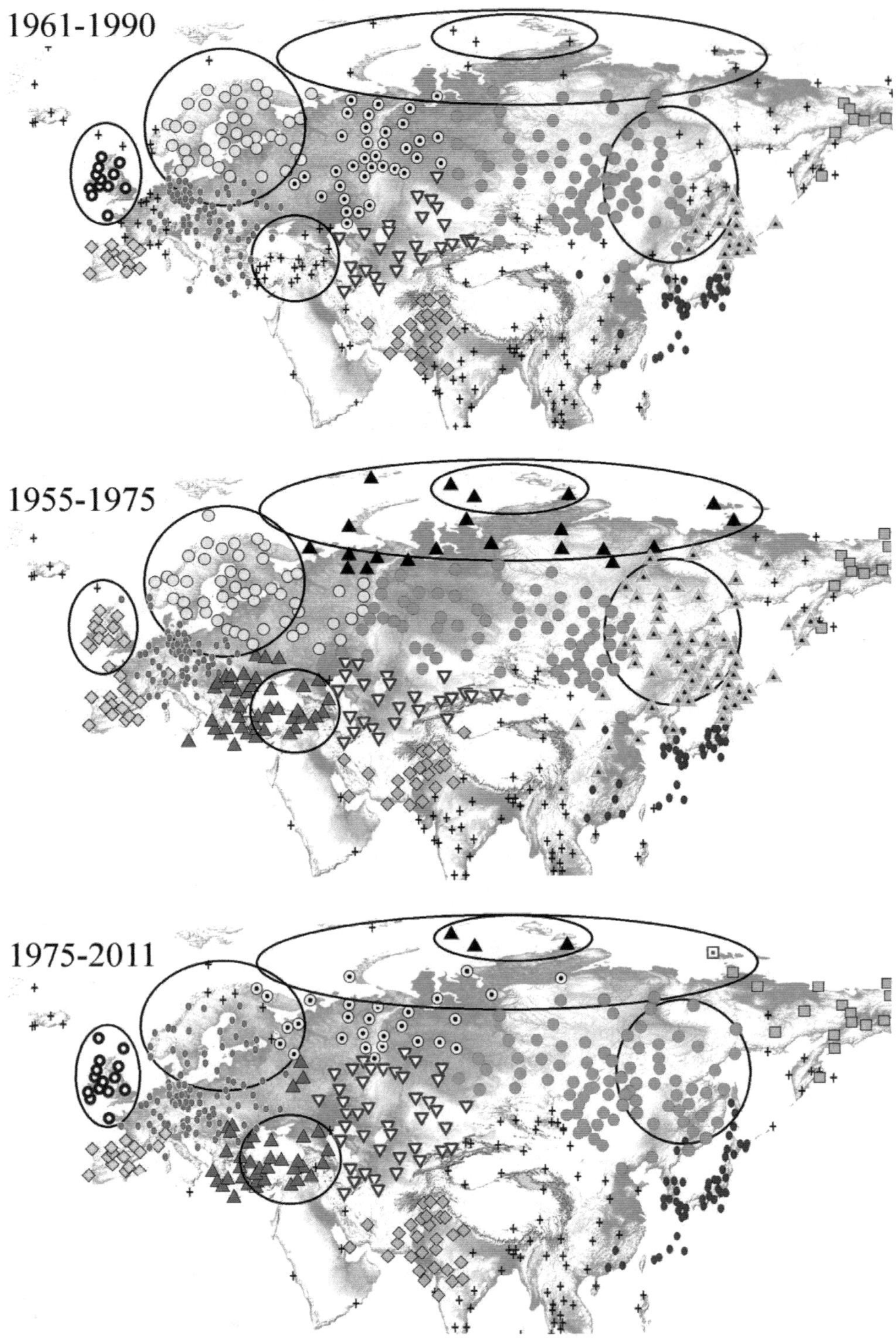

Figure 1. Classification of surface temperature field of Eurasia at different time intervals. Stations not included in any of the classes are marked with a "+". Different classes of stations are indicated by different symbols. The ovals are areas where changes occurred in the field structure.

IOP Conf. Series: Earth and Environmental Science **386** (2019) 012009 doi:10.1088/1755-1315/386/1/012009

Some stations (Figure 1) have a special type of the phase; hence, they are not included in any of the classes and are marked "+". The annual variation in the areas where these stations are located is not the dominant temperature fluctuation and, therefore, the phase changes are quite small. A common feature of most of these stations is their location in the areas of the monsoon circulation effect, which has various degrees of severity in different regions. Within the period of the most pronounced global warming the number of stations not belonging to any of the classes increased (Figure 1). This may, on average, indicate a downward trend in the consistency of the dynamics of the temperature field compared to 1955-1975. However, there are areas where synchronization of the temperature changes increased. For example, the stations of the Sakhalin and Japan Islands which had been distributed in two classes formed one class in 1976-2011. The latter can be associated with a decrease in the activity and duration of the winter monsoon and an increase in the summer monsoon in this area within the period, as a result of which the synchronizing effect of the ocean in the temperature field of this region may increase [8].

There is a well-known phenomenon of increasing warming in the Arctic at the end of the 20th and the beginning of the 21st centuries [9]. The obtained classification results confirmed that the Russian Arctic and Subarctic are regions which are sensitive to changes in the global temperature. On average, in the field structure the stations of these regions of Eurasia were combined into one class for the period from 1955 to 1975 (Figure 1). During the period of pronounced global warming the temperature dynamics at these stations became consistent with the type of temperature fluctuations of the stations belonging to more southern classes (Figure 1).

The inland continental regions of Eurasia are also sensitive to fluctuations in the global temperature. In the period from 1955 to 1975 the stations of the Central Siberian Plateau formed one extensive class with the stations of the West Siberian Plain, the mountains of Southern Siberia and Mongolia. Against the background of the global climate warming the stations of the West Siberian Lowland were divided into three classes in spite of a vast uniform inland lowland landscape. It is likely that the activation of the western transfer contributed to an increase in the consistency of the temperature fluctuations in the central and northern parts of Western Siberia and over the East European Plain. The Middle Urals which previously served as a natural climatic boundary, being the lowest part of the Ural Mountains, lost this function. The stations of the south of the West Siberian Plain were united in one class with the stations of the plains of Inner Kazakhstan, Caspian and Central Asia, as well as the mountains and highlands of Central Asia.

During the period of intensified warming the stations of the Central Siberian Plateau, the mountains of Southern Siberia and Mongolia became more consistent with the climate dynamics at the stations of Inner Yakutia and the Continental East of Asia. In the period from 1955 to 1975 the stations of Inner Yakutia, Mainland East Asia, Kamchatka, and the northern part of the Japanese Islands formed a separate class. Such redistribution can be associated with an increased influence of the Asian maximum. According to [10], in the late 20th – early 21st centuries it is possible to observe a tendency of the Asian anticyclone centre to shift to the north and west. At the same time, above we noted a tendency associated with a decrease in the activity and duration of the winter monsoon here in the recent decades [9].

The stations located in the regions of the mountainous terrain of Eurasia confirmed their high sensitivity to the changing climate. 24% of the stations located above 800 m proved to be sensitive to changes in the global temperature. Of the 12 considered stations located above 2,000 metres, 4 stations changed their classes depending on the period: the alpine stations Villacheralpe, Saentis, Sonnblick and the Himalayan station Mukteswar.

Throughout the periods of the greatest warming rates the temperature fluctuations at the stations of Fennoscandia synchronized with the stations of more southern areas: the Atlantic and Central European ones. The stations of the British Isles formed a class separately from the stations of the Iberian region, which had been included in the class before the warming period.

As a result of the analysis it can be concluded that the consistency structure of the temperature fluctuations appears to be most stable in the European Mediterranean territories, West and South-West

Asia, the Indo-Gangetic lowland, and in South Asia. The influence of such a large-scale mode of climatic variability as the North Atlantic Oscillation is not determinative in this case. Identifying other causes of such structure stability requires additional research.

Using only the parameter of temperature for the classification, namely, the phase of its annual course, allows simplifying the dynamic analysis of transformation of the climate system structure. The results have shown significant changes in the consistency structure of the temperature fluctuations in most parts of Eurasia.

4. Conclusions

The advantages of the above-proposed classification method are its compactness and objectivity, which are expressed in obtaining a natural set of structures reflecting the consistency of temperature fluctuations, as well as using accurately measured data not taking into account bioclimatic criteria whose database is insufficient.

Using this method of objective classification, we identified areas where temperature synchronization is most sensitive to global climate change: the Russian Arctic and Subarctic regions, Fennoskandia, the British Isles, the Central European Region, South Western Siberia, as well as stations located in mountainous regions of Eurasia.

Using only one parameter, temperature, for the classification significantly simplifies the dynamic analysis of climate system transformations. The above-detected spatial synchronization in the nonlinear chaotic climate system introduces an element of order and may become one of the effective ways to optimize both the development of an instrumental observation network and the creation of regional forecast models of climate system dynamics.

Acknowledgements

This research was supported by the Russian Academy of Sciences (project IX.138.2.4).

References

[1] Cheredko N N, Tartakovsky V A, Krutikov V A and Volkov Yu V 2017 Climate Classification in the Northern Hemisphere Using Phases of Temperature Signals *Atmos. Ocean. Opt.* **30** (1) 63–9 DOI: 10.1134/S1024856017010043

[2] Hurrell J W and Loon H V 1997 Decadal variations in climate associated with the North Atlantic Oscillation *Cl. Change* **36** 301–26

[3] Popova V V and Shmakin A B 2006 Circulation Mechanisms of Large-Scale Winter Air Temperature Anomalies in Northern Eurasia at the End of the 20th Century *Meteorologiya i Gidrologiya* **12** 15–25

[4] Popova V V 2018 Present-day changes in climate in the north of Eurasia as a manifestation of variation of the large-scale atmospheric circulation *Fund. and appl. climatology* **1** 84–111

[5] Hurrell J W *et al* 2006 Atlantic Climate Variability and Predictability: A CLIVAR Perspective *J. Clim.* **19** 5100–21

[6] IPCC 2013. *Climate Change 2013: The Physical Science Basis. Contribution of Working Group I to the Fifth Assessment Report of the Intergovernmental Panel on Climate Change* ed. T F Stocker *et al* (Cambridge: Cambridge University Press)

[7] http://www.metoffice.gov.uk; https://crudata.uea.ac.uk/cru/data/temperature (01.12.2018)

[8] Rostov I D, Rudykh N I, Rostov V I and Vorontsov A A 2016 Expressions of global climatic changes in coastal waters of Northern part of the Sea of Japan *Vestnik FEB RAS* **5** 100–12

[9] Walsh J E 2014 Intensified warming of the Arctic: Causes and impacts on middle latitudes *Glob.and Pl. Change* **117** 52–63

[10] Latysheva I V, Loshchenko K A and Shahaeva E V 2011 The study dynamics of the Asian High and cold periods of circulation of on the territory of Irkutsk Range *The Bull. of Irkutsk St. Univ.* **2** 161–71

Dynamics of the Arctic polar vortex during the 1984/1985 sudden stratospheric warming

V V Zuev and E S Savelieva

Institute of Monitoring of Climatic and Ecological Systems of the Siberian Branch of the Russian Academy of Sciences, 10/3, Academichesky Ave., Tomsk, Russia

vzuev@list.ru, ekat.savelieva@gmail.com

Abstract. The Arctic stratospheric polar vortex usually forms in October, reaches its peak intensity in December, and decays from February to April. In most cases, the breakdown of the Arctic polar vortex occurs under the influence of vertically propagating planetary Rossby waves. The influence of planetary wave activity on the polar vortex leads to a displacement or breakdown accompanied by a sudden stratospheric warming (SSW). One of the earliest SSWs was observed on December 29, 1984, after which the polar vortex in the lower stratosphere had not fully recovered and existed for less than 3 months since October. In this study, we analyze the dynamics of the stratospheric polar vortex under the influence of planetary wave activity in the winter of 1984-1985.

1. Introduction

The Arctic stratospheric polar vortex usually exists for 5–7 months, from autumn to spring, whereas the Antarctic polar vortex usually exists more than 8 months. The earlier breakdown of the Arctic polar vortex occurs due to the influence of vertically propagating planetary Rossby waves [1]. Vertical fluxes of Rossby wave activity (Eliassen–Palm (EP) fluxes), moving from the troposphere, are destroyed in the polar stratosphere as a result of interaction with the eastward zonal flow (i.e. the stratospheric polar vortex) [2–4]. The sudden stratospheric warmings (SSWs), which occur due to adiabatic heating of the middle stratosphere during the planetary wave breaking, are accompanied by a displacement or splitting of the polar vortex [5, 6]. SSWs are characterized by a sudden increase in the latitudinal mean temperature poleward from 60° latitude at the 10 hPa level (~ 30 km) or below [2]. SSWs are divided into major and minor ones. A strong displacement of the polar vortex occurs during minor SSWs, whereas major SSWs are accompanied, as a rule, by the polar vortex splitting into two smaller vortices.

The dynamics of the polar vortex plays an important role in the stratospheric ozone distribution, the movement of air masses in the polar stratosphere and the change in the stratospheric temperature over the polar region [7, 8]. The winter and spring stratospheric circulation is determined by the activity of the polar vortex, and its changes are associated with variations in the shape and location of the vortex. The area and extent of the stratospheric ozone depletion over the polar region depend on the polar vortex strength and persistence in spring [9]. The polar vortices act as containment vessels and prevent the inflow of warm and ozone-rich air masses from the midlatitude stratosphere into the polar region [10, 11]. At the same time, polar stratospheric clouds (PSCs) form at extremely low temperatures (lower than –78 °C) inside the polar vortex. Heterogeneous reactions, releasing photochemically active molecular chlorine, occur on and in PSC particles [12]. The stratospheric ozone depletion over the polar region occurs as a result of catalytic cycles involving chlorine species in the presence of weak solar radiation from late winter to spring [13].

The influence of planetary wave activity on the polar vortex does not always lead to its breakdown. Short-term splitting of the polar vortex is sometimes observed for several days, after which the polar vortex strengthens again and PSCs form inside the vortex. Such a recovery of the polar vortex is most probable in winter. Major SSWs appearing due to the polar vortex splitting usually occur in February or in late January [2]. However, there are 2 cases when major SSWs which appeared in the lower stratosphere were observed in late December: the 1984/1985 SSW and the 1998/1999 SSW. Some studies [14, 15] investigated the dynamics of the stratospheric circulation in the winter of 1984-1985 before and after the SSW. The preSSW state was characterized by the strong polar vortex and deep penetration of planetary waves in the upper stratosphere [14]. In contrast, the postSSW state showed the weak polar vortex and higher wave activity in the lower stratosphere [14]. The aim of this study is to investigate the dynamics of the Arctic polar vortex under the influence of planetary wave activity in the winter of 1984-1985 using the ERA-Interim reanalysis data and the NASA GSFC satellite data.

2. Data and methods

The EP fluxes data were taken from the NOAA Earth System Research Laboratory's Physical Sciences Division (ESRL PSD) [16]. The potential vorticity and temperature data for 30°–90° N on the 50, 10, and 1 hPa pressure surfaces were retrieved from the European Centre for Medium-Range Weather Forecasts (ECMWF) ERA-Interim reanalysis data [17; http://apps.ecmwf.int/datasets]. The data on zonal mean zonal wind for 60° N and zonal mean temperature for 60°–90° N on the 50, 10, and 1 hPa pressure surfaces, zonal mean eddy heat flux for 45°–75° N on the 10 hPa pressure surface, the area of the polar vortex on the 460 K isentropic surface and PSC volume for 60°–90° N were retrieved from the NASA Goddard Space Flight Center (GSFC) online database [18; http://ozonewatch.gsfc.nasa.gov].

To explore variations of meteorological data from November 1984 to March 1985, we obtained 38-year climatological mean seasonal cycles of zonal wind and temperature in the Arctic stratosphere over the 1979–2016 period and 40-year climatological mean seasonal cycles of eddy heat flux, vortex area and PSC volume in the Arctic stratosphere over 1979–2018 with standard deviations (σ). The climatological means and standard deviations were smoothed by a 15-point fast Fourier transform (FFT) filter.

3. Results

3.1. 1984/1985 sudden stratospheric warming

SSWs appear due to the breaking and dissipation of planetary waves (in particular, the EP fluxes) as a result of interaction with the stratospheric polar vortex [2]. Figure 1 shows 1-day averaged EP flux vectors from 5 December 1984 to 30 March 1985 with a time step of 2–8 days, taken from the NOAA ESRL PSD. An increase in the EP flux during the winter-spring 1984/1985 was observed from 1 to 7 December, from 25 December to 4 January, from 9 to 11 January, and from 9 to 10 March. The maximum activity of the EP flux was observed on 28 December (Figure 1), after which the SSW was recorded on 29 December. Thus, as seen from Figure 1, in the 1984/1985 winter-spring period an increase in the EP flux was observed in December, and an unusual decrease in February and March.

Potential vorticity and temperature are usually used to describe SSWs. Enhanced values of potential vorticity in the winter-spring period usually determine the location of the polar vortex, whereas lower and higher temperatures are observed inside and outside the polar vortex, respectively. Figure 2 shows potential vorticity and temperature distributions at the 50, 10, and 1 hPa pressure levels (~ 20, 30, and 46 km) over the Arctic from December 1984 to March 1985 retrieved from the ERA-Interim reanalysis. As seen from Figure 2, the polar vortex had a slightly elongated shape and was displaced relative to the pole on 20 December. Then the polar vortex elongated, and its splitting in the lower, middle, and upper stratosphere was observed on 29 December. T2wo regions of high values of potential vorticity on 29 December 1984 indicate the polar vortex splitting and the occurrence of the SSW.

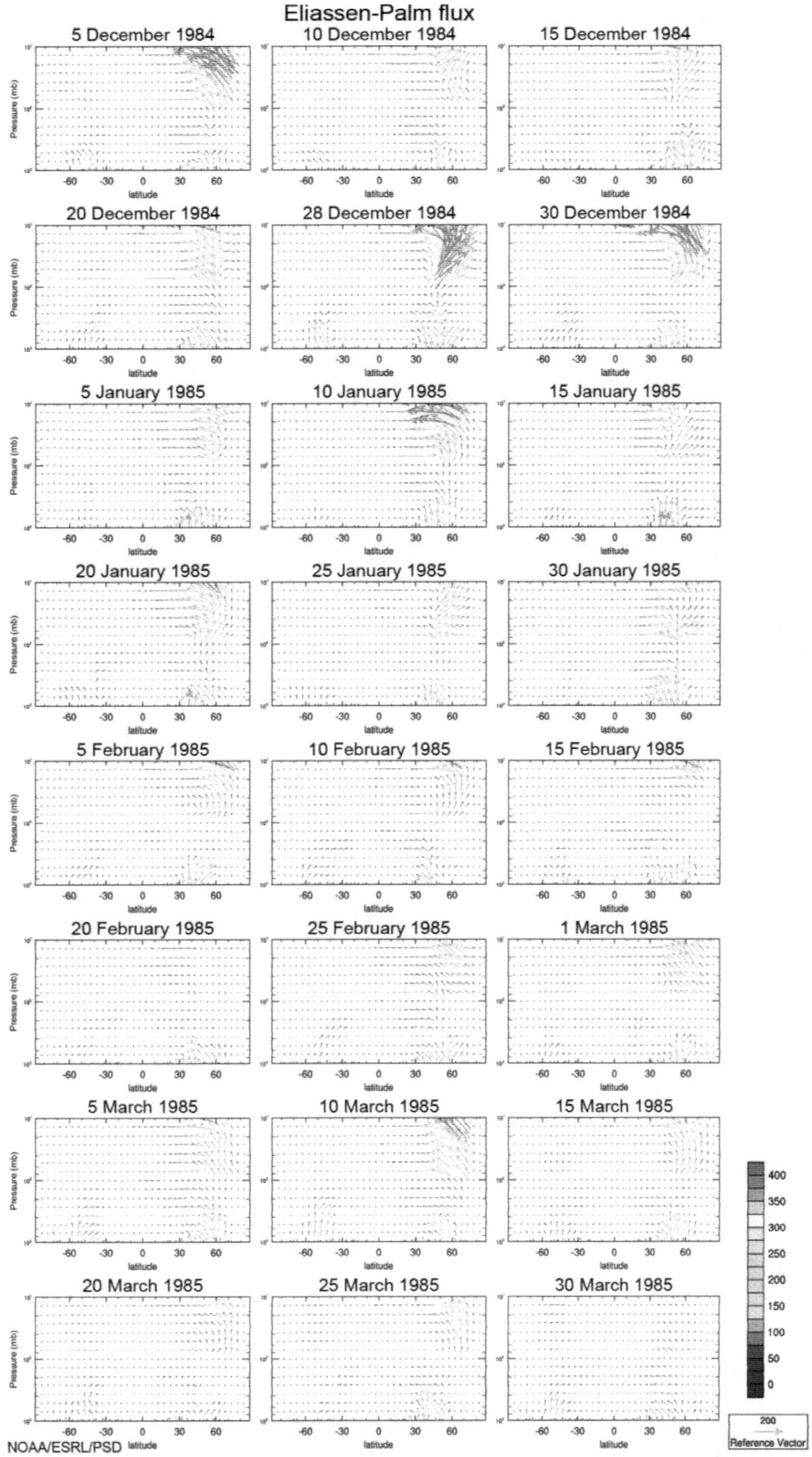

Figure 1. EP fluxes from December 1984 to March 1985.

IOP Conf. Series: Earth and Environmental Science **386** (2019) 012010 doi:10.1088/1755-1315/386/1/012010

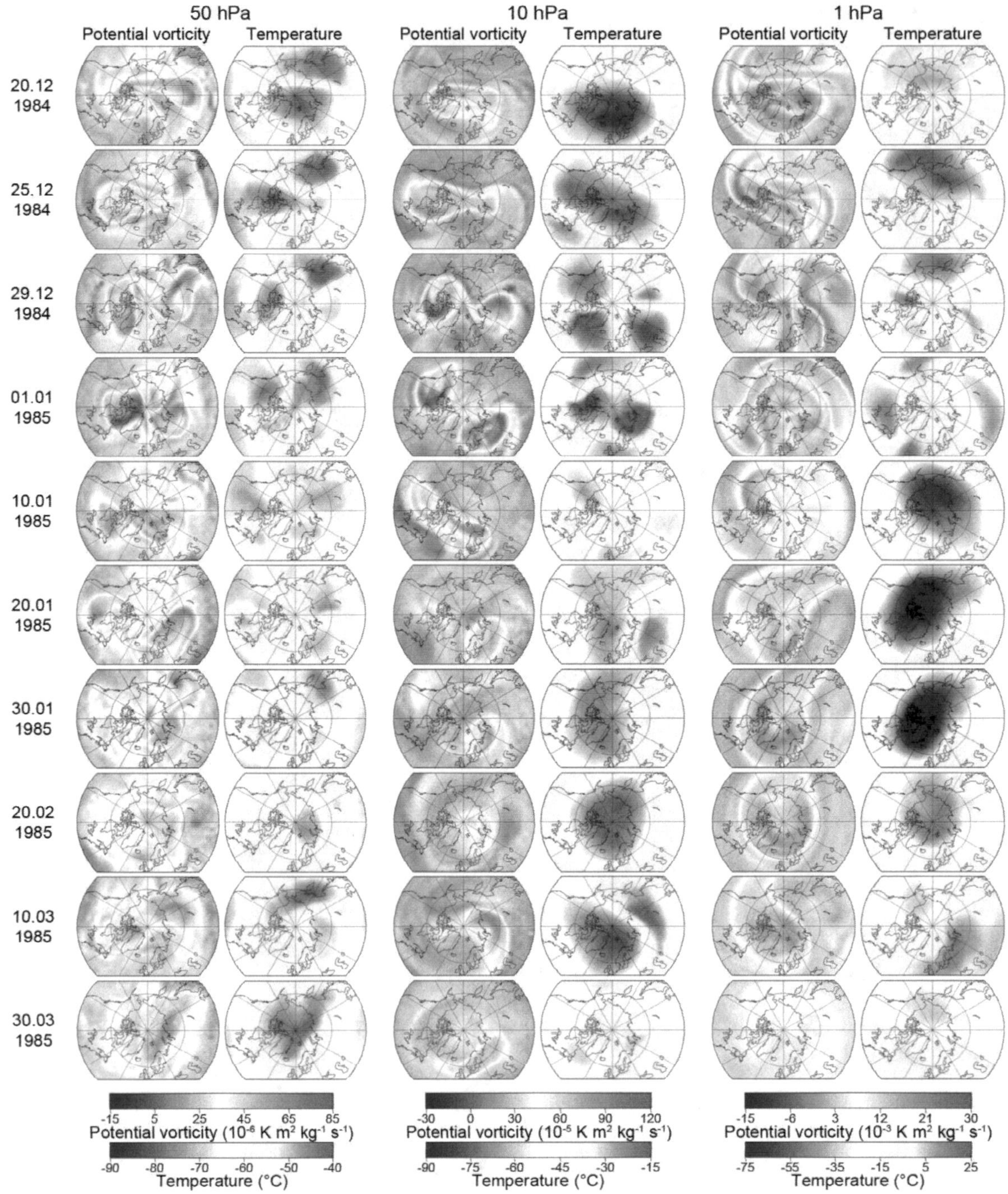

Figure 2. Potential vorticity and temperature distributions at the 50, 10 and 1 hPa pressure levels over the Arctic from December 1984 to March 1985.

The SSW that began on December 29 was most clearly pronounced in the middle stratosphere on January 1, 1985 [15] (Figure 2). After such an early SSW the process of polar vortex recovery started, and then it was interrupted on January 10 as a result of intensification of the EP fluxes (Figures 1 and 2). A constant change in the recovery and elongation (preceding the splitting) of the polar vortex was observed during January. The second splitting was observed in the lower stratosphere on January 20, after which at the 50 hPa pressure level the polar vortex no longer appeared in the winter-spring 1984/1985. While there is still an area of high potential vorticity values over the Arctic, the polar vortex is not well defined [15] (Figure 2). The potential vorticity values are lower than those in early winter and are spread more thinly over the Arctic [15]. In the middle and upper stratosphere, a weak polar vortex appeared in the temperature distributions (Figure 2) in February and in early March. Then, after a small increase in the EP fluxes on March 10, the eastward zonal flow was replaced by the westward one in mid-March about 20 days earlier than usual.

3.2. Arctic polar vortex dynamics during the winter 1984/1985

Figure 3 shows the time series of the zonal wind for 60° N and the temperature for 60°–90° N in the lower, middle, and upper stratosphere (at the 50, 10 and 1 hPa pressure levels) from November 1984 to March 1985 compared to its 38-year climatological mean seasonal cycles over 1979–2016 with ±1 standard deviations. All time series and seasonal cycles are retrieved from the ERA-Interim reanalysis. As seen in Figure 3, after the impact of planetary waves in late December 1984 the polar vortex recovered in the middle and upper stratosphere in February and was destroyed in mid-March. The zonal wind in the upper stratosphere even exceeded the climatological mean in February. It is important to note that in the upper stratosphere high temperatures are observed within the strong polar vortex, while the weakening of the vortex leads to their decrease. For this reason, during the polar vortex splitting in late December 1984, in contrast to the temperature increase in the lower and middle polar stratosphere, in the upper stratosphere an abrupt temperature decrease was observed (Figure 3).

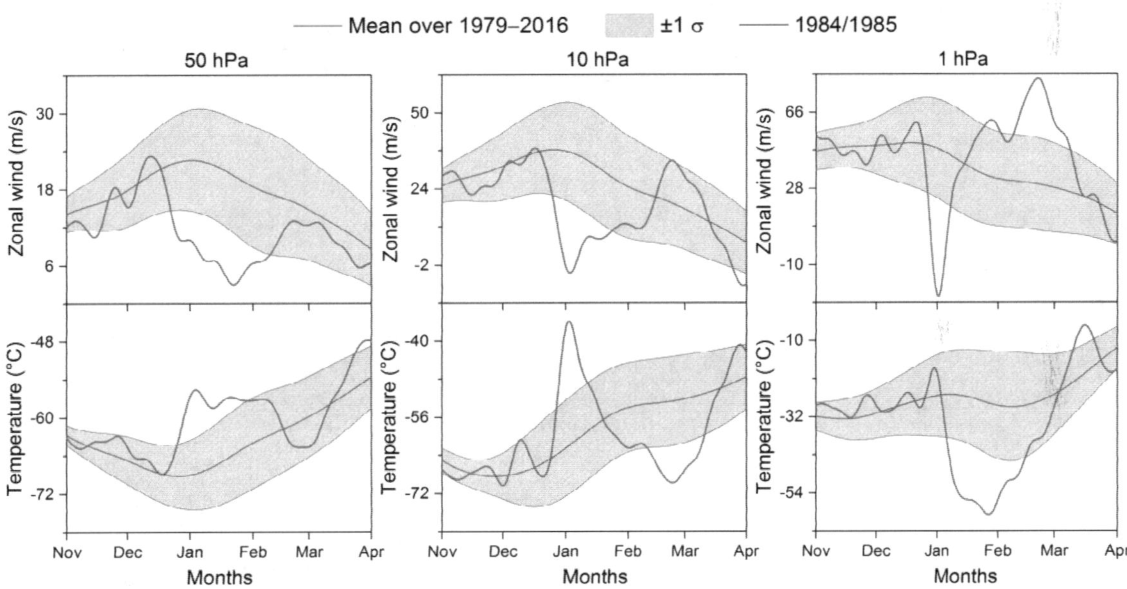

Figure 3. Time series of zonal mean zonal wind for 60° N and zonal mean temperature for 60°–90° N at the 50, 10 and 1 hPa pressure levels from November 1984 to March 1985 (red lines) and 1979–2016 climatological means (green lines) with ±1 standard deviations (green areas).

Figure 4 shows the time series of an eddy heat flux for 45°–75° N at the 10 hPa pressure level, the polar vortex area at the 460 K isentropic level (~ 20 km), and the PSC volume for 60°–90° N from November 1984 to March 1985 compared to its 40-year climatological mean seasonal cycles over 1979–2018 with ±1 standard deviations. All time series and seasonal cycles are retrieved from the NASA GSFC database. In the time series of the heat flux, extremes are observed on 5 and 28 December 1984 and 10 March 1985 (Figure 4), which is in good agreement with the NOAA ESRL PSD data (Figure 1). At the same time, the polar vortex splitting in the lower stratosphere was observed on January 20 under the conditions of low activity of planetary waves in January. It is assumed that the high influence of the planetary wave activity played a key role in the development of the Arctic polar vortex in the winter-spring of 1984-1985. After the impact of planetary waves on December 28, 1984, even a relatively small increase in the wave activity in January led to the second splitting of the polar vortex in the lower stratosphere. The dynamics of the polar vortex described above can be traced in the time series of the vortex area in the lower stratosphere (Figure 4). Finally, the absence of PSCs from January to March 1985, which are formed in the lower stratosphere, confirms that after the splitting on December 29, 1984 the polar vortex did not fully recover during the winter-spring of 1984-1985.

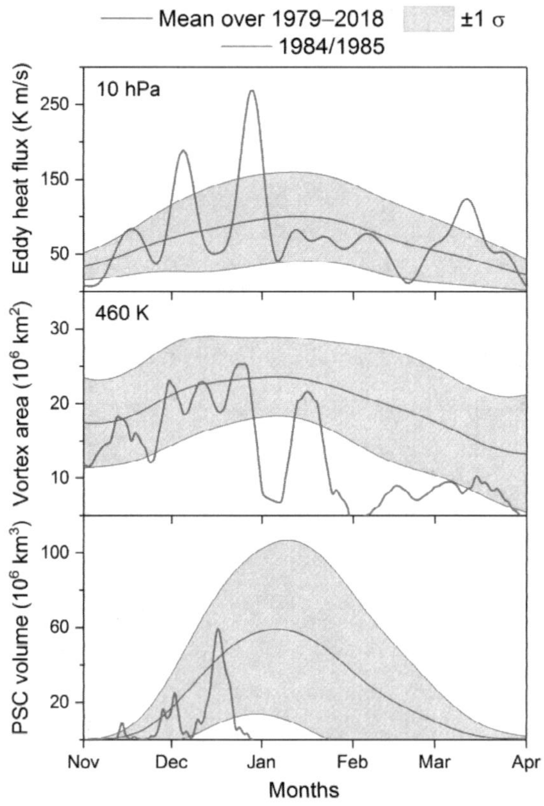

Figure 4. Time series of zonal mean eddy heat flux for 45°–75° N at the 10 hPa pressure level, polar vortex area at the 460 K isentropic level, and PSC volume for 60°–90° N from November 1984 to March 1985 (red lines) and 1979–2018 climatological means (green lines) with ±1 standard deviations (green areas).

4. Conclusions

In most cases, the effect of planetary waves on the polar vortex from December to early January is limited to a displacement or short-term splitting observed for several days, after which the polar vortex recovers. An increased activity of EP fluxes was recorded on December 28, 1984. The polar vortex splitting which occurred on December 29 was so strong that the relatively small subsequent activity of the EP fluxes did not allow the vortex to fully recover. The newly formed vortex was characterized by temperatures that were insufficiently low for the formation of PSCs, and after the polar vortex splitting in December no PSCs appeared until late spring. It is assumed that the high-intensity impact of planetary waves on the polar vortex on December 28, 1984 determined the further dynamics of the vortex. As a result, even the short-term small intensification of the EP fluxes, which, as a rule, does not have a significant effect on the polar vortex, prevented its recovery this time. Thus, the case of the 1984/1985 SSW is one of the exceptional ones because, as a result of the impact of planetary waves on December 28, the polar vortex in the lower stratosphere had not fully recovered and existed for less than 3 months since October.

References

[1] Waugh D W, Sobel A H and Polvani L M 2017 What is the polar vortex and how does it influence weather? *B. Am. Meteorol. Soc.* **98** 37–44

[2] Kuttippurath J and Nikulin G 2012 A comparative study of the major sudden stratospheric warmings in the Arctic winters 2003/2004–2009/2010 *Atmos. Chem. Phys.* **12** 8115–29

[3] Flury T, Hocke K, Haefele A, Kämpfer N and Lehmann R 2009 Ozone depletion, water vapor increase, and PSC generation at midlatitudes by the 2008 major stratospheric warming *J. Geophys. Res.* D **114** D18302

[4] Limpasuvan V, Thompson D W J and Hartmann D L 2004 The life cycle of the Northern Hemisphere sudden stratospheric warmings *J. Climate* **17** 2584–96

[5] Torre L, Garcia R R, Barriopedro D and Chandran A 2012 Climatology and characteristics of stratospheric sudden warmings in the Whole Atmosphere Community Climate Model *J. Geophys. Res.* D **117** D04110

[6] Butler A H, Seidel D J, Hardiman S C, Butchart N, Birner T and Match A 2015 Defining Sudden Stratospheric Warmings *B. Am. Meteorol. Soc.* **96** 1913–28

[7] Waugh D W and Randel W J 1999 Climatology of Arctic and Antarctic polar vortices using elliptical diagnostics *J. Atmos. Sci.* **56** 1594–613

[8] Waugh D W and Polvani L M 2010 Stratospheric polar vortices *The Stratosphere: Dynamics, Transport, and Chemistry (Geophys. Monogr. Ser.)* vol 190 (Washington: AGU) pp 43–57

[9] Newman P A 2010 Chemistry and dynamics of the Antarctic ozone hole *The Stratosphere: Dynamics, Transport, and Chemistry (Geophys. Monogr. Ser.)* vol 190 (Washington: AGU) pp 157–171

[10] Manney G L and Zurek R W 1994 On the motion of air through the stratospheric polar vortex *J. Atmos. Sci.* **51** 2973–94

[11] Sobel A H, Plumb R A and Waugh D W 1997 Methods of calculating transport across the polar vortex edge *J. Atmos. Sci.* **54** 2241–60

[12] Finlayson-Pitts B J and Pitts J N 2000 *Chemistry of the Upper and Lower Atmosphere: Theory, Experiments, and Applications* (California: Academic Press) pp 657–726

[13] Solomon S, Garcia R R, Rowland F S and Wuebbles D J 1986 On the depletion of Antarctic ozone *Nature* **321** 755–8

[14] Randel W J and Boville B A 1987 Observations of a Major Stratospheric Warming during December 1984 *J. Atmos. Sci.* **44** 2179–86

[15] Kwasniok F, Beaumont R and Thuburn J 2019 Vortex dynamics of stratospheric sudden warmings: A reanalysis data study using PV contour integral diagnostics *Q. J. Roy. Meteor. Soc.* **145** 1013–33

[16] The NOAA Earth System Research Laboratory's Physical Sciences Division (NOAA ESRL PSD). Plotting & Analysis. https://www.esrl.noaa.gov/psd/data/epflux (retrieved 03/2019)

[17] Dee D P *et al* 2011 The ERA-Interim reanalysis: configuration and performance of the data assimilation system *Q. J. Roy. Meteor. Soc.* **137** 553–97

[18] Goddard Space Flight Center (GSFC). NASA's Ozone Hole Watch Web Site (online database). http://ozonewatch.gsfc.nasa.gov/NH.html (retrieved 03/2019)

IOP Conf. Series: Earth and Environmental Science **386** (2019) 012011 doi:10.1088/1755-1315/386/1/012011

A study of the mutual influence of global circulation objects by a method of dispersive analysis

S V Morozova[1], E A Polyanskaya[1], N K Kononova[2], N P Molchanova[3] and A V Letuchiy[3]

[1] Saratov State University, Saratov, Russia
[2] Institute of Geography, Russian Academy of Sciences, Moscow Russia
[3] Vavilov Saratov State Agrarian University, Russia

E-mail: swetwl@yandex.ru

Abstract. Three structural elements of the general circulation of the atmosphere on a global scale are considered: the Arctic oscillation, the Southern Oscillation, and the planetary tall frontal zone. The degree of influence of the Arctic Oscillation and the Southern Oscillation on the dynamics of the planetary tall frontal zone is estimated. The assessment is carried out by a method of analysis of the variance, which allows assessing the significance of the influence of one or more meteorological quantities of the studied value. A comparison of the variances has shown that the degree of influence of these objects on the dynamics of the planetary high altitude zone is 6% from the Arctic Oscillation and 7% from the Southern Oscillation. A comprehensive analysis has shown that the planetary high-altitude frontal zone has the highest intensity with negative anomalies of the Arctic Oscillation and the Southern Oscillation, and the lowest one with negative anomalies of the Arctic Oscillation and positive anomalies of the Southern Oscillation. A comparison with the Fisher criterion has not revealed any statistical significance. An important conclusion is formulated about the predominance of the influence of inherent internal processes in the dynamics of the planetary high-altitude frontal zone in comparison with the circulation structures adjacent to it.

1. Introduction

It is well-known that the greatest climatic variability is typical for the temperate latitudes, whose main circulation feature is active cyclonic activity. The directions of the air flow are determined by the deformation of the planetary tall frontal zone (PTFZ) - a three-dimensional object of circulation on a global scale encircling the entire globe. The curvature of the PTFZ and strong development of wave processes in it lead to an increase in the degree of meridionality of the flows, so that sharp invasions of cold air masses into the southern latitudes occur in the rear parts of the baric hollows, and powerful heat removal to the Arctic is observed in the rear parts of the high-altitude baric ridges.

This process is accompanied by the activation of the Arctic center of action, which is reflected in a more active penetration into the moderate latitudes of the Arctic anticyclones. At the same time, a cyclonic curvature is formed in the deformation region of the PTFZ. When the ridges of subtropical anticyclones propagate northward, an anticyclonic curvature arises in the PTFZ.

Thus, the PTFZ functions not isolated from other structures of the general circulation of the atmosphere, and it is directly affected by their influence. Circulating objects of the arctic and tropical latitudes are adjacent to it. The main circulation structures in them are the Arctic Oscillation and the Southern Oscillation, respectively.

It remains an open question: to what extent do the circulation objects adjacent to it have a strong influence on the planetary tall frontal zone, do the wave processes in the PTFZ develop under the influence of smaller-scale structures (centers of action of the atmosphere), or the curvature of fluxes in the PTFZ results from its own internal dynamics? It should be noted that parametric estimates of the interaction and mutual influence of the structural elements of the general atmospheric circulation of various space-time scales are practically absent.

The purpose of this paper is to assess the influence of the Arctic Oscillation and the Southern Oscillation on the dynamics of the planetary altitude frontal zone.

In this work, the task is to assess the influence of circulation structures of high and low latitude on the planetary high-altitude frontal zone.

Assessment of the influence of one or more factors on the object being studied can be carried out by different methods. Thus, the authors of [1] evaluated the contribution of the radiation exposure of greenhouse gases and multidecadal oscillations to the near-surface temperature trends using an autoregressive model.

The contribution of a single factor to the studied value is estimated much more simply. The authors of [2] obtained a quantitative assessment of the influence of one variable on another by a determination coefficient R2. The predictor is the second variable in the case of an obvious causal relationship.

In papers [3, 4], the contribution of the influence of one quantity on another is also estimated using the coefficient of determination, but in this case the causal relationship does not seem obvious.

According to the authors of this study, in cases where a causal relationship is not obvious, it is better to use the analysis of variance (factor analysis). Factor analysis allows us to assess the significance of the influence of one quantity on another. This technique makes it possible to evaluate individually the effect of the first investigated quantity on the second one, and vice versa, the effect of the second quantity on the first one. This method of factor analysis also allows you to evaluate the influence of a set of factors on the value being studied.

2. Material and methods

The source materials for the study were the data on the Arctic Oscillation and the Southern Oscillation taken from the site (origin.cpc.ncep.noaa.gov).

Quite a few parameters are used to characterize the planetary height frontal zone: the area, length, tortuosity, intensity, width of the location of the axial isohypse, etc. [5]. In the present study, for the analysis of the dynamics of PTFZ, its characteristic was chosen as intensity.

The rate of geostrophic wind Vg at the average level of the troposphere is taken as the intensity of the PTFZ. The geostrophic wind speed is calculated by the formula

$$Vg = \frac{9.8}{l}\frac{\partial H}{\partial n}, \tag{1}$$

where l is the Coriolis parameter (l = 2ωsinφ),

ω is the angular velocity of the Earth's rotation,

φ is the latitude of the axial isohypse,

H is the height of the geopotential of the isobaric surface of 500 hPa,

$\frac{\partial H}{\partial n}$ is the derivative of the geopotential height in the direction perpendicular to the tangent to the isohypse.

In the numerical estimation of Vg, the derivative was approximated by the ratio $\Delta H/\Delta n$, where ΔH is the difference in the geopotential heights, and Δn is the distance along the normal to the tangent drawn at a given point of the isohypse. Given that in monthly (and even more so in annual) averaging, isohypses on an AT-500 map are usually parallel, the change in the distance normal to the tangent was considered identical to the change in the distance along the meridian. In this case $\Delta n = \Delta\varphi \cdot 111 \cdot 10^3$ *m.*

By formula (1), the average annual geostrophic wind speeds from 1957 to 2015 were calculated. The obtained values are taken as the intensity of the PTFZ.

The problem posed in this study was solved by using the methods of analysis of variance, allowing one to evaluate the significance of the influence of one value (or several values) on another. The essence of the analysis of variance consists in comparing the factor dispersion generated by the effect of a factor and the residual dispersion due to random causes. If the difference between these variances is significant, the factor has a significant impact on the quantity *X* being studied.

Let *Sx* denote the total sum of squares of deviations of the observed values from the total average; S_f is the factor sum of the squared deviations of group means; *Sz* is the residual sum of the squares of deviations of the observed values of the group from the group average. Then *Sx* characterizes the scattering of values relative to a common average; S_f is scattering between the groups; and *Sz* is scattering within the group.

Thus, we find that *Sx* characterizes the overall variability of *X*; S_f is the variability of the magnitude under the action of the studied factor *f*. If the factor has a significant impact on the value of X, the group of observed values of the trait at one particular level will differ from the groups of observations at other levels. Consequently, the group averages will also differ, and the more they are scattered around a common average, the stronger will be the influence of the factor. The value of *Sz* characterizes the variability due to the influence of random or unrecorded factors [6,7].

Since the magnitude of *X*, in addition to the factor *f*, is also affected by random causes, the observations of the same group are different and scattered around their group average. Therefore, to estimate random causes, the deviations of the observed values of each group from their group average are calculated. In this case, the average of the observed values at each level (in each gradation) will also differ significantly.

Hence, $S_x = S_f + S_z$.

In practice, the sum of the deviations Sx and Sf is usually calculated, and the residual sum of the deviations is found as $S_z = S_x - S_f$ [1].

For calculations we use the formula [1]

$$S_x = \sum^n x_i^2 - \frac{1}{n}\left(\sum^n x_i\right)^2 \tag{2}$$

$$S_f = \sum^{q} \frac{\sum^{m_i} (x_i)^2}{m_i} - \frac{1}{n}\left(\sum^{n} x_i\right)^2; \tag{3}$$

Dispersions are found, respectively, by the expressions:

$$\sigma_f^2 = \frac{S_f}{K_f}, \sigma_z^2 = \frac{S_z}{K_z}, \sigma_x^2 = \frac{S_x}{K_x}, \tag{4}$$

In this case, $Kf = q-1$, $Kz = Kx - Kf$, $Kx = N-1$ are the numbers of degrees of freedom,
where *q* is the number of levels (gradations) of the studied factor;
N is the total number of observations.

In multivariate analysis of variance, the study breaks down into two stages. At the first stage, the significance of the influence of all complexes on the studied value is revealed. This is done according to the single-factor analysis formulas. Since the uniformity (randomness) of each complex of factors is important in multivariate analysis, the formulas change.

$$S_x = \sum^{q}\sum^{p} x_i^2 - \frac{1}{pq}\left(\sum^{q}\sum^{p} x_i\right)^2, \tag{5}$$

$$S_f = \frac{1}{p}\sum^{q}\left(\sum^{p} x_i\right)^2 - \frac{1}{pq}\left(\sum^{q}\sum^{p} x_i\right)^2, \tag{6}$$

where q is the number of complexes, and

p is the number of cases in each complex.

If the first stage is established, then we proceed to the second stage. At the second stage, the significance of the influence $F > F_p$, of each combination of factors on the studied value x is established. This is done based on the calculation of the Fisher criterion according to the formula [7]:

$$F = \frac{pqp}{(pq+p)\sigma_z^2}\left(\bar{x}_q - \bar{x}_{pq}\right)^2, \tag{7}$$

where $\bar{x}_q$ is the average value for this complex; and

$\bar{x}_{pq}$ is the total average.

Complexes for which $F > F_p$ have a significant impact on the change in x. The value F_p is found based on the level of significance (usually 5%) and degrees of freedom $k_f = 1$ and $k_z = k_{z1}$, where k_{z1} is the number of degrees of freedom found at the first stage.

3. Results and discussion

Since the influence of the characteristics of the Arctic Oscillation and the Southern Oscillation on the intensity of the PTFZ is being studied, it is these characteristics that are divided into gradations. For splitting into gradations, we apply the standard statistical procedure: by the formula $k = 5ln\ N$ we determine the number of gradations, and by the formula $l = (x_{max} - x_{min}) / k$ we determine the width of the gradation.

The values of the PTFZ intensity are distributed according to the gradations [2]. The table shows the distribution of the intensity of the PTFZ from the intensity of the Arctic oscillations.

The total sum of squared deviations calculated over the entire series was 1888.3, while the factor sum of squares was 233.3. The ratio of the calculated variances according to Fisher's criterion at the 95% significance level did not show the statistical significance of the contribution of the Arctic Oscillation to the change in the PTFZ intensity. The contribution of the Arctic Oscillation to the variability of PTFZ is estimated at 6%. Similar results on the absence of the influence significance were obtained for the Southern Oscillation. The contribution of the Southern Oscillation to the PTFZ intensity is estimated at 7%.

Table. Interval values of PTFZ intensity depending on Arctic oscillations.

Characteristic	Gradations of intensity of the Arctic Oscillation						
	1.04;-0.78	-0.78;-0.53	-0.5;-0.27	-0.27;-0.01	-0.01;0.25	0.25;0.51	0.51;1.02
Amounts	22.6	27.9	80.0	87.2	73.6	22.2	22.5
Number of cases	4	5	14	16	13	4	4
Average value	5.7	5.6	5.7	5.5	5.7	5.6	5.6

The calculated correlation coefficients between the intensity of the planetary tall frontal zone and the indices of the Arctic Oscillation and the Southern Oscillation turned out to be 0.02 and -0.03, respectively.

In addition to one-factor analysis, the authors performed a multivariate analysis. The simultaneous influence of two factors — the Arctic Oscillation and the Southern Oscillation — on the variability of the planetary altitudinal frontal zone was estimated. The calculations were carried out according to formulas (5) – (7).

Arctic oscillation

South wobble	-1.0482 -0.7844	-0.7844 -0.5260	-0.5260 -0.2676	-0.2676 -0.0092	-0.0092 0.2491	0.2491 0.5075	0.5075 1.0243
-1.233 -0.8948			4.8 4.9		5.8 5.5		
-0,8947 -0.5562			6.2	4.9 4.8 5.9 5.2 5.4	5.4 5.2 5.7	5.5	5.9
-0.5561 -0.2176	6.4		6.1 6.1 6.3	6.1 5.8			5.6
-0.2175 0.1210	5.2		5.8 5.9 5.8	4.9 5.1 5.8 5.7	5.3		
0.1211 0.4596		4.9 5.9 6.3 6.1	5.6 5.7	5.1 5.8 5.8	5.3 5.7 6.3 6.0		5.7
0.4597 0.7982	6.5		5.3 5.7 5.8		5.8 5.8	5.5	5.3
0.7983 1.1368	4.7				5.9		
1,1368 1.4753		5.3		5.9			

Figure 1. Dependence of the intensity of the planetary high-altitude frontal zone on the intensity of the Arctic Oscillation and the Southern Oscillation.

The figure shows a graphical interpretation of the influence of two factors - the Arctic Oscillation and the Southern Oscillation - on the PTFZ intensity. One can see in the figure that the lowest values of the PTFZ intensity are observed for negative anomalies of the Arctic Oscillation and positive anomalies of the Southern Oscillation. That is, when the Arctic anticyclone is weakened and the subtropical one intensifies, the PTFZ intensity is weakened. The highest intensity of the PTPZ is observed with negative anomalies of both the Arctic oscillations and the Southern Oscillation. That is, in the case of weakening of both the Arctic and subtropical anticyclones, the PTFZ exhibits the highest intensity.
Thus, in this study we obtained parametric estimates of the impact of adjacent circulating structures – the Arctic and the Southern Oscillations - on the planetary frontal zone. However, these estimates did not show the statistical significance of such an impact. Therefore, it is possible to make an important conclusion that the dynamics of the planetary high-altitude frontal zone is determined more by the processes occurring in the region of its localization than by the effect of circulating structures adjacent to it.

Conclusions

1. The planetary tall frontal zone is affected by objects of circulation of high and low latitudes adjacent to it: the Arctic Oscillation and the Southern Oscillation. The degree of influence is estimated as six and seven percent, respectively.

2. The intensity of the planetary tall frontal zone depends on the intensity of the Arctic Oscillation and the Southern Oscillation. The PTFZ has the highest intensity with negative anomalies of the Arctic Oscillation and the Southern Oscillation, and the lowest one with negative anomalies of the Arctic Oscillation and positive anomalies of the Southern Oscillation. However, this effect is statistically insignificant according to the Fisher criterion.

Acknowledgements
This work was performed under Fundamental Research Program no. 0148-2019-0009, Institute of Geography RAS.

References

[1] Mokhov I I, Smirnov D A 2018 Contribution of Greenhouse Gas Radiative Forcing and Atlantic Multidecadal Oscillation to Surface Air Temperature Trends *Meteorology and hydrology* **9** 5-13
[2] Sporyshev P V, Govorkova V A 2013 Temperature changes according to observation and model simulations with a separate account of antropogenic and natural external impacts *Proceeding of MGO* **568** 51-79
[3] Perevedentsev Y P , Vilfand R M, Shantalinskii K M, Guryanov V V, Nikolaev A A, Ismagilov N V 2019 Monitoring and prediction of climate variability in the Volga Region *Proceedings of Hydrometcentre of Russia* **371** 67-94
[4] Perevedentsev Yu P and Shantalinsky K M 2014 Estimation of Observed Variations of Air Temperature and Wind Speed in the Troposphere *Meteorology and hydrology* **10** 19-31
[5] Neushkin A I, Sidorenkov N S, Sanina A T, Ivanova T B, Berezhnaya T V, Pankratenko N V, Makarova M Ye 2013 Monitoring obshchey tsirkulyatsii atmosfery. Severnoye polushariye *Obninsk VNIGMIMTSD Press* 200
[6] Gmurman V E 2003 Probability Theory and Mathematical Statistics *Moscow Higher School Press* 480
[7] Panovfsky H A, Brier G W 1967 Some Applications of statistics to Meteorology *Leningdad Hydrometeorology Press* 241
[8] htpp:www.origin.cpc.ncep.noaa.gov

Influence of various parameters of INM RAS climate model on the results of extreme precipitation simulation

M A Tarasevich[1] and E M Volodin[2]

[1] Moscow Institute of Physics and Technology (MIPT NRU), Institutskiy per., 9, Dolgoprudny, Moscow Region, 141701, Russian Federation
[2] Marchuk Institute of Numerical Mathematics of the Russian Academy of Sciences (INM RAS), Gubkin str., 8, Moscow, 119333, Russian Federation

Email: mashatarasevich@gmail.com

Abstract. The results of simulation of extreme weather indices of modern climate by two versions of the INM RAS climate model (INMCM4, INMCM5) are considered. It is shown that the INMCM5 model simulates most of the temperature and average precipitation related indices better, but the simulation results of extreme precipitation related indices are worse. Various precipitation related physical factors in the model are analyzed. A parameterization of vertical mixing of the horizontal velocity components due to large-scale condensation, as well as air resistance acting on falling precipitation, is implemented in the INMCM5. The corresponding model parameters are adjusted. The effect of the adjustments on the simulation results of extreme indices is presented.

1. Introduction

Modern models of the Earth's climate can reproduce not only the average climate condition, but also extreme weather and climate phenomena. Therefore, there arises the problem of comparing climate models for observable extreme weather events.

In [1, 2], various extreme weather and climatic situations are considered. According to the paper, 27 extreme indices are defined, characterizing different situations with high and low temperatures, with heavy precipitation or with absence of precipitation.

The results of simulation of the extreme indices with the INMCM4 [3] climate model were compared with the results of other models which took part in the CMIP5 project (Coupled Model Intercomparison Project, Phase 5) [2]. The comparison demonstrates that this model performs well for most indices except for those related to daily minimum temperature. For those indices the model shows one of the worst results.

The parameterizations of physical processes in the next model version, INMCM5, were replaced or tuned [4, 5], so that changes in the extreme indices simulation are expected.

The simulation results were compared to the ERA-Interim [6] reanalysis data, which were considered as the observational data for this study. Indices averaged for the 1981–2010 year range were compared. Mann-Whitney test with 1% significance level was used to examine where changes are significant.

To evaluate the quality of simulation of extreme weather phenomena, the extreme indices were calculated [7] using the results of computations performed by two versions of the INM RAS climate model (INMCM4 and INMCM5) and the ERA Interim reanalysis. We took the root mean square deviation of the index value computed from the modeled and reanalysis data as the measure of simulation quality. The mean is averaged over the land.

Table 1. Root mean square deviation of temperature indices for models: INMCM4, INMCM5, INMCM5VM ($A_0 = 7.5$, $C_{INC} = 3$), INMCM5AR ($W_L = 5$ m/s, $W_S = 0.5$ m/s, $A_0 = 8$, $C_{INC} = 2$) compared to ERA-Interim reanalysis.

Index name	Label, unit	INM CM4	INM CM5	INM CM5VM	INM CM5AR
Max T_{max}	TXx, K	4.6	3.7	3.5	3.5
Min T_{max}	TXn, K	5.4	4.8	5.4	5.4
Max T_{min}	TNx, K	3.0	2.3	2.3	2.4
Min T_{min}	TNn, K	13.4	6.7	7.4	7.7
Diurnal temperature range	DTR, K	6.4	2.8	2.7	2.8
Growing season length	GSL, days	24	15	15	15
Frost days	FD, days	58	20	21	22
Ice days	ID, days	15	11	11	12
Summer days	SU, days	45	40	37	37
Tropical nights	TR, days	99	48	51	57
Cold days	TX10p, %	1.2	1.0	1.2	1.1
Cold nights	TN10p, %	1.1	0.9	1.2	1.1
Warm days	TX90p, %	1.7	1.7	1.9	1.8
Warm nights	TN90p, %	1.7	1.7	1.7	1.7
Warm spell duration	WSDI, days	4.7	4.9	4.7	4.4
Cold spell duration	CSDI, days	3.2	3.1	3.4	3.1

Table 2. Root mean square deviation of precipitation indices for models: INMCM4, INMCM5, INMCM5VM ($A_0 = 7.5$, $C_{INC} = 3$), INMCM5AR ($W_L = 5$ m/s, $W_S = 0.5$ m/s, $A_0 = 8$, $C_{INC} = 2$) compared to ERA-Interim reanalysis.

Index name	Label, unit	INM CM4	INM CM5	INM CM5VM	INM CM5AR
Number of wet days	R1mm, days	41	40	39	39
Heavy precipitation days	R10mm, days	32	26	24	22
Very heavy precipitation days	R20mm, days	9.8	6.2	6.5	6.3
Consecutive dry days	CDD, days	35	37	38	36
Consecutive wet days	CWD, days	46	30	27	27
Simple daily intensity	SDII, mm/day	1.7	1.5	1.5	1.4
Max 1 day precipitation	RX1day, mm	19	19	17	15
Max 5 day precipitation	RX5day, mm	28	43	40	36
Very wet days precipitation	R95p, mm	120	130	130	120
Extremely wet days precipitation	R99p, mm	57	53	49	48
Total wet-day precipitation	PRCPTOT, mm	550	480	470	450

Tables 1 and 2 present the names of extreme indices related to temperature and precipitation, their labels and measurement units, as well as the land only averaged standard deviations for these indices between the ERA-Interim reanalysis and different versions of the INM RAS climate model.

Table 1 shows that the simulation of almost all temperature indices has improved in the INMCM5 compared to INMCM4. In particular, the simulation of the following extreme indices related to the minimum daily temperature improved significantly (by 37–56%): the annual daily minimum temperature (TNn), the number of frost days (FD) and tropical nights (TR), the diurnal temperature range (DTR), and the growing season length (GSL).

Table 2 shows that the simulation of the number of heavy (R10mm) and very heavy (R20mm) precipitation days, consecutive wet days (CWD), simple daily intensity (SDII), and total wet-day precipitation (PRCPTOT) noticeably improved in INMCM5. At the same time, the simulation of indices related to the intensity (RX5day) and the amount (R95p) of precipitation on very rainy days became worse.

Figures 1a,b–3a,b show the spatial distribution of max 1 day (RX1day) and 5 day precipitation (RX5day) according to ERA-Interim reanalysis and their deviations from reanalysis according to INMCM5 data.

Figures 1a,b–3a,b show that INMCM5 overestimates RX1day and RX5day in Central and Eastern North America, the Amazon basin, South Africa, South, East, Southeast Asia and Australia, as well as underestimates max 1 day and 5 day precipitation in West Africa. Thus, INMCM5 overestimates the intensity of precipitation on very rainy days, especially in those regions where it is large according to the ERA-Interim reanalysis data.

To improve the simulation of extreme precipitation by the INMCM5 model, the following physical processes were considered: evaporation of precipitation in the upper atmosphere; mixing of horizontal velocity components due to large-scale condensation and deep convection; air resistance acting on falling precipitation particles.

2. Evaporation of precipitation in the atmosphere

In the INMCM5 model precipitation is formed as a result of balancing of two processes: the conversion of cloudy moisture into precipitation and the evaporation of falling precipitation [8].

The evaporation rate of precipitation in the model cell is determined by

$$E_{\mathrm{p}} = 5.44 \cdot 10^{-4}(1-a)(q_{\max} - q)\left(\frac{P\sqrt{p/p_{\mathrm{s}}}}{5.9 \cdot 10^{-3}}\right)^{1/2}, \qquad (1)$$

where a is the fraction of the cell occupied with clouds, $q_{\max}$ is the maximal specific humidity at the current cell temperature, q is the specific humidity in the cell, P is the total precipitation flux from above (kg m^{-2} s^{-1}), p is the pressure in the cell and p_{s} is the pressure at the surface level. Here E_{p} cannot exceed the $E_{\mathrm{p}}^{\max}$ value at which all water falling as precipitation from cloud moisture evaporates.

To estimate the effect of evaporation in the upper atmosphere on the formation of extreme precipitation, the dependence (1) was simplified to proportional one

$$E_{\mathrm{p}} = KE_{\mathrm{p}}^{\max}, \quad K = \mathrm{const}. \qquad (2)$$

With this simplified formula and the proportionality coefficient taken to be K = 0.2 and K = 1, computations for three model years were carried out. This was done to test the sensitivity of the model simulation of max 1 day precipitation (RX1day) to changes in the precipitation evaporation rate in the upper atmosphere. Figures 1c, 1d show deviations of RX1day for the INM RAS climate model with changed evaporation rate for two different values of K compared to the original INMCM5 model.

Note that for K = 0.2 (Figure 1c) there is both an overestimation of RX1day in the Amazon basin, South Africa, and East Asia and its underestimation in Central North America, South America, and South Asia. In Australia, significant overestimation of max 1 day precipitation in some areas is compensated by an equally significant understatement in the others. In general, it can be noted that at K = 0.2 the quality of reproducing the RX1day index has not changed.

Figure 1d for K = 1 illustrates significant overestimation of RX1day in the Amazon basin, South Africa, South and Southeast Asia and Australia, that is, in those regions where the max 1 day precipitation according to INMCM5 is overestimated compared to the ERA-Interim reanalysis data. Thus, with K = 1 the positive RX1day simulation error only increased.

Thus, changing the precipitation evaporation rate in the upper atmosphere only led to an increase in max 1 day precipitation in regions where, according to INMCM5, it was already too high compared to the ERA-Interim.

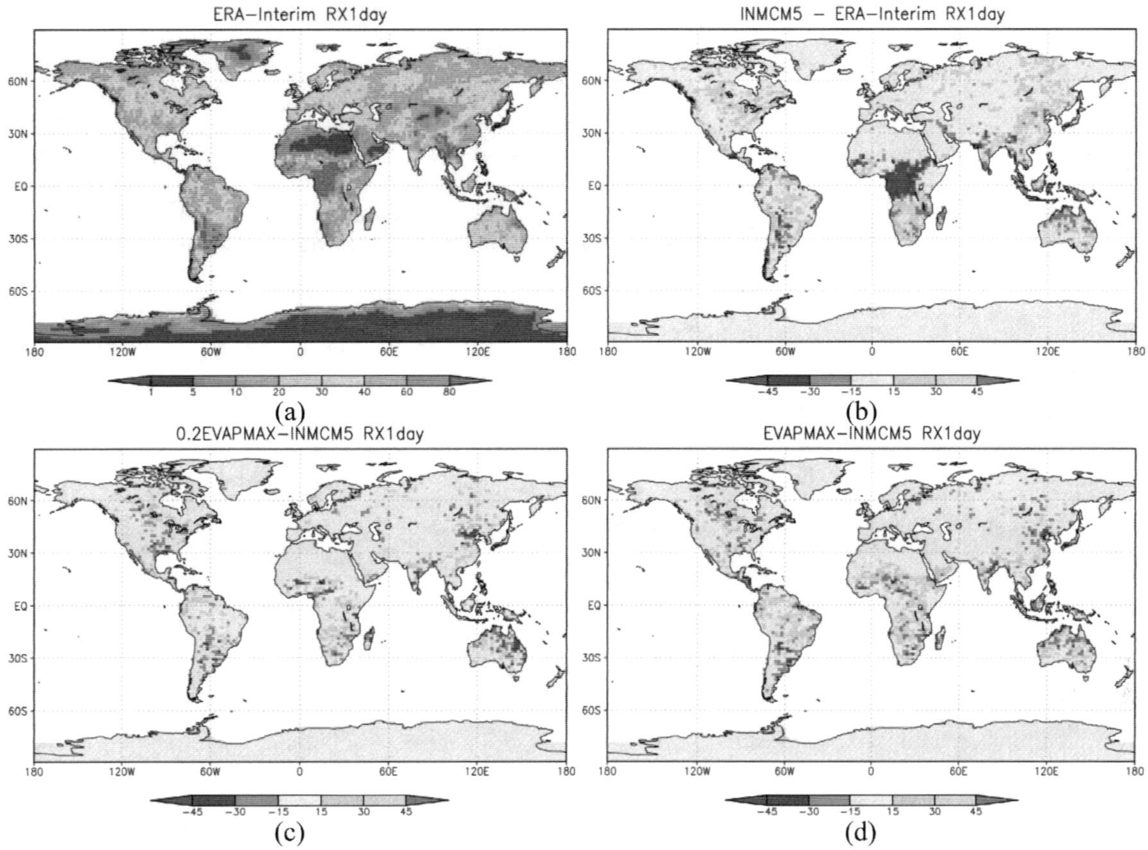

Figure 1. (a) RX1day index values averaged over 1979–1981 according to ERA-Interim data. (b-d) Deviations of the same average computed from data by INMCM5 model and its modification with $K = 0.2$ and $K = 1$.

3. Vertical mixing of horizontal velocity components

Both large-scale condensation and deep convection cause vertical motion, which redistributes the horizontal momentum between the nearby air layers.

The implementation of mixing due to large-scale condensation was added to the model. For two vertically adjacent cells the mixing is produced by the formulas

$$\frac{\tilde{\phi}_k - \phi_k}{\Delta t} = \alpha_{k+1/2}(\phi^{\text{avg}}_{k+1/2} - \phi_k),$$

$$\frac{\tilde{\phi}_{k+1} - \phi_{k+1}}{\Delta t} = \alpha_{k+1/2}(\phi^{\text{avg}}_{k+1/2} - \phi_{k+1}),$$

where ϕ is either of the horizontal velocity components, k, $k+1$ are adjacent vertical level numbers, $\phi^{\text{avg}}_{k+1/2}$ is the value between ϕ_{k} and $\phi_{\text{k+1}}$ averaged over the sigma-coordinate, $\alpha_{\text{k+1/2}} = \min(\Delta t^{-1}, A_0 P_{\text{k+1/2}})$ is the mixing speed, Δt is the time step, and $P_{\text{k+1/2}}$ is the precipitation flux from above. The mixing speed is artificially limited by Δt^{-1} to guarantee numerical stability of the procedure. The parameter A_0 is a fixed value chosen from the $1 \div 10$ m^2 kg^{-1} range.

A similar procedure is carried out to take into account mixing associated with deep convection. The differences are that it is performed in the region between the lower and upper convection boundaries, and the mixing rate β is calculated by the formula

$$\beta = \min\left(\frac{1}{\Delta t}, \frac{C_{\text{INC}}}{\tau_{\text{HIG}}}\frac{P_{\text{CV}}}{Q_{\text{TOT}}}\right).$$

Here τ_{HIG} = 12 h is the characteristic convection time, C_{INC} is a dimensionless constant parameter in the 1 ÷ 3 range, P_{CV} is the precipitation amount due to deep convection, Q_{TOT} is the total amount of moisture in the convective region.

To adjust the parameter A_0 (which is responsible for mixing horizontal velocity components due to large-scale condensation), a number of short-term computations with different A_0 values in the [1; 10] interval were carried out. The optimal A_0 value which minimizes the standard deviation from the reanalysis data for the RX1day and RX5day indices was found to be A_0 = 7.5. The parameter C_{INC}, which is responsible for mixing horizontal velocity components due to deep convection, was optimized with A_0 being fixed at 7.5. The optimal value for C_{INC} was found to be 3.

For short we will refer to the INMCM5 version with these changes as INMCM5VM (INMCM5 Velocities Mixing).

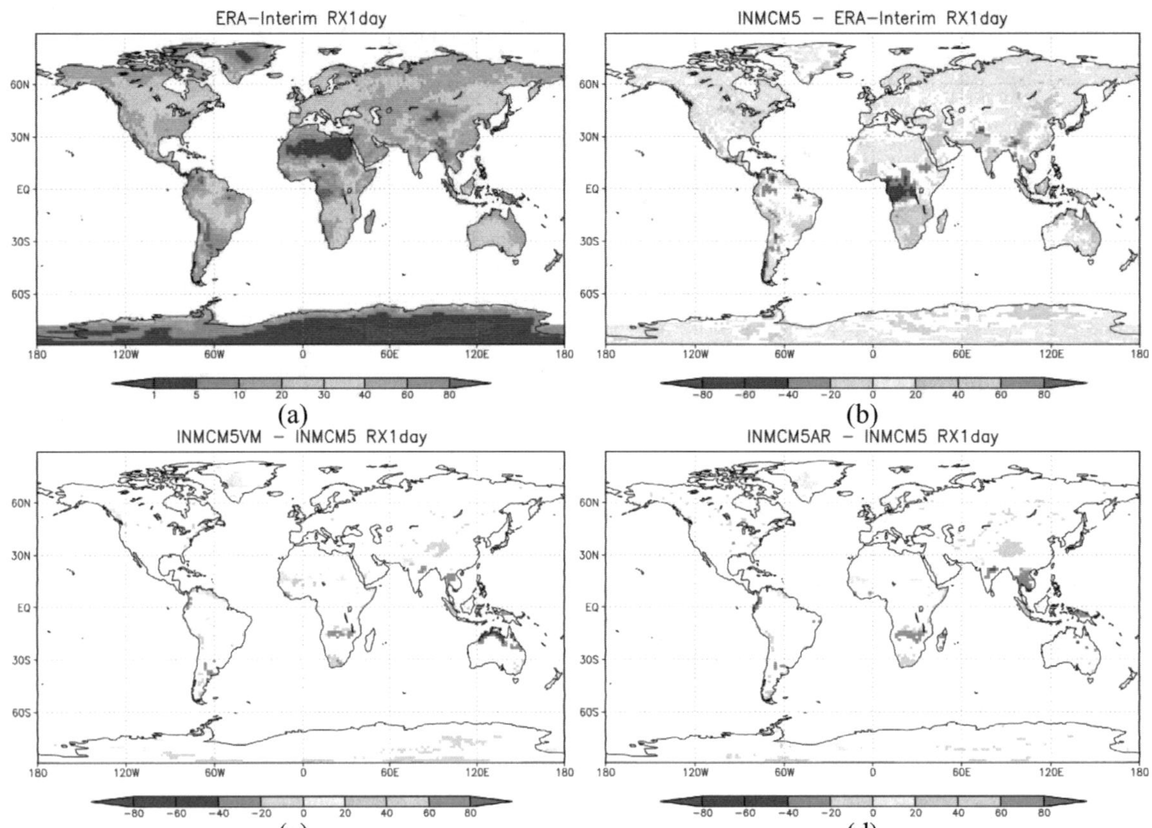

Figure 2. (a) RX1day index values averaged over 1981–2010 according to ERA-Interim data. (b-d) Deviations of the same average obtained from INMCM5, INMCM5VM, and INMCM5AR data. Statistically insignificant deviations are presented as white.

After these parameters were tuned, a long-term computation was performed. The computed data were used to calculate the extreme indices. Table 1 shows that in the model with vertical mixing of the horizontal velocity components the quality of simulation of most of the temperature extreme indices either improved slightly or did not change in comparison with INMCM5. Exceptions are the annual minimum of the maximum daily temperature (TXn), the annual minimum of the minimum daily

IOP Conf. Series: Earth and Environmental Science **386** (2019) 012012 doi:10.1088/1755-1315/386/1/012012

temperature (TNn), the number of frost days (FD) and tropical nights (TR), and the cold spell duration (CSDI): the simulation of these indices in INMCM5VM worsened by 5–13 % compared to INMCM5.

Table 2 shows that the quality of simulation of the R1mm, R10mm, CWD, RX1day, RX5day and R99p indices improved in INMCM5VM by 2–10 % compared to INMCM5. However, for the R20mm and CDD indices the standard deviation from the reanalysis data increased by 3–5 %.

Figures 2c, 3c show the spatial distribution of deviations for max 1 day (RX1day) and 5 day (RX5day) precipitation between INMCM5VM from INMCM5. The model with the adjusted vertical mixing of horizontal velocity components compared to INMCM5 significantly underestimates the RX1day in Australia and slightly underestimates the RX1day in South America, South Africa, South, East and South-East Asia. Also INMCM5VM significantly underestimates RX5day in East Asia and Australia compared to INMCM5, and also slightly underestimates RX5day in South America, South Africa, South and Southeast Asia.

Therefore, adjusting the vertical mixing parameters significantly reduces the overestimation of RX1day and RX5day observed in INMCM5 in South Africa, East Asia, and Australia and, in general, leads to an improvement in the quality of extreme indices related to the precipitation amount and intensity on very rainy days simulation, although not as significant as necessary. At the same time, a significant overestimation of the RX1day, RX5day indices in the Amazon basin, South and Southeast Asia, as well as their underestimation in West Africa remains.

4. Air resistance acting on falling precipitation particles

Since precipitation particles (water droplets or ice crystals) move in the surrounding air, a drag force arises that carries the air along with the particles. This resistance force can be included in the right-hand side of the momentum balance equation, which is part of the atmosphere hydrothermodynamic system of equations. Accurate accounting for the effect of this force requires numerical solving of an additional Poisson-type equation.

On the other hand, if we consider the air-precipitation system, in it the drag force is internal and does not contribute to the total momentum of the system. The atmosphere hydrothermodynamic equations remain valid if the air density is replaced by the density of the air-precipitation mixture.

In the model, this additional precipitation density is obtained as the ratio of the precipitation flux to the steady velocity of falling of precipitation particles:

$$\Delta\rho = \frac{P}{W_{\mathrm{prec}}}.$$

The range of the possible falling velocities is quite large and, therefore, in the calculation they were fixed with some constant values, different for water droplets and ice crystals.

The atmosphere dynamic equations used in the INMCM5 model do not contain the density in explicit form; instead it is expressed in terms of temperature and pressure from the ideal gas law:

$$\rho = \frac{p}{R_s T},$$

Thus, the density in the equations is uniquely determined by the temperature. Therefore, before calculating the solution of the atmosphere dynamic equations at the next step, the temperature in the model is corrected by

$$\Delta T = -T\frac{\Delta\rho}{\rho} = -\frac{R_s T^2}{p}\Delta\rho,$$

and after calculating the solution at this step, the inverse correction for the same ΔT value is made.

Since the relation

$$\frac{\Delta T}{T} = -\frac{\Delta\rho}{\rho}$$

is valid only for small $\Delta\rho$, the following restriction is also added to the model:

$$0 \le -\frac{\Delta T}{T} \le 0.05.$$

Since the temperature correction interferes with other processes in the model, it is only used in the atmosphere dynamic equations to determine the divergence of the horizontal velocity.

For short, we will refer to the INMCM5 model version with the air resistance and vertical mixing of the horizontal velocity components as INMCM5AR (INMCM5 Air Resistance).

During the computations the precipitation falling velocities were fixed at $W_L = 5$ m/s for liquid precipitation (water droplets) and $W_S = 0.5$ m/s for solid precipitation (ice crystals). The parameters A_0 and C_{INC} were retuned again for this model, and the optimal values for them were found to be $A_0 = 8$ and $C_{INC} = 2$.

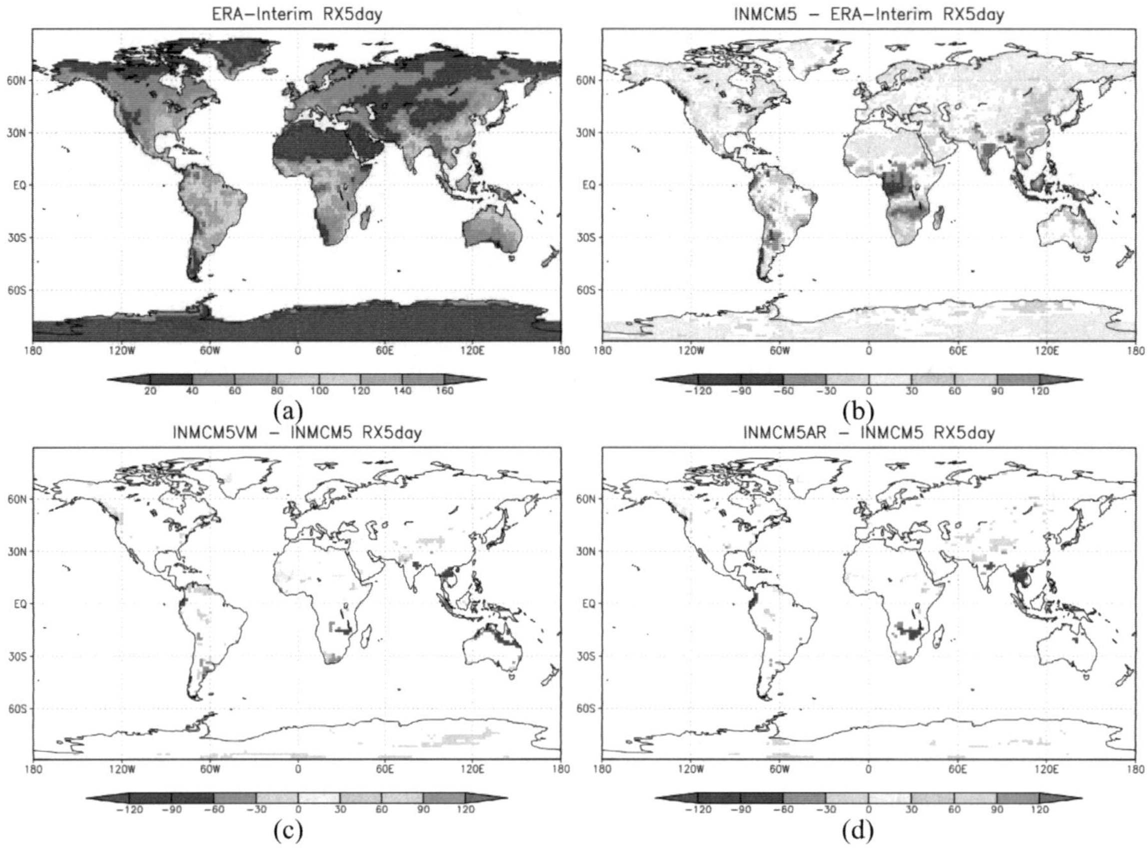

Figure 3. (a) RX5day index values averaged over 1981–2010 according to ERA-Interim data. (b-d) Deviations of the same average obtained from INMCM5, INMCM5VM, and INMCM5AR data. Statistically insignificant deviations are presented as white.

Table 1 shows that in the model with air resistance acting on falling precipitation particles the quality of simulation of most temperature-related extreme indices became worse by 9–19 % compared to INMCM5. The exception is the annual maximum of daily maximum temperature (TXx), the number of summer days (SU), and the warm spell duration (WSDI): the error in reproducing these indices decreased by 5–10 %. The reason for the increase in the error of simulation of the temperature indices is probably the inaccurate temperature correction used to adjust the density in the model. This error can be decreased by applying a temperature correction only for those equations where temperature was substituted instead of density using the ideal gas state equation.

Table 2 shows that the quality of simulation of all precipitation-related extreme indices in INMCM5AR either improved by 3–21 % compared to INMCM5 or remained unchanged. Figures 2d, 3d show the spatial distribution of the deviations for max 1 day (RX1day) and 5 day (RX5day) precipitation according to INMCM5AR compared to INMCM5. The model with air resistance acting on falling precipitation particles compared to INMCM5 significantly underestimates RX1day and RX5day in South Africa, South and East Asia, and slightly underestimates the indicated extreme indices in Tibet.

Taking into account the air resistance acting on falling precipitation particles significantly reduces the overestimation of RX1day and RX5day observed in INMCM5 in South Africa, South and East Asia, and leads to an improvement in the quality of extreme indices associated with the precipitation amount on very rainy days and their intensity simulation by 9–21 %. At the same time, a significant overestimation of the RX1day and RX5day indices in the Amazon basin and Southeast Asia, as well as their underestimation in West Africa, still remain.

5. Conclusions

To improve the results of simulations of extreme precipitation, the effect of evaporation in the upper atmosphere was studied first. Attempts to increase the evaporation rate led to a redistribution of precipitation over an area where the model extreme precipitation intensity was already too high compared to the ERA-Interim values. As a result, this change had a negative impact on the quality of simulations of problematic indices and was rejected.

Thereafter the vertical mixing of air masses due to large-scale condensation and deep convection was implemented in the model. The A_0 and C_{INC} model parameters, which are responsible for the intensity of these processes, were optimized. These changes have led to improvements in the simulation results of the precipitation related extreme indices by 2–10 % compared to the INMCM5.

Next, air resistance acting on falling precipitation particles was added to the model. This significantly (by 3–21 %) improved the quality of all precipitation related indices simulation results, but worsened the simulation results of some temperature indices. This degradation of the temperature indices may be due to the temperature correction which is used as a way to introduce additional density into the model.

At the same time, the above changes have not significantly improved the extreme precipitation simulation results in Southeast Asia. A peculiarity of this region is an unsatisfactory resolution of the computational grid. It is also possible that in this area the ERA-Interim reanalysis does contain significant deviations from the observational data. These issues require additional research involving, possibly, data from other reanalyses, more detailed observations, and a more detailed analysis of the hydrological cycle of the model.

Acknowledgements
This study was supported by the RAS under Program on «Climate Changes: Causes, Risks, Consequences, Problems of Adaptation and Regulation». The authors are grateful to R. Yu. Fadeev for valuable advice given during the course of the study.

References

[1] Flato G, Marotzke J, Abiodun B et al 2014 Evaluation of climate models *Climate Change 2013. The physical Science Basis* (Cambridge University press) p 1535
[2] Sillmann J, Kharin V V, Zhang X *et al* 2013 Climate extremes indices in the CMIP5 multimodel ensemble: Part 1. Model evaluation in the present climate *J. Geophys. Res. Atmospheres* **118** 1716–33
[3] Volodin E M, Dianskii N A and Gusev A V 2010 Simulating present-day climate with the INMCM4.0 coupled model of the atmospheric and oceanic general circulations *Izv. Atmos. Ocean. Phys.* **46** 414–31
[4] Volodin E M, Mortikov E V, Kostrykin S V *et al* 2017 Simulation of the present day climate

with the new version of INM RAS climate model (in Russian) *Izv. Atmos. Ocean. Phys.* **53** 164–78

[5] Volodin E M, Mortikov E V, Kostrykin S V *et al* 2017 Simulation of the present day climate with the climate model INMCM5 *Clim. Dyn.* **49** 3715–34

[6] Dee D P *et al* 2011 The ERA-Interim reanalysis: Configuration and performance of the data assimilation system *Quart. Journal. Roy. Meteorol. Soc* **137** 553–97

[7] Volodin E M and Tarasevich M A 2018 Simulation of climate and weather extreme indices with the INM-CM5 climate model *Russ. Meteorol. Hydrol.* **43** 756–62

[8] Volodin E 2017 Representation of heat, moisture and momentum fluxes in climate models. Convection and condensation (In Russian) *Fund. and Appl. Climatol.* **2** 26–41

IOP Conf. Series: Earth and Environmental Science **386** (2019) 012013 doi:10.1088/1755-1315/386/1/012013

Numerical simulation of breeze circulation in Crimea using WRF-ARW model

M V Shokurov and N Yu Kraevskaya

Department of Atmosphere Ocean Interaction, Marine Hydrophysical Institute, Russian Academy of Sciences, Kapitanskaya Street 2, Sevastopol 299011, Russia

shokurov.m@gmail.com, krayevskaja@gmail.com

Abstract. Breeze circulation is an important atmospheric process that affects the weather and climate of coastal regions. Crimea is one of such regions. One of the methods of breeze research is numerical simulation, which has been quite successfully used to obtain information on breeze circulation in various geographical regions. The purpose of this work is to describe the regional features of sea breeze in Crimea, specifically, a breeze circulation near the Crimean Mountains. We use an array of satellite images and data of global operational analysis, NCEP GDAS, to choose a day with a pronounced breeze. A numerical experiment is performed for the day using a mesoscale model, WRF-ARW. Based on the results of the numerical simulation, the structure of the sea breeze circulation in Crimea is described. Some classical elements of the breeze characteristics of the flat part of Crimea are observed. A unique phenomenon of collision of breeze gravity currents has been revealed. It is a characteristic of the rather small land areas. Much attention is given to the features of breeze in the mountains associated with the complex topography.

1. Introduction

Sea breeze circulation is an important atmospheric process that affects the weather and climate of coastal regions [1, 2]. Weak background pressure gradients are needed for the development of breeze circulation, which is typical of the middle latitudes in summer.

Various methods are used to study the breeze circulation: analytical theory (linear and nonlinear), *measurements* (*in situ* and remote), laboratory experiments, and numerical simulation (ideal and realistic).

The breeze of weak intensity is described by a linear theory, in which breeze circulation is represented as a forced internal gravity wave localized near the coast. This wave is a reaction of the atmosphere to a heat source with a daily frequency [3, 4].

For a high intensity breeze, the lower branch of the breeze circulation cell is a propagating gravity current. It is a substantially nonlinear phenomenon that has been the subject of a number of studies [2, 5 – 7].

Breeze field observations include the data from standard weather stations, the results of specialized programs and expeditions [8–11], as well as numerous satellite data in different spectral ranges [12 – 13].

In laboratory experiments, the features of the breeze circulation and gravity currents are studied under controlled conditions, such as stratification, velocity profile, etc. [2, 14].

IOP Conf. Series: Earth and Environmental Science **386** (2019) 012013 doi:10.1088/1755-1315/386/1/012013

In the ideal numerical simulation, the problem is usually posed with maximum simplifications, such as a straight endless coast, constant buoyancy frequency, no complex topography, and simple parameterization of viscosity and thermal conductivity [7, 15, 16]. Ideal simulation makes it possible to determine and investigate the influence of each factor that is involved in the formation and modification of the breeze.

Realistic numerical simulation of the breeze in a particular region, taking into account real geography, mountain topography, etc., using mesoscale atmospheric models and reliable input data from global forecasts and analyses, makes it possible to describe in detail the real breeze circulation with high spatial resolution, taking into account all other processes, for example, synoptic ones.

Realistic numerical simulation together with the data from weather stations and satellite data has been successfully used to obtain reliable information on breeze circulation in various geographical areas with complex topography, such as Spain [8, 10], Greece [9], Mississippi area [11], etc.

To describe the breeze circulation of the Crimean Peninsula, a number of studies were performed using numerical simulation [17 – 18]. It was noted that the features of breeze in Crimea are explained by its limited size, a complex shape of the coastline, and the presence of mountains. A significant difference between the breeze in the flat terrain and in the mountains was also noted.

Two diagnostic methods can be used in realistic numerical simulation of the breeze:

1. Numerical simulation and subsequent averaging over some period (months, years) of time, in order to get an idea of the average characteristics of the breeze in a given area. In this case, background synoptic and large-scale processes should be taken into account and properly filtered [17 – 18].

2. Case study: diagnosing the structure of a breeze for one particular day makes it possible to investigate the development and dynamics of the breeze in a specific synoptic situation, to highlight the influence of geographical features of a territory.

Due to the influence of numerous atmospheric factors, such as the vertical profile of the synoptic wind, the vertical profile of the buoyancy frequency, the distribution of humidity, etc., the structure and dynamics of the breeze in a given area greatly vary from day to day. Therefore, the use of the averaging method can largely distort the picture and, therefore, it is preferable to consider the breeze circulation on a particular day when there is no such distortion.

In order to consider the breeze circulation in its pure form, it is necessary to choose the simplest state of the surrounding atmosphere characterized by the absence of strong synoptic processes, large-scale clouds, and enhanced cumulus convection. For this purpose, a day with a well pronounced breeze over the Crimean Peninsula was selected from the archived data of satellite images and global operational analysis data.

For the selected day, a numerical experiment was performed using the mesoscale atmospheric model WRF-ARW, and a detailed analysis of the breeze circulation was made.

The work consists of four sections. The following section describes the area under study, the model, the input data, and the model options used for the calculation. The third section presents the simulation results, describes the structure and development of the breeze during the day in the flat and mountainous parts of Crimea. The last section is devoted to conclusions.

2. Numerical experiment

2.1. Description of the study area

The study area includes the Crimean Peninsula with the adjacent Black and Azov Seas. Most of the peninsula is flat terrain. In the south and southeast of the peninsula a mountain ridge stretches along the coast; its height does not exceed 1600 m. On all sides Crimea is surrounded by seas. In the north, the peninsula is connected with the continent by an isthmus. From the West and South of Crimea, the Black Sea is located, and in the north-east – the Sea of Azov.

2.2. Choice of the day for the study

The day for the study was selected using the data of the global operational analysis NCEP GDAS. We considered the wind speed field at levels from the surface to 850 mb and estimated the maximum wind speed and the general synoptic situation. Days with the lowest wind speed and the absence of cyclones and anticyclones in the vicinity of the Crimean Peninsula were selected.

We also used the data in the visible range from MODIS Terra and MODIS Aqua satellites, with a resolution of 250 m. The breeze indicator on the satellite imagery was characteristic cumulus clouds that formed in front of the breeze front. Since these satellites pass over the same point of the surface with an interval of 3 hours, this allows one to see the development of cumulus clouds during the day.

Days without strong displacement of the cumulus clouds of the breeze front by the synoptic wind were selected for numerical modeling. The days were also selected without deep convection and the formation of cumulonimbus clouds strongly influencing the development of the breeze.

Figure 1. MODIS Aqua satellite image, 29 July 2018.

Three summer seasons of 2016-2018 were analyzed and, as a result, the day of 29 July 2018 was chosen for the simulation. Figure 1 shows the distribution of clouds over the Crimean Peninsula according to the MODIS Aqua satellite data for this day.

An analysis of the GDAS global operational analysis data for the Crimean Peninsula for this day showed that in the lower part of the troposphere the wind speed, wind shift, and horizontal temperature gradients were small. At the lower levels, a weak northwest wind was observed with a speed of 2-3 m/s, and at the level of 500 hPa the wind speed was 5 m/s. In the upper part of the troposphere, above the level of 350 hPa, a southwestern flow was observed with a vertical wind shift of 24 m/s between the levels of 350 and 250 hPa, and the corresponding horizontal temperature gradient. However, the motion in the upper troposphere does not affect the breeze circulation.

2.3. Description of the model and input data

For the numerical simulation, the mesoscale model WRF-ARW of version 3.7 [19] was used, based on a system of nonhydrostatic equations written for a compressible atmosphere. The model vertical coordinate is a terrain-following eta coordinate, with stretching permitted by the vertical grid. The horizontal grid is presented by a staggered Arakawa C-grid. Time-split integration is performed using a 2-nd- or 3-rd-order Runge-Kutta scheme. For spatial discretization, 2-nd- to 6-th-order advection schemes in the horizontal and vertical directions are used.

The model has a wide range of options for describing subgrid processes, such as turbulence in the boundary planetary layer, radiation transfer, cloud microphysics, cumulus convection, heat and moisture transfer in the soil (Table 1).

Data from the global operational analysis NCEP GDAS / FNL 0.25 were taken for the numerical simulation of the breeze circulation as initial conditions in the internal area of both computational domains, and then in the process of calculating the boundary conditions at the boundary of the external domain. The data had a resolution of 0.25 degrees in space, 6 hours in time. Data on the topography were taken from the GEOG array with a spatial resolution of 30 seconds.

2.4. Numerical simulation

Table 1 provides information on the numerical experiment for the selected day and the model options used.

For the internal domain, a resolution of 2 km was chosen, which is enough to describe the features of the breeze structure.

32 vertical levels were selected to have a sufficiently high resolution in the lower troposphere, in which the breeze circulation develops.

The time step is determined by the spatial resolution.

The listed parameterizations are traditionally used in the numerical simulation of a breeze under conditions of complex topography [8, 9, 11, 17, 18].

Table 1. Details of the domain grids and options for physical parameterizations used in the numerical simulation.

Domains	Domain 1	Domain 2
Horizontal resolution	6 km	2 km
Domain coordinates	41.028 — 49.433 N, 27.968 — 40.632 E	44.132 — 46.939 N, 32.227 — 37.108 E
Time step	36 s	12 s
Vertical resolution	32 levels	
Simulation time	1 day	
Physical parameterizations		
microphysics	WSM 3-class simple ice scheme	
longwave radiation	RRTM scheme	
shortwave radiation	Dudhia scheme	
surface-layer	Revised MM5 Monin-Obukhov scheme	
land-surface	thermal diffusion scheme	
boundary-layer	YSU scheme	
cumulus	Kain-Fritsch (new Eta) scheme	

3. Results

To analyze the breeze circulation, we used the results of numerical simulation in the internal domain with a resolution of 2 km.

Below we consider the structure of the breeze over the Crimean Peninsula and provide an interpretation of its features.

3.1. Sea breeze over Crimea

A detailed analysis of the results of the numerical simulation showed that the breeze circulation over Crimea has standard characteristics inherent in classical concepts. These include: 1) formation of the daytime convective boundary layer; 2) development and propagation of breeze gravity currents from the coast; 3) formation of the breeze front, head, and tail of the breeze gravity current; 4) intense vertical lift in front of the gravity current; 5) reverse branch of the breeze circulation cell.

To highlight these characteristics for different moments in time, maps were constructed at different levels and various vertical sections of the horizontal and vertical components of wind velocity, temperature and potential temperature, humidity, the planetary boundary layer height, cloudiness, etc.

Due to a limited volume of this article, of all these maps only a map of direction and module of the horizontal velocity component is shown in Figure 2. This figure clearly depicts the spread of gravity currents.

In addition to these classical characteristics, the breeze circulation in Crimea has its own features. One of them is the propagation of several gravity currents from different shores to the center of the peninsula and subsequent collision of these currents, which can lead to the development of deep convection. The Crimean Mountains also have a significant impact on the breeze, especially on the southern slope.

Consider the horizontal and vertical structure of the breeze over Crimea (Figure 2). To obtain information about the vertical structure, consider two vertical sections marked in Figure 2 as black straight lines.

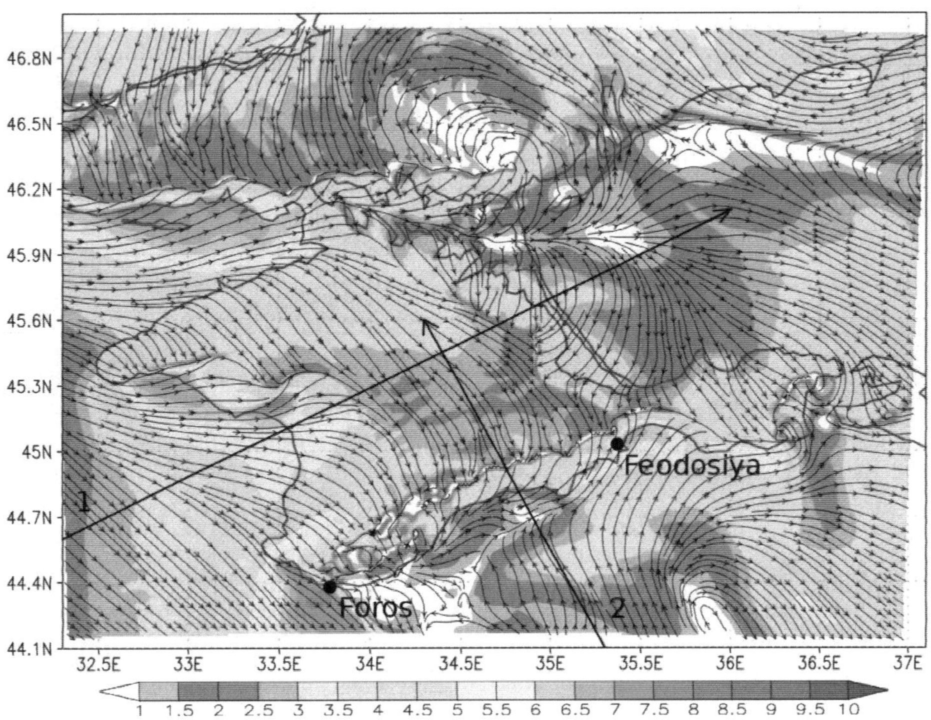

Figure 2. Wind speed (color) (m/s) and direction (streamlines) at 15:00 local time (LT) (UTC + 3) at 2.5-d eta level (73,5 m) for velocity. 1 – cross section along which the vertical structure of sea breeze in the flat part of Crimea is considered. 2 – cross section along which the vertical structure of sea breeze in the Crimean Mountains is studied.

IOP Conf. Series: Earth and Environmental Science **386** (2019) 012013 doi:10.1088/1755-1315/386/1/012013

3.2. Sea breeze on the plain

Consider the sea breeze circulation over the central plain part of Crimea. In the morning, a convective atmospheric boundary layer is formed over most of the peninsula, in which there is a uniform vertical profile of potential temperature. A heated convective boundary layer over land is necessary for the intrusion of cold sea air to the land, the formation of breeze gravity currents, and their subsequent propagation from the coast to the land.

The Crimean Peninsula is surrounded by seas on all sides, but breeze does not develop on each side. Only two fronts are clearly visible in Figure 2: from the west (Black Sea) and from the east (the Sea of Azov).

The third breeze front from the northwest, from the Karkinitsky Bay, is absent on the study day. This is because on this day there was a weak northwest synoptic wind over the entire territory of Crimea; it prevented the development of a breeze on the north-western coast of Crimea.

The fourth front from the side of the Southern Coast of Crimea is being formed, but the Crimean Mountains prevent its propagation. Its features are described in more detail in the next section.

A more detailed analysis of the breeze gravity currents at different moments shows that during the day, as their structure develops, the feature elements of sea breeze form: the front, the head of the current (with a height of about 600-800 m), and the tail (with a height of about 300-500 m). In front of and above the fronts of breeze gravity currents, intense vertical movements (about 0.5-0.7 m/s) occur in rather narrow areas (about 10 km), which leads to enhanced cumulus convection seen in satellite images (Figure 1). When advancing further on land, the fronts of gravity currents follow the shorelines of the coasts where they formed.

Above the near-surface currents, the wind direction changes to the opposite one, and reverse branches of the sea breeze circulation cells form.

The time of occurrence and the distance that the breeze fronts cover during the day differ at different shores. On the west coast, a breeze front occurs at 10:00 LT and on the east coast, at 12:00 LT. During the day the western front covers a distance of about 100 km, and the eastern front, of about 80 km.

In the evening at 19:00 – 20:00 LT there is a collision of breeze currents. Figure 3 shows a vertical section of the longitudinal component of wind velocity at the time before the collision at 15:00 LT. The direct and reverse branches of the circulation of both fronts, their heads, and tails are clearly visible. After the collision, the current extending from the Sea of Azov propagates above the current from the Black Sea, which continues to spread on the surface. This is due to the fact that the Black Sea breeze front is stronger (it has a greater thickness and propagation speed) than the Azov one.

The collision of sea breeze gravity currents in the center of the Crimean Peninsula is a unique phenomenon associated with the small size of the peninsula. The breeze fronts manage to travel distances (approximately 100 km) from the coast to the center of the peninsula within a few hours (8 hours) with a characteristic speed of about 5 m/s.

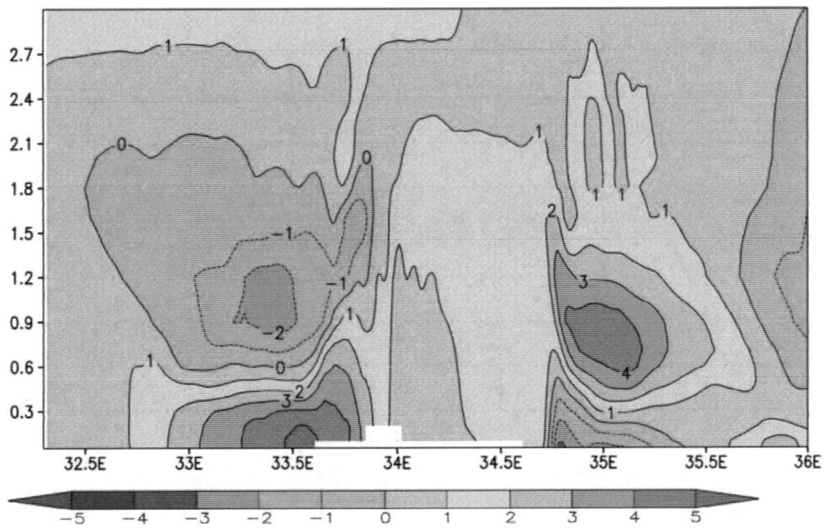

Figure 3. Velocity component (m/s) along cross-section 1 at 15:00 LT. The position of cross-section 1 is shown in Figure 2.

3.3. Sea breeze in the mountains

The wind in the mountains occurs due to the breeze circulation and the anabatic slope wind, caused by one reason — the daytime heating of the (inclined) land surface. On the northern slope, there is a slope wind, a northwest background wind which coincides in direction with the slope wind, and breezes from the western and eastern shores of Crimea. Pure breeze is not observed on the northern slope, but it occurs on the southern slope of the mountains.

Consider the sea breeze circulation on the southern slope of the Crimean Mountains, where the picture of motion is less complex than on the northern slope.

On the southern coast of Crimea, the breeze from the adjacent Black Sea propagates up the slope of the whole chain of the Crimean Mountains.

In this part of the peninsula, the circulation develops earlier (at about 8 am local time) compared with the other shores of Crimea (at about 10:00 – 12:00 LT).

The breeze covers the entire southern coast from Foros to Feodosiya (Figure 2), and represents a narrow band with a wind of the southern direction, at a speed of up to 4 m/s.

During the day, the width of the area occupied by breeze changes from 2 – 4 km at 11:00 LT to 20 km during the maximum breeze development at 15:00 LT, and then decreases by 19:00 LT, that is, by the end of the sea breeze circulation.

The visible difference in the distribution of the breeze in the western and eastern parts of the mountains (Figure 2) is due to different heights of the mountains and different steepness of the slopes. The western part of the Crimean Mountains is higher than the eastern part, and has steeper southern slopes, which significantly affects the characteristics of the sea breeze. First, the coastal band width on the Southern Coast covered with the sea breeze circulation in the west (5 km) is less than in the east (15 - 20 km). Second, the intensity of the breeze circulation in the west is less (3 m/s) than in the east (5 m/s). Third, the duration of the sea breeze in the east (11:00 – 20:00 LT) is longer than in the west (11:00 – 18:00 LT).

In particular places of the Southern Coast of Crimea, along the mountain passes, large values of the horizontal velocity are observed during the day.

Consider now the vertical structure of the breeze in the mountains. To do this, we choose a cross-section that runs perpendicular to the ridge of the Crimean Mountains, and consider the distribution of the velocity component perpendicular to the ridge (Figure 4).

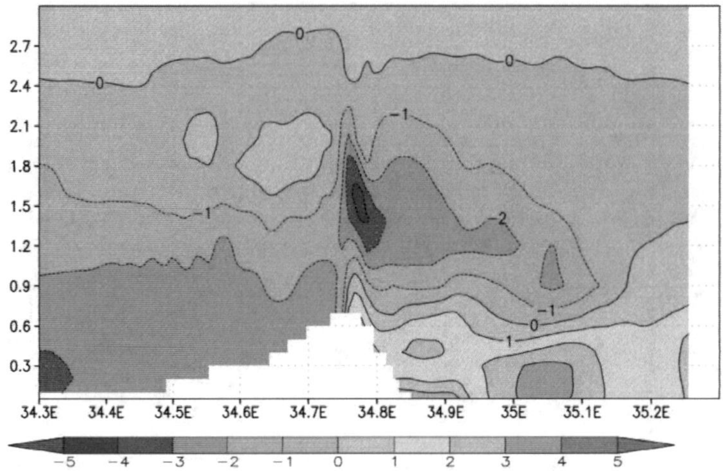

Figure 4. Velocity component (m/s) along cross-section 2 at 15:00 LT. The position of cross-section 2 is shown in Figure 2.

The breeze structure in the mountains also has characteristic elements of a classical breeze, in particular, the lower branch of the breeze circulation, which after the breeze commences (at about 8:00 LT) propagates up the slope. Its advance is accompanied by an intense vertical motion.

The breeze front extends only to the top of the ridge, and after 14:00 LT it is observed along the southern slope of the mountain, not moving further north. The current cannot pass further due to a northwestern synoptic headwind on the northern slope of the mountain ridge, which flows on the breeze gravity current from above. Due to the collision of the breeze and the background wind, convergence areas are observed along the entire chain of the Crimean Mountains.

After 20:00 LT, the breeze structure collapses. Maximum speeds in the surface gravity current were observed during the day (from 12:00 to 15:00 LT) and reached 3 m/s.

4. Conclusions

Based on the results of a numerical simulation, the structure of a sea breeze circulation in Crimea has been described. Some classical elements of the sea breeze of the flat part of Crimea were observed. A unique phenomenon of collision of breeze gravity currents has been revealed. It is a characteristic of small-size peninsulas and islands (see, for example, [20]).

Much attention was given to the features of the breeze associated with the influence of the mountains. On the Southern Coast of Crimea, in the daytime, two mesoscale atmospheric phenomena occur simultaneously — a breeze circulation and an anabatic slope wind caused by daytime heating of the (inclined) land surface.

The numerical modeling allowed us to describe in detail the sea breeze circulation of the southern coast of the peninsula. Here the sea breeze develops earlier compared to the other shores. The breeze of the southern coast is a narrow band with a southern wind, and during the day the width of this band changes greatly. There is a difference in the propagation of the breeze in the western and eastern parts of the mountains, due to the difference in height between the mountains and the steepness of the slopes.

Some elements typical of a classical breeze have been observed in the vertical structure of the breeze in the mountains, specifically, a gravity current near the surface, which propagates up the southern slope of the mountain.

Acknowledgments
This work was carried out under state assignment no. 0827-2018-0001 "Interaction of the Ocean and the Atmosphere".

References
[1] Simpson J E 1994 *Sea breeze and local winds* (Cambridge: Cambridge University Press) p 234
[2] Simpson J E 1997 *Gravity Currents in the environment and the laboratory* (Cambridge: Cambridge University Press) p 244
[3] Rotunno R 1983 *J. Atmos. Sci.* **40** 1999–2009
[4] Jiang Q 2012 *J. Atmos. Sci.* **69** 1890–909
[5] Benjamin T B 1968 *J. Fluid Mech.* **31** 209-48
[6] Haase S P and Smith R K 1989 *Geophys. Astrophys. Fluid Dyn.* **46** 1-33
[7] Robinson F J, Patterson M D and Sherwood S C 2013 *J. Atmos. Sci.* **70** 653-68
[8] Arrilaga J A, Yagüe C, Sastre M and Román-Cascón C 2016 *Atmospheric Research* **181** 265 – 80
[9] Papanastasiou D K, Melas D and Lissaridis I 2010 *Atmospheric Research* **98** 102-17
[10] Soler M R, Arasa R, Merino M, Olid M and Ortega S 2011 *Bound.-Layer Meteorol.* **140** 37–56
[11] Challa V S, Indracanti J, Rabarison M K, Patrick C, Baham J M, Young J, Hughes R, Hardy M G, Swanier S J and Yerramilli A 2009 *Atmospheric Research* **91** 9-25
[12] Wakimoto R M and Atkins N T 1993 Mon. Wea. Rev. **122** 1092 – 114
[13] Lensky I M and Dayan U 2012 *Atmos. Chem. Phys.* **12** 6505–13
[14] Shin J O, Dalziel S B and Linden P F 2004 *J. Fluid Mech.* **521** 1-34
[15] Du Y and Rotunno R 2015 *J. Atmos. Sci.* **72** 2806 – 21
[16] Crosman E T and Horel J D 2010 *Boundary-Layer Meteorol* **137** 1-29
[17] Efimov V V and Komarovskaya O I 2015 *Physical oceanography* **6** 69 – 78
[18] Efimov V V 2017 *Izvestiya, Atmospheric and Oceanic Physics* **53**(1) 84–94
[19] Skamarock W C, Klemp J B, Dudhia J, Gill D O, Barker D M, Duda M G, Huang X and Wang W 2008 *A Description of The Advanced Research WRF Version 3*. NCAR Tech. Note NCAR/TN – 475+STR p 113
[20] Ramis C, Jansa A and Alonso S 1990 Meteorol. Atmos. Phys. **42** 249-58

Interannual variability of wind stress curl in the Black Sea and its response to changes in prevailing wind frequency

I G Shokurova

Oceanography Department, Marine Hydrophysical Institute RAS, 2, Kapitanskaya str., Sevastopol 299011, Russia

e-mail: igshokurova@ mail.ru

Abstract. The purpose of this study is to identify trends in the long-term variability of the wind stress curl in the Black Sea and to analyze its response to changes in the direction of the prevailing wind flow over the sea. NCEP/NCAR reanalysis (1948–2016) and ERA Interim reanalysis (1979–2016) wind data at 10-m height are used. The temporal and spatial characteristics of the wind stress curl are compared by using these reanalyses with various spatial resolutions. Interannual changes in the wind stress curl are examined with long-term time series data of NCEP/NCAR reanalysis. The wind stress curl seasonal cycle, the dependence of its magnitude on wind direction, and the frequency of wind direction are in good agreement for the two reanalyses. There are multi-year periods of predominant positive and negative anomalies of the curl according to annual mean and winter data. The cyclonic wind stress curl increases in the late 1960s and early 1970s and weakens at the end of the 1990s. There is a weak positive trend in the curl time series in summer. The long-term variations of the wind stress curl are related to changes in the wind frequency of certain directions. High values of positive correlation coefficients have been obtained between the time series of the basin-averaged wind stress curl and the frequency of northeastern, eastern winds, and negative coefficients for the frequency of southwestern, western winds.

1. Introduction

The Black Sea is the inland sea of the Mediterranean Basin. The main feature of the sea is weak mixing between the surface and deep waters because of a positive water balance at the sea surface – the excess flow of rivers and precipitation over evaporation [1]. Wind impact is an important factor determining the nature and intensity of currents in the sea and affecting the vertical mixing in the upper layer [2]. Spatial heterogeneity in the wind field leads to the appearance of a vertical velocity at the lower boundary of the Ekman layer. The magnitude and direction of the velocity depend on the magnitude and sign of the wind stress curl. On average, a positive wind curl prevails over the Black Sea [3–5]. This leads to the upwelling of water in the deep water area of the basin, its divergence to the coast, the emergence of a horizontal pressure gradient, and the formation of gradient currents. Along with the river flow, the positive wind curl supports the cyclonic character of large-scale circulation in the sea [2, 6]. All this makes it important to study the variability of the wind stress curl.

Numerical estimates and analysis of the temporal and spatial variability of the wind curl and the wind stress curl were performed based on reanalyses [3, 5, 7–9], regional models [4], and satellite measurements [10] data to study the impact of wind on the circulation in the Black Sea. The seasonal cycle demonstrates an increase in the cyclonic wind stress curl in winter and its weakening in summer

[3–5, 9, 11]. The interannual changes of the wind stress curl were studied using reanalyses data from 1950s to 2000s in [8, 12] and reanalyses and satellite measurements data over the past decades since 1980s in [9, 10].

The large-scale atmospheric circulation, together with local orographic effects in the Black Sea region, determines the direction of the wind above the sea [13] and the spatial structure of the wind curl [3, 10, 11, 14]. Statistical analysis of the magnitude of the wind stress curl and its spatial distribution showed that they depend on the direction of the wind prevailing over the sea [11, 15]. In this research it is examined whether these results are valid for the interannual changes of the wind stress curl.

The purpose of this study is to detect long-term trends of the wind stress curl in the Black Sea and to analyze the response of the wind stress curl magnitude to changes in the direction of the prevailing winds using long-term time series data. The study of the relationship between the wind stress curl and the wind directions is one of the intermediate stages in the research of the relationship between the long-term variability of the Black Sea currents and changes in the atmospheric synoptic situation in the region.

High temporal resolution of data is needed to accurately determine the frequency of each wind direction. NCEP/NCAR reanalysis wind data since 1948 (4 times daily) [16] satisfy such requirements. However, as a rule, long-term series data have insufficient spatial resolution. Therefore, a comparison between the spatial-temporal variability of the wind stress curl based on wind data of NCEP \ NCAR reanalysis and of ERA Interim reanalysis was performed. ERA Interim reanalysis has a shorter time series (since 1979), but a higher spatial resolution [17].

In Section 2, the data and research methods are described. The seasonal cycle and interannual changes of the basin-averaged wind stress curl and its correlation with the frequency of winds with different directions are presented in Section 3. The main conclusions and a discussion are presented in Section 4.

2. Data and methods

Data sets of 4 times daily wind at 10-m height of NCEP/NCAR reanalysis for 1948–2016 (1.905° × 1.875°) [16] and of ERA-Interim reanalysis for 1979–2016 (0.75° × 0.75°) [17] are used. For each 6-hourly time interval, the wind stress, the wind stress curl, and the prevailing wind direction were calculated.

The wind stress was calculated using the bulk formula

$$\tau = \rho_{\text{air}} C_d |V| V, \ (\text{N m}^{-2}),$$

where (kg m^{-3}) is the air density; C_d is a dimensionless drag coefficient, and $V = (u, v)$ is the 10-m height wind velocity (ms^{-1}). The drag coefficient was taken to be 1.2×10^{-3} for a wind speed <11 m s^{-1} and $(0.49+0.065\ V) \times 10^{-3}$ for a wind speed >11 m s^{-1} [18].

The wind stress curl was calculated using the formula

$$\text{rot}_z \tau = \partial \tau_y / \partial x - \partial \tau_x / \partial y, \ (\text{N m}^{-3}).$$

The curl values were calculated at the centers of cells of the reanalyses data (Figure 1a).

The correspondence between the magnitude of the wind stress curl and the direction of the wind flow prevailing over the sea is examined. The prevailing wind direction is determined for each 6-hourly time interval by averaging the wind vector components $V = (u, v)$ using the grid points located in the sea area. The resulting direction refers to one of the geographical directions: northern (N), northeastern (NE), eastern (E), southeastern (SE), southern (S), southwestern (SW), western (W), and northwestern (NW). Thus, there were the prevailing wind direction and the corresponding curl field for each time interval.

Composites of the wind stress curl fields were formed for events with each wind direction. The annual mean wind stress curl corresponding to an individual wind direction was calculated by averaging over a set of situations with this wind direction. The annual mean frequency of the wind directions was calculated in percent of the total number of situations.

In order to examine the relationship between the long-term time series data of the wind stress curl and the frequency of the winds, the correlation between them was calculated. The two-monthly mean frequency of the wind directions and the two-monthly mean basin-averaged wind stress curl were calculated for every year and every season in this case.

All average quantities were calculated by averaging the data located in the area bounded by the 50-meter isobath (Figure 1a).

3. Results

3.1. Seasonal variability

3.1.1. Spatial distribution. The spatial distributions of the annual mean and January-February wind stress curl have a similar structure according to both reanalyses for 1979-2016 (Figure 1, b, c). Positive values (the cyclonic curl) of the wind stress curl are located in the northeastern and southwestern parts of the sea; negative values (the anticyclonic curl) are located near the western and southeastern coasts. In July-August and in January-February, the main differences between the reanalyses are located in the southeastern part of the sea (Figure 1, c, d). These differences are probably caused by the insufficient spatial resolution of the long-term NCEP/NCAR reanalysis.

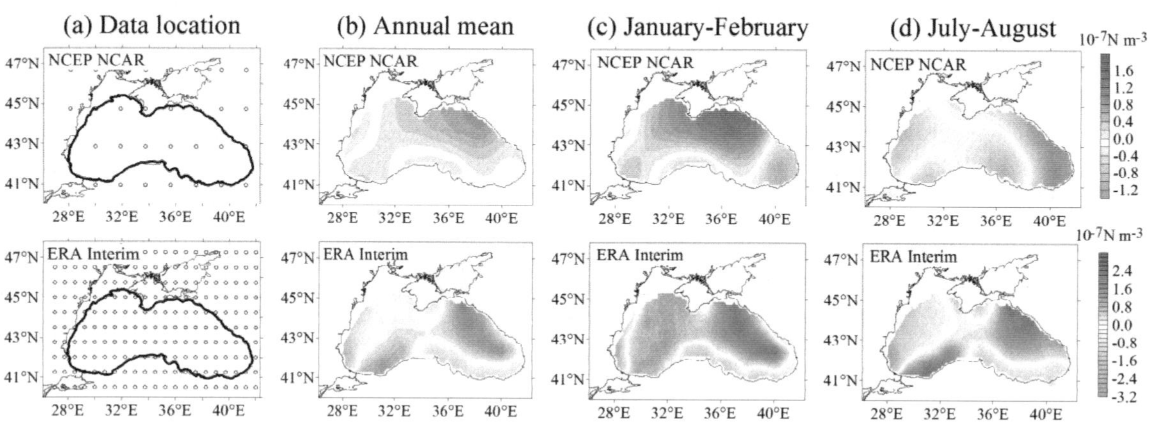

Figure 1. Spatial grid of reanalysis data (the bold line marks the border of the region with depths exceeding 50 m) (a); annual mean of the wind stress curl ($\times 10^{-7}$ Nm^{-3}) (b), in January-February (c), in July-August (d) according to the data of reanalysis NCEP/NCAR for 1979-2016 (top) and ERA Interim (bottom).

3.1.2. Basin-averaged wind stress curl. The seasonal cycle of the basin-averaged wind stress curl is presented in Figure 2a. The curl magnitude according to ERA Interim reanalysis is higher throughout the year with the predominance of the cyclonic curl compared with NCEP/NCAR reanalysis. A maximum positive value of the wind stress curl is in February. In summer, the anticyclonic curl prevails over the sea only in June according to ERA Interim data and during all summer months according to NCEP/NCAR.

The magnitude of the wind stress curl varies with change in the direction of the wind prevailing over the sea area (Figure 2b). On average over the year, the maximum cyclonic wind stress curl is observed during synoptic situations with northeastern and eastern winds prevailing over the sea, and the maximum anticyclonic curl occurs during southwestern and western winds. High values of the curl are associated with high heterogeneity of the wind velocity field. The northeastern, southwestern, and western winds have the highest speeds [11] and, therefore, they have the greatest speed shifts. Land-sea surface change and tall mountains on the eastern and southern coasts contribute to an increase in the heterogeneity of the wind speed field [3, 14]. A high cyclonic curl occurs in the eastern and southwestern parts of the sea during the northeastern winds flow around the Caucasus Mountains and around the Pontic Mountains (Figure 3a). A high anticyclonic curl was found in the southern part of the sea during the western or southwestern winds flow around the Pontic Mountains (Figure 3b).

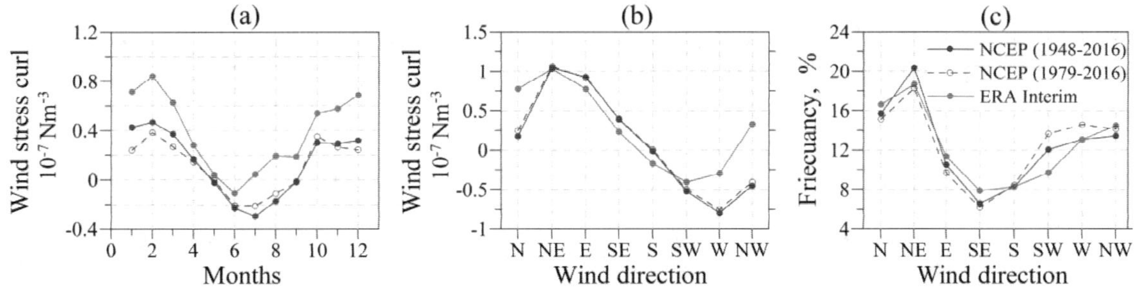

Figure 2. Annual cycle of the basin-averaged wind stress curl ($\times 10^{-7}$ Nm^{-3}) (a); wind stress curl at different wind directions (b); the wind direction frequency (%) (c). NCEP/NCAR: 1948-2016 – blue solid line, 1979-2016 – blue dashed line; ERA Interim – red line.

The northeastern winds have maximum frequency, and the southeast winds have minimum frequency (Figure 2c). The frequency of the wind of each individual direction changes throughout the year, but the northeastern wind has a high frequency in all months [11, 13]. Therefore, there are high values of the cyclonic curl in the eastern part of the sea in all seasons (Figure 1, b–d).

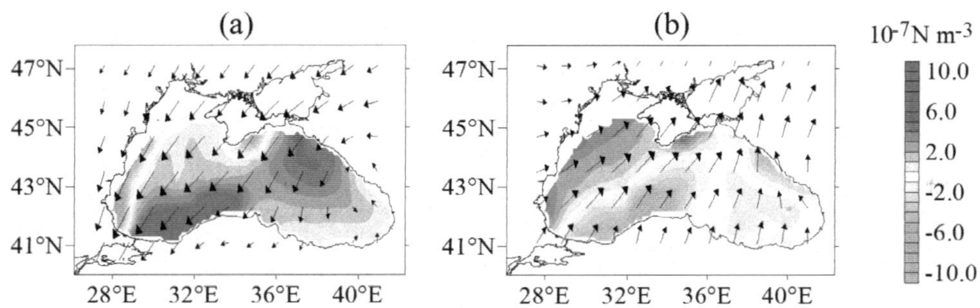

Figure 3. Mean wind stress curl ($\times 10^{-7}$ Nm^{-3}) under northeastern winds (a) and southwestern winds (b) in January-February according to ERA Interim data.

3.2. Interannual variability

3.2.1. Comparison of reanalyses. A comparison of the time series of the basin-averaged wind stress curl anomalies calculated on the basis of different reanalyses wind data shows good agreement between them for 1979-2016, which is common for two reanalyses (Figure 4, top row). The correlation coefficients between them have high significant values in all seasons (Table 1). Thus, despite the difference in the absolute values of the wind stress curl, the character of its interannual variability is similarly reproduced by reanalyses with high and low spatial resolution.

Next, we will examine the interannual variability of the wind stress curl using the long-term wind data of NCEP/NCAR reanalysis.

Table 1. Correlation coefficients between annual mean time series (1979-2016) of basin averaged wind stress curl in the Black Sea from NCEP/NCAR and ERA Interim wind data.

	Wind stress curl (ERA-Interim)				
	Year	Jan- Feb	Apr-May	Jul-Aug	Oct-Nov
Wind stress curl (NCEP/NCAR)	0.81	0.87	0.74	0.88	0.89

3.2.2. Long-term trends. There are evident periods of positive and negative anomalies in the basin-averaged wind stress curl time series (Figure 4, top row). According to the annual mean data, the cyclonic curl increased in the late 60s – early 70s and weakened in the late 90s (Figure 4a). In winter, the same trends are observed, but the changes have a larger amplitude (Figure 4b). In the summer months, there is a weak positive trend of the curl (Figure 4c). Similar results were obtained in earlier studies using shorter data series. The decrease of the wind stress curl from the 1960s to the 2000s was revealed earlier in [12], and its increase since 2000s was obtained in [10].

3.2.3. Connection with winds directions frequency. To analyze the relationship between the long-term changes of the wind stress curl and the synoptic situation over the sea, the correlation between the curl and the prevailing winds frequency time series was calculated. The largest positive correlation coefficients were obtained for the north-eastern wind, and a negative one for the south-western and western winds (Table 2). The values of the correlation coefficients with winds of other directions are smaller. Thus, the dependence of the curl on the wind direction obtained from the annual mean data (Figure 2b) is also detected in long-term data series.

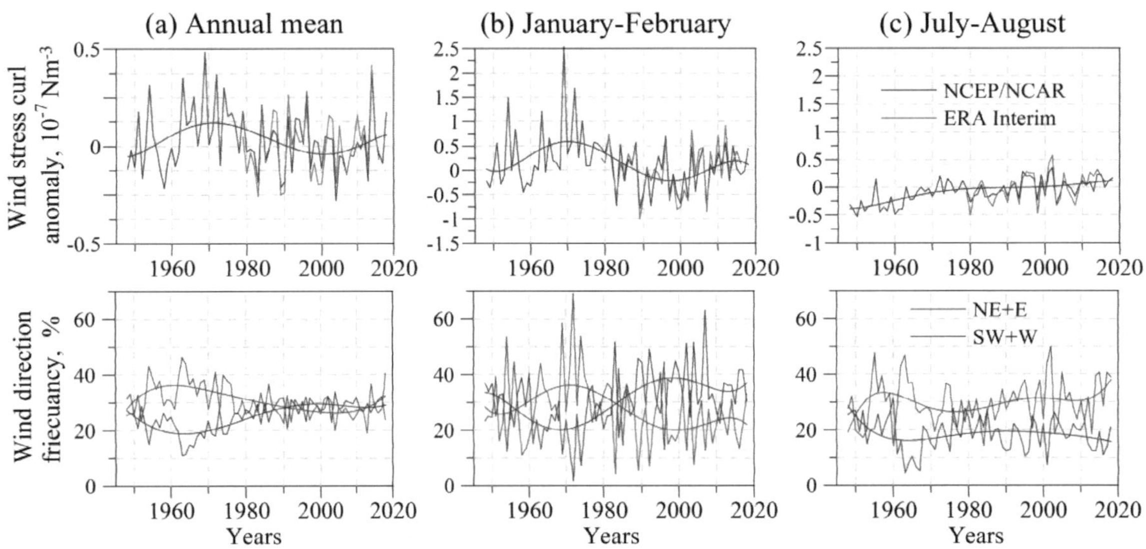

Figure 4. Top: basin-averaged wind stress curl ($\times 10^{-7}$ Nm^{-3}) anomalies from NCEP/NCAR (blue), ERA Interim (red); bottom: frequency of north-eastern and eastern (NE+E - green) and south-western and western (SW+W - brown) winds in annual mean (a), in January-February (b) and in July-August (c) in the Black sea. Smoothing curves are polynomials of order 5.

The maximum significant correlation coefficients are in winter and autumn, when the wind velocity is high. The correlation with the total frequency of northeastern and eastern winds (NE+E) and southwestern and western winds (SW+W) has higher values than for individual wind directions for winter, autumn, and the annual mean (Table 2). The interannual variability of the total frequency of these winds on average for the year, in January-February, and in July-August is shown in Figure 4 (bottom row).

The relationship between the frequency of winds and the wind stress curl is well expressed in the annual means (Figure 4a) and January-February (Figure 4b) data. Long-term changes in the frequency of northeastern and eastern winds occur simultaneously with changes in the wind stress curl. High values of the cyclonic curl occur when the frequency of these winds increases. High values of the anticyclonic curl take place when the frequency of the western and southwestern winds grows. The variability of the total frequency of SW+W winds and NE+E winds has opposite trends.

Table 2. Correlation coefficients between time series of the basin-averaged wind stress curl and the frequency of prevailing winds (NCEP/NCAR) in the Black Sea (maximum positive and negative correlation coefficients for individual wind directions are shown in bold).

		Frequency of wind								Total frequency	
		N	NE	E	SE	S	SW	W	NW	NE+E	SW+W
Wind stress curl	Annual	0.23	**0.68**	0.42	0.11	-0.41	**-0.60**	-0.58	-0.44	0.72	-0.64
	Jan-Feb	0.04	**0.77**	0.75	0.37	0.01	-0.57	**-0.70**	-0.54	0.87	-0.77
	Apr-May	-0.02	**0.74**	0.16	-0.06	-0.40	-0.41	**-0.44**	-0.24	0.72	-0.50
	Jul-Aug	0.02	**0.49**	0.08	-0.20	-0.04	-0.47	-0.20	**-0.51**	0.43	-0.35
	Oct-Nov	-0.11	**0.73**	0.44	0.15	-0.35	-0.59	**-0.65**	-0.54	0.75	-0.73

The total frequency of northeastern and eastern winds was high from the 1950s to the early 1980s, and has been low since the 1990s in January-February (Figure 4b). The period from 1960s to 2000s was the time of decreasing frequency of northeastern and eastern winds. The weakening of the frequency of strong winds and wind speed in this period was obtained earlier using data of meteorological measurements [19, 20]. The western and southwestern winds dominated from the mid-1980s in January-February over the Black Sea.

The relationship between long-term variations of the wind stress curl and changes in the wind direction is less expressed in summer, when the wind velocity is low (Figure 4c, Table 2).

4. Discussion and conclusions

Analysis of the seasonal and interannual variability of the wind stress curl in the Black Sea and its dependence on the frequency of prevailing wind directions was performed using NCEP/NCAR and ERA Interim reanalyses wind data with various spatial resolutions and lengths of data series.

The seasonal variability of the basin average wind stress curl, the dependence of the curl magnitude on wind direction and the frequency of wind directions are in good agreement for the two reanalyses. However, according to the ERA Interim reanalysis with a higher spatial resolution, the magnitude of the wind stress curl has a larger value throughout the year with a predominant cyclonic curl.

The long-term variations of the wind stress curl magnitude depend on changes in the frequency of wind directions. The wind stress curl time series has a positive correlation with the frequency of northeastern and eastern winds, and a negative correlation with the frequency of western and southwestern winds. The increase in the frequency of northeastern end eastern winds is accompanied by intensification of the cyclonic wind stress curl and of the anticyclonic one in the western and southwest winds.

The direction of the prevailing wind is determined by the large-scale atmospheric synoptic situation, the position of large-scale atmospheric cyclones and anticyclones in particular [13]. Changing the synoptic situation leads to a change in the wind direction and a change in the position and sizes of the areas with cyclonic and anticyclonic wind stress curls [11].

The orographic effects due to the presence of high and elongated Caucasus and Pontic Mountains make an additional contribution to the creation of the wind stress curl [3, 4]. The seasonal change in the temperature contrasts between the sea and the surrounding land makes a great contribution to the seasonal variability of the wind curl [3, 4, 21]. In winter, when the sea is warmer than the land, a cyclonic circulation is formed over the sea. In summer, the situation is reversed, and an anticyclonic circulation is formed.

The interannual variability of the wind stress curl and the frequency of individual direction winds are probably related to long-term changes in the trajectories and position of large-scale atmospheric cyclones and anticyclones [13].

Acknowledgements
This study was carried out under state assignment no. 0827-2019-0001

References

[1] Ivanov V A and Belokopytov V N 2013 *Oceanography of the Black Sea* (Sevastopol: ECOSY-Gidrofizika) p 210
[2] Stanev E V 2005 Understanding Black Sea Dynamics: Overview of Recent Numerical Modeling *Oceanography* **18**(2) 56-75
[3] Efimov V V, Shokurov M V and Barabanov V S 2002 Physical Mechanisms of Wind Circulation Forcing over the Inland Seas *Izv. Atmos. Oceanic Phys.* **38**(2) 217-27
[4] Efimov V V and Anisimov A E 2011 Climatic parameters of wind-field variability in the Black Sea region: numerical reanalysis of regional atmospheric circulation *Izv. Atmos. Oceanic Phys.* **47**(3) 350-61
[5] Kubryakov A A and Stanichny S V 2015 Seasonal and Interannual Variability of the Black Sea Eddies and its Dependence on Characteristics of the Large-Scale Circulation *Deep Sea Research Part I: Oceanographic Research Papers* **97** 80-91
[6] Korotaev G, Oguz T, Nikiforov A and Koblinsky C 2003 Seasonal, interannual, and mesoscale variability of the Black Sea upper layer circulation derived from altimeter data *Journal of Geophysical Research: Oceans* **108** (C4)
[7] Kara A B, Hurlburt H E, Wallcraft A J and Bourassa M A 2005 Black Sea Mixed Layer Sensitivity to Various Wind and Thermal Forcing Products on Climatological Time Scales *J. Climate* **18**(24) 5266-93
[8] Polonskii A B and Shokurova I G 2009 Decadal variability of characteristics of the Black Sea pycnocline and geostrophic circulation in the wintertime *Russian Meteorology and Hydrology* **34**(4) 243-55
[9] Kubryakov A A, Stanichny S V, Zatsepin A G and Kremenetskiy V V 2016 Long-term Variations of the Black Sea Dynamics and Their Impact on the Marine Ecosystem. *J. Mar. Syst.* **163** 80-94
[10] Kubryakov A, Stanichny S, Shokurov M and Garmashov A 2019 Wind velocity and wind curl variability over the Black Sea from QuikScat and ASCAT satellite measurements *Remote sensing of environment* **224** 236-58
[11] Shokurov M V and Shokurova I G 2017 Wind Stress Curl over the Black Sea under Different Wind Regimes *Physical Oceanography* **6** 12-23
[12] Polonskii A B and Shokurova I G 2011 Long-term variability of the wind stress curl over the Black Sea according to reanalysis data *Ecological Safety of the Coastal and Shelf Zones and Complex Use of Shelf Resources* **24** 182-89 (In Russian)
[13] Sorkina A I 1974 *Reference book on the Black Sea climate* (Moscow: Gidrometeoizdat) (In Russian)
[14] Efimov V V and Yurovsky A V 2017 Formation of vorticity of the wind speed field in the atmosphere over the Black Sea *Physical Oceanography* **6** 3-11
[15] Capet A, Barth A, Beckers J M and Marilaure G 2012 Interannual variability of Black Sea's hydrodynamics and connection to atmospheric patterns *Deep Sea Research Part II: Topical Studies in Oceanography* **77** 128-42

[16] Kalnay E *et al* 1996 The NCEP/NCAR 40-year reanalysis project *Bulletin of the American meteorological Society* **77**(3) 437-72
[17] Dee D *et al* 2011 The ERA-Interim reanalysis: Configuration and performance of the data assimilation system *Quarterly Journal of the royal meteorological society* **137**(656) 553-97
[18] Large W G and Pond S 1981 Open ocean momentum flux measurements in moderate to strong winds *Journal of physical oceanography* **11**(3) 324-36
[19] Repetin L N, Belokopytov V N and Lipchenko M M 2003 Winds and waves in the coastal zone of the southwest part of Crimea *Ecological Safety of the Coastal and Shelf Zones and Complex Use of Shelf Resources* **9** 13-28 (In Russian)
[20] Repetin L N and Belokopytov V N 2009 The wind regime of the north-western part of the Black Sea and its climatic change *Ecological Safety of Coastal and Shelf Zones and Complex Use of Shelf Resources* **17** 225–43 (In Russian)
[21] Korotaev G K 2001 On the causes of the seasonal course of circulation in the Black Sea *Morsk. Gidrofiz. Zh.* **6** 14-20 (In Russian)

IOP Conf. Series: Earth and Environmental Science **386** (2019) 012015 doi:10.1088/1755-1315/386/1/012015

Investigation of eddy and advective heat fluxes in the southeastern part of Western Siberia

E I Usova[1], S V Loginov[1], E V Kharyutkina[1] and Yu V Martynova[1,2]

[1]Institute of Monitoring of Climatic and Ecological Systems SB RAS, 10/3 Academichesky ave., Tomsk, 634055, Russia
[2]Siberian Research Hydrometeorological Institute of Roshydromet, 30 Sovetskaya str., Novosibirsk, 630099, Russia

E-mail: kh_ev@imces.ru

Abstract. Estimates of seasonally averaged advective and eddy heat fluxes and relative vorticity at the earth's surface are obtained for the southeastern part of Western Siberia and, in particular, for the Tomsk region (55°-62°N, 74°-90°E) based on daily ERA-Interim reanalysis data with a high spatial resolution. It has been found that, in general, in the early 21st century in the whole region a decrease in advective air transfer is observed in autumn and winter, and an increase in spring and summer. At the same time, the heat fluxes coming to the region from the southern border play a significant role in regulating these changes. For the eddy heat fluxes, the situation is quite similar, in general. In the past few years, there has been a decrease in cyclonic activity and an increase in anticyclonic circulation. Besides, there was a great fluctuation in relative vorticity at the beginning of the 21st century. However, in general, a tendency towards negative values is observed.

1. Introduction

During the last decades the climate change has led to increase in the number of unfavorable and dangerous hydrometeorological events [1]. Thus, in Russia the number of dangerous phenomena is growing annually by 6.3–7% [2]. The greatest number of dangerous events in Russia in the recent years has been observed in Western Siberia [3]. Moreover, the most frequently occurring dangerous phenomena are associated with strong winds (24.5%), as well as heavy precipitation (15.3%) [4]. The most frequent and long-lasting strong winds are in the valley of the river Ob and in the southern parts of Tomsk region. These events have a convective origin. Their frequency of formation and intensity describe the variability of atmospheric circulation processes.

One of the reasons of such changes could be the variability in advective and eddy heat and moisture fluxes coming to the territory, due to changes in the large-scale atmospheric circulation over the recent decades. Experiments with the ocean general circulation model using variations in the atmospheric influence [5] indicate that meridional heat transfer has increased during the recent decades.

The main goal of this study is to estimate the dynamics of advective and eddy heat fluxes in Tomsk region at the end of the 20th and at the beginning of the 21st centuries

2. Materials and Methods

The study of heat transfer in Tomsk region (55°-62°N, 74°-90°E) in 1979 – 2018 is based on the daily ERA-Interim reanalysis data with a spatial resolution of 1.25° × 1.25° for calculating advective heat fluxes and 0.125° × 0.125° for the analysis of relative vorticity.

The calculation method of advective heat fluxes was described in more detail in [6], and that of eddy fluxes, in [7]. The temporal variability of heat fluxes was obtained directly for the whole territory of Tomsk region and separately for its borders (northern, southern, western, and eastern ones). This approach made it possible to identify and estimate the dominant direction of heat transfer to the territory in each season. Analysis of advective and eddy heat transfer was carried out in the layer of 1000 - 850 hPa.

Therefore, to estimate the eddy activity in the region, the relative vorticity was also calculated at the levels of 1000 hPa and 850 hPa.

To reveal the tendencies and to reduce the signal noise in time series of climatic parameters, low-pass filtering (LPF) was applied [8]. The LPF characteristics were derived using the Fourier spectrum of a time series of climatic values with a 10-year window width. The method of fluctuation calculation was described in [7].

The error of the calculations was determined through the standard deviation. Verification of the null hypothesis when estimating the linear trend coefficients was performed for $\alpha = 0.05$ [9].

3. Results and Discussion

The interannual variability of seasonal values of the total advective heat fluxes coming to Tomsk region in 1979 – 2018 is presented in Figure 1. The dotted line shows the variability of fluxes smoothed by the low-pass filter. The total heat flux represents the sum of all incoming and outcoming heat fluxes across the borders of the region. At the beginning of the 21st century over the whole region a decrease in the advective flux was observed in autumn and in winter, and an increase in spring and in summer (Figure 1). In addition, the maximum value of the incoming advective heat flux for 1979 – 2018 was observed in the winter season ($5.52*10^{12}$ W), and the minimum one in the summer season ($1.65*10^{12}$ W).

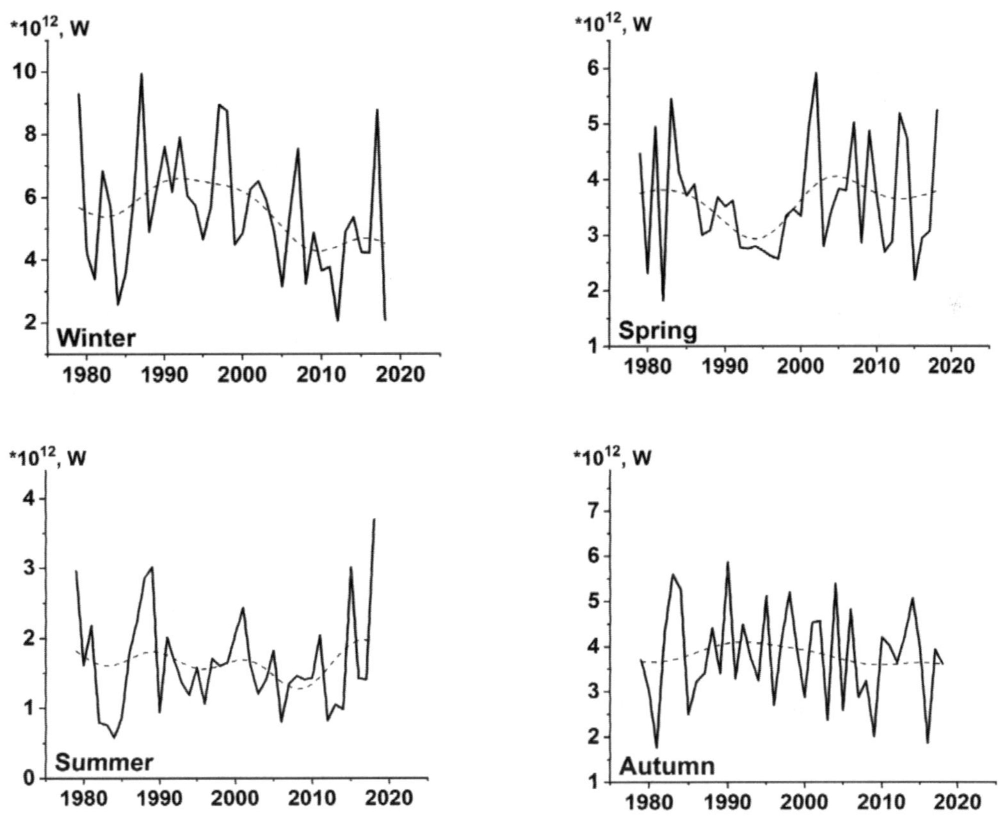

Figure 1. Temporal variability of the total advective heat flux coming to Tomsk region.

As for the eddy heat fluxes, the temporal tendencies are quite similar to those described above, but there are differences in some seasons. Thus, in winter while the advective heat transfer decreased, the magnitude of the eddy heat flux did not change. Significant fluctuations of the eddy heat flux were not observed in spring but, at the same time, the advective flux coming from Tomsk region increased. In contrast to the advective fluxes, the maximum of the eddy ones was revealed in spring months.

The value of the advective flux exceeds the corresponding total value of the eddy flux 3 - 5 times. However, they have the same order, except for the fluxes coming from the territory in summer and in autumn (here the advective flux value is much higher).

To investigate the distribution of heat fluxes in Tomsk region in more detail, analyses of the fluxes incoming and outcoming through each border of the region were carried out. Table 1 shows the percentage of advective heat flux in Tomsk region across its different borders for 1979 – 2018. The advective flux values that exceeded 40% are in bold. As a result, it was shown that in winter and autumn the heat flux coming to the region from the southern border dominated (up to 80 %). In spring, the value reduced down to 50 % due to the growth of the heat flux coming from the northern border. In summer, the opposite situation was observed: the heat incoming from the northern border was 50%; and the heat incoming from the southern border was 30 %. The outcoming advective heat flux had maximal values for the northern border in winter (61%) and for the southern one in summer (57%).

Table 1. Percentage of advective heat flux in Tomsk region for its different borders over 1979 – 2018.

Flux	Season	Total flux, 10^{12} W	West, %	East, %	North, %	South, %
Incoming	Winter	**5.52**	18	0	2	**80**
	Spring	3.60	19	0	31	**50**
	Summer	1.65	15	2	**50**	33
	Autumn	3.80	18	0	5	**77**
Outcoming	Winter	**4.23**	0	37	**61**	1
	Spring	2.43	0	28	**54**	17
	Summer	1.50	3	18	22	**57**
	Autumn	2.37	0	35	**60**	5

An analysis of the interannual variability of the heat fluxes incoming to and outcoming from the dominant directions to Tomsk region was carried out (Figure 2). The dotted line shows the variability of fluxes smoothed by a low-pass filter. It can be noted that there was a significant decrease in the heat fluxes coming to the region from south (Figure 2a).

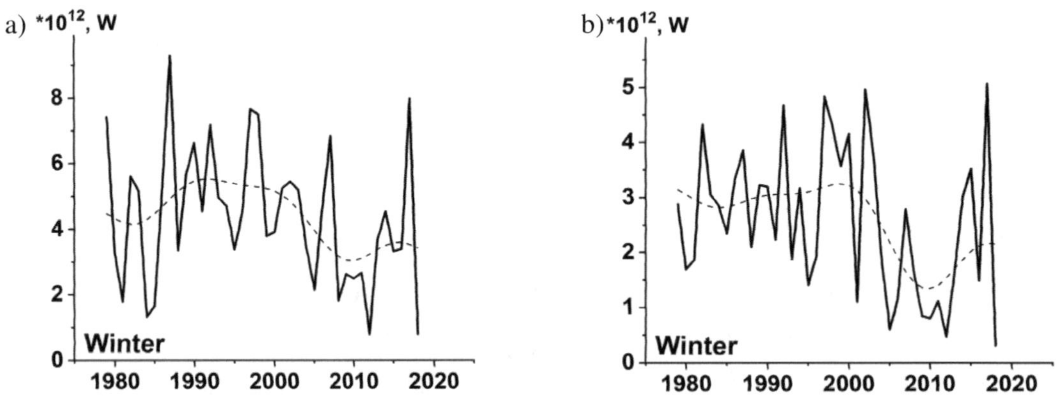

Figure 2. Temporal variability of the advective heat flux in winter: a) coming to Tomsk region from south; b) coming from Tomsk region from north.

It was established that, in general, at the beginning of the 21st century a decrease in the outcoming advective hear flux through the northern border (Figure 2b) was observed over the whole region. However, during the last years the value of the advective flux on these borders had increased significantly (~2 times). It is expressed mainly in the winter season. In spring, the situation was reversed: the incoming heat flux to the region from south and the outcoming one from north increased. In autumn, the incoming advective heat flux to the territory from the northern border increased, and the outcoming one through the eastern border decreased. In summer, the heat flux comes from Tomsk region through almost all borders, while cold air comes from the eastern border, and warm air comes from north. The incoming heat in summer to the territory was observed from both south and west.

In general, the advective heat flux incoming through the western border decreased, which confirms the conclusions on the strengthening of the meridional type of circulation [5, 6].

The distribution of the eddy heat fluxes at the borders of Tomsk region is similar to that of the advective heat fluxes. Table 2 shows the percentage of eddy heat flux in Tomsk region across its different borders for 1979 – 2018. It was shown that in winter and autumn the heat flux coming to the region from the southern border dominated (up to 73 %). The outcoming advective heat flux was maximal for the northern border in winter (52 %) and for the southern border in summer (48 %). However, unlike the advective flux, the maximum eddy heat flux for 1979 – 2018 was observed in spring ($1.12*10^{12}$ W).

Table 2. Percentage of eddy heat flux in Tomsk region for its different borders over 1979 – 2018.

Flux	Season	Total flux, 10^{12} W	West, %	East, %	North, %	South, %
Incoming	Winter	0.74	17	2	8	**73**
	Spring	**1.12**	15	0	34	**51**
	Summer	0.67	15	5	**57**	23
	Autumn	0.81	14	1	13	**72**
Outcoming	Winter	0.79	1	42	**52**	5
	Spring	**1.05**	0	26	**44**	30
	Summer	0.58	5	15	34	**46**
	Autumn	0.73	1	41	**48**	10

The temporal variability of the eddy flux from season to season was similar to the variability of the advective heat flux (Figure 3). However, their tendency did not always match. For example, in autumn against the background of advective flux decrease, the eddy heat flux coming from the southern border practically did not change. The eddy flux was associated with the growth of the incoming heat from the northern border of Tomsk region in winter, and from the southern one in spring and summer. Also, the eddy flux was related to the decrease of the outcoming heat from the eastern border in winter and summer.

The eddy activity in Tomsk region can be described by using the relative vorticity parameter. This parameter can characterize both the cyclonic (positive values) and anticyclonic (negative values) types of circulation. Thus, the study of this parameter allows one to better understand and describe the dynamics of individual elements of the regional climatic system, in particular, the features of the atmospheric circulation processes. The interannual variability of the relative vorticity at the levels of 1000 hPa and 850 hPa for winter and summer in Tomsk region is shown in Figure 4. It was found that over the last decades a significant decrease in the relative vorticity was observed, which is well pronounced in the winter season. This was typical for the levels of 1000 hPa and 500 hPa.

It was established that, in general, in Tomsk region the vorticity values in the troposphere were 2 times higher than that at the surface. There was a great fluctuation in this parameter at the beginning of the 21st century. However, a tendency towards the appearance of negative values was observed. This indicates a decrease in the cyclonic activity and an increase in the anticyclonic circulation type. It was noted that the tendency of the relative vorticity was similar to that in Western Siberia [10].

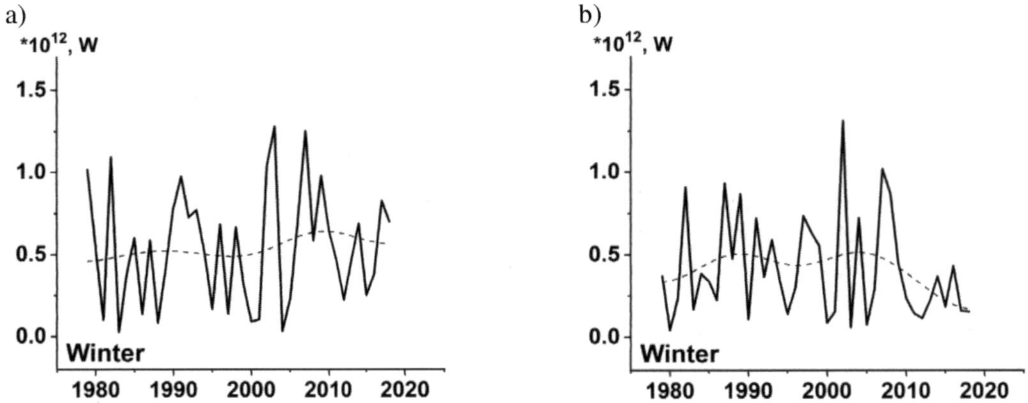

Figure 3. Temporal variability of eddy heat flux in winter: a) coming to Tomsk region from south; b) coming from north.

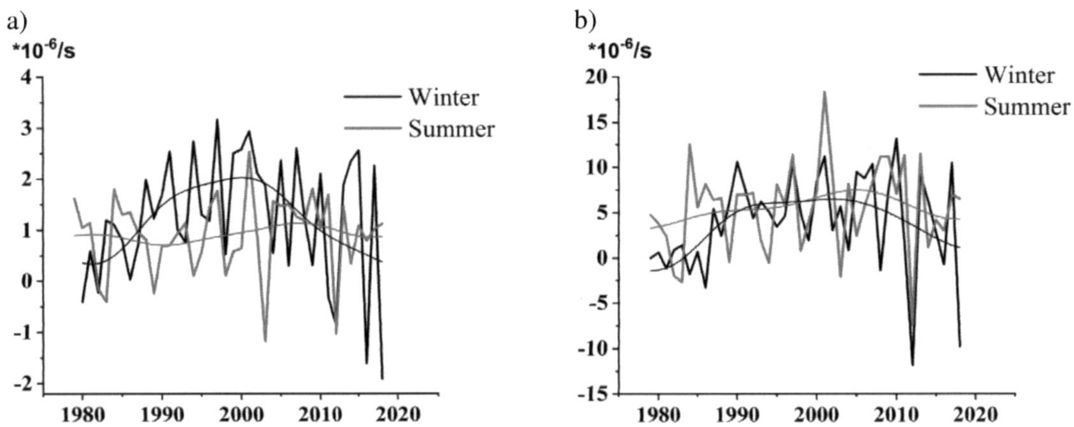

Figure 4. Temporal variability of relative vorticity ($*10^{-6}$/s) in Tomsk region: a) 1000 hPa; b) 500 hPa.

Since the above-mentioned changes are associated with the atmospheric circulation variability, the fluctuation values in the processes of synoptic scale (2-7 days) were calculated for the air temperature (t), air humidity (q), the zonal (u), meridional (v), and vertical (w) components of wind speed in Tomsk region at the level of 850 hPa over two time intervals: 1979-1998 and 1999-2018. The derived estimates revealed areas with significant fluctuations ($\frac{x-\bar{x}}{\sigma} \geq 3$) of the temperature, humidity, and meridional component of wind speed in the winter season at the beginning of the 21st century. However, in the presence of nonlinearity in climatic systems many signals cannot be analyzed adequately by second-order statistical methods. For this reason, higher order statistical methods have been used [11]. Thus, the values of these parameters were calculated based on cumulants κ_{14} (Table 3). For this relationship, maximal estimates of the 5th-order cumulant were derived.

Maximal values of the cumulants correspond to the maximal contribution to the estimation of the probability of extreme events. As for Tomsk region, in the winter season maximal values for a combination of the two parameters were obtained for the air temperature and air humidity.

Table 3. Cumulant values of air temperature (t), air humidity (q), meridional component of wind speed (v) in Tomsk region. Statistically significant values (α=0.05) are in bold.

Time interval	1979 – 1998			1999 – 2018		
	t – q	v – q	v – t	t – q	v – q	v – t
Winter	**0.76**	**0.37**	**0.38**	**0.47**	-0.02	-0.10
Spring	**-0.29**	-0.18	-0.18	0.14	0.00	-0.04
Summer	0.11	-0.10	0.20	0.20	0.09	0.08
Autumn	**-0.51**	-0.13	0.23	0.21	-0.03	0.09

It should be noted that these values decreased at the beginning of the 21st century, but the values remained statistically significant. The approach being used specified the sources of nonlinear relationships and their significance for prognostic models.

4. Conclusions

Thus, estimates of seasonally averaged advective and eddy heat fluxes and relative vorticity at the earth's surface were obtained for the southeastern part of Western Siberia and, in particular, for the Tomsk region based on daily ERA-Interim reanalysis data with a high spatial resolution.

It was found that, in general, in the early 21st century in the whole region a decrease in advective flux was observed in autumn and winter, and an increase in spring and summer. At the same time, the heat fluxes coming to the region from the southern border play a significant role in regulating these changes. For example, in winter and autumn the heat flux coming to the region from the southern border was dominant (up to 80 %). As for the outcoming advective heat flux, the maximal values were observed at the northern border also in the winter season (61 %).

For the eddy heat fluxes, in general, the situation is quite similar. In the past few years there was a decrease in cyclonic activity and an increase in anticyclonic circulation. Also, it was established that, in general, in the Tomsk region the vorticity value in the troposphere was 2 times higher than that at the surface. There was a great fluctuation in this parameter at the beginning of the 21st century.

It was assumed that, to describe the changes in the meteorological parameters and in the atmospheric circulation processes, nonlinear statistical methods could be applied. As for the Tomsk region, in winter the maximal values of the two parameters were obtained for air temperature and air humidity.

Thus, during these periods the weather and climatic conditions were formed not only by the processes of large-scale atmospheric circulation. Probably, the influence of some local factors prevailed: some local forms of the circulation (cyclones and anticyclones) can contribute to the intensification of convective processes and, as a result, to the associated weather conditions. Therefore, when investigating regional climatic changes it is very important to take into account the local scale circulation processes.

Acknowledgements

This work was partially supported under PFSR SB RAS research project AAAA-A17-117013050031-8 and by the RFBR and the Tomsk region administration under grant 18-45-703014 r_mol_a.

References

[1] Sillmann J, Donat M G, Fyfe J C and Zwiers F W 2014 Observed and simulated temperature extremes during the recent warming hiatus *Environ. Res. Lett.* 9 1-8

[2] *Strategic forecast of climate change in the Russian Federation for the period 2010–2015*

yy. and their impact on industries 2005 (Moscow) p 30

[3] *A report on climate features in the territory of the Russian Federation in 2017* 2018 (Moscow: Federal service for hydrometeorology and environmental monitoring) p 70

[4] Evseeva N S, Romashova T V 2011 Hazardous meteorological phenomena as an integral part of natural risk (using the example of the south of the Tomsk region) *Bulletin of Tomsk State University* **353** 199-204 [in Russian]

[5] Boning C W, Scheiner M, Dengg J, Biastoch A, Funk A 2006 Decadal variability of subpolar gyre transport and its reverberation in the North Atlantic overturning *Geophys Res Lett* **33** L21S01

[6] Usova E I, Loginov S V, Kharyutkina E V 2018 The influence of heat fluxes in the Barents Sea on the temperature regime of West Siberia in winter season *Atmos. Oceanic Opt.* **31** 8 652-59 [in Russian]

[7] Loginov S V, Eliseev A V, Mokhov I I 2017 Impact of non-Gaussian statistics of atmospheric variables on extreme intramonth anomalies *Izvestiya. Atmospheric and Oceanic Physics.* **53** 3 269-278.

[8] Jenkins G M, Watts D G 1968 Spectral analysis and its applications (San Francisco: Holden-Day) p 525

[9] von Storch H, Zwiers F W 2003 Statistical Analysis in Climate Research (Cambridge: Cambridge University Press) p 484 ISBN 0-521-45071-3

[10] Kharyutkina E V, Loginov S V 2018 Spatio-temporal changes in energetic characteristics of eddy Atmospheric circulation in the troposphere of West Siberia in the end of XX – the beginning of XXI centuries *IOP Conf.Series: Earth and Environmental Science* **190** 012030 doi:10.1088/1755-1315/190/1/012030

[11] Sanaullah M 2013 A Review of Higher Order Statistics and Spectra in Communication Systems *Global Journal of Science Frontier Research Physics and Space Science* **13** 4 30-50

Nonlinear interactions of stationary planetary waves during February 2016 sudden stratospheric warming

K A Didenko[1,2], A I Pogoreltsev[1,2], T S Ermakova[2] and G M Shved[1]

[1] Department of Atmospheric Physics, St. Petersburg State University, St. Petersburg-Petrodvorets, Ulyanovskaya street, 1, 198504, Russia
[2] Department of Meteorological Forecasting, Russian State Hydrometeorological University, St. Petersburg, Maloohtinsky Prospekt, 98, 195196, Russia

E-mail: didenko.xeniya@yandex.ru, apogor@rshu.ru, taalika@mail.ru and shved1936@gmail.com

Abstract. An analysis of nonlinear wave-wave and wave-mean flow interactions has been carried out during a sudden stratospheric warming (SSW) event in February 2016. This approach is based on a study of the perturbed potential enstrophy (Ertel's potential vorticity squared) balance equation. The results obtained by using UK Met Office reanalysis data are presented. It has been shown that an increase in the nonlinear interactions occurs at higher-middle latitudes of the boreal stratosphere during the SSW. It is noted that it is necessary to include the stationary planetary wave with zonal wave number m=3 in the analysis of major warmings.

1. Introduction

In the stratosphere, there is an anti-correlation between changes in the amplitudes of stationary planetary waves with zonal wave numbers 1 and 2 (SPW1 and SPW2) due to nonlinear wave-wave interactions. To interpret the observed behavior of SPW amplitudes, it is useful to consider the nonlinear interactions of SPWs with the zonal mean flow and between the SPWs with different zonal wave numbers. Using this approach, the conservation of the perturbed potential enstrophy is investigated. In this case the terms responsible for nonlinear interactions in the balance equation of the potential enstrophy (Ertel's potential vorticity squared) are calculated [1].

To obtain the balance equation of perturbed potential enstrophy, we use the conservation equations of Ertel's potential vorticity (EPV) perturbation for SPW1 and SPW2 and multiply these equations by the perturbed EPV for each SPW. The result is a general form of the eddy enstrophy balance in a log-pressure coordinate system:

$$\underbrace{\frac{\partial}{\partial t}(\overline{P'^2}/2)}_{A} + \underbrace{\overline{P'\left(\vec{V'}\cdot\vec{\nabla}P'\right)}}_{B} + \underbrace{\overline{P'\left(\vec{V'}\cdot\vec{\nabla}\bar{P}\right)}}_{C} + \underbrace{\overline{P'\left(\vec{\bar{V}}\cdot\vec{\nabla}P'\right)}}_{D} = \underbrace{\overline{P'Q'}}_{E}, \qquad (1)$$

where P' and $\bar{P}$ are the perturbed and zonally averaged components of Ertel's potential vorticity; $\vec{V'}$ and $\vec{\bar{V}}$ are the perturbed and zonally averaged components of the wind vector; Q' represents the perturbation of diabatic sources and the sinks and terms describing the subscale contributions to the momentum equation [2]. The expression for Ertel's potential vorticity is as follows:

$$P = \overrightarrow{\omega_a} \cdot \vec{\nabla}\theta / \rho_0, \quad (2)$$

where $\overrightarrow{\omega_a}$ is the absolute vorticity, θ is the potential temperature, and ρ_0 is the background density.

The first term, A, denotes the wave transience and can be defined as a measure of wave activity variability [3]. The next terms, B and C, describe the wave-wave and wave-mean flow interactions, respectively. The term D is responsible for the advective transport of potential enstrophy. The right-hand side term E describes the changes in perturbed enstrophy due to diabatic heating. This term involves subscale contributions to the momentum equation including momentum deposition by gravity and inertial-gravity waves.

Equations for SPW1 and SPW2 similar to equation (1) may be obtained using the secondary wave generation method. If a signal consisting of two cosine waves with zonal wavenumbers and frequencies (m1, ω1) and (m2, ω2) passes through some quadratic system, the output of this system will contain the secondary waves (2m1, 2ω1), (2m2, 2ω2), (m1–m2, ω1–ω2), and (m1+m2, ω1+ω2) [4, 5]. Thus, secondary SPWs with zonal wave number 2 (SPW2) are generated as a result of the nonlinear SPW1-SPW3 interaction and SPW1 self-interaction. Secondary SPW1 arise due to SPW1-SPW2 and SPW2-SPW3 interactions.

2. Data

To show the behavior of nonlinear terms characterizing the wave-wave and wave-mean flow interactions in the balance equations of perturbed potential enstrophy for SPW1 and SPW2, considering secondary wave generation the UK Met Office assimilated fields were used. Since the activity of SPWs increases before and decreases after the SSW, the major SSW in February 2016 was chosen for analysis. The UK Met Office data are currently unique in terms of the upper boundary height. In November 2009 the upper boundary of the UK Met Office model was extended to 0.01 hPa, i. e., practically into the mesosphere. To investigate the dynamic processes observed during 2015-2016 winter months in the Northern Hemisphere, the assimilated meteorological fields were decomposed at each latitude and altitude into Fourier-series taking into account zonal harmonics with wave number m = 0-3. Figure 1 demonstrates the time evolution of the amplitude of a planetary wave with m=1 in geopotential height (the upper panel), the mean zonal wind at latitude 62.5°N (the middle panel), and the change in the zonal mean temperature at latitude 87.5°N (the lower panel). One can see two strong increases in the amplitude of the SPW1 on the 28th of January and 8th of February, which was accompanied by the reversal of the zonal mean flow in the stratosphere. As a consequence, a sudden stratospheric warming was observed at the beginning of February. This SSW event was attended by significant increases in the amplitudes of SPW2 and SPW3 (see Figure 2).

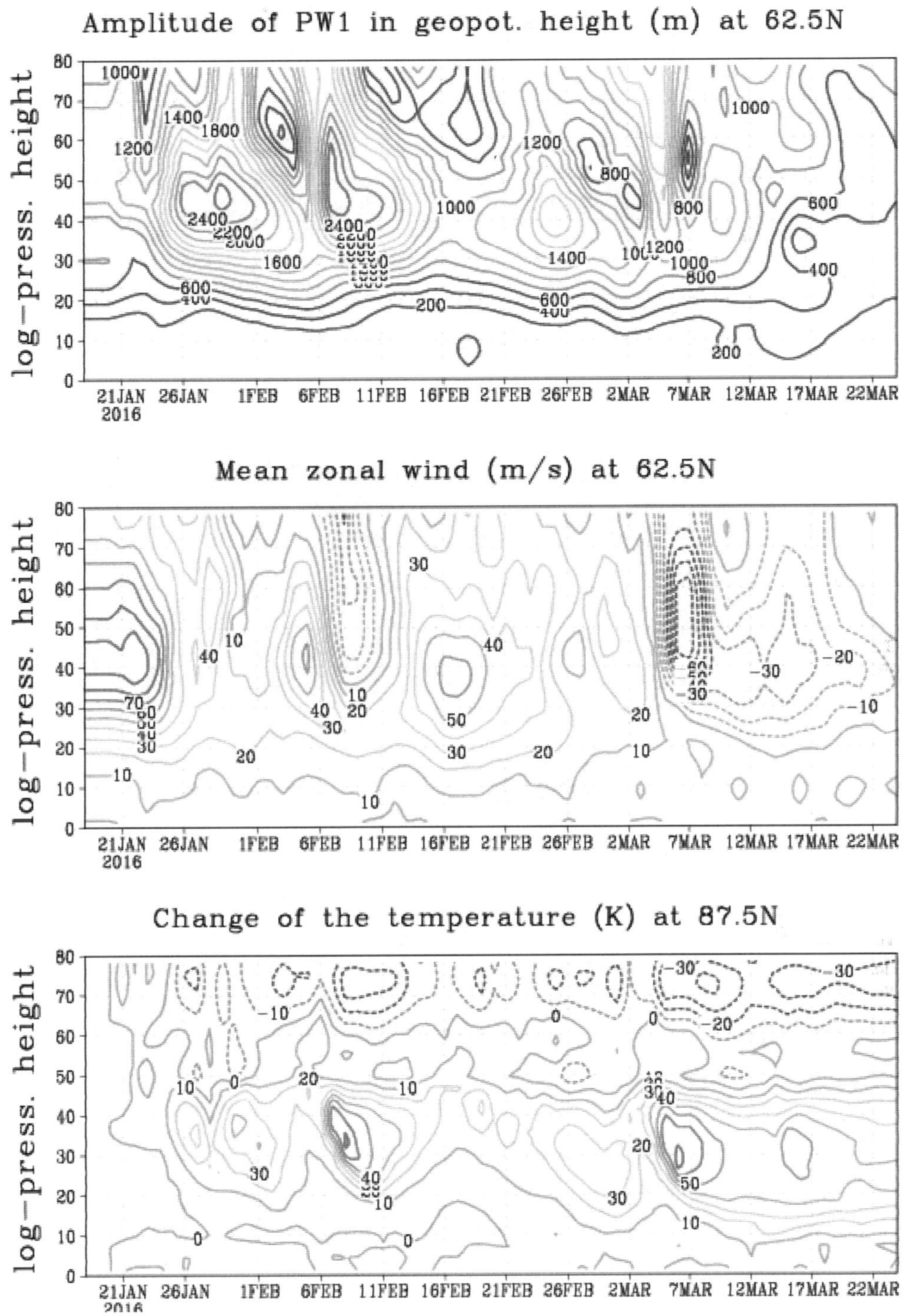

Figure 1. Time-altitude cross-sections of the amplitude of zonal harmonic with m = 1 in geopotential height and the mean zonal wind at latitude 62.5°N (upper and middle panels) for January-March 2016; change of the zonal mean temperature at 87.5°N (lower panel). UK Met Office data.

IOP Conf. Series: Earth and Environmental Science **386** (2019) 012016 doi:10.1088/1755-1315/386/1/012016

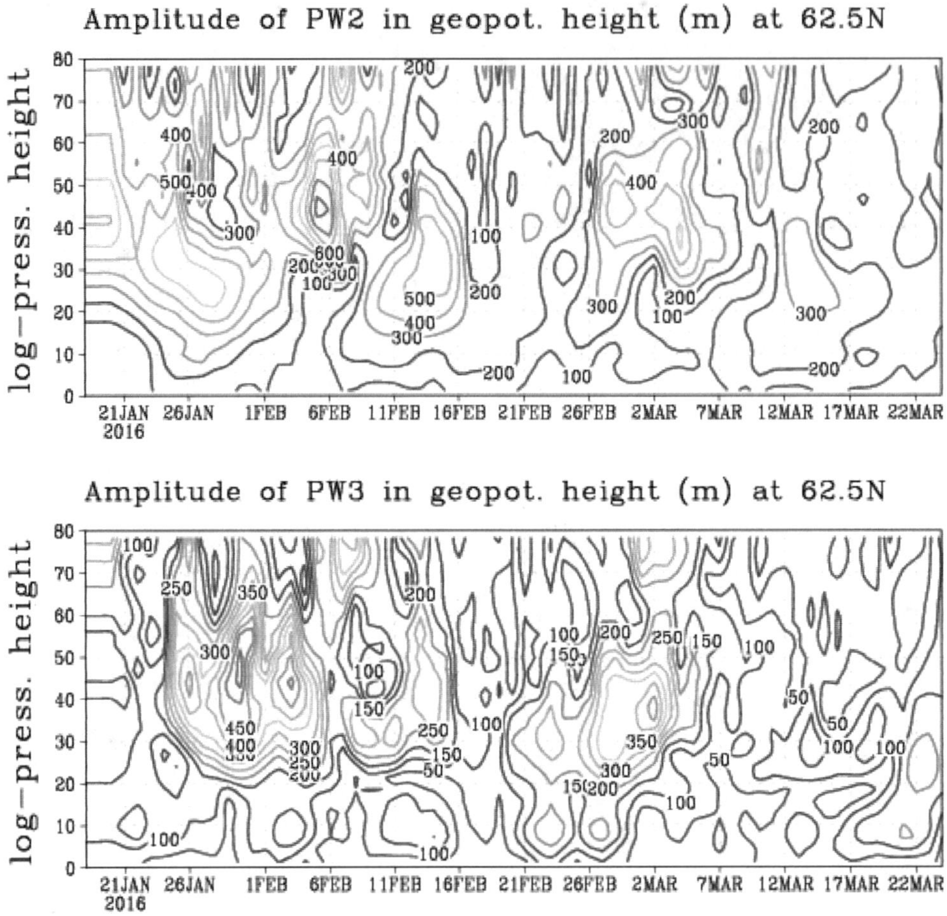

Figure 2. Time-altitude cross-sections of the amplitude of zonal harmonics with m = 2 (upper panel) and m = 3 (lower panel) in geopotential height at latitude 62.5°N for January-March 2016. UK Met Office data.

The winter of 2015-2016 is also characterized by a very strong final warming at the beginning of March. This event was preceded by increases in the SPW2 and SPW3 amplitudes. At the same time, there was no a substantial rise in the amplitude of SPW1, which is usually observed during such warming events. The analysis of this SSW is out of scope of the present paper.

3. Nonlinear interactions during the SSW event in February 2016

Using UK Met Office data, the terms of equation (1) were calculated and averaged over 2-6 of February (Figures 3 and 4).

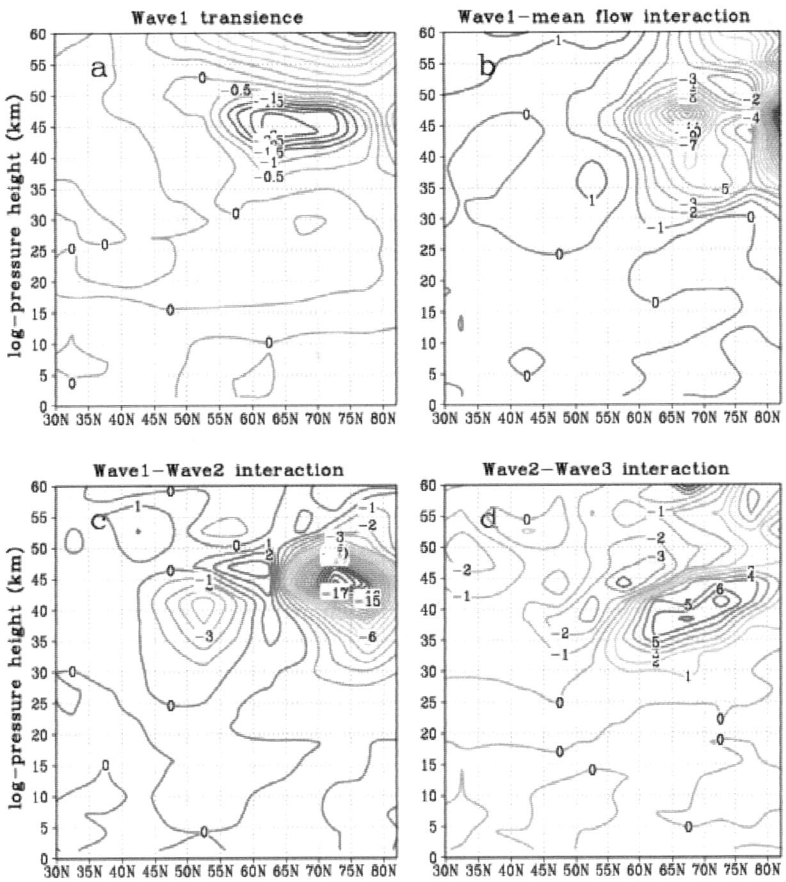

Figure 3. Latitude-height cross sections of terms in the potential enstrophy balance for SPW1: transience (a), SPW1-mean flow interaction (b), SPW1-SPW2 (c) and SPW2-SPW3 (d) interactions. Units: $10^{12}(kg\cdot m^{-3})^2\cdot PVU^2/day$, where $1PVU=10^{-6}\cdot K\cdot m^2\cdot kg^{-1}\cdot s^{-1}$.

IOP Conf. Series: Earth and Environmental Science **386** (2019) 012016 doi:10.1088/1755-1315/386/1/012016

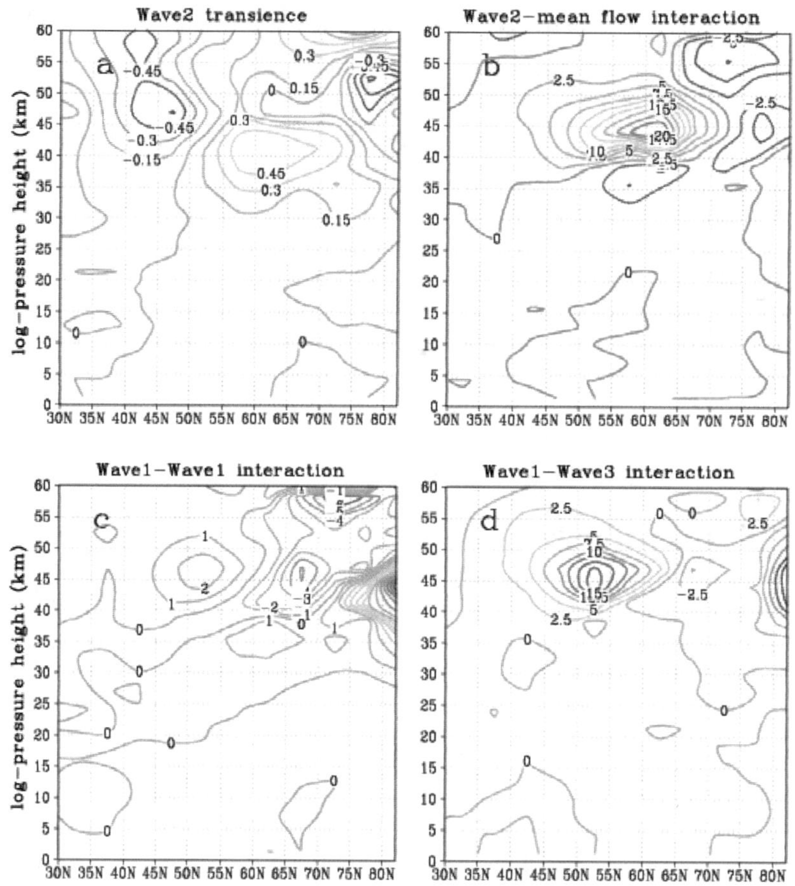

Figure 4. Latitude-height cross-sections of terms in the potential enstrophy balance for SPW2: transience (a), SPW2-mean flow interaction (b), self-interaction of SPW1 (c) and SPW1-SPW3 interaction (d). Units: $10^{12}(kg\cdot m^{-3})^2\cdot PVU^2/day$, where $1PVU=10^{-6}\cdot K\cdot m^2\cdot kg^{-1}\cdot s^{-1}$.

The calculation results show that the warming event in February 2016 was accompanied by an extremely strong increase in the wave-wave and wave-mean flow interactions in the stratosphere in higher-middle latitudes. During the considered time interval a decrease in the SPW1 wave activity was attended by a rise in the SPW2 wave activity (see Figures 3 (a) and 4 (a).

The wave-mean flow interaction (Figures 3 (b) and 4 (b)) requires special attention. The greatest influence on the mean flow is exerted by SPW2 with a strong maximum in the region where SPW1 has a minimum in the higher-middle latitudes. Considering the results for the wave-wave interaction (Figures 3 (c, d) and 4 (c, d)), it can be noted that the order of magnitude indicates a significant SPW3 contribution to the generation of secondary waves during the SSW in February 2016. Strengthening of the wave-wave interaction in the case of SPW1 generation takes place in higher-middle latitudes and in the case of SPW2 generation, throughout all middle latitudes.

4. Summary and conclusions

In the present paper, nonlinear interactions during a SSW event at the beginning of February 2016 were analyzed by using the perturbed potential enstrophy balance equation. The above-obtained results showed the contributions of the transience, wave-wave, and wave-mean flow interaction terms to this equation. It is noted that the SSW event is described by an increase in all types of interactions in the stratosphere. The possibilities for further study, both analytical and numerical, are immense. For example, SPW3 should be involved in the analysis in the case of strong sudden stratospheric warmings. This implies the derivation of an equation (1) taking into account the secondary planetary wave generation method for SPW3. Additionally, a similar approach can be used to investigate the final warmings, especially during their anomalous development (e.g., as observed in March 2016). An identical approach can be used to analyze the nonlinear interactions between higher-frequency planetary waves (for instance, atmospheric tides) themselves or/and with stationary planetary waves during a SSW.

Acknowledgments
This research was supported by the Russian Science Foundation under scientific project no. 19-17-00198.

References

[1] Smith A K 1983 Observation of wave-wave interactions in the stratosphere *J. Atmos. Sci.* **40** 2484–93
[2] Pogoreltsev A I, Savenkova E N, Aniskina O G, Ermakova T S, Chen W and Wei K 2015 Interannual and intraseasonal variability of stratospheric dynamics and stratosphere-troposphere coupling during northern winter *J. Atmos. Sol. Terr. Phys.* **136** 187–200
[3] White I P, Hua L, Mitchell N J and Phillips T 2015 Dynamical response to the QBO in the northern winter stratosphere: signatures in wave forcing and eddy fluxes of potential vorticity *J. Atmos. Sci.* **72** 4487–407
[4] Spizzichino A 1969 Etude des interactions entre les differentes composantes du vent dans la haute atmosphere *Ann. Geophys. 3e. Partie* **25**(4) 773–83
[5] Pogoreltsev A I 2001 Numerical simulation of secondary planetary waves arising from the nonlinear interaction of the normal atmospheric modes *Phys. Chem. Earth (Part C)* **26**(6) 395–403

Modeling the dynamics of comfort thermal conditions in Arctic cities under regional climate change

A A Semenova[1], P I Konstantinov[1], M I Varentsov[2] and T E Samsonov[1]

[1] Lomonosov Moscow State University, Faculty of Geography, Russia, Moscow, GSP-1, Leninskiye gory, 119991

[2] Lomonosov Moscow State University, Research Computing Center, Russia, Moscow, Leninskiye gory, 119991

E-mail: mos7kit@mail.ru , kostadini@mail.ru

Abstract: This work is devoted to modeling of characteristics of comfort thermal conditions in some Arctic cities under climate change. In this paper, the authors try to solve the problem of scarce meteorological observing stations in the cities by modeling thermal comfort indices in the Arctic urban landscape. For this, seven Arctic cities are classified according to a universally accepted classification of local climatic zones (LCZs) to identify the zones that best characterize each of the cities. Building data and data from the weather stations closest to the cities are included in a diagnostic microclimatic model RayMan. As a result of the experiments, air temperature trends and bioclimatic environmental comfort indices for the Arctic cities from 1966 to 2017 have been obtained. An assessment of the evolution of the indices has shown that the transitional seasons (April-May and October-November) make the greatest contribution to the change in comfort of the Arctic cities, and the number of days with severe cold stress decreases in all these cities of the Russian Arctic.

1. Introduction & motivation

The Arctic region stands out especially brightly against the background of global climate change. It is well-known that the rate of temperature increase in this region is higher than the average for the globe[1]. Temperature as a climatic resource largely determines the conditions of human life and the possibility of economic development for the territories. However, the well-being of each individual is affected not only by temperature, but by all meteorological parameters as a whole. Therefore, thermal comfort indices are used to assess its impact on a person, his health, and life expectancy. These are climatic indicators characterizing a state of comfort in which an optimal level of physiological functions of the body is formed, while a person is neither cold nor hot [2].

In Russia, about 74% of the population lives in cities. In the Arctic zone, due to the specifics of its development, this figure reaches 85% [3]. Therefore, the assessment of thermal comfort conditions in applied bioclimatology can be considered an urgent issue not only on average for the territory, but also for urbanized areas separately. In these territories the effect of urban heat islands is strongly affected [4],[5] , and the majority of the region's population lives in addition to general climate warming.

In studying the climate of urbanized territories, there appears the problem that weather stations are located outside the city limits almost in all cities. These weather stations cannot characterize the weather conditions of the whole city. Therefore, there are almost no data on the climate of urban areas in the Arctic. This work provides an example of a solution to this problem. For the study, the territory that officially entered the Arctic zone of the Russian Federation by the decree of the President of the Russian Federation dated May 2, 2014 was chosen, and the following cities were selected: Murmansk, Kandalaksha, Arkhangelsk, Naryan-Mar, Salekhard, Anadyr, and Nadym, also included in the Arctic zone [6]. These cities were selected on the basis of the presence of a weather station within the city limits or within a 50-km radius with a complete climatic range of data for 1966-2017 (Figure 1).

Figure 1 Map of the Arctic zone of Russia with selected cities.

2. Data and methods

2.1 Assessment of temperature trends

At the beginning of the work, an analysis of the average annual temperature trends in different regions of the Arctic zone of the Russian Federation was made. To do this, we used data from the reanalysis database of the National Centers for Environmental Information, NCEI NOAA [7] and data from the weather stations of the Roshydromet network closest to the cities under the study (the data were obtained from the VNIIGMI-WDC [8] resource). According to the reanalysis data and the data from weather stations, the comparison between the rates of temperature growth showed that the rates are of the same order: about 0.3 ° - 0.55 ° C / 10 years over 52 years in all areas of the Russian Arctic zone.

It can be concluded that there are no representative data on what is happening directly in the cities: first, urban weather stations do not capture what is happening with the temperature in each particular city and, second, the spatial scale of almost all bases existing today does not allow one to capture the climatic changes occurring specifically in the cities (the size of the city is much smaller than the step of the reanalysis grid).

2.2 Modeling the parameters of the urban environment

To solve the problem of lack of meteorological data within urban landscapes and for further modeling of climate and comfort, an attempt to model the urban environment was made. For this, zoning of 7 Arctic cities by local climatic zones (LCZs) [9] was carried out.

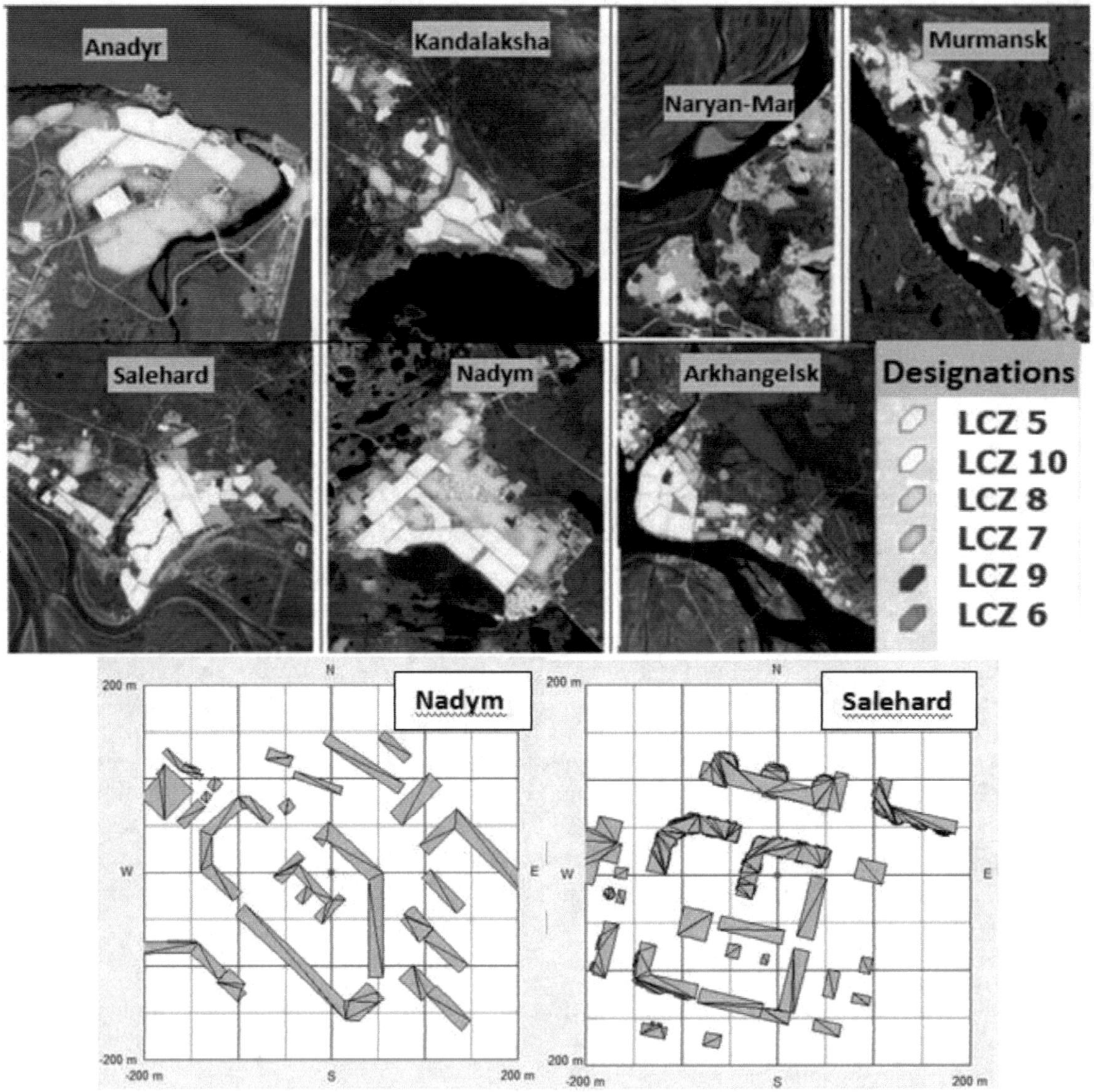

Figure 2. a) Zoning of cities by LCZs. From right to left: Anadyr, Kandalaksha, Naryan-Mar, Murmansk, Salekhard, Nadym, Arkhangelsk. b) Typical building area for microscale modeling.

Local climatic zones (LCZs) are areas with a relatively homogeneous surface coverage: density and height of buildings, the number of green spaces, building materials and a certain nature of human activity, which have a special type of interaction with the surface layer of the atmosphere and, in aggregate, have a specific effect on the local climatic characteristics. In total, 17 types of zones are distinguished in the LCZ system on a local scale from 100 to 10 000 m: 10 zones with urban development and 7 natural zones [10], [11]. For all 17 zones, the characteristics of the underlying surface (for example, albedo) and the thermophysical parameters of the environment, as well as the degree of human transformation were calculated [9]. The zoning was carried out manually, according to the instructions of the WUDAPT project (www.wudapt.org). For manual zoning, Google Earth software, Landsat 7 images, and SAGA-GIS were used (Figure 2a).

During the zoning process (Figure 2), it was found that in all selected cities LCZ 5 and LCZ 6 climatic zones are predominant. The most distinctive zone for most cities is the LCZ 5 zone (wide streets with houses about 10-25 meters high). That is ordinary city blocks. The exception is Naryan-Mar, for which the most characteristic is "LCZ 6" (wide streets with dense buildings and houses 1-3 floors high).

2.3 Modeling the conditions of thermal comfort in the Arctic cities

To calculate the biometeorological indices, the diagnostic model RayManPro 3.1 [12] was used. This model calculates radiant temperature, average radiation fluxes, and biometeorological indices (PET, PMV, and UTCI) based on the equation of the energy balance of a human body in a particular place at a specific moment in time for a particular person, as well as short-wave and long-wave radiation fluxes. The model is compatible with Microsoft Windows and can analyze complex urban structures and other environments.

Data on the thermophysical properties of the surface were obtained on the basis of information about the distinct type of development for each city with the type of development most characteristic of each city. Data on the closed horizon and building altitude were obtained from the OpenStreetMap city database. Using a script in the QGIS environment, it was possible to create files with a simulated urban area characteristic for each particular city — "typical building areas" (Figure 2b).

This information, as well as regular data from meteorological observations with 3-hour resolution in 1966-2017, were used for modeling in a three-dimensional block model RayMan Pro 3.1[13]. In this work, we used the most modern bioclimatic indices that take into account the individual characteristics of a person as much as possible. These indices are based on the heat balance equation of a person and take into account physiological processes (metabolism, sweating, etc.) under the influence of different meteorological factors.

1) PET - Physiologically Equivalent Temperature. This is a measure of the thermal sensation of a person at rest conditions. This is an indicator characterizing the complex effect of temperature, air humidity, wind speed, radiation flows on a person and take into account the degree of physiological activity of a person and the degree of thermoregulatory properties of clothes [14]. This index can detect levels of thermal exposure to the environment (heat stress), if they exist [15].

2) UTCI is a universal index of thermal comfort. It was developed as a concept of "equivalent temperature": it includes the definition of a basic condition against which all other climatic conditions will be compared [16].

To assess the dynamics of cold stress separately, the Wind chill index was calculated [17]. This is an indicator that reflects the sensations of a person with simultaneous exposure to wind and low temperature. The subjective sensation by a person of the air temperature in an open area of the skin is estimated taking into account the wind speed. This is a cold index and it is not physiologically founded.

3. Results and discussion

As a result of the work, it became possible to simulate the dynamics of thermal comfort conditions in the cities of the Arctic zone for the first time ever (Figure 3).

The graphs show that, despite a significant interannual dynamics, there is an increase in the temperature and an increase in the simulated indices. The indices in all studied cities have a growth rate comparable to the growth rate of the surface air temperature.

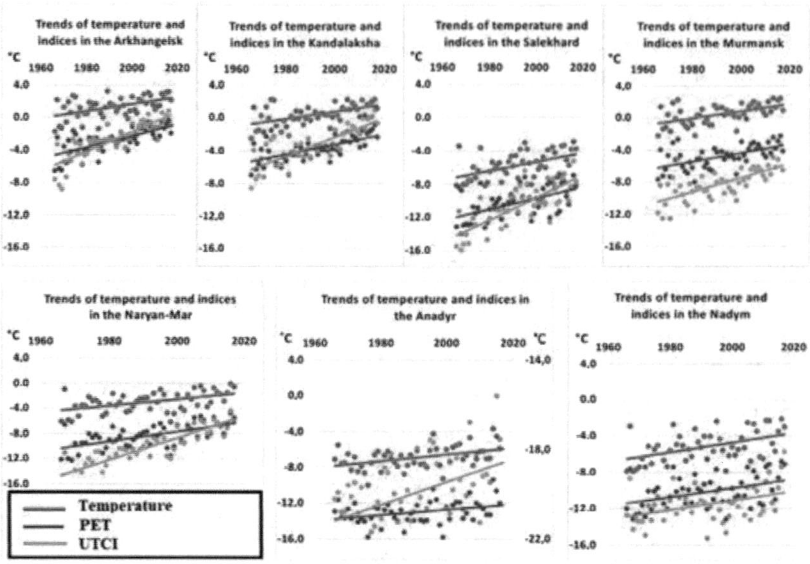

Figure 3. Average annual trends in temperature and simulated indices of thermal comfort for Arctic cities.

The results can be divided into two groups. The reason for this heterogeneity is probably both regional and microclimatic (urban landscape) features. The first group is Arkhangelsk, Kandalaksha, Salekhard, and Naryan-Mar, where the trends of the PET and UTCI indices grow faster than the air temperature. This suggests that in these cities the level of comfort, on average, improves faster than the air temperature rises.

The second group is the cities of Murmansk, Nadym, and Anadyr, where the temperature graph and the PET and UTCI index graphs are almost parallel. This means that changes in the temperature and thermal comfort conditions in combination with other parameters embedded in these indices occur at the same rate. For convenience of evaluating the results, a table of the rates of temperature change and the comfort indicators for 10 and 50 years was compiled (Table 1).

Table 1. The growth rate of temperature and indices for 10 years and for 50 years.

City	ΔT	ΔPET		ΔUTCI		ΔWind Chill	
	For 10 years	For 10 years	For 50 years	For 10 years	For 50 years	For 10 years	For 50 years
Murmansk	0.45	0.54	2.71	0.68	3.42	0.60	2.99
Kandalaksha	0.45	0.61	3.05	1.10	5.48	0.65	3.25
Arkhangelsk	0.45	0.69	3.47	1.22	6.10	0.72	3.60
Naryan-Mar	0.52	0.78	3.90	1.68	8.41	0.93	4.65
Salekhard	0.53	0.67	3.33	0.82	4.08	0.41	2.04
Nadym	0.54	0.50	2.51	0.47	2.36	0.60	2.99
Anadyr	0.36	0.30	1.48	0.29	1.47	0.56	2.79

The average annual temperature for all cities changes at a rate of 0.47 ° C / 10 years over 52 years (Table 1). The indices change, on average, at a rate of 0.57 ° C / 10 years (PET) and 0.9 ° C / 10 years (UTCI). It can be assumed that these are the average rates of change of the comfort indicators in the Arctic region. It is worth noting that an increase in the trends of the calculated indices can be dangerous, since the spread of values (variance) also increases, that is, the number of very cold and very warm days that leads to thermal stress. In all respects, comfort in the city of Naryan-Mar is changing most rapidly. An analysis of the trends in the wind speed and relative humidity in Naryan-Mar has shown a steady and significant trend towards a decrease in the wind speed (the determination coefficient is 0.86) and a slightly less significant trend towards a decrease in the relative humidity. A change in these meteorological parameters included in the calculated indicators influenced the change in comfort in this city. In other cities of significant trends, changes in the wind speed and relative humidity are practically not observed (Figure 4).

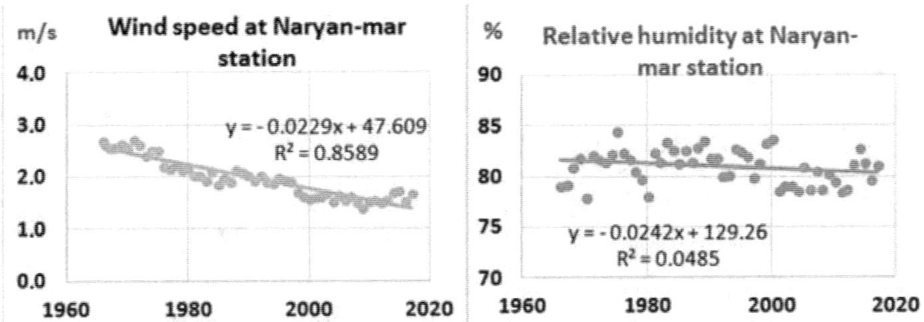

Figure 4. Change in wind speed and relative humidity at Naryan-Mar station.

According to the Wind Chill index, steady growth of this indicator is also observed in all cities (Table 1). These changes are comparable to the rate of change of the air temperature. Salekhard stands out separately, where this indicator grows more slowly than the temperature. The highest growth rates for changes in this index are in Arkhangelsk and Naryan-Mar: the growth is about 1.5 times faster than that of the temperature.

The results obtained may indicate that in the average annual climate of cities in the Arctic zone is becoming less rigid and more comfortable for living in this region.

3.1 Assessment of the dynamics of heat and cold stress

In order to assess which season of the year and which level of discomfort affects the change in comfort in the Arctic region, we analyzed the dynamics of the number of days with thermal stress according to UTCI and PET, as well as the number of days with cold stress according to UTCI and Wind Chill separately. Moreover, an analysis of the annual progress of these indicators was made. Assessment of the dynamics of the comfort indicators in the Arctic cities showed:

- A decrease in the number of days with severe cold stress in all cities of the Russian Arctic. The risk of hypothermia (according to the Wind Chill index and UTCI) and cold stress are most pronounced in Anadyr and Nadym, and least of all in Arkhangelsk, Kandalaksha, and Murmansk. On average, cold stress lasts 8 months a year with a maximum in January (Figure 5).

IOP Conf. Series: Earth and Environmental Science **386** (2019) 012017 doi:10.1088/1755-1315/386/1/012017

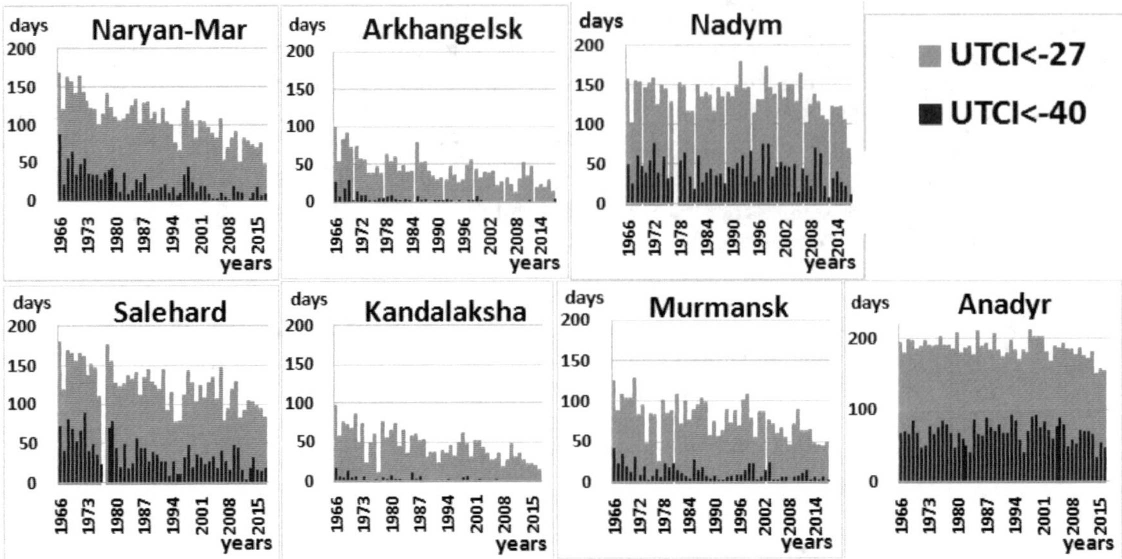

Figure 5. The dynamics of the number of days with very strong and extreme cold stress (UTCI <-27 and < - 40) in the cities of the Arctic zone.

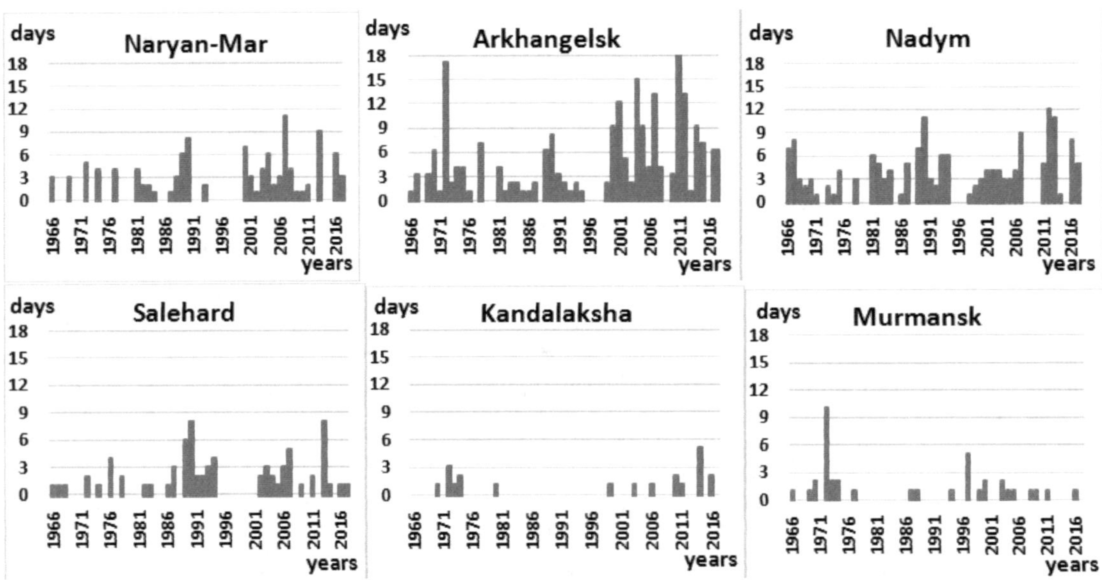

Figure 6. The number of days with strong heat exposure (PET> 35).

- An increase in the number of days with severe heat stress in Nadym, Salekhard, Naryan-Mar, and Arkhangelsk (PET > 35). In Kandalaksha and Murmansk, such days are almost completely absent, while in Anadyr there are none at all. During the year, heat stress in most cities can be observed only in the summer months, with a maximum in July (Figure 6).
- Transitional seasons (April-May and October-November) make the largest contribution to the change in comfort in the Arctic cities (Salekhard) (Figure7).

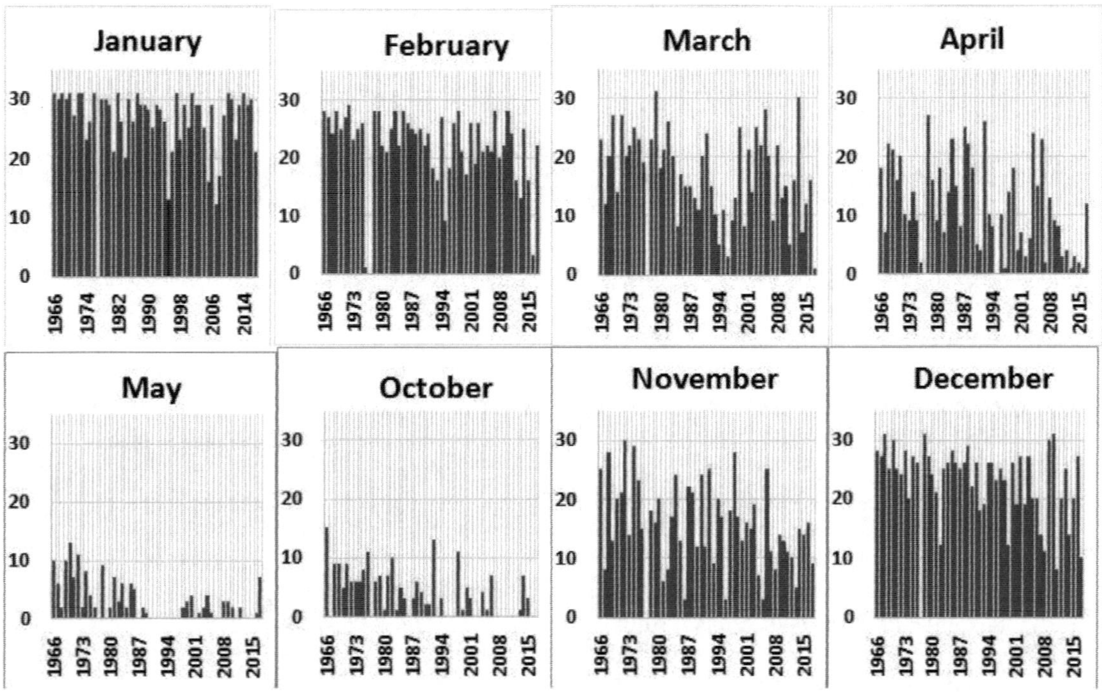

Figure 7. The number of days with very severe cold stress in Salekhard (UTCI<-27).

4. Conclusions

For the first time, a classification of local climatic zones (LCZs) was carried out for some cities of the Arctic zone of the Russian Federation. It was determined that the best characteristic for most of the cities is a "LCZ 5" zone (wide streets with houses about 10-25 meters high).

Using building data in each of the cities in a three-dimensional block of the RayMan diagnostic model, most common bioclimatic environmental comfort indices were calculated for 7 cities of the Arctic zone of the Russian Federation for 52 years: from 1966 to 2017. As a result, it was found that there are trends in an increase in the average annual temperature at a rate of 0.3-0.55 ° C / 10 years. Despite this fact, almost throughout the whole Arctic zone of the Russian Federation the indices of thermal comfort in the cities of the Arctic are changing differently. Thus, in Arkhangelsk, Kandalaksha, Salekhard, and Naryan-Mar the PET and UTCI indices grow 1.5-3 times faster than the air temperature. This suggests that comfort in these cities, on average, improves faster than the air temperature rises. In Murmansk, Nadym, and Anadyr the trends in temperature and all three indices are almost parallel. This means that the changes in temperature and thermal comfort conditions, in combination with the other parameters embedded in these indices, take place at the same rate. This heterogeneity is caused by regional and microclimatic (urban landscape) features. The results obtained indicate that, on average, climate of the cities in the Arctic zone is becoming less rigid and more comfortable.

Acknowledgements

Most of this research was supported by the Russian Science Foundation (project no. 17-77-20070 on "Preliminary Assessment and Projection of Bioclimatic Comfort in Russian Cities in the 21st Century under Climate Change"). The thermal comfort indices assessment for Archangelsk region (performed by P.I. Konstantinov) was supported by the Russian Foundation for Basic Research (project no. 18-05-60146_Arctica).

IOP Conf. Series: Earth and Environmental Science **386** (2019) 012017 doi:10.1088/1755-1315/386/1/012017

References

[1] Agassiz A *Climate change 2001: impacts, adaptation, and vulnerability*

[2] Isaev A A 2003 *Jekologicheskaja klimatologija* (Nauch. mir)

[3] http://www.gks.ru/

[4] Konstantinov P, Varentsov M and Esau I 2018 A high density urban temperature network deployed in several cities of Eurasian Arctic *Environ. Res. Lett.* **13** *075007*

[5] Varentsov M, Konstantinov P, Baklanov A, Esau I, Miles V and Davy R 2018 Anthropogenic and natural drivers of a strong winter urban heat island in a typical Arctic city *Atmos. Chem. Phys.* **18** 17573–87

[6] http://www.kremlin.ru/acts/bank/38377

[7] https://www.noaa.gov/

[8] http://meteo.ru/

[9] Bechtel B, Alexander P J, Böhner J, Ching J, Conrad O, Feddema J, Mills G, See L and Stewart I 2015 Mapping local climate zones for a worldwide database of the form and function of cities *ISPRS Int. J. Geo-Information* **4** 199–219

[10] Stewart I D 2011 A systematic review and scientific critique of methodology in modern urban heat island literature *Int. J. Climatol.* **31** 200–17

[11] Stewart I D and Oke T R 2012 Local climate zones for urban temperature studies *Bull. Am. Meteorol. Soc.* **93** 1879–900

[12] Matzarakis A, Rutz F and Mayer H 2010 Modelling radiation fluxes in simple and complex environments: Basics of the RayMan model *Int. J. Biometeorol.* **54** 131–9

[13] Matzarakis A and Rutz F 2005 Application of RayMan for tourism and climate investigations *Annalen der Meteorologie 41* **2** 631-6

[14] Ajzenshtadt B.A. 1987 Ocenki radiacionnogo vlijanija razlichnyh jelementov gorodskoj sredy na teplovoe sostojanie cheloveka dlja celej gradostroitel'stva . *Stroit. klimatologija* **6** 66–71

[15] Chen Y C and Matzarakis A 2018 Modified physiologically equivalent temperature—basics and applications for western European climate *Theor. Appl. Climatol.* **132** 1275–89

[16] Jendritzky, Gerd, George Havenith, Philipp Weihs E B and R D 2007 *The Universal Thermal Climate Index UTCI Goal and state of COST Action 730*

[17] Bluestein M and Zecher J 1999 A New Approach to an Accurate Wind Chill Factor *Bull. Am. Meteorol. Soc.* **80** 1893–9

IOP Conf. Series: Earth and Environmental Science **386** (2019) 012018 doi:10.1088/1755-1315/386/1/012018

Assessment of the response of Russian forest ecosystems to different climatic conditions from model data

Yu V Martynova[1,2] **and V N Krupchatnikov**[1,2,3]

[1] Institute of Monitoring of Climatic and Ecological Systems SB RAS, Tomsk, 634055, Russia
[2] Siberian Regional Hydrometeorological Research Institute, Novosibirsk, 630099, Russia
[3] Novosibirsk State University, Novosibirsk, 630090, Russia

E-mail: FoxyJ13@gmail.com, vkrupchatnikov@yandex.ru

Abstract. The response of extra-tropical evergreen and deciduous Russian forest ecosystems to a wide range of atmospheric background conditions is considered. This study is performed by using a land-surface model, JSBACH. Two numerical experiments are carried out: with hotter and more humid (model PLASIM) and more moderate (model INMCM4) atmospheric background conditions. A perturbation in the climate system is set by using the RCP 8.5 climate scenario. The study has shown a geographical redistribution of the extra-tropical evergreen and deciduous forests across Russia as well as an increase in their fractions in some already forested regions. An increase in the gross and net CO_2 assimilation and respiration of these types of forest vegetation, as their response to the climatic disturbances, is obtained. However, even under the conditions of extremely intense climatic disturbances the Russian forest ecosystems will continue to act as CO_2 sinks.

1. Introduction

Due to its geographical location Russia has a wide variety of surface characteristics, such as orography, vegetation types, and greenhouse gas fluxes (carbon dioxide, methane). Forests in general and Russian forests in particular act as natural greenhouse gas sinks. Assessing the carbon budget of forests and carbon dioxide fluxes for Russia is particularly important in the context of the Paris Agreement [1].

The biosphere is an essential component of the Earth climate system in terms of the global cycle of main greenhouse gases, the hydrological and energy cycles. The interaction of the surface and the atmosphere is one of the key characteristics for an understanding of the climate system dynamics, which determines the energy, water, and greenhouse gases (such as CO_2, CH_4) fluxes [2,3]. The surface plays an essential role in climate modeling. It affects biogeophysical and biogeochemical processes [4,5], controls the distribution of energy. The energy cycle is associated with the hydrological cycle, which redistributes the precipitation to surface runoff, drainage and infiltration, determining the soil moisture and, as a result, the total evaporation. The biogeochemical processes of the carbon cycle are mainly represented by the carbon sink on the surface, which is closely related with the hydrological cycle through the stomatal control of the leaves during photosynthesis and transpiration [6-8].

Climatic conditions are a governing factor in the growth of various types of vegetation. At the same time, vegetation affects climate by controlling energy balance, surface water, and CO_2 concentration budget [9-13]. In the context of global climate change, it is important to understand what changes are possible for a given territory, how energy and greenhouse gas fluxes and the distribution limits of different types of vegetation and surface types can change. This knowledge will allow assessing the risks associated with these changes and developing the most effective ways to adapt the population, industry, and agriculture to the predicted changes.

In an earlier paper the authors have already studied the response of plant ecosystems (extra-tropical forests, shrubs and grass vegetation) to climatic changes for Siberia [14]. In the current study, the area

IOP Conf. Series: Earth and Environmental Science **386** (2019) 012018 doi:10.1088/1755-1315/386/1/012018

of consideration has been increased, and the entire territory of Russia is considered. Attention is focused only on the forest ecosystems (extra-tropical evergreen and deciduous forests). A wide range of atmospheric background conditions is considered, that allow us to estimate the possible variety of the ecosystem response to the applied climatic disturbances.

2. Data and method

In the study, two numerical experiments were carried out, differing in the atmospheric background conditions. Each of the experiments consisted of two parts: simulation for the historical period and the climate scenario RCP 8.5 [15].

2.1 Experimental design

In this study, for the numerical experiments a technique used in a previous paper was applied [14]. As in that paper, the land surface model JSBACH [16-18] was used. The JSBACH gives a complete description of the processes in the soil-vegetation-atmosphere system and can work offline with external climate data as forcing. The model reproduces the soil hydrology, the heat transfer in soils and the energy balance on the soil surface [19], dynamic processes of carbon absorption and emissions from soils and plants, photosynthetic and phenological processes, variation of the surface radiation and background albedo [20, 21], changes in the vegetation cover (including damage by wind and fire) [18, 22]. For the experiments of this study, the atmospheric background conditions required for the JSBACH model were set using two data sets previously obtained using the global large-scale model of intermediate complexity PLASIM [23] and the global climate model INMCM4 [24].

The numerical experiment with each of the atmospheric datasets consisted of two parts: the historical part and the part (until 2005) of the climatic scenario RCP 8.5 (until 2100). The climatic conditions for the historical pre-industrial period (until 2005) were set in the case of the PLASIM model by fixing the global atmospheric CO_2 concentration at 360 ppm, and in the case of the INMCM4 model according to the CMIP5 protocol "Historical simulation". Further, following the RCP 8.5 climate scenario (until 2100), an exponential increase in the global atmospheric CO_2 concentration was established from the current value to a level of 936 ppm. This final CO_2 concentration value is almost 2.3 times higher than its pre-industrial value.

2.2 Background atmospheric conditions

The climatic scenario RCP 8.5 used in the work allowed us to set conditions for extreme climate warming for Russia (Table 1). For a given increase in the carbon dioxide concentration, the model PLASIM demonstrated an increase in the time-period average of the minimum (T_{min}) and maximum (T_{max}) daily temperature. In the north of the territory under consideration, the difference reached 8°C. A slight increase in the average precipitation (Pr_{sum}) was obtained over most of Siberia, and a slight decrease was along the southern border of Russia. The surface specific humidity (Q_{air}) increased with anthropogenic load rising throughout the territory. The wind speed (*Wind*) increased by 0.3-0.6 m/s in the north of Eastern Siberia and the Far East and decreased by 0.3-1.2 m / s in the rest of the territory. As a result of an increase in the global atmospheric CO_2 concentration rise, the radiation parameters were changed as well. There was an increase in the surface downward longwave radiation (R_{lw}) and a decrease in the surface balance of solar radiation with a cloudless sky throughout Russia. The surface downward shortwave radiation (R_{sw}) increased in southwestern Siberia and decreased in the rest of the Russian territory.

For the Russian territory, the climate reproduced by the INMCM4 model in the same conditions of the global atmospheric CO_2 concentration growth turned out to be significantly less hot, less humid, and less windy than the climate of the PLASIM model. Moreover, in comparison with the PLASIM model, the INMCM4 model demonstrated, under the given conditions, a stronger increase in the precipitation, a more intensive growth of minimum and less intensive growth of maximum daily temperatures (Table 1). The intensity of changes of the other mean territory atmospheric parameters is similar to each other for these models. Comparing the models INMCM4 and PLASIM, the patterns of the geographical distribution over Russia, changes of all used atmospheric background parameters are similar.

Table 1. Background atmospheric parameters mean for 20-year periods averaged over Russia.

		T_{min}, ^oC	T_{max}, ^oC	Pr_{sum}, *mm*	Q_{air}, 10^{-3} *kg/kg*	*Wind, m/s*	R_{lw}, W/m^2	R_{sw}, W/m^2
PLASIM	1981-2000	4.45	6.87	956.33	5.03	7.71	298.13	104.07
	2081-2100	9.79	11.89	1029.86	6.82	7.47	328.82	99.72
	Difference	5.35	5.02	73.54	1.78	-0.24	30.69	-4.35
INMCM4	1981-2000	-11.30	-0.13	563.32	3.48	3.61	257.85	126.95
	2081-2100	-5.87	4.10	670.67	4.59	3.63	281.42	123.54
	Difference	5.44	4.23	107.34	1.11	0.03	23.57	-3.41

3. Results

Extra-tropical evergreen and extra-tropical deciduous forests were considered. Here deciduous means summergreen forests (leaf-bearing forests including larches), and evergreen means forests which are green all year (including pines and fir-trees). Estimates of the response of the spatial distribution of the above forest types, as well as their gross and net CO_2 assimilation, to assigned climatic perturbations are presented. The total CO_2 flux from the surface of the Russian territory to the atmosphere, in whose formation all vegetation types (including shrubs and grass, not just forests) simulated by the JSBACH model are involved was also assessed.

3.1. Forest vegetation

The numerical experiments performed for different background atmospheric conditions demonstrate similar, but not identical, nature of the spatial distribution of the forest vegetation throughout Russia (Figures 1 and 2).

At the end of the 20th century extra-tropical evergreen forests became more widespread throughout the territory under the atmospheric conditions specified by the INMCM4 model than in the climate conditions specified by the PLASIM model (Figure 1). It is especially clearly seen in the Far East of Russia and the Kamchatka Peninsula, where the fraction of this type of vegetation is substantially larger for the INMCM4 climate conditions than for the PLASIM.

By the end of the 21st century, after injecting climate perturbations according to the RCP 8.5, extra-tropical evergreen forests shift to the north with a decrease of their fraction at the southern Russian borders and an increase in the North of Russia for the atmospheric conditions specified by the PLASIM model. The milder (in general) atmospheric conditions set by the INMCM4 model with a similar intensity of climatic disturbances cause significantly less dramatic changes in the spatial distribution of extra-tropical evergreen forests than the conditions of the PLASIM model. They cause only a small increase in the fraction in the north of the European part of Russia and Western Siberia, and also on the eastern coasts of Russia.

For the distribution of the extra-tropical deciduous forests, as compared with the extra-tropical evergreen, a different picture was obtained at the end of the 20th century. The deciduous forests have become more widespread in a milder climate defined by the INMCM4 model. Under these conditions, the extra-tropical evergreen forests fraction more or less remains throughout Russia. Unlike the INMCM4 conditions, under the conditions specified by the PLASIM this vegetation type is almost absent in the Far East of Russia.

IOP Conf. Series: Earth and Environmental Science **386** (2019) 012018 doi:10.1088/1755-1315/386/1/012018

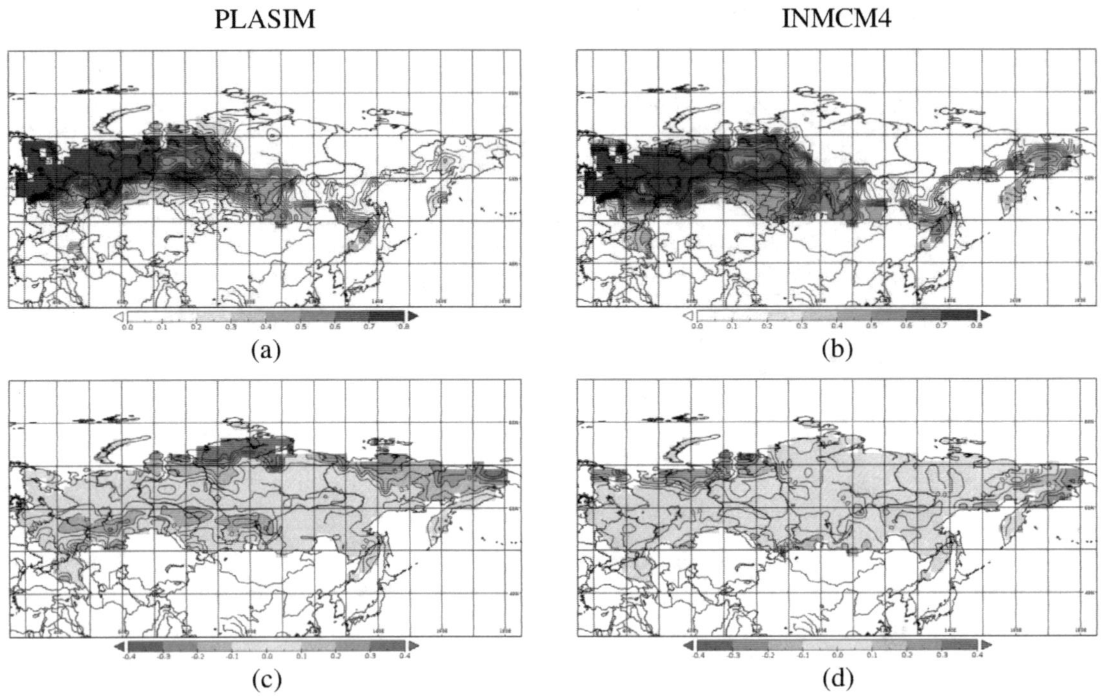

Figure 1. Extra-tropical evergreen forests, vegetation fraction (a, b) and difference between the end of 21st century (2081-2100) and 20th century (1981-2000) (c, d) for atmospheric conditions from PLASIM (a, c) and INMCM4 (b, d) model output.

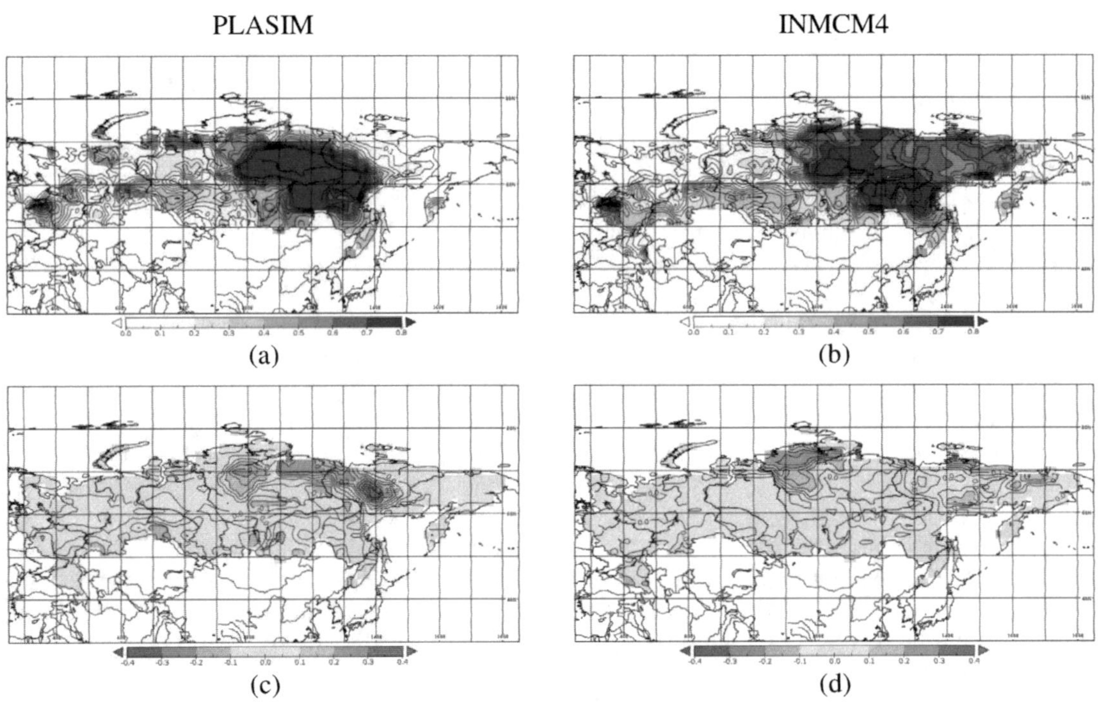

Figure 2. The same as Figure 1 but for extra-tropical deciduous forests.

As in the case of extra-tropical evergreen forests, the response of extra-tropical deciduous forests to injected climatic disturbances differs for the atmospheric conditions given by the INMCM4 and PLASIM models. By the end of the 21st century, under the PLASIM atmospheric conditions the fraction of this vegetation type increases in the north of the central part of Russia, especially in the North Siberian Lowland, along the Lena River and in Transbaikal. However, in the area of the Vilyui River and at the southern borders of Russia a focal reduction of the extra-tropical deciduous forests fraction was obtained. Under the conditions specified by the INMCM4 model changes in the distribution of the vegetation fraction of this type are less intense. These conditions cause a slight increase of the fraction in the Taymyr Peninsula and the north of Kolyma and Chukotka.

3.2. Canopy CO_2 assimilation

When the geographic distribution and density (fraction) of vegetation cover change, it is quite natural to expect a change in the amount of carbon dioxide absorption by it from the atmosphere.

In the whole territory, an increase in the total gross and net CO_2 assimilation was obtained by the end of the 21st century compared to the end of the 20th century for both types of forest vegetation considered (Table 2). Within each numerical experiment for both the end of the 20th century and the end of the 21st century, comparable values between the evergreen and deciduous forest vegetation types gross and net CO_2 assimilation were obtained. In the warmer and more humid atmospheric conditions specified by the PLASIM model, as compared with the INMCM4, for both types of forest vegetation under consideration there is more intensive absorption of CO_2 from the atmosphere, accumulation of CO_2 in the biomass, and more intense respiration (which is the difference between the gross and net CO_2 assimilation). However, the evergreen and deciduous forest vegetation shows various responses to climatic disturbances under different background atmospheric conditions. The gross CO_2 assimilation of the evergreen forest vegetation demonstrates a higher percentage of the response to the disturbances injected to the background conditions set by the INMCM4 model than those set by the PLASIM. For the deciduous forest vegetation, the situation is opposite: a higher gross CO_2 assimilation response was obtained for the atmospheric conditions specified by the PLASIM model. The respiration of these vegetation types demonstrates a similar response. However, the response of the net CO_2 assimilation to the injected disturbances differs from the gross one. By the end of the 21st century, as compared with the end of the 20th century, for both types of forest vegetation under consideration there is a more intensive increase in the CO_2 accumulation in vegetation under the conditions specified by the INMCM4 model than by the PLASIM.

Table 2. Mean annual gross and net CO_2 assimilation by canopy and canopy respiration for Russia, [Gt C yr^{-1}].

		Gross assim.		Net assim.		Respiration	
		Evergreen	Deciduous	Evergreen	Deciduous	Evergreen	Deciduous
PLASIM atm.	1981-2000	4.881	4.385	4.538	4.080	0.343	0.305
	2081-2100	5.876	5.785	5.409	5.286	0.467	0.499
	Difference	0.995 (20.39 %)	1.399 (31.90 %)	0.872 (19.21 %)	1.206 (29.57 %)	0.124 (36.15 %)	0.194 (63.61 %)
INMCM4 atm.	1981-2000	3.759	3.248	3.529	3.035	0.230	0.213
	2081-2100	4.987	4.263	4.653	3.956	0.334	0.307
	Difference	1.228 (32.66 %)	1.016 (31.27 %)	1.125 (31.87 %)	0.921 (30.35 %)	0.104 (45.22 %)	0.094 (44.13 %)

The spatial distribution of changes in the gross and net CO_2 assimilation by the forest vegetation was also affected by different climatic conditions from the numerical experiments (Figure 3).

Under the PLASIM atmospheric conditions the spatial distribution pattern of the change in the gross CO_2 assimilation by extra-tropical evergreen forests is similar to the pattern of the spatial distribution of this vegetation type. A reduction in the gross CO_2 assimilation was obtained in those areas where a reduction in the fraction of this vegetation type occurred, and its increase in areas where the extra-tropical evergreen forests fraction increased (Figures 1c and 3a). A slightly different picture was obtained for the extra-tropical deciduous forests. An increase in the gross CO_2 assimilation was obtained not only in areas where the deciduous forests fraction increase was present, but also in already forested areas where the fraction increase did not occur (Figures 2c and 3c).

Similar changes were obtained for the atmospheric background conditions specified by the INMCM4 model. However, in this case the vegetation types seemed to reverse roles. By the end of the 21st century for the extra-tropical deciduous forests the gross CO_2 assimilation increase was obtained only in regions with an increased fraction of this vegetation type (Figures 2d and 3d). For the extra-tropical evergreen forests growth was additionally shown in already forested regions where the vegetation fraction has not changed (Figures 1d and 3b).

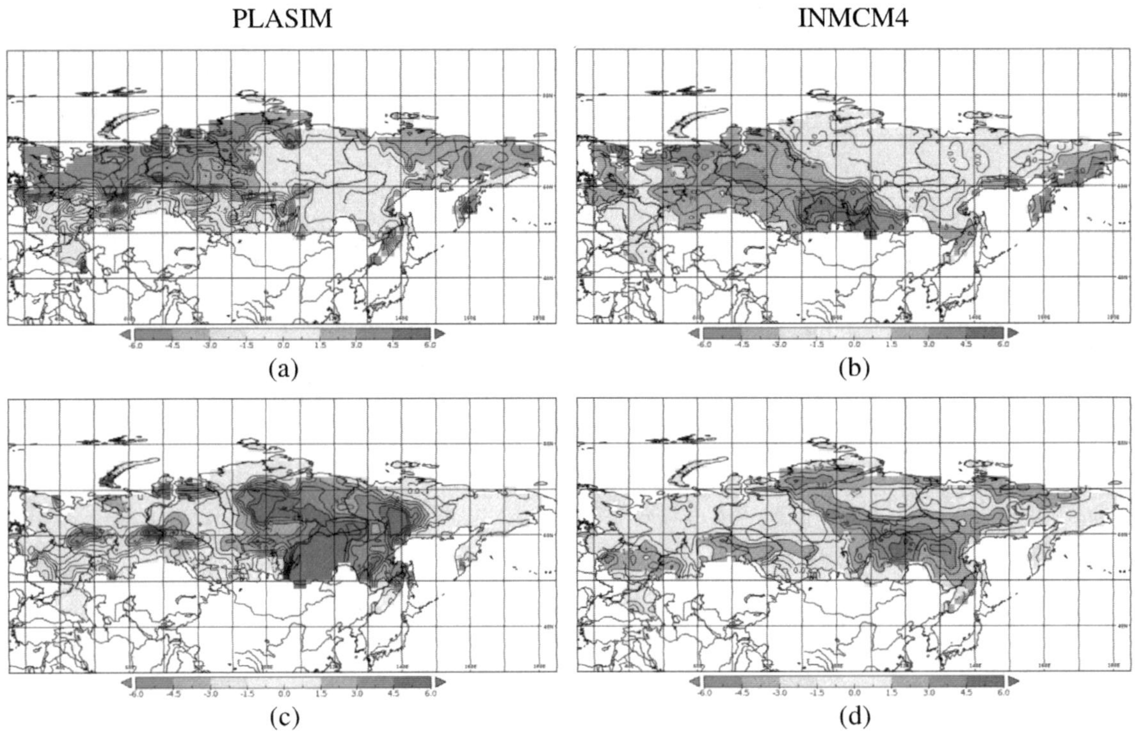

Figure 3. Difference between the end of 21st century (2081-2100) and 20th century (1981-2000) for annual gross CO_2 assimilation by extra-tropical evergreen (a, b) and deciduous (c, d) forests in atmospheric conditions from PLASIM (a, c) and INMCM4 (b, d) model output, [Mt C m^{-2} yr^{-1}].

The spatial distribution of the net CO_2 assimilation response to the climatic forcing in the given atmospheric conditions is similar to the gross CO_2 assimilation response (not shown in the figures).

3.3. CO_2 flux to the atmosphere

Forests cover a substantial part of the territory of Russia. Their spatial redistribution, as well as a change in their CO_2 absorption properties, will contribute significantly to the geographical distribution of CO_2 fluxes to the atmosphere.

For the Russian territory as a whole, by the end of the 21st century the numerical experiments showed a decrease of CO_2 absorption from the atmosphere. Moreover, for some regions the CO_2 fluxes to the atmosphere even change their sign (Figure 4). The nature of the spatial distribution of the CO_2 fluxes changes similarly to the nature of the changes obtained for the gross CO_2 assimilation by extra-tropical evergreen and deciduous forests (Figure 3). In particular, for the atmospheric conditions specified by the PLASIM model the reduction in the annual CO_2 flux to the atmosphere was obtained in the same regions where the decrease in the gross CO_2 assimilation was shown (Figure 4 e and Figure 3 a, c). For the conditions specified by the INMCM4 model the result was the same but with a less magnitude of values.

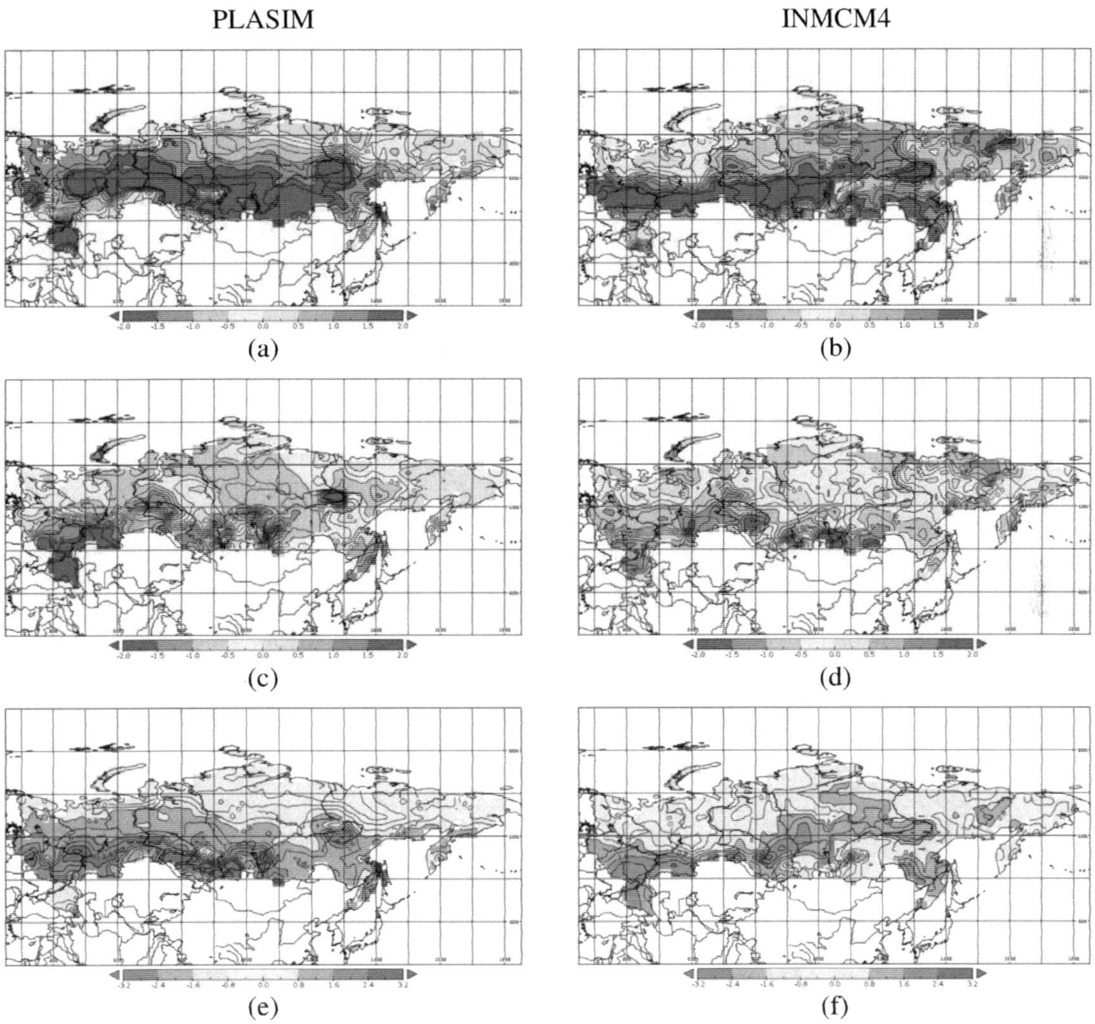

Figure 4. Annual CO_2 flux to the atmosphere for 1981–2000 (a, b) and 2081–2100 (c, d), and the difference between the ends of the 21st century (2081–2100) and 20th century (1981–2000) (e, f) for atmospheric conditions from PLASIM (a, c, e) and INMCM4 (b, d, f) model output, [Mt C m^{-2} yr^{-1}].

The total CO_2 flux to the atmosphere for the Russian territory for both time periods considered varies for different background atmospheric conditions (Table 3). For warmer and more humid conditions set by the PLASIM model lower values of the CO_2 flux to the atmosphere were obtained than for milder conditions set by the INMCM4 model. The response of the CO_2 flux to the atmosphere integral over the territory to the injected climatic disturbances (RCP 8.5) also depends on the background atmospheric conditions. By the end of the 21st century, a higher percentage of the response to the climatic disturbance was obtained for the PLASIM atmospheric conditions than for the conditions set by the INMCM4 model. However, despite high intensity of the injected disturbances the sign of the annual CO_2 flux to the atmosphere accumulated over Russia has not changed and remains negative.

Table 3. Annual CO_2 flux to the atmosphere accumulated over Russia, [Gt C yr^{-1}].

	PLASIM	INMCM4
1981-2000	-0.763	-0.849
2081-2100	-0.121	-0.360
Difference	0.641 (-84.08 %)	0.489 (-57.64 %)

4. Conclusions

The above numerical experiments allowed us to estimate the sensitivity of Russian forest ecosystems to climatic changes. This research has shown that the sensitivity of the extra-tropical evergreen forests is higher than that of the deciduous forests, to global climate changes. A reduction in the vegetation fraction and growing areas near the southern borders of Russia and an increase in the northern regions of the territory were obtained. Thus, under warmer and more humid climate conditions on the territory of Russia, a significant geographical redistribution of the growing forest vegetation types is to be expected. A logical consequence of this, for each of the above-considered plant types, is a redistribution over the territory of the gross and net CO_2 assimilation. The increase in vegetation amount also contributes to the increase in CO_2 absorption from the atmosphere, which, in turn, contributes to the increase in the vegetation biomass. Of course, these mutually reinforcing processes cannot be infinite. A natural limiter is the climatic conditions formed in the territory (temperature and humidity conditions, in particular).

It should be noted that in some areas of Russia, due to injected extreme climate changes and the decrease in the vegetation amount, a change in CO_2 absorption up to emission was observed. However, the total CO_2 flux to the atmosphere for the Russian territory, even under extreme disturbances (RCP 8.5), remains negative by the end of the 21st century. It indicates that the vegetation cover of Russia provides CO_2 absorption from the atmosphere even in highly aggressive climatic conditions.

Forest vegetation, due to its biological properties, is quite an inert component of the climate system. These studies have shown that, no matter how extreme the climatic changes are, the Russian forest ecosystems will act as CO_2 sinks from the atmosphere. The Russian forests will play this role as long as the climatic conditions are suitable for the growth of biomass, and the CO_2 emissions to the atmosphere will not exceed the absorptive capacity of the forest ecosystems.

Acknowledgments

This study was supported under state assignment AAAA-A17-117022250098-1. The archive of global georeferenced land surface data was made under state assignment AAAA-A17-117013050037-0. The Siberian Branch of the Russian Academy of Sciences (SB RAS) Siberian Supercomputer Center is gratefully acknowledged for providing supercomputer facilities (http://www.sscc.icmmg.nsc.ru/).

IOP Conf. Series: Earth and Environmental Science **386** (2019) 012018 doi:10.1088/1755-1315/386/1/012018

References

[1] Kokorin A and Lugovaya D 2018 Pogloshchenie CO_2 lesami Rossii v kontekste Parizhskogo soglasheniya *Ustojchivoe lesopol'zovanie* **54** (2) 13–8 *(in Russian)*

[2] Sellers P J, Mintz Y, Sud Y C and Dalcher A 1986 A simple biosphere model (Si B) for use within General Circulation Models *J. Atmos. Sci.* **43** 505–31

[3] Bonan G B 1995 Land-atmosphere CO_2 exchange simulated by a land surface process model coupled to an atmospheric general circulation model *J. Geophys. Res.* **D 100** 2817–31

[4] Farquhar G D, von Caemmerer S V and Berry J A 1980 A biochemical model of photosynthetic CO_2 assimilation in leaves of C3 species *Planta* **149** (1) 78–90

[5] Collatz G J, Ball J T, Grivet C and Berry J A 1991 Physiological and environmental regulation of stomatal conductance, photosynthesis and transpiration: a model that includes a laminar boundary layer *Agricultural and Forest meteorology* **54** (2-4) 107–36.

[6] Krupchatnikov V 1998 Simulation of CO_2 exchange in the atmosphere-surface biomes system by climate model ECSib *Russian J. Num. Anal. and Math. Modelling.* **13** (6) 479–92

[7] Krupchatnikov V N, Volodin E M, Galin V Ya and Lykossov V N 2000 Climatology of the near-the-surface fluxes of CO2 in the coupled model of atmospheric general circulation, vegetation and soil: a case with the prescribed architectonics of vegetation *Numerical Mathematics and Mathematical Modelling vol II (Moscow: Institute of Numerica Mathematics)* 97–112.

[8] Lykosov V and Krupchatnikov V 2006 Sovremennye modeli poverhnosti sushi s detal'nym opisaniem gidrologii, processov v biosfere i v pochve *Bioraznoobrazie i dinamika ekosistem (in Russian)*

[9] Bonan G B, Pollard D and Thompson S L 1992 Effects of boreal forest vegetation on global climate *Nature* **359** (6397) 716

[10] Claussen M 1998 On multiple solutions of the atmosphere–vegetation system in present-day climate *Global Change Biology* **4** (5) 549–59

[11] Ganopolski A, Kubatzki C, Claussen M, Brovkin V and Petoukhov V 1998 The influence of vegetation-atmosphere-ocean interaction on climate during the mid-Holocene *Science* **280** (5371) 1916–9

[12] Bounoua L et al. 1999 Interactions between vegetation and climate: radiative and physiological effects of doubled atmospheric CO_2 *J. Climate* **12** (2) 309–24

[13] Foley J A, Costa M H, Delire C, Ramankutty N and Snyder P 2003 Green surprise? How terrestrial ecosystems could affect earth's climate *Frontiers in Ecology and the Environment* **1** (1) 38–44

[14] Martynova Yu V and Krupchatnikov V N 2018 Siberian vegetation cover response to projected future climate change *IOP Conf. Series: Earth and Environmental Science* **211** 012013 DOI:10.1088/1755-1315/211/1/012013

[15] Moss R et al. 2010 The next generation of scenarios for climate change research and assessment *Nature* **463** (7282) 747–56 DOI:10.1038/nature08823

[16] Raddatz T J, Reick C H, Knorr W, Kattge J, Roeckner E, Schnur R, Schnitzler K G, Wetzel P and Jungclaus J 2007 Will the tropical land biosphere dominate the climate-carbon cycle feedback during the twenty-first century? *Clim. Dyn.* **29** (6) 565–74

[17] Reick C H, Gayler V, Raddatz T, Schnur R and Wilkenskjeld S 2012 JSBACH–The new land component of ECHAM *Max Planck Insitute for Meteorology (Hamburg, Germany)* 1–167.

[18] Reick C, Raddatz T, Brovkin V, and Gayler V 2013 The representation of natural and anthropogenic land cover change in MPI-ESM *Journal of Advances in Modeling Earth Systems* **5** 459–82

[19] Roeckner E et al. 2003 *The atmospheric general circulation model ECHAM 5. PART I: Model description*

[20] Knorr W 1997 Satellite remote sensing and modelling of the global CO_2 exchange of land vegetation: a synthesis study *Max-Planck-Institut fur Meteorologie Examensarbeit* **49** 185

[21] Vamborg F S E, Brovkin V, and Claussen M 2011 The effect of a dynamic background albedo scheme on Sahel/Sahara precipitation during the mid-Holocene *Climate of the Past* **7** 117–31 DOI:10.5194/cp-7-117-2011
[22] Brovkin V, Boysen L, Raddatz T, Gayler V, Loew A and Claussen M 2013 Evaluation of vegetation cover and landsurface albedo in MPIESM CMIP5 simulations *Journal of Advances in Modeling Earth Systems* **5** (1) 48–57
[23] Fraedrich K, Jansen H, Kirk E, Luksch U and Lunkeit F 2005 The Planet Simulator: Towards a user friendly model *Meteorol. Zeitschrift* **14** 299–304
[24] Volodin E M, Dianskii N A and Gusev A V 2010 Simulating present-day climate with the INMCM4.0 coupled model of the atmospheric and oceanic general circulations *Izv. - Atm. Ocean. Phys.* **46** (4) 414–31

IOP Conf. Series: Earth and Environmental Science **386** (2019) 012019 doi:10.1088/1755-1315/386/1/012019

Stability of relic methane hydrates under climatic changes in the Holocene

M M Arzhanov[1], V V Malakhova[2], I I Mokhov[1, 3] and M R Parfenova[1]

[1]A.M. Obukhov Institute of Atmospheric Physics RAS, 119017, Moscow, Russia
[2]The Institute of Computational Mathematics and Mathematical Geophysics SB RAS, 630090, Novosibirsk, Russia
[3]M.V. Lomonosov Moscow State University, 119991, Moscow, Russia

Abstract. The results of numerical modeling of the thermal regime of high-latitude permafrost regions of Western Siberia (the Yamal Peninsula) for the last 6 thousand years are presented. Thermobaric conditions for stability and dissociation of continental methane hydrates are defined. It is shown that at the present time relict methane hydrates can exist at depths of up to 150 m in the strata of frozen soil of Yamal above the modern boundary of the stability zone, having "survived" warming in the Mid-Holocene warm period (about 6,000-5,000 years ago) at negative temperatures in the permafrost. According to the results obtained, the current warming in Yamal region exceeds the warming of the Mid-Holocene. The increase of the temperature of the permafrost layer can reduce the strength of the soil and lead to dissociation of the near-surface gas hydrates.

1. Introduction

The most significant growth of the near-surface temperature occurs in the high and middle latitudes of the Northern Hemisphere and may continue in the 21st century and the next few centuries [1, 2]. According to observational data and simulation results, the spatial structure of the permafrost temperature response to near-surface warming has regional maxima in the north of Western Siberia [3-5]. The maximum increase in the active layer depth in the 21st century is also associated with this region, according to calculations with global climate models [6–8].

The raise of the soil temperature reduces the strength of frozen soil and can lead to dissociation of relict methane hydrates located above the modern boundary of the hydrate stability zone, which are sensitive to changes in the soil temperature. A probable reason for the formation of hydrates in the upper horizons of the cryolithozone is the baric factor that might be explained by the pressure of overlying ice sheet or by marine transgression. The results of calculations with a dynamic model of thermophysical processes in the glacier-soil system show that the upper boundary of the stability zone of methane hydrates could reach the soil surface [12 -14] in Yamal region during glaciation periods about 95-80 and 70-60 thousand years ago [9-11].

The warming that began more than 10 thousand years ago with a maximum about 6 thousand years ago led to the fact that the southern boundary of the permafrost in Western Siberia moved northward towards about 68th degree north [5, 15], but Yamal region remained in the permafrost. Due to the soil temperature increase, gas hydrates located in the frozen soil could have existed above the modern boundary of the stability zone (at depths of up to 100-150 m), and remain in a metastable state as a result of the self-preservation effect at negative temperatures. In particular, experimental data indicate that methane hydrates can exist in a metastable state at a negative temperature of about -6 ° C in the pore space of frozen soil for a considerable time [16, 17].

According to one of the hypotheses, the current warming can lead to dissociation of the underlying surface gas hydrates and can be one of the causes of gas emissions into the atmosphere. This hypothesis is supported by the fact that since 2014 craters with a diameter of about 10–20 m with traces of discarded rocks have been found on the Yamal Peninsula and in neighboring regions. This may be a consequence of an increase in the near-surface temperature in the recent years with dissociation of shallow deposits of relict methane hydrates [18–20]. The absence of traces of external exposure and combustion, as well as an increased concentration of methane in the funnel, serve as an evidence of gas release into the atmosphere [18]. Gas emissions in this region are not related to the gas flow from the underlying lithospheric layers because, according to isotope analysis, the gas taken from sediments to depths of about 120 m on Yamal contains nitrogen, which is of biochemical origin and is formed as a result of microbial processing of the organic matter contained in soil [21]. Large gas flows, as well as the composition, structure, and properties of the frozen soil in Yamal region indicate possible existence of gas accumulations in hydrate form [21]. Small depths and location above the boundary of the stability zone indicate that the gas emissions may be a result of dissociation of relict methane hydrates.

The aim of this study is to assess the response of the permafrost thermal state to the most significant climatic changes of the Holocene, in particular, in the Mid-Holocene, and at the end of the 20th - beginning of the 21st centuries.

2. Numerical simulations

To obtain estimates of the soil temperature regime and the thermobaric conditions for the existence of methane hydrates in the permafrost of the Yamal Peninsula in the Holocene (during the last 6 thousand years), calculations with the model of heat transfer in the soil were performed [22, 23]. The results of the calculations with the global climate model CLIMBER-2 [24] in the period from 6 thousand years ago to the middle of the 20th century (1950) were used as external atmospheric forcing. For 1950-2010, CERA-20C reanalysis data were used [25] for Yamal region (66N–73N; 65E–75E). This reanalysis reproduces the temperature anomalies in the Northern Hemisphere most realistically [25]. It reproduces the spatial structure of the temperature changes, as well as the periods of warming and cooling, in particular, in the 1970s [26].

In the numerical experiments, the depth of the computational area in the soil was taken to be 1500 m. The soil thermal characteristics were chosen to correspond to clay loam in accordance with the drilling data from Yamal region [27]. The concentration of salts in the pore solution, lowering the water freezing point in the permafrost were also taken into account, and the temperature of the water-ice phase transitions during the modeling of the soil thermal regime was assumed to be -1°C. At the initial time, the soil temperature profile was determined by solving the stationary heat transfer equation for the average annual surface air temperature at the specified heat flux at the lower boundary of the computational domain. For the upper boundary of the region, the temperature of the surface soil (or snow, if any) was taken to be equal to the surface temperature for this region, according to the CLIMBER-2 model calculation results and reanalysis data. The geothermal flux at the lower boundary was assumed to be 50 mW/sq.m, which corresponds to the average heat flux for this region [28]. The snow accumulation dynamics was taken into account according to the CLIMBER-2 model precipitation data and from the CERA-20C reanalysis.

The equilibrium equations for the existence of methane hydrates in thawed and frozen environments were used during the simulations of the stability zone of methane hydrates [29]. In the model, the soil pressure is calculated as hydrostatic.

3. Results and discussion

Analysis of the results of calculations with CLIMBER-2 and reanalysis data revealed a negative trend in the surface temperature in Yamal region starting from the Mid-Holocene to the modern period, which is consistent with paleo reconstructions based on palynological data [30]. The average monthly surface temperatures on the Yamal Peninsula in the Mid-Holocene (5,500 years ago) and in the mid-20th century (1950-1960) according to the CLIMBER-2 model results and in the first decade of the 21st century using CERA-20C reanalysis are presented in Figure 1a. The temperature in the middle of the 20th century is lower than the temperature in the Mid-Holocene in summer, as well as in winter in January and February. In the spring and autumn periods in the middle of the 20th century the surface temperature is higher compared to the Mid-Holocene.

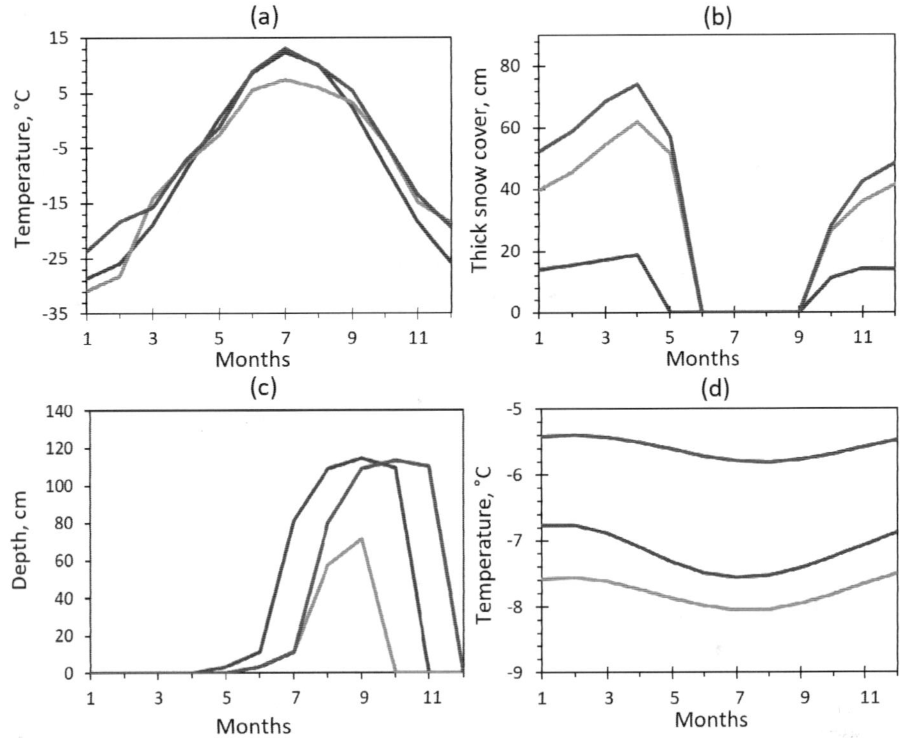

Figure 1. CLIMBER-2 model surface temperature and CERA-20C reanalysis data (a), snow depth (b), seasonal thawing depth (c), and soil temperature at a depth of 10 m (d) for Mid-Holocene (blue curve), mid-20th century (green curve), and the modern period (red curve).

A comparison of the results of calculations with the CLIMBER-2 model and reanalysis data showed that the current near-surface warming exceeds the warming in the Mid-Holocene. The greatest difference (about 5°C) was obtained for the winter period. In the autumn seasons the values of the near-surface temperature at the beginning of the 21st century are also higher by 2-3°C compared to the Mid-Holocene. The temperatures for the summer periods for the beginning of the 21st century and for the Mid-Holocene are almost equal.

The trend of surface temperature increase in the region for 1950-2010 is 0.03°C/year according to CERA-20C reanalysis, which is consistent with the observational data [5] (0.02-0.03°C/year). The results of calculations obtained in [31] demonstrate that the global surface temperature at the beginning of the 21st century (2000–2009) could also exceed the Mid-Holocene temperatures by 0.3°C.

The calculated snow depth for the beginning of the 21st century exceeds that of the middle of the 20th century and the Mid-Holocene. The underestimation of the snow depth in the Mid-Holocene in CLIMBER-2 calculations is associated with an underestimation of precipitation by the model. This underestimation leads to the disappearance of the snow cover in May in the Mid-Holocene and, therefore, to earlier permafrost thawing, according to the simulations.

The maximum thawing depth is dependent on the air temperature in the warmest season and is estimated to be about 1.1 m at the beginning of the 21st century and in the Mid-Holocene. Higher autumn temperatures at the beginning of the 21st century compared to the Mid-Holocene lead to later soil freezing during this period. The model estimates of the maximum thaw depth do not correlate with the snow cover thickness. This is due to the numerical scheme specifications, which consider the major amount of heat loss during the thawing of ice-containing soil to be consumed by the ice-water phase transition. The energy amount needed for heating the ice-saturated soils depends on the initial soil temperature, but even for low-temperature permafrost it does not exceed 5-10% of the energy amount required for ice melting. The snow cover has a heat insulating effect on the permafrost in winter.

The soil temperature of the upper horizons at the beginning of the 21st century exceeds the temperature in the middle of the 20th century and in the Mid-Holocene (Figure 1d) due to the higher average annual air temperature and the greater snow cover thickness in the 21st century. Thus, the obtained estimates of the soil temperature at a depth of 10 m are about -5°C for the beginning of the 21st century, which is consistent with observations [32]. For all three periods simulated, the minimum annual temperature values for 10-m depth are reached in summers due to thermal inertia, which creates a phase shift between the temperature fluctuations at different depths.

Figure 2 presents the results of the calculation of the average annual soil temperature for Yamal region over the past 6 thousand years. The soil temperature in the upper 100-150 m of the permafrost in the Mid-Holocene was about -6°C – -7°C. Starting from about 4000 thousand years ago, the temperature of the soil drops to -7°C – -8°C in the upper 20 m and to -6°C at a depth of 100 m.

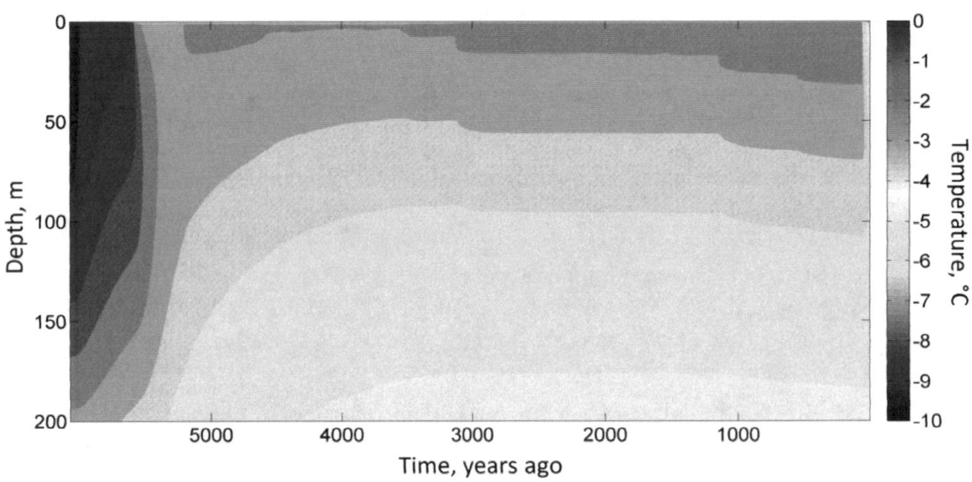

Figure 2. Model average annual temperature of the soil in the upper 200 m over the past 6 thousand years for Yamal region.

At the same time, the trend of soil temperature decreases with depth. A significant increase in the soil temperatures in the upper layers of 50-60 m was obtained for the modern climatic conditions (Figure 2).

Figure 3 shows the soil temperature profiles for the Mid-Holocene, mid-20th and early 21st centuries. According to the results obtained, the current near-surface warming in the north of Western Siberia exceeds the warming in the Mid-Holocene about 6 thousand years ago. The current soil temperature in the upper few tens of meters exceeds the soil temperature in the Mid-Holocene, and may exceed the threshold temperature for the dissociation of methane hydrates. In addition, near-surface warming, particularly in Yamal region, also contributes to the permafrost degradation. As a result, this can lead to the dissociation of gas hydrates and gas emissions with the destruction of frozen soil [13]. One can also see that the permafrost thickness in the region simulated decreases from 500 m about 6 thousand years ago to 430 m under the current climatic conditions.

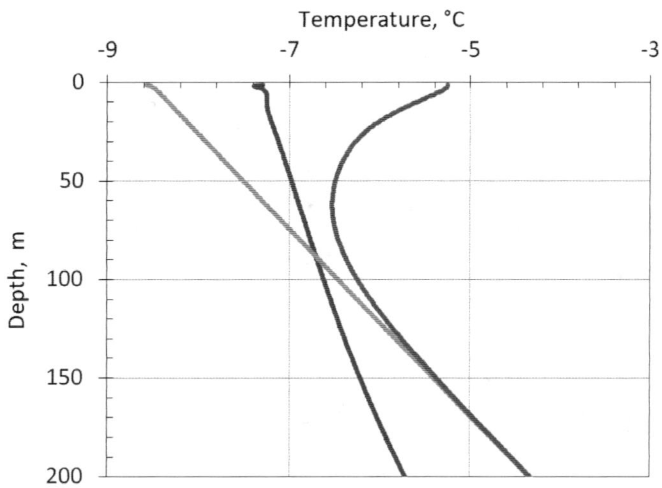

Figure 3. Model profiles of the average annual soil temperature for Mid-Holocene (blue curve), mid-20th century (green curve), and the beginning of the 21st century (red curve).

Based on the model estimates for the vertical temperature and pressure distribution in the permafrost, the temperature and pressure conditions for the equilibrium state of methane hydrates were calculated, and the depth of the upper and lower boundaries of the stability zone was determined. It was found that the upper boundary depth of the stability zone of methane hydrates in the Mid-Holocene drops from 194 m to 220 m as a result of warming during this period. With a subsequent decrease in the air temperature, the model depth of the stability zone boundary remains almost unchanged and is located within the frozen strata. The results of the calculation of the thermobaric conditions show that at the present time stable methane hydrates may exist at depths of more than 200-250 m in the frozen soil masses of Yamal, together with relict methane hydrates that persist at depths of up to 150 m [9] at negative temperatures below -7°C over the past 6 thousand years, including the Mid-Holocene. The disruptions of stability of methane hydrates near the surface may be a result of the current climate change.

4. Conclusions

The above analysis of the calculation results of the global climate model CLIMBER-2, the CERA-20C reanalysis and the calculation results of the permafrost thermal regime revealed that the current warming may exceed the warming in the Mid-Holocene, particularly in Yamal region. The average annual near-surface temperature in this region at the beginning of the 21st century can be higher than the temperature in the Mid-Holocene due to warmer winters and autumns. According to the results of the calculations, modern climatic changes lead to an increase in permafrost temperatures. At the same time, soil temperatures in the upper horizons (several tens of meters below the surface) can exceed the equilibrium temperature of metastable methane hydrates. This affects the stability of the relict hydrates of the continental cryolithozone gases which are located above the present-day stability zone boundary. Near-surface warming also contributes to reducing the strength of the frozen soil. As a result, this leads to frozen soil destruction and dissociation of the gas hydrates which "survived", in particular, the Mid-Holocene warming.

Acknowledgments
This study was supported by the Ministry of Science and Higher Education of the Russian Federation, agreement no. 14.616.21.0082 (RFMEFI61617X0082).

References
[1] Mokhov I I and Eliseev A V 2012 *Dokl. Earth Sci.* **443** 532–36
[2] *Climate Change 2013: The Physical Science Basis. Contribution of Working Group I to the Fifth Assessment Report of the Intergovernmental Panel on Climate Change* ed T Stocker *et al* (New York: Cambridge University Press) pp 659–740
[3] Mokhov I I 2015 *Herald Russ. Acad. Sci.* **85** 265-71
[4] Arzhanov M M and Mokhov I I 2013 *Dokl. Earth Sci.* **449** 319–23
[5] Pavlov A V 2008 *Permafrost monitoring (*Novosibirsk: Geo) p 230 (In Russian)
[6] Kitover D C, Van Balen R T, Vandenberghe J, Roche D M and Renssen H 2016 *Permafr. Periglac. Process.* **27** 31-42
[7] Slater A G and Lawrence D M 2013 *J. Climate* **26** 5608-23
[8] Arzhanov M M, Eliseev A V, Klimenko V V, Mokhov I I and Tereshin A G 2012 *Izv. Atmospheric and Oceanic Physics.* **48** 573-84
[9] Ingolfsson O, Moller P and Lokrantz H 2008 *Polar Res.* **27** 152-61
[10] Kleman J, Fastook J, Ebert K, Nilsson J and Caballero R 2013 *Clim. Past.* **9** 2365-78
[11] Lambeck K, Purcell A, Funder S, Kjær K H, Larsen E and Poller P 2006 *Boreas* **35** 539-75
[12] Arzhanov M M, Mokhov I I and Denisov S N 2016 *Dokl. Earth Sci.* **468** 616-18
[13] Arzhanov M M and Mokhov I I 2017 *Dokl. Earth Sci.* **476** 1163-67
[14] Arzhanov M M, Malakhova V V and Mokhov I I 2018 *Dokl. Earth Sci.* **480** 826-30
[15] Flint R F 1957 *Glacial and Pleistocene Geology* (New York: John Wiley and Son) p 553
[16] Chuvilin E M and Davletshina D 2018 *Geosciences* **8** 467
[17] Fray N, Marboeuf U, Brissaud O and Schmitt B 2010 *J. Chem. Eng. Data* **55** 5101–08
[18] Leibman M O, Kizyakov A I, Plekhanov A V and Streletskaya I D 2014 *Geogr., Environ., Sustainabil.* **7** 68-79
[19] Bogoyavlensky V I 2014 *Oil Drill.* **9** 13–8 (In Russian)
[20] Kizyakov A, Zimin M,Sonyushkin A, Dvornikov Yu, Khomutov A and Leibman M 2017 *Remote Sens.* **9** 1023
[21] Yakushev V S, Perlova E V, Makhonina N A, Chuvilin E M and Kozlova E V 2003 *Khimicheskiy Zhurnal* **3** 80-90 (In Russian)
[22] Malakhova V V and Eliseev A V 2017 *Glob. Planet. Change* **157** 18–25
[23] Malakhova V V and Eliseev A V 2018 *Ice and Snow* **58** 231-42
[24] Ganopolski A, Calov R and Claussen M 2010 *Clim. Past.* **6** 229-44
[25] Hersbach H, Poli P and Dee D P 2015 *The observation feedback archive for the ICOADS and ISPD data sets* (ECMWF: Shinfield Park)
[26] Bokuchava D and Semenov V A 2018 *Fundamental and Applied Climatology* **1** 28-51
[27] Chuvilin E M, Yakushev V S and Perlova E V 1998 *Polarforschung* **68** 215-19
[28] Davies J H 2013 *Geochem. Geophyst. Geosyst.* **14** 4608-22
[29] Moridis G J 2003 *Society of Petroleum Engineers Journal* **32** 359-70
[30] Arslanov Kh A, Savelieva L, Gey N A, Klimanov V A, Chernov S B, Chernova G M, Kuzmin G V, Tertychnaya T V, Subetto D and Denisenkov V P 1999 *Radiocarbon* **41** 25-45
[31] Marcott S A, Shakun J D, Clark P U and Mix A C 2013 *Science* **339** 1198-1201

[32] Babkina E A, Leibman M O, Dvornikov Yu A, Fakashchuk N Yu, Khairullin R R and Khomutov A V 2019 *Russ. Meteorol. Hydrol.* **44** 283-90

IOP Conf. Series: Earth and Environmental Science **386** (2019) 012020 doi:10.1088/1755-1315/386/1/012020

Numerical simulation of modern methane emissions and their influence on climate variability and gas composition of the Arctic atmosphere

Maria Cherepova and Sergei Smyshlyaev

Russian State Hydrometeorological University

79 Voronezhskaya str., Saint Petersburg 192007, Russia

Abstract. The impact of Arctic methane hydrate emissions on the gas composition and climate is studied by using a chemistry-climate model. Model runs are carried out using two methane emission scenarios (with Arctic methane emissions increased by 5 and 10 times, respectively, relative to the reference level of emissions of 10 Tg/year). The methane hydrate emissions have a strong impact on the concentrations of methane and other gases. Temperature distributions in the polar and nonpolar regions in the troposphere and stratosphere are obtained. The influence on the climate variability and gas composition of the atmosphere is estimated.

1. Introduction

Now much attention has been given to the impact of non-carbon dioxide greenhouse gases on climate change. Atmospheric methane, being one of the most important greenhouse gases, on the other hand, plays an important role in the chemical balance of the lower and middle atmosphere. Methane determines tropospheric ozone formation and affects hydroxyl radical content, which is the main oxidizer of the troposphere [1-5]. In the stratosphere, methane is a major source of water vapor and, therefore, affects the catalytic ozone destruction [1-2].

An important feature of CH_4 is that it is not formed in the atmosphere and the only source of atmospheric methane is its surface emissions [6]. At the same time, methane surface emissions are determined by both natural processes and anthropogenic factors. Considering the climatic effects of methane, it is necessary to take into account that the intensity of natural biogenic methane emissions greatly depends on climate and promptly responds to its changes, which leads to the formation of feedbacks between climate changes and the methane content in the atmosphere [7]. In particular, in the Arctic zone, where accumulated a huge amount of methane in the permafrost, the continental shelf and the gas hydrates and where climate change is faster than in the other regions, the melting of permafrost and the destruction of gas hydrates due to climate change can lead to release of large amounts of methane. This can lead to further strengthening of climate change which, in turn, can affect the increase in methane emissions from melting permafrost and the destruction of gas hydrates [8].

Gas hydrates are unstable with increasing temperature and lowering pressure, which can lead to the release of a significant part of the methane accumulated in them during climate warming [9]. As a result, the amount of methane emitted from gas hydrates may become comparable to emissions from other globally significant sources. In the future, with increasing temperature, the role of Arctic gas hydrates as a source of methane on a global scale will only increase, which necessitates theoretical and experimental studies to assess the power of methane emissions from gas hydrates and the consequences of their impact on climate change and atmospheric composition. At the same time, due to the long lifetime in the atmosphere (on average, about 10 years), methane variability affects the radiative and chemical properties of the atmosphere not only in the source zone, but also in other regions and altitudes, up to the impact on global effects.

Volodin E.M. [10] used a general circulation model developed at the Institute of Numerical Mathematics of the Russian Academy of Sciences (INM RAS) without chemistry included to study the influence of an additional source of Arctic methane on the interhemispheric asymmetry of the CH_4 concentration and climate [10]. The results of numerical experiments with additional methane emissions of 50 or 100 Tg/year in the northern high latitudes demonstrate that the differences in the methane concentrations between the Arctic and the Southern Hemisphere are 14 and 17%, respectively, a value that is higher than the observed 10% [10]. The spatial distribution of the warming generated by an additional methane emission is determined only by the average amount of methane contained in the atmosphere and has almost no dependence on where the source of methane lies. An additional methane emission of 4000 Tg/year causes the global mean surface temperature to rise by about 1.5° [10]. This work is an extension of [10] with the chemistry-climate interaction added.

In addition to the chemical and climatic feedbacks associated with methane emissions increase into the atmosphere in the Arctic zone as a result of climate change, there are also purely chemical feedbacks formed due to the fact that the intensive melting of methane hydrates and the growth of their content in these zones can greatly reduce hydroxyl radicals, because methane is one of its main destroyers. As a result, the methane content may increase even more, since the main destruction of atmospheric methane occurs as a result of chemical interaction with OH [1,2]. On the other hand, the effect of reduction of hydroxyl radicals on methane itself can have an impact on other gases, since the self-cleaning ability of the atmosphere is mainly determined by the content of hydroxyl radicals, OH.

Numerical simulations using chemistry-climate models (CCMs) that take into account the interaction of physical and chemical processes in the troposphere and stratosphere are best suited to assess the impact of potential increases in methane emissions on changes in the temperature regime and gas composition of the atmosphere. In this paper, a chemistry-climate model of the lower and middle atmosphere is used to estimate the effect of possible enhanced methane emissions from Arctic gas hydrates on changes in ozone and other atmospheric gases and the temperature of the troposphere and stratosphere.

2. Model description and numerical experiment structure

This study was carried out using a numerical chemistry-climate model of INM RAS and Russian State Hydrometeorological University (RSHU) [11]. The INM RAS – RSHU CCM covers the entire globe with a horizontal resolution of 4 × 5 degrees of latitude-longitude and from the earth's surface up to the mesopause. The model chemical scheme takes into account the variability of 74 minor gases, among which there are both organic and inorganic compounds, long-lived or short-lived. The model takes into account advection processes, chemical transformations, turbulence, dry and wet deposition, and phase transitions. The chemical scheme includes 174 chemical reactions and 51 photodissociation processes. Surface emissions of modeled gases are based on EDGAR, GISS NASA, GEIA, and MEGAN databases. In particular, the biogenic (termites, wetlands, decomposition of organic matter in oceans, forest fires) and anthropogenic (industrial emissions, transport, oil and gas production, energy, agricultural activity and its waste, biomass burning) methane surface emissions are used to set the model lower boundary condition.

To estimate the impact of possible increased methane emissions from Arctic gas hydrates on the gas composition and structure of the atmosphere, the INM RAS – RSHU CCM was run for 15 years starting in 2015. At the same time, methane emissions from Arctic gas hydrates were increased by 5

IOP Conf. Series: Earth and Environmental Science **386** (2019) 012020 doi:10.1088/1755-1315/386/1/012020

and 10 times compared to a background level of 10 Tg/year. The results of the calculations averaged over the period from 11 to 15 years of the model calculations were compared with the calculations based on the baseline scenario in which methane emissions from gas hydrates remained at the present level.

3. The results

Figure 1 demonstrates the percentage change in the methane concentration due to increased Arctic methane emissions by 5 and 10 times. Tropospheric increases in methane are consistent with enhanced Arctic emissions. At the same time, the largest increase, as expected, is revealed in the Arctic, i.e. near the zone of enhanced emissions. In the upper stratosphere, there is also an increase in the methane content, in order of magnitude corresponding to the tropospheric increase in the methane content. However, in the lower and middle stratosphere methane does not increase as much as below and above. Moreover, in the polar regions of the middle stratosphere there is the smallest increase in the methane content, and in the Antarctic in spring there is even a decrease in the methane content. The latter effect points to complex nonlinear effects of methane emissions that are observed across the globe, and the most unusual results are recorded in the opposite part of Earth's atmosphere.

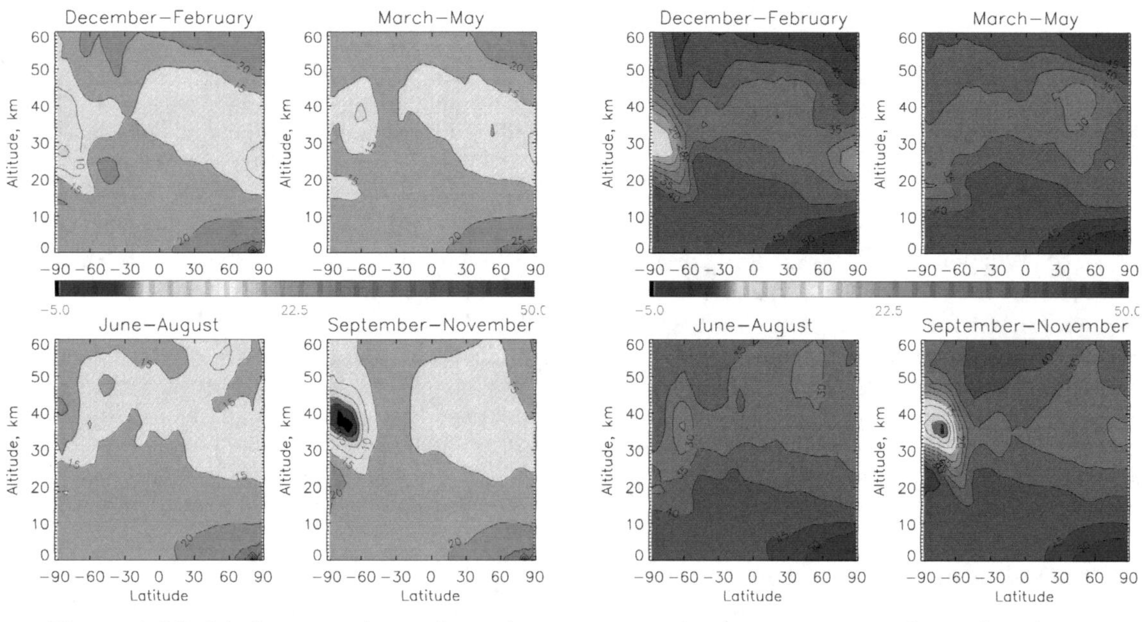

Figure 1. Modeled percent change in methane concentration in response to enhanced methane emission from Arctic gas hydrates by (a) 5 times, (b) 10 times.

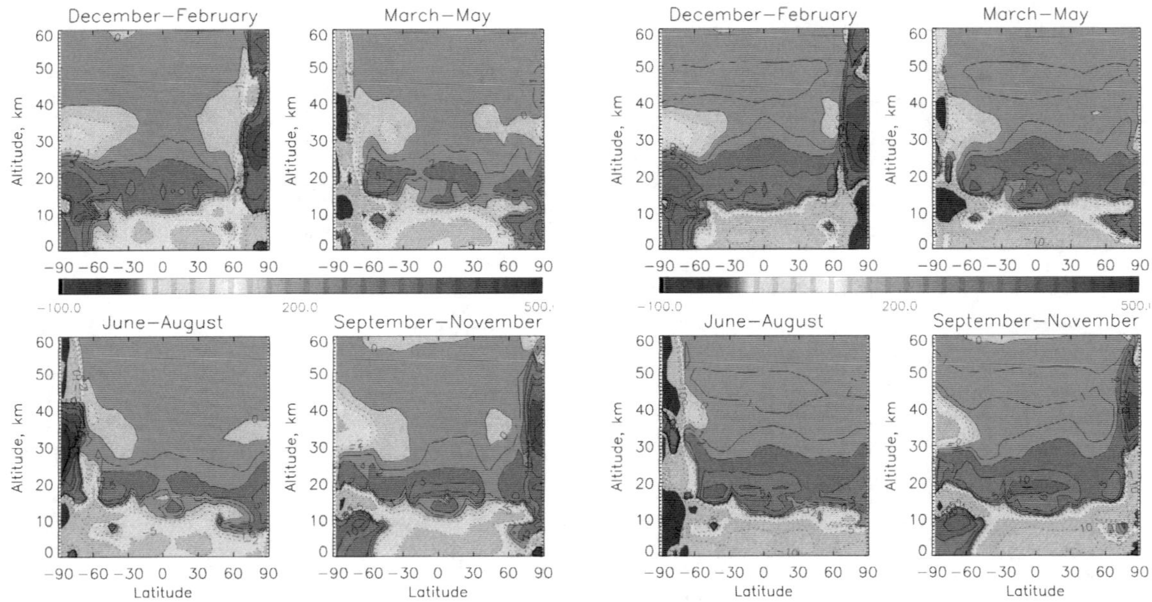

Figure 2. Modeled percent change of hydroxil radical OH concentration in response to enhanced methane emission from Arctic gas hydrates by (a) 5 times, (b) 10 times.

Hydroxyl radical OH is the main chemical methane loss in the atmosphere and, on the other hand, OH is largely destroyed by methane. As shown in Fig.2, with a rise in Arctic methane emissions as a result of possible melting of gas hydrates due to climate changes, the tropospheric hydroxyl radical OH significantly reduced in the troposphere, which corresponds to the known fact of destruction of OH molecules by methane molecules. At the same time, in the stratosphere, on the contrary, there is an increase in the hydroxyl content, which is probably a consequence of an increase in the stratospheric water vapor, which is a source of hydroxyl radicals. In the troposphere, the influence of methane on the water vapor content is insignificant due to the dominant role of other sources of water vapor.

In the polar regions during the polar night, when there is no formation of hydroxyl radicals under the influence of solar radiation, the influence of methane on the OH content is very different from other latitudes. At the same time, the tropospheric effects of reducing hydroxyl radicals are amplified during the polar night, as the destruction of methane is preserved, and in the stratosphere the effects differ in the Northern and Southern Hemispheres. In the Southern Hemisphere the content of hydroxyl radicals is mainly reduced during the polar night both in the troposphere and stratosphere, which can be explained by the destruction of hydroxyl methane and in the absence of solar radiation, and lack of water vapor production from methane, whereas in the Northern Hemisphere in the stratosphere an increase in methane is often observed, indicating the influence of complex chemical-dynamic feedbacks. It should be noted, however, that the concentration of hydroxyl radicals during the polar night is negligible and OH has little effect on the atmospheric chemistry in the absence of the sun.

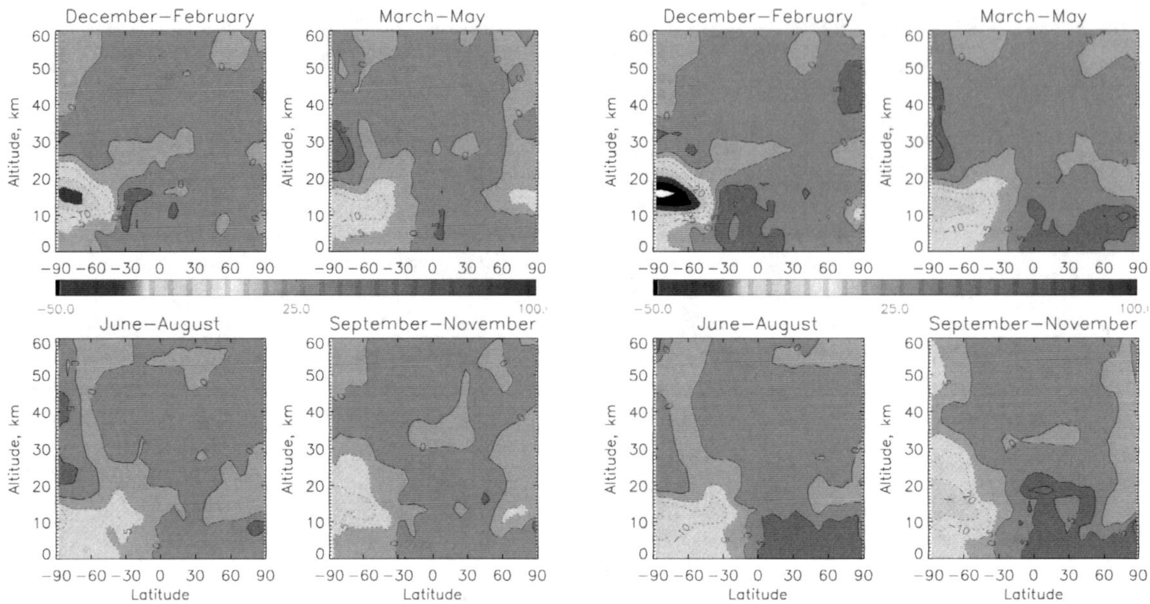

Figure 3. Modeled percent change in ozone concentration in response to enhanced methane emission from Arctic gas hydrates by (a) 5 times, (b)10 times.

Methane contributes to the production of tropospheric ozone. Thus, its increase as a result of the melting of Arctic hydrates should contribute to an increase in the ozone content in the troposphere. The results of the model experiments presented in Fig.3 demonstrate that this effect is available, and it becomes most noticeable with a tenfold increase in the methane emissions from Arctic gas hydrates. However, it is noticeable that the effect of methane increase on the tropospheric ozone is not linear. It is expressed, on the one hand, in the absence of geographical consistence between the regions of methane and ozone increase (compare to Fig.1), and, on the other hand, in the absence of a significant change in the ozone reaction between scenarios with fifth and tenth increase in the methane emissions. This may be due to the influence of other gases, in particular, nitrogen oxides, on the ozone production due to methane oxidation. The content of nitrogen oxides in the model experiments varies little with the increase in methane fluxes from Arctic gas hydrates. Therefore, the effect of methane on the tropospheric ozone is less than would be expected on the basis of a significant increase in the Arctic methane emissions.

In the stratosphere, the impact of Arctic methane emissions on ozone is negligible, with the exception of the lower Antarctic stratosphere. In most of the stratosphere, methane has little effect on ozone, despite increases in the water vapor and hydroxyl radicals (Fig.2). This effect is a consequence of the determining role of nitrogen catalytic cycles of the stratospheric ozone depletion, and the increase in hydroxyl radicals leads more to the binding of nitrogen radicals than to an increase in ozone depletion in hydrogen catalytic cycles. In the lower Antarctic stratosphere, where the formation of Antarctic "ozone holes" is developing, the increase in methane content is likely to contribute to even greater denitrification of the polar stratosphere by binding nitrogen gases, which increases the intensity of ozone depletion in chlorine and bromine cycles. Accordingly, unlike other regions of the stratosphere, the ozone content in Antarctica is significantly reduced with enhanced methane emissions in the Arctic.

Figure 4 shows the temperature percentage change due to an increase in the Arctic methane emissions by 5 and 10 times. There are some areas where there is temperature decrease or increase. In Antarctic spring and summer in the lower stratosphere (between altitudes of 15 km and 45 km and between altitudes of 11 km and 21 km) Arctic methane emissions increase by 5 times is observed. The

IOP Conf. Series: Earth and Environmental Science **386** (2019) 012020 doi:10.1088/1755-1315/386/1/012020

other scenario when the Arctic emission increase is by 10 times shows more intense and wide temperature decrease. A temperature decrease by 2 % is observed from ~ 10 km to 27 km in winter and from 12 to 44 km in spring. Besides, it is important to note that in spring and summer a temperature increase is observed. It is situated about 6-7 km higher than the temperature decrease.

In the Arctic stratosphere, the temperature is increasing from September to November. In the troposphere, the temperature is stable (no decrease and no increase). Thus, when setting scenarios for increasing methane hydrate emissions, an increase in temperature is observed in relation to the current state (at the current level of emissions). It should also be noted that warming in the stratosphere in winter with an increase in methane hydrate emissions is 10 times more intense than the increase in methane hydrate emissions by 5 times - both in terms of the temperature change and spatial resolution - the focus of a positive temperature increase; the summer period is characterized by a decrease in the size and intensity of the tropospheric focus of warming.

In the Antarctic, the temperature decrease is more intense when the Arctic methane emissions increase by 10 times than by 5 times. The Arctic decreasing temperature is observed from September to February with the Arctic methane emissions increase by 10 times. If the Arctic methane emissions increase by 5 times, the decrease is observed from September to November.

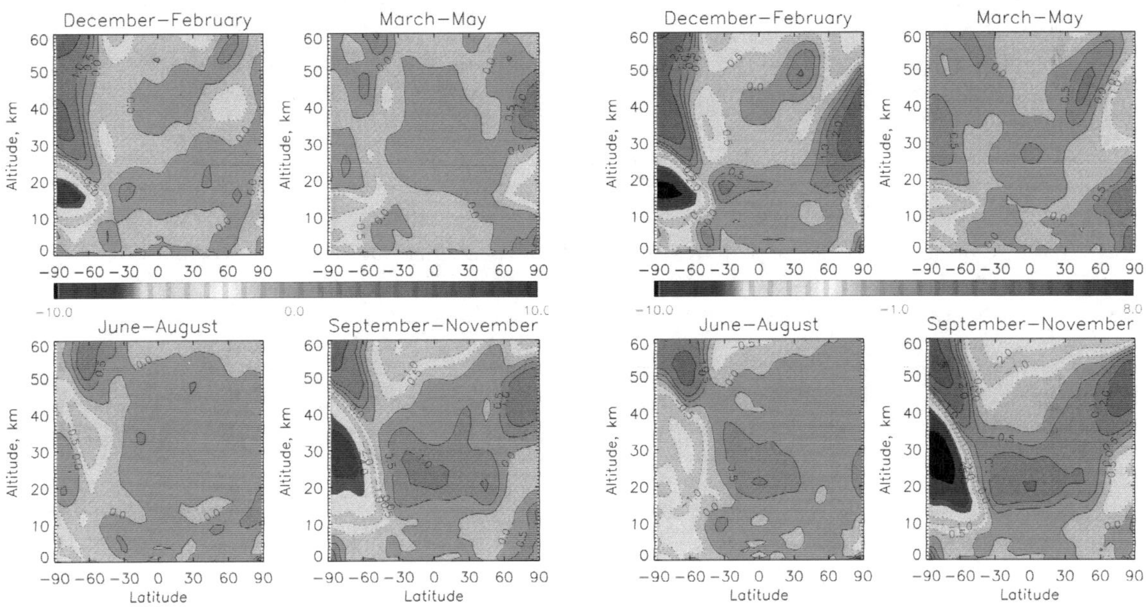

Figure 4. Modeled percent change of temperature in response to enhanced methane emission from Arctic gas hydrates by (a)5 times, (b) 10 times.

4. Conclusions

Numerical experiments with a chemistry-climatic model of the lower and middle atmosphere were carried out to assess the effect of a probable increase in methane emissions from the melting of Arctic gas hydrates due to climate warming on the gas composition and temperature of the atmosphere.

The results of the numerical experiments have shown that the increase in methane emissions affects the composition and structure of the atmosphere not only in the emission zone, i.e. in the Arctic, but also in some other regions of the Globe. The effects in the troposphere and stratosphere are different, and the methane emissions may have a great impact on the atmospheric gas composition and temperature, especially in the polar regions. Moreover, in Antarctica, i.e. in the opposite region of the Globe, changes in both the gas composition and the temperature of the atmosphere, which are most significant and different from those in the other regions, are recoded. The results of the model experiments demonstrate the nonlinearity of the effect of methane emissions from the Arctic gas hydrates on the gas composition and temperature regime of the atmosphere.

Acknowledgments
This research was supported by the Russian Science Foundation under scientific project 19-17-00198 and by the Russian Foundation of Basic Research (projects 17-05-01277).

References

[1] Jacob J D 1999 *Introduction to atmospheric chemistry* (Princeton: Princeton University Press) p 264
[2] Murry L Salby 2012 *Physics of the atmosphere and climate* (Cambrige University Press) p 717
[3] Seinfeld J H and Pandis S N 1998 *Atmos1pheric Chemistry and Physics. From Air Pollution to Climate Change* (New Jersey: John Wiley and Sons) p 1326
[4] Isaksen I S A, Gauss M, Myhre G et al 2011 Strong atmospheric chemistry feedback to climate warming from Arctic methane emissions *Global Biogeochem. Cycles* 25 GB2002 doi:10.1029/2010GB003845)
[5] Brasseur G P, Orlando J J, Tyndal G S 1999 Atmospheric Chemistry and Global Change (Oxford: Oxford University Press) p 688
[6] Bazhin N M 2000 Methane in the atmosphere *Soros Educational J.* **6** 3 52–7
[7] Ginzburg A S, Vinogradova A A, Fedorova E I 2011 Some features of seasonal variations in the methane content in the atmosphere over northern Eurasia *Izvestiya. Atmospheric and Oceanic Physics* **47** 1 45-58
[8] Yakutseni V P 2013 Gas hydrates – unconventional gas sourses, their formation, propetis, distribution ang geological resourses *Neftegazovaya Geologiya. Teoriya I Praktika* **8** 4 http://www.ngtp.ru/rub/9/50_2013.pdf
[9] A Dickens G R, Quinby-Hunt M S 1994 Methane hydrate stability in seawater *Geophys. Res. Lett.* **21** 19 2115–18
[10] Volodin E M 2015 Influence of methane sources in Northern hemisphere high latitudes on the interhemispheric asymmetry of its atmospheric concentration and climate *Izvestiya. Atmospheric and oceanic physic* **51** 3 287–294
[11] Galin V Ya, Smyshlyaev S P, Volodin E M 2007 Combined chemistry-climate model of the atmosphere *Izvestiya. Atmospheric and Oceanic Physics* **43** 4 399-412

IOP Conf. Series: Earth and Environmental Science **386** (2019) 012021 doi:10.1088/1755-1315/386/1/012021

Simulation of the influence of the ocean and the El-Niño – Southern Oscillation phenomenon on the structure and composition of the atmosphere

A R Jakovlev and S P Smyshlyaev

Russian State Hydrometeorology University, Voronejskaya, 79, St. Petersburg, 192007, Russia
E-mail: endrick@yandex.ru; endrusj@rambler.ru

Abstract. The influence of El-Niño and La-Niña on the Arctic stratosphere is studied. Sea surface temperature (SST), potential vorticity, air temperature, ozone mixing ratio, and column ozone ERA Interim data are analyzed for 1997, 1999, and 2016. It is shown that El-Niño leads to a series of sudden stratospheric warmings and, consequently, to instability of the polar vortex and to an increase in column ozone, whereas La-Niña leads to a stable polar vortex and low temperatures in the stratosphere and, therefore, to a decrease in column ozone. The influence of SST and CO_2 levels on the air temperature in the troposphere and stratosphere, and ozone content from 1980 to 2015 is also studied. It is shown that there is warming of the ocean and the troposphere, as well as cooling of the stratosphere. It is also shown that the changes in air temperature in the troposphere are mainly due to changes in SST, whereas for the stratosphere the impact of CO2 variability prevails. It is also shown that ozone content varies little.

1. Introduction

In the Earth's atmosphere there are a lot of processes caused by the interaction of the atmosphere and ocean. Among them, one of the most known events is the phenomenon of El-Niño, representing the current in the eastern part of the tropical region of the Pacific Ocean. The El-Niño is a regional phenomenon, and it leads to a rise in the temperature of the ocean top layer, but the influence of this phenomenon is revealed on a global scale [4]. The El-Niño atmospheric component is the Southern Oscillation [1,4] at which there are fluctuations of atmospheric pressure between the eastern and western parts of the Pacific Ocean. Because of low pressure in the eastern part of the Pacific Ocean, there begins deep convection and easing of trade winds.

La-Niña represents a cold phase of this phenomenon, at which there is a temperature downturn in the ocean top layer [4]. Thus, the atmospheric pressure increases, and the convective processes stop.

An important dynamic process occurring in a polar stratosphere is the circumpolar vortex, which has an essential influence on the temperature and gas structure of the stratosphere of polar areas. The stability of a circumpolar vortex is defined by the stability of the carry zone of air weights, whereas the meridional carry leads to instability of this vortex [5]. The stability of a vortex also influences the maintenance of ozone and temperature: under the theory a circumpolar vortex exists longer, there will be an ozone hole, and below there will be temperature in a polar stratosphere [7].

The temperature of a polar bottom stratosphere is an important factor of reduction in the maintenance of ozone. The cold winter season with 1996 for 1997, which was a consequence of lowered distribution of the wave activity, led to strengthening of the ozone destruction due to heterogeneous activation of the chloric and bromic gases and to easing movement of the planetary waves, whereas the warm winter season with 1998 for 1999 was characterized by a strengthened distribution of the wave activity [6]. The temperature in a stratospheric polar vortex defines the volume of stratospheric polar clouds, on which the force of destruction of ozone also depends. Easing of planetary waves can lead to the downturn of polar temperatures and easing of ozone carry, which can lead to tropical tropopause heating [6, 8].

Changes in the sea surface temperature (SST) leading to the meridional gradient of SST can have an essential influence on the stratosphere [7.] The SST meridional gradient has a stronger influence on the zone circulation than the SST global variability. This means that the SST gradient has a strong influence on the circumpolar vortex and the maintenance of ozone in the northern hemisphere. Also, it has been revealed with the assumption that reduction in the maintenance of ozone can cause the phenomenon of El-Niño - southern fluctuation, which causes redistribution of the heat streams in the ocean and atmosphere, whereas increase in the maintenance of ozone can provoke La-Niña [9]. It can lead to the transition of energy of ultra-violet radiation in the ocean, increasing its heat maintenance. Simulation has shown that the influence of the El-Niño on the parameters of the Europe ground climate is carried out through the stratosphere [10].

Increase in the CO_2 level leads to the fact that infrared radiation of the Earth is late an atmosphere, which leads to increase of average by the ground temperatures of the air [2,3]. As a result of the ocean surface heating, there are gradients of temperature and density which lead to various ocean currents. These currents have an essential influence on the weather and a climate of separate regions, and of the Earth as a whole [1].

In this article, results of two studies connected with the influence of oceanic processes on the atmosphere are presented. In the first study, data of reanalysis ERA Interim for 1997, 1999, 2010, 2011, 2015, and 2016 are analyzed. In the second one, results of numerical experiments by means of a chemistry-climate model (CCM) and data of reanalysis for the period with 1980 for 2015 are analyzed. The purpose of this work is revealing the interrelation of oceanic processes (the El-Niño and La-Niña) and structures of the bottom and average atmosphere.

2 Materials and Methods

The CCM model represents a combination of two parts: a dynamic part developed at the Institute of Numerical Mathematics of the Russian Academy of Sciences (INM RAS) and a chemical part developed at the Russian State Hydrometeorological University (RSHU) [11,12].

The resolution of this model is -5^0 longitude and 4^0 latitude. The scope on longitude is from 185^0 W to 180^0 E, and on latitude, from 88^0 S to 88^0 N. On the vertical the number of levels of the model is equal to 39 (from 0 to 88 km or from the surface to a level of 0.003 hPa). Thus, the results of the simulation are interpolated in an isobaric system of coordinates on 31 standard isobaric surface. The quantity of units in the model is 72 on longitude and 45 on latitude. The time step is 6 minutes. In order to account for the temporal variability of the influencing factors, the external parameters are updated every six hours [11,12].

The chemical part of the model describes changes in the concentration of 74 basic gases which influence the maintenance of ozone in the atmosphere. The model solves the system of equations of carry of gas impurity in view of the photochemical interaction for a long-living gas impurity. For short-lived gases, the equations are solved not taking into account the atmospheric carry [11,12].

The model considers the interaction between the chemical and physical processes at each time step. The solar radiation streams first pay off. Thus, the current maintenance of ozone and the molecular dispersion of light are considered. These streams are used for the calculation of the photodissociation speeds of gases on which count the speeds of photochemical formation of gases. Thus, for the calculation the temperature values calculated in the dynamic part of the model are used. These speeds of chemical reactions allow one to model the evolution of ozone and other gases in the atmosphere [11,12].

The basic method of the research is numerical simulation of atmospheric processes. As mentioned earlier, changes in the basic meteorological parameters (temperature and humidity of the air, atmospheric pressure, density of the air and speed of the wind) are described by the equations of thermodynamics, preservation water pair, the conditions of air, the indissolubility and movement of the atmosphere. These equations are solved only by numerical methods [11.] The non-adiabatic processes connected with the solar radiation, turbulence, convection and the influence of a spreading surface on the atmosphere, are considered by means of parameterization. The evolution of gas impurities is defined by means of the equations of their carry, which are also, certainly, solved by different methods. Thus, to solve the given problem the speeds of chemical reactions [11,12] are defined. Between the dynamic and chemical parts of the model there is an exchange of settlement data, which allows considering the influence of all these processes on each other [11,12].

The initial data for the meteorological parameters were taken from CIRA-86 and NCEP [14,15]. The ozone distribution data were taken from TOMS satellite data [16]. To study the influence of the ocean on the atmosphere, various experiments were performed. In each of these experiments, different data on the ocean surface temperature and the ice cover area, namely, data Met Office [17], data ERA Interim [13], and ERA20C and ERA5 [13] were used.

These results of the simulation, which could be compared with processes observed during simulation in the atmosphere, were used given reanalysis data about the temperature, wind speeds, the concentration and total column of ozone: MERRA, MERRA2 [18], ERA Interim, ERA20Century, ERA5 [19], JRA [20], with which the simulation results were compared. Thus, in MERRA, ERA, and JRA there are 25, 21, and 21 vertical levels, respectively. Also, data on the total column of ozone were used. Reanalysis data led to a simulation grid (72 units on longitude, 45 units on breadth, and 31 level on the vertical) for a comparison with the simulation results. The simulation was carried out for the period with 1979 on 2015. For reanalysis data MERRA the period with 1979 for 2015, for MERRA2 with 1980 on 2016, for ERA Interim and JRA with 1979 on 2015, for ERA20Century with 1979 on 2010, and for ERA5 with 2010 on 2016 were chosen.

3. Results and Discussion

Figure 1 demonstrates the SST anomalies for December-March of 1996-1997, 1998-1999, and 2015-2016 when different phases of the Southern Oscillation [21,22] took place. Polar projections of potential vorticity (PV) at a potential temperature level of 500 K were also presented for these years according to ERA Interim reanalysis data [19]. In 1997 and 1999 there is an area of negative SST anomalies (up to-2 K) in the equatorial part of the Pacific Ocean, whereas in 2016 there is an area of positive SST anomalies (up to 2 K). For all three years, the polar vortex is stable in the first half of the winter of the Northern Hemisphere, but in the second half of the winter the polar vortex continues to be stable for the years with negative SST anomalies, which corresponds to the La Niña phenomenon, and is getting unstable for the years with positive temperature anomalies, which corresponds to the El Niño phenomenon. In general, it can be noted that the years with El Niño correspond to the unstable polar vortex in the second half of the winter of the Northern Hemisphere and the years with La Niña facilitate the formation of a stable polar vortex.

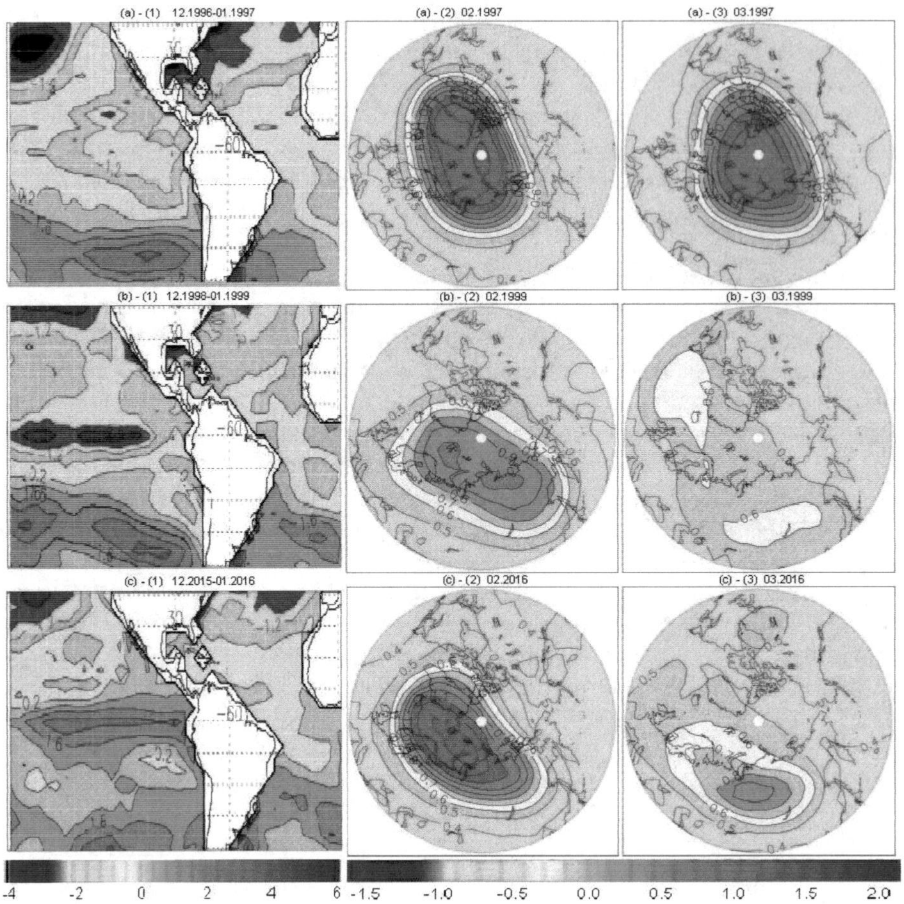

Figure 1. ERA Interim reanalysis data of SST anomalies (K) at 40^0 S - 40^0 N and 180^0 W - 0^0 for December-January of 1996-1997 (a - 1), 1998-1999 (b - 1), 2015-2016 (c - 1), and monthly average potential vorticity (10^4 K m^2 kg^{-1} s^{-1}) for northern polar region (40^0 N - 90^0 N) at 24.5 km (30 hPa) for February and March of 1997 (a - 2 and 3), 1999 (b - 2 and 3), and 2016 (c - 2 and 3).

In Figure 2, polar projections of the air temperature at the level of 30 hPa and vertical zonal average annual profiles of temperature for 1997, 1999, and 2016 according to ERA Interim reanalysis [19] are shown. In 1997, when there was the La Niña phenomenon, the low temperature region in the Arctic stratosphere was larger and lasted longer than in the years with the neutral phase (1999) and the El Niño phenomenon (2016). Thus, in these years low temperatures interrupt a series of sudden stratospheric warming (SSW), which leads to instability of this area, apparently on vertical profiles [22].

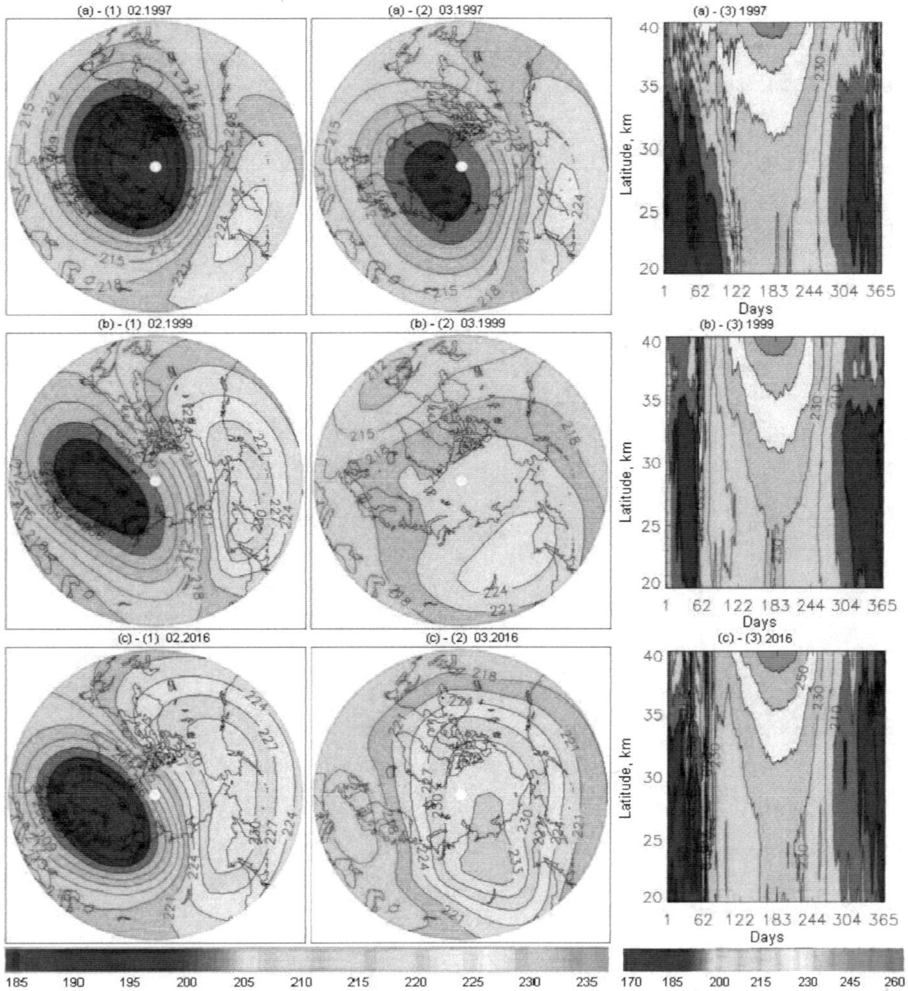

Figure 2. ERA Interim reanalysis data of monthly average air temperature (K) for northern polar region (40^0 N - 90^0 N) at 24.5 km (30 hPa) for February and March, 1997 (a - 1 and 2), 1999 (b - 1 and 2), 2016 (c - 1 and 2), and annual vertical profiles of daily average air temperature (*K*) at latitude 84^0 N for 1997 (a - 3), 1999 (b - 3), 2016 (c - 3).

In Figure 3 polar projections of the ozone mixed ratio in ppm (the ozone density divided by the air density and multiplied by 10^6) at a height of 30 hPa and the total column of ozone in Dobson units (DU) are shown. In 1997, following the La Niña phenomenon, the ozone content in the lower stratosphere in February and March was significantly lower than in 1999 and 2016. At the same time, in 2016, when the El Niño phenomenon was observed, in February the ozone content was even lower than in 1997, but a sudden stratospheric warming in March 2016 led to the destruction of the polar vortex and an increase in the ozone content.

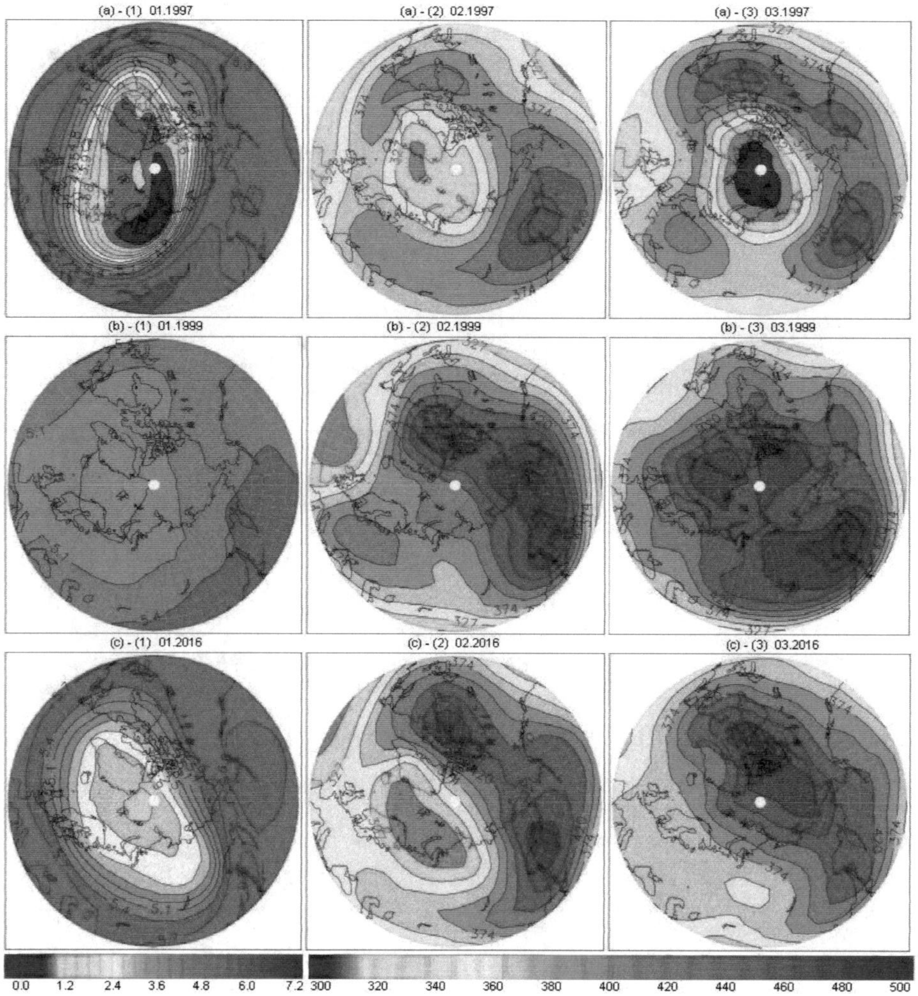

Figure 3. Monthly average distribution of the ozone mixed ratio in Northern Hemisphere in ppm in an area of 40^0 N - 90^0 N and at a height of 24.5 km (30 hPa) according to ERA Interim for February 1997 (a - 1), January 1999 (b - 2), March 2016 (c - 3), monthly average distribution of the total column of ozone in Northern Hemisphere in Dobson units (DU) in an area of 40^0 N - 90^0 N according to ERA Interim for February and March 1997 (a - 2 and 3), 1999 (b - 2 and 3) and 2016 (c - 2 and 3).

The results of the model experiments with the CCM under different scenarios of the influence of SST variability and CO_2 content during the climatic period from 1980 to 2015 are shown in Figure 4. A comparison of the ozone and temperature at the beginning of the given climate period (1980-1985) and at the end of the period (2010-2015) compares the effects of SST variability and CO_2 increase. Differences of the SST values, air temperatures at levels of 925 and 20 hPa, and the ozone mixed ratio at a level of 20 hPa for the period with 1980 for 2015 are presented. During this period, warming of the ocean on average on the planet (Figure 4, 1, 2 and 3) at 0.7-0.8 K is observed though about Antarctica and South America downturn of temperature at 0.6-0.8 K. By the results of the experiments in view of the SST changes (Figure 4, G and J) and ERA Interim data (Figure 4, M), warming troposphere on average on the planet at 0.5-0.7 K is observed. Thus, the character of distribution of the temperature differences corresponds to the distribution of the SST differences (the correlation factor is equal to 0.7-0.9). In the experiments not taking into account the SST changes (Figure 4, A and D) warming on the planet at 0.2-0.4 K is observed. In the stratosphere, in the experiments in view of CO_2 changes (Figure 4, E and K) a cold snap at 2.2-3.0 K is observed. In the experiments with a fixed CO_2

IOP Conf. Series: Earth and Environmental Science **386** (2019) 012021 doi:10.1088/1755-1315/386/1/012021

level (Figure 4, B and H) and according to ERA Interim data (Figure 4, N) the air temperature in the stratosphere varies little. The ozone mixed ratio in all experiments (Figure 4 C, F, I, L) varies little, although in the experiment with fixed SST values and CO_2 (Figure 4, C) a downturn of the mixed ratio at 4 ppm above Antarctica is observed. According to Era Interim (Figure 4, O) there is an increase in the mixed ratio at 15 ppm in the tropical belt and a reduction by 4 ppm in latitudes 40^0-60^0.

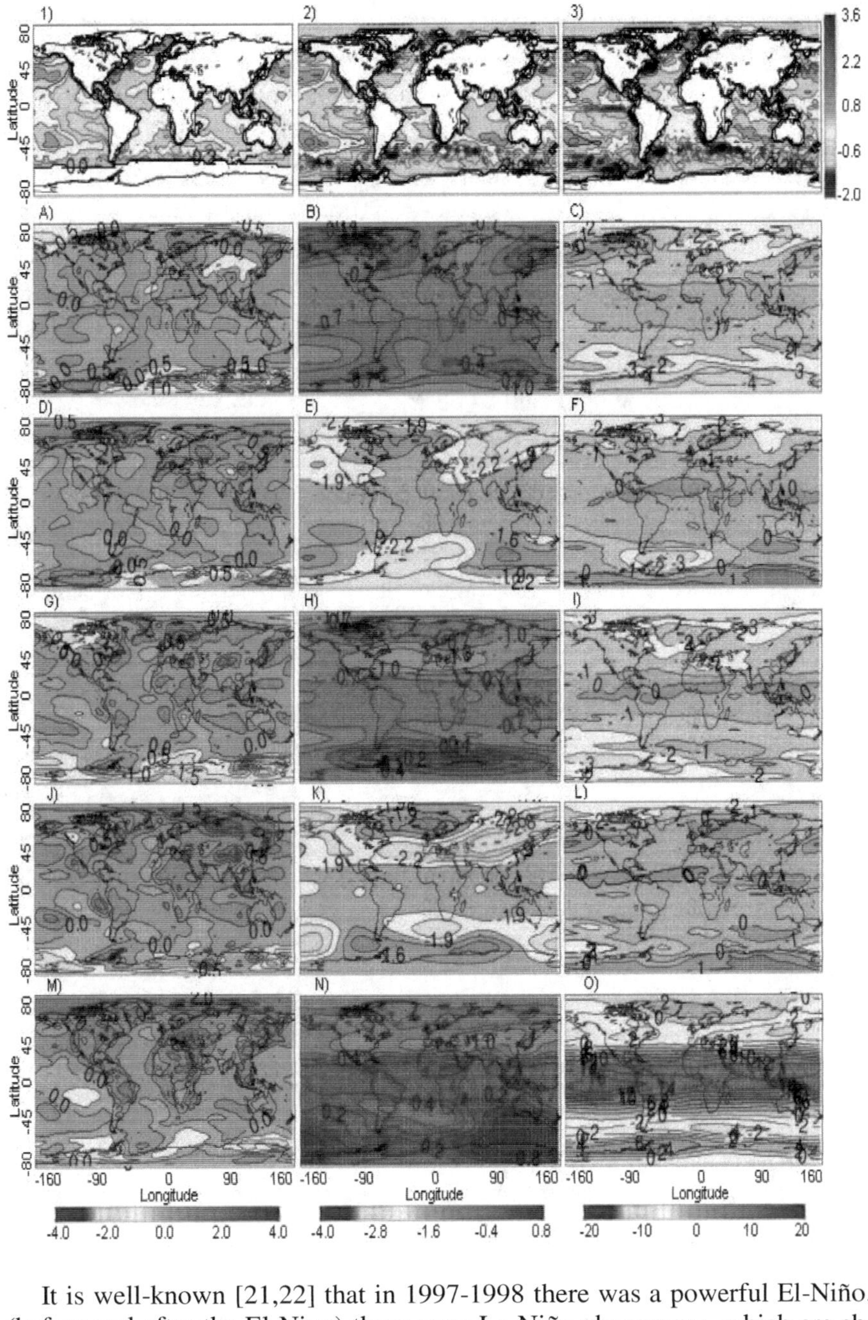

Figure 4. Difference of averaged sizes for 2010-2015 and 1980-1985: sea surface temperatures (K) according to Met Office (1), ERA Interim (2), ERA20Century and ERA5 (3), differences of simulation results averaged for the periods with 2010 on 2015 and with 1980 on 1985: in experiments with fixed SST, SIC, and CO_2 values at a level of 1979 (2 line), with fixed SST and SIC at a level of 1979 (3 line), with fixed CO_2 at a level of 1979 (4 line), base experiment (5 line) and ERA Interim reanalysis data (6 line); (from 2 line) left column: air temperature (K) at a level 925 hPa, average column: air temperature (K) at a level of 20 hPa, right column: ozone mixed ratio (ppm) at a level of 20 hPa.

It is well-known [21,22] that in 1997-1998 there was a powerful El-Niño. Thus, in 1997 and 1999 (before and after the El-Nino) there were La-Niña phenomena, which are shown in Figure 1. In 2016 there was also an El-Niño phenomenon. Besides, 2010, 2011, and 2015 were also analyzed. Thus, in 2011 there were La-Niña, and in 2010 and 2015, El-Niño.

IOP Conf. Series: Earth and Environmental Science **386** (2019) 012021 doi:10.1088/1755-1315/386/1/012021

As a result of study [22] it is found that the instability of CPV can be a consequence of El-Niño, whereas its stability can be a consequence of La-Niña. Within La-Niña (1997 and 2011) the CPV was steady and it kept till March. Also, steady CPV is characteristic for years of weak El-Niño (2010, 2015 and 2016) but, thus, it breaks up in February. In 1999, after a powerful El-Niño the CPV was unstable. Thus, the El-Niño phenomenon reduces the stability of CPV.

The CPV and El-Niño also influence the temperature in the stratosphere. Within a steady CPV (1997 and 2011) when a La-Niña phenomenon took place there was a steady area of low temperatures (150-180 K) till April. Within the El-Niño (2010, 2015, and 2016) and also after the El-Niño 1997-1998 (in 1999) the period of low temperatures interrupted the SSW series, which led to the fact that the area of low temperatures broke up already in the middle of March. It means that the SSWs can be a consequence of El-Niño and are the reason of instability of the CPV.

As for the concentration and maintenance of ozone in the atmosphere, its steady reduction was observed in 1997 and 2011 till April when a La-Niña phenomenon took place, there were a steady CPV and an area of low temperatures, although the given reduction is also caused by the chemical processes in the atmosphere. In 2010, 2015, and 2016 from February to March there was an increase in the maintenance of ozone that can be caused by a SSW series and, hence, by the El-Niño phenomenon. In 1999 the maintenance of ozone varied a little and remained high, which caused unstable CPV and SSW series, and, probably, means a phenomenon of El-Niño in 1997-1998. It means that the El-Niño can lead to increase in the maintenance of ozone by means of SSW.

Besides the El-Niño, the ocean also has a general global influence on the atmosphere by means of SST. It is shown that the distribution of temperature changes in the troposphere corresponds to the distribution of SST changes, which is apparent from Figure 4, with a correlation at 0.7-0.9. Also, it is shown that there is a warming of the ocean at 0.8 K and, thus, there is also a warming at 0.7 K in the troposphere according to the reanalysis and experiments in view of the SST changes. It means that the SST is a significant factor in the troposphere. Changes in the CO_2 level are poorly significant for the troposphere. However, they indirectly influence the SST (increase in CO_2 leads to heating of the atmosphere and thawing of ices, which leads to increase in the open ocean area and the area of absorption of solar beams, and, hence, to increase in the SST).

In the stratosphere, according to the results of the experiments considering changes in CO_2, there is a downturn of temperature on average on the planet at 2.5-3 K. This corresponds to the theory of global warming, whereas not taking into account changes in CO_2 the temperature varies but little. It specifies that changes in CO_2 are of great value for the stratospheric temperature, whereas the SST is poorly significant.

As for the concentration of ozone, CO_2 level also matters for it. However, the processes connected with ozone are of great value. In Figure 4, the results of the experiments in view of the return influence of ozone on the atmosphere are presented. In this case changes in CO_2 influence the maintenance of ozone a little, and it varies little during time. In the experiments not taking into account the return influence of ozone, the CO_2 level becomes a significant factor: at a fixed CO_2 there is a reduction of the concentration of ozone on average on the planet, whereas at a varying one there is an increase.

Thus, it is assumed that the El-Niño leads to the SSW, CPV instability, and increase in the maintenance of ozone, and La-Niña - on the contrary. Also, it is supposed that the increase in the CO_2 level leads to cooling of the stratosphere and increase in the concentration of ozone on the planet, whereas heating of the troposphere occurs, mainly, due to increase in the SST on the planet.

4. Conclusions

Research on the influence of the El-Niño-La-Niña phenomenon on the stratosphere of the Arctic regions and the ozone layer was carried out. It was shown that El-Niño events are connected with SSW series, instability of the CPV, and increase in the maintenance of ozone. La-Niña events are connected with a steady CPV, low temperatures, and also a reduction in the maintenance of ozone. Also, a research was carried out on the influence of the changes in the SST and CO_2 level on the troposphere and stratosphere. It was shown that there is warming of the ocean and the troposphere on average on the planet, and also cooling of the stratosphere. The results of this research may be of great value in studying the influence of the ocean, the El-Niño phenomenon, and the CO_2 level on the atmosphere.

Acknowledgments
This research was supported by the Russian Science Foundation under scientific project no. 19-17-00198 and by the Russian Foundation of Basic Research (project 17-05-01277).

Reference

[1] Arnaud Czaja 3-7 September 2012 Ocean-atmosphere coupling in midlatitudes: does it invigorate or damp the stormtrack? *ECMWF Seminar on Seasonal Prediction* pp 35-46
[2] Solomon S, Dahe Q, Manning M et al 2007 Climate Change 2007 *Intergovernmental Panel on Climate Change (IPCC), Cambridge Univ Press: New York* p 996
[3] Hansen J, Sato M, Ruedy R, Lacis A and Oinas V 2000 Global warming in the twenty-first century: An alternative scenario *Proc Natl Acad Sci.* **97** 9875-80
[4] Bondarenko A L and Zhmur V V 2004 Nature and forecasting possibility of the El-Niño/La-Niña phenomenon *Russian Meteorology and Hydrology* **11** 26-35
[5] Smyshlyaev S P, Pogoreltsev A I, Galin V J *et al* 2016 Influenve of wave activity on gaseous composition of stratosphere of polar regions *Geomagnetism and Aeronomia* **56**(1) 102-16
[6] Newman Paul A, Nash Eric R and Rosenfield Joan E 2001 What controls the temperature of the Arctic stratosphere during the spring? *Journal of geophysical research* **106**(17) 19999-20010
[7] Dingzhu Hu, Wenshou Tian, Fei Xie, Jianchuan Shu et al july 2014 Effects of meridional sea surface temperature changes on stratospheric temperature and circulation *Advances in atmospheric science* **31** 888-900
[8] Chipperfield M P and Jones R L 1999 Relative influences of atmospheric chemistry and transport on Arctic ozone trends *Nature* **400** 551-54
[9] Manatsa D and Mukwada G 2017 A connection from stratospheric ozone to El Niño-Southern Oscillation *Scientific Reports* **7** 5558
[10] Bell C, Gray L, Charlton-Perez A, Joshi M and Scaife A 2009 Stratospheric Communication of El Ninõ Teleconnections to European Winter *J. Climate* **22** 4083-96
[11] Galin V Ya, Smyshlyaev S P and Volodin E M 2007 Combined Chemistry–Climate Model of the Atmosphere *Izvestiya. Atmos. Ocean. Phys.* **43**(4) 399-412
[12] Vargin P N and Volodin E M 2016 Analyze of reproduction of dynamical processes in stratosphere by climate model ICM RAS *Izvestiya. Atmospheric and Oceanic Physics* **52**(1) 3-18
[13] Taylor K, Williamson D and Zwiers F September 2000 The sea surface temperature and sea-ice concentracion boundary conditions for AMIP II simulations *PCMDI Report* **60**
[14] Fleming E L, Chandra S, Barnett J J and Corney M 1990 Zonal Mean Temperature, Pressure, Zonal Wind and Geopotential Height as Functions of Latitude *Adv. Space Res.* **10**(12) 11-59
[15] Kalnay E, Kanamitsu M *et al* 1996 The NCEP/NCAR 40-Year Reanalysis Project Bull *Amer. Met. Soci.* **77** 437-71
[16] McPeters R D, Bharita P K, Krueger A J and Herman J R 1996 Nimbus-7 Total Ozone Mapping Spectrometer (TOMS) Data Products User's Guide *NASA Reference Publication* p 1384
[17] Twain M and McCurley K United Kingdom Meteorological Office *TOP500 Supercomputing Sites Top500.org*
[18] Rienecker M M, Suarez M J *et al* 2011 MERRA: NASA's Modern-Era Retrospective Analysis for Research and Applications *Journal of Climate* **24**(4)

[19] Dee D P *et al* 2011 The ERA-Interim reanalysis: Configuration and performance of the data assimilation system *Quart. J. Roy. Meteor. Soc.* **137** 553–97
[20] Kobayashi S *et al* 2015 The JRA-55 Reanalysis: General Specifications and Basic Characteristics *J. Met. Soc. Jap.* **93(1)** 5-48
[21] Jakovlev A R and Smyshlyaev S P 2017 The numerical simulations of global influence of ocean and El-Nino – La-Nina on structure and composition of atmosphere *Uchenie zapiski RSHU* **49** 58-72
[22] Jakovlev A R and Smyshlyaev S P 2019 Impact of the Southern Oscillation on Arctic Stratospheric Dynamics and Ozone Layer Izvestiya *Atmospheric and Oceanic Physics* **55(1)** 86-99

A simulated distribution of Siberian river runoff in the Arctic Ocean

E Golubeva[1,2], G Platov[1,2], D Iakshina[1,2] and M Kraineva[1]

[1] Institute of Computational Mathematics and Mathematical Geophysics SB RAS, 6 Prospekt Akademika Lavrentieva, Novosibirsk, 630090, Russia
[2] Novosibirsk State University, 1, Pirogova str., Novosibirsk, 630090, Russia

Email: elen@ommfao.sscc.ru

Abstract. Continental runoff is one of the major sources of the Arctic freshwater budget. As is generally known, it influences water column stratification and maintains Arctic halocline, which isolates the sea ice and the cold, fresh upper layer from the warmer, saltier Atlantic waters of the Arctic Ocean. An increase in river runoff was observed in recent years. It is suggested that this will have an impact on Arctic water mass transformations. However, few details are known regarding river freshwater export to the Central Arctic Basin. It is assumed that river water pathways in vast shelf seas and deep basins are closely related to atmospheric variability. In this study, we use three-dimensional coupled regional ocean-ice model simulations forced by atmospheric reanalysis data to investigate the change in Siberian rivers freshwater pathways in the Arctic Ocean due to the variability of atmospheric dynamics. A numerical experiment with an increasing runoff of the largest Siberian rivers is carried out. The consequences of adding freshwater to particular regions of the Arctic Ocean are analysed.

1. Introduction

The Arctic river discharge is one of the major sources for the Arctic freshwater budget. The Arctic Ocean accounts for about 1% of the global ocean volume but receives about 11% of the global river discharge [1, 2]. Eight largest Arctic rivers, namely, Mackenzie, Yukon, Kolyma, Lena, Yenisey, Ob, Pechora, and Severnaya Dvina together, cover approximately two-thirds of the pan-Arctic drainage area and account for over 60% of the river water inflow to the Arctic Ocean [3]. A long-term increasing trend for the Arctic river discharge, which is the greatest for the rivers of the Eurasian Arctic, has been well documented [4, 5, 6, 3]. Peterson et al. [4] show a 7% increase in the Eurasian runoff over the period from 1936 to 1999. Overeem et al. [6] analyzed the observational records for 19 large rivers encompassing the Arctic region and found a consistent increase in the annual discharge (+9.8%) over the entire region from 1977 to 2007. The authors show that a combined change in the river water discharge over 30 years exceeds the previous values obtained for the Canadian Arctic (+2%) over 1964–2000 and for Eurasia (+7%) over 1936-1999. An observation data analysis performed by Holmes et.al. [3] over 1976-2018 has shown that long-term trends in the annual discharge were 3.3±1.6% per decade for the Eurasian rivers and 2.0±1.8% per decade for the North American Arctic rivers.

Fresh water supplied by the Arctic rivers influences the Arctic Ocean stratification and the sea ice formation. The Arctic stratification is characterized by a cold and fresh surface layer, a relatively warm and salty Atlantic layer formed by Atlantic Waters entering the Arctic Ocean through the

Barents Sea and the Fram Strait, and an intermediate layer of cold water with a gradual increase in the salinity with depth known as the cold halocline [7, 8, 9]. Moving from the shelves to the Central Arctic Ocean, the river waters maintain the Arctic halocline, which isolates the sea ice from the warmer and saltier Atlantic layer. Changes in the river discharge can cause notable changes both inside and outside the Arctic Ocean as perturbations in the freshwater export from the Arctic Ocean have a potential for changing the dense water formation rate and the ocean convection in the North Atlantic [10].

It is generally known that the river water pathways over the vast shelf seas and the residence time are closely linked to the atmospheric dynamics [11, 12, 13]. To date not much is known regarding the river freshwater export into the Central Arctic Basin, and the distribution of the river water in the Arctic Ocean and beyond it. It is assumed that the variability of the Eurasian river runoff pathways is largely governed by the Arctic Oscillation (AO), the leading mode of atmospheric variability in the Northern Hemisphere [14]. Since the beginning of 2000 this issue has been debated in scientific papers.

The numerical modeling carried out by Harms et al. [15] investigates the role of the Siberian river runoff for transporting possible river contaminants in the Arctic Ocean. Based on the numerical results the authors have concluded that the Kara Sea river water dominates in the Siberian branch of the Transpolar Drift, whereas the Lena River water dominates in the Canadian branch. Analysis of the numerical results obtained has shown that the dissolved contaminants supplied by the river water would be able to reach the Canadian Archipelago in about 15 years after releasing into the Siberian estuaries. In a paper by Jahn et al. [16] it was shown that the fresh water exported through the western Canadian Arctic Archipelago mainly comes from the Pacific and North American runoff. The Eurasian runoff export through the Fram Strait occurs in the years with an anticyclonic circulation anomaly and it takes three years to reach the Fram Strait after leaving the shelf. Newton et.al. [17] used a high-resolution numerical simulation of the Arctic Ocean to study the river runoff distribution in the Arctic Ocean in the 1990s characterized by an increased AO index. In the numerical results obtained the freshwater plumes have shifted eastward over the both shelves and the deep basin. The tracers migrate away from the Eurasian Basin; first to the Canadian side of the Lomonosov Ridge over the Makarov Basin, and then even further into the Beaufort Sea, north of the Canadian continental slope. Morison et al [18] argued that due to a cyclonic (anticlockwise) shift in the ocean pathway of the Eurasian runoff forced by strengthening of the west-to-east Northern Hemisphere atmospheric circulation in the1990s, a large amount of the Eurasian river runoff was transported mainly into the Canada Basin, and caused an increase in the freshwater content in the Canada basin balanced by a decrease in the Eurasian basin.

In this study we aim at tracing the Siberian river runoff distribution over the Arctic Ocean and try to identify modifications in the river freshwater pathways due to variability of the atmospheric dynamics. We discuss the numerical results based on a three-dimensional coupled regional ocean-ice model forced by atmospheric reanalysis data. In our first experiment we analyzed the spatio-temporal scales of the distribution of particles periodically emitted at the shelves from the Siberian river mouths. In our second experiment we increased the discharge values of the Siberian Rivers and assumed that this has an impact on the Arctic water mass transformation. By analyzing the numerical results obtained we tried to understand the consequences of adding freshwater for particular regions of the Arctic Ocean.

2. Model

2.1. Coupled Ice-Ocean model

Our study is based on a three-dimensional regional coupled ocean-ice model called SibCIOM (Siberian Coupled Ice-Ocean Model) [19, 20] developed at the Institute of Computational Mathematics and Mathematical Geophysics (Siberian Branch of the Russian Academy of Sciences). The ocean model is based on the conservation laws for heat, salt, and momentum, as well as on conventional approximations: the Boussinesq, hydrostatic, and "rigid lid" ones. The QUICKEST scheme is employed to approximate advection [21]. The multidimensional extension uses the COSMIS approach [22]. The vertical adjustment is considered to be a mixed layer parameterization based on the Richardson number [23].

The ocean circulation model has been coupled with the CICE v3 model of the thermodynamics of elastic viscous-plastic ice [24] and the multi-category sea ice thermodynamics [25]. The sea ice advection utilizes a semi-Lagrangian scheme [26]. The fast ice parameterization is the most simplified approach, and the ice velocity was set to zero in the shallowest part of the Laptev and East Siberian Seas (a depth of < 30 m) for the period from October 30 to June 1.

2.2. Particle Tracer Model

In order to numerically track the river water distribution we used the method of Lagrangian particles. Particles were individually and periodically emitted in the region of a certain source and moved within the numerical domain with a model velocity. The advective motion of particles was also accompanied by diffusion, which is considered to be a stochastic process. The position of a particle caught in the layer of convective or wind mixing was also stochastically determined based on a uniform distribution in the mixed layer. A particle of any river runoff was deployed in a way that it represents the volume $V_0 = 0.6859$ km^3. This means that the time interval between two successive particle releases is determined as $\Delta t = V_0/R(t)$, where $R(t)$ is the current river discharge rate in km^3/s.

2.3. Model domain

The model domain includes the Arctic and the Atlantic Ocean north of 20° S. The grid resolution for the North Atlantic is chosen to be 0.5° x 0.5°. At 65° N the North Atlantic spherical coordinate grid is merged with the displaced poles of the Arctic grid. The horizontal grid size in the Arctic varies from 10 to 25 km with an average grid spacing of about 18 km. The model version used here has 38 unevenly spaced vertical levels with a maximum resolution of 5 m in the upper 20-meter layer. The minimum depth of the shelf zone is taken to be 20 m.

2.4. Forcing

The model is forced by the CORE-II Reanalysis data [27]. The model takes into account the inflow of Bering Strait and the 52 largest rivers in the region, among which are the Siberian rivers Yenisei, Ob, Lena, Indigirka, Olenek, Yana, and Kolyma. Data on the average seasonal runoff from these rivers were obtained from hydrological station measurements [28]. In addition, according to the estimates by Aagaard and Carmack [1] the total runoff of the continental waters in the Arctic is approximately 1.3 times greater than that of the main rivers. Therefore, to obtain the whole picture the discharge of the above-mentioned rivers was increased 1.3 times, including those of the Atlantic basin. The rivers' fresh water flux was calculated on the basis of an assumption that the river water has zero salinity. The specified mass transports at the open boundaries and river inflows are compensated by transports through the outflow boundary at 20° S.

3. Simulation

3.1. Variability of Siberian river runoff pathways in the Arctic Ocean

In our basic experiment (hereafter E1) we have simulated large-scale variability of the Arctic Ocean circulation and the sea ice state caused by the variability of the atmosphere from 1985 to 2014. We started our experiment using the ocean and sea ice state previously obtained in numerical experiments discussed in our published works [20, 29] and showed how the numerical model restores the variability of the thermohaline fields of the Arctic Ocean known from the observational data. In this paper we analyze possible paths of the river water distribution of the most full-flowing Siberian rivers: Ob, Yenisei, and Lena based on a consideration of the trajectories of tracers entering the shelf zone of the Siberian seas with the river waters.

Our simulation forced by the atmospheric reanalysis has shown that a significant part of the tracers arriving with the waters of the Ob, Yenisei, and Lena rivers in different periods of the atmospheric circulation remains within the shelf zone. Earlier in paper [30] we discussed the results of modeling the water circulation and the distribution of river waters on the shelf of the Laptev Sea. In the paper in question it was shown that the circulation pattern of the passive tracers injected from the Lena River mouth is fairly complicated, being subject to the wind forcing and the ice edge position. The river tracers changed their direction several times per year and, eventually, a lot of the tracers remained within the eastern part of the Laptev Sea shelf longer than one year.

Having reached the mainland slope, the river tracers are involved in the large-scale circulation system of the Arctic basin. Two main modes of the surface circulation of the Arctic Ocean, cyclonic and anticyclonic, determined by the atmospheric dynamics, have been repeatedly discussed in the literature [31]. By analyzing the tracer propagation paths, we tried to identify these two periods. In Figures 1-3 the distribution of the river tracers outgoing from the three main Siberian rivers (Lena, Ob, and Yenisei) is presented. Figure 1 shows the distribution of tracers that arrived in 1989-1996. The Arctic Ocean Oscillation index (AOO) [32, 31] was defined on the basis of a wind-driven simulated sea surface height field across the Arctic and represents a measure of the intensity and sense (clockwise/anticyclonic or counterclockwise/cyclonic) of the Arctic Ocean wind-driven upper oceanic circulation. These years, 1989-1996, are characterized by an increased negative AOO index, which corresponds to the formation of a cyclonic mode of the surface water circulation. Figure 1 shows mainly the north-east distribution of the river tracers.

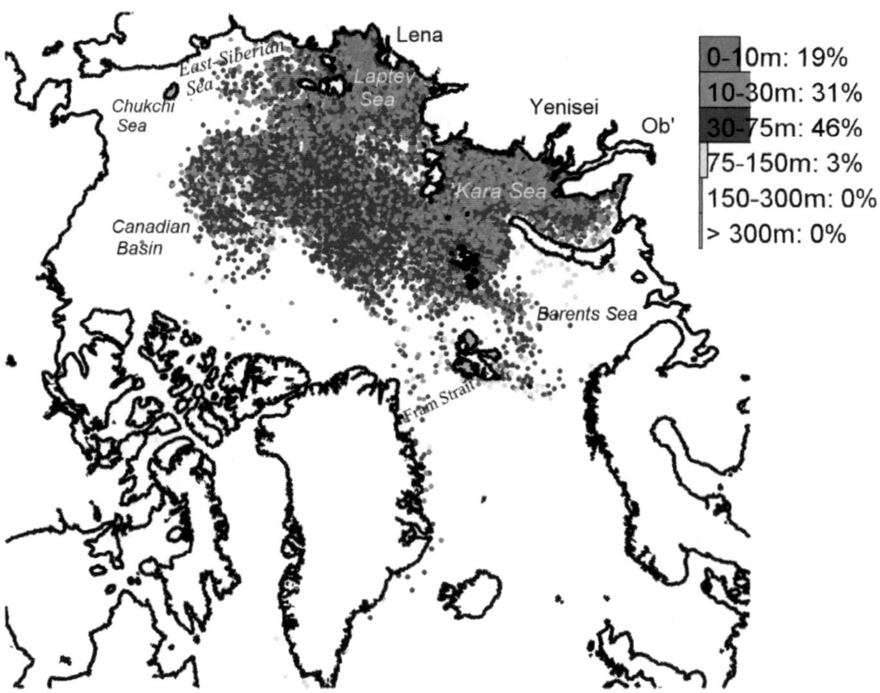

Figure1. Distribution of the sum of all tracers released from largest Siberian rivers in 1989-1996.

The transition to the anticyclonic circulation mode in the subsequent years [31] is shown in Figure 2. A significantly smaller number of tracers released during these years have appeared in the central part of the basin. The figure shows that the tracers are distributed towards the Fram Strait. The percentage of tracers at various depths is also shown in the figure.

To complete the picture described, we have also presented the distribution of all tracers released between 1989 and 2006. Examination of Figures 1-3 shows that the part of the tracers moving to the central basin during the cyclonic circulation period (Figure 1) was redistributed according to the changed water circulation system. Some of the tracers were involved in the anticyclonic cycle and moved towards the Canadian Straits. A significant number of the tracers were located in the Eurasian basin, and some tracers went beyond the Fram Strait and joined the subpolar cycle. A small number of the tracers (less than 5%) went down to the deep layers of the Canadian basin.

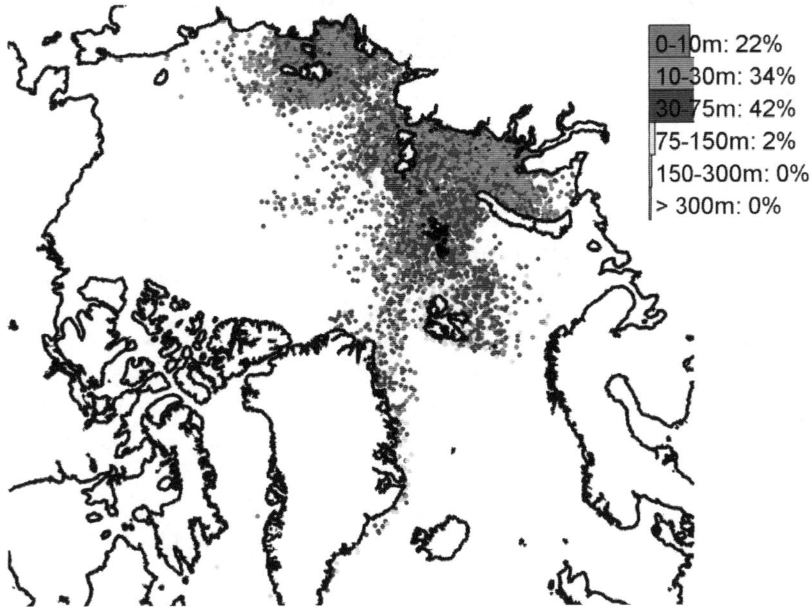

Figure 2. The same as Figure 1, but over 2000-2006.

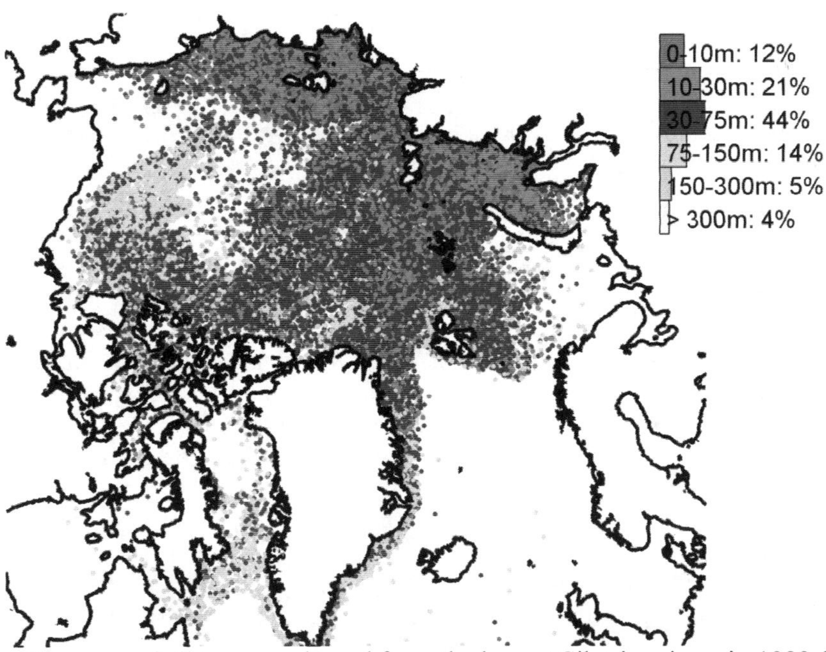

Figure 3. Distribution of the sum of all tracers released from the largest Siberian rivers in 1989-2006.

3.2. Sensitivity study to an increase in the largest Siberian rivers discharge

Given the tendency of an increase in the runoff of Siberian rivers in the recent decades [3], we have carried out a numerical experiment to study the sensitivity of the model temperature and salinity distributions in different regions of the Arctic Ocean to an increase in the discharge of the rivers. Using the ocean and sea ice state obtained in the reference experiment E1 for 2000 as the initial one, we have increased the runoff of the Siberian rivers one and a half times and repeated the numerical experiment for the period from 2000 to 2014.

A discussion of our results will be based on an analysis of hydrological characteristics averaged over individual regions. As such characteristics we will consider the difference in the temperature and salinity obtained from the results of the present (E2) and reference (E1) experiments.

An analysis of the comparison of the calculated fields for the two experiments has shown that due to an increase in the flow of rivers into the Siberian seas the most significant decrease in the salinity was in the upper 25-meter layer. Regional averaging over 15 years of the model calculation of the salinity of the surface layer has shown a decrease of this value by 2.5 psu in the Laptev Sea, 1.75 psu in the East Siberian Sea, and 2 psu in the Kara Sea. In the Kara Sea, the most dramatic changes in the salinity are sharply limited to the upper 20-meter layer. Below the surface layer the Atlantic waters penetrating into the region make a significant contribution to the water stratification. In the East Siberian Sea anomalies of salinity penetrate deeper than in the Kara Sea and in the Laptev Sea (up to 60 m at the end of the experiment). This can be explained by remoteness of the region from the mouths of the most powerful rivers. The salinity reduction in the region occurs due to the fresh water transport from the Lena and Kolyma rivers along the coast of eastern Siberia.

An analysis of the temporal variability of the temperature averaged over the regions also makes it possible to highlight some features of its distribution in the Siberian seas caused by an increase in the river inflow. By analyzing the deviation field in the Laptev Sea and in the East Siberian Sea, we have detected positive temperature anomalies (up to 0.3 ° C in the Laptev Sea and up to 1.15 ° C in the East Siberian Sea). These anomalies regularly occur in the summer in the surface layer, propagate into deeper layers, and cause an increase in the bottom layer temperature. A similar character of the anomalies was obtained in numerical tests [30] when the thermal runoff of the Lena River was taken into account. The measurement data obtained from 1940 to 2011 confirm the process of increasing the bottom layer temperature of the Laptev Sea. The results obtained for the Kara Sea do not allow us to say anything definite about the temperature anomalies that occasionally appeared and were localized in the bottom layer.

According to our experiment, visible changes in the salinity field begin to appear over the continental slope of Eurasia after about 2 years (Figure 4b). The strongest anomalies were recorded in the Nansen basin region (1-3 psu), first at the continental slope of the Kara and the Laptev Seas (R4) and near the St. Anna trough (region R3), and then the salinity anomaly moved toward the Fram Strait (regions R2 and R1). The salinity anomalies in the eastern part of the continental slope are much weaker and do not exceed 1 psu at the sea surface. In the central part of the basin and along the American coast anomalies not exceeding 0.5 psu at the sea surface appear after 6-8 years.

It should be noted that in some cases, in areas adjacent to the Fram Strait, a decrease in the surface salinity is accompanied by an increase in the salinity and temperature in deeper layers (Figure 4c). Such a temperature anomaly has been detected in regions R1 and R2 in 2009 – 2011 (0.1-0.2C) and in 2014 (0.55C) in region R1. This means that the appearance of fresh water at the surface led to the suppression of the mixing processes and to a more prolonged conservation of the heat in the layer of Atlantic waters flowing through the Fram Strait. Positive temperature anomalies are observed in the layer of Atlantic waters in the regions located along the continental slope (in R4). An analysis of the simulation results shows that the depth of the upper mixed layer in this period is 10 m smaller than that in the reference experiment E1. A negative temperature anomaly in the region R3 over the past three years indicates that intense mixing has occurred. This fact underlines the ambiguity of the Arctic Ocean response to possible climate changes.

In the areas adjacent to the continental slope in the eastern part of the basin, small positive temperature anomalies were detected in the uppermost layer. Their values do not exceed 0.06 C. After 7 years of calculation, a positive temperature anomaly in the layer from 100 to 200 m is constantly present in the Canadian basin. This depth corresponds to the distribution of the Pacific waters layer in the Arctic basin. The magnitude of this anomaly (maximum: 0.04 C) is too small, but the fact of the appearance of this anomaly seemed of interest to us. It may mean a decrease in the Pacific water heat loss on the Chukchi Sea shelf due to more stable stratification.

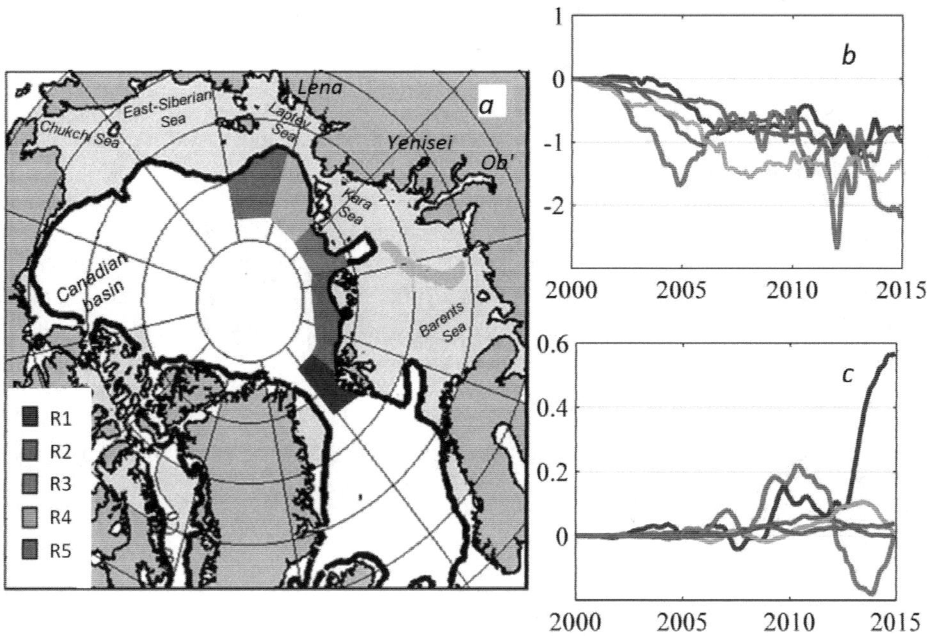

Figure 4. Left panel: Regions of analyzing temperature and salinity anomalies, Upper-right panel: Salinity anomalies (psu) averaged over a certain region and in 25-meter layer. Lower-right panel: Temperature anomalies (°C) averaged over a certain region at a depth of 200 m. The line color corresponds to the region of averaging.

The response of the Arctic Ocean ice cover to a decrease in the surface layer salinity is ambiguous depending on certain regions and time periods. The largest ice volume reduction caused by an increase in the fresh water amount occasionally reaches 10% in the Nansen basin region. In most regions it does not exceed 1-2%. In the central part of the basin the numerical model shows a trend for an increase in the ice volume.

4. Summary

Our model results support the hypotheses that the atmospheric changes reflected in the AOO play an important role in the distribution of fresh water in the Arctic Ocean. We have simulated paths of tracers representing distributions of the Siberian river runoffs in different modes of the Arctic Ocean circulation. Analyses of the simulation results have shown that a lot of tracers injected from river mouths remain trapped near the shelves and tend to exit from the shelves after several years. In a cyclonic mode of the circulation (1989-1996) the tracers, crossing the shelves, moved north-eastward into the central Arctic, and part of them reached the northern part of the Canadian basin. In response to a change in the wind patterns associated with a trend toward positive values of the AOO in 2000-2006, the tracers shifted westward into the Eurasian basin and then moved to the Fram Strait. Some of the tracers exiting in 1989-1996 were involved into an anticyclonic circulation and reached the Canadian Straits. Only a small number of the tracers reached the Beaufort Sea in 2006.

To understand the consequences of an increase in the runoff of the Siberian rivers known from observational data, we carried out a numerical experiment in which we increased the runoff of the Lena, Yenisei, and Ob Siberian rivers by one and a half times. The experiment was performed for the period from 2000 to 2014. The results of our experiment have shown that different thermohaline characteristics of the Siberian seas respond differently to the input of additional fresh water. Their behavior in the Kara Sea is quite different from that in the Laptev Sea and the East Siberian Sea.

Outside the Arctic shelf, some regions of the Eurasian basin are most sensitive to the increased runoff of the Siberian Rivers. This is probably due to the fact that during this period there was a positive mode in the AO index. A predominantly anticyclonic circulation was established during this period, which contributed to the transfer of excessive fresh water to the Fram Strait. Temperature anomalies of various intensities were detected not only in the surface layer, but also in the layer of Atlantic waters entering the Arctic Ocean through the Fram Strait. These anomalies correspond to a decrease in the mixing depth of the upper layer.

Small positive temperature anomalies were found in the Pacific Water layer in the Canadian Basin. We believe that, despite the fact that these are small anomalies, this result deserves attention and should be verified with a higher-resolution model. Our arguments in favor of this conclusion are as follows: It is the contribution of the Lena River waters in the experiment that caused the appearance of these anomalies, since these waters can be transferred to the East Siberian Sea and then carried off the shelf of the Chukchi Sea, where they can affect the mixing processes.

Acknowledgments

This work was supported by the Russian Foundation for Basic Research (grants 17-05-00396 and 17-05-00382). Some parts of the research were performed under Government Order 0315-2019-0004. We gratefully acknowledge the support of the staff of the Siberian Supercomputer Center of the Siberian Branch of the Russian Academy of Sciences for providing us with their supercomputer facilities.

References

[1] Aagaard K and Carmack E C 1989 The role of sea ice and other fresh water in the Arctic circulation. *J. Geophys. Res.* **94** (C10) 14485– 98 doi:10.1029/JC094iC10p14485

[2] McClelland J W, Holmes R M, Dunton K H *et al* 2012 *Estuaries and Coasts* **35** 353 https://doi.org/10.1007/s12237-010-9357-3

[3] Holmes R M, Shiklomanov A I, Suslova A, Tretiakov M, McClelland J W, Spencer R J M and Tank S E 2018 River Discharge *Arctic Report Card* (https://www.arctic.noaa.gov/Report-Card/Report-Card-2018/ArtMID/7878/ArticleID/786 /River-Discharge)

[4] Peterson B J, Holmes R M, McClelland J W, Vörösmarty C J, Lammers R B, Shiklomanov A I, Shiklomanov I A and Rahmstorf S 2002 Increasing river discharge to the Arctic Ocean *Science* **298** (5601) 2171–73 doi:10.1126/science.1077445

[5] McClelland J W, Déry S J, Peterson B J, Holmes R M and Wood E F 2006 A pan-arctic evaluation of changes in river discharge during the latter half of the 20th century *Geophys.Res. Lett.* **33** L06715 doi:10.1029/2006GL025753

[6] Overeem I and Syvitski J P M 2010 Shifting discharge peaks in Arctic rivers, 1977–2007 *Geogr. Ann.* **92**A 285–96 https://doi.org/10.1111/j.1468-0459.2010.00395.x

[7] Aagaard K, Coachman L K and Carmack E C 1981 On the halocline of the Arctic Ocean *Deep Sea Res. Part A* **28** (6) 529–45

[8] Rudels B, Anderson L G and Jones E P 1996 Formation and evolution of the surface mixed layer and halocline of the Arctic Ocean. *J. Geophys. Res.* **101** (C4) 8807–21

[9] Rudels B, Jones E P, Schauer U and Eriksson P B 2004 Atlantic sources of the Arctic Ocean surface and halocline waters *Polar Res.* **23** (2) 181–208 doi:10.1111/j.1751-8369.2004.tb00007.x

[10] Manabe S and Stouffer R 1995 Simulation of abrupt climate change induced by freshwater input to the North Atlantic Ocean *Nature* **378** 165–67 doi:10.1038/378165a0

[11] Guay C K, Falkner R D, Muench R D, Mensch M, Frank M and Bayer R 2001 Wind-driven transport pathways for Eurasian Arctic river discharge *J. Geophys. Res.* **106** (11) 469–80
[12] Dmitrenko I, Kirillov S, Tremblay L, Kassens H, Anisimov O, Lavrov S, Razumov S and Grigoriev M 2011 Recent changes in shelf hydrography in the Siberian Arctic: Potential for subsea permafrost instability. *J. Geophys. Res.* **116** C10027
[13] Bauch D, Dmitrenko I A, Wegner C, Hölemann J, Kirillov S A, Timokhov L A and Kassens H 2009 Exchange of Laptev Sea and Arctic Ocean halocline waters in response to atmospheric forcing. *J. Geophys. Res.* **114** C05008
[14] Thompson D W J, Lee S and Baldwin M P 2002 Atmospheric Processes Governing the Northern Hemisphere Annular Mode/North Atlantic Oscillation. *AGU monograph on the North Atlantic Oscillation* pp 85-89
[15] Harms I H, Karcher M J and Dethleff D 2000 Modelling Siberian river runoff — implications for contaminant transport in the Arctic Ocean *Journal of Marine Systems* **27** 95-115
[16] Jahn A, Tremblay L B, Newton R, Holland M M, Mysak L A and Dmitrenko I A 2010 A tracer study of theArctic Ocean's liquid freshwater export variability *J. Geophys. Res.* **115** C07015 doi:10.1029/2009JC005873
[17] Newton R, Schlosser P, Martinson D G and Maslowski W 2008 Freshwater distribution in the Arctic Ocean:Simulation with a high-resolution model and model-data comparison, *J. Geophys. Res.* **113** C05024 doi:10.1029/2007JC004111
[18] Morison J, Kwok R, Peralta-Ferriz C, Alkire M, Rigor I, Andersen R and Steele M 2012 Changing Arctic Ocean freshwater pathways *Nature* **481** (7379) 66–70 doi:10.1038/nature10705
[19] Golubeva E N and Platov G A 2007 On improving the simulation of Atlantic Water circulation in the Arctic Ocean. *J. Geophys. Res.* **112** C04S05
[20] Golubeva E N and Platov G A 2009 Numerical Modeling of the Arctic Ocean Ice System Response to Variations in the Atmospheric Circulation from 1948 to 2007 *Izvestiya. Atmospheric and Oceanic Physics* **45** (1) 137–51
[21] Leonard B P 1979 A stable and accurate convective modeling procedure based on quadratic upstream interpolation *Computer Methods in Applied Mechanics and Engineering* **19** 59–98
[22] Leonard B P, Lock A P and MacVean M K 1996 Conservative explicit unrestricted-timestep multidimensional constancy-preserving advection schemes *Monthly Weather Review* **124** 2588–606
[23] Golubeva E N, Ivanov Ju A, Kuzin V I and Platov G A 1992 Numerical modeling of the World Ocean circulation including upper ocean mixed layer *Oceanology* **32**(3) 395-405
[24] Hunke E C and Dukowicz J K 1997 An elastic-viscous-plastic model for ice dynamics. *J. Phys. Oceanogr.* **27**(9) 1849–67
[25] Bitz C M and Lipscomb W H 1999 An energy-conserving thermodynamic model of sea ice *J. Geophys. Res.* **104** (15) 669-77
[26] Lipscomb W H and Hunke E C 2004 Modeling Sea Ice Transport Using Incremental Remapping. *Mon. Wea. Rev.* **132**(6) 1341–54
[27] Large W G and Yeager S G 2004 Diurnal to decadal global forcing for ocean and sea–ice models: The data sets and flux climatologies *NCAR Technical note NCAR/TN-4601STR* doi:10.5065/D6KK98Q6
[28] Vorosmarty C J, Fekete B M and Tucker B A 1998 Global River Discharge, 1807-1991, Version. 1.1 (RivDIS) Data set *Oak Ridge National Laboratory Distributed Active Archive Center* Oak Ridge (Tennessee, U.S.A.) (http://www.daac.ornl.gov)
[29] Kuzin V I, Platov G A, Golubeva E N and Malakhova V V 2012 Certain results of numerical simulation of processes in the Arctic Ocean *Izvestiya, Atmospheric and Oceanic Physics* **48** (102) https://doi.org/10.1134/S0001433812010069
[30] Golubeva E, Platov G, Malakhova V, Kraineva M and Iakshina D 2018 Modelling the Long-Term and Inter-Annual Variability in the Laptev Sea Hydrography and Subsea Permafrost

State *Polarforschung* **87** (2)
[31] Proshutinsky A, Dukhovskoy D, Timmermans M-L, Krishfield R and Bamber J L 2015 Arctic circulation regimes. *Phil. Trans. R.Soc.A* **373** 20140160 doi: 10.1098/rsta.2014.0160
[32] Proshutinsky A and Johnson M 1997 Two circulation regimes of the wind-driven Arctic Ocean. *J. Geophys. Res.* **102** 12493–12514 doi:10.1029/97JC00738

IOP Conf. Series: Earth and Environmental Science **386** (2019) 012023 doi:10.1088/1755-1315/386/1/012023

Sea ice modeling in the Sea of Azov for a study of long-term variability

A I Mizyuk and O S Puzina

Marine Hydrophysical Institute, RAS, 2 Kapitanskaya str., Sevastopol, 299011, Russia

E-mail: artem.mizyuk@mhi-ras.ru, oksana_puzina@mhi-ras.ru

Abstract. This study deals with an attempt to use the NEMO modeling framework to investigate the long-term variability of the sea-ice fields in the Sea of Azov. A coupled ocean-ice model is used to perform a simulation by using an earlier developed configuration of cascading sea basins (Azov, Black, and Marmara). Various features of geographical location cause the formation of sea ice in the waters of the Sea of Azov and in the coastal area of the north-western Black Sea shelf. For a preliminary accuracy assessment, the simulation results are compared with the data based on satellite observations. The restricted amount of the data for the rather small area under study, as well as the features of the regional weather conditions, produce rather low accuracy of the available data products based on the satellite data. At the same time, the available data from coastal weather stations, as well as archived observations, produce a fairly adequate pattern of the simulated variability of the sea-ice conditions in the basin.

1. Introduction

The Sea of Azov is a semi-enclosed inland basin located rather far from the ocean. Therefore, the variability of its physical parameters is highly dominated by the atmospheric conditions formed over the continent. Despite the very small area and depth (with a maximum as large as 14.5 m), the basin can be considered as an important transport hub. In addition, a large amount of river runoff affects its high fish productivity.

The Sea of Azov is characterized by significant spatial inhomogeneity of physical parameters. Interannual, seasonal, and storm-surge induced variations in the thermohaline regime are well pronounced [1]. A significant number of abnormal phenomena have been observed here recently under the stable anthropogenic impact, e.g. [2]. An important role both for the basin itself and for the neighboring Black Sea is played by the water exchange through the Kerch Strait. The inflow of more saline water masses and circulation features significantly affect the salinity of the basin. At the same time, the outflow and further propagation of nutrient-enriched Azov Sea waters can be manifested in the Black Sea from satellite observations [3]. Therefore, to produce forecasts of the thermohaline and circulation properties in the region, one should consider the cascading basins of the adjacent seas.

This problem is rather important within the framework of establishing regional marine nowcasting/ forecasting systems, incorporating available observations and general circulation models to detail information about the sea state, e.g. the Black Sea Marine Forecasting Centre (BSMFC, http://mis.bsmfc.net), which is still operating and open to end users. Note that there are no systems at the moment able to provide similar information for the Sea of Azov with sufficient resolution. In the global ocean forecasting centers, e.g. Copernicus Marine Environment Monitoring Service, CMEMS (http://marine.copernicus.eu/) or HYCOM (https://www.hycom.org), the spatial resolution of the provided products is rather low for studying regional circulation features and variability of thermohaline parameters of the Sea of Azov. Note that the operational forecast of the basin produced by the Hydrometeorological Centre of Russia (HMC) (http://hmc.meteorf.ru) is based on a model from [5]. However, it reproduces only circulation and storm surges not taking into account the temperature, salinity, and exchange with the Black Sea.

One of the difficulties for the solution of such problems is the need for sea ice modeling that is formed in the basin and characterized by significant interannual and seasonal variability. This results from the geographical position of the basin, its shallowness, and rather low sea water salinity [6]. Prediction of the sea ice conditions is complicated by continuous freezing of the new ice cover, together with its melting in certain areas of the basin in conjunction with a constant ice regime, as well as the atmospheric conditions. For example, a difficult navigation situation occurred in the winter of 2011 – 2012 [7]. The whole basin of the Sea of Azov was completely covered with ice and, due to the stable impact on the formed sea ice, its concentration and layering increased, leading to the formation of hummocks.

The need to implement a coupled model for sea and ice thermodynamics complicates the solution of the problem. A number of important results on numerical modeling of the ice conditions in the basin were obtained earlier in Russian works. For example, in [8], for the region of the Kerch Strait for 5 months, the period of 2011-2012, a locally one-dimensional model of the thermodynamics of ice formation and melting was used. The results obtained were compared only with data from coastal stations. The paper [8] is devoted to a study of the dependence of the evolution of the ice drift velocity field in the Sea of Azov on the direction and time of the wind exposure. A more sophisticated two-dimensional numerical model was used here considering, however, the idealized cases of uniform stable wind forcing.

Improving the quality of numerical modeling is usually achieved through the observational data assimilation. Regular measurements are available from coastal weather stations (CWSs). However, directly over the sea surface such information can be obtained from satellite images. An example of monitoring the emerging sea-ice extent situation and vessel positioning according to the COSMO-SkyMed radar satellites (E-GEOS, Italy) is presented in [10]. In [11], a set of visible and infrared NOAA satellite images made at the Japan station "Furuno Su-8" installed at "YugNIRO" (Kerch) for the winter of 2015–2016 was used as initial data. The main objective of these studies was to construct maps describing the main phases of the ice formation regime, and a description of the main ice characteristics. In [12], a series of optical images was investigated obtained from different satellites by using the satellite service "See The Sea" and the community sharing center "IKI-Monitoring". Based on such observations, the ice drift was analyzed in the Kerch Strait before and after the construction of the Crimean Bridge.

A more complex approach was implemented in [13]. Here MODIS sensor data from Terra and Aqua for 2006 – 2016 as well as *in-situ* measurements from coastal stations were used. They allowed one to obtain quality estimates of the proposed multi-component balance hydrological model, based on which long-term variability of sea-ice cover, sea-ice thickness, and seawater temperature of the Sea of Azov were obtained.

In addition to CWS data, regular observations of sea-ice characteristics in the delta of the Don River are carried out by the South Scientific Center, RAS [14]. Note that other available sources of sea-ice observations (ship and aircraft surveys) collected by the State Oceanographic Institute (SOI) are not open. Nevertheless, the results of a comprehensive analysis of these data taking into account satellite images and CWS data are usually published as an atlas, e.g. [15]. It allows one to obtain a representation of the long-term and spatial variability of the sea-ice characteristics in the basin.

The present paper demonstrates the results of using the NEMO ocean modeling framework [16] to reproduce the long-term variability of sea-ice cover in the basin of the Sea of Azov for further development of the nowcasting/forecasting system for the cascading basins of the Azov, Black, and Marmara Seas. To assess the quality of the simulation results, we used various types of products obtained from satellite observations, the available CWS data, as well as sea-ice modeling results obtained in other works. The paper is organized as follows. A brief description of the LIM3 sea-ice model being used is presented in Section 2. The parameters of the cascading basins configuration developed for the NEMO are given together with a description of the numerical experiments and surface boundary conditions. In addition, the observational data used for preliminary validation and their brief characteristics are presented. Section 3 presents some results of variability analysis of simulated sea-ice fields in 2007 – 2016 and their comparison with CWS observations and satellite data products. Conclusions and discussion of the results are presented in Section 4. Inaccuracy of the used satellite data reprocessing is discussed.

2. Materials and methods

2.1. Brief description of used configuration

In this study, numerical simulation of sea-ice evolution was performed using the NEMO modeling framework [16], i.e. the ocean general circulation model OPA coupled with the LIM3 sea-ice model, whose detailed description can be found in [18]. The model simulates the thermodynamics of snow and ice on the sea surface taking into account ice-drift. In the LIM3 model, global (extensive) thermodynamic state variables are considered: sea-ice concentration, sea-ice volume, snow volume, ice enthalpy, snow enthalpy, sea-ice salt content, and sea-ice age.

In our simulations we use an eddy-resolving (1/24° in the meridional direction ≈ 4.6 km) configuration for the cascading basins of the Azov, Black, and Marmara Seas developed earlier. The domain bottom topography was prepared using an EMODNet bathymetry array [19] (Fig. 1(a), only the area of the Sea of Azov is presented).

Atmospheric fields were obtained using the CORE bulk formula from the ERA5 reanalysis product provided by the European Centre for Medium-Range Weather Forecasts (ECMWF) [20] with a spatial resolution of 1/4°. The time resolution of 1 hour can be important for the reproduction of short-period processes and the diurnal cycle.

The climatic river runoff is prescribed in the Black Sea basin. The monthly data on the volume discharge for the largest rivers in the Azov Sea (Kuban and Don) [21] were used for the simulations. It was shown that this allows better reproduction of an unusual positive tendency of the Azov Sea water salinity observed in recent years. Initial conditions for the Sea of Azov were prepared using objective analysis of *in-situ* observations provided by CMEMS and the oceanographic database of the SeaDataNet project (https://www.seadatanet.org/).

Using the presented configuration, we perform two free run simulations of the cascade general circulation for the period from August 2007 to December 2016. The first one was carried out not taking into account the sea-ice model (exp. 1) using a special correction of the surface boundary conditions for the heat flux. This approach allows only obtaining information about the expected area of the sea-ice cover based on simple criteria for the sea water temperature on the first model level: it should be below the freezing point for the Azov waters (near –0.06 °C).

The second numerical experiment (exp. 2) was carried out in the coupling mode. Simulation allows one to reveal a large number of parameters of the sea state in the basins of the cascade and the sea-ice cover with an output frequency of 1 day^{-1}. Here we analyze only the characteristics related directly to the sea-ice fields. A comparison of the simulation results showed that the correction boundary conditions approach allowed reproducing the position of the boundaries and the size of the sea-ice cover area quite close to the results of exp. 2. Thus, our analysis will be focused more on the latter.

2.2. *Products of reprocessing of satellite measurements*

To validate the simulation results, daily data on the sea-ice concentration from OSTIA [22] with a spatial resolution of 0.05° were used. The product is based on sea surface radiation brightness data from a passive microwave radiometer Special Sensor Microwave Imager Sounder (SSMIS). It uses a combination of modern algorithms and atmospheric correction. The spatial resolution of the data is 0.05°. The data are provided by CMEMS (http://marine.copernicus.eu).

(a)

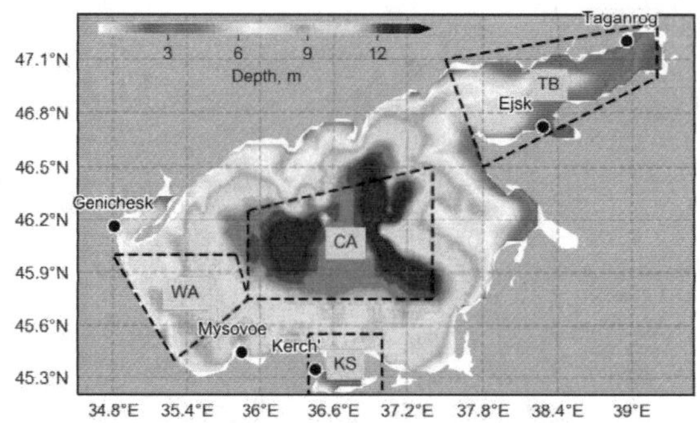

(b)

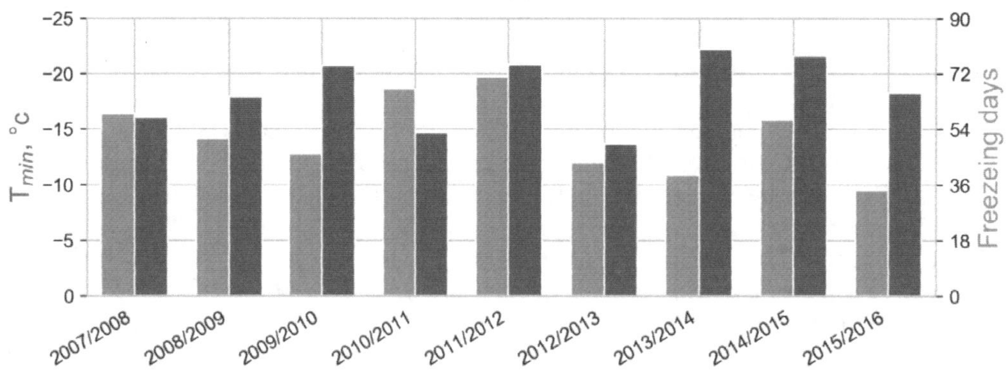

Figure 1. Initial and boundary conditions for numerical simulation: (a) bathymetry (in color) [m] for the Azov Sea basin and a simplified scheme for areas (TB – Taganrog bay, CA – Central Azov, WA – Western Azov, KS – Kerch Strait);
(b) - minimum value of 2 m air temperature (°C) over the Azov Sea in winter (blue) and the number of freezing days (i.e. with an air temperature below 0°C) (orange) according to ERA5 reanalysis product.

On the other hand, the sea surface temperature (SST) and the sea-ice temperature (SIT) data from MODIS / Aqua (https://modis.gsfc.nasa.gov) were obtained using a technique from [23] (the spatial resolution is 0.05°). For certain dates we use a subjective (visual) classification of the area covered by sea-ice from visual imagery with a resolution of 250 m obtained by the same sensor.

3. Results

A simple classification of winters by atmospheric conditions was proposed in [15]. The simulation results showed a rather close classification. As noted above, according to [15] 2007/2008, 2010/2011, and 2011/2012 were classified as mild winters; and 2008/2009, 2009/2010, 2012/2013 and 2013/2014, as warm ones. Figure 1(b) shows the agreement of the reanalysis data from ERA5 with the proposed classification, i.e. one can identify mild winters with a greater number of freezing days than for the warm one. For the latter, rather low air temperatures can be observed, but with a smaller number of such days.

The analysis of the sea-ice cover area (not presented here) and sea-ice concentration according to the results of numerical modeling (Figure 2) allows us to distinguish mild and warm winters presented in Table 1. During mild winters the sea-ice cover is general for the entire Azov Sea basin, while during warm ones sea ice covers about 60% of the basin. The mild winter of 2013/2014 according to [15] can be mentioned here particularly. According to NEMO simulations, the period of sea-ice covering the area of the Azov Sea is quite lasting, over 2 months. This fact is in better agreement with the MODIS / Aqua observations. The same can be revealed from OSTIA reprocessing and MODIS / Aqua images.

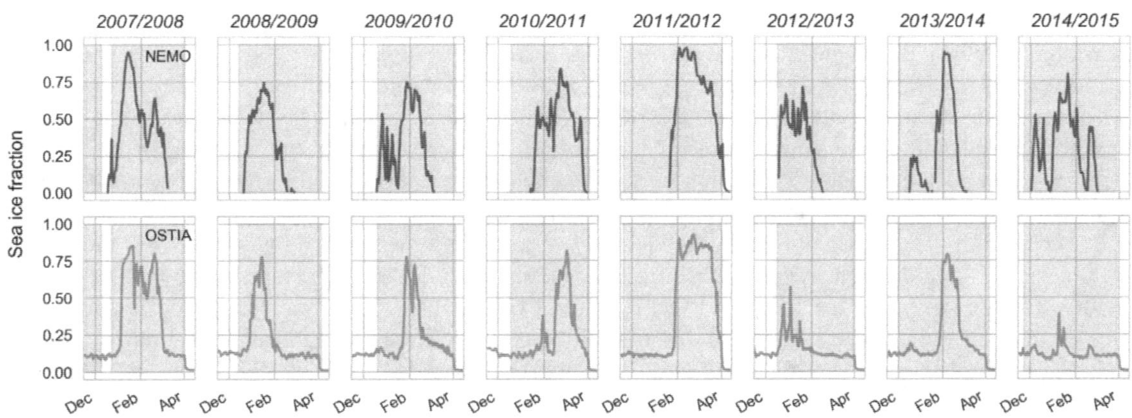

Figure 2. Interannual variability of the basin-averaged sea-ice concentration according to the results of numerical modeling (blue) and to OSTIA (orange); gray indicates periods of sea-ice cover according to the results of MODIS / Aqua treatment.

Table 1. Classification of sea-ice conditions in the Sea of Azov basin according to the results of numerical modeling.

Years	Winter type	Ice creation	Ice vanishing	Lasting, days
2007/2008	mild	2007-12-17	2008-03-09	83
2008/2009	warm	2008-12-20	2009-02-28	66
2009/2010	warm	2009-12-18	2010-03-05	78
2010/2011	mild	2011-01-13	2011-03-30	77
2011/2012	mild	2012-01-20	2012-04-11	82
2012/2013	warm	2012-12-17	2013-02-16	61
2013/2014	mild	2013-12-15	2014-03-06	74
2014/2015	warm	2014-12-02	2015-03-03	91
2015/2016	warm	2016-01-01	2016-02-22	53

When comparing the numerical simulation results and observed data OSTIA, a fairly good agreement is found between the average sea-ice concentrations during the ice formation. In addition, the phase of the local maxima winter matches sufficiently exactly. It is important to note here that according to OSTIA sea ice is observed in December and March. Visual analysis of MODIS / Aqua RGB images is more consistent with exp. 2 throughout the period under study. Moreover, the analysis procedure implemented in OSTIA introduces noticeable artifacts for the coastal areas of the Sea of Azov. This leads to the fact that sea ice in some parts of the water area is observed throughout the year, which, obviously, is not true. In addition, this significantly spoils the representation of the array data in 2016. Therefore, for this year the results of the comparison are not presented.

Figure 3 shows the spatial distribution of the sea-ice thickness. For a mild winter of 2007/2008 the characteristic sea-ice thickness does not exceed 30 cm, except for coastal regions where ice rafting is observed. The water area of the basin was completely covered with ice only two weeks in January. The average sea-ice concentration in this period reaches 95%. Then its area decreases, and by the end of February the sea-ice cover occupies about half of the basin. Complete cleansing of the sea-ice occurs in early March. In the mild winter of 2011/2012 the sea-ice cover begins to form somewhat later (at the end of December). By the end of January the sea-ice thickness reaches a large value (up to 120 cm, off the Crimean coast). A feature of this winter, together with long and cold atmospheric conditions, is a stable northeastern wind, which forms a layer of sea ice in the western part of the sea. The average sea-ice concentration in this period reaches almost 100% and is maintained for almost a month. Complete cleansing of the sea ice occurs in early April.

For a preliminary assessment of the adequacy, we compared the sea-ice thickness from the results of numerical modeling with field measurements taken from [13]. It presents data on the sea-ice thickness of the Azov Sea ice on the CWS (Taganrog, Genichesk, Yeisk) according to the simulation results and measurements from the HMC Russia (hmc.meteorf.ru). For Taganrog, maximum values of thickness were obtained in 2013, 2015, and 2016 and are equal to 50-60 cm. However, according to measurements the maximum thickness reaches 50 cm in 2011/2012. In other years the simulation results correspond to measurements. In Genichesk, differences are observed only for the winter of 2010/2011 where the sea-ice thickness by modeling is more than 50 cm and by measurements – 30 cm. In Yeisk, according to the calculation the maxima were observed in 2010/2011 and 2013/2014 and from measurements in 2011/2012. At station Mysovoye 2011/2012 stands out, where the maximum thickness is 75 cm, and according to HMC data it is only 30 cm. Qualitatively, the simulation results correspond to measurements, except for several events.

The location of the sea-ice edge is compared according to the results of the numerical simulation and satellite data. To do this, we found the areas of intersection and the difference between the areas covered with ice and normalized to the total area of the Sea of Azov. This characteristic allows one to get an idea of the spatial coordination of sea-ice fields. The analysis showed that most of the sea-ice coverage according to MODIS / Aqua data lies within the area occupied by ice obtained in exp. 2.

The results of a comparison of the numerical simulation and OSTIA data are shown in Figure 4. One can see that during the period of ice formation the areas of sea-ice coverage are more than 85% consistent. This is especially true for the mild type of winters: 2007/2008, 2010/2011, 2011/2012, and 2013/2014. In warm winters low values of intersection result from a small area of simulated sea-ice cover. For December and March the difference between the analyzed arrays is about 50%. It can be explained by the presence of false sea ice in OSTIA described above.

The position of the proposed ice coating region obtained from the numerical experiment, exp. 1, quite accurately corresponds to the calculation of exp. 2 (not shown in the Figure 4).

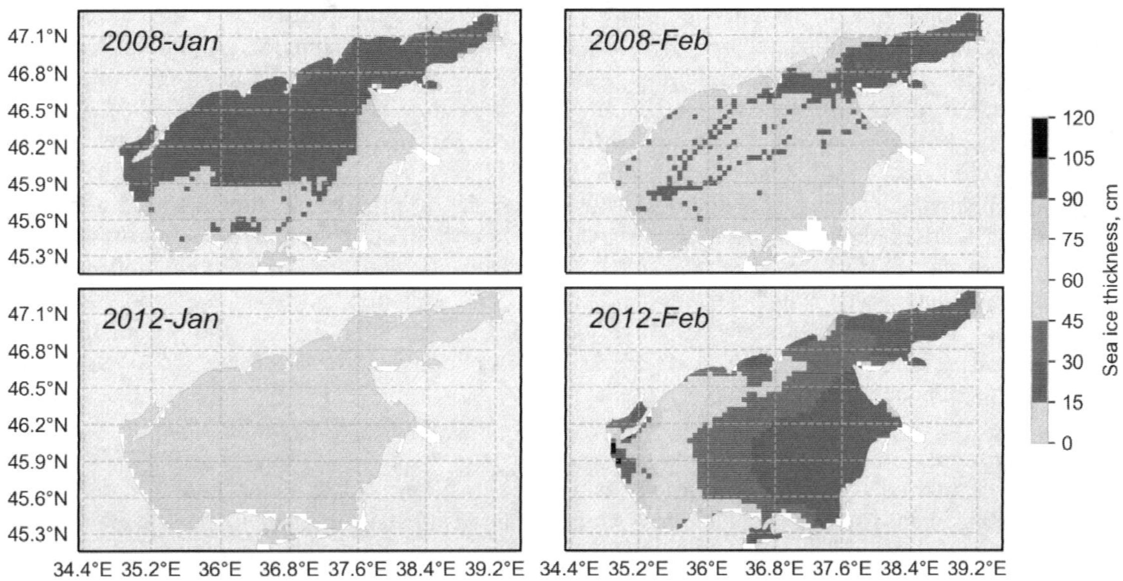

Figure 3. Average monthly sea-ice thickness (cm) according to the results of numerical modeling (exp 2) in January (left) and February (right) for warm (top) and mild (bottom) types of winters.

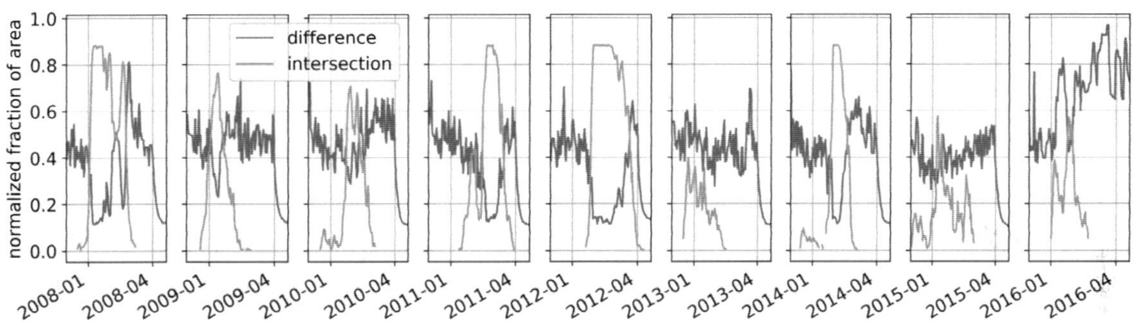

Figure 4. Areas of intersection and difference of sea-ice cover between the simulation results and OSTIA data normalized to the total sea area.

4. Conclusions and discussion

This study presented the results of an attempt to reproduce the spatial and temporal variability of the ice cover of the Sea of Azov obtained by using NEMO modeling framework. A regional configuration of cascading seas was adapted to achieve the objectives. The thermodynamics and dynamics of the sea ice were reproduced by the LIM3 NEMO component. The simulated sea-ice cover parameters from 2007 to 2016, the concentration and thickness, in particular, were verified with satellite observational data products and data from coastal weather stations. The model reproduces a typical pattern of formation of a sea-ice field which does not contradict the previously obtained results and observations. The features of this scheme for the period under study agrees with a classification of winter atmospheric conditions based on the available field data.

The sea-ice concentration averaged over the basin according to the simulation results is in good agreement with the OSTIA satellite measurements reprocessing, despite the fact that the time of presence of sea ice according to the modeling results is shorter. The phase and amplitude of the local maxima of sea-ice concentration during this period seem to be simulated quite well. The noted

disagreement with the OSTIA data may be caused by the inaccuracy of the mapping procedure used to obtain the gridded array, since an analysis of another data from the MODIS / Aqua showed better consistency with the numerical modeling results. The preliminary numerical experiment was performed only with an ocean circulation model but, nevertheless, it allows one to determine the expected areas of sea-ice formation. The size of the sea-ice coverage area in this case is rather close to the results of numerical simulations coupled with the LIM3.
The presented simulation, however, was performed by means of a model developed for the sea ice of Artic regions, and it should be used after accurate tuning by taking into account the regional peculiarities of the basin. The thickness of the modeled sea ice does not agree well with the measurements from coastal weather stations. This can be explained by using a too coarse spatial resolution of the configuration, which was insufficient for reproducing the circulation and thermodynamics of the coastal zone. Nevertheless, the location of the sea-ice edge agrees rather well with the results of numerical modeling.

Acknowledgments
This study was carried out at the Marine Hydrophysical Institute, RAS, under state assignment no. 0555-2019-0002 "Interdisciplinary Numerical Modeling and High-Performance Computing."

References
[1] D'yakov N N, Ivanov V A and Gorbach S B 2002 Sezonnaya i mezhgodovaya izmenchivost' gidrologicheskikh kharakteristik pribrezhnoy zony Azovskogo morya *Ekologicheskaya bezopasnost' pribrezhnoy i shel'fovoy zon morya* **6** 39-46
[2] Dashkevich L V, Kulygin V V and Berdnikov S V 2017 Many-year variations of the average salinity of the sea of Azov *WATER RESOURCES* **44** 749-57
[3] Aleskerova A A, Kubryakov A A, Goryachkin Yu N and Stanichny S V 2017 Propagation of waters from the Kerch Strait in the Black Sea *Physical Oceanography* **4** 47-57
[4] Korotaev G K, Oguz T, Dorofeyev V L, Demyshev S G and Kubryakov A I 2011 Development of Black Sea nowcasting and forecasting system *Ocean Sci.* **7** 629–49
[5] Popov S K and Lobov A L 2017 Modelirovaniye izmeneniy urovnya Azovskogo morya v 2015–2016 godakh *Trudy Gidrometeorologicheskogo nauchno-issledovatel'skogo tsentra Rossiyskoy Federatsii* **364** 131-43
[6] Dobrovol'skiy A D and Zalogin B S 1982 Morya SSSR (Moscow:Izd-vo MGU) p 192
[7] Dumanskaya I O 2014 Ledovyye usloviya morey yevropeyskoy chasti Rossii (Moskow, Obninsk: IG–SOTSIN) p 608
[8] Ivanov V A and Ryabtsev Y N 2018 Simulation of formation and melting of ice in the Kerch Strait *Russian Meteorology and Hydrology* **43** 36-42
[9] Bukatov A E, Zav'yalov D D and Solomakha T A 2012 Numerical modeling of ice drift in the Sea of Azov *Russian Meteorology and Hydrology* **37** 6 379–85
[10] Pietranera L, Cezarano L and Kantemirov YU I 2012 Primer monitoringa ledovoy obstanovki i sudokhodstva v zamerzshey akvatorii Azovskogo morya i Kerchenskom prolive po dannym COSMO-SkyMed *GEOMATIKA* 1 72–76
[11] Borovskaya R V 2017 Issledovaniye ledovykh usloviy Azovskogo morya i kerchenskogo proliva v zimniy period 2015-2016 gg. I otsenka ikh vliyaniya na promyslovuyu obstanovku i protsess migratsii i nagul ryby *Trudy Yuzhnogo nauchno-issledovatel'skogo instituta rybnogo khozyaystva i okeanografii* **54** 35-41
[12] Lupyan Ye A, Lavrova O Yu, Mityagina M I and Kostyanoy A G 2017 Ledovaya obstanovka v rayone stroitel'stva Krymskogo mosta v fevrale 2017 g *Sovremennyye problem distantsionnogo zondirovaniya Zemli iz kosmosa* **14** 247–51
[13] Dashkevich L V, Nemtseva L D and Berdnikov S V 2016 Otsenka ledovitosti Azovskogo morya v XXI veke po sputnikovym snimkam Terra/Aqua MODIS i rezul'tatam matematicheskogo modelirovaniya *Sovremennyye problemy distantsionnogo zondirovaniya*

Zemli iz kosmosa **13** 91–100

[14] Matishov G G, Berdnikov S V, Sheverdyaev I V and Chikin A L 2014 The extreme flood in the Don river delta, March 23-24, 2013, and determining factors *Doklady Earth Sciences* **455** 360-3

[15] D'yakov N N, Timoshenko T Yu, Belogudov A A and Gorbach S B 2015 Atlas l'dov Chernogo i Azovskogo morey *Sevastopol'skoye otdeleniye federal'nogo gosudarstvennogo byudzhetnogo uchrezhdeniya "Gosudarstvennyy okeanograficheskiy institut imeni N.N.Zubova" Sevastopol'* p 219

[16] Madec G and NEMO System Team 2016 NEMO ocean engine (Scientific Notes of Climate Modelling Center, Institut Pierre-Simon Laplace IPSL) pp 1288-619

[17] Mizyuk A I, Grigoriev A V, Puzina O S and Lishaev P N 2019 Long-term variability of the thermohaline characteristics of the Azov Sea by means of numerical eddy-resolving model *Physical Oceanography* **5** (in press)

[18] Rousset C *et al* 2015 The Louvain-La-Neuve sea ice model LIM3.6: global and regional capabilities *Geosci. Model Dev.* **8** 2991– 3005

[19] EMODnet Bathymetry Consortium EMODnet Digital Bathymetry (DTM) 2018 http://www.emodnet-bathymetry.eu

[20] Copernicus Climate Change Service (C3S) 2017 ERA5: Fifth generation of ECMWF atmospheric reanalyses of the global climate. Copernicus Climate Change Service Climate Data Store (CDS) (lass access: 22.08.2018) https://cds.climate.copernicus.eu/cdsapp#!/home

[21] Database «Osnovnyye gidrologicheskiye kharakteristiki morskikh ust'yev rek yevropeyskoy territorii Rossii» Certificate of state registration of the database No. 2012020681 July 18 2012

[22] McLaren A, Fiedler E, Roberts-Jones J, Martin M, Mao C and Good S 2016 Global Ocean OSTIA Near Real Time Level 4 Sea Surface Temperature Product SST-GLO-SST-L4-NRT-OBSERVATIONS-010-001 p 23

[23] Hall D K, Riggs G A and Salomonson V V 2006 MODIS snow and sea ice products. *Earth science satellite remote sensing* (Berlin Heidelberg: Springer) pp 154-81

The impact of various types of open boundary conditions in numerical simulation of the Black Sea coastal circulation

M V Senderov

Marine Hydrophysical Institute, RAS, Kapitanskaya str. 4, Sevastopol, 299011, Russia

E-mail: senderovmaxim@gmail.com

Abstract. An important problem is to obtain a short-term forecast in the Black Sea coastal zone with a high resolution. It is solved on the basis of numerical modeling of ocean circulation. Carrying out simulations that allow reproducing the meso- and submesoscale processes in a reasonable machine time requires a considerable computational power. One of the common ways to solve this problem is to create a regional configuration (a nested domain). This work is performed to study the influence of methods of specifying open boundary conditions. To conduct such a study, a regional configuration corresponding to the southern coast of Crimea with a resolution of 1.5 km is developed. Four short-term numerical experiments are carried out to investigate the influence of various types of open boundary conditions on the simulation results. The results of the simulations have shown that for a correct reproduction of the hydrophysical parameters, barotropic velocities at the open border should not be taken into account. At the open boundary, in the fields of temperature and salinity, there is no numerical noise when using a flow relaxation scheme. The sea level is reproduced correctly when using a Flather-type scheme.

1. Introduction

Since 2012, Marine Forecasting Center has been established at the Marine Hydrophysical Institute of RAS (MHI). It is still operating, allowing one to obtain information on the general circulation in the Black Sea in a real-time mode with quite good resolution (about 5 km). This system rather accurately describes the main features of the general circulation and the vertical structure of the basin: the Rim Current, its meandering, the location of mesoscale eddies, the depth of the cold intermediate layer (CIL), etc. To increase the quality of marine forecasts, reproduction of the large-scale processes is not sufficient. One should increase the resolution of the forecasting model to meso- and submesoscales. Processes of such scales are usually observed in the coastal and shelf areas. In these areas, a spatial resolution of ~ 5 km is not sufficient to adequately reproduce dynamic structures. In the coastal and shelf areas, both quasi-persistent and moving eddies of various scales and intensities are constantly observed. Therefore, reconstruction of the dynamics in the coastal areas should be carried out with a finer spatial resolution. Carrying out high-resolution simulations of the entire basin for an adequate machine time requires large computing capabilities. One of the solutions is modeling with an approach of nested grids. This method allows one to perform model runs of the restricted area of investigated domain using a fairly small grid step.

Due to small scales of the coastal processes, simulation results need to be validated against different types of observations obtained from a sub-satellite polygon. For the Black Sea, such studies are possible using data from the Black Sea Branch of MHI.

Considering the Black Sea basin, such studies are carried out at MHI by means of a numerical circulation model developed in [1]. Most of them are devoted to the reconstruction of currents of the northwestern shelf of the Black Sea [2]. This area, as the shallowest part of the basin, is of great interest for study. Another part of the studies is devoted to the reproduction of thermohaline fields and circulation in the coastal area of Crimea [3]. Upwelling phenomena are observed in this region. The vicinity of the Rim Current to the coast affects the generation of a large number of mesoscale and submesoscale eddies. Also, less salty and warmer waters of the Azov Sea flowing through the Kerch Strait spread along this coast. The results of numerical simulations in the described areas are verified with in-situ data of ARGO buys, field data of Black Sea polygons, and during cruises of the R/V Professor Vodyanitsky.

When creating a nested domain or a regional configuration, the main impact on the internal region is carried out by setting open boundary conditions. A problem arises of setting open boundary conditions in the computational domain. A wide variety of methods for setting open boundary conditions is associated with the features that arise during their implementation [4],[5],[6]. The circuits described in [4] accumulate energy and cause a recirculation problem along the open boundary. To correctly reproduce the tidal impact in [5], a modification of the tidal scheme for the ROMS model is proposed. Some features of the implementation of the Flather scheme for barotropic variables and the scheme proposed by Orlanski for baroclinic velocities are given in [6].

Within the framework of the RSF project, MHI plans to create a nowcast/forecast system of the entire oceans [7]. To obtain a forecast in the Black Sea basin, the NEMO (Nucleus for European Modeling of the Ocean) [8] model will be used. It is well parallelized and well established in European forecast centers. Therefore, it is used in this work, whose purpose is to reproduce thermohaline fields in the coastal zone off the southern coast of Crimea. The main goal of the work is to study the influence of methods for setting open boundary conditions on the results of numerical modeling.

In this paper, the first part describes the model used to achieve the goals and the boundary conditions used in the calculations. The second part is devoted to ae description of the configurations used in the calculations. The third part is devoted to an analysis of the influence of boundary conditions and a comparison of the results with the results of the basin configuration.

2. Methods and materials

2.1. Brief description of the OGCM and open boundary conditions

This study was carried out mainly using the NEMO (Nucleus for European Modeling of the Ocean) modeling framework [8]. The ocean component of NEMO (OPA) is based on primitive equations. Choosing an orthogonal set of unit vectors **i, j, k** (**k** is the local upward vector, and **i, j** are two vectors orthogonal to **k**), the equations are as follows:

$$\frac{\partial U_h}{\partial t} = -\left[(\nabla \times U) \times \mathrm{U} + \frac{1}{2}\nabla(U^2)\right]_h - \mathrm{fk} \times U_h - \frac{1}{\rho_0}\nabla_h \mathrm{p} + \mathrm{D}^U + \mathrm{F}^U, \tag{1}$$

$$\frac{\partial p}{\partial z} = -\rho g, \tag{2}$$

$$\nabla \cdot \mathrm{U} = 0, \tag{3}$$

$$\frac{\partial T}{\partial t} = -\nabla(\mathrm{TU}) + \mathrm{D}^T, \tag{4}$$

IOP Conf. Series: Earth and Environmental Science **386** (2019) 012024 doi:10.1088/1755-1315/386/1/012024

$$\frac{\partial S}{\partial t} = -\nabla(\mathrm{SU})+\mathrm{D}^S, \quad (5)$$

$$\rho=\rho(\mathrm{T,S,p}), \quad (6)$$

$$\mathrm{w}=\frac{\partial \eta}{\partial t}+\mathrm{U}_{h|z=\eta}\cdot \nabla_h(\eta)+\mathrm{P}+\mathrm{R}-E, \quad (7)$$

where **U** is the current velocity vector; $\mathbf{U}_h$ is the current velocity in the (**i,j**) - plane;
$f = 2\,\mathbf{\Omega}\,\mathbf{k}$ is the Coriolis acceleration;
Ω is the Earth's angular velocity vector,
η is the sea surface height,
P is the precipitation,
R is the river runoff,
E is the evaporation;
$\mathbf{F}^{\mathrm{U}}$ are the surface forcing terms;
$\mathbf{D}^{\mathrm{U}}$, D^T, D^S are the parameterizations of small-scale physics for the momentum, temperature, and salinity.

The equation of state (6) is the *UNESCO* formula [9]. The time splitting scheme is used for the calculation of the sea-surface height [10].

The system of equations is discretized using Arakawa's "C" grid [11]. We use the vector invariant form of the momentum equation (1) with the energy and enstrophy conservation scheme [12]. The *TVD* scheme [13] is used for the discretization of the nonlinear terms in the transport–diffusion equations (4) and (5). The vertical mixing is described using a *k-ε* turbulent closure model [14].

In NEMO, for each open boundary set a boundary condition has to be chosen for the sea-surface height, barotropic and baroclinic velocities, temperature, and salinity. We can choose a scheme for each variable. In this research, three of them are considered.

The Flow Relaxation Scheme (FRS) applies a simple relaxation of the model fields to externally-specified values over a zone next to the edge of the model domain [8],[15]. Given a model prognostic variable Φ,

$$\Phi(d)=\alpha(d)\Phi_e(d) + \big(1-\alpha(d)\big)\Phi_m(d);\mathrm{d}=1,\mathrm{N},$$

where Φ_e is the specified external field;
Φ_m is a model solution;
d is the discrete distance from the boundary of the computational domain, and α is a parameter that varies from 1 at $d = 1$ to a small value at $d = N$. The function α (d) is specified as a *tanh* function:

$$\alpha(d) = 1-\tanh\left(\frac{d-1}{2}\right);\mathrm{d}=1,N$$

Thus, the model solution is completely prescribed by the external conditions at the edge of the model domain and is relaxed towards the external conditions over the rest of the FRS zone. The application of a relaxation zone helps to prevent spurious reflection of outgoing signals from the model boundary. The parameter d usually ranges from 8 to 10.

The Flather scheme is a radiation condition on the normal, depth-mean transport across the open boundary [8,16]. It takes the form:

$$\mathrm{U}=\mathrm{U}_e+\frac{c}{h}(\eta-\eta_e),$$

where U is the depth-mean velocity normal to the boundary and η is the sea surface height, both from the model. The subscript e indicates the same fields from external sources. The speed of external gravity waves is given by $c=\sqrt{gh}$, and h is the depth of the water column. The depth-mean normal velocity along the edge of the model domain is set equal to the external depth-mean normal velocity, plus a correction term that allows gravity waves generated internally to exit the model boundary.

When using Dirichlet boundary conditions the values at the boundary of the region are equal to the values from the external (basin) configuration.

2.2. Model setup and numerical experiments

A regional configuration with a spatial resolution of ≈ 4.6 km ($^1/_{24}°$) was developed recently as the starting point for the new version of the Black Sea Marine Forecasting Center (BS MFC). However, the model domain covers cascading basins of the Azov, Black, and Marmara Seas (Figure 1). It can be considered as an extension of a configuration of the Black and Azov Seas proposed in [17]. The NEMO cascading configuration is being continuously adapted to the operational mode and validated against observations showing rather adequate results of the Black Sea general circulation and vertical structure, exchange through the Bosporus and Kerch Straits, the long-term salinity tendency in the Azov Sea [18].

Bearing in mind the location of the experimental polygon of the Black Sea Branch of MHI RAS, we choose an area covering the south-eastern coast of Crimea (Figure 1). As the starting point, the factor of 3 was used to produce the nested domain inferring the grid cell size is ≈1.5 km. The bottom topography for the parent and child domains was prepared using an EMODnet bathymetry set [19] (Figure 1). The same setup of z-levels for the domains was used expecting further simulations using the nested grids technique.

The surface boundary conditions for the numerical experiments are based on the ERA5 product provided by the European Center for Medium-Range Weather Forecasts (ECMWF) [20]. The product has a spatial resolution of 0.25° resolving the atmosphere mesoscale processes, and can be fairly satisfactory for the proposed study. Heat, mass, and momentum fluxes on the surface are obtained using a CORE bulk-formula [21].

The bilaplacian formulation is used for the representation of the lateral mixing ($\mathbf{D}^U$, D^T, and D^S in equations (1), (4) and (5). The long-term simulations of the general circulation in the cascading basins were made using values of $-\ 5\times10^9$ m^4/s and $-\ 4\times10^8$ m^4/s for the lateral turbulent viscosity and diffusion. Based on the results of preliminary simulations using a nested domain configuration, their values were chosen as large as -5×10^8 m^4/s and -5×10^7 m^4/s, respectively.

The free-slip boundary condition is used for the solid boundaries with a zero flux across the boundaries. On the bottom we use quadratic friction. No heat and salt flux condition is set for the temperature and salinity.

A preliminary free-run simulation for 1/1/2008 – 12/31/2008 was carried out using a configuration of cascading basins (basin). Conditions for temperature, salinity, baroclinic and barotropic velocities and sea-surface height prescribed on the liquid open boundaries for the nested domain configuration were obtained from respective results of a preliminary basin model run. After that, four model runs were performed using different combinations for prescribing open boundary conditions (Table 1). All experiments were carried out for the period of a month. The results of the experiments were compared with each other and with those obtained from the basin simulation.

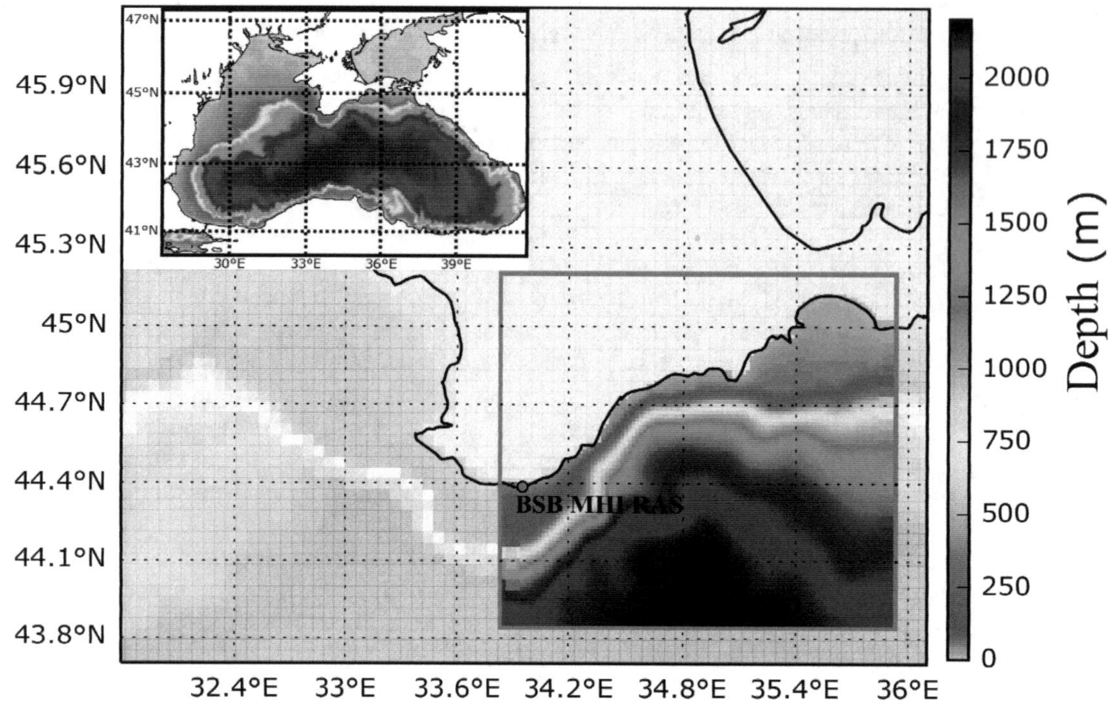

Figure 1. Geographical location and bathymetry in the area under study: nested domain (dark-grey lines) within the domain of cascade configuration.

The Flather radiation scheme is prescribed for sea-surface height for all experiments. In the first experiment, the FRS was used for baroclinic velocities, and temperature and salinity. In the second experiment, the Dirichlet conditions were used for temperature and salinity, and baroclinic velocities were prescribed by the FRS (Table 1).

Table 1. Model runs and types of used open boundary conditions.

	Open boundary conditions			
	Sea-surface height	Temperature and salinity	Baroclinic velocity	Barotropic velocity
Exp. 1	Flather	FRS	FRS	none
Exp. 2	Flather	Dirichlet	FRS	none
Exp. 3	Flather	Dirichlet	none	Flather
Exp. 4	Flather	FRS	none	Flather

3. Results

Analyzing the spatial structure of simulated temperature and salinity, one can note their similarity. One can see that the sea water temperature on the first model level from exp. 1 and exp. 2 are rather close to the results obtained using the basin configuration. We can observe numerical noise at the lateral boundary in exp. 2, which is similar to two-step waves. They could appear due to the reflected wave. In the first experiment this is not observed, because a flow relaxation scheme was used for temperature and salinity, which prevents their appearance. The vertical temperature distribution in these experiments is also close to the basin one. In experiment 2, at an open boundary numerical noise is observed to a depth of 100 m.

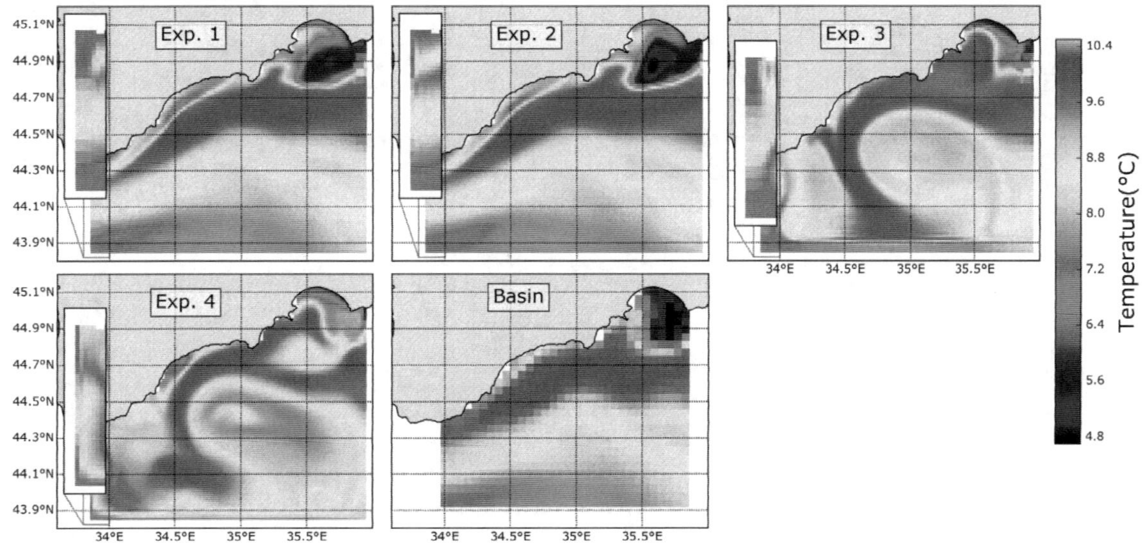

Figure 2. Sea water temperature (°C) on the first z-level for January 26, 2008.

The temperature and salinity fields in this simulation have different structures from the fields obtained in the basin configuration and in the first experiment (Figure 2).

The last, fourth experiment was carried out with the schemes used in the second experiment (sea-surface height - Flather scheme, barotropic velocities - flow relaxation scheme, temperature and salinity - Dirichlet type).

The results obtained in the last two experiments in the fields of temperature and salinity (Figure 2) have significant differences in the results of the basin configuration. The average temperature for the regional domain is close to experiments 1, 2, and 4. In the third experiment starting from January 8, the average temperature increases by 0.2 °C. Cyclone formation is observed in the center of the computational domain, which prevents the spread of warmer water from the Sea of Azov along the coast of Feodosia (Figure 2).

The formation of cyclone eddies can be more clearly observed in the field of flow velocities and in changes in the sea-surface height (Figure 3). The diameter of the cyclone is about 100 km. The reproducible cyclone exerts its influence on all levels. There is a significant decrease in the sea level in the center of the cyclone. In the first two experiments, the average sea-surface height is qualitatively and quantitatively close to each other and to the sea-surface height of the basin configuration.

The effect of using barotropic velocities in experiments 3 and 4 can be estimated by analyzing the average kinetic energy over the domain (Figure 4). The cyclone formation was directly reflected in the kinetic energy values. For the latest simulations it increased compared with the basin and experiments 1 and 2 by 10 times.

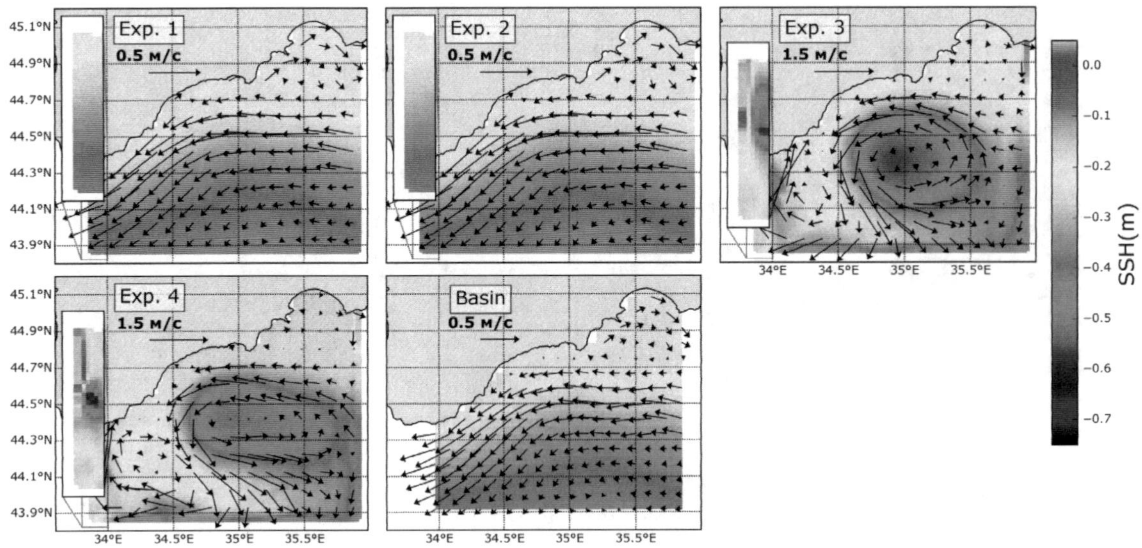

Figure 3. Sea-surface height and surface current for January 26, 2008.

Since the last two simulations qualitatively poorly reproduce the thermodynamic parameters, it should be noted that barotropic velocities at the open lateral boundary should not be taken into account.

Analyzing the first part of the experiments, the following can be noted. In the first two simulations, the temperature and salinity fields are reproduced qualitatively and quantitatively close to the data obtained in the global configuration. In the first experiment, a flow relaxation scheme was used for these fields; numerical noise does not occur at the open lateral boundary, because this scheme prevents the appearance of waves reflected from the boundary. The use of barotropic velocities (experiments 3 and 4) leads to the appearance of a cyclone, which is not in the basin configuration. Its formation prevents the spreading of warm waters along the coast. It should be noted that the method of setting boundary conditions in the first experiment makes it possible to most faithfully reproduce the hydrophysical parameters being studied.

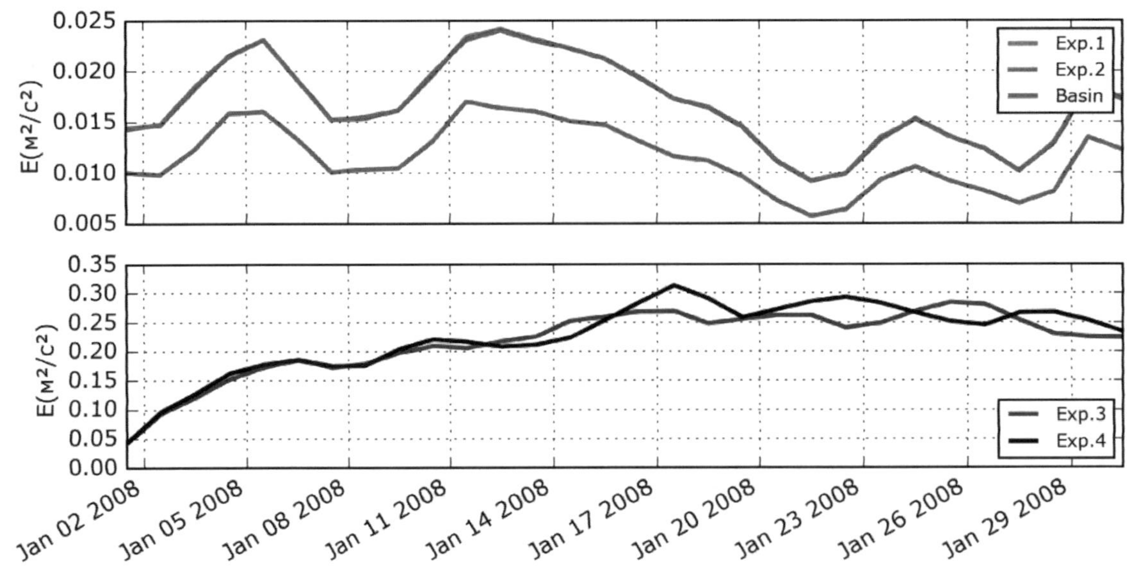

Figure 4. Average kinetic energy in the domain.

IOP Conf. Series: Earth and Environmental Science **386** (2019) 012024 doi:10.1088/1755-1315/386/1/012024

Conclusions

In the present study, a numerical simulation of physical parameters in the coastal area near the southeastern part of the Crimean Peninsula was performed. A nested grid configuration was developed with a spatial resolution of 1.5 km. Four short-term experiments were carried out with various types of open boundary conditions. In the fields of temperature and salinity at the open boundary there is no numerical noise when using the flow relaxation scheme. This scheme reduces or prevents the reflection of waves from the border. The sea-surface height is reproduced correctly by the Flather scheme. Analysis of the spatial structure of the simulated parameters has shown that for a correct reproduction of the hydrophysical parameters the barotropic velocities at the open boundary should not be taken into account. From this we conclude that the schemes selected in the first calculation most faithfully reproduce the hydrophysical parameters.

Acknowledgments

This work was performed under state assignment no. 0555-2019-0002 on "Interdisciplinary Numerical Modeling and High-Performance Computing".

References

[1] Demyshev S and Korotaev G 1992 Chislenaya energosbalansirovanaya model baroklinich techenij okeana s nerovnym dnom na setke C Chislenye modely i rezultaty kalibrovochnich raschetov techenij v Atlanticheskom okeane Moscow INM RAS p 163 – 231

[2] Demyshev S and Evstigneeva N 2013 Analiz gidrofizicheskich polej na severo-zapadnom shelfe Chernogo morya *Oceanology* **53** 585–95

[3] Demyshev S and Evstigneeva N 2016 Analiz gidrofizicheskich polej v rayone uznogo berega Kryma na osnove rascheta s visokim prostranstvenum razresheniem (Sevastopol: Ecologicheskaya bezopasnost pribrejnoj i shelfovoj zon morya) **3** p 21-5

[4] Xinping Chen , Chuanyu Liu, Kieran O'Driscoll, Bernhard Mayer, Jian Su and Thomas Pohlmann 2013 On the nudging terms at open boundaries in regional ocean models *Ocean modeling* **66** 14-25

[5] Zhiqiang Liu and Jianping Gan 2016 Open boundary conditions for tidally and subtidally forced circulation in a limited-area coastal model using the Regional Ocean Modeling System (ROMS) *Journal of Geophysical Research: Oceans* **121** 6184-203

[6] Marsaleix P, Auclair F and Estournel C 2006 Considerations on open boundary conditions for regional and coastal ocean models *Journal of atmospheric and oceanic technology* **23** 1604 – 13

[7] Korotaev G 2018 Operational Oceanography: A New Branch of Modern Oceanological Science *Herald of the Russian Academy of Sciences* **88** 272–80

[8] Madec G and NEMO System Team 2016 NEMO ocean engine *Scientific Notes of Climate Modelling Center, Institut Pierre-Simon Laplace IPSL* **1288-1619**

[9] UNESCO 1983 Algorithms for computation of fundamental property of sea water *Techn. Paper in Mar. Sci* **44**

[10] Roullet G and Madec G 2000 Salt conservation, free surface, and varying levels: a new formulation for ocean general circulation models *J. Geophys. Res.* **23**

[11] Mesinger F and Arakawa A 1976 Numerical methods used in Atmospheric models *GARP Publication Series* **17**

[12] Arakawa A and Hsu Y-J G 1990 Energy conserving and potential-enstrophy dissipating schemes for the shallow water equations *Mon. Wea. Rev.* **118** 1960–69

[13] Zalesak S 1979 Fully multidimensional flux corrected transport algorithms for fluids *J. Comput. Phys.* **31**

[14] Rodi W 1987 Examples of calculation methods for flow and mixing in stratified fluids *J. Geophys. Res.* 5305-28

[15] Davies H 1976 A lateral boundary formulation for multi-level prediction models. *Quart. J. Roy. Meteor. Soc.* **102** 405–18

[16] Flather R 1991 A storm surge prediction model for the northern bay of Bengal with application to the cyclone disaster in april *J. Phys. Oceanogr.* **24** 172–90

[17] Mizyuk A, Senderov M, Korotaev G and Sarkisyan A 2016 Features of the horizontal variability of the sea surface temperature in the Western Black Sea from high resolution modeling *Izvestiya Atmospheric And Oceanic Physics* **52**

[18] Mizyuk A I, Grigoriev A V, Puzina O S and Lishaev P N 2019 Long-term variability of thermohaline characteristics of the Azov Sea based on eddy-resolving ocean model *Physical oceanography* **5** (*Preprint*)

[19] EMODnet Bathymetry Consortium EMODnet Digital Bathymetry (DTM) 2018 *http://www.emodnet-bathymetry.eu*

[20] Copernicus Climate Change Service (C3S) 2017 ERA5: Fifth generation of ECMWF atmospheric reanalyses of the global climate *Copernicus Climate Change Service Climate Data Store (CDS)[lass access: 22.08.2018]*

[21] Large W G and Yeager S 2004 Diurnal to decadal global forcing for ocean and sea-ice models: the data sets and flux climatologies *NCAR Technical Note CGD Division of the National Center for Atmospheric Research*

IOP Conf. Series: Earth and Environmental Science **386** (2019) 012025 doi:10.1088/1755-1315/386/1/012025

Deterministic and stochastic parameterizations of kinetic energy backscatter in the NEMO ocean model in Double-Gyre configuration

P Perezhogin

Marchuk Institute of Numerical Mathematics of the Russian Academy of Sciences, Moscow, Russia

E-mail: pperezhogin@gmail.com

Abstract. Eddy-permitting numerical ocean models often resolve mesoscale turbulence only partly, which leads to underestimation of eddy kinetic energy (EKE). Mesoscale dynamics can be amplified by using kinetic energy backscatter (KEB) parameterizations returning energy from the unresolved scales. We consider two types of KEB: stochastic and negative viscosity ones. The tuning of their amplitudes is based on a local budget of kinetic energy, thus, they are "energetically-consistent" KEBs. In this work, KEB parameterizations are applied to the NEMO ocean model in Double-Gyre configuration with an eddy-permitting resolution (1/4 degree). To evaluate the results, we compare this model with an eddy-resolving one (1/9 degree). We show that with the KEBs the meridional overturning circulation (MOC), meridional heat flux, and sea surface temperature (SST) can be significantly improved. In addition, a better match has been found between the time power spectra of the eddy-permitting and the eddy-resolving model solutions.

1. Introduction

Numerical ocean models used in climate research [1] have a relatively coarse resolution to properly resolve mesoscale eddies, which contribute largely to the transport of tracers and momentum. Mesoscale eddies emerge on the length scale of the Rossby radius of deformation due to baroclinic instability, and in the midlatitude ocean their qualitative length scale is 30 km. Considering the ability to resolve mesoscale eddies, the ocean general circulation models (OGCMs) can be divided into three groups: "non-eddy-resolving" models have an approximate resolution of 1° and cannot simulate the generation of mesoscale eddies; "eddy-resolving" models have several mesh points for a mesoscale eddy and their resolution is 10 times finer. Finally, there are models named "eddy-permitting" which have an intermediate resolution. In the eddy-permitting models eddies are represented in a computational grid, but their dynamics and generation are damped. Currently the OGCMs in climate models are moving from non-eddy-resolving to eddy-permitting resolutions.

In the non-eddy-resolving models there are two main parameterizations of subgrid eddies (i.e. unrepresented in the grid): usual turbulent diffusion/viscosity and Gent-McWilliams (GM) diffusion of layer thickness [2]. The former models turbulent mixing of resolved fields by subgrid eddies, while the latter reduces the slope of isopycnals, thereby introducing a sink of available potential energy (APE) required to mimic baroclinic instability occurring on unresolved scales. Turbulent diffusion at non-eddy-resolving resolutions is usually designed with a Laplace operator acting along isopycnal surfaces

[3]. Subgrid parameterizations at eddy-permitting resolution are substantially different. Diffusion and viscosity operators are usually changed to a squared Laplace (biharmonic) one acting along the horizontal direction. Thus, its purpose is not to perform mixing, but to ensure stability of the simulation to reduce damping of mesoscale eddies. Also, the GM parameterization is not recommended for the eddy-permitting models [4].

Kinetic energy backscatter (KEB) is a parameterization which returns energy from unresolved turbulent scales to resolved ones, contrary to eddy viscosity. This parameterization models poorly resolved inverse energy cascade inherent to 2D turbulence [5]. The first turbulence closures of this type were tested in a simple barotropic model [6] and later in an atmosphere ensemble prediction system [7]. Recently it was proposed to use KEB parameterizations to improve eddy-permitting ocean simulations [8], [9]. The simplest KEB closures are based on the Laplace operator with a negative viscosity coefficient [8], [9] or on a stochastic tendency [8], [10]. Both types of the KEB were shown to be effective in restoration of the barotropic eddy kinetic energy spectrum [8], [9], [11], and they can improve the mean flow [8], [9]. In spite of the success of these parameterizations in simulations with simplified quasi-geostrophic equations, we can mention only one paper where the KEB was applied to a primitive-equation ocean model [12].

In this work we apply two KEB parameterizations, negative viscosity and stochastic ones, to improve the NEMO ocean model [3] at coarse eddy-permitting resolution ($1/4^o$) in Double-Gyre configuration [13]. Intending to improve the mesoscale physics, we compare coarse models with respect to an eddy-resolving one ($1/9^o$), which allows for mesoscales eddies, but not submesoscale ones. Both KEBs are "energetically consistent", which means that we tune the KEB amplitude to compensate the energy loss due to eddy viscosity. Similarly to [9] and [12], we define the negative viscosity KEB as a Laplace operator with a negative viscosity coefficient depending on the amount of subgrid energy. The computation stability is ensured by biharmonic eddy viscosity damping. The stochastic KEB has the functional form of a random streamfunction weighted by the local dissipation rate, as proposed in [7]. The spatial correlation is controlled by multiple applications of a simple spatial filter to the generated spatial white noise, as proposed in [10]. Both KEB parameterizations increase the eddy kinetic energy (EKE) in an eddy-permitting model. It makes the meridional eddy heat flux close in eddy-permitting and eddy-resolving models. As a result, the sea surface temperature (SST) and the meridional overturning circulation (MOC) were restored with the use of KEB. Finally, we report improvements in the time power density spectrum.

2. Double Gyre setting

We use an open access NEMO model (version 3.6) [14] in Double-Gyre configuration described in [13]. Note that there is some discrepancy in the parameters between the open access Double Gyre code and paper [13]. We hope that this is the reason why we have some quantitative differences. If it is not mentioned, we do not change the source code. The ocean model solves primitive equations (the vectors are in bold):

$$\frac{dT}{dt} = F_T, \frac{dS}{dt} = F_S, \quad (1)$$

$$\frac{\partial \boldsymbol{U}_h}{\partial t} + adv_h + cor_h = -\frac{1}{\rho_0}\nabla_h p + \boldsymbol{F}_{\boldsymbol{U}_h}, \quad (2)$$

$$\frac{\partial \eta}{\partial t} = -H\nabla_h \overline{\boldsymbol{U}}_h, \quad (3)$$

$$\frac{\partial p}{\partial z} = -\rho g, \nabla \cdot \boldsymbol{U} = 0, \quad (4)$$

$$\rho = \rho_0\big(1 - a(T - T_0) + b(S - S_0)\big), \quad (5)$$

where $T, S, \boldsymbol{U}, \eta, \rho, p$ are the potential temperature, salinity, velocity, free surface height, density, and pressure; $\boldsymbol{U}_h$ is the horizontal part of the velocity, and $\overline{\boldsymbol{U}}_h$ is its vertical average. $F_T, F_S, \boldsymbol{F}_{\boldsymbol{U}_h}$ are the external forcings and physical parameterizations. We introduce a Lagrangian derivative $\frac{d}{dt} = \frac{\partial}{\partial t} +$

$(\boldsymbol{U}\cdot\nabla)$, a nabla operator $\nabla=(\partial_x,\partial_y,\partial_z)$, and its horizontal part $\nabla_h=(\partial_x,\partial_y)$. Exact expressions for the advective and Coriolis terms adv_h and cor_h are given in [14]. In our configuration the free surface equation (3) and the equation of state (EOS (5)) are linear. The EOS parameters are as follows: $\rho_0=1026\,kg\,m^{-3}$, $a=2\cdot10^{-4}K^{-1}$, $b=7.7\cdot10^{-4}psu^{-1}$, $T_0=10C^o$, $S_0=35psu$. The computational domain is a flat-bottomed rectangular box $L_x\times L_y\times L_z=3180\,km\times2120\,km\times 4\,km$ in the β-plane approximation with free-slip and no heat flux, no salt flux spatial boundary conditions, and quadratic bottom drag. The basin is centered at $\sim30^oN$ and rotated by 45° to the zonal direction, in lat-lon coordinates; it is shown in Figure 1. Note that the metric terms inherent to spherical geometry are excluded.

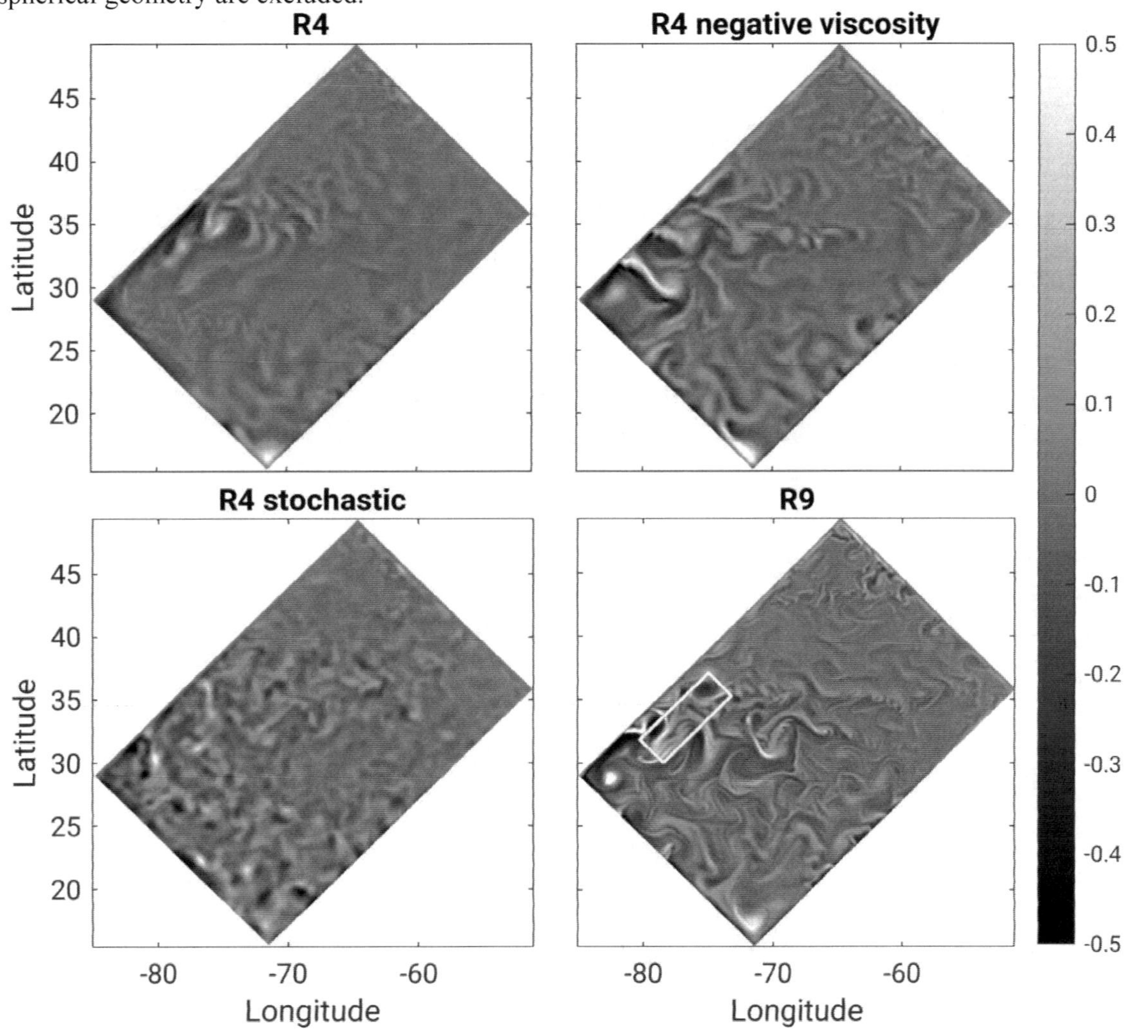

Figure 1. Snapshot of surface relative vorticity in f (Coriolis parameter) units. 30 March after spin-up. R4 – eddy-permitting model, R9 – eddy-resolving model, R4 negative viscosity and R4 stochastic – eddy-permitting models with KEB parameterizations. Colorbar is saturated at ±0.5. The white rectangle is explained in Figure 4.

The free surface is stressed by zonal wind having its maximum eastward speed at 36^oN and maximum westward speed at 22^oN . The following three surface buoyancy fluxes sustain a south-north decline of the surface temperature and the growth of surface salinity. The conductive heat flux from the atmosphere is given by $Q=\gamma(T_{atm}-SST)$, SST and T_{atm} – the surface ocean temperature and the

atmosphere temperature, $\gamma = 40Wm^{-2}K^{-1}$. Also, solar radiation and fresh water fluxes are prescribed at the surface. The listed forcings vary with latitude and seasonally and are given in [13].
We consider models with three uniform spatial resolutions. The parameters are given in Table 1. R1 is initialized at rest with vertical profiles of temperature and salinity uniformly applied to the whole basin [13]. The number of vertical layers is 30 for all simulations. We accomplish spin-up of R4 and R9 models, first running R1 model for 1000 years and then continuing the computations with R4 or R9 models for 120 years. The last 20 years are stored for analysis. The Rossby deformation radius for the resulting stratification varies from 40 km in the south to 5 km in the north [13]. The diffusion/viscosity coefficients for models R1 and R9 are taken from [13]. The viscosity in model R4 is almost the smallest possible. Further decreasing leads to wrong reproduction of the meridional eddy heat flux.

Table 1. Model experiments (R1, R4, R9) and their parameters. Unless otherwise mentioned, diffusivity/viscosity act in horizontal direction.

	R1, non-eddy-resolving	R4, eddy-permitting	R9, eddy-resolving
$n_x \times n_y \times n_z$	$30 \times 20 \times 30$	$120 \times 80 \times 30$	$270 \times 180 \times 30$
mesh step	$1^o, 106\ km$	$1/4^o, 26.5\ km$	$1/9^o, 11.7\ km$
time step	120 min	30 min	800 sec
eddy diffusivity	Isoneutral ∇_h^2, $10^3 m^2 s^{-1}$	∇_h^4, $-10^{10} m^4 s^{-1}$	∇_h^4, $-10^9 m^4 s^{-1}$
eddy viscosity	∇_h^2, $10^5 m^2 s^{-1}$	∇_h^4, $-5 \cdot 10^{11} m^4 s^{-1}$	∇_h^4, $-5 \cdot 10^{10} m^4 s^{-1}$

3. Kinetic energy backscatter (KEB) parameterizations

3.1. *Negative viscosity KEB*

This parameterization supplements horizontal biharmonic momentum damping in the momentum equation with (2) additional term returning the energy in the form of a Laplace operator, as proposed in [8]:

$$\frac{\partial \boldsymbol{U}_h}{\partial t} = \cdots \nu_4 \nabla_h^4 \boldsymbol{U}_h + \nabla_h(\nu_2 \nabla_h \boldsymbol{U}_h). \qquad (6)$$

Here ∇_h is assumed to act on the vector componentwise. When $\nu_4 < 0$, the grid-scale numerical noise is effectively dissipated. The viscosity $\nu_2 \leq 0$ must be negative to return the energy. The choice of a Laplace operator for KEB is natural, since it is the simplest linear operator with a characteristic length scale larger than that of the biharmonic operator. Moreover, it is shown by R. Kraichnan in [15] that subgrid 2D turbulence produces a tendency represented by the Laplace operator in the middle and large scales with a negative viscosity coefficient. To take into consideration spatial non-homogeneity of the eddy field and weak dynamics near the bottom, we follow works [9], [16], and [12] and introduce dependence on the coordinates $\nu_2(x, y, z, t) \leq 0$. ν_2 is found based on the "energetically consistent" property: the joint energy flux to subgrid scales corresponding to the Laplace and biharmonic operators must be zero. This property is consistent with the idea of inverse energy cascade: the energy shall not pass to subgrid scales. The energy fluxes to subgrid scales related to the biharmonic and Laplace operators are expressed in Galilean invariant form following [16]:

$$\dot{E}_{diss} = \nu_4 \nabla_h \boldsymbol{U}_h \cdot \nabla_h (\nabla_h^2 \boldsymbol{U}_h), \qquad (7)$$

$$\dot{E}_{back} = \nu_2 \nabla_h \boldsymbol{U}_h \cdot \nabla_h \boldsymbol{U}_h. \qquad (8)$$

The signs of the fluxes are as follows: $\int \dot{E}_{diss}\, dxdydz > 0, \int \dot{E}_{back}\, dxdydz \leq 0$. Local equalization of these fluxes $\dot{E}_{diss}(x, y, z, t) + \dot{E}_{back}(x, y, z, t) = 0$ leads to an ill-posed problem for finding ν_2 if

$|\nabla_h \boldsymbol{U}_h| = 0$. To overcome it, papers [9] and [16] introduced an equation for subgrid energy ($e \equiv e(x,y,z,t)$), which is produced by dissipation ($\dot{E}_{diss}$) and lost to the resolved scales ($\dot{E}_{back}$):

$$\frac{de}{dt} = c_{diss}\dot{E}_{diss} + \dot{E}_{back} + \nu_e \nabla_h^2 e. \qquad (9)$$

Here d/dt is the Lagrangian derivative (advection), which is implemented in the code with the simplest upwind scheme [14], $\nu_e = 1000 m^2 s^{-1}$ is equal to the diffusivity of eddies in model R1. Contrary to [16], we apply the Lagrangian derivative instead of a partial derivative, which seems to be more physical and was originally proposed for the subgrid energy equation in [17]. As in [12], $c_{diss} \in (0,1)$ allows one to reduce the backscatter power. We found $c_{diss} = 0.8$ to be satisfactory. Larger values produce overestimated strong vortices and a strong meridional eddy heat flux near the surface at the WBC separation latitude. The subgrid energy e defines the negative viscosity:

$$\nu_2 = -c_{back}\Delta x\sqrt{\max(e,0)}, \qquad (10)$$

where Δx is the grid spacing and $c_{back} = 0.4\sqrt{2}$, as in [9]. The max-function allows us to treat situations when subgrid energy is negative, which can occur since the biharmonic operator is not positive definite. For the terms $\nabla_h(\nu_2 \nabla_h \boldsymbol{U}_h)$ in (6) and $\nu_e \nabla_h^2 e$ in (9), Neumann boundary conditions are applied: $(\nabla_h \boldsymbol{U}_h)\cdot \boldsymbol{n} = 0$ and $(\nabla_h e)\cdot \boldsymbol{n} = 0$, $\boldsymbol{n}$ is the normal vector to the horizontal boundary. As in [9], we report weak sensitivity of the results on the model for the e used. Our understanding is that an additional equation for subgrid energy is needed only to make the problem for finding ν_2 well-posed. In the next section it will be shown that the stochastic KEB can be tuned without an additional equation for subgrid energy.

3.2. *Stochastic KEB*

Let us construct a quasi-barotropic (quasi-2D) stochastic streamfunction as proposed in [7]:

$$\psi(x,y,z,t^n) = \phi(x,y,t^n)\cdot A(x,y,z,t^n), \qquad (11)$$

where $\phi(x,y,t^n)$ is a discrete random field. ϕ is independent at each mesh point and time layer t^n (i.e. discrete-space-time white noise) and has distribution $N(0,1)$ with zeros at the boundary. A is the amplitude controlling the energy input. The streamfunction modifies the momentum equation (2) as follows:

$$\frac{\partial \boldsymbol{U}_h}{\partial t} = \cdots + \alpha \nabla_h^{\perp} S^n(\psi), \qquad (12)$$

where $\nabla_h^{\perp} = (-\partial_y, \partial_x)$. α will be defined later. The function $S^n(\cdot)$ introduces n applications of a spatial discrete filter:

$$S(\psi) = \psi + \frac{(\Delta x)^2}{8}\nabla_h^2 \psi. \qquad (13)$$

This filter is based on the Laplace operator already contained in the model and nullifies the checkerboard grid noise ($(-1)^{i+j}$; i and j are indexes along the $x-$ and $y-$ directions) if a second order approximation is used. Zero Dirichlet boundary conditions are applied. The physical reasoning for stochastic parameterization with spatial white noise is again taken from [15], where subgrid turbulence was shown to produce stochastic forcing in small resolved scales. In contrast to [10], we generate random streamfunction, instead of random Reynolds stress components. Streamfunction approach gives analogous wavenumber spectrum, but doesn't require removing divergence of the resulting forces in momentum equation [18]. According to [10], the filter $S^n(\cdot)$ can be considered as a correction of a sharp wavenumber spectrum near the grid scale to account for numerical effects of non-spectral advection schemes. Also, the filter defines the correlation radius of the resulting parameterization. We use $n = 6$, which gives a correlation radius of several mesh points.

Contrary to the negative viscosity KEB, local equalization of the energy fluxes $\dot{E}_{diss}$ and $\dot{E}_{back}$ is possible without an additional equation for subgrid energy. According to [19], energy generation by the white noise process is proportional to the squared amplitude ($\dot{E}_{back} \sim -A^2$) and, hence, we choose:

$$A(x,y,z,t^n) = \sqrt{\max(\dot{E}_{diss},0)}. \qquad (14)$$

Finally, the energy generation and dissipation fluxes integrated over the domain should be equal. Again, using [19] we compute the energy generation for white noise and obtain a relation for finding α:

$$\frac{\alpha^2 \Delta t}{2} \int \left\langle \left| \nabla_h^\perp S^n(\psi) \right|^2 \right\rangle dxdydz = \int \dot{E}_{diss} dxdydz. \qquad (15)$$

Here Δt is the time step, and the angle brackets $\langle \cdot \rangle$ denote averaging over realizations of the random field ϕ. The left-hand side of the above equation can be estimated analytically taking into consideration knowledge of the wavenumber spectra [7], but still with simplifications. Our novelty is to drop averaging over an ensemble:

$$\int \left\langle \left| \nabla_h^\perp S^n(\psi) \right|^2 \right\rangle dxdydz \approx \int \left| \nabla_h^\perp S^n(\psi) \right|^2 dxdydz, \qquad (16)$$

where the right-hand side is computed directly for the current realization of φ. This method is based on the following facts: 1) the ensemble-mean of RHS equals LHS 2) the standard deviation of RHS is 7% of its mean value.

4. Results

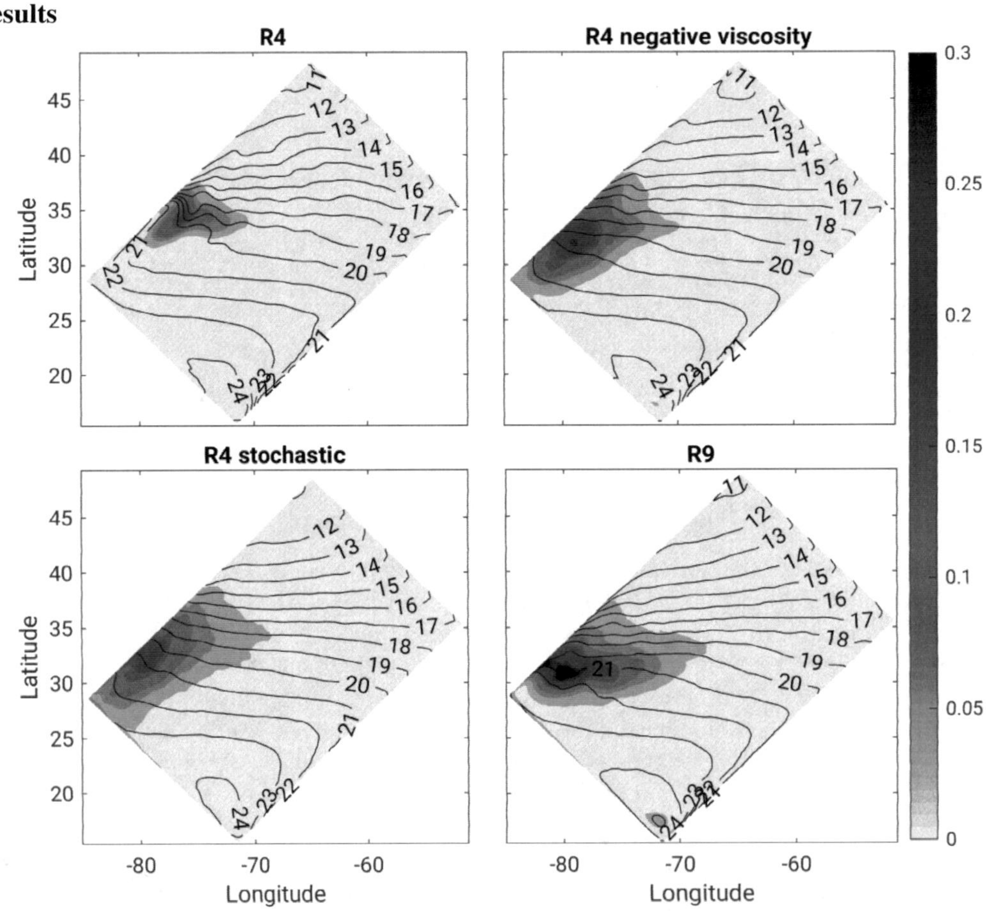

Figure 2. In colour: 20-year mean surface eddy kinetic energy (EKE), m^2/s^2. The colorbar is saturated at 0.3. In contours: 20-year mean sea surface temperature (SST), C^o. R4 – eddy-permitting model, R9 – eddy-resolving model, R4 negative viscosity and R4 stochastic – eddy-permitting models with KEB parameterizations.

We consider 4 models: an eddy-permitting model R4, an eddy-resolving model R9, and R4 with two KEB parameterizations: negative viscosity and stochastic ones. They are spun up as described in Section 2. Unless otherwise mentioned, we show the results averaged over the last 20 years. The model in Double Gyre setting simulates the western boundary current (WBC) similar to Gulfstream or Kuroshio. At some latitude, the WBC separates from the boundary forming, in the zonal direction, an offshore extension which divides the north and south gyres (see Figure 1). The separation latitude is strongly dependent on the resolution, and it moves to the south as the mesoscale and submesoscale eddies become resolved [13]. The WBC separation point and its extension strength are strongly associated with the sea surface temperature (SST) and the sea surface height (SSH), since their isolines pass along the jet current. Figure 2 shows the 20-year mean SST in contours for 4 models. The major discrepancy between models R4 and R9 is in isotherms 21 and 22. Applying the KEB straightens these isotherms, which leads to a significant fall of the errors in the mean surface fields (see Table 2). Improvements in the mean fields results from the amplification of the eddy activity, whose strength can be attributed to the eddy kinetic energy (EKE) level. Applying the KEB parameterizations allows us to increase the surface-averaged EKE (see Table 2) and – concerning its spatial distribution – to shift its maximum southward (see Figure 2 in colour). However, we point out that the surface EKE in models R4 with KEBs is elongated along the boundary, contrary to elongation in the zonal direction in model R9. It means that long jet extension cannot be simulated at a coarse grid even with KEB parameterizations.

Table 2. 20-year mean surface-averaged EKE and norm of errors in SST (sea surface temperature) and SSH (sea surface height) fields in models R4 with respect to reference R9. Two norms for any $\varphi(x,y)$ are $(|\varphi_{R4} - \varphi_{R9}|)$and $mean(|\varphi_{R4} - \varphi_{R9}|)$.

	R4	R4 negative visc.	R4 stochastic	R9
surf. EKE, m^2s^{-2}	0.01	0.023	0.027	0.025
error SST, C^o	7.0; 0.40	3.1; 0.30	4.3; 0.27	
error SSH, m	0.68; 0.062	0.38; 0.039	0.40; 0.040	

One of the most apparent indicators of the eddy activity is the transport of tracers. Under the prescribed surface heat fluxes, ocean flows produce meridional heat transport (MHT) towards north: $Q = \rho_0 C_p TV$, $C_p \approx 3992 JK^{-1}kg^{-1}$ is the heat capacity, and V is the meridional velocity. Note that we neglect the heat transport corresponding to diffusion, since its action is minimized with the use of a biharmonic operator. The MHT consists of two parts: mean-flow meridional heat transport (MMHT, $\rho_0 C_p \bar{T}\bar{V}$, the overline stands for time-averaging over the last 20 years) and eddy meridional heat transport (EMHT). The EMHT strongly depends on the resolution, and it is almost zero for model R1. Models R4 and R9 have a significant EMHT: of the order of the MMHT (see [13] for details). The distribution of the EMHT in depth for models R4 and R9 is shown in Figure 3 by colour. In model R9, the most significant EMHT (note the logarithmic depth-scale) corresponds to a region $200m - 500m$ in depth and $23^oN - 30^oN$. The heat in this region fluxes southward, which is consistent with the northward gradient of potential temperature at this depth (not shown). Both KEB parameterizations amplify this flux in model R4 and break the wrong negative heat flux in the region of $32^oN - 35^oN$.

The meridional overturning circulation (MOC) is shown by contours in Figure 3. The MOC is described by the following streamfunction: $\Psi_{MOC}(y',z) = \int_{-H}^{-z} \bar{V}(x',y',z',t)dx'dz'$, where x' and y' are the coordinates along longitude and latitude, $z \in (-H,0)$ is the depth, the overline is time averaging over 20 years, and V is the meridional velocity. The circulation in all models consists of 4 cells, and the largest one is at the bottom (note logarithmic depth-scale). This cell is shifted northward in model R4 as compared to R9. Both KEB parameterizations restore the correct position of the bottom cell at about 30^oN. The upper northern cell acquires correct shape as a result of applying the

KEBs. The northern bottom cell is better reproduced by the stochastic model R4 (see -1.5 isoline in Figure 3).

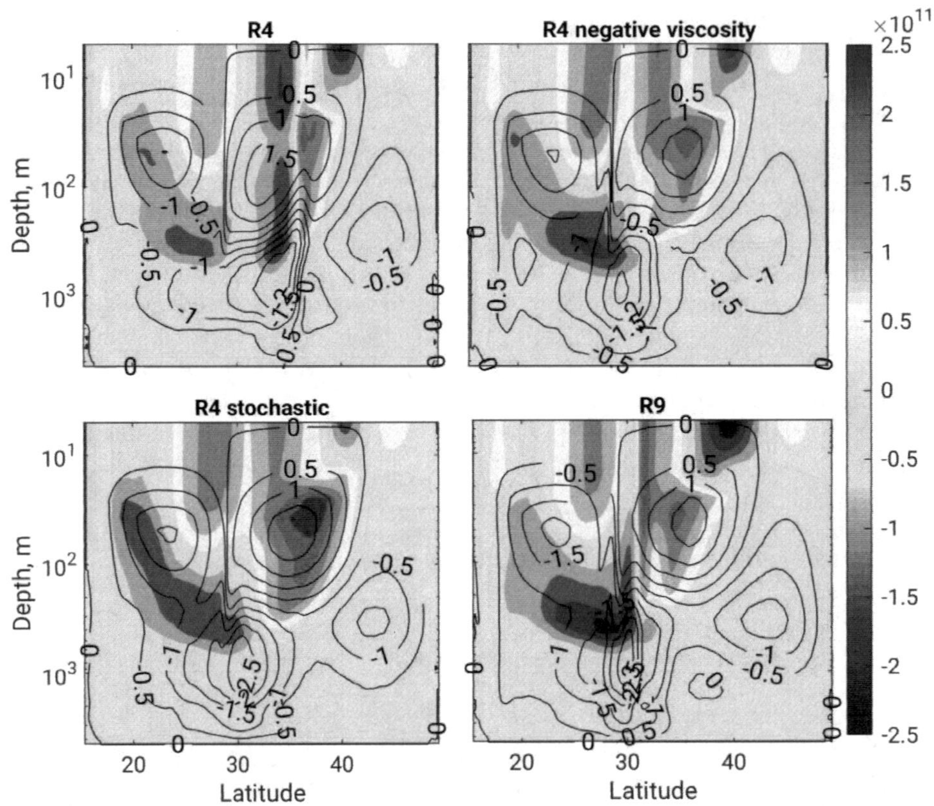

Figure 3. In colour: 20-year mean eddy meridional heat transport integrated zonally, W/m. In contours: 20-year mean meridional overturning streamfunction Ψ_{MOC} in Sverdrups. R4 – eddy-permitting model, R9 – eddy-resolving model, R4 negative viscosity and R4 stochastic – eddy-permitting models with KEB parameterizations.

We study the time variability of the models by computing the time power density of the surface EKE, which is shown in Figure 4. The power density is averaged over a white rectangle shown in Figure 1 to reduce its oscillations and to exclude its strong dependence on the WBC separation point. The spectra consist of two power-law intervals and a long-period tail. The short-period power law ν^{-4} possibly corresponds to the wave motions, while the middle-period ν^{-2} possibly corresponds to the mesoscale turbulence. Both power-law intervals are underestimated in model R4. The KEB parameterizations allow one to increase the power density in the middle-period interval up to the level of model R9. The negative viscosity KEB also improves the power density in the short-period interval. The stochastic KEB gives discrepancy with model R9 in the short-period interval and has a singularity near 1-day oscillation. We decided that the white-noise parameterization excites inertial waves which have a similar period. Earlier it was not reported anywhere that KEB with the white-noise stochastic process excites inertial waves, possibly since most works consider quasi-geostrophic equations (as [8] and [10]), where inertial oscillations are filtered out. We have tried a simple modification of the stochastic KEB, including a temporal correlation with a lag of 10 days to force only the middle-period interval. Correlated noise was generated using an autoregressive model of order 1 (AR-1), as [7] suggests. The time-correlated stochastic KEB does not excite inertial oscillations (not shown), but it has one moderate drawback: there is no more exact formula for the energy generation like (15), and it introduces one free parameter to be tuned by hand.

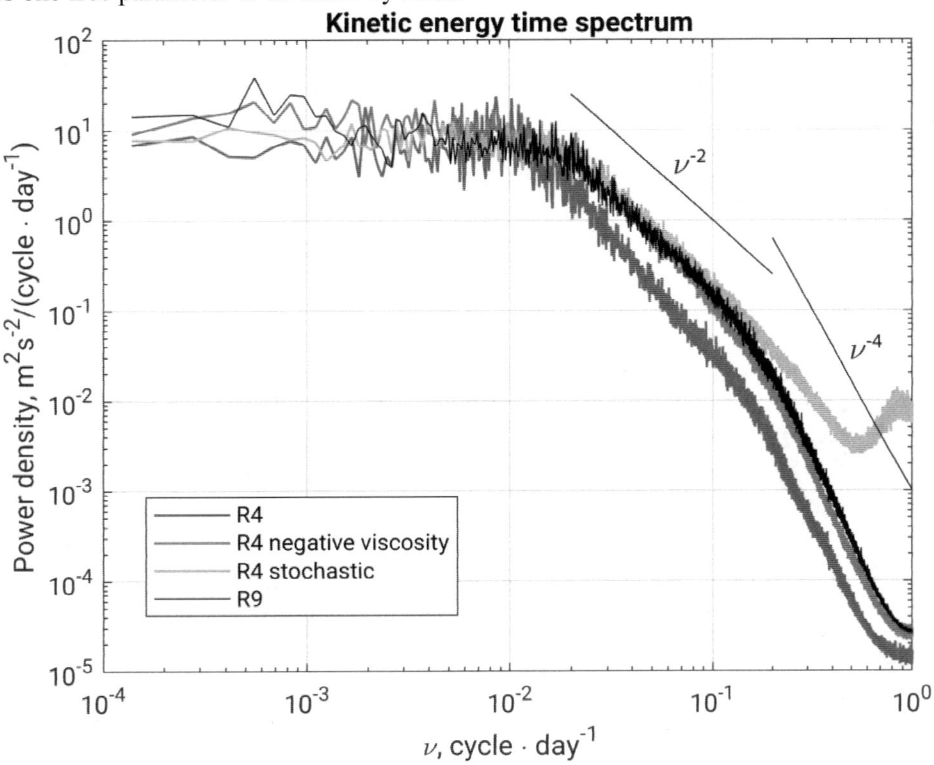

Figure 4. Time power density of surface EKE. Density is averaged over a white rectangle shown in Figure 1. R4 – eddy-permitting model, R9 – eddy-resolving model, R4 negative viscosity and R4 stochastic – eddy-permitting models with KEB parameterizations. Time series correspond to the last 20 years.

Finally, let us consider snapshots of relative vorticity (see Figure 1). The R4 and R9 solutions are highly different in the number of turbulent eddies and filaments. Both KEB parameterizations induce eddy activity in model R4, but in different way. The stochastic KEB solution looks like "synthetic" turbulence which consists of turbulent eddies without filaments and it similar in the shape of its eddies to the stochastic KEB tendency itself. The amount of undesirable noise in the solution can be reduced if the time-correlated stochastic process is used (not shown). The negative viscosity KEB, on the contrary, amplifies the existing filaments and large eddies, but the solution suffers from the absence of small-scale features.

5. Conclusions

In this work, we have demonstrated that simple kinetic energy backscatter (KEB) parameterizations accounting for a badly resolved barotropic inverse energy cascade are able to improve general ocean circulation models with an eddy-permitting resolution. The novelty of this work is the use and comparison of the two KEBs for the primitive equations of the ocean. Both KEBs give similar improvements in the mean characteristics. Specifically, the mesoscale eddies are amplified and, consequently, eddy kinetic energy (EKE) rises. As a result, the eddy meridional heat transport near the surface and in a moderately deep ocean was restored. Large errors in sea surface temperature and height were reduced in the region of jet separation. Also, the bottom cell of the meridional overturning circulation (MOC) was shifted southward, as in the high resolution model. A moderate difference concerns the bottom northern MOC cell, which is better reproduced with the stochastic KEB. It has been shown that the KEBs effectively improve the time power spectra of EKE, but a white-noise process can lead to an undesirable generation of inertial waves, which can be overcome by introducing a time-correlated stochastic process. The major difference between the KEBs is in the solution type. The stochastic KEB seems to introduce too much small eddies without filaments, but the negative viscosity one amplifies the existing filament-eddy field. It is possible that they can be effectively used together. Future work will be devoted to studying more realistic ocean configurations with varying ocean depth and spherical geometry.

Acknowledgments
The work was supported by the Russian Science Foundation (grant 17-17-01210, development of subgrid turbulence models) and the Russian Foundation for Basic Research (grant 18-05-60184, calculations with idealized ocean model and analysis of results).

References

[1] Haarsma R *et al* 2016 *Geosci. Model Dev.* **9** 4185-4208
[2] Gent P and McWilliams J 1990 *J. of Phys. Oceanography* **20** 150-55
[3] Madec J *et al* 1997 Ocean general circulation model reference manual. *Note du Pôle de modélisation*
[4] Bernard B *et al* 2006 *Ocean dynamics* **56** 543-67
[5] Kraichnan R 1967 *The Phys. of Fluids* **10** 1417-23
[6] Frederiksen J and Davies A 1997 *J. of the Atmos. Sci.* **54** 2475-92
[7] Berner J, Shutts G, Leutbecher M and Palmer T 2009 *J. of the Atmos. Sci.* **66** 603-26
[8] Jansen M and Held I 2014 *Ocean Modelling* **80** 36-48
[9] Jansen M, Held I, Adcroft A and Hallberg R 2015 *Ocean Modelling* **94** 15-26
[10] Grooms I, Lee Y and Majda A 2015 *Multiscale Model.&Sim.* **13** 1001-21
[11] Perezhogin P, Glazunov A and Gritsun A 2019 *Rus. J. of Num. Anal. and Math. Model.* **34**(4) 197-213
[12] Juricke S, Danilov S, Kutsenko A and Oliver M 2019 *Ocean Model.* **138** 51-67
[13] Levy M *et al* 2010 *Ocean Model.* **34** 1-15
[14] Madec G *et al* 2015 NEMO ocean engine
[15] Kraichnan R 1976 *J. of Atm. Sci.* **33** 1521-36
[16] Zurita-Gotor P, Held I and Jansen M 2015 *J. of Adv. In Modeling Earth Sys.* **7** 1117-35
[17] Eden C and Greatbatch R 2008 *Ocean Modelling* **20** 223-39
[18] Carati D, Ghosal S and Moin P 1995 *Phys. Of Fluids* **7** 606-16
[19] Alvelius K 1999 *Phys. Of Fluids* **11** 1880-89

Simulation of the Pacific equatorial thermocline with an ocean general circulation model

K V Ushakov[1,2,3] **and M N Kaurkin**[1,2]

[1] Marine Hydrophysical Institute of RAS, Sevastopol, Russia
[2] Shirshov Institute of Oceanology, Russian Academy of Sciences, Moscow, Russia
[3] Marchuk Institute of Numerical Mathematics, Russian Academy of Sciences, Moscow, Russia

Email: ushakovkv@mail.ru

Abstract. The reproduction of equatorial thermocline characteristics by numerical models is of great importance for seasonal forecasts, and it is also a key factor in operational ocean forecasting. Due to difficulties in the numerical description of adiabatic eddy processes, the model thermocline is usually diffused and shifted in depth. In this work, we perform a numerical experimental study of the sensitivity of the Pacific equatorial thermocline to horizontal and vertical mixing parameterizations in the INMIO ocean general circulation model. It has been shown that the sharpness and location of the thermocline can be improved by a combined parameterization of eddy mixing and isopycnal diffusion, as well as by proper tuning of the background coefficient of vertical diffusion.

1. Introduction

The ocean thermocline is one of the key objects whose reproduction largely determines the adequacy of an ocean general circulation model. In the middle latitudes, its physics has been studied in sufficient detail and generally does not cause problems for modelers. On the other hand, the equatorial thermocline is more difficult to reproduce. At the same time, its parameters significantly affect the structure of horizontal and vertical heat fluxes and, consequently, are key to the model climatic system characteristics. This primarily applies to the intensity and duration of the El Niño – Southern Oscillation phases [1]. In [2], an analysis of the equatorial heat budget was carried out, which showed that, depending on the sharpness of the model thermocline, the *westward* or *vertical* heat transfer becomes the dominant factor in the development of El Niño phases. Knowledge of this development is of great importance both as a source of long-term predictability for seasonal forecasts, and as a factor of short-term operational ocean forecasting. It directly influences the system of equatorial currents and upwellings that affect the state of the ocean, including the mid-latitudes [3,4].

In nature, the equatorial thermocline of the Pacific Ocean represents a very sharp temperature jump. Its depth, usually defined by the position of the 20°C isotherm, gradually increases from the eastern to the western side of the ocean. A peculiar feature of the Pacific equatorial region is the dominant role of adiabatic processes in the ocean circulation [1]. Thus, even small values of model diffusion, both induced by numerical schemes and explicitly specified, can introduce significant distortions into the structure of the thermocline. Models often experience a complex of problems in this area, which is called the “diffuse thermocline”. First of all, this is expressed in a decrease of the vertical temperature gradient and, accordingly, in an increase of the interval of depths at which the temperature jump occurs. In addition, the model solution may manifest an excessively sharp immersion of the thermocline when moving along the equator to the west. In the meridional section, the isotherms can strongly bend upward, which is manifested in a surface cold bias and in the corresponding distortion of the heat fluxes balance. In [5], this was interpreted as the model underestimation of the intensity of tropical instability waves, which are responsible for the flattening of isotherms through heat transfer into the region of the East Pacific cold equatorial "tongue". With the INMIO model used in that work, this bias appeared despite the high spatial resolution of the model, generally sufficient for the description of the equatorial eddy dynamics.

Many works studied the formation mechanisms of the equatorial thermocline by means of idealized examples, e.g. [6,7]. With the development of numerical modeling, a number of studies have appeared that improve the characteristics of full three-dimensional numerical models in the equatorial Pacific, mainly concerning the parameterizations of vertical mixing [2,8]. The goal of the present work is to experimentally study the sensitivity of the equatorial thermocline to the settings of the horizontal and vertical mixing schemes in a numerical model with fixed advection schemes. For this, we use the non-global model configuration of the isolated Pacific Ocean that has realistic boundary conditions and allows us to save computer resources during a series of experiments.

2. Model and experiment configuration

The INMIO model [5,9] approximates the 3DPEM system of three-dimensional ocean dynamics equations with the Boussinesq and hydrostatics approximations by the finite volume method on a B-type grid in vertical z-coordinates with an explicit description of surface fluxes and a nonlinear kinematic free surface condition. When approximating horizontal differential operators, only explicit time schemes are applied, which has made it possible to efficiently parallelize the model under the control of the CMF software package [10].

The Pacific Ocean model calculation domain is confined by solid boundaries from the north in the Bering Strait and from the west on the 116°E meridian. In the east, the periodic boundary condition is set in the Drake Strait, allowing the waters flowing into the strait to appear from a "window" in the western boundary at the same latitudes, thus mimicking the Antarctic Circumpolar Current. A regular latitude-longitude grid is used with a horizontal resolution of 0.25 degrees, which allows partial resolving of mesoscale eddies. The vertical discretization includes 49 horizons with a step from 6 m in the upper layer of the ocean to 250 m in deep waters.

One of the key ideas of the modern general circulation model development is the most possible explicit description of the eddy dynamics. In particular, this implies the deprecation of the second-order viscosity operator ("Laplace viscosity"), which was previously used mainly to ensure numerical stability and to simulate non-resolved lateral boundary layers. The transition to such low-viscosity mode of calculations requires providing self-consistency of the model numerical schemes at the difference level. This was performed for the INMIO model in previous works [11,5]. In the present study, this allowed us to take settings of the turbulent exchange coefficients similar to the eddy-permitting configuration ORCA025 [12], which has the same horizontal resolution.

In the control experiment, the horizontal diffusion coefficient for heat and salt is taken equal to 300 m^2/s at the equator and scaled to the poles in proportion to the square root of the grid cell area. In the momentum equations, the Laplace viscosity is not used. To ensure numerical stability, the biharmonic filter is used with a coefficient of $-1.5 \cdot 10^{11}$ m^4/s at the equator, scaled in proportion to the cell area to the power of 3/2 and with the adaptive addition according to the Smagorinsky method (in the formulation [13] with coefficient $C = 1$).

The background coefficients of vertical viscosity and diffusion are 10^{-4} and 10^{-5} m^2/s. The shear vertical mixing is parameterized according to the Munk – Anderson scheme with maximum values equal to 10^{-2} and 10^{-3} m^2/s, respectively. The time step is 10 minutes, while the barotropic velocities are calculated in a nested loop with a step of 20 seconds. To approximate the advection of momentum and tracers, we use the centered difference scheme in space and the leapfrog scheme in time. On solid lateral boundaries, the conditions of no normal fluxes of mass, heat, salt, and momentum are used.

All experiments were performed for 20 years of model time, during which the fluxes at the atmosphere–ocean interface were defined by the atmospheric conditions and bulk formulas of the CORE-I protocol [14]. The model starts from the state of rest and the annual mean temperature and salinity fields of the World Ocean Atlas 2009 (WOA09) climatology [15,16]. The focus of this work is on the cold tongue of the East Pacific equatorial waters, which spreads westward from the coast of South America. Therefore, in the polar regions to simulate the sea ice we use a simple thermodynamic model [17], which does not require a large amount of computation.

3. Experiment results

In this section, we present the results of the series of numerical experiments. Continuing to each subsequent experiment, we will change one model parameter each time. All presented model fields will be annual means for the last (20th) year of experiments.

Figure 1 shows the equatorial distribution of the temperature averaged in the band from 1°S to 1°N. Plate 1(a) presents the annual mean data of WOA2009. The thermocline is pronounced over most of the considered interval of 140°E – 80°W. The vertical temperature gradient (hereinafter, the average between the isotherms of 16 – 24°C at 160°W) is 0.129 °C/m. The depth of the 20°C isotherm gradually increases from 40 m in the east to 180 m in the west, while at 160°W it is 150 m.

The control experiment described above is presented in plate 1(b). It can be seen that the thermocline is substantially diffused, especially in the eastern part of the region. The vertical gradient is 0.072 °C/m. In the course of the experiment the upper layer of the ocean warms excessively and, hence, the depth of the thermocline is much higher than that observed in nature. From 110°W to 85°W, the surface temperature is overestimated by 1-1.5°C. The isotherms of the upper part of the thermocline sink more sharply than in the observational data. The depth of the thermocline (isotherm of 20°C at 160°W) is 207 m.

The configuration of the control experiment is generally similar to the configuration of the calculations carried out in [5]. In that work, similar biases of the diffuse thermocline were encountered. Now, to overcome them, let us pay attention primarily to the mechanism of model heat diffusion. Since the adiabatic heat transport processes dominate in the region under consideration, the parameterization of their explicitly not resolved component is of great importance. It is necessary to describe the process of the available potential energy (APE) transfer into the kinetic energy of subgrid eddies, in particular, into the subgrid component of tropical instability waves. As mentioned in [18], the baroclinic instability plays a significant role in their generation. Typically, these processes are described by using one or another modification of the Gent–McWilliams closure [19]. In the present work, we use this scheme in the skew formulation of [20] thanks to its numerical stability, as well as to the ease of its implementation together with the isopycnal diffusion scheme [21], which is acknowledged as more physically justified than the simple horizontal Laplace diffusion. In our case, the procedure [22] is applied to taper the components of the mixing tensor in regions where the slope of isoneutral surfaces exceeds the value of $S_{max} = 0.01$.

Thus, experiment (c) differs from experiment (b) in that the horizontal Laplace diffusion scheme is replaced by the parameterization [20] with the same diffusion coefficient. As a result, the vertical temperature gradient of the thermocline increased to 0.075 °C/m. Its depth decreased to 201 m, while the improvement in the position of the isotherms of 14–16°C in the lower part of the considered depth interval is more noticeable. In (c) and subsequent plates, the solid lines show the results of the current experiment, while the dashed ones repeat the previous experiment for convenience of comparison.

In experiment (d), compared to the one (c), the Smagorinsky coefficient C of [13] was increased from 1 to 1.5. This led to a slight increase in the thermocline depth (by 1-2 meters), while the vertical gradient change was negligible. The plate does not allow one to clearly see these changes, but to preserve the sequence of experiments and for consistency with Figure 3, we will keep it.

Further, in experiment (e), the nominal (equatorial) value of the diffusion coefficient was increased from 300 to 500 m^2/s. This led to a decrease in the slope of the isotherms in the upper part of the thermocline, which better corresponds to the observations. Besides, it is consistent with a qualitative

IOP Conf. Series: Earth and Environmental Science **386** (2019) 012026 doi:10.1088/1755-1315/386/1/012026

example from [20], where it was shown how the Gent–McWilliams parameterization rotates sloping isotherms in a clockwise manner. The vertical temperature gradient increased to 0.077 °C/m, and the thermocline depth decreased to 199 m. Note that in case of a decrease in the diffusion coefficient isotherms rotate in the opposite direction (not shown).

Finally, in experiment (f), following paper [2], the background coefficient of vertical diffusion was reduced from 10^{-5} m^2/s to 10^{-6} m^2/s. As in the above-mentioned work, this made the thermocline sharper. The vertical temperature gradient increased to 0.086 °C/m, and its depth decreased to 196 m. Note that under the same conditions an increase in the *maximum* diffusion coefficient (the value achieved at low Richardson numbers) leads to a noticeable decrease in the thermocline depth, but also to an unrealistic increase in the slope of isotherms (not shown). Also, a small positive effect in the form of a decrease in the slope of isotherms is achieved by an increase in the limiting parameter of the isoneutral surfaces slope S_{max} up to 0.05.

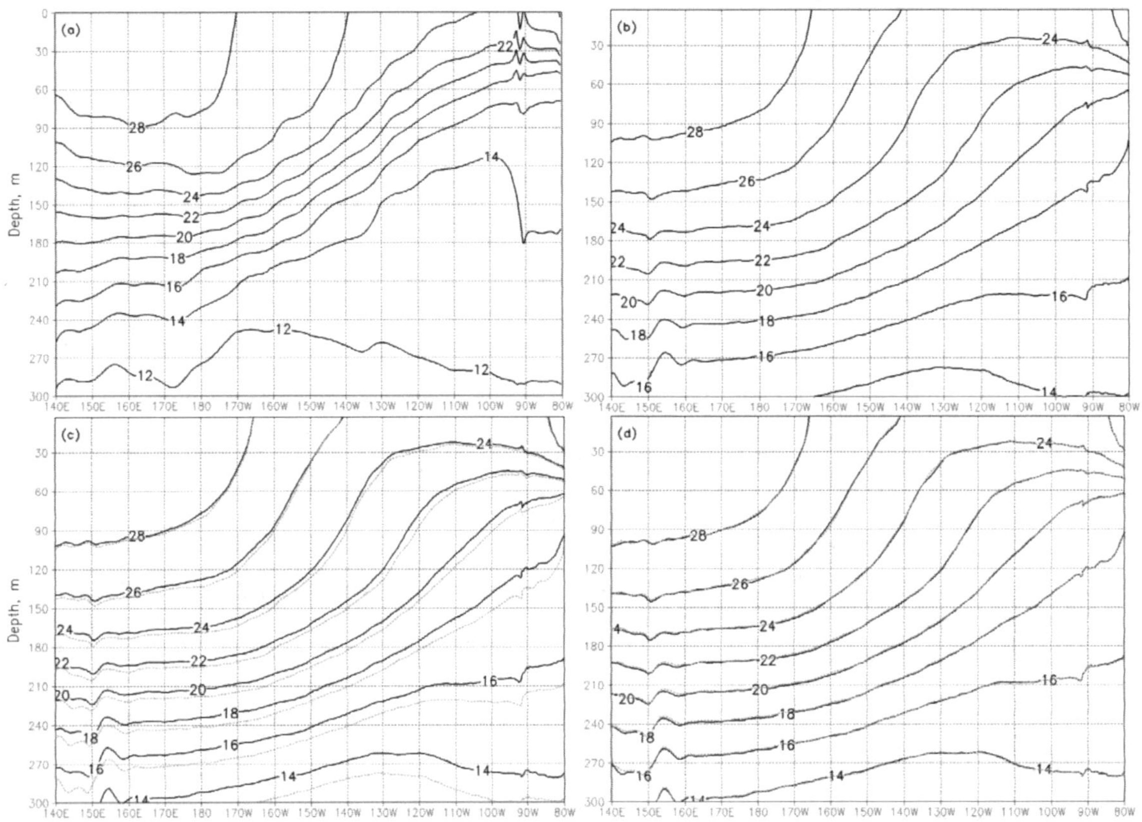

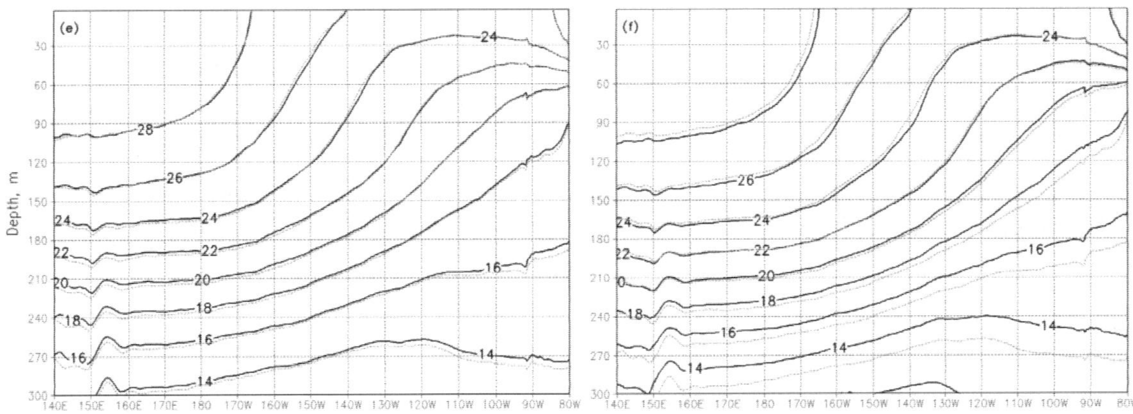

Figure 1. Annual mean temperature in 1°S – 1°N band by WOA09 climatology (a) and the 20th year of model experiments (b-f, details in text). In each of (c) – (f) plates, the dashed lines show the isotherms of the previous plate for comparison.

In Figure 2, the same experiments are illustrated by the zonal temperature distribution near the equator on the 140°W meridian. One can see an overestimation of the thermocline depth by the model, as well as the convexity of the upper isotherms, which apparently indicates an underestimated intensity of the heat exchange of the cold tongue with the surrounding waters. The most noticeable effect in correcting these biases under the conditions of our experiments was exerted by the transition to the parameterization [20] and by the decrease in the background vertical diffusion coefficient. The performed tuning of the model made it possible to bring the surface temperature at this point of the equator closer to the observational data, thereby improving the temperature contrast between the waters above and below the thermocline, which is important for the reproduction of the ENSO phases.

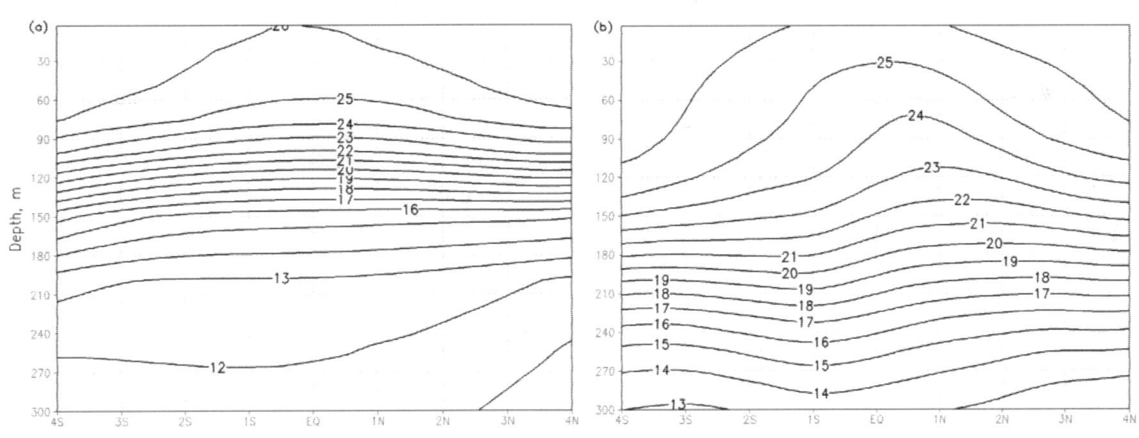

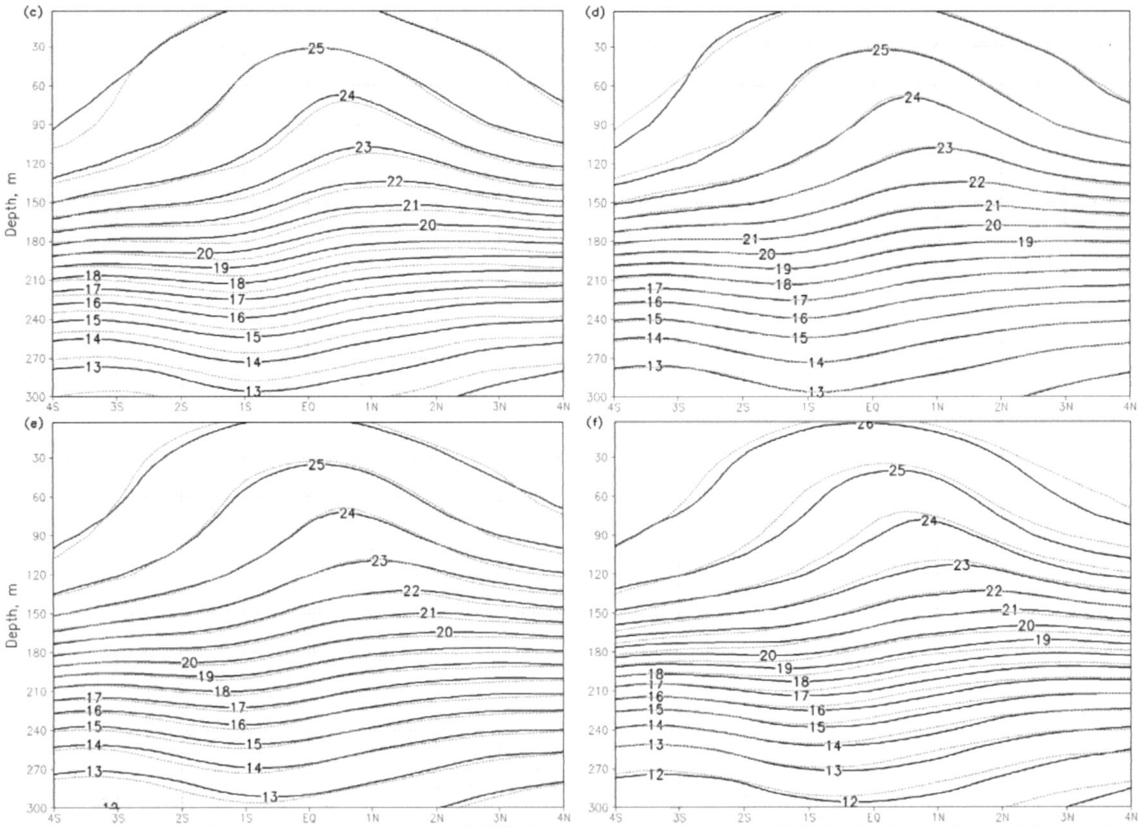

Figure 2. Annual mean temperature on 140°W near the equator by WOA09 climatology (a) and the 20th year of model experiments (b-f, details in text). On each of (c) – (f) plates, the dashed lines show the isotherms of the previous plate for comparison.

4. Conclusions

We performed a series of numerical experiments with the eddy-permitting INMIO model to study the sensitivity of the Pacific equatorial thermocline to the settings and characteristics of turbulent mixing parameterizations. Although the problem of a diffused thermocline is still open, the results of this work may help us better understand the mechanisms of this bias and possible ways to overcome them, as well as directions for improving the ocean research and forecasting model systems. The results of our calculations have shown that the utilization of the Gent – McWilliams eddy mixing parameterization and the Redi isopycnal diffusion in a combined formulation [20] reduces the thermocline reproduction errors by increasing its sharpness and decreasing the depth. It also allows one to adjust the slope of the isotherms. Calculations with a small background vertical diffusion have made it possible to improve the adiabaticity of the model solution and, as a result, to reduce the near-surface stratification biases. At the same time, a decrease in the maximum values of the diffusion coefficient has led to an improvement in the characteristics of the equatorial thermocline in the western part of the Pacific Ocean, but caused an unacceptable increase in the slope of the isotherms. Also, it was found that the solution is somewhat sensitive to the background viscosity parameters and the isopycnals slope tapering.

Acknowledgments
This work was supported by the Russian Science Foundation (project no. 17-77-30001) and performed at the Marine Hydrophysical Institute of RAS.

IOP Conf. Series: Earth and Environmental Science **386** (2019) 012026 doi:10.1088/1755-1315/386/1/012026

References

[1] Tatebe H and Hasumi H 2010 Formation mechanism of the Pacific equatorial thermocline revealed by a general circulation model with a high accuracy tracer advection scheme *Ocean Modelling* **35** 245–52

[2] Imada Y and Kimoto M 2006 Improvement of thermocline structure that affect ENSO performance in a coupled GCM *SOLA* **2** 164–67

[3] Webb D J 2018 On the role of the North Equatorial Counter Current during a strong El Niño *Ocean Sci.* **14** 633–60

[4] National Academies of Sciences, Engineering, and Medicine 2016 *Next Generation Earth System Prediction: Strategies for Subseasonal to Seasonal Forecasts, Chapter 4* (Washington, DC: The National Academies Press) p 350

[5] Ushakov K V and Ibrayev R A 2018 Investigation of interannual variability and budget of heat in an eddy-resolving numerical model of tropical instability waves in the Pacific Ocean *Russian Meteorology and Hydrology* **43** 787–94

[6] Boccaletti G, Pacanowski R, Philander S and Fedorov A V 2004 The thermal structure of the upper ocean *J. Phys. Oceanogr.* **34** 888–902

[7] Pedlosky J 1998 *Ocean Circulation Theory* (Berlin Heidelberg: Springer-Verlag) p 455

[8] Meehl G *et al* 2001 Factors that affect the amplitude of El Niño in global coupled climate models *Clim. Dinam.* **17** 515–26

[9] Ibrayev R A, Khabeev R N and Ushakov K V 2012 Eddy-resolving 1/10° model of the World Ocean *Izv. Atmos. Ocean. Phys.* **48** 37–46

[10] Kalmykov V V, Ibrayev R A, Kaurkin M N and Ushakov K V 2018 Compact Modeling Framework v3.0 for high-resolution global ocean–ice–atmosphere models *Geosci. Model Dev.* **11** 3983–97

[11] Ushakov K V, Ibrayev R A and Kalmykov V V 2015 Simulation of the World Ocean climate with a massively parallel numerical model *Izv. Atmos. Ocean. Phys.* **51** 362–80

[12] Barnier B *et al* 2006 Impact of partial steps and momentum advection schemes in a global ocean circulation model at eddy permitting resolution. *Ocean Dynamics* **56** 543–67

[13] Griffies S M, Hallberg R W 2000 Biharmonic friction with a Smagorinsky-like viscosity for use in large-scale eddy-permitting ocean models *Mon. Wea. Rev.* **128** 2935–46

[14] Griffies S M *et al* 2006 Coordinated ocean-ice reference experiments (COREs) *Ocean modelling* **26** 1–46

[15] Locarnini R A *et al* 2010 *World Ocean Atlas 2009, Volume 1: Temperature* ed S Levitus (Washington, D.C.: U.S. Government Printing Office) p 184

[16] Antonov J I *et al* 2010 *World Ocean Atlas 2009, Volume 2: Salinity* ed S Levitus (Washington, D.C.: U.S. Government Printing Office) p 184

[17] Schrum C and Backhaus J 1999 Sensitivity of atmosphere–ocean heat exchange and heat content in the North Sea and the Baltic Sea *Tellus A: Dynamic Meteorology and Oceanography* **51** 526–49

[18] Imada Y and Kimoto M 2012 Parameterization of tropical instability waves and examination of their impact on ENSO characteristics *J. Climate* **25** 4568–81

[19] Gent P R and McWilliams J C 1990 Isopycnal mixing in ocean circulation models *J. Phys. Oceanogr.* **20** 150–155

[20] Griffies S M 1998 The Gent–McWilliams skew flux *J. Phys. Oceanogr.* **28** 831–41

[21] Redi M H 1982 Oceanic isopycnal mixing by coordinate rotation *J. Phys. Oceanogr.* **12** 1154–58

[22] Griffies S M 2012 *Elements of the Modular Ocean Model (MOM) (2012 release with updates) GFDL Ocean Group technical report No. 7* (Princeton: NOAA Geophysical Fluid Dynamics Laboratory) p 645

Sensitivity of results of Black Sea level modeling to the choice of boundary conditions on the free surface

S G Demyshev, O A Dymova and N A Miklashevskaya

Wave Theory Department, Marine Hydrophysical Institute RAS, 2, Kapitanskaya st., Sevastopol 299011, Russia

Email: olgadym@yahoo.com

Abstract. This work is devoted to an analysis of the influence of sea surface boundary conditions on results of Black Sea level modeling. Two numerical experiments are carried out by using a z-coordinate model of the Marine Hydrophysical Institute, RAS. A linear kinematic condition on the free surface is used for calculating a level rise in the first experiment and a nonlinear one – in the second experiment. The analysis shows that both experiments reconstruct the Black Sea surface dynamics in accordance with altimetry data. It is found that the sea level near the western boundary of the basin decreases under the nonlinear kinematic condition in cases of storm winds. The greatest differences in the structure of the sea level fields are observed during weakening of the RIM Current in the evolution zones of Sevastopol and Batumi anticyclones. The minimal and maximal values of the level are almost identical in both experiments, but the locations do not match. The discrepancies in the spatial positions of raising/lowering levels are due to a mismatch in the motion phases of anticyclones and mesoscale eddies along the periphery of the RIM Current.

1. Introduction

Level height variations in coastal zones of the enclosed seas can pose a serious threat to the population and objects of state and private business. As shown in the observational data, catastrophic level rises in the Black Sea and the Azov Sea reached heights of about 7 m [1], and in the Caspian Sea, of more than 20 m [2], which caused flooding of the territory and significant damage. Such extreme situations were induced by meteorological events. On the other hand, over the second part of the twentieth century a trend to increase of the Black Sea level was observed, which was mainly connected with the growth of the surface freshwater flux calculated as the difference between the precipitation and evaporation [3]. The tide-gage and satellite altimetry data of 1992 – 2005 also show that the rising rate of the mean sea level is 7.6 ± 0.3 mm/year, which is 2 – 3 times higher than the estimates for the previous periods [4]. In addition to a positive trend of the Black Sea level into a basin scale, the spatial heterogeneity of the energy distribution of the level oscillations was also noted. According to [5], the shallow area of the North-Western Shelf exposed stronger variability in the synoptic and mesoscale ranges than the abyssal zones near the Crimean and eastern coasts.

Therefore, accurate prediction of the sea level changes is important from the point of view of risk assessment for the coastal areas. One of the main ways to obtain information about the past, present, and possible future state of ocean systems is numerical modeling. The quality of the diagnosis and prognosis depends on correct consideration of numerous factors. In particular, the accuracy of the sea level reconstruction in the hydrodynamic models is determined on the basis of adequate formulation of boundary conditions on the sea surface. This work is devoted to the analysis of modeling results of the sea level fields for different surface boundary conditions.

Section 2 describes the problem formulation. The numerical model being used and the parameters of the experiment are presented in Section 3. A comparison of the modeling results of the Black Sea level fields is made in Section 4. Analysis of the eddy kinetic energy is performed in Section 5. The main conclusions are presented in Section 6.

2. Problem formulation

There are two main approaches to constructing circulation models. In the first case, the solution of the ocean thermodynamics equations is constructed in terms of a stream function (for example, the MICOM [6]) and in the second one, in terms of a reference ocean level (for example, POM [7], NEMO [8]). The second approach allows one to calibrate models by taking into account in-situ observations of the sea level. Earlier the rigid-lid condition was used in works on modeling circulation in the Black Sea [9]. This condition filters fast inertial-gravity waves. As shown in paper [10], in the shelf areas of the Black Sea the dynamics of the currents during the adaptation period after the initial perturbation differs significantly from the data obtained with the rigid-lid condition. Consequently, inertial-gravity waves make a considerable contribution to the process of adaptation of density fields to the bottom relief and shore orography. Thus, it is necessary to specify more rigid conditions on the sea surface.

The free surface equation is constructed under the assumption that the kinematic condition is satisfied, indicating that the free surface is impenetrable for the liquid:

$$z = -\zeta(x, y, t)\,.$$

Since the vertical velocity of the surface motion is defined as the total time derivative from z, the free surface equation has the following form for case of absence of a mass flux from the atmosphere:

$$w + \frac{\partial\zeta}{\partial t} + u\frac{\partial\zeta}{\partial x} + v\frac{\partial\zeta}{\partial y} = 0\,, \tag{1}$$

where u, v, w are the current velocity components on the free surface; ζ is the sea level elevation. If the ratio $\zeta/H << 1$ (where H is depth), the third and fourth terms in the left-hand side of equation (1) are neglected (the linear condition). The maximum depth of the Black Sea exceeds 2 km, while more than 20% of the sea area is occupied by domains with depths less than 100 m. The North-Western part of the sea is an extensive shelf zone with average depths of about 20 m. As shown in book [1], extreme level rises can reach here several meters and, therefore, the condition $\zeta/H << 1$ becomes incorrect in some cases. The aim of this work is to estimate the accuracy of the modeling results of the Black Sea level fields by using the linear and nonlinear kinematic conditions on the free surface.

3. Model and experimental setting

An eddy-resolving z-model of the Black Sea circulation developed at the Marine Hydrophysical Institute of the Russian Academy of Sciences (MHI RAS) [11] was used in our study. The model is based on the primitive equations of ocean taking into account the approximations of Boussinesq, hydrostatics, and incompressibility of the ocean water:

$$u_t - (\xi + f)v + wu_z = -g\zeta_x - \frac{1}{\rho_0}(p' + E)_x + (\nu_V u_z)_z - \nu_H \nabla^4 u\,, \tag{2}$$

$$v_t + (\xi + f)u + wv_z = -g\zeta_y - \frac{1}{\rho_0}(p' + E)_y + (\nu_V v_z)_z - \nu_H \nabla^4 v\,, \tag{3}$$

$$u_x + v_y + w_z = 0, \tag{4}$$

$$T_t + (uT)_x + (vT)_y + (wT)_z = -\kappa_H \nabla^4 T + (\kappa_V T_z)_z, \tag{5}$$

$$S_t + (uS)_x + (vS)_y + (wS)_z = -\kappa_H \nabla^4 S + (\kappa_V S_z)_z, \tag{6}$$

$$\rho = \varphi(T, S), \tag{7}$$

$$p = g\rho_0\zeta + g\int_0^z \rho d\mu = g\rho_0\zeta + p', \quad \xi = \frac{\partial v}{\partial x} - \frac{\partial u}{\partial y}, \quad E = \rho_0 \frac{u^2 + v^2}{2}.$$

Here f is the Coriolis parameter; g is the gravity acceleration; p is the pressure; T is the temperature; S is the salinity; ρ is the density; $\rho_0 = 1$ g cm^{-3}; ρ_1 is the mean density in the first model horizon; ν_V and ν_H are the vertical and horizontal viscosity coefficients; κ_V and κ_H are the vertical and horizontal diffusion coefficients. The density nonlinearly depends on the temperature and salinity (equation (7)). The density change due to pressure was neglected according to [12]. The wind stress (τ^x, τ^y), the heat flux Q, the precipitation Pr , and the evaporation Ev are set on the sea surface as boundary conditions:

$$\rho_1 \nu_V u_z = -\tau^x, \quad \rho_1 \nu_V v_z = -\tau^y, \quad \rho_1 \kappa_V T_z = Q, \quad \kappa_V S_z = \frac{Pr - Ev}{\rho_1} S_0 + \beta(S_0 - S^{cl}),$$

where S_0 is the surface salinity; S^{cl} is the surface climatic salinity; and β is the relaxation parameter. The relaxation parameter β is equal to 2/1728 m day^{-1} and it denotes the salinization velocity. We set the non-flow condition on the solid boundary. The water exchange through the Bosporus and the Kerch Strait and the runoff of the main Black Sea rivers are taken into account by the Dirichlet boundary condition. The impermeability condition and the absence of heat and salt fluxes are set on the bottom, and the bottom friction is not considered. The slip condition is satisfied, and the normal derivatives for temperature and salinity are zero on the solid lateral boundaries.

The model is implemented on a C-grid [13]. The advective terms in equations (5), (6) are approximated by the total variation diminishing (TVD) schemes [14]. Biharmonic operators for the calculation of the horizontal viscosity and diffusion in equations (2), (3), (5), (6) were used to provide filtering of the computational noise and stabilizing of the numerical solution. The vertical turbulent mixing is parameterized by the Mellor-Yamada turbulent closure model of level 2.5 [15]. Full description of the model, mathematical assumptions, and numerical approximations are presented in [11].

We carried out two numerical experiments on Black Sea circulation modeling taking into account real atmospheric forcing. The difference between the experiments was in a form of the free surface equation. The linear kinematic condition (equation (8)) was set in the first experiment, and we used the nonlinear kinematic condition (equation (9)) in the second one:

$$w\big|_{z=0} + \frac{\partial \zeta}{\partial t} = \frac{Pr - Ev}{\rho_1}, \tag{8}$$

$$w\big|_{z=0} + \frac{\partial \zeta}{\partial t} + u\frac{\partial \zeta}{\partial x} + v\frac{\partial \zeta}{\partial y} = \frac{Pr - Ev}{\rho_1}. \tag{9}$$

The experiments were performed on a uniform grid with a horizontal step of 1.6 km, 27 z-horizons were set vertically, and the time step was 96 sec. The biharmonic operator coefficients in terms describing the horizontal turbulence in equations (2), (3), (5), (6) are $\nu_H = 10^{16}$ cm^4 s^{-1} and $\kappa_H = 10^{16}$ cm^4 s^{-1}. The bathymetry was constructed by using the Black Sea depths from the MHI oceanographic data bank. The sea level, temperature, salinity, and horizontal velocity fields were set at the initial moment of time. These data correspond to January 1 of a climatic year. The data for 2011 obtained by the SKIRON/Eta model with a horizontal resolution of 0.1° in latitude and longitude were used as atmospheric forcing. Observation data are not assimilated in the model. All boundary and initial fields were linearly interpolated to the model grid nodes. Both experiments were carried out for one year. The daily-mean fields of all hydrophysical parameters were returned as output data. Validation of the modeling results by in-situ data is presented in [16].

4. Modeling results

Let us consider the calculated sea level elevation ζ for two experiments. Subscripts L and N in the text below will be used to denote the results of experiment 1 (linear case) and experiment 2 (nonlinear case), respectively. The difference $\Delta\zeta = \zeta_L - \zeta_N$ was calculated at each grid node. Maps of the spatial distribution $\Delta\zeta$ were created to quantify the estimation of the discrepancy between the results of the two calculations. The analysis of maps of the level difference fields for each day showed that the values $\Delta\zeta$ did not exceed an average of 0.6 cm during the first three months after the beginning of simulation. A fairly smooth structure of the currents field was observed in the upper layer in the winter of 2011. Therefore, the sea level fields were reconstructed almost identically in both experiments. The accuracy of the modeling results was assessed based on data of the operational system for diagnosis and forecast of hydrophysical characteristics of the Black Sea [17]. Note that these materials were calculated with the assimilation of available observational data (including altimetry). To estimate the correspondence of our results to reanalysis data, the differences between the year-mean sea level obtained in [17] and the year-mean values of ζ_L and ζ_N were compared. Before that the experimental outputs were interpolated to the reanalysis model grid (resolution: 5 km). Spatial distributions of the difference are presented in Figure 1. The largest deviations (3.4 and 2.7 cm for experiments 1 and 2, respectively) were observed in the abyssal part of the sea, and they are associated with the initial conditions in the experiments. Recall that we used, as the starting point, climatic hydrophysical fields which are smoother and lower than real data. As we see, the maximum absolute magnitude of the sea level difference decreased for experiment 2, especially in the northern and south-western parts. A good agreement between the daily-mean sea level fields and reanalysis data was obtained for situations when the level height increased two and more times compared with the year-mean maximum absolute values.

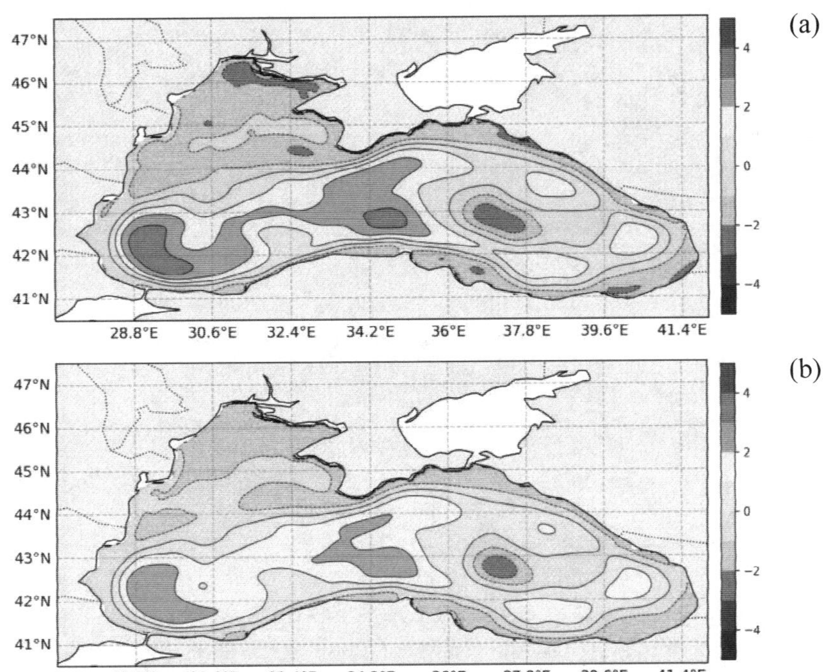

Figure 1. Spatial distribution of the difference (cm) between year-mean sea level heights: (a) reanalysis and experiment 1; (b) reanalysis and experiment 2.

IOP Conf. Series: Earth and Environmental Science **386** (2019) 012027 doi:10.1088/1755-1315/386/1/012027

A comparison of the sea level fields with SKIRON wind stress maps allowed us to conclude that the significant increase in the level height (more than 25 cm) on the shelf near the western boundary of the basin was the result of the impact of storm winds. Here the value of $\Delta\zeta$ was positive, which means that ζ_N is less than ζ_L. The most indicative situations were on April 11 – 15 and October 17 – 18, 2011 when the difference between the modeling data reached 3 cm. Maps of the sea level fields on April 12, 2011 by the data of the system [17] and experiments 1 and 2 are presented in Figure 2. One can see that the structure of the fields qualitatively corresponds to each other. The maximum values of the level height in Figure 2 are located near the mouth of the Danube and are equal to 24 cm, 25.1 cm, and 22.9 cm for Figure 2 (a), (b), and (c), respectively. The minimum values of the sea level (about -15 cm) were observed in the area of the Dnieper-Bug estuary. It was found that the modeling results have a delay of one day compared to the data of the system [17] for these situations.

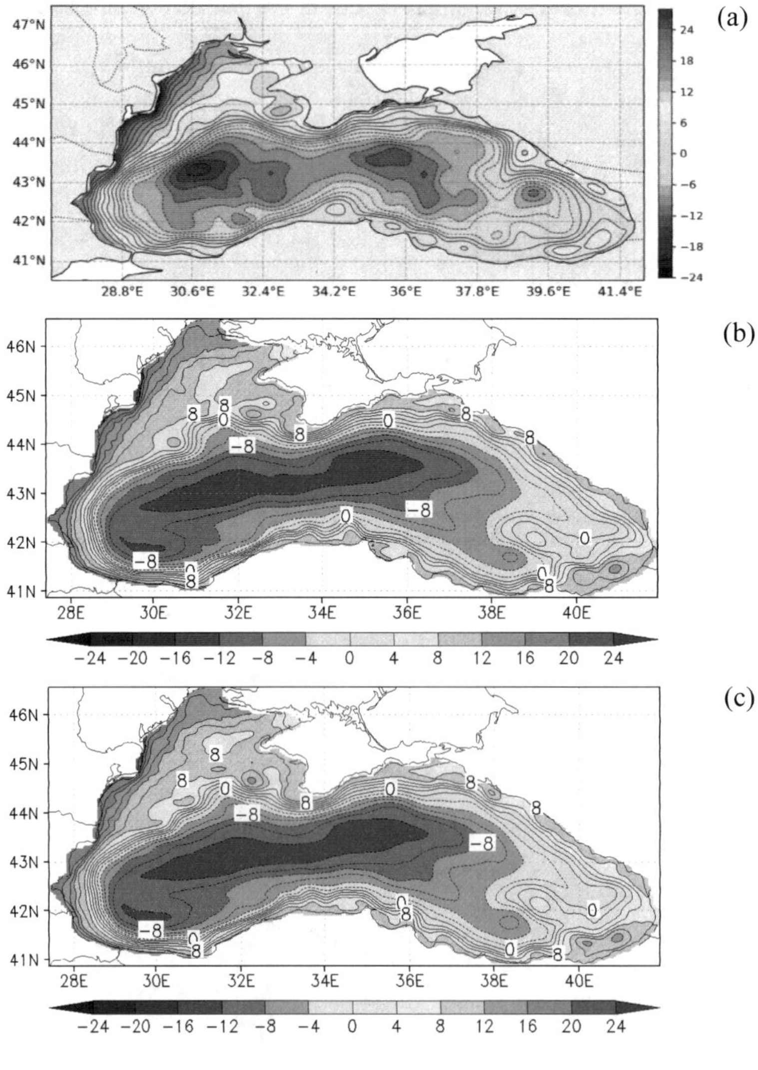

Figure 2. Maps of the sea level field (cm) on April 12, 2011: (a) by the data of [17]; (b) by the results of experiment 1; (c) by the results of experiment 2.

Significant differences in the level field structure were revealed approximately 7 months after the start of the simulation. Spatial distributions of the time-averaged $\Delta\zeta$ calculated for January-July 2011 and August-December 2011 are presented in Figure 3. One can see that extremum values of the time-mean $\Delta\zeta$ increased by an order in the second part of the year compared with the first half. For the daily-mean magnitudes the maximum $\Delta\zeta$ did not exceed 10% of the maximum values ζ_L and ζ_N in the first half of the year. However, occasionally in the second part for both experiments $\Delta\zeta$ reached 15 cm and was commensurate with the maximum level height.

The Black Sea circulation is characterized by a sub-basin cyclonic gyre, the so-called RIM Current propagating along the continental slope. The RIM Current weakens in the warm season and intensifies in autumn-winter. Anticyclonic eddies are mainly formed between the RIM Current and the shore. The most intensive and long-lived eddies in the Black Sea are the Sevastopol and Batumi anticyclonic eddies. Their size is of the order of 100 km and their life-time varies from 4 to 8 months. Relatively the Rossby deformation radius for the Black Sea (about 25 km), these eddies are mesoscale and quasigeostrophic (the Rossby number << 1) in accordance with a classification of the oceanic processes of [18]. Figure 3 (b) illustrates that the greatest differences in the level are located in the center of the sea eastern part and in domains of the Sevastopol and Batumi anticyclones. A comparison of the daily-mean sea level and surface current fields revealed that the displacement phases of the mesoscale eddies moving along the periphery of the RIM Current did not coincide with the experimental results. It was found that the Sevastopol anticyclonic eddy began to move to the south-west about 40 days earlier in experiment 2 than by data of experiment 1 in the second part of the year. A cyclonic eddy located to the north-east of the Batumi anticyclone was more intense and occupied a large area according to the results of the first experiment.

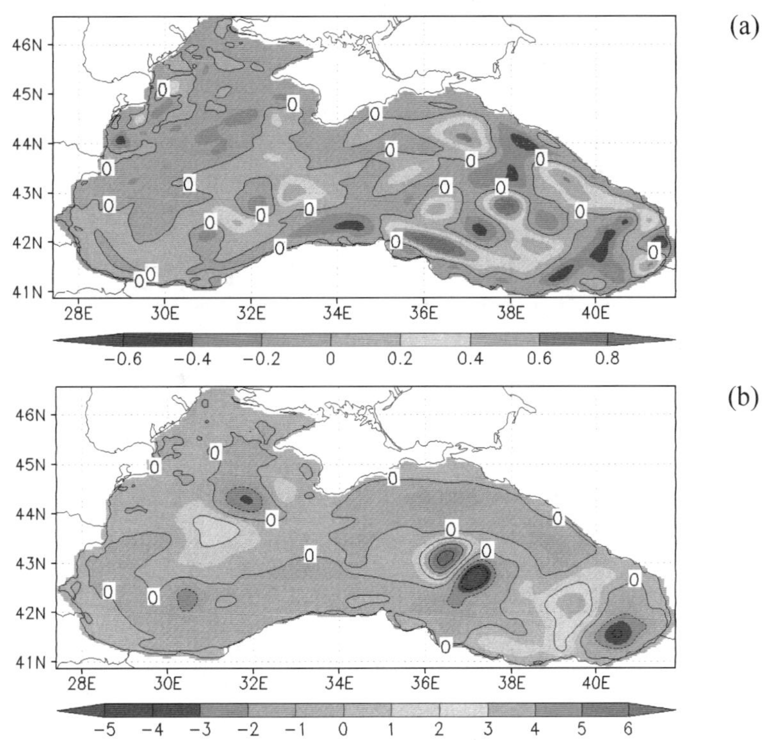

Figure 3. Maps of the time-mean $\Delta\zeta$ (cm): (a) averaged over January-July 2011; (b) averaged over August-December 2011.

An interesting situation was observed in autumn 2011. A storm wind with a velocity of about 20 m/s was blowing over the Black Sea on October 17 – 18, 2011 according to satellite observations (http://dvs.net.ru/mp/data/201110vw_ru.shtml). A level surge near the western coast was the result of such rapid and intense inflow of the kinetic energy from the wind. The response of the sea to the wind impact was manifested in a sequential increase of the sea level along the basin periphery, at first along the Anatolian coast, the Caucasus coast, and then the Crimea. This wave rounded the whole sea in about one week gradually attenuating. Thus, the maximum level height was 33 cm near Burgas (Bulgaria) on October 18 and 22 cm in the Feodosia area (Russia) on October 26. The results are approximately the same in both experiments. The dynamics of the mesoscale eddies in the basin was different according to the results of the two experiments after the wave propagation.

In experiment 1, the Sevastopol anticyclone was practically not observed on October 27 – 30. Then it was again visualized on November 1, greatly weakened. The eddy existed until November 13 moving along the RIM Current to the south-west, where it dissipated to the south of 44°N. At the same time, on November 1, another anticyclone formed near the southern tip of the Crimea. Its size increased to about 100 km, and the shape stretched in the north-western direction along the RIM Current by the end of November. In early December, a cyclone was formed to the north-east of the anticyclone. This one gradually increased in size and, together with the RIM Current meander, prevented displacement of the anticyclone until the second decade of December. The anticyclone size decreased. By the end of December, a new cyclonic eddy was generated from the meander of the RIM Current. It shifted to the north and, together with the already existing cyclone, deformed the Sevastopol anticyclone along the RIM Current in the zonal direction. The anticyclone significantly intensified, and the length of the large axis of the ellipse exceeded 200 km at the end of the year.

In experiment 2, the Sevastopol anticyclone did not disappear completely after the storm, but only slightly weakened on October 27 – 30, in contrast to experiment 1. It shifted along the core of the RIM Current to the south-west and dissipated in mid-November. A new anticyclone generated to the south-west of Sevastopol in October occupied its place in early November. The eddy increased in size and moved along the RIM Current till the end of the year. During December a chain of anticyclonic eddies (rather than individual cyclones, as in experiment 1) was formed near the Crimean coast. This structure was moving to the west not enhancing the Sevastopol anticyclone. It was found that the Sevastopol anticyclone was less intense than in experiment 1 and its size varied in the range of 100 – 150 km.

The Batumi anticyclone had almost the same size and intensity by the beginning of the storm on October 17 – 18 in experiments 1 and 2. It dissipated a few days after the storm, the RIM Current being pressed to the shore in both cases. Since November, the formation of coastal anticyclones in the south-eastern part was reconstructed in the results of the numerical experiments differently. The coastal eddies were more intense and larger in area for experiment 2. During December, the number of anticyclonic formations between the RIM Current and the shore in both experiments was different. Their number remained unchanged and was equal to two in the second experiment, whereas the number of eddies varied from two to four and their sizes were smaller in the first experiment. Cyclonic eddies were formed between the anticyclonic ones in December, but they were more extensive in experiment 2.

5. Energetics

As shown above, the dynamics of mesoscale eddies changed when applying the nonlinear kinematic condition on the free surface. The mesoscale variability of circulation is usually associated with the eddy kinetic energy determined by the fluctuations of the velocity of currents from the mean value [19]. The Lorenz Energy Cycle analysis [20] is generally used to study the mechanisms of circulation variability. The energy cycle is defined by the balance of four components: the mean kinetic energy (MKE), the eddy kinetic energy (EKE), the available mean potential energy (MPE), and the available eddy potential energy (EPE). To investigate the possible reasons of distinctions in the eddy displacement phase, the EKE variability for both experiments was considered. The potential energy components were not investigated here. The year-mean (time-averaged over 2011) and time-varying (deviations from the year-mean for each day of the year) zonal and meridional velocity components were calculated. Based on these data, the volume-integrated values of the EKE and the rate of energy conversion from the EKE to the MKE through barotropic instability (denoted by *EKE* and *C*(*EKE*, *MKE*), respectively) were calculated for each day by using the following formulas:

$$EKE = \frac{1}{2}\int_V \rho_0 \left(u'^2 + v'^2\right) dV, \quad C(EKE, MKE) = -\int_V \rho_0 \left[u'^2 \frac{\partial u}{\partial x} + v'^2 \frac{\partial v}{\partial y} + u'v' \left(\frac{\partial v}{\partial x} + \frac{\partial u}{\partial y} \right) \right] dV,$$

where $\int_V dV$ is the integral over the basin volume; the apostrophe denotes the deviation from the year-mean value. If *C*(*EKE*, *MKE*) is positive, the energy is converted from the EKE to the MKE. The year-mean value of the volume-integrated kinetic energy (*MKE*) was equal to 0.563 PJ (petajoules = 10^{15} joules) for both experiments. Also, it was found that the mean value of EKE_L was equal to 0.228 PJ and the mean value of EKE_N was equal to 0.217 PJ. Thus, in 2011 the year-mean contribution of the EKE to the kinetic energy was 40.5% for experiment 1 and 38.5% for experiment 2.

Changes in the wind stress maximum over the Black Sea, *EKE* and *C*(*EKE*, *MKE*) during 2011 for both experiments are shown in Figure 4. One can see in Figure 4 (b) and (c) that the energy characteristics were close in the first half of the year. Later EKE_N was less than EKE_L by about 15 – 20% from August to October 2011 (Figure 4 (b)). At the same time, $C(EKE, MKE)_N$ increased: two peaks of the red line are clearly defined in August and September in Figure 4 (c). The positive value of $C(EKE, MKE)_N$ indicates that the eddy energy transformed to the kinetic energy of the basin-scale circulation in the second experiment. $C(EKE, MKE)_L$ was mainly negative from May to September 2011 in the first experiment, which means that EKE increased due to MKE. For both experiments, a significant surge of *EKE* was observed on October 18 – 19, 2011, and it was caused by the storm wind (Figure 4 (a)). The maximum value of EKE_L was equal to 0.4 PJ, in contrast to max EKE_N = 0.34 PJ. In November and December, the range of the $C(EKE, MKE)_N$ oscillation was from -0.16 to 0.2 GW (gigawatt = 10^9 watts), and the range of the $C(EKE, MKE)_L$ oscillation, from -0.04 to 0.12 GW. For the year-mean values it was found that $C(EKE, MKE)_L$ = -0.0166 GW and $C(EKE, MKE)_N$ = -0.0157 GW. Consequently, the decrease of EKE_N compared with EKE_L coinciding with a reduction of the energy inflow from the MKE to the EKE for experiment 2 (recall the sign of *C*(*EKE*, *MKE*) indicates the direction of energy conversion). Since *MKE* is the same for both experiments, a possible reason for the restructuring of the mesoscale circulation in the second half of 2011 is the redistribution of kinetic energy into faster processes such as submesoscale.

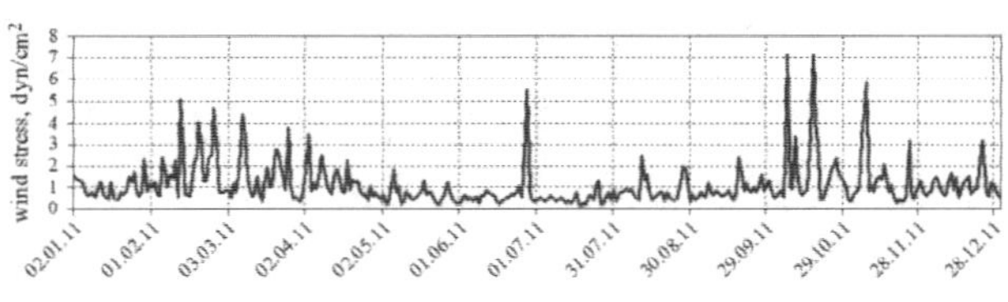

(a)

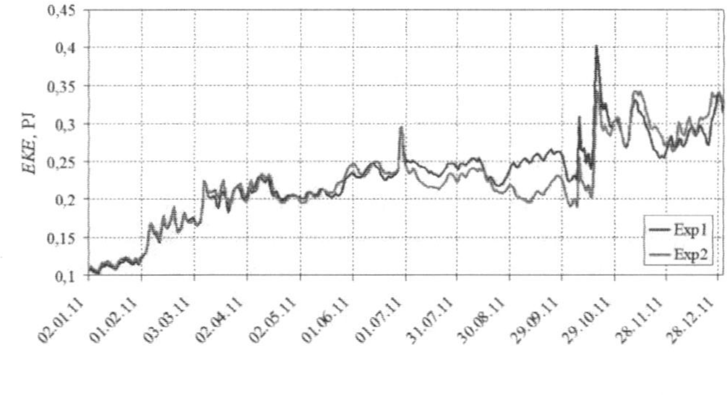

(b)

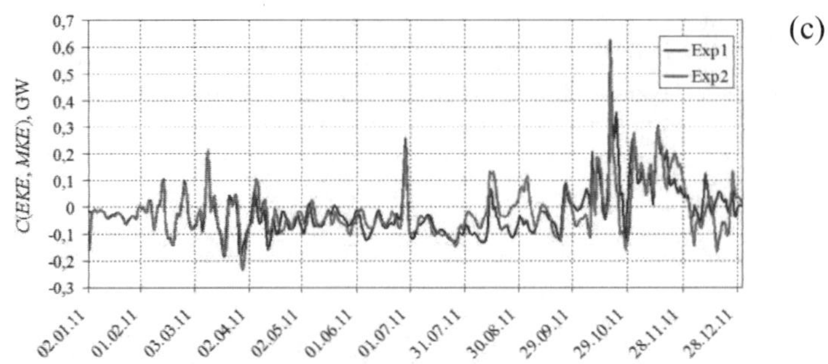

Figure 4. Time variability of the wind stress maximum (a), volume-integrated values of EKE (b) and conversion rate between EKE and MKE (c) for experiments 1 and 2.

6. Conclusions

A study of results of Black Sea level modeling calculated by using linear and nonlinear kinematic conditions on the free surface was performed. Experimental data were obtained by an eddy-resolving numerical MHI-model with a resolution of 1.6 km with real atmospheric forcing for 2011. The daily-mean and year-mean values of the sea level were analyzed. The differences between the year-mean reanalysis sea level and year-mean values of ζ_L and ζ_N were compared to estimate the correspondence of the modeling results with the reanalysis data (with altimetry assimilation). It has been revealed that the deviation of the sea level height from the reanalysis data decreased in the experiment with the nonlinear kinematic condition. It has been found that the daily-mean height of the sea level near the western boundary of the basin was decreased by using the nonlinear kinematic condition for cases of significant level rises (more than 25 cm). The reduction of the maximum level height in the coastal zones can be explained by the fact that the account of all terms in the free surface equation allows us to describe a wider class of level oscillations. As a result, in the second experiment the eddy kinetic energy was decreased due to a more correct redistribution between the movements on different spatial and temporal scales. To construe the types of level oscillations reconstructed by using the nonlinear kinematic condition, we plan to perform a spectral analysis of the modeling results.

The comparison of the experimental results demonstrated that the greatest differences in the structure of the sea level fields were observed during the period of weakening of the RIM Current (the warm season) in the evolution areas of Sevastopol and Batumi anticyclones. Note that the minimum and maximum values of level heights are almost identical in both experiments, but their locations do not match each other. The discrepancies in the spatial positions of raising/lowering levels are due to a mismatch in the displacement phases of anticyclonic eddies along the periphery of the RIM Current. The Sevastopol and Batumi anticyclones are components of mesoscale variability in the Black Sea. In the second part of 2011, changes in the dynamics of mesoscale eddies were observed. This process was displayed in the behavior of the volume-integrated eddy kinetic energy when the discrepancy between EKE_L and EKE_N became significant after a storm in July and reached a maximum after a storm in October 2011. The restructuring of the Black Sea mesoscale circulation in 2011 in the second experiment is associated with the decrease in the eddy kinetic energy due to a reduction in the inflow from the mean current kinetic energy. It is probably related to a redistribution of kinetic energy towards the submesoscale processes.

The above-obtained results show that the account of nonlinear terms in the equation for calculating the free surface height led to a qualitative change in the response of the Black Sea current field to an intense wind impact in 2011. Thus, the deviations of the level heights from the reanalysis data were reduced. These conclusions should be taken into account for improving the accuracy of forecasting the hydrophysical fields in the Black Sea, especially in the submesoscale ranges of variability. Now the operational system for diagnosis and forecast uses a version of the MHI-model where the level height is calculated by using the linear condition. However, the nonlinear kinematic condition is used in most of the modern models of ocean circulation.

Acknowledgements
The authors are grateful to a Reviewer for helpful comments. Experiment 1 was carried out under support of the Russian Foundation for Basic Research (project no. 18-45-920019), experiment 2 and comparative analysis were performed under state assignment 0827-2019-0003.

References
[1] Dotsenko S F and Ivanov V A 2010 *Natural disasters of the Black Sea and Azov Sea region* (Sevastopol) p 174 (in Russian)
[2] Dyakonov G S and Ibraev R A 2018 Reproduction of interannual variability of the Caspian Sea level in a high-resolution hydrodynamic model *Oceanology* **58**(1) 8-18
[3] Tsimplis M N, Josey S A, Rixen M and Stanev E V 2004 On the forcing of sea level in the Black Sea *J. Geophys. Res. Oceans* **109**(C8) C08015
[4] Kubryakov A A and Stanichnyi S V 2013 The Black Sea level trends from tide gages and satellite altimetry *Russ. Meteorol. Hydrol.* **38**(5) 329-33
[5] Medvedev I 2018 Analysis of Variance of the Black Sea Level Oscillations in a Wide Range of Frequencies *Phys. Oceanogr.* **25**(6) 448-58
[6] Haza A C, Mariano A, Chin T M, Olson D B 2007 The mean flow and variability of the Gulf stream-slopewater system from MICOM *Ocean Modelling* **17**(3) 239-76
[7] Mellor G L 2002 *A three-dimensional, primitive equation, numerical ocean model* (Princeton University, Princeton, NJ 08544-0710) p 42
[8] Madec G and the NEMO team 2016 *NEMO ocean engine* (Note du Pôle de modélisation, Institut Pierre-Simon Laplace (IPSL), France) p 396
[9] Demyshev S G 1996 The improvement of calculation accuracy of the Black Sea currents using a numerical model with a reference level *Russ. Meteorol. Hydrol.* **1**(9) 53-60
[10] Ibraev R A 2001 A study of the sensitivity of the model of the Black Sea current dynamics to the surface boundary conditions *Oceanology* **41**(5) 615-21
[11] Demyshev S G 2012 A numerical model of online forecasting Black Sea currents *Izv. Atmos. Ocean. Phys.* **48**(1) 120-32
[12] Mamaev O N 1975 *Temperature-salinity Analysis of World Ocean Waters* (Elsevier Scientific Publishing) p 374
[13] Arakawa A and Lamb V R 1981 A potential enstrophy and energy conserving scheme for the shallow water equation *Monthly Whether Rev.* **109**(1) 18-36
[14] Harten A 1983 High resolution schemes for hyperbolic conservation laws *J. Comput. Phys.* **49**(3) 357-93
[15] Mellor G L and Yamada T 1982 Development of a turbulence closure model for geophysical fluid problems *Rev. Geophys.* **20**(4) 851-75
[16] Markova N V, Dymova O A and Demyshev S G 2019 Numerical Simulations of the Black Sea Hydrophysical Fields Below the Main Pycnocline: Validation by ARGO Data *Physical and Mathematical Modeling of Earth and Environment Processes* (Switzerland: Springer, Cham) pp 15-21
[17] Korotaev G K, Ratner Y B, Ivanchik M V, Kholod A L and Ivanchik A M 2016 Operational system for diagnosis and forecast of hydrophysical characteristics of the Black Sea *Izv. Atmos. Ocean. Phys.* **52**(5) 542-49
[18] Thomas L N, Tandon A, Mahadevan A 2008 Submesoscale processes and dynamics *Ocean Modeling in an Eddying Regime. Geophysical Monograph Series* **177** 17-38
[19] Holland W R and Lin L B 1975 On the generation of mesoscale eddies and their contribution to the oceanic general circulation. I. A preliminary numerical experiment *J. Phys. Oceanogr.* **5** 642-69

[20] Von Storch J-S, Eden C, Fast I, Haak H, Hernández-Deckers D, Maier-Reimer E, Marotzke J and Stammer D 2012 An Estimate of the Lorenz Energy Cycle for the World Ocean Based on the 1/10° STORM/NCEP Simulation *J. Phys. Oceanogr.* **42**(12) 2185-205

IOP Conf. Series: Earth and Environmental Science **386** (2019) 012028 doi:10.1088/1755-1315/386/1/012028

Assessment of the carbon balance of treed bogs under climate change with observation and modelling data

E A Dyukarev[1,2], Yu V Martynova[1] and E A Golovatskaya[1]

[1] Institute of Monitoring of Climatic and Ecological System of the Siberian Branch Russian Academy of Sciences, 10/3 Akademicheskii ave., Tomsk, 634055, Russia
[2] Yugra State University, 16 Chekhov st., Khanty-Mansiysk, 628012, Russia

E-mail: egor@imces.ru

Abstract. Carbon dioxide fluxes in treed bogs in the South Taiga Zone of Western Siberia are estimated by using field data and a mathematical model calibrated against observation data. Forecasts of carbon balance under climate change are made by using the mathematical model.

1. Introduction

Peatland ecosystems play a significant role in the global carbon cycle, being sources and sinks of greenhouse gases (GHGs) [1,2]. Despite covering a relatively small part of the Earth surface (about 3%), peatlands store a large amount of organic matter ranging between 500 and 700 billion tons of C [3, 4]. In Western Siberia peatlands occupy over 30% of the area [5-7]. According to IPCC estimates [1], the contribution of natural mires into the total natural methane emissions ranged between 61 and 82%. The intensity of GHG fluxes is controlled by different factors including the hydrological and thermal regime of the peat deposit [8-15].

The gaseous exchange between the atmosphere and peatlands is governed by photosynthetic fixation of CO_2 from the atmosphere and by soil and vegetation respiration losses of CO_2. The balance between them is known as the net ecosystem exchange (NEE) of CO_2 [10, 16, 17]. The rise in surface air temperature [18] and the lowering of water levels causes peat drying, increase of temperature and aeration, which contributes to the intensity of greenhouse gas emissions [19, 20]. Peatland ecosystems in different years can also serve as both a source and a sink of carbon [21, 22]. The variety of direct and inverse relationships existing between the components of peatlands and the surrounding areas indicates a complex nonlinear impact of the peatlands on the environment in different geographic, climatic, and geomorphological conditions [23-26]. Quantitative estimation of the rate of carbon exchange between peatlands and the atmosphere, as well as the revealing of environmental factors affecting carbon exchange, is an important scientific issue [5, 27].

The main purpose of this study is to asses CO_2 exchange fluxes in treed bogs in the South Taiga Zone in Western Siberia and to forecast carbon balance disturbances under climate change using a mathematical model.

2. Observation sites

Field studies of net primary production (NPP) and CO_2 emissions were carried out in the eastern part of the Ob-Irtysh interfluve, within the eastern margin of the Vasyugan Plateau in the Bakchar bog in 2008-2011 [21]. Oligotrophic peatland ecosystems with a narrow set of forming species have a dominant position in the study area. Pine-shrub-sphagnum bogs (ryams) is the most widespread type of bogs in the south taiga subzone. They are confined to margins or well-drained slopes of bogs with surface slopes of 0.001 - 0.006 degrees. Pine-shrub-sphagnum bogs are also found in the most convex central parts of the peatlands. Observation points were organized in the tall ryam, low ryam, and drained low ryam.

The tall ryam is located at the margin of the Bakchar bog massif. The microrelief is presented by moss hummocks up to 50 cm high, occupying about 50% of the surface. The vegetation of tall ryam belongs to the pine-shrub-sedge-sphagnum association. The tree layer consists of pine (*Pinus silvestris L.*) with single trees of *Pinus sybirica Mayr.* and *Betula pubescens Ehrh.* The average height of the tree stand is 18 m, and the average diameter of the stems is about 20 cm. The projective cover of the tree cover is 90%. The shrub layer is well developed, the shrubs reach a height of 50 cm, and the projective cover is 90%. The dominants are *Ledum palustre L., Chamaedaphne calyculata L., Vaccinium vitis-idea L.*, and cranberries (*Oxycoccus microcarpus*). *Carex globularis L., Eriophorum vaginatum L.,* and *Rubus chamaemorus L.* are found in the grass cover. The dominants in the moss cover (96%) are sphagnum mosses (*Sphagnum angustifolium*). A small amount of green moss is found on the tree trunks, and lichens are present on the trunks of trees and shrubs. The peat deposit of the tall ryam is 1 m.

The tall ryam towards the center of the bog turns by the low ryam. The low ryam microrelief is wavy due to a large number of large moss hummocks about 30 cm high and up to 3 m in diameter. The low ryam vegetation belongs to the pine-shrub-sphagnum association with low pine trees. The oppressed tree layer is presented by *Pinus silvestris f. Litwinowii.* The average height of the trees is 2-3 m, and the average diameter of the stems is 3 cm. The shrub layer is developed abundantly on hummocks, and the total projective cover is 60-70%. The grass cover is composed of *Ledum palustre L., Chamaedaphne calyculata L., Andromeda polifolia L.*, and *Vaccinium uliginosum L.* On the tops of the hummocks there grow small-fruited cranberries (*Oxycoccus microcarpus Turcz.*). The grass protective cover is less than 5%, and it is presented by *Eriophorum vaginatum L., Rubus chamaemorus L.,* and *Drosera rotundifolia L. Sphagnum fuscum Klinggr.* dominates in the moss cover. The peat thickness in the low ryam is 2 – 3 m.

To reveal the effect of peatland drainage on the carbon balance of oligotrophic bogs, studies were performed at the drained low ryam at Vasyuganskoye peatland on the watershed plateau between the Bakchar and Iksa Rivers. The distance between the draining channels is 150 m, and the design drainage rate is 0.6 m. The phytocenosis of the drained area is also represented by pine-shrub-sphagnum vegetation and is characterized as a low ryam. The tree layer is formed by pine trees up to 2-3 m high, with a trunk diameter of 2-5 cm; the projective cover is 40%. In the surface cover there are abundant wild rosemary, cassandra, and blueberries (80%). Cotton grass is involved in the grass layer (5%). Sphagnum fuscum Klinggr. prevails in the moss cover. There are lichen stains occupying up to 20% of the surface. The thickness of the peat deposit is 2.5 m. The water content at the site is low, and the level of bog waters is 0.2 m and below.

3. Mathematical modelling

To obtain continuous data records, to extrapolate them to other periods when experimental data are missing, and to calculate the annual carbon budget of the ecosystem, a model of total ecosystem carbon exchange was proposed [28, 29]. The measured total NEE was partitioned into the incoming (GPP) and expenditure (ER) components [30-33]:

$$NEE = ER - GPP; \quad (1)$$

$$ER = HR + AR; \quad (2)$$

GPP is defined as the total amount of carbon fixed in the process of photosynthesis by plants in an ecosystem, while NEE refers to GPP minus ER. ER is the sum of plant respiration, or autotrophic respiration (AR) and heterotrophic respiration (HR).

It is well-known that the photosynthetic response under low light intensities is characterized by a linear response, and photosynthetic saturation is observed at high light intensities [34]. A rectangular hyperbolic function (3) is used for the light response of the NEE in the daytime [30, 31]:

$$GPP = LAI \times \alpha \times PAR \times Gm / (\alpha \times PAR + Gm), \tag{3}$$

where α is the initial slope of the light response curve at low light (photosynthetic efficiency) (mg μmol^{-1}), and Gm is the theoretical maximum rate of photosynthesis at infinite PAR (photosynthetic capacity) (mg m^{-2} h^{-1}). Carbon dioxide fluxes are given in mg of CO_2 per m^2 per hour. The PAR is measured in μmol m^{-2} s^{-1}. Possible GPP limitation at high air temperatures was not accounted for.

The respiration was modelled using an exponential equation (4) widely used for explanation of the ER variation [32, 33]:

$$HR = E_0 \times exp(k_T \times Ta); \tag{4}$$

$$AR = LAI \times k_A \times HR, \tag{5}$$

where Ta is the air temperature (°C), E_0 is the reference heterotrophic respiration (mg m^{-2} h^{-1}) at Ta = 0°C; k_T is a coefficient describing the respiration temperature response (K^{-1}); k_A is the ratio of autotrophic respiration to heterotrophic one at LAI = 1.

The NEE is negative when the GPP value exceeds the ER value and there is a net removal of carbon dioxide from the atmosphere. The NEE is positive when the ER value exceeds the GPP value and the carbon dioxide is released from the ecosystem into the atmosphere.

A simple model was proposed for the seasonal changes in green phytomass (or LAI). The LAI is following variations in the mean decadal air temperature (Tm). At the beginning of the growing season (at the transition of Tm through 5 °C) the index is LAI = 0.5. The same LAI value is enforced at the end of the growing season. In the first half of the growing season (until reaching a maximum of Tm), the LAI increases:

$$LAI_{k+1} = LAI_k + L_1 \times (Tm - 5),\ if\ Tm > 5, \tag{6}$$

After the seasonal maximum temperature Tm, the green vegetation dies off, and the LAI decreases exponentially:

$$LAI_{k+1} = LAI_k - L_2 \times exp(1 - Tm),\ if\ Tm > 5, \tag{7}$$

where L_1 and L_2 are the parameters for growth and dying off of vegetation, and k is the number of the time step. To account for the slower growth of vegetation at high air temperatures, L_1 is reduced by 25% for Tm above 15 ° C.

The model has been calibrated using all available data set on field NPP data in 1999-2011. The temperature sensitivity coefficient of the heterotrophic respiration was calculated from measurements of CO_2 emission fluxes from the bog surface [35]. It was found that for high and low ryam the temperature sensitivity of respiration is the same and k_T = 0.045 g m^{-2} K^{-1}. For a drained ryam the temperature sensitivity of respiration is approximately twice as high and k_T = 0.109 g m^{-2} K^{-1}. The accumulated phytomass in vegetation in the model is taken into account through the increase in the LAI during the season. The multi-objective optimization procedure was performed in the Matlab software using the fmin search function. The minimum of the unconstrained multivariable function was found using a derivative-free optimization method [36]. The root-mean-square error was used as a minimizing function.

4. Simulation results

The proposed model allows calculating carbon dioxide fluxes for the ecosystems being studied on the hourly time scale. The results of carbon flux estimations for each ecosystem were summarized annually for the period from May 1 to September 30.

The gross carbon accumulation by vegetation (GPP) in the model is completely controlled by the amount of incoming photosynthetically active radiation and biomass reserves accumulated during the growing season (via the LAI). Therefore, the interannual fluctuations of the GPP at different sites are almost synchronous (r> 0.76). However, the GPP value differs significantly between the sites. The greatest accumulation of carbon by vegetation is observed in the tall ryam, and the smallest one is in the drained ryam.

The total ecosystem respiration (ER) varies depending on the weather conditions and takes the maximum values in 2012 for all bogs, when the average summer temperature exceeded the multi-year average by 2 °C. Also abnormally large emission fluxes associated with an increase in the air temperature were observed in 2016. On average, over the study period, the ecosystem respiration decreases in the row high ryam – low ryam - drained ryam. The expected increase in the emission fluxes during the drainage of the peat deposit is not observed in the model estimates. Some increase in the heterotrophic respiration due to an increase in the capacity of the aerated layer is fully compensated by a decrease in the dark respiration of plants.

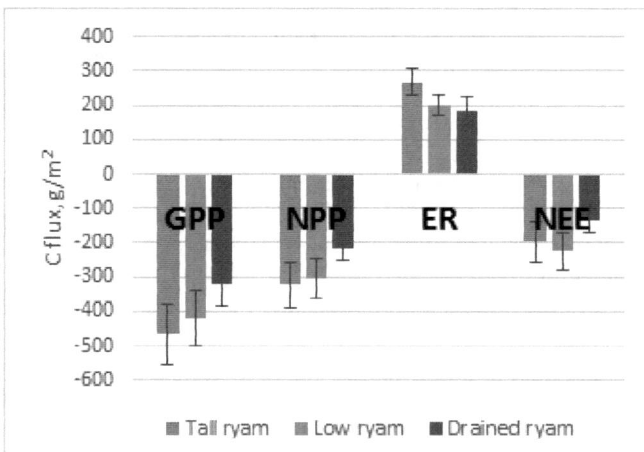

Figure 1. Carbon balance of pine-shrub-sphagnum bogs of the Bakchar peatland and its components (mean ± STD) for 1999 -2017.

The net ecosystem exchange, which is considered as an estimate of the carbon balance (Fig. 1.), has the lowest values for the drained ryam (134 g m^{-2}). For the undisturbed high and low ryam ecosystems, the NEE values turned out to be close (197 and 223 g m^{-2}). The net primary production, whose observations were used to calibrate the model, decreases from 323 g m^{-2} for the high ryam to 306 g m^{-2} for the low ryam, and then to 214 g m^{-2} for the drained ryam.

Thus, forest reclamation after peatland drainage in the long term leads to a decrease in both the input and expenditure components of the carbon balance, but the total annual carbon absorption in the drained ecosystems decreases by 20-224 g m^{-2} (or 16-70%) depending on the weather conditions. However, this decline does not shift the ecosystem from a sink of atmospheric carbon to its source.

5. Long-term changes

The forecast of changes in the carbon balance was made using the model described above. As control parameters of the model (air temperature and incoming photosynthetically active radiation), the results of calculations using the PLASIM model [36] for 1901-2100 were used, with an increase in the carbon dioxide concentration according to the RCP 8.5 scenario [37] (http: //climate.uvic.ca/EMICAR5). The PASIM model consists of a series of modules of varying complexity: atmospheric, oceanic, soil ones, modules for sea ice and land surface. Being a model of intermediate complexity, the PASIM allows modeling for a sufficiently long period in a reasonable estimated time and, at the same time, makes it possible to estimate the magnitude and overall dynamics of the studied processes.

According to the RCP 8.5 scenario, the average global concentration of CO_2 increases from 296 to 936 ppm, and according to the model estimates, the surface air temperature also increases (Figure 2). Thus, the average annual temperature averaged over the first and last 20 years of the study period increases from -0.4 to 5.0 °C, and the average temperature of the growing season (May-September), from 9.7 to 15.8 °C. The average PAR during the growing season for two centuries does not undergo significant directional changes, increasing from 278 to 330 μmol m^{-2} s^{-1} (Figure 2). However, in the twenty-first century the interannual variability of the PAR increases.

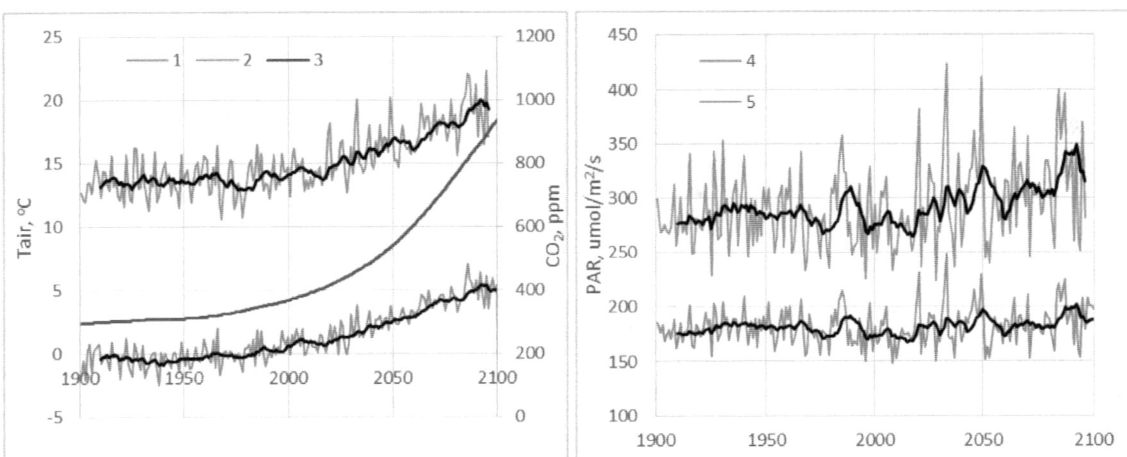

Figure 2. Changes of meteorological parameters in XX-XXI centuries according to model estimations in scenario RCP 8.5. 1 - average annual air temperature, 2 - average air temperature in May-September, 3 - scenario of growth of average CO2 concentration, 4 - average annual PAR, 5 - average PAR for May-September (black lines: 10-year running average).

Model (1-7) allows one to extrapolate data on carbon fluxes both in the past (from 1901) and into the future (until 2100). The model parameters were determined by the results of field studies of the net primary production of bog vegetation. Since the end of the 21st century, according to the model calculations the duration of the period with air temperatures above 5 °C increases. Figure 4 shows the changes in the net annual carbon fluxes in the 20th – 21st centuries. Since the model does not support the transformation of the vegetation cover, changes in the hydrological regime, the existing estimates of the fluxes are simply extrapolated to the future in accordance with the predicted changes in the temperature and the carbon fluxes in different ecosystems are synchronous. The linear correlation coefficient of annual carbon fluxes in various ecosystems is 0.99 - 0.92. The exception is the correlation coefficient between the NEE in the drained and the native ryams (r = 0.58 and 0.64). Like the air temperature, the carbon fluxes over the 20th century change slightly, and the main changes in the magnitudes of the fluxes are predicted during the 21st century (Figure 3).

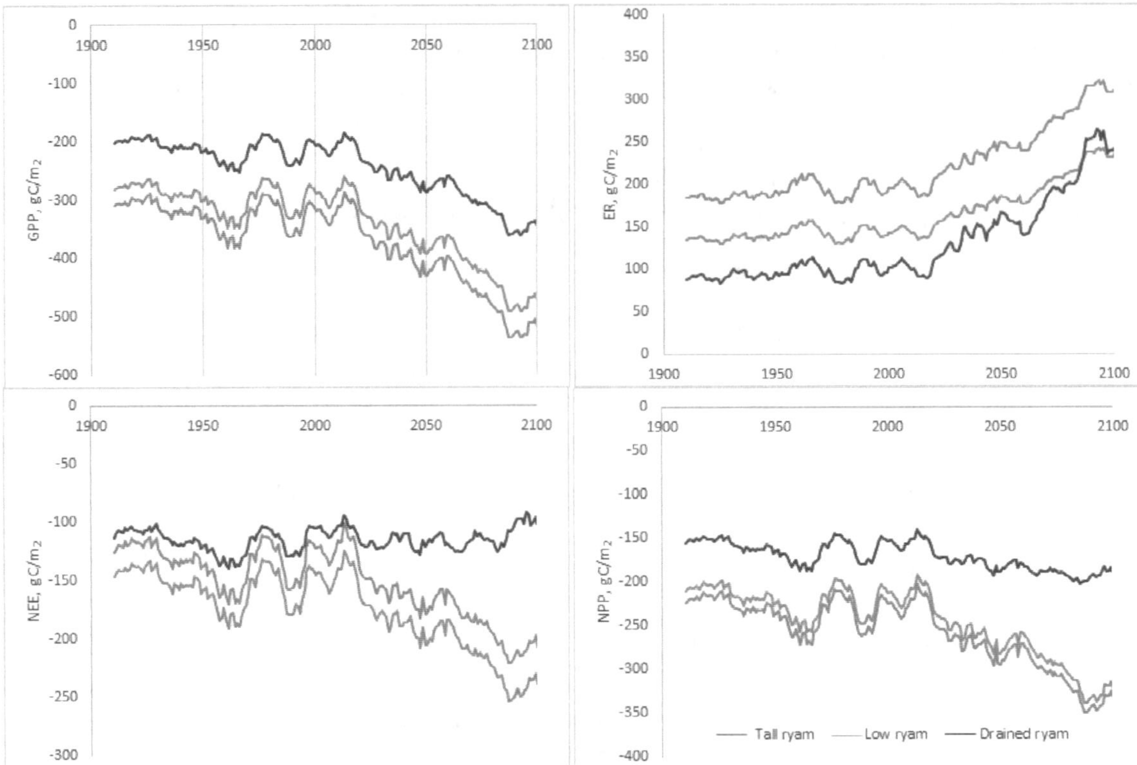

Figure 3. Forecast of annual carbon fluxes (GPP, ER, NEE, NPP) in XX and XXI centuries (points - model, lines - 10-year running average).

The model incorporates the exponential sensitivity of the ecosystem respiration to the temperature, and the ER value increases in accordance with the increase in the air temperature over two centuries. In the low and tall ryams, the annual value of ER increases by 75 and 71% (Table 1). The coefficient of respiration sensitivity to temperature for the drained ryam is approximately 2 times higher, which leads to a significant increase in the ER by 182%.

Table 1. Forecast estimates of carbon fluxes (g m-2) in pine-shrub-sphagnum bogs during the first 20 years of XX century and the last 20 years of XXI century, Δ is the difference between fluxes (%).

		GPP	NPP	ER	NEE
Tall ryam	1901-1920	-305	-220	183	-122
	2081-2100	-523	-339	313	-210
	Δ	71%	54%	71%	73%
Low ryam	1901-1920	-277	-208	134	-143
	2081-2100	-478	-328	235	-243
	Δ	73%	58%	75%	70%
Drained low ryam	1901-1920	-198	-154	88	-110
	2081-2100	-349	-192	248	-101
	Δ	76%	25%	182%	**-8%**

The gross carbon accumulation by plants (GPP) also increases in all ecosystems. The GPP growth is partly due to minor increases in the PAR and, to a greater extent, due to an increase in the maximum values of biomass storages accumulated during the growing season. However, it is most likely that with significant warming there will be a violation of the hydrological regime of the wetlands and a change in the vegetation type. Thus, according to the present model estimates in all studied ecosystems the GPP growth is projected at 71 - 76%.

6. Conclusions

The net primary production of the above-studied treed bogs increases over two centuries under climatic changes. However, the NPP growth in a drained treed bog is projected to be approximately two times less (a NPP growth of 25% over the study period) than in the native high and low treed bogs (the NPP rises by 54 and 58%). Although in the above scenario of warming and increasing carbon dioxide concentration, both the expenditure (ecosystem respiration) and input (gross primary production) parts of carbon balance are predicted to grow, the net ecosystem exchange (carbon accumulation in the ecosystem) also increases in the native treed bogs. The increase in the NEE for a high treed bog is 73% and for a low treed bog 70%. However, in a drained treed bog with warming the ecosystem respiration increases faster than the gross primary production, which results in a certain (of 8%) decrease in the NEE. Nevertheless, even with a significant warming scenario it was predicted that even a drained treed bog will continue to act as an absorber of atmospheric carbon. Some additional forecast estimates have shown that a drained treed bog will transit to a net carbon source in an atmosphere with additional two degrees of warming, according to the RCP 8.5 scenario.

Acknowledgments

This study was supported by project AAAA-A17-117013050031-8 and SB RAS integration project AAAA-A18-118012490022-2.

References

[1] Ciais P, Sabine C, Bala G *et al* 2013 *Carbon and Other Biogeochemical Cycles. In: Climate Change 2013: The Physical Science Basis. Contribution of Working Group I to the Fifth Assessment Report of the Intergovernmental Panel on Climate Change* (Cambridge, United Kingdom and New York, NY, USA: Cambridge University Press) pp 465-570
[2] Rydin H, Jeglum J 2015 *The Biology of Peatlands* (Oxford. Univ. Press) p 400
[3] Page S, Baird A 2016 *Ann. Rev. Environ. Resour.* **41** 35–57
[4] Leifeld J, Menichetti L 2018 *Nat. Commun.* **9** 1071
[5] Sheng Y, Smith L C, MacDonald G M *et al* 2004 *Global Biogeochemical Cycles* **18** GB3004
[6] Terentieva I E, Glagolev M V, Lapshina E D *et al* 2016 *Biogeosciences* **13** 4615–26
[7] Dyukarev E A, Pologova N N, Dyukarev A G, Golovatskaya E A 2011 *Environ. Res. Lett.* **6**(3) 035203 9 doi: 10.1088/1748-9326/6/3/035203
[8] Naumov A V 2009 *Soil respiration* (Novosibirsk, Izd SO RAN) p 208
[9] Sasakawa M, Ito A, Machida T *et al* 2012 *Tellus, Ser. B: Chem. and Phys. Meteorology* **64**(1) 1–10
[10] Helfter C, Campbell C, Dinsmore K J *et al* 2015 *Biogeosciences* **12**(6) 1799–811
[11] Molchanov A G 2015 *Russian J. of Ecol.* **46**(3) 230–35
[12] Walker T N, Garnett M H, Ward S E *et al* 2016 *Glob. Change Biol.* **22**(5) 1880–9
[13] Glagolev M V, Ilyasov D V, Terentyeva I E *et al* 2017 *Atmospheric and Oceanic Optics* **30**(4) 301–9
[14] Veretennikova E E, Dyukarev E A 2017 *Russian Meteorol. and Hydrol.* **42**(5) 319–26 doi:10.3103/S1068373917050077
[15] Leroy F, Gogo S, Guimbaud C *et al* 2017 *Soil Biol. and Biochem.* **107** 164–7
[16] Bubier J L, Crill P M, Mosedale A *et al* 2003 *Glob. Biogeochem. Cycles* **17**(2) 1066
[17] Olchev A, Novenko E, Desherevskaya O *et al* 2009 *Environ. Research Lett.* **4**(4) 045007
[18] Zhaojun B, Joosten H, Hongkai L *et al* 2011 *Acta Ecologica Sinica* **31**(3) 157–62

[19] Baird A, Belyea L, Comas X *et al* 2013 *Carbon Cycling in Northern Peatlands* (Geophysical Monograph Series. AGU) p 297
[20] *The second assessment report of Roshydromet on climate change and their consequences on the territory of the Russian Federation* 2014 (Federal service for Hydrometeorology and environmental monitoring) p 605
[21] Golovatskaya E A, Dyukarev E A, Ippolitov I I, Kabanov M V 2008 *Doklady Earth Sci.* **4** 1–4
[22] Parazoo N C, Koven C D, Lawrence D M *et al* 2018 *The Cryosphere* **12** 123–44
[23] *Peatlands of West Siberia. Their composition and hydrological regime* 1976 (Leningrad: Hydrometeoizdat) p 615
[24] Vomperskiy S E 1994 The role of peatlands in carbon cycling, in *Biogeocenotic features of peatlands and their rational exploitation* (Moscow: Nauka) pp 5-37
[25] Ratcliffe J L, Creevy A, Andersen R *et al* 2017 *Sci. of the Total Env.* **607-608** 816–28
[26] Webster K L, Bhatti J S, Thompson D K *et al* 2018 *Carbon Balance and Management* **13**(1) 16
[27] Kabanov M V 2015 *Geogr. and Natural Res.* **3** 207–13
[28] Dyukarev E A 2017 *Agricult. and Forest Meteorol.* **239** 236–48 doi:10.1016/j.agrformet.2017.03.011
[29] Dyukarev E, Godovnikov E, Karpov D *et al* 2019 *Geography, Environment, Sustainability* **12**(2) 227–44 doi:10.24057/2071-9388-2018-77
[30] Mäkelä A, Hari P, Berninger F *et al* 2004 *Tree Physiol.* **24**(4) 369–76
[31] Laine A, Riutta T, Juutinen S *et al* 2009 *Ecol. Modell.* **220** 2646–55
[32] Kandel T P, Elsgaard L, Larke P E 2013 *Bioenergy* **5**(5) 548–61
[33] Campbell D I, Wall A M, Nieveen J P *et al* 2015 *Agric. Ecosyst. Environ.* **202** 68–78 doi:10.1016/j.agee.2014.12.019
[34] Pessarakli M 2005 *Handbook of Photosynthesis* (Taylor & Francis Group) p 928
[35] Golovatskaya E A and Dyukarev E A 2012 *Eurasian Soil Sci. J.* **45**(6) 588–97 doi:10.1134/S106422931206004X
[36] Lagarias J C, Reeds J A, Wright M H, Wright P E 1998 *SIAM Journal of Optimization* **9**(1) 112–47
[37] Fraedrich K., Jansen H., Kirk E *et al* 2005 *Meteorol. Zeitschrift* **14** 299–304
[38] Moss R *et al* 2010 *Nature* **463**(7282) 747–56 doi: 10.1038/nature08823

IOP Conf. Series: Earth and Environmental Science **386** (2019) 012029 doi:10.1088/1755-1315/386/1/012029

Temperature regimes of drained and natural peatlands in arid and water-logged years

M V Kiselev[1], N N Voropay[1,2], E A Dyukarev[1,3] and Yu I Preis[1]

[1]Institute of Monitoring of Climatic and Ecological Systems SB RAS, Russia, Tomsk, Akademicheskii 10/3
[2]V B Sochava Institute of Geography SB RAS, Russia, Irkutsk, Ulan-Batorskaya 1
[3]Yugra State University, Russia, Khanty-Mansiysk, Chekhova 16

Abstract. The results of a study of temperature regimes of oligotrophic bog ecosystems in the south taiga zone of Western Siberia in 2011-2018 are presented. Soil temperature regimes are studied using an atmospheric-soil measuring complex at several depths from the surface to 320 cm. Waterlogged (excessive moisture) and arid years are determined by calculating the aridity index of Ped (S). The years with most severe hydrothermal conditions over the past 10 years have been found to be 2012 and 2018. In 2012, a drought is observed from April to July, with extreme aridity in June (S = 4.07), and a severe drought in July (S = 3.42). It has been found that the native treed bog in a wet year warms up much more than in a dry year. On average, the temperature values are higher by 1.5°C for all depths. On average, the drained bog is colder than the natural one at depths up to 80 cm by 4.0°C, and at depths of 120 to 240 cm by 2.5°C. The maximum temperature differences of the native bog during the warm period of a waterlogged (excessive moisture) year are 8.0°C (at a depth of 60 cm) and, at depths from 80 to 240 cm they are reduced from 5.0 to 2.5°C. The maximum differences of the native bog in the waterlogged (excessive moisture) 2018 and dry 2012 in the warm period at a depth of 60 cm are 8.0-9.0°C, and from 80 to 240 cm they are reduced from 6.0 to 1.5°C.

1. Introduction

A peat deposit is a complex organo-mineral system with specific properties: high water content and porosity, with a large amount of slightly decomposed organic matter [1]. The temperature of the peat soil affects the growth of terrestrial vegetation (mosses, herbs. shrubs, etc.), the formation of a microclimate, and is a key factor controlling many biotic and abiotic processes in soils (peat or mineral): the decomposition and mineralization of soil organic matter, greenhouse gas emissions (CO_2, CH_4, N_2O), and the release of dissolved organic carbon [2].

The main world reserves of peat (and carbon) are located in Western Siberia. According to various estimates, the peatlands of Western Siberia contain up to 70 billion tons of carbon [3]. Such a significant reservoir of carbon due to climate change or anthropogenic impact can potentially transfer to the atmosphere in the form of CO_2 or CH_4 and make a significant contribution to the carbon balance of the atmosphere.

Studies of the temperature regime of soils were carried out in the territory of the Vasyuganie geophysical station [4]. Observations were made at two sites with significantly different hydrological regimes and the levels of bog waters.

2. Study area

The first site is located 200 m from the edge of the bog massif in the basin of the Key creek. The vegetation is represented by an oligotrophic low ryam, a pine-shrub-sphagnum community with a depressed tree stand (*Pinus sylvestris f. Willkommii*) of 2-m height (h); the stem diameter (D) is 3-7 cm, and the closure of crowns (P) is 0.4. *Chamaedaphne calyculata* dominates in the dense grass-shrub layer, and *Ledum palustre, Oxycoccus microcarpus*, and *Eriophorum vaginatum* are less abundant. Large moss hummocks (h – 30-70 cm, d – 100-150 cm) are formed by *Sphagnum fuscum. S. angustifolium* prevails between the hummocks. The average bog water table level (WTL) is about 25 cm below the surface. The peat deposit has a depth of 2 m. The transitional and lowland woody-moss and woody-peat deposits are under the layer of high-grade *Fuscum* and *Magellanicum* peats.

The other site is located on the drained, burned out site at Iksinskoe bog, 50 m from the main drainage channel. At the place of the former tall ryam, single tall (h – up to 6.5 m, D – 10-20 cm) pines were preserved, abundant (P – 0.7) undergrowth from pine and birch was formed (h – 2-2.3 m; D – 1.5-5 cm), a dense (P – up to 85%) shrub layer from *Ledum palustre* mixed with *Chamaedaphne calyculata, Vaccinium myrtillus, Andromeda polifolia, Carex globularis*. Small moss hummocks formed by *S. magellanicum* mixed with *S. angustifolium* and *Polytrichum strictum* occupy 60% of the surface. The average WTL is 50 cm. The peat deposit has a depth of 1.2 m, and is a pine-cotton grass peat. Above the middle-laid upland pine and cotton grass peats lies 20 cm of transitional woody and *Magellanicum* peats.

At the sites, measurements of the soil temperature were made at 14 depths (0, 2, 5, 10, 15, 20, 30, 40, 60, 80, 120, 160, 240, 320 cm) and the air temperature at a height of 2 m using an atmospheric soil measuring complex [5] in the period from September 11, 2011 to September 10, 2018 (2384 days) with a time step of 1 hour [6, 7]. The work also used data on the air temperature and rainfall sums from observations of the nearest Bakchar weather station (30 km to the west of the experimental sites) obtained from the RIHMI-WDC archive [8]. Data on the precipitation amount in the bog were obtained using a HOBO Rain Gauge 3M precipitation gauge (Onset Computer Corp., USA) installed at the first site.

The differences between the average monthly values of the air temperature measured at the meteorological station and in the bog have a seasonal variation: from March to August they are positive (the temperature in the bog is higher, on average, by 0.2-0.7°C), and in the other months they are negative (the bog temperature is lower, on average, by 0.1-0.9°C). The differences in the precipitation amount in the winter months are negative: on average, the bog falls by 10–20 mm / month less than at the Bakchar meteorological station. During the warm months (May-September), both positive deviations (up to 78 mm, July 2017) and negative ones (up to -68 mm, August, 2011) are observed, with the average difference of -3 ... - 7 mm / month.

3. Hydrothermal conditions of the territory

The specific features of the temperature regime of soils in certain years are determined not only by the characteristics of the soils, but also by the hydrothermal conditions of a particular year. The complex characteristic of temperature and atmospheric humidification are hydrothermal coefficients, which are based on the relationship between the air temperature and precipitation amount [9, 10]. It is most correct to use normalized indicators when comparing different landscapes for revealing droughts and periods of waterlogging [11, 12]. In this paper, the determination of waterlogged and dry years was carried out by calculating the aridity index of Ped [13],

$$S_i = \frac{\Delta t_i}{\sigma_t} - \frac{\Delta r_i}{\sigma_r}, \tag{1}$$

where $\Delta t_i = t_i - t_{norm}$ is the air temperature anomaly for the *i* time interval, σ_t is its standard deviation, $\Delta r_i = r_i - r_{norm}$ is the anomaly of the precipitation amount over the *i* time period, and σ_r is its standard deviation. The S index takes positive values during droughts and negative values during periods of overwetting. According to the classification, there are weak ($1 \leq |S| < 2$), moderate ($2 \leq |S| < 3$), strong ($3 \leq |S| < 4$), and extreme events ($|S| \geq 4$).

Despite significant differences in the monthly precipitation data from observations at the Bakchar meteorological station and the bog observation site, due to the localization of convective precipitation (especially during the summer period), the aridity / overwetting characteristics of individual months using the Ped index are identical for the two meteorological observation sites.

In relation to the long-term period (1936-2018) during the years of microclimatic observations at the meteorological station during the growing season, both dry and overwetting periods of varying intensity were observed, ranging from 1 to 4 months. The most contrasting hydrothermal conditions over the past 10 years were 2012 and 2018. In 2012, droughts were observed from April to July, with extreme aridity in June (S = 4.07) and a drought of strong intensity in July (S = 3.42). The beginning of the warm season of 2018 was characterized by a prolonged over-wetting conditions of the territory with a maximum in May (S = –4.05) (table 1).

Table 1. Aridity index of Ped.

Year	Months						
	April	May	June	July	August	September	October
2011	2.60	0.64	1.75	**-3.20**	**-3.49**	1.83	1.34
2012	1.80	0.44	**4.07**	**3.42**	-1.10	1.56	0.85
2013	0.39	**-4.31**	-1.30	1.33	0.23	-0.89	0.77
2014	0.57	-1.26	0.54	1.30	1.71	-2.25	-2.53
2015	-1.66	0.16	1.11	0.86	0.16	-2.06	-2.04
2016	0.38	0.55	1.54	-1.54	1.85	**3.58**	-1.64
2017	0.32	-0.33	-2.14	-0.29	1.33	-0.41	0.54
2018	**-2.51**	**-4.05**	**-1.74**	**-1.63**	0.56	1.64	1.01

4. Temperature regime of natural and anthropogenically disturbed bogs

It was found that the natural low ryam in the wet year (2018) warms up much better than in the dry year (2012). On average, for all depths the temperature in 2018 at a low ryam is higher by 1.5°C. The drained low ryam has small maximum temperatures and generally warms worse than an undisturbed low ryam under the same conditions of atmospheric drought. On average, the drained bog is colder than the natural ones at depths of up to 80 cm by 4.0°C, and at depths of 120-240 cm by 2.5°C.

At the studied sites in different years, the number of days with temperatures above + 15°C was calculated. These temperatures were observed in June-August, and the duration of the period varies depending on the depth.

The maximum number of days with a peat temperature above 15°C was observed at a low ryam in the overwetted year 2018. The duration was 57 days on the surface, 42 days at a depth of 10 cm, and 19 days at 20 cm. In the dry 2012, despite the fact that the air temperature was higher than in 2018, the number of days with temperatures above + 15°C at the low ryam was less at all depths except for the surface. The duration was 57 days on the surface and 29 days per 10 cm. At a depth of 20 cm, the peat warmed up to + 14.8°C. Such differences show that atmospheric aridity can reduce the duration of a period with temperatures above + 15°C for 10–20 days during a warm period. This decrease is mainly due to changes in the level of bog waters. Its increase leads to an increase in the amount of heat absorbed during the daytime heating of the peat deposit and to a decrease in the rate of cooling of the peat deposit during the night period.

The number of days with a peat deposit temperature of more than + 15°C at the drained low ryam in the overwetted year of 2018 is much lower than in the undisturbed bog. The duration was 16 days on the surface, 4 days at 10 cm, and below it was not observed. Such a short duration of the warm period is associated with the initially lower level of the bog waters (60 cm on average) and with a higher rate of reduction of the WTL after intense rains in comparison with the low ryam. Such differences lead to the fact that an increase in the WTL at the drained low ryam in an overmoistened year occurs for a shorter time period, which averages 5-7 days.

To assess the degree of influence of the hydrothermal conditions on the heating of the peat deposit, we calculated the differences between the soil temperature at selected sites at depths from the surface to 320 cm (Figure 1). The results obtained show that the peat deposit under the natural pine-shrub sphagnum bog in the overmoistened year 2018 warms up better than the similar peat deposit under the drained pine shrub sphagnum bog. The maximum differences for the warm period of the overwetted year (from April to the end of September) to a depth of 60 cm are 8.0°C, and at depths from 80 to 240 cm they decrease from 5.0 to 2.5°C.

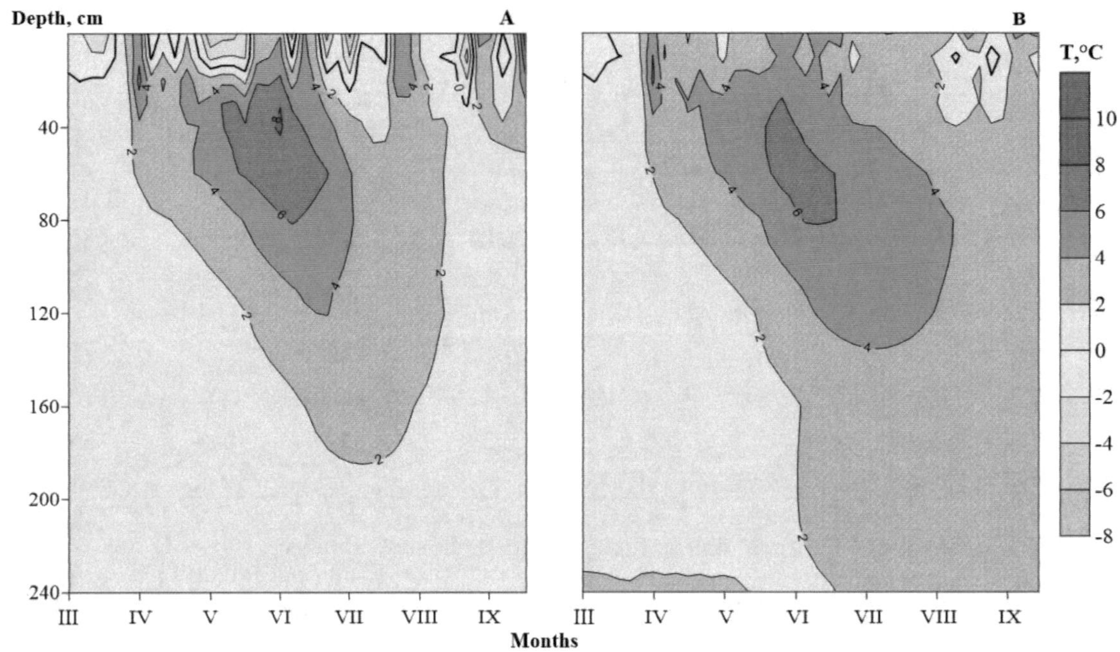

Figure 1. Temperature differences for peat soils. A – natural low ryam ($T_{2018} - T_{2012}$), B – natural and drained ryam ($T_{\text{low ryam 2018}} - T_{\text{drained low ryam 2018}}$).

A similar result was observed when comparing the temperature of the peat deposit at the undisturbed site in the overmoistened 2018 and in the dry 2012. The maximum differences for the warm period (from April to the end of September) to a depth of 60 cm are 8.0–9.0°C, and from 80 to 240 cm they decrease from 6.0 to 1.5°C.

5. One-dimensional heat flux model

In order to investigate the effect of periods of droughts and overmoistening on the intensity of the heat exchange processes inside the soil [14, 15], we used a one-dimensional model of heat transfer (2). The heat diffusion model describes the nonstationary temperature distribution in a homogeneous solid medium within which there is no bulk motion (advection or convection). We use the one-dimensional approach due to neglectable small heat fluxes along the horizontal x and y coordinates. The term $-\partial G/\partial z$ is related to the net conduction heat flux into the control volume for the z-coordinate direction. Within the medium there may also be an energy source term associated with the rate of thermal energy generation. This term is represented as dF.

$$C\frac{\partial T}{\partial t} = -\frac{\partial G}{\partial z} + dF, \tag{2}$$

where *G* is the heat flux due to conductive (or molecular) heat transfer, *dF* is the change of heat not associated with molecular thermal conductivity, but with phase transitions of water, ice, and water vapor, an explicit heat transfer with water migration, heat release during wetting, heat sources from decomposition of plant remains, etc., $\partial T/\partial t$ is the temperature gradient over time, and z is the depth, cm.

Actually, the model is very simplified and does not take into account vertical heat transfer with percolated water, turbulent and radiative heat transfer within soil pores, capillary water movement, and subsequent heat effects, and also possible horizontal heat transfer related to water movement. However, the model includes two basic soil heat fluxes: conductive heat transfer and water phase transition. The model (2) is applicable for preliminary estimation of the relation between molecular (diffusive, conductive) heat flux and heat storage changes due to other reasons.

For the layer of soil located between the depths of z0 and z1, the following was calculated:

1. The total change in the heat content (ΔS) during Δt:

$$\Delta S = \int_{z0}^{z1} C\frac{\partial T}{\partial t}dz \tag{3}$$

2. The heat flux due to molecular heat transfer:

$$G = -k\frac{\partial T}{\partial z} \tag{4}$$

Having determined these two values from the observation data, it is possible to calculate the heat flux due to other reasons:

$$F = \Delta S - G \tag{5}$$

The temperature and temperature gradients were determined by the available observation data using finite-difference schemes:

$$\frac{\partial T}{\partial z} \approx \frac{T_k{}^i - T_{k-1}{}^i}{z_k - z_{k-1}}, \quad \frac{\partial T}{\partial t} \approx \frac{T_k{}^{i+1} - T_k{}^i}{\Delta t}. \tag{6}$$

The coefficients of heat capacity (*C*) and thermal conductivity (*k*) depend on the characteristics of the soil, where the soil is represented as a three-phase medium consisting (Figure 2) of a solid phase, the gaps between which are filled with water and air [16].

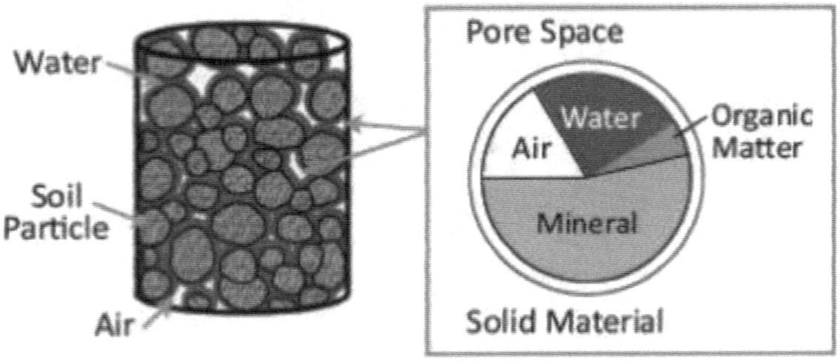

Figure 2. Composition of an Unsaturated Soil Sample [16].

The heat capacity coefficient is calculated by the equation

$$C = c_A\rho_A A + c_W\rho_W W + c_I\rho_I I + c_{S1}\rho_{S1}(1 - \Pi)S_1 + c_{S2}(1 - \Pi)(1 - S_1), \quad (7)$$

where c_A, c_W, c_I, c_{S1}, c_{S2} are the heat capacities of the air, water, ice, peat and clay, respectively, ρ_A, ρ_W, ρ_I, ρ_{S1}, ρ_{S2} are the densities of the air, water, ice, peat and clay, respectively, Π is porosity.

The thermal conductivity coefficient is calculated by the following equation:

$$k = k_A A + k_W W + k_I I + k_S(1 - \Pi), \quad (8)$$

where k_A, k_W, k_I, k_S are the thermal conductivities of the air, water, ice, and soil, respectively, A is the number of pores filled with air, W is the number of pores filled with water, and I is the number of pores filled with ice in the peat deposit.

At negative temperatures all water is replaced by ice ($I = W$). Below the level of bog waters, air (A) is absent, and all pores are occupied by water (W) or ice (I). The thermophysical characteristics of the substances are tabulated constants (Table 2) [http://thermalinfo.ru/].

Table 2. Thermophysical characteristics.

Characteristic	Heat capacity, J/kg/K	Density, kg/m^3	Thermal conductivity, W m^{-1} K
Air	1005	1.247	0.025
Water	4182	1000	0.59
Ice	2150	918	2.25
Peat	1880	1000	0.05
Clay	750	1700	1.6

6. Results of one-dimensional modeling of heat flux

The dynamics of daily heat fluxes at some depths over six years are presented in Figure 3. The conductive (G) heat flux in summer takes positive values, which corresponds to the transfer of heat from the warm surface layers into the soil (Figure 3). In winter, the sign of molecular heat transfer changes, the flux values are usually lower than in summer.

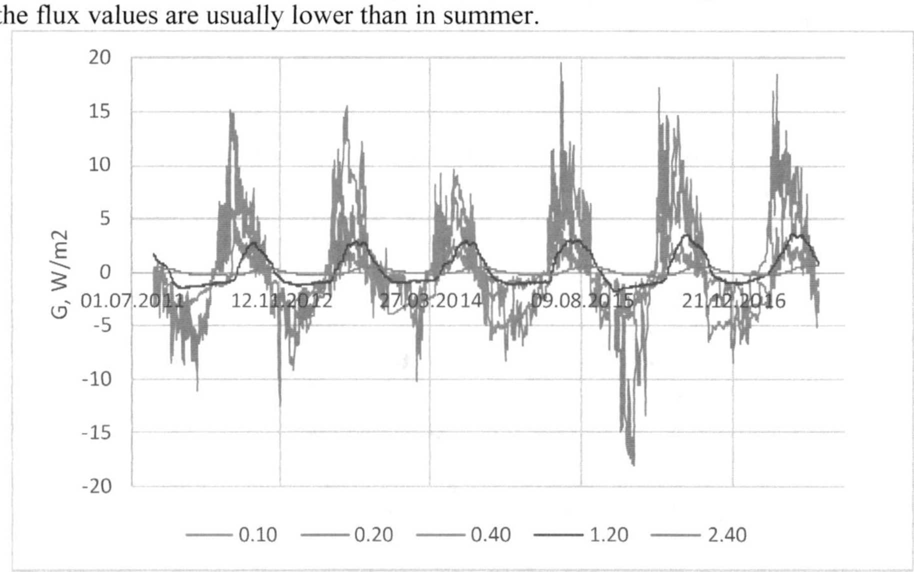

Figure 3. Mean daily soil conductive heat fluxes in low ryam.

The change in the heat content (dS) of the soil in summer at a depth of more than 20 cm is positive, which means the accumulation of heat in the soil. In the surface layer, the change in the heat content has large values, while their sign changes from positive to negative during summer. Such fluctuations mean rapid heat accumulation and sink.

The dynamics of the non-conductive heat flux (*F*), defined as the difference between the change in the heat content and the conductive heat flux, has a more complex structure. A large negative flux F was observed in the upper layers up to 40 cm in summer (heat consumption for evaporation) and positive fluxes, in winter (heat generation when water freezes).

The maximum conductive heat transfer (*G*) averaged over certain months in the low ryam (Figure 4) is observed in summer not at the surface, but in a layer of 10-30 cm (at the boundary of the bog water level). In water saturated peat layers, the temperature gradients are less and, as a result, the molecular heat transfer is not so intense. However, even on the lower layer (240 cm) the heat flux is positive from June to November.

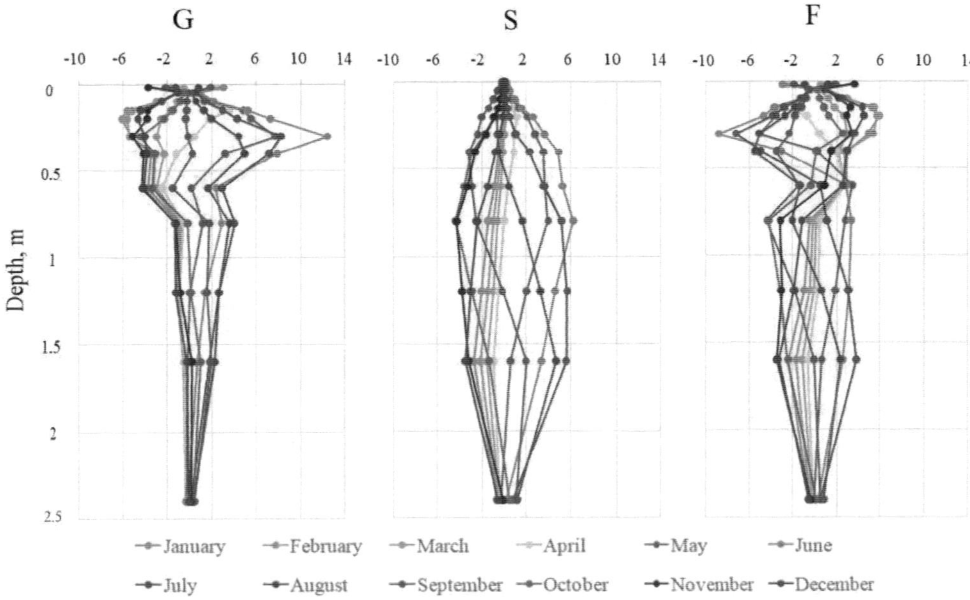

Figure 4. Soil heat fluxes in low ryam peat deposit (*G* – conductive heat transfer, *dS* – heat storage change, *F* – non-conductive heat flux).

In the cold season (from October to March), the values of conductive heat transfer (*G*) are negative. The largest negative, i.e., upward heat flux was obtained in December-February in a layer of 15-30 cm (inside the frozen layer). The thermal conductivity of ice is higher than that of water, and the frozen layer conducts heat better.

The change in the heat storages (*dS*) at monthly averaging is maximum inside the saturated water peat deposit (30-160 cm). In the upper layers with pores filled with air and in the lower layer with dense clay and less water, the heat storages vary slightly.

The distribution profiles of the non-conductive heat flux (*F*) are similar to the conductive flux profiles, but in the reflected (left to right) projection. For example, the largest heat fluxes entering the soil by molecular diffusion in June in a layer of 20-40 cm practically do not change in the heat storages, but they are completely spent on the evaporation of water. In winter, relatively large fluxes of molecular heat from the soil to the surface also do not cause a significant reduction in the heat storage and are generated from heat release during the formation of ice.

Inside the water-saturated peat layers (80-160 cm) there are significant fluxes of non-conductive heat transfer (*F*). The molecular heat transfer here is small, and the change in the heat content is significant. Interpretation of the calculation results of the fluxes is impossible within the framework of the one-dimensional heat transfer model. Possible causes of the appearance of these fluxes are the heat release during the decomposition of organic matter or the influx / removal of heat associated with the migration of internal bog waters.

In the peat deposits with a disturbed hydrological regime due to drainage, all above-described patterns of heat flux distribution are preserved (Figure 5).

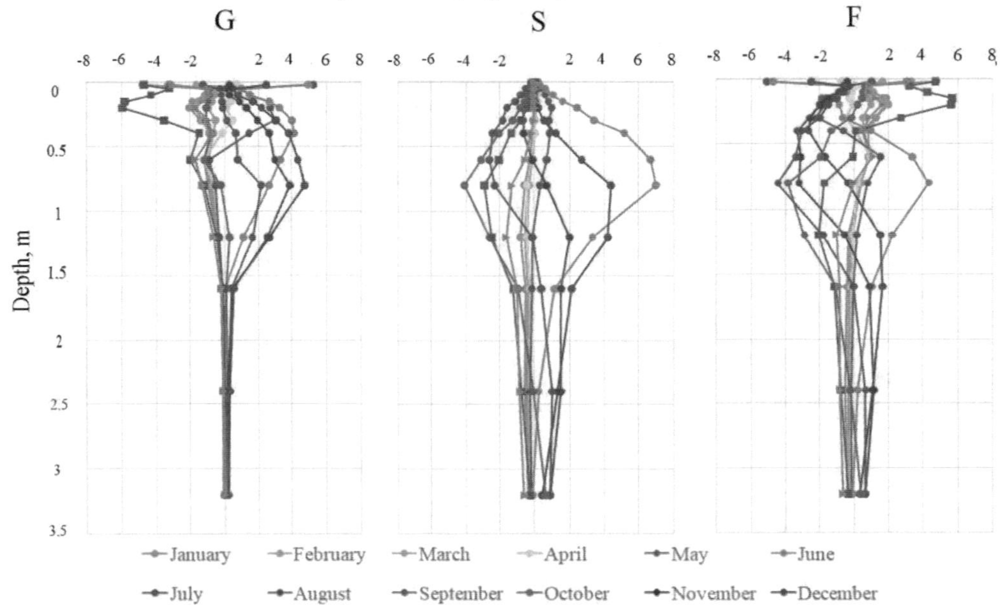

Figure 5. Soil heat fluxes in drained low ryam (*G* – conductive heat transfer, *dS* – heat storages change, *F* – non-conductive heat flux).

However, due to lower water levels the maximal molecular (*G*) transfer layer shifts deeper and to later periods (June, depth: 80 cm). The change in the heat storages (*dS*) is also maximal within the peat layer, which is 110 cm here.

Peat freezing in the drained area begins in November. Since the moisture in the peat is low, within only a few days the frozen layer reaches a considerable depth, which subsequently varies only slightly during winter and spring. For example, from November 8 to November 24 in 2015, the soil froze to 33 cm, and a maximum freezing depth of 45 cm was observed on May 5, 2016. Intense freezing is reflected in the non-conductive (F) heat transfer profile in November.

In general, in the natural ryam the intensity of heat exchange is higher compared to the drained ryam. This can be traced both in fluxes of molecular heat transfer and latent heat, and is explained by a high moisture content in the soil profile of the natural ryam both in the aerated part and in the water-saturated layers. In turn, in the drained ryam the moisture content is lower as the peat density is higher, the porosity is lower and, accordingly, the peat deposit water content is lower.

7. Conclusions

The temperature and moisture of peat deposits play an important role in emission of methane and carbon dioxide. The latest results show that temperatures from + 15°C to + 20°C initiate the most active period of methane emission [17, 18], provided that the peat deposit is filled with water. Thus, waterlogging of the peat deposit contributes to both increasing temperatures and increasing methane emissions.

Also, an increase in moisture and temperature leads to an increase in carbon dioxide emission from the surface of treed sphagnum bogs [19]. However, the peat deposit must remain unsaturated with water. Due to an initially small WTL and relatively fast filtration of water into a drained low ryam peat deposit, a significant warming up of the peat occurs in the overmoistened year. Thus, it is possible to increase the emissions of carbon dioxide the main part of which is emitted in the upper layers of the peat deposit [20].

The results obtained above give an assessment of the influence of atmospheric aridity on the temperature regime of forested wetlands. They show some possible negative consequences of anthropogenic or natural drainage of peat deposits, such as an increase in carbon dioxide emissions from taiga bogs in wet years. One of the possible positive consequences is a slight decrease in methane and carbon dioxide emissions in dry years. However, drainage of peat deposits leads to an increase in fire danger in dry years. The above-obtained results of one-dimensional modeling of the heat fluxes have shown that this approach has insufficient efficiency in estimating the latent heat flux.

Acknowledgements
The study was carried out with the financial support of RFBR in the framework of scientific projects №18-45-700015, №18-05-00306 and basic research project №AAAA-A17-117013050030-1.

References

1. Romanov V V 1961 *Hydrophysics of bogs* (Leningrad: Gidrometeoizdat) p 359
2. Golovatskaya E A, Dyukarev E A, Ippolitov I I, Kabanov M V 2008 The influence of landscape and hydrometeorological conditions on CO2 emission in peatland ecosystems *Reports of the Russian Academy of Sciences* **4** 1-4
3. Ciais P *et al* 2013 Carbon and Other Biogeochemical Cycles *Climate Change 2013: The Physical Science Basis. Contribution of Working Group I to the Fifth Assessment Report of the Intergovernmental Panel on Climate Change* ed Stocker T F *et al* (Cambridge, United Kingdom and New York, NY, USA: Cambridge University Press) p 570
4. Kiselev M V, Voropai N N, Dyukarev E A 2016 Features of temperature regime of soils upland of the wetland massif *Reports of higher educational institutions. Physics* **7 2** 93-7
5. Kiselev M V, Voropay N N, Dyukarev E A, Kurakov S A, Kurakova P S, Makeev E A 2018 Automatic meteorological measuring systems for microclimate monitoring *IOP Conference Series: Earth and Environmental Science* **190**(1)
6. Dyukarev E A 2015 Influences of air temperature and snow cover on the seasonally frozen soil layer characteristics *Earth's Cryosphere* **19**(3) 39-45
7. Kiselev M V, Voropay N N and Dyukarev E A 2017 The temperature regime of the soil sedge-sphagnum fen in the southern taiga of Western Siberia *Geography and Natural Resources* **1** 110-7
8. Bulygina O N Description of the array of urgent data on the main meteorological parameters at the stations of Russia (Electronic resource) Certificate of state registration of the database No. 2014620549 Access mode: http://meteo.ru/data/163-basic-parameters#data- array of data
9. Ryazanova A A, Voropay N N 2017 Droughts and Excessive Moisture Events in Southern Siberia in the Late XXth - Early XXIst Centuries *IOP Conference Series: Earth and Environmental Science* **96**(1)
10. Voropay N N, Maksyutova E V and Riazanova A A 2016 Hydrothermal conditions at the south of East Siberia during the ongoing warming *IOP Conference Series: Earth and Environmental Science* **48**(1)
11. Voropay N N, Ryazanova A A 2018 A comparative assessment of the aridity indices for analysis of the hydrothermal conditions *IOP Conf. Series: Earth and Environmental Science* **190**
12. Voropay N N and Ryazanova A A 2018 Atmospheric droughts in Southern Siberia in the late 20th and early 21st centuries *IOP Conf. Series: Earth and Environmental Science* **211**
13. Ped D A 1975 On the indicator of drought and excess moisture *Transactions of the Hydrometeorological Center of the USSR* **156** 19-38
14. Chudnovsky A F 1976 *Thermophysics of soils* (Moscow: Nauka) p 353

IOP Conf. Series: Earth and Environmental Science **386** (2019) 012029 doi:10.1088/1755-1315/386/1/012029

15. Rode A A 1965 *Soil Moisture Basics 1* (Leningrad: Gidrometeoizdat) p 664
16. Hillel D 1980 *Fundamentals of Soil Physics* (New York: Academic Press) p 288
17. Glagolev M V, Ilyasov D V, Sabrenkov A F, Goncharov V M 2018 Response of Methane Emission to Temperature Anomalies of Mires: Case Study of the Southern Taiga in Western Siberia *Water resources* **45**(2) 44-52
18. Veretennikova E E and Dyukarev E A 2017 Diurnal variations in methane emissions from West Siberia peatlands in summer *Russian Meteorology and Hydrology* **42**(5) 319-26
19. Molchanov A G 2015 Sphagnum gas exchange at various levels of surface groundwater *Ecology* **3** 182
20. Dyukarev E, Godovnikov E, Karpov D, Kurakov S, Lapshina E, Filippov I, Filippova N, Zarov E 2019 Net Ecosystem Exchange, Gross Primary Production And Ecosystem Respiration In Ridge-Hollow Complex At Mukhrino Bog *Geography, Environment, Sustainability* **12**(2) 227-44

Modeling of near-surface flows over an aeolian relief

E A Malinovskaya[1] and O G Chkhetiani[1]

A M Obukhov Institute of Atmospheric Physics RAS, Pyzhevskii 3, Moscow, 119017 Russia

elen_am@inbox.ru

Abstract. As a result of interaction of an air flow with the elements of an aeolian relief, the character of the flow changes. It affects both the conditions of dust blowing from the surface and the surface shape itself. A model is developed for the formation of the windward slope under the influence of wind. It makes it possible to estimate the dynamic speed when moving up the windward slope on the basis of data on height and movement of the dunes. The coefficient of linear change in the dynamic velocity and its decrease with increasing structure are obtained. This property is probably due to a change in the nature of the flow and circulation at the surface. To study such problems, dome methods are applied to simulate the air flow near a complex 3D surface using the open OpenFOAM package with a 4-level spatial grid. A technology is also proposed for converting the digital terrain data into a format understandable to mathematical packages for modelling of air flow around a 3D object. Based on the simulation results, the coefficient of linear change in the dynamic velocity for aeolian structures with a height of about 100 m has been calculated, which generally corresponds to values obtained from theoretical estimates and real data on changes in height and movement of dunes.

1. Introduction

Modern changes in the ecological situation and global warming [1] lead to increase in desertification and soil degradation [2]. Remote sensing data, such as near-surface temperature, albedo, and vegetation indices, are intensively used in the analysis and forecasts of desertification [3, 4] and dust storms modeling [4, 5, 6]. It is well-known that for moving sands involved in the formation and movement of the dune-type aeolian structures wind removal is more intensive [5]. There is a clear relationship between the ripple motion velocity and the wind speed [7]. The relief on a small scale [8], [9] affects the regimes of wind removal [10] and the circulation in the boundary layer over desert territories [11]. The structure of the wind-sand flow has an impact on the formation of the aeolian relief [9, 12].

In this regard, the study of air circulations and motions in the vicinity of a heated surface in conjunction with the transformations of aeolian relief forms is an especially important problem. The influence of changes in the shape of dunes on the features of their flow is of interest when refining paleogeographic scenarios. The relief data obtained on the basis of geoinformation models and software tools for constructing 3D surface models are used after conversion as a flow object in a numerical experiment [13].

In contrast to the consideration of mesoscale atmospheric dynamics with the relief [13], the analyses of aeolian processes must include the study of the relationship between two media: air and sand. In our study, the relief data were used to solve two problems: 1) estimation of the air flow deceleration when moving up the windward slope based on the analysis of the principle of formation of the windward slope and the dynamic processes on the leeward side; 2) numerical simulation of the airflow around a complex 3D surface.

2. Estimation of air flow deceleration when moving up the windward slope based on SRTM data

Evaluation of the air flow deceleration over aeolian structures is possible using SRTM data [14]. The SRTM (Shuttle Radar Topographic Mission) is a radar topographic survey of most of the globe, which is used, in particular, in Google Earth [13].

To assess the air flow deceleration when moving up the windward slope, we used a model of the windward slope formation under the assumption of a layered ripple structure [10]. The ripple structure is characterized by the arrangement of particles in layers at different horizontal levels. A certain fraction of the particles of the upper layer is blown out by the air flow. The remaining sand grains roll along the surface of the lower layer to an obstacle in the form of a platform formed from stationary particles. This formation is an extended element of the layer. Considering the linear approximation for $k = k_0 + \mu z$, we can obtain the equation for the length of the elements layer x:

$$\frac{\partial x}{\partial t} = -(k_0 + \mu z)\frac{\langle r \rangle}{t_p}\frac{\partial x}{\partial z}. \qquad (1)$$

Here $\langle r \rangle$ is the average particle radius, and t_p is the average value of the detachment time.

The displacement of the leeward slope can be estimated from [10] as

$$\Delta x = \frac{1}{2}\frac{\Delta z^2 \Delta k^2}{h_{str}} \cdot \cos\beta. \qquad (2)$$

Here $\Delta k = k_L - k > 0$ is the difference between k and k_L, k, k_L are the fractions of detached particles at a given point and at a point at a distance equal to the saltation length L, β is the critical angle of the leeward to the horizon, Δz is the increase in the aeolian structure height, and h_{str} is the structure height. Since the fraction of detached particles at each point on the windward slope is determined as a relative value, taking into account the fraction of the delivered particles. Thus, $\Delta k = k_0 - k$. Here k_0 is the fraction of detached particles at the base of the windward slope.

The linear dependence in (1) for $k = k_0 + \mu z$ was obtained by estimating the fraction of lifting particles taking into account the peculiarities of the particle size distribution and critical wind friction velocities:

$$1 - k(u_*) = \begin{cases} A_1 \cdot u_* + B_1 \; at \; u_* \in (0.26 - 0.28), \\ A_2 \cdot u_* + B_2 \; at \; u_* \in (0.28 - 0.3). \end{cases}$$

Here $A_1 = 46.0$, $A_2 = 7.85, B_1 = -12.0$, $B_2 = -1.35$ obtained from estimates of the fraction of separated particles taking into account their characteristic sizes when exceeding the critical wind speed [15]. These values correspond to a linear approximation for the friction velocity $u_* = u_{*0} + \gamma z$. The fraction of detached particles in a given area, $1 - k$, depends on the fraction of particles deposited and transferred from another one for an extended surface $1 - k_L$. For a fixed value of the particle saltation length, we obtain $\Delta k = 1 - k - (1 - k_L) = k_L - k$. For successive areas related to long saltation, we obtain values k_0, k_1, k_2 … k. Then on top $\Delta k = k_0 - k$. For Δk we have

$$\Delta k = A_i(u_{*0} - u_*) = A_i\gamma(z_0 - z) = A_i\gamma\, h_{str}\,. \tag{3}$$

Here

$$\gamma = \frac{\Delta k}{A_i\, h_{str}}\,. \tag{4}$$

We used images of the Earth's surface included in Google Earth (https://www.google.com/earth/ https://www.google.com/intl/ru_ALL/earth/) to estimate the numerical values of the coefficients used in formulas (1)-(4). These data also allow one

1) to obtain historical photographs of the relief of the territories for a period of time up to 30 years,
2) to save broken lines that determine the position of the dune ridges in different years,
3) to measure the distances and the angles on the relief under consideration.

We find that according to the data taken for the four above-listed regions using the geometric characteristics of the aeolian forms from (3), the changes of the friction velocity when moving up the windward slope correspond to 0.015 m·s^{-1} for aeolian structures 10 m high, 0.02 m / s for 70 m, 0.17 m·s^{-1} if above 80 m [15]. The change depends on the structure height. The estimated linear friction velocity deceleration coefficient γ in (4) matches the values 0.0005 s^{-1} for $h_{str} = 10$ m, 0.0002 s^{-1} for $h_{str} = 20$ m, about 0.0001 s^{-1} for $h_{str} > 40$ m (Figure 1).

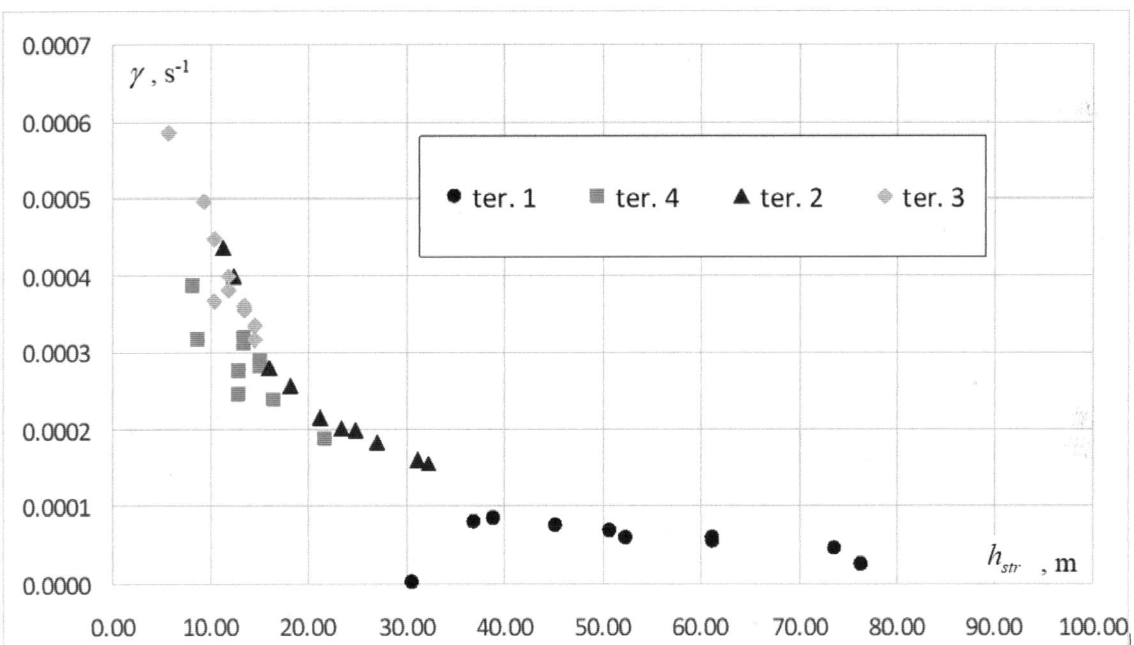

Figure 1. Results of calculating the coefficient γ for structures of different heights from 10 to 80 m. Here ter1, ter2, ter3, and ter4 are data obtained for several structures of different heights for each of the four selected territories.

The velocities on the ridge become higher with an increase in the aeolian structure height. This property leads to an increase in the coefficient of linear change. Moreover, for structures with different heights, almost equal increments in the values of the friction velocity between the lower and upper points above the windward slope are observed. A decrease in the coefficient of linear friction velocity change with increasing dune height may be a result of the arising vertical convective flows from a heated surface.

Later we used the capabilities of numerical simulation of the flow around the aeolian structure using the OpenFoam free software to find out the reasons for the change in the air flow velocity. An important point in this approach is the ability to integrate geographic digital data, 3D modeling programs, and a package for mathematical calculations. Therefore, further we will consider the process of preparing relief data in the required format.

3. The use of terrain data in the design of 3D surfaces for numerical modeling in flow problems

The technology for preparing a 3D surface for numerical modeling of the flow processes taking into account terrain features includes several stages: selecting a territory, converting the source file with elevation data for the terrain to the required format, and presenting the data in the STL file format. The next step is to prepare the surface to obtain a sustainable solution and start solving the problem.

The preparation of the relief data as a file in text format for meshing and applying iterative solution methods in a computational experiment involves the following steps:

Stage 1: work with altitude data for the selected area for combining 3D modeling file formats with programs for the numerical solution of the flow problem:

1) receive a file with a map element of the studied area in geotif format using USGS GIS data (US Geological Survey) from the EarthExplorer service (https://earthexplorer.usgs.gov, Google Maps);

2) trim using the 3DEM program to the required size in connection with the task;

3) create a height file in dew format (Autodesk Civil 3D), since the height file obtained by unloading from 3DEM is not accepted by mathematical packages;

Stage 2: preparation for the numerical solution of the flow problem:

4) to export the drawing (Autodesk Civil 3D, 2013), import the drawing file in Multiphysics or Salome programs in the 3D geometry section, set the correct orientation (rotation) and scale (by transforming the surface mesh);

5) use the tools of geometry in the Multiphysics or Salome programs to build a volume bounded above by the exported surface shape, below and on the sides by planes, defining it as a streamlined element with the relief upper part;

6) perform flattening, that is, reduce the level of heights at the edges of the volume to level 0 in the volume under consideration where the flow is simulated, to speed up and improve the quality of calculations;

7) by means of an open package OpenFoam, perform grid construction and numerical simulation of the flow.

Exporting real data on the earth's surface relief makes it possible to simplify the process of constructing a 3D-model of the surface, which is used in detailed numerical modeling of the boundary layer flow. Such data give a realistic picture of the surface structure and can be taken, including for other planets, where geodetic measurements are not possible.

Let us dwell on point 7 in more detail.

4. Surface preparation for solving flow problems

At this stage, it is important to use a good version of the spatial grid for calculations. The result is a multi-level grid using snappyHexMesh, which has a smaller scale of elements near the surface with a minimum grid pitch of 0.2 m, which allows us to study the effect of the surface topography on the features of the air flow circulation. Since the surface has a complex shape, the VTK format was used to represent the grid data; this allows us to improve its quality and reduce the number of defects (Figure 2).

An object is created for a streamlined surface where boundary conditions and a surface grid are specified. Patches for each surface element are registered separately:

```
"cone_patch.+"
{ type        calculated;
  value       uniform 0;}
```

Figure 2. Structure of a multi-level grid using the utility snappyHexMesh and FoamToVTK.

As a result, a flow object is placed in the cube region, for which boundary conditions are set as for the surface. The adapted multi-level grid allows one to detail small-scale processes at the surface and abstract from them at some distance. The grid parameters can be configured and refined.

At the next step, we configure the selected solver.

5. Steady flow

In the numerical experiment, an airflow with neutral stratification around a hill is simulated not taking into account the effect of saltating particles. The region of the flowing medium is represented by the volume of a parallelepiped with a height of 500 m and a length of the faces of the base of 1000 m. The transverse and vertical dimensions of the computational domain are chosen so as to exclude the influence of the boundary conditions. The hill length corresponds to 350 m. We assume that the air flow generally moves horizontally, at the upper boundary a flow is set with a velocity of 7 m/s characteristic of this height. On the left and right walls there is a free flow with a zero gradient. On the bottom wall and for the surface modeling the shape of the hill, the adhesion conditions are used. The solver SimpleFoam with $k-\varepsilon$ model for a stationary incompressible RANS flow from the OpenFoam open-source was used to evaluate the effects of changes in the friction velocity. The pressure-correction scheme is associated with employing the contravariant velocity. Semi-explicit calculation methods use nonorthogonal, bodyfined coordinates, which is better for irregular geometry [16, 17]. In [18], a number of issues are discussed regarding the use of turbulent models in OpenFoam in considering the air flow above a single dune. As initial parameters we used:

$$E = \frac{3}{2}\left(u_{ref}T_i\right)^2,$$

E is the kinetic energy of turbulent mixing, u_{ref} is the velocity at the upper boundary, and T_i is the turbulence intensity,

$$\varepsilon = 0.09^{3/4}\frac{E^{3/2}}{l}\ \text{m}^2\cdot\text{s}^{-3},\ l = 0.07L.$$

The streamlined surface is a hill about 100-120 m high. The length of the windward slope is about 150-190 m. The size of the modeling area is 760 m x 1000 m x 1300 m. Since the height of the streamlined object is, on average, 110 m, we can use this value as an estimate for L ($L = 110$ m). Depending on the conditions, the turbulence intensity T_i values vary from 0.05 to 0.25. For a given scale the turbulence model used provides stable solutions with the ratio $\frac{E}{\varepsilon} \approx 43$ (s).

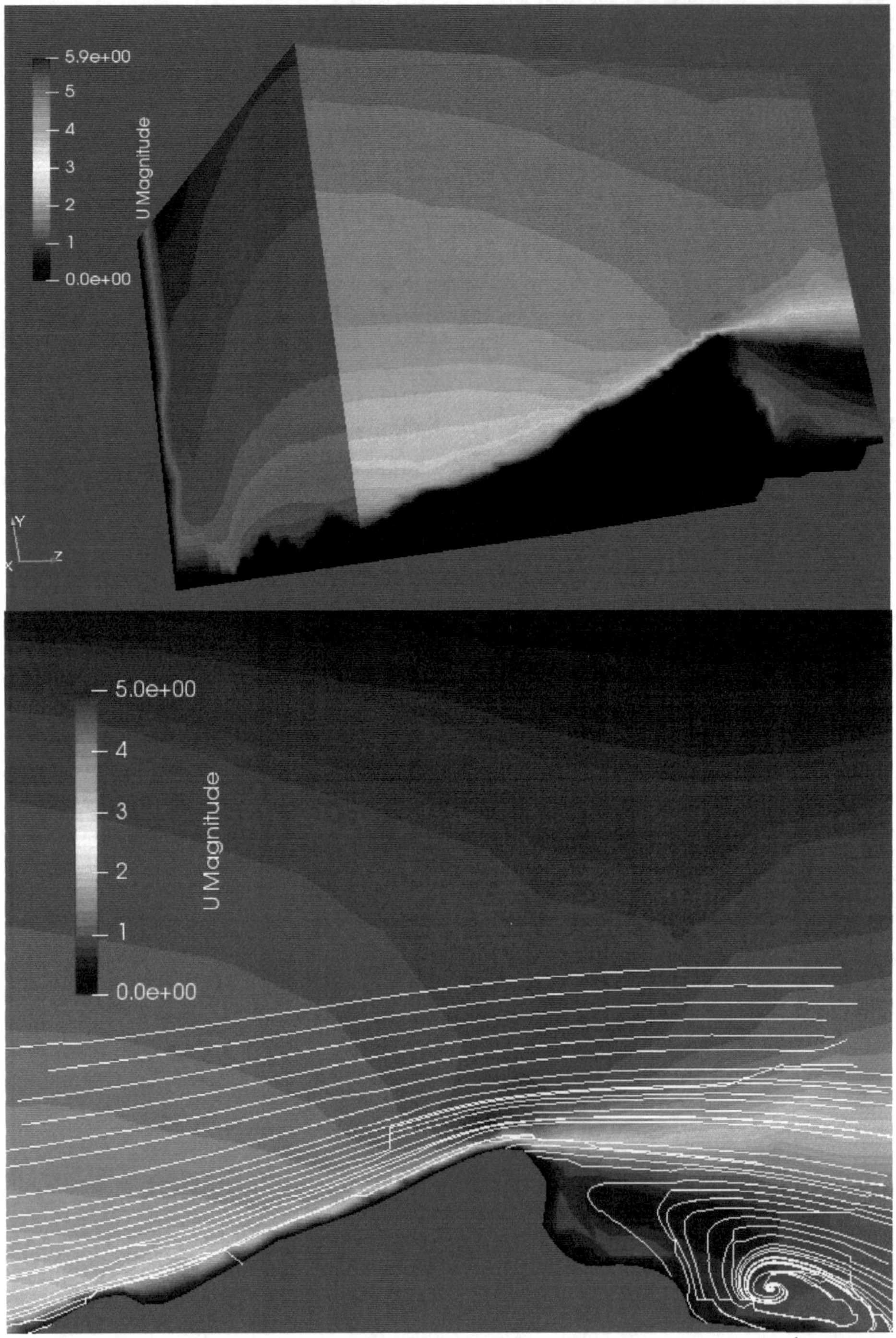

Figure 3. Wind velocity field above a hill on the windward slope for $u_{ref} = 7$ m·s^{-1}, $k = 1.22$ m^2·s^{-2}, $\varepsilon = 0.029$ m^2·s^{-3}.

In Table 1 we present the values of T_i approximately corresponding to this value of $\frac{E}{\varepsilon}$. The minimal number of iterations for obtaining a stable solution has been reached at $u_{ref} = 7$ m·s^{-1} (Figure 3).

Table 1. Selection of T_i values with a stable solution.

L (m)	u_{ref} (m·s^{-1})	E (m^2·s^{-2})	ε (m^2·s^{-3})	$\frac{E}{\varepsilon}$ (s)	T_i
110	9	1.22	0.029	42.51	0.1
110	8	1.16	0.027	43.48	0.11
110	7	1.24	0.030	42.05	0.13
110	5	1.22	0.029	42.51	0.18

Since the linear variation of the friction velocity with height was estimated in Section 1, the difference derivative along the contour line repeating the surface profile at a certain height was determined for the results of a numerical experiment,

$$\Delta u = \frac{u(z, x_2) - u(z, x_1)}{x_2 - x_1}.$$

Here x_1 and x_2 are on the contour.

Along the lines of the surface contours at heights of 2 and 8 m, the velocity varies along the slope by 1 m·s^{-1}and 1.5 m·s^{-1}. The differential derivative γ has values of 0.01 s^{-1} and 0.016 s^{-1}, respectively.

The wind velocity profile $u(z)$ over a sand surface can be represented as [1-3]

$$u(z) = \frac{u_*}{\chi} \ln\left(\frac{9u_* z}{\nu}\right). \tag{5}$$

Here $\chi = 0.4$ is the Karman constant, u_* is the friction velocity, and ν is the kinematic viscosity. From Equation (5) we obtain the velocity change along the slope according to the data of a numerical experiment: γ=0.0001 s^{-1} for $h_{str} = 100$ m. This value of γ corresponds to the estimates using the relief data and the computational model given in Section 1. The coefficients of the velocity change along the surface at various heights are determined by the shift of the slope line to a given height (Figure 4).

6. Conclusions

The use of SRTM images taking into account the flow processes can combine the results of studies of dynamic processes on different time scales.

A technology was proposed for converting digital terrain data into a format applicable to mathematical packages for numerically solving the problem of flow around an object with a 3D surface.

The technology for the preparation of digital terrain data includes the creation of a 3D surface, its extension, cropping, orientation, flattening, file preparation in the required format for calculations in numerical modeling on high-performance computers.

A test of turbulent flow modeling implemented using the OpenFOAM open-source package and snappyHexMesh utility, as well as estimates based on changes in the geometric characteristics of aeolian landforms, have shown close results for changing friction velocity when moving up the windward slope.

This study was supported by a major project of the Presidium of the Russian Academy of Sciences, P12/KP19-270, and by a joint project of the Russian Foundation for Basic Research and the Russian Geographical Society (No. 17-05-41121).

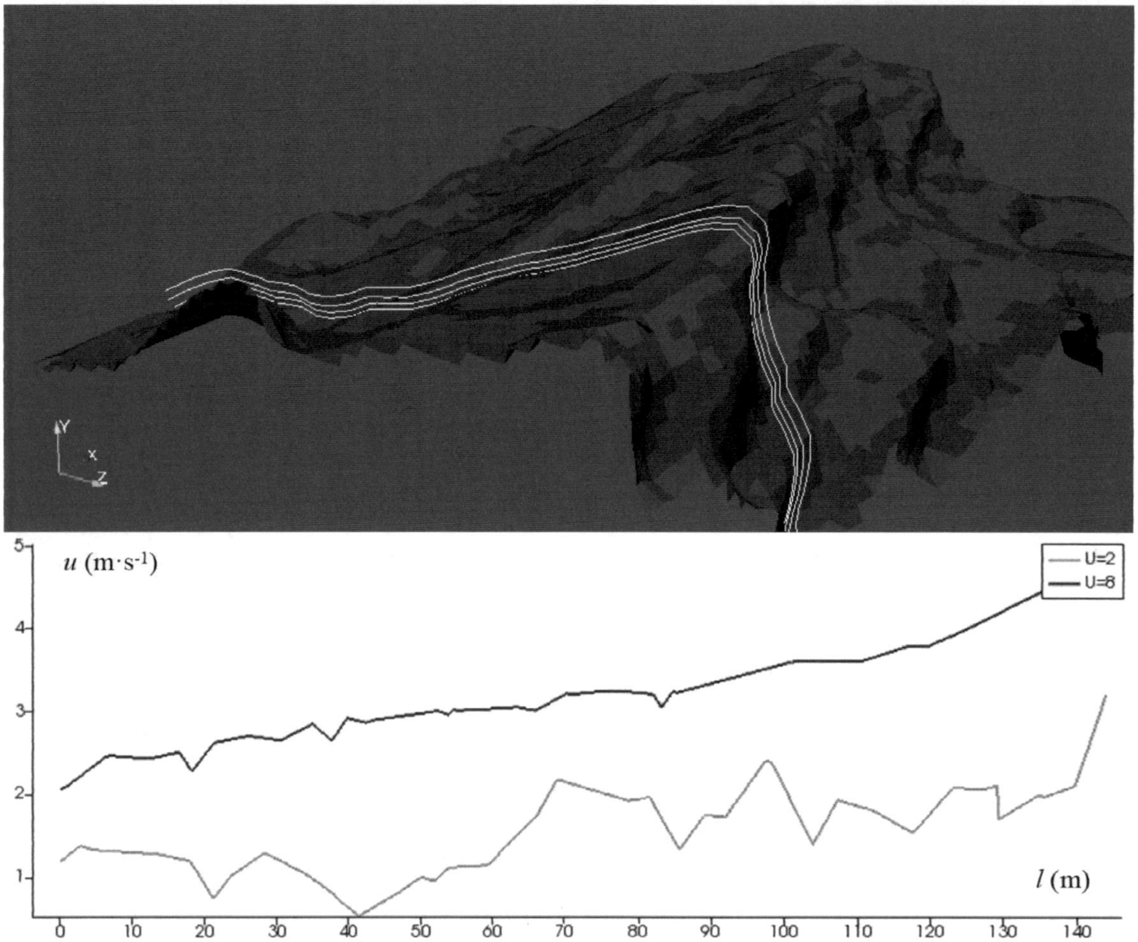

Figure 4. Change in the air flow velocity above the slope. The orange line is at 2 m. The blue line is at 6 m.

References

[1] *Vtoroi otsenochnyi doklad Rosgidrometa ob izmeneniyakh klimata i ikh posledstviyakh na territorii Rossiiskoi Federatsii.* (*Rosgidromet 2nd evaluation report on climate change and its aftermaths for Russia*) 2014 (Moscow: NITs Planeta) p 1008

[2] Baraev A I and Gossen E F 1980 *Vetrovaya eroziya i bor'ba s nej v aziatskoj chasti SSSR* (*Wind erosion and the fight against it in the Asian part of the USSR*) (Moscow: Kolos) 7-47

[3] Zolotokrylin A N Global Warming, Desertification/Degradation, and Droughts in Arid Regions *Izvestiya Rossiiskoi Akademii Nauk. Seriya Geograficheskaya* **1** 3-13 https://doi.org/10.31857/s2587-5566201913-13

[4] Groisman P *et al* Dryland belt of Northern Eurasia: contemporary environmental changes and their consequences *Environmental Research Letters* **13(11)** 115008 https://doi.org/10.1088/1748-9326/aae43c

[5] Kaipov I V *et al* 2012 Ocenka vynosa massy alevritovyh chastic s osushennogo dna Bol'shogo Arala s ispol'zovaniem dannyh distancionnogo sputnikovogo zondirovaniya i

mezomasshtabnoj gidrodinamicheskoj modeli (Evaluation of the removal of the mass of silt particles from the drained bottom of the Great Aral Sea using data from satellite remote sensing and a mesoscale hydrodynamic model) *Gidrometeorologiya i ekologiya* **1** 7-40

[6] Zakarin E A *et al* 2010 GIS-tekhnologiya monitoringa i modelirovaniya pyl'nyh bur' (GIS technology for monitoring and modeling dust storms) *Gidrometeorologiya i ekologiya* **3** 8-20

[7] Semenov O E 2014 Eolovye formy rel'efa i ih peremeshchenie vetrom (Aeolian landforms and their movement by the wind) *Gidrometeorologiya i ekologiya* **3** 7-33

[8] Zilitinkevich S S, Vager B G and Utina Z M 1972 Topography impact on the wind velocity profile *Tr. Gl. Geofiz. Obs.* **276** 168-77

[9] Manukyan E, Prigozhin L 2009 Formation of aeolian ripples and sand sorting *Physical Review* E **79(3)** 031303 https://doi.org/10.1103/physreve.79.031303

[10] Gorchakov G I, Karpov A V, Kuznetsov G A, Buntov D V 2016 Quasiperiodic saltation in the windsand flux over desertified areas *Atmospheric and Oceanic Optics* **29** (6) 501-06 https://doi.org/10.1134/S102485601606004X

[11] Chkhetiani O G, Kalashnik M V, Ingel L K 2013 Generation of thermal wind over a nonuniformly heated wavy surface *Izvestiya, Atmospheric and Oceanic Physics* **49** (2) 121-27 https://doi.org/10.1134/S0001433813020059

[12] Yizhaq H, Kok J F, Katra I 2014 Basaltic sand ripples at Eagle Crater as indirect evidence for the hysteresis effect in martian saltation *Icarus* **230** 143-50 https://doi.org/10.1016/j.icarus.2013.08.006

[13] Wang Y, Huynh G, Williamson C 2013 Integration of Google Maps/Earth with microscale meteorology models and data visualization *Computers & Geosciences* **61** 23-31 http://dx.doi.org/10.1016/j.cageo.2013.07.016

[14] Farr T G *et al* 2007 The shuttle radar topography mission *Reviews of Geophysics* **45** (2) RG2004 https://doi.org/10.1029/2005RG000183

[15] Popov A Yu 2011 Ob ispol'zovanii svobodno rasprostranyaemogo programmnogo obespecheniya dlya resheniya zadach vychislitel'noj gidroaerodinamiki (On using freeware for solving problems of computational fluid dynamics) *Vestnik Moskovskogo gosudarstvennogo tekhnicheskogo universiteta im N E Baumana Seriya «Estestvennye nauki»* **S1** 96-108

[16] Malinovskaya E A 2019 Windward aeolian slope formation model Izvestiya *Atmospheric and Oceanic Physics* **55** (2) 218-28 https://doi.org/10.1134/S0001433819020129

[17] Choi S K, Nam H. Y, Cho M 1993 Use of the momentum interpolation method for numerical solution of incompressible flows in complex geometries: choosing cell face velocities *Numerical Heat Transfer* **23** (1) 21-41 https://doi.org/10.1080/10407799308914888

[18] Moukalled F, Mangani L, Darwish M 2016 The finite volume method in computational fluid dynamics:. An advanced introduction with OpenFoam® and Matlab® *Springer International Publishing* https://doi.org/10.1007/978-3-319-16874-6

[19] Michelsen B, Strobl S, Parteli E J, Pöschel T 2015 Two-dimensional airflow modeling underpredicts the wind velocity over dunes *Scientific reports* **5** 16572 https://doi.org/10.1038/srep16572

IOP Conf. Series: Earth and Environmental Science **386** (2019) 012031 doi:10.1088/1755-1315/386/1/012031

Verification of the INM RAS-MSU land surface scheme using temperature and moisture measurements in peat and mineral soils

V Bogomolov[1,2], E Dyukarev [1,3], V Stepanenko [2,4] and E Volodin[5]

[1] Institute of Monitoring of Climatic and Ecological Systems, Siberian Branch of the Russian Academy of Science, Russia, Tomsk
[2] Research Computing Center, Moscow State University, Russia, Moscow
[3]Yugra State University, Russia, Khanty-Mansijsk
[4] Faculty of Geography, Moscow State University, Russia, Moscow
[5] Institute of Numerical Mathematics, Russian Academy of Sciences, Russia, Moscow

Bogomolov@scert.ru

Abstract. Detailed monitoring of soil temperature provides a unique experimental material for studying the complex processes of heat transfer from the surface layer of the atmosphere to the soil. According to air temperature monitoring data, within each of the key areas there are no significant differences between the data of the observation sites. According to annual (2011-2018) observations of soil temperature, it has been found that the microclimatic properties of bog ecosystems clearly manifest themselves in the characteristics of daily and annual variations of soil temperature. The thickness of the seasonally frozen layer at all sites is 20-60 cm, and maximum freezing of the peat layer is reached in February-March. There is evidence of degradation of the seasonally frozen layer that occurs both from above and from below. It has been found that similar bog ecosystems in different bog massifs may have significantly different temperature regimes. The peat stratum of northern bogs may be both warmer (in winter) and colder (in summer), in comparison with some bogs located 520 km to the south and 860 km to the west.

1. Introduction

An important uncertainty in predicting climate change is caused by the lack of knowledge about the state of the land surface. This issue is mainly related to the determination of energy, moisture, and greenhouse gas fluxes from the surface, temperature and humidity in the atmospheric surface layer, and the state of vegetation. In addition, high-quality maps of the distribution of ecosystems of various types are often lacking. The surface fluxes mentioned above are characterized by complex feedbacks between the atmosphere and terrestrial ecosystems. These relations have pronounced regional features in the wetlands of Western Siberia. Wetland ecosystems are included in some modern climate models as a specific type of underlying surface.

The thermal regimes of peat and mineral soil differ significantly. According to long-term (2011-2018) observations of the soil temperature regime of the oligotrophic bog "Bakcharskoye", it was found that the microclimatic specificity of bog ecosystems is clearly manifested in daily and annual variations in the soil temperature. Peat soils are colder than mineral ones by 8-10 °C in warm time and warmer in the cold season by 1-3 ° C [1].

The peat deposit is a complex organic-mineral system with specific properties: high water content and porosity; it keeps a large amount of poorly decomposed organic matter [2]. In low humid places with a large amount of organic matter, low thermal conductivity often promotes freezing on the soil surface in spring and autumn, and thick peaty soils in the northern latitudes contribute to the elevation of permafrost and its advance to more southern areas. The thermal transport in the peat deposits of wetlands, especially during the period of freezing and thawing, depends on the processes of heat production or absorption by phase transitions of water. In other words, the heat propagation velocity in a peat deposit is determined by the effective thermal diffusivity, which is established experimentally [3]. A number of works have been devoted to model estimates of the thermophysical characteristics of soil in the tundra zone based on temperature profile measurements.

Studying the temperature regime of peat soils is an important task, since the highest rates of modern warming are in the northern latitudes [4], where the main reserves of peat (and carbon) are located. According to various estimates, peatland ecosystems contain about 120–455 billion tons of carbon [5],[6]. In Russia, carbon stocks in the form of peat are estimated as 215 billion tons [7]. Peatlands of Western Siberia contain up to 70 billion tons of carbon [8]. Such a significant amount of carbon as a result of climate change or anthropogenic impact can potentially partially go into the atmosphere in the form of CO_2 or CH_4 and make a significant contribution to the carbon balance of the atmosphere [9].

It should also be emphasized that detailed monitoring of the temperature of the soil layer provides a unique experimental material for studying the complex processes of heat transfer from the surface layer of the atmosphere to the soil.

In the state-of-the-art models of the active land layer, such as the CLM [10] and JSBACH [11], the thermodynamic regime and moisture transfer in wetlands are reproduced taking into account the porosity and thermal conductivity characteristic of the peat soil types. Both models use the TOPMODEL, which calculates the level of groundwater taking into account the subgrid scale orography, surface runoff, and precipitation.

In these land surface schemes, only one type of wetland soil with the properties averaged over a variety of wetlands is taken into account, while, for example, the CLASS model [12] accounts for the difference between different types of wetlands, in particular, in moisture diffusivity. Also, different types of wetlands have different thermodynamic regimes and physical properties, both during autumn freezing and spring melting.

In the model of the active soil layer of INM RAS (Institute of Numerical Mathematics, Russian Academy of Sciences) and Moscow State University, the land is divided into 5 types of surface: vegetation, open soil, water bodies, snow, and a layer of intercepted moisture (raindrops remaining on the leaves, surface moisture film on the soil, etc.). Land cells contain different fractions of these types, which depend on time. The differences in surface types are taken into account in the surface heat balance equation through the following characteristics: roughness lengths, air humidity at the surface, surface stomatal resistance, geographical differences in the model for land cells, the fractions of sand and clay in soil, depth of lakes. The distribution of surface types, the ocean surface temperature is represented for each of the 12 months, and the albedo of the land surface which is free from snow - for January, April, July, and October. The state of the active layer in the INM RAS-MSU model is characterized by ten values. As initial conditions of the variables, value fields are used, which are external parameters of the active layer model.

In the soil, solid particles of three fractions are considered: sand, silt, and clay. The data obtained characterize the texture of the soil stratum as a whole. However, it is assumed that the clay fraction corresponds to the organic matter of the soil and is concentrated in the upper layers of the model to a level of 70 cm [13].

In the model, there is no such soil type as peat or another type close to it in the hydrodynamic properties. Within the framework of this work, it is necessary to understand whether it is possible, with the land surface model formulation and the physical parameterizations, to reproduce the temperature and moisture conditions, in both mineral and peat soils, differing only the external soil parameters that are responsible for its hydrodynamic properties.

By peat soils we mean soils of oligotrophic bogs in which the groundwater level depends more directly on precipitation falling on the bog ecosystem itself, and not on the surface runoff from the territories surrounding the bog, as in the case of eutrophic bogs [3]. That is, remaining within the framework of the equations of heat and moisture transfer of the soil, we can estimate the correctness of the reproduction of the groundwater level in oligotrophic bogs depending on precipitation and mechanical properties of the soil.

2. Measurement data

Field studies of the soil temperature regime were carried out in the eastern part of the Ob-Irtysh interfluve, within the eastern margin of the Vasyugan Plateau in 2011-2018 [14]. The mean annual air temperature for 1936–2017 is –0.3 °C, with July ranking warmest (18.1 °C), and January (–19.2 °C) coldest on the records. The total annual precipitation amounted to 468 mm, of which 45 % fell in the summer months and 12 % during winter [15].

Observation sites were organized at two contrast soil types. The first site was located within the territory of the Bakchar weather station with clay loam soil [15]. The Bakchar weather station is situated on the Bakchar river terrace, where the soil is described as sod-gley heavy-textured clay loam. The surface vegetation cover is short grass mowed several times in summer.

The second site was at a typical oligotrophic treed bog with peat soil [16]. Oligotrophic peatlands with a narrow set of forming species occupy a dominant position in the study area. The treed pine-shrub-sphagnum bogs (ryams) is the most widespread type of bogs in the south taiga subzone. The low ryam microorography is wavy due to a large number of large moss hummocks about 30 cm high and up to 3 m in diameter. The peat thickness at the observation site is 2.1 m.

The measurements of the soil temperature were performed at several depths from the surface to 240 cm (0, 2, 5, 10, 15, 20, 30, 40, 60, 80, 120, 160, 240 cm), and the air temperature was measured at a height of 2 m using an atmospheric-soil measuring complex [17]. The depth-time distribution of the soil temperature for the two studied sites is shown in Figure 1.

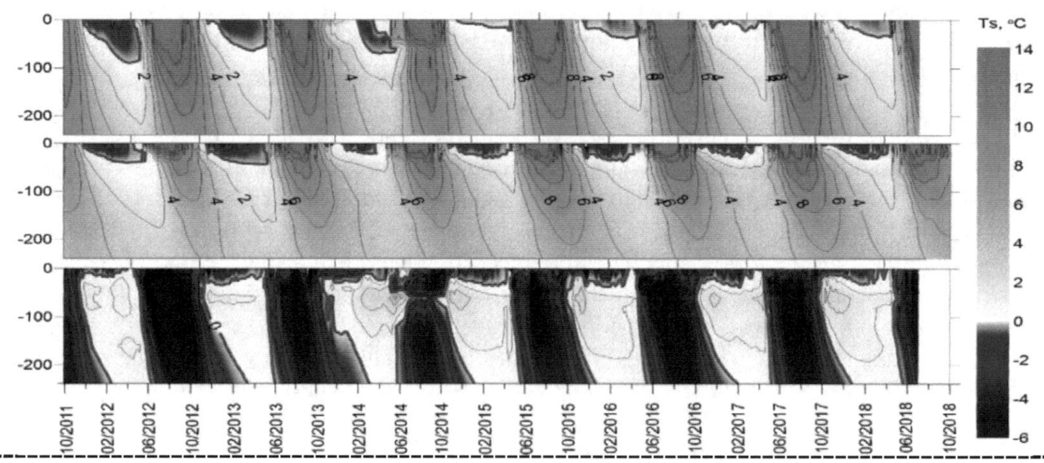

Figure 1. Soil temperature for clay loam soil (top), peat soil (middle), and soil temperature difference (bottom) (Tpeat – Tclay).

The maximum of long-term average surface temperature at the weather station was observed in July (20.3 °C), and the minimum in February (-0.9 °C) beneath a thick snow cover. The surface temperature at the bog is lower than at the weather station reaching 17.8 °C in July and – 2.4 °C in January. The difference in the soil temperatures between the sites is negligible in winter with – 0.6 °C at a 20-cm depth in February at both sites, and it raises in summer with 18.4 °C and 13,8 °C at the clay and peat soils, and subsequently in July at 20 cm. The maximal positive temperature difference between the sites is observed in July at a depth of 60 cm with the soil temperatures reaching 15.3 °C and 10,6 °C at the clay and peat soils, subsequently. The most pronounced negative temperature difference (– 2.1 °C) was registered in November at a depth of 60 cm with clay and peat soil temperatures of 2.7 °C and 4.8 °C. At a depth of 240 cm, the annual temperature variations in the clay soil are in the range from 2.9 °C (April) to 9.5 °C September), while in the peat soil the temperature varies from 3.3 °C (June) to 5.0 °C (November).

The top 20-cm layer of peat soil is always colder than in clay soil, but deeper peat soil layers are warmer in winter and colder in summer. The observed temperature differences are related to essentially different soil composition and extremely high soil moisture of the peat soil.

3. Modeling results

The modified INM RAS-MSU land surface scheme (LSM) was used to simulate the temperature regime of mineral and peat soils. This version of the model is a software package decoupled from the Earth System model and capable of working both in single-column mode for a specific point of interest regional and global modes, the latter simulating the whole land surface of the Earth. Atmospheric forcing can be taken from both measured data sets and simulation results, produced by climate models or reanalyses. This allows to configure the model for a specific study area and verify the result using local measurement data.

We used the single-column mode of INM RAS-MSU LSM to perform numerical experiments. As atmospheric boundary conditions, we used the observation data on the basic meteorological variables (air temperature, humidity, precipitation between the observation periods, surface wind speed and atmospheric pressure) with increments of 3 hours, as well as the intensity of the incoming short-wave and long-wave radiation. Measurements of the above data were carried out near the installation points of soil stations.

External input parameters for INM RAS-MSU LSM related to the thermohydrodynamic properties of soils were set from experimental data obtained for the two studied sites, with clay and peat soils (Table 1), namely: b – Clapp-Hornberger dimensionless parameter; $\Psi_{\max}$ - moisture potential at saturation, m; П – porosity; γ_{max} - maximum hydraulic conductivity, m/s; λ_{max} - maximum values of the moisture diffusion coefficient, m^2/s ; W_0 - the amount of water remaining unfrozen at 0 °C, kg/kg; W_m - the amount of water remaining unfrozen at very low temperature, kg/kg.

Table 1. Parameters of two types of soil, clay and peat.

Physical parameter	Clay	Peat
b	5.30	11.40
$\boldsymbol{\Psi_{\max} * 100}$	56.6	18.6
П	0.31	0.850
$\boldsymbol{\gamma_{max} * 100}$	0.00072	0.0001
$\boldsymbol{\lambda_{max} * 10000}$	0.20400	0.00926
$\boldsymbol{W_0}$	0.18	0.40
$\boldsymbol{W_m}$	0.07	0.20

During the numerical experiments, in contrast to the basic version of the LSM_INMCM, one hundred percent content of the same type with properties from Table 1 that did not change depth was set at each of the calculated depth levels. Also, peat density values were introduced for peat, which were the same for all types in the basic version (Table 2).

Table 2. Density parameters of dry skeleton and saturated for clay and peat.

	Clay	Peat
ρ (dry soil), kg/m^3	1200	100
ρ (solids), kg/m^3	2650	1550

IOP Conf. Series: Earth and Environmental Science **386** (2019) 012031 doi:10.1088/1755-1315/386/1/012031

Of course, in reality the property of the soil, in particular, the porosity of peat, varies greatly with depth and it is necessary to set the hydrodynamic properties of the soil as a function of depth, but for this model it cannot be done.

In order to simulate heat and moisture transfer in soil, in addition to boundary conditions, initial conditions are also necessary. The initial distribution of the moisture and thermal characteristics of the soil can have a decisive influence on the reproduction of the thermal regime of the soil cover. Therefore, at the initial time it is necessary to know with sufficient accuracy the depth distribution of the soil temperature, as well as the amount of liquid and frozen moisture in fractions of dry soil. The initial temperature and humidity profile can be determined from the measurement data, and the ice content in the pore space of the soil in July can be set to zero.

The calculation period was 8 years, 2011-2018, identical to the measurement period.

4. Results

As seen in Figure 2 and Figure 3, in the case of clay soil the temperature has a similar amplitude and close values, both on the surface and to a depth of 160, and for the deeper layer the correlation coefficient between the modelled and observed data decreases.

In peat soils we see a clear shift in the amplitude with increasing depth (Figure 2) and an increase in the absolute error (Figure 3), while on the surface, up to 40 cm, the temperature is reproduced quite correctly.

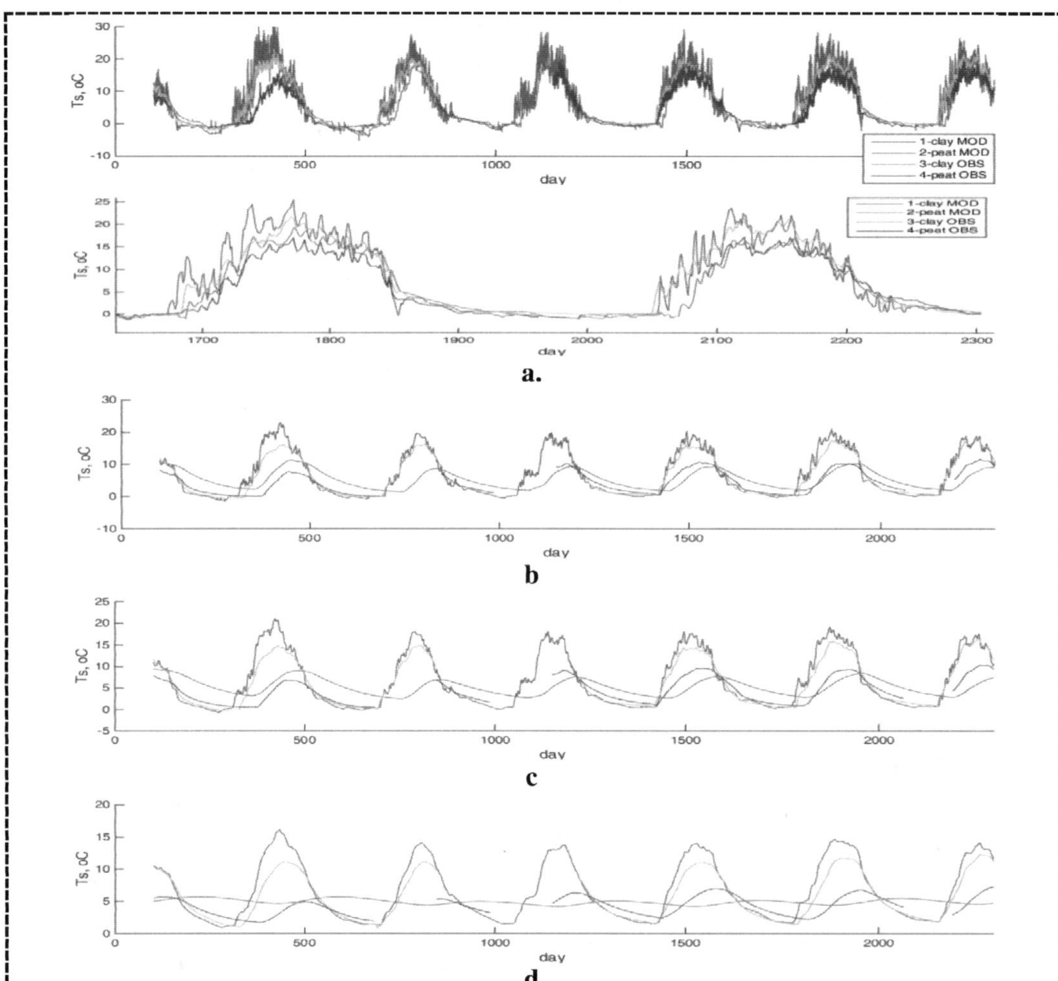

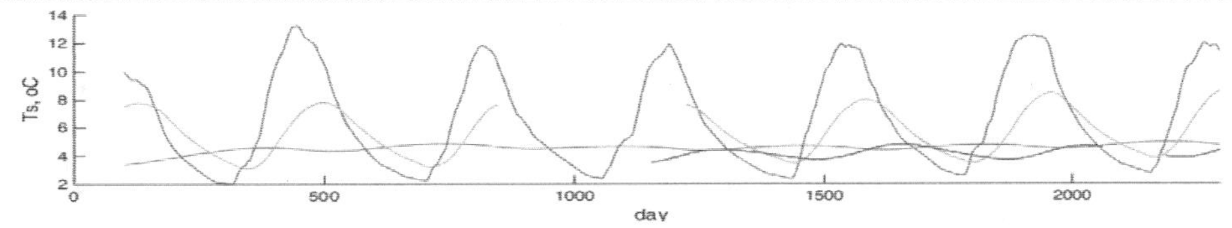

Figure 2. Daily average temperature modelled and observed data for 2011 to 2018: a. 20 cm; b. 60 cm.; c 80 cm.: 160 cm.: 320 cm depth.

The surface temperature in the active layer is completely determined by the atmospheric exposure and radiation forcing. In deep soil there are discrepancies associated with improperly described hydrodynamic properties which vary quite a lot with depth. The not enough correct reproduction of the groundwater level affects inertia in the processes of thawing and freezing as well.

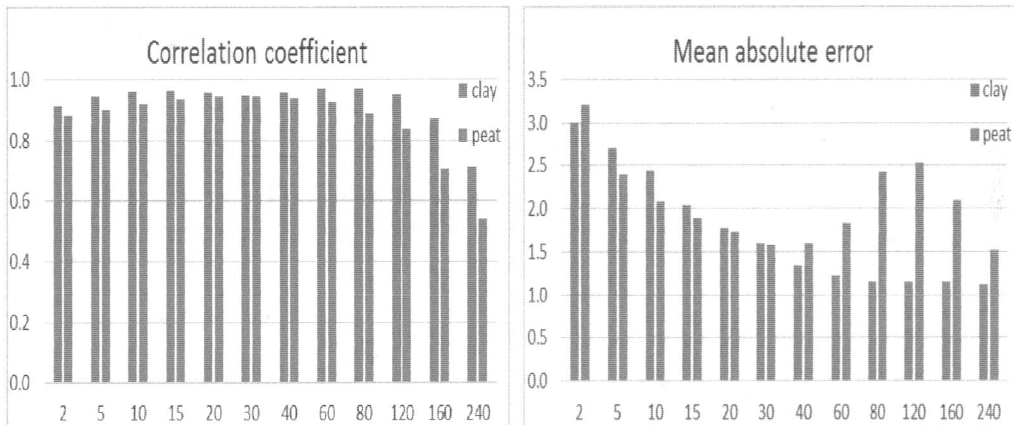

Figure 3. Correlation coefficient and mean absolute error for modelled and observed data. Average for 2012 to 2018.

In cases of moisture on the surface (Figures 4 and 5), both clay and peat reproduce seasonal freezing quite correctly. For clay this is about 90 cm, and for peat 30 cm. With depth, both types are represented as saturated with water, which is a characteristic for the West Siberian lowland. In the case of peat, the surface processes have no influence on humidity below 150 cm for all months, which is correctly reproduced (all lines collapse (Figure 5)). This, of course, in reality is connected with the presence of groundwater. Moreover, for peat the volume of moisture below 150 cm is reproduced as being of significantly lesser value, 0.5 m^3/m^3, than in reality, 0.8-0.9 m^3/m^3. This is due to the fact that the ground water flow is at the lower boundary in the model, which is not correct for this type of soil in which there is hydraulic resistance in the form of a clay soil layer under a peat deposit at a depth of 1.5 -3 meters.

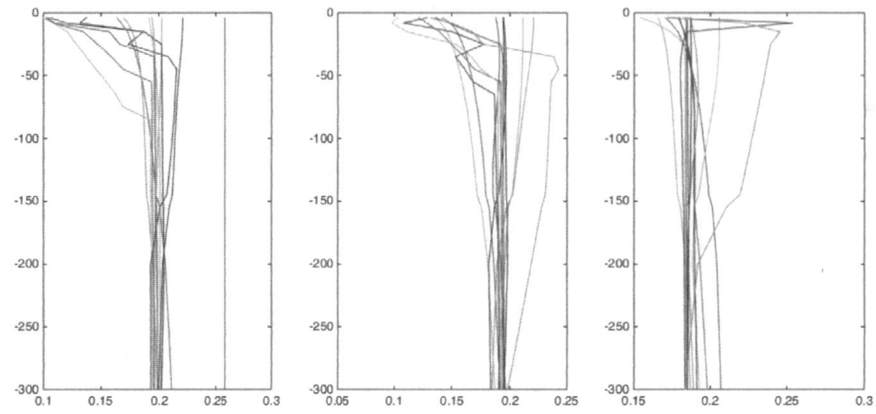

Figure 4. Monthly-averaged soil moisture profile for clay, three different years (2012, 2013, 2014).

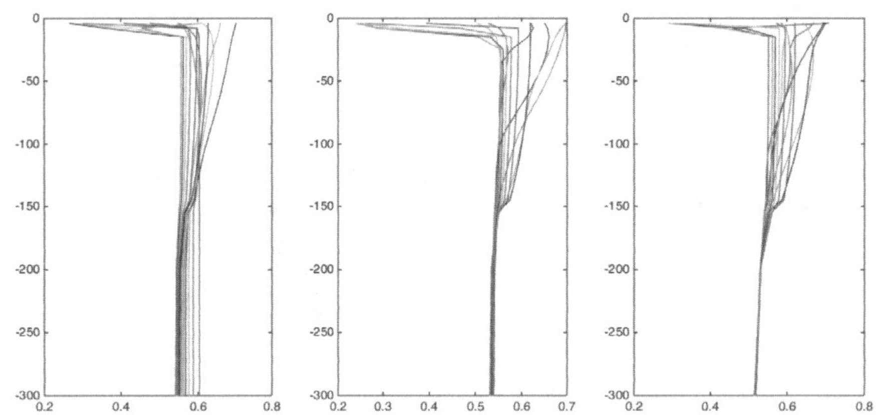

Figure 5. Monthly-averaged soil moisture profile for peat, three different years (2012, 2013, 2014).

The above analysis of the results suggests that to reproduce the temperature regime of bog ecosystems it is not enough just to introduce a new type of hydrodynamic properties, it is necessary to switch to the hydrodynamic properties of a soil as a function of depth for the correct reproduction of bogs at any depth of a peat deposit. Moreover, the above-obtained results allow us to state that in oligotrophic bogs the groundwater level can be reproduced correctly without adding any of the parameters related to the runoff.

The study was carried out with the financial support of RFBR in the framework of scientific project 18-05-00306 and basic research project AAAA-A17-117013050037-0.

References

[1] Dyukarev E A 2017 *Agricult. and For. Meteorol.* **239** 236–48
[2] Romanov V V 1961 *Hydrophysics of mires* (Leningrad: Gidrometeoizdat) p 365
[3] Ivanov K E, Novikov S M 1976 *Mires of Western Siberia, their structure and hydrological regime* (Leningrad: Gidrometeoizdat) p 447

[4] *IPCC, 2013: Climate Change 2013: The Physical Science Basis. Contribution of Working Group I to the Fifth Assessment Report of the Intergovernmental Panel on Climate Change* ed Stocker T F, Qin D, Plattner G-K, Tignor M, Allen S K, Boschung J, Nauels A, Xia Y, Bex V and Midgley P M (Cambridge, United Kingdom and New York, NY, USA: Cambridge University Press) p 1535
[5] Gorham E 1988 *Trans. Royal Soc. Canada* **3** 21-3
[6] Vompersky S E 1994 *Biogeocenotic features of bogs and their rational use* (Moscow: Nauka) 5-37
[7] Botch M S, Kobak K I, Vinson T S, Kolchugina T P 1995 *Glob. Biogeochem. Cycles* **9**(1) 37-46
[8] Sheng Y, Smith L C, MacDonald G M, Kremenetski K V, Frey K E, Velichko A A, Lee M, Beilman D W, Dubinin 2004 *Glob. Biogeochem. Cycles* **18** GB3004
[9] Karelin D V, Zamolodchikov D G 2008 *Carbon metabolism in cryogenic ecosystems* (Moscow: Nauka) p 344
[10] Oleson K W, Lawrence D M, Bonan G B et al. 2010 *Technical Description of version 4.0 of the Community Land Model (CLM)* (Boulder: NCAR) http://www.cesm.ucar.edu/models/cesm1.0/clm/CLM4_Tech_Note.pdf
[11] Roeckner E *et al* 2003 *The atmospheric general circulation model ECHAM 5. PART I: Model description (*MPI Report) p 349
[12] Letts M G, Roulet N T, Comer N T, Skarupa M R and Verseghy D L 2000 *Atmosphere-Ocean* **38**(1) 141-60
[13] Volodin E M and Lykosov V N 1998 *Izvestiya. Atmospheric and Oceanic Physics* **34** (4) 405-16
[14] Dyukarev E A, Pologova N N, Dyukarev A G and Golovatskaya E A 2011 *Environ. Res. Lett.* **6(3)** 035203
[15] Dyukarev E A 2015 *Earth's Cryosphere* **19**(3) 45 – 51
[16] Dyukarev E A, Golovatskaya E A, Duchkov A D and Kazantsev S A 2009 *Russian Geology and Geophysics* **50** (6) 579–86
[17] Kiselev M V, Voropay N N, Dyukarev E A, Kurakov S A, Kurakova P S, Makeev E A 2018 *IOP Conference Series: Earth and Environmental Science* **190**(1) 012031

IOP Conf. Series: Earth and Environmental Science **386** (2019) 012032 doi:10.1088/1755-1315/386/1/012032

Estimation of methane emissions from wetlands and most extreme fires in Western Siberia based on MACC / CAMS and GFED data

O Yu Antokhina[1], P N Antokhin[1] and Yu V Martynova [2,3]

[1]V.E. Zuev Institute of Atmospheric Optics of the Siberian Branch of the Russian Academy of Sciences, 1, Academician Zuev Square, Tomsk, Russia
[2]Institute of Monitoring of Climatic and Ecological Systems of the Siberian Branch of the Russian Academy of Sciences, 10/3, Academichesky ave., Tomsk, Russia
[3]Siberian Regional Hydrometeorological Research Institute, 30 Sovetskaya st., Novosibirsk, Russia

olgayumarchenko@gmail.com

Abstract. Methane is an important greenhouse gas, and its atmospheric concentration has nearly tripled since pre-industrial times. This paper presents estimates of the strength of two primary sources of methane emissions in Western Siberia: wetlands emissions (WEs) and methane emissions from biomass burning (BBEs). The atmospheric circulation data used are from ECMWF Era-Interim, BBE – GFED (Global Fire Emissions Database) and CAMS GFAS (Global Fire Assimilation System), WE – MACC-III greenhouse gases inversions, v10_an. We have revealed that the highest BBE in Western Siberia (summer periods of 2005, 2006, 2007, 2010, 2011, 2012, and 2013) are accompanied by blocking in the atmosphere. We have also showed that there is a lag of 3-6 days between the beginning of the blocking and the periods of maximum BBE. Features of Rossby wave dynamics and blocking highs and ridges are examined as a possible reason for the lag. It turned out that the maximum BBE is commonly observed when a blocking high (or ridge) is shifted from Western to Eastern Siberia. The GFAS data demonstrate higher BBE than GFED (except for 2010). According to GFED, the average emission for all years is only 14.2% from WEs (the maximum emission is 35% (2012), and the minimum one, 2.2% (2007) from WEs). According to GFAS, the average emission for all years is 38.1% from WEs (the maximum emission is 127,4% (2012), and the minimum one, 6.1 (2010) % from WEs).

1. Introduction

Methane is an important greenhouse gas, and its atmospheric concentration has nearly tripled since pre-industrial times. The growth rate of atmospheric methane is determined by the balance between the surface emissions and the photochemical destruction by the hydroxyl radical [1]. Emission sources comprise anthropogenic activity: fossil fuel combustion, rice agriculture, livestock, landfill, and waste treatment, and some biomass burning, and natural sources such as wetlands, termites and the ocean. The past three decades have seen prolonged periods of increasing atmospheric methane, but the growth rate slowed in the 1990s, and from 1999 to 2006 the methane burden (that is, the total amount of methane in the air) was nearly constant. Yet, strong growth resumed in 2007. The reasons for these changes remain poorly understood because of limited knowledge of what controls the global methane budget [2-4]. Earlier, in [5] it was shown that the increased growth rate during 1998 corresponds to an increase in the imbalance between CH_4 sources and sinks equal to -24 Tg CH_4, the largest perturbation observed in 16 years of measurements. The authors of [5] suggest that wetland and boreal biomass burning sources may have contributed to the anomaly. Later in [6], the authors demonstrated an anomalously high flux observed in June and July 2007 in the West Siberian Lowland. Thus, the north part of Eurasia is an important source of methane, due to extensive wetlands and wildfires. Estimation of the power of both sources has always been necessary for compiling a global methane balance. For example, in paper [7] the authors estimated the methane wildfire emissions from Northeast Eurasia for 2000-2011 and compared them with the methane wetland emissions reported in the literature; it turns out that the wildfire adds about 5-20%. We suppose that the maxima of methane emissions from biomass burning in the northern part of Eurasia are linked to atmospheric blocking [8]. Therefore, a critical factor is the formation of blocking in the most important regions in terms of biomass density in the summer season. Depending on the development scenario of blocking during the summer period, the ratio of methane emissions from wetlands (WEs) and biomass burning (BBEs) can vary greatly. Of course, the blocking changes in the future may have an ambiguous effect on the ratio of WEs and BBEs. An analysis of some recent blocking events shows [9] that not only the change in the frequency of blockings is essential, but also their configuration. However, an analysis of such features is still to be made. In the present work, our goal is to demonstrate an important role of atmospheric blocking in Western Siberia (WS) as a driving force of wildfires and calculate the ratio of methane emission from wildfires accompanied by the blocking events and wetland emission.

2. Data and methods

Atmospheric data used in this study are from the European Center for Medium-Range Weather Forecasts ECMWF Era-Interim [10]. The spatial and temporal resolution is 2.5×2.5 and 12 UTC. In this work, we used the blocking periods obtained in [8] for 2005-2013. To clarify the period and position of blocking events, we use a GHGS (geopotential height – gradient south) criterion which was developed in [11-13] (equation (1)). We use the GHGS with fix blocking latitude (φ_{fix}) and flexible blocking latitude (φ_{flex}) based on [14]. The flexible latitude was used only for some of the periods. The GHGS criterion was used to determine the blocking dates in 2016.

$$GHGS = \frac{Z(\varphi_0)-Z(\varphi_S)}{\varphi_0-\varphi_S}, \quad (1)$$

where Z is the 500 hPa geopotential height, for φ_{fix}: φ_0=60° N ±Δ, φ_s=40° N ±Δ, for φ_{flex}= φ_0=70° N ±Δ, φ_s=50° N ±Δ (for Western Siberia).

In contrast to [11-12], we took the following values for Δ: -5°, -2.5°, 0°, 2.5° or 5°, which were first proposed for use in [13]. To clarify the blocking dates, we also used the potential temperature on the dynamic tropopause (PV–θ). According to [15], PV–θ is a perfect candidate for studying the development of blocking, since it is materially conserved in time, providing an excellent tracer for the air masses contributing to blocking formation, and can be inverted to give a balanced component of the flow. Also, the reversal of the meridional gradient PV–θ is associated with Rossby wave-breaking [15]. The determined blocking dates are shown in Table 1.

We used daily data of methane emission from GFED (Global Fire Emissions Database) [16, available at https://www.globalfiredata.org/] and CAMS GFAS (Global Fire Assimilation System) [17, available at https://apps.ecmwf.int/datasets/data/cams-gfas/]. The Global Fire Assimilation System (GFASv1.0) calculates biomass burning emissions (BBEs) by assimilating Fire Radiative Power (FRP) observations from the MODIS instruments onboard the Terra and Aqua satellites (resolution: 0.1×0.1). The Global Fire Emissions Database (GFED4) calculated emission by burned area from the MODIS. For an estimate of the methane emission from wetland (WE), we used the data from MACC-III greenhouse gases inversions, v10_an based on delayed-mode analysis of the GOSAT satellite and surface observations. Regular updates of global CH4 flux inversions are provided every

IOP Conf. Series: Earth and Environmental Science **386** (2019) 012032 doi:10.1088/1755-1315/386/1/012032

six months (for a period of ~7-13 months before real-time). These 'Delayed-Mode' CH4 inversions generally use both satellite and surface observations. Since the beginning of 2012 the GOSAT Remote C PROXY v2.0 XCH4 retrievals are used. The model output from these GOSAT based delayed-mode inversions is available under the MACC version ID v10_an. Further details of the 'Delayed-mode' CH4 flux inversions are described in [18].

Table 1. Blocking dates for the summer period with the maximum blocking frequency in WS (the dates are shown for the longitude where the blocking conditions had the longest time).

Y, m[a]	lon, lat[b]	data	Y,m	lon, lat	data
2005/7	75E, 70-50N	22-28 July	2011/06	75E, 70-50N	1-5 July
	75E, 60-40N	22-28 July		75E, 60-40N	3 July
2006/7	70E, 70-50N	10-16 July	2012/6-7	75E, 60-40N	9-16, 21-26 June,
	70E, 60-40N	10-17 July			1-5, 18-22 July
2007/7	75E, 70-50N	28 June-7 July,	2013/7-8	75E, 70-50N	16-25 July
		12-16 July		75E, 60-40N	16-28 July
	75E, 60-40N	2-7 July 12-16 July			
2010/7-8	65E, 60-40N	26 July- 6 Aug.	2016/7	75E, 70-50N	14-22 July
				75E, 60-40N	17-22 July

[a] Year, month
[b] longitude, latitude

3. Results and discussion

Figure 1 shows a comparison of BBEs based on two databases for summertime characterized by blocking events (Table 1). It can be seen (figure 1) that the periods of blocking and the periods of maximum BBEs coincide, although there is a lag in the data of the maxima of emissions relative to the beginning of blocking. We explored all events, and have concluded that this lag was mainly due to the peculiarities of blocking in Western and Eastern Siberia (except for 2010, when the blocking bordered the WS region from the west).

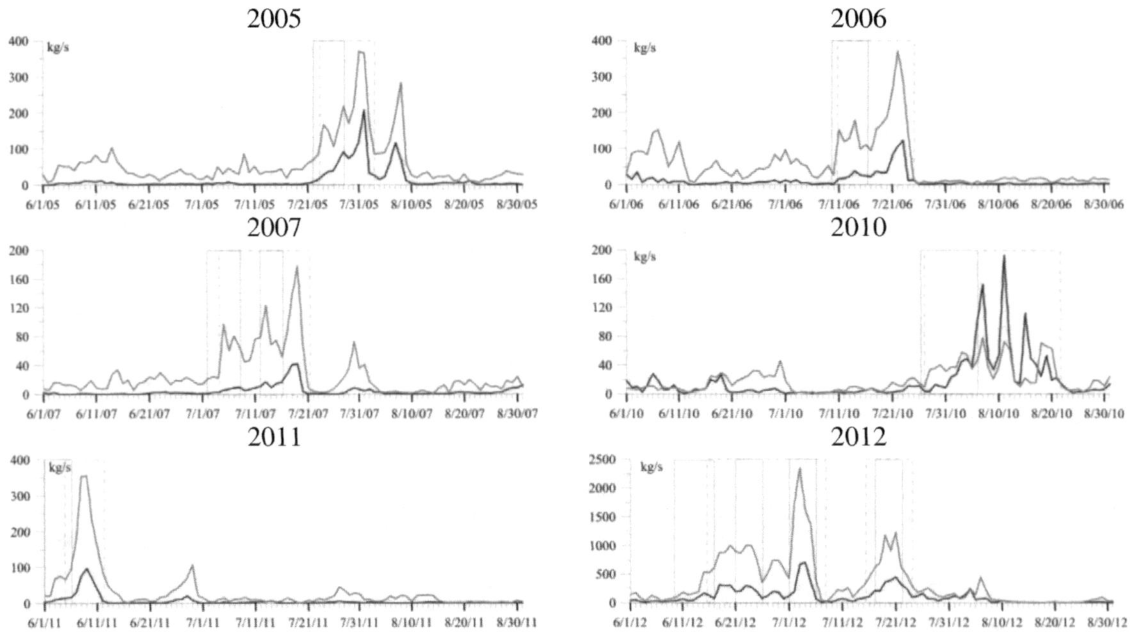

2013 2016

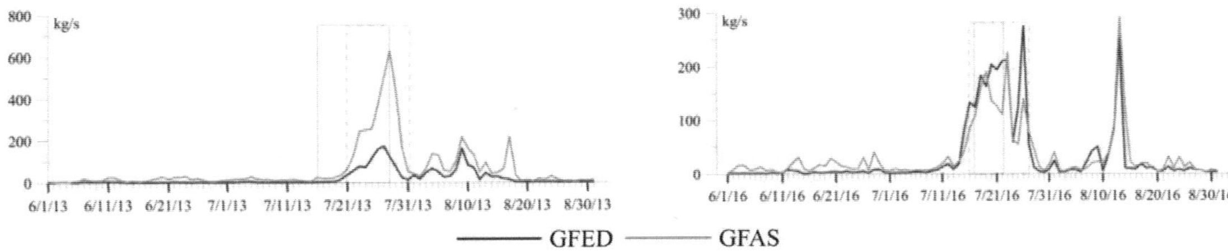

Figure 1. Biomass burning emission in summertime according to GFED and GFAS for an area of 50-70N, 60-90E. *Solid line: blocking periods, dash line: maximum methane emission.*

To demonstrate the reason for the lag, we present Figure 2, which shows the methane emission according to GFAS, PV-Θ, and the streamflow at 500 hPa. The case of July 2006 (Figure 2) is typical in terms of the development of blocking and high-pressure ridge in Siberia. Figure 3 shows the time-longitude diagram of GHGS for July 2006. In Figures 2 and 3 from 10 to 23 July we see the following blocking scenario caused by the increasing BBEs. The moving of the blocking ridge takes place: first we see a Ural-WS blocking (10-16 July), then an eastward shifting of the blocking ridge (17-19 July) and, finally (20-23 July), an Eastern Siberia-Russian Far East blocking. The maximum BBEs were observed for the third stage. This eastward movement of the high-pressure ridge depending on the growth amplitude can be as follows:

1. WS (WS-Urals) ridge – eastward moving – ES (ES-RFE) blocking
2. WS (WS-Urals) blocking – eastward moving – ES (ES-RFE) ridge
3. WS (WS-Urals) blocking – eastward moving – ES (ES-RFE) blocking

Also, we should note that the inertia of formation of the BBEs maximum is caused by increase in the anomalous dryness by the end of blocking.

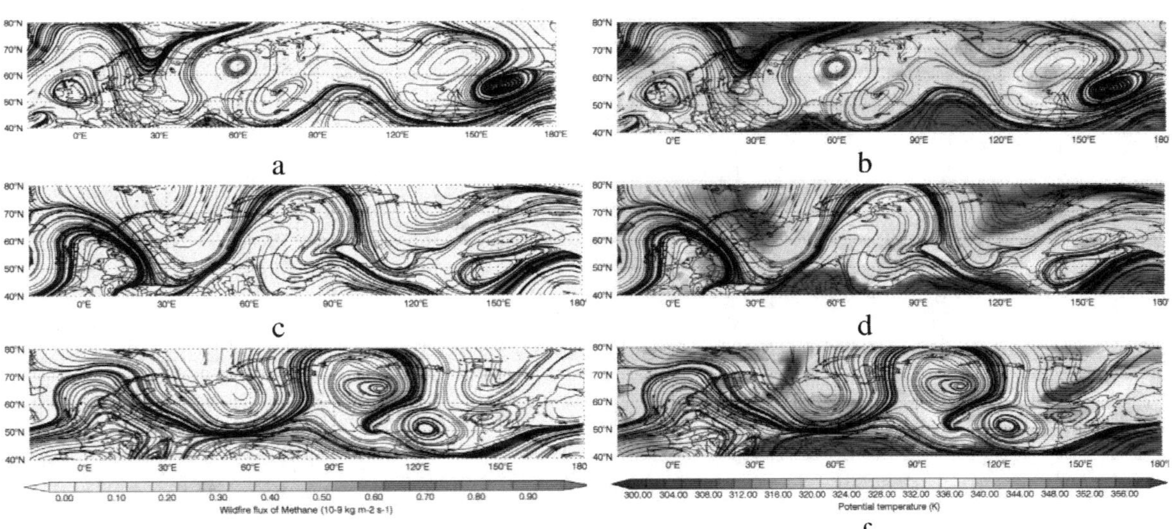

Figure 2. Maps of BBE (a, c, e), and PV–θ (b, d, f) for 15 (a, b), 19 (c, d), 22 (e, f) July 2006 (for both panels, the streamlines at 500 hPa are shown).

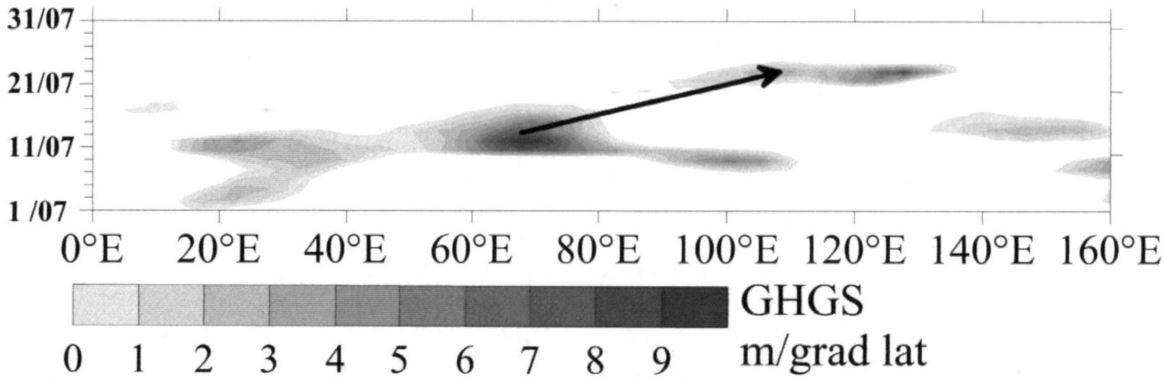

Figure 3. Time-longitude diagram of GHGS (fix latitude) for July 2006.

In Table 2, we demonstrate a comparison of the BBEs for two databases and the WEs. First of all, we pay attention to the difference between the GFAS and GFED data for WS. Except for 2010 and 2016, the GFAS shows much higher values than the GFED. Depending on which of these values will be compared with the WEs, we will have different outcomes. According to the GFED, the average emission for all years is only 14.2% from WEs (the maximum emission is 35% (2012), and the minimum emission is 2.2% (2007) from WEs). According to the GFAS, the average emission for all years is 38.1% from WEs (the maximum emission is 127.4% (2012), and the minimum emission is 6.1 (2010) % from WEs).

Table 2. The ratio of methane emission from wetland and biomass burning.

Summer (June-August) average total for WS emission from wetland calculated for 2012 and 2013.	BBEs averaged for the period of maximum emission (Figure 1).				
		GFAS		GFED	
	Year	kg/s	% of wetland	kg/s	% of wetland
	2005	197.6	28.8	72.4	10.6
	2006	175.9	25.6	42.6	6.2
	2007	83.4	12.1	14.8	2.2
	2010	42.0	6.10	54.0	7.9
	2011	155.6	22.7	34.0	5.0
	2012/6	872.9	127.4	239.7	35
	2012/7	469.0	68.5	183.5	26.8
685 kg/s – 100%	2013	229.7	33.5	73.6	10.7
	2016	123.1	18.0	155.3	22.7
	Mean	**261**	**38.08**	**96.66**	**14.12**

4. Summary

Methane is an important greenhouse gas, and its atmospheric concentration has nearly tripled since pre-industrial times. The growth rate of atmospheric methane is determined by the balance between the surface emissions and the photochemical destruction by the hydroxyl radical. The northern part of Eurasia is an important source of methane due to its extensive wetlands and wildfires. Estimation of the strength of both sources has always been of interest for calculating the global methane balance. We supposed that the maximum methane emissions from biomass burning in the northern part of Eurasia are related to atmospheric blocking.

The atmospheric data used in this study are from the European Center for Medium-Range Weather Forecasts ECMWF Era-Interim [10]. The spatial and temporal resolution is 2.5×2.5 and 12 UTC. In this work we used blocking periods obtained in [8] for 2005-2013. Also, we used a GHGS (geopotential height – gradient south) criterion developed in [11-13]. To clarify the blocking dates, we also used the potential temperature at the dynamic tropopause (PV–θ).

We used daily data of methane emission from GFED (Global Fire Emissions Database) and CAMS GFAS (Global Fire Assimilation System). To estimate the methane emission from wetland (WE), we used data from MACC-III greenhouse gases inversions, v10_an based on delayed-mode analysis of the GOSAT satellite and surface observations.

We have revealed that the maximum BBE in Western Siberia (summer periods of 2005, 2006, 2007, 2010, 2011, 2012, and 2013) are accompanied by blocking in the atmosphere. We have also shown that there is a lag of 3-6 days between the beginning of the blocking and the periods of maxima BBE. Features of Rossby wave dynamics, as well as blocking highs and ridges, were examined as a possible reason for the lag. It turned out that the maximum BBE is commonly observed when a blocking high (or ridge) is shifted from Western to Eastern Siberia. The GFAS data demonstrated higher BBE than GFED (except for 2010). According to GFED, the average emission for all years is only 14.2% from WE's (the maximum emission is 35% (2012), and the minimum one, 2.2% (2007) from WEs). According to GFAS, the average emission for all years is 38.1% from WEs (the maximum emission is 127,4% (2012), and the minimum one, 6.1 (2010) % from WEs).

Acknowledgments
This study was supported by the Russian Science Foundation under project 17-17-01095.

References

[1] Bousquet P *et al* 2006 Contribution of anthropogenic and natural sources to atmospheric methane variability *Nature* **443** 439–43
[2] Nisbet E G, Dlugokencky E J and Bousquet P 2014 Methane on the Rise--Again *Science* **343** 493–95
[3] Nisbet E G *et al* 2019 Very Strong Atmospheric Methane Growth in the 4 Years 2014–2017: *Implications for the Paris Agreement Global Biogeochemical Cycles* **33** 318–42
[4] Rakitin V S *et al* 2017 Study of trends of total CO and CH4 contents over Eurasia through analysis of ground-based and satellite spectroscopic measurements *Atmospheric and Oceanic Optics* **30** 517–26
[5] Dlugokencky E J, Walter B P, Masarie K A, Lang P M and Kasischke E S 2001 Measurements of an anomalous global methane increase during 1998 *Geophysical Research Letters* **28** 499–502
[6] Sasakawa M, Ito A, Machida T, Tsuda N, Niwa Y, Davydov D, Fofonov A and Arshinov M 2012 Annual variation of CH4 emissions from the middle taiga in West Siberian Lowland (2005–2009): a case of high CH4 flux and precipitation rate in the summer of 2007 *Tellus B: Chemical and Physical Meteorology* **64** 17514
[7] Vasileva A and Moiseenko K 2013 Methane emissions from 2000 to 2011 wildfires in Northeast Eurasia estimated with MODIS burned area data *Atmospheric Environment* **71** 115–21
[8] Antokhin P N, Antokhina O Yu, Arshinov M Yu, Belan B D, Davydov D K, Sklyadneva T K, Fofonov A V, Sasakawa M, Machida T 2017 Impact of atmospheric blocking in Western Siberia on the change in methane concentration in the summer period *Atmos. Oceanic Opt.* **30**(5) 393-403 (in Russian with English abstract)
[9] Sitnov S A and Mokhov I I 2018 Anomalies in the Atmospheric Methane Content over Northern Eurasia in the Summer of 2016 *Doklady Earth Sciences* **480** 637–41
[10] Dee D P *et al* 2011 The ERA-Interim reanalysis: configuration and performance of the data assimilation system *Quarterly Journal of the Royal Meteorological Society* **137** 553–97
[11] Lejenäs H and Økland H 1983 Characteristics of northern hemisphere blocking as determined from a long time series of observational data *Tellus* **35**A 350–62
[12] Tibaldi S and Molteni F 1990 On the operational predictability of blocking *Tellus* **42**A 343–65
[13] Barriopedro D, García-Herrera R, Lupo A R and Hernández E 2006 A Climatology of Northern Hemisphere Blocking *Journal of Climate* **19** 1042–63

[14] Antokhina O Y, Antokhin P N, Zorkal'Tseva O S and Devyatova E V 2017 Atmospheric blockings in Western Siberia. Part 1. Detection features, objective criteria, and their comparison *Russian Meteorology and Hydrology* **42** 644–52

[15] Masato G, Hoskins B J and Woollings T J 2011 Wave-breaking characteristics of midlatitude blocking *Quarterly Journal of the Royal Meteorological Society* **138** 1285–96

[16] Giglio L, Randerson J T and Werf G R V D 2013 Analysis of daily, monthly, and annual burned area using the fourth-generation global fire emissions database (GFED4) *Journal of Geophysical Research: Biogeosciences* **118** 317–28

[17] Kaiser J W, Heil A, Andreae M O, Benedetti A, Chubarova N, Jones L, Morcrette J-J, Razinger M, Schultz M G, Suttie M and Werf G R V D 2011 Biomass burning emissions estimated with a global fire assimilation system based on observed fire radiative power *Biogeosciences Discussions* **8** 7339–98

[18] Bergamaschi P *et al* Report on the quality of the inverted CH4 fluxes, MACC-II Deliverable D_43.3, 2013b *Tech. Rep., Joint Research Center, European Commission,* available at https://atmosphere.copernicus.eu/documents/maccii/deliverables/ghg/

IOP Conf. Series: Earth and Environmental Science **386** (2019) 012033 doi:10.1088/1755-1315/386/1/012033

Calculating scheme of ground freezing depth on the basis of data on seasonal snowfall deposition, snow cower accumulation and temperature variation

D Frolov

Geographical Department, Lomonosov Moscow State University, Moscow, 119991, Russia

E-mail: denisfrolovm@mail.ru

Abstract. Soil thermal regime remains today an insufficiently studied field of environmental computing science. For the determination of the influence of air-temperature and snow cover thickness absolute values and dynamics on the ground freezing depth, and the development of root systems of trees and plants, a calculating scheme on the basis of data on seasonal snowfall deposition, snow cover accumulation, and temperature variation is considered. The calculating scheme is based on a three-layer media heat conduction problem (snow cover, frozen and thawed ground) with phase transition on the boundary between frozen and unfrozen grounds. The heat balance equation includes phase transition energy, inflow of heat from the unfrozen ground and outflow to the frozen ground, snow cover and the atmosphere. The heat flux is calculated on the basis of the Fourier law as the product of heat conductivity and temperature gradient. It is supposed that the temperature changes in each of the media linearly. An assumption is also made that snow cover consists of different layers deposited by different snowfalls and having different structure, density, and heat conductivity depending on its density. The density and heat conductivity of each layer and the whole thickness of snow cover are determined, a regional stratigraphic column for snow cover is compiled, and the ground freezing intensity and freezing depth are calculated. A comparison of the calculated and observed values of ground freezing depth for Narayan-Mar is performed, and a correlation of 0.76-0.77 is obtained.

1. Introduction

Earth's climatic system can be described by a nonlinear dynamic system of differential equations of variables varying in time. Newton's laws of mechanics are used for the composition of this system of differential equations on the basis of a description of the interaction in the five main components of the Earth: atmosphere, ocean, land, cryosphere, and biosphere. This system is not conservative; there is dissipation of energy. Forcing (the energy income) of the system is the income of solar radiation on the Earth's surface.

On the basis of all these mentioned objectives for forecasts of the weather and climate, different climatic models are launched (for example, one national SL-AV model (which is semi-Lagrangian, based on Absolute Vortices, authored by M.A. Tolstykh from the Marchuk Institute of Numerical Mathematics RAS) and another one from the Main Geophysical Observatory in Saint-Petersburg). The models are launched with perturbed initial conditions, and an ensemble forecast is obtained.

This procedure is made in a similar way as for other natural systems (simplified physical models constructed from systems of differential equations). These systems of differential equations often do not have explicit solutions but are solved numerically. The solutions of these systems may present stable trajectories, between which switching at some values of variables and parameters may take place. In the vicinity of the switching (tipping) points (bifurcation) the moment of switching is determined by "noise". In the solution behavior there may also appear "chaos". J.M.T. Thompson and J. Sieber [1] proposed considering the tipping points of the Earth climatic system as stable points, and trajectories and switching between them similarly to the theory of stable state and forced oscillation in a nonlinear dynamic system of dumped oscillator developed by J.M.T. Thompson and H.B. Stewart [2].

A description of the tipping elements of the Earth's climate system is taken from the paper of T.M. Lenton *et al.* [3] (see Figure 1).

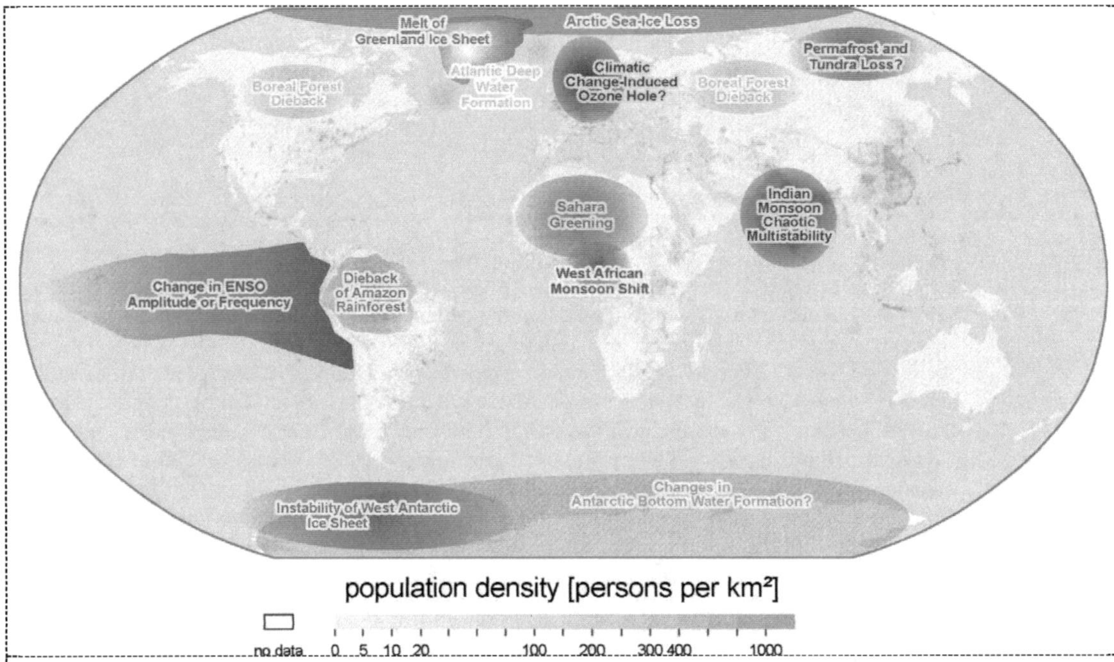

Figure 1. Tipping elements of Earth's climate system (as in paper of T.M. Lenton *et al* [3]).

T.M. Lenton described the properties of possible sources of the Earth's climate system tipping points: their character and mechanism of origin and the influence on the Earth's climate alteration. T.M. Lenton has also started to describe the early prediction of the occurrence of Earth's climate tipping points in his work with Valeria Livina [4].

Under the main Earth's climate tipping point sources, T.M. Lenton considered the Arctic sea-ice, the Greenland ice sheet, the west Antarctic ice sheet, the Atlantic thermohaline circulation, the El Nin˜o–southern oscillation, the Indian summer monsoon, the Sahara/Sahel and west African monsoon, the Amazon rainforest, and the boreal forest (see Figure 1).

For example, as the Arctic sea ice melts, it exposes a much darker ocean surface, which absorbs more radiation. This amplifies the warming. This area covered by ice decreases and less energy from insolation is reflected, resulting in the increasing temperature and, thus, it decreases the ice coverage. This positive feedback (it is assumed that this ice-albedo decrease is a positive feedback) may lead to the destruction of the energy balance and to instability or another stable state.

Such instability is not expected for the Southern Ocean sea ice, because the Antarctic continent covers that entire region. Some researchers think that a summer Arctic ice-loss point, if not already passed, may be very close and a transition could occur within this century. However, Lindsay & Zhang in [5] are not so confident about that, and Eisenman & Wettlaufer in [6] argue that there is probably no bifurcation for the loss of seasonal (summer) sea-ice cover: but there may be one for the year-round loss of ice cover.

The Atlantic thermohaline circulation is a conveyer, where cold and salty (more dense) waters sink to the bottom on the north of the Atlantic and fresher and warmer waters come in its place. The increased income of these fresh (and/or warm) waters on the north of the Atlantic (for example, from the Greenland glacier melt) may shutoff this convection. According to the results of numerous models, this shutoff is reversible but some hysteresis is necessary. In some models this sea water convection is similar to the convection in the atmosphere, where in 1963 Lorenz observed some chaotic attractor (see [7]).

The boreal system is a complex interaction between the tree physiology, permafrost, and fire. Under climate alteration there is increase in the water stress, peak summer heat stress and increase in the mortality, vulnerability to diseases, and subsequent fire, as well as decrease in the reproduction rates, which could lead to large-scale dieback of the boreal forests [8,9], with transitions to open woodlands or grasslands. Newly unfrozen soils that drain well regionally and reduction in the amount of snow also support the drying, cause more fire and, hence, less biomass. In contrast, increased thaw depth and increased water-use efficiency under elevated CO_2 will tend to increase the available soil moisture, decreasing the frequency of fires and increasing the woody biomass.

So are the examples of tipping points of Earth's climate system and the objectives of their origin. The soil thermal regime of boreal forests, despite the simplicity of measuring the soil temperature, remains today an insufficiently studied field. To determine the influence of the air-temperature and snow cover thickness absolute values and the dynamics on the ground freezing and thawing depth, develop microbiota and root systems of trees and plants, a number of models and calculation schemes have been developed. For example, V.A. Kudryavtsev [10] characterized the warming and cooling action of the snow cover on the ground depending on the snow accumulation regime and its duration, and proposed an equation for the estimation of the ground freezing depth including snow cover thickness, its thermal properties, and the amplitude of yearly air temperature oscillations. In our case the calculation scheme for ground freezing is constructed on the basis of three layer media heat conductivity problem (snow cover, frozen and thawed ground) with a phase transition on the boundary of frozen and unfrozen ground (see Figure 2).

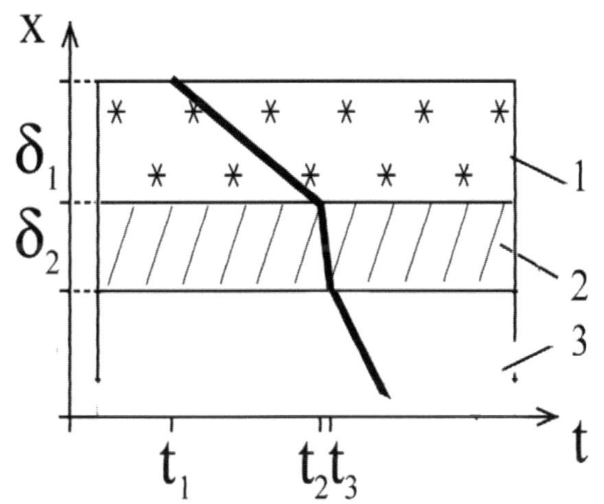

Figure 2. Temperature distribution in the media consisting of snow thickness (1), layer of frozen (2) and thawed (3) soil.

IOP Conf. Series: Earth and Environmental Science **386** (2019) 012033 doi:10.1088/1755-1315/386/1/012033

The heat balance equation includes phase transition energy, the inflow of heat from unfrozen ground, and the outflow to frozen ground, snow cover, and atmosphere. The heat flux is calculated on basis of the Fourier law as a product of the heat conductivity and temperature gradient. It is assumed that temperature changes in each medium linearly. An assumption is also made that the snow cover consists of different layers deposited by different snowfalls and having different structure and density and heat conductivity depending on its density. Consideration of meteorological data of the air temperature, precipitation and snow thickness and snowfall intensity at the nearest meteorological station makes it possible to define the density and heat conductivity of each layer of snow cover, compile a regional stratigraphic column for the snow cover, and make more precise calculation of the ground freezing intensity and freezing depth.

2. Materials and methods

On the basis of meteorological data on the air temperature, precipitation, and snow cover thickness, we extracted data on the deposition and intensity of snowfalls at the nearest meteorological station Narayan-Mar [11] (see the distribution in Figure 3,a), and generalized stratigraphy columns were compiled for this region for the winter seasons of 1990/91-2015/16 (as in [12, 13]) (see Figure 3, b).

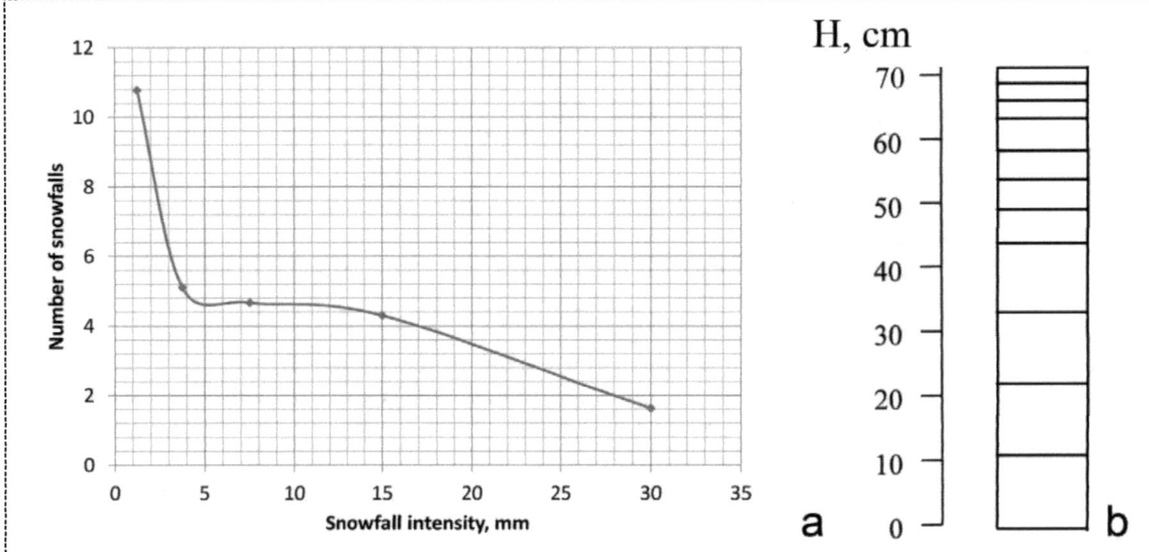

Figure 3. a – average number of snowfalls of particular intensity according data of meteorological station Narayan-Mar for 1988-2008, b - generalized stratigraphic column of snow cover for meteorological station Narayan-Mar for 1988-2008.

On the basis of the relation of the snow heat conductivity on density according to A.V. Pavlov's [14] formula, estimation of the heat conductivity of separate snow layers is performed. According to the formula of heat conductivity of multilayer media, as in equation (1),

$$\frac{\lambda}{\Delta x} = \frac{1}{\frac{\Delta x_1}{\lambda_1} + \cdots + \frac{\Delta x_n}{\lambda_n}}, \quad (1)$$

and on basis of information on the snow layers, the ground freezing depth is calculated.

The calculations of freezing of the snow-covered ground in the winter period on the basis of daily data on the air temperature, snow thickness, and heat conductivity of the snow cover allows estimating the rate of movement of the ground freezing interface during this winter period. The rate of movement of the ground freezing interface can be expressed by the formulas:

The equation of heat balance can be written as in equation (2):

$$F_1 = cLV + F_2. \tag{2}$$

F_1 is the heat outflow through the snow cover and frozen ground from the ground freezing interface (W/m^2);

cLV is the heat value for the phase transition in the ground, c is the ground moisture content (1-4 kg/cm*m^2), (the last value corresponds to full filling of porous medium by water for light clay with a density of 2000 kg/m^3 and a porosity coefficient of 0.617 [15]);

L is the energy of H_2O phase transition (335 kJ/kg), and V is the rate of movement of the ground freezing interface (cm/s);

F_2 is the heat inflow from cooling of thawed ground in front of the ground freezing interface (W/m^2).

The heat flux through the snow cover and frozen ground from the ground freezing interface is expressed according to the Fourier law by means of the temperature gradient and heat conductivity as $F=\lambda\ (grad\ T)$. The heat conductivity and heat flux through a combination of two media (snow and frozen ground) can be expressed as in equation (3):

$$F_1 = \lambda \frac{\Delta T}{\Delta x} = \frac{\Delta T}{\left(\frac{\Delta x_s}{\lambda_s} + \frac{\Delta x_{fg}}{\lambda_{fg}}\right)} = \frac{T_{air}}{\left(\frac{h_s}{\lambda_s} + \frac{l_{fg}}{\lambda_{fg}}\right)}. \tag{3}$$

Here T_{air} is the air temperature, h_s and l_{fg} are the snow cover thickness and ground freezing depth, and λ_s and λ_{fg} are the heat conductivity of snow and frozen ground.

It is assumed that at a depth of 10 m in the ground there is a point of zero annual temperature oscillation with a temperature T_0 of about 3°C. That is why the heat inflow can be expressed as in equation (4):

$$F_2 = \lambda_{thg} \frac{\Delta T}{\Delta x} = \lambda_{thg} \frac{T_0}{10 - l_{fg}}. \tag{4}$$

To validate the three-layers-calculating-scheme, an experiment of one-direction freezing of a snow-covered sand sample is performed in a refrigerated chamber under the action of negative temperatures (as in [16]). The intensity of freezing and the rate of movement of the phase transition front are determined and compared with scheme values obtained by calculation. For this, a dry sand sample with a mass of 5.2 kg, a density of 1.35 g/cm^3, and less than a millimeter of grain size is placed in a plastic volume of 14*14*30 cm. One liter of distillated water is also added to the sand for its moisture content to become maximal and equal to 20%. The volume with wet sand is placed into a heat isolating cover for one direction freezing, and thermal probes at levels of -20, -10, and 0 cm from the upper sand surface level are also placed. The installation is placed in the refrigerated chamber with a negative temperature of -5°C. As the upper sand surface is cooled down, the coarse grained snow with an initial density of 0.3 g/cm^3 is placed on the top of the sand, and a thermal probe at a level of +5 cm from the upper sand surface is placed close to the upper outer snow surface.

While cooling of the installation, a phase transition on the lower snow - upper ground surface takes place. Further, at the assumed heat losses of the sample at one direction freezing (1 J/s) the calculated time of freezing of the whole ground sample is 90 hours. In reality the phase transition in the ground sample finished in 84 hours. This could be explained by the heat losses through the sidewalls of the installation that were not accounted for. Therefore, the three-layers-calculating-scheme estimations are validated with experimental observations.

3. Results

The calculations of ground freezing depth were made according to the above-constructed calculating scheme on the basis of a heat conduction Stephan problem with multilayer heat conductivity. For this reason, on the basis of knowledge about winter snowfall frequency and intensity, generalized snow stratigraphy columns for each winter season were compiled, and the heat conductivity of the multilayer system was determined.

The calculations were made with a step-size of one day. For initial conditions, it was supposed that the frozen ground thickness l_{fg} was equal to 0.5 cm. For each time step (for each day) the rate of movement of the freezing interface V and the frozen ground thickness l_{fg} for the next day (time-step) were calculated.

According to [7], the averaged heat conductivity of thawed and frozen ground was assumed to be equal to 1.5 and 1.8 W/m °C, respectively.

The results of the calculations by the estimation scheme have shown general consistency of the calculated ground freezing depth values with the observed ones. The correlation coefficient is equal to 0.76-0.77, which is quite good for such a simple calculation scheme (see Figure 4).

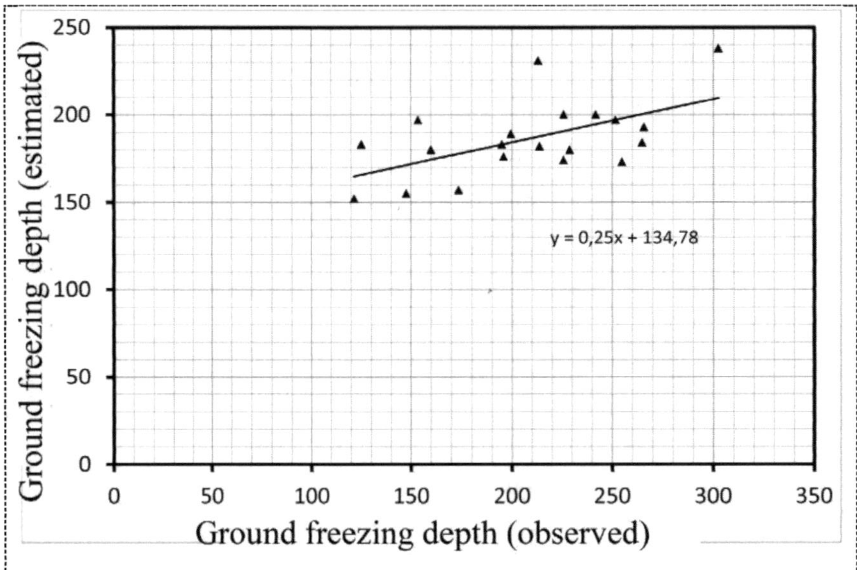

Figure 4. Correlation of observed and estimated ground freezing depth in Narayan-Mar in 1988-2008.

4. Acknowledgments

The work was done under state program AAAA-A16-116032810093-2 on «Mapping, Modelling and Assessment of Risk of Hazardous Natural Processes».

5. References

[1] Thompson J M T and Sieber J 2010 Predicting climate tipping points *Chapter in book: Geo-Engineering Climate Change: Environmental Necessity or Pandora's Box?* Ed B Launder and J M T Thompson (Cambridge University Press)

[2] Thompson J M T and Stewart H B 2002 *Nonlinear Dynamics and Chaos* (2nd edition) (Chicheste: Wiley) p 340

[3] Lenton T M, Held H, Kriegler E, Hall J W, Lucht W, Rahmstorf S and Schellnhuber H J 2008 Tipping elements in the Earth's climate system *Proceedings of the National Academy of Sciences* **105** 6 1786–93 doi:10.1073/pnas.0705414105

[4] Lenton T M, Livina V N, Dakos V, Van Nes E H and Scheffer M 2012 Early warning of

climate tipping points from critical slowing down: comparing methods to improve robustness *Philosophical Transactions. Series A, Mathematical, Physical, and Engineering Sciences* **370** 1962 1185–204 doi:10.1098/rsta.2011.0304

[5] Lindsay R W and Zhang J 2005 The thinning of arctic sea ice, 1988-2003: have we passed a tipping point? *J Climate* 18 4879-94

[6] Eisenman I and Wettlaufer J S 2009 Nonlinear threshold behaviour during the loss of Arctic sea ice *Proc. Nat. Acad. Sci. USA* 106 28-32

[7] Lorenz E N 1963 Deterministic nonperiodic flow *J. Atmos. Sci.* **20** 130-41

[8] Lucht W, Schaphoff S, Erbrecht T, Heyder U and Cramer W 2006 Terrestrial vegetation redistribution and carbon balance under climate change *Carbon Balance & Management* **1** 6 doi:10.1186/1750-0680-1-6

[9] Joos F, Prentice I C, Sitch S, Meyer R, Hooss G, Plattner G-K, Gerber S. and Hasselmann K 2001 Global warming feedbacks on terrestrial carbon uptake under the Intergovernmental Panel on Climate change (IPCC) Emission Scenarios *Global Biogeochem Cycles* **15** 891-907

[10] Kudryavtsev V A 1954. *Temperature of the upper horizons of permafrost thickness within the USSR* (Leningrad: Akad. Nauk SSSR) p 183 [in Russian]

[11] Specialized data for climatic research http://aisori.meteo.ru/ClimateR

[12] Golubev V N, Petrushina M N and Frolov D M 2017 Snowfall events as a factor of snow cover's stratigraphy formation *2nd Int. Electron. Conf. Atmos. Sci.* vol. 2 ed A Lupo doi: 10.3390/ecas2017-04135

[13] Golubev V N, Petrushina M N and Frolov D M 2008 Winter regime of temperature and precipitation as a factor of snow-cover distribution and its stratigraphy *Annals of Glaciology* **49** 179-86

[14] Pavlov A V 2008 *Cryolithozone monitoring* (Novosibirsk:Academician publishing house) p 231 [In Russian]

[15] Trofimov V T 2005 *Soil science* (Moscow: Moscow State University Press) p 1024 [In Russian]

[16] Golubev V N and Frolov D M 2015 Water vapor flows across snow–air and snow–soil interfaces *Earth's Cryosphere* **19** 1 20–26

IOP Conf. Series: Earth and Environmental Science **386** (2019) 012034 doi:10.1088/1755-1315/386/1/012034

Analysis of the effects of climatic factors on flood peak formation

M E Bukovskiy and A V Semenova

Derzhavin Tambov State University, 33 Internatsional'naya Street, 392000 Tambov, Russia
E-mail: asv273@mail.ru

Abstract. Spring flooding is an annual significant overflow of water in the central Russia rivers. As a rule, the maximum flood levels in the Tambov region rivers are reached in the late March - early April. However, this parameter is very variable. Moreover, the maximum flood water level varies from year to year. In this paper, we study the effects of various climatic parameters on the spring flood peak formation. We consider the Tsna River (the Oka basin) near Tambov as the model object of our study. Using methods of multiple correlation analysis, we determine the most significant climatic factors which affect the maximum flood level. In the research, 16 climatic and hydrological parameters from 1970 to 2018 are analyzed. We assess the contribution of the various factors to flood wave formation by using multiple regression analysis. Using the analysis, we have identified the most significant flooding factors.

1. Introduction

Human development is closely linked to the state of water resources. Water land plays a key role in people's lives. It is widely used for industrial and domestic water supply, shipping, timber rafting, energy, irrigationб and other purposes [1].

Every year more and more specialists in climatology, meteorology, and hydrology note, among the problems which threaten humanity, the problem of flooding of the inhabited territories [2].

European researchers believe that floods cause great economic losses and people's death in many parts of Europe. Among the 100 most expensive natural disasters of the 20th century, flood damage amounts to 40% of the total losses (including earthquakes, typhoons, storms, fires, etc.), which are evaluated as 680 billion euros. Operative flood management and flood protection planning require detailed spatial and temporal information about flooding when the flood is active or in other states [3].

Local protection of buildings and infrastructure, early evacuation can be very effective measures to protect the population against the flood consequences. Reliable hydrological forecasts are necessary for all these measures [4].

Domestic researchers also consider flood as one of the most dangerous natural phenomena that cause emergencies and material damage to industrial and agricultural facilities and to human health and lead to their death [5]. Flood can be caused by high water, freshet and ice jams on rivers.

The causes of floods in the Tambov region are mostly spring snow melting and the water level rising in the rivers connected with it.

In this paper, one of the flood types – spring flood (high water) – is considered. It is a multifactorial phenomenon that is affected by a whole complex of natural factors: water reserves in snow, precipitation, temperature values, etc. [6].

The spring flood is a characteristic feature of the lowland rivers of Russia, including the Volga River basin [7]. The city of Tambov is located on the banks of the Tsna River, which relates to Volga basin. High floods near Tambov is a source of hydrological emergencies. It causes flooding of summer cottages, houses, and the infrastructure.

Many researchers have been studying the problem of flood forecasting in different aspects.

Over a long period of time, physical modeling was used to forecast hydrological phenomena, as well as to create hydraulic models of atmospheric, oceanic, and other flows [8,9,10,11]. Undoubtedly, physical-type models have shown high efficiency in forecasting various flooding options. However, the need for hydro-geomorphological monitoring data obtained during intensive calculations makes them unsuitable for short-term forecasts [12]. According to Kim et al., the creation of physical models requires fundamental knowledge of various hydrological parameters and experience in the designing of such models that complicate the forecasting activity [13]. In addition, many researchers note drawbacks of the short-term forecasting capabilities of physical-type models [11].

Besides the numerical and physical models, data-driven models are gaining more and more popularity in flood modeling. These methods combine fixed climatic indexes and hydrometeorological parameters for increasing the modeling capability.

The Russian scientists D.A. Burakov, V.F. Kosmakova, and I.N. Gordeev use the regression dependencies for creating a physical and mathematical model of long-term forecast for large Siberian rivers [14,15,16]. Their methodology is based on multiple linear regression, which makes it possible to study the interconnected multifactorial natural processes [14].

The most popular methods of flood frequency analysis (FFA) for the modelling of flood forecasts are statistical models of autoregressive moving average (ARMA), multiple linear regression (MLR), and autoregressive integrated moving average (ARIMA) [17,18,19].

The FFA is one of the first methods of statistics in flood forecasting [20]. In comparison with physical forecasting models; a more developed and, at the same time, a more efficient one is regional flood frequency analysis (RFFA) including the computation cost and generalization [21].

B.M. Belyaev and N.A. Varentsova developed a methodology for long-term forecasts of maximum flood level based on hydrometeorological information. They also considered its features, effectiveness, and prospects of the statistical approach in hydrological forecasts, especially for small rivers [6].

At the same time, A.N. Arzhakov and K.I. Kusatov tested the methodology for forecasting the maximum levels of spring floods based on integrated comparative geographic and statistical analysis taking into account satellite information [22].

Because of the computing power development, flood forecasting models, such as machine learning (ML) models, are becoming increasingly common. Another driving force for the widespread use of such models is their ability to numerically formulate flood nonlinearity based only on historical data without information on the basic physical processes. These models are really prospective for forecasting because of their fast development with minimal cost of it [23]. Continuous development of the ML methods over the past two centuries showed that these methods are better suited for forecasts with optimal speed than conventional methods, although it requires high technical equipment [24].

In the sphere of effective preparedness and prompt preventive measures, one of the most effective systems for reducing damage is a well-functioning warning system [25]. That is why long-, medium-, and short-term forecasting is an important scientific objective whose results can be demanded by the regional services of the Ministry of Emergency Situations [26].

Some foreign researchers indicate a significant role of precipitation and the spatial distribution of the hydrological cycle in flood modeling [27]. However, as it turned out later, for more accurate flood forecasting one cannot rely on a precipitation forecast. Thus, in the long-term forecast of a river it is advisable to take into account not only the precipitation, but also the estimates of the soil moisture in the catchment [28].

Analysis of the formation conditions of spring runoff performed by A. M. Shustov showed that «high floods are formed with large water reserves in cold winter snow and late united snow melting» [29].

IOP Conf. Series: Earth and Environmental Science **386** (2019) 012034 doi:10.1088/1755-1315/386/1/012034

At the same time, A.M. Vladimirov notes that significant factors which form a flood peak are climatic [30]. First, it is precipitation which can be solid and liquid. The solid one is represented by snow reserves, which are measured in a catchment at the flood start, and snowfalls during the flood. It determines the volume and the peak of the flood. The liquid one is rainfalls and drizzle, which are able to form new peaks on the rise and on the decline of the flood.

Temperature is the climatic factor that is second in significance. Its course strongly affects the intensity of snow melting. If the sum of positive temperatures during the snowmelt increases, it intensifies and the period of the rising flood level is reduced, and the peak is reached faster.

The goal of this paper is to assess the significance of various climatic factors in flood formation using the statistical methods.

2. Data and methods

The basis of our research is the data of the Tambov Center for Hydrometeorology and Environmental Monitoring.

Our model object is the Tsna River (the Oka basin) near Tambov.

A review of papers of other researchers makes it possible to highlight the main factors of flood formation, which include the water reserves in snow cover, the soil freezing depth, the temperature regime, and the precipitation before and during the flood.

The factors which affect the maximum level in a river during the flood were formalized in the following parameters:

- snow melting starting dates;
- dates of stable transition of the average daily air temperature above 0°C upwards (also below – "dates of transition above 0°C");
- dates of complete snow melting;
- flood starting dates;
- maximum flood level dates;
- water reserves in the snow cover (also below – just "snow") on the snow melting starting date;
- water reserves in the snow cover on the date of stable transition of the average daily air temperature above 0°C upwards;
- water reserves in the snow cover on the flood starting date;
- maximum flood levels;
- sums of average daily air temperatures above 0°C (also below – "sums of positive temperatures") from the snow melting starting date to the maximum flood level date;
- sums of average daily air temperatures above 0°C from the dates of stable transition of the average daily air temperature above 0°C upwards to maximum flood level date;
- sums of average daily air temperatures above 0°C from the flood starting date to the maximum flood level date;
- sums of precipitation from the snow melting starting date to the maximum flood level date;
- sums of precipitation on the dates of stable transition of the average daily air temperature above 0°C upwards to the maximum flood level date;
- sums of precipitation from the flood starting date to the maximum flood level date;
- soil freezing depth on the snow melting starting date;
- soil freezing depth on the date of stable transition of the average daily air temperature above 0°C upwards;
- soil freezing depth on the flood starting date.

The analysis was performed for the period from 1970 to 2018.

At the first stage, we have created a data base containing values of the parameter mentioned above for each year.

Then we analyzed the degree of linking of the maximum flood level with each factor being considered. The partial Pearson's correlation coefficients were verified by Student's t-test.

The calculations were carried out by MS Excel. To characterize the correlation we used the Cheddock's scale.

After that we sorted the values of all parameters by the most significant criterion, which is the snow melting staring date. Thus, we got the periods of 15 days (at the end of February it is 13 (14) days, and at the end of March, 16 days).

To assess the contribution of climatic factors to the formation of flood wave and determine significant ones, we used multiple regression analysis.

Further analysis of significance of the factors which affect the flood peak formation was already carried out within these periods.

3. Results and discussion

During the correlation analysis of parameters which affect the maximum flood level, we determined the ones that have weak and middle direct links with the flood peak. The t-statistic value modulus in all cases exceeded the critical tabular values. Figure 1 shows the parameters whose correlation coefficients are higher than 0.4 at a significance level of 0.01.

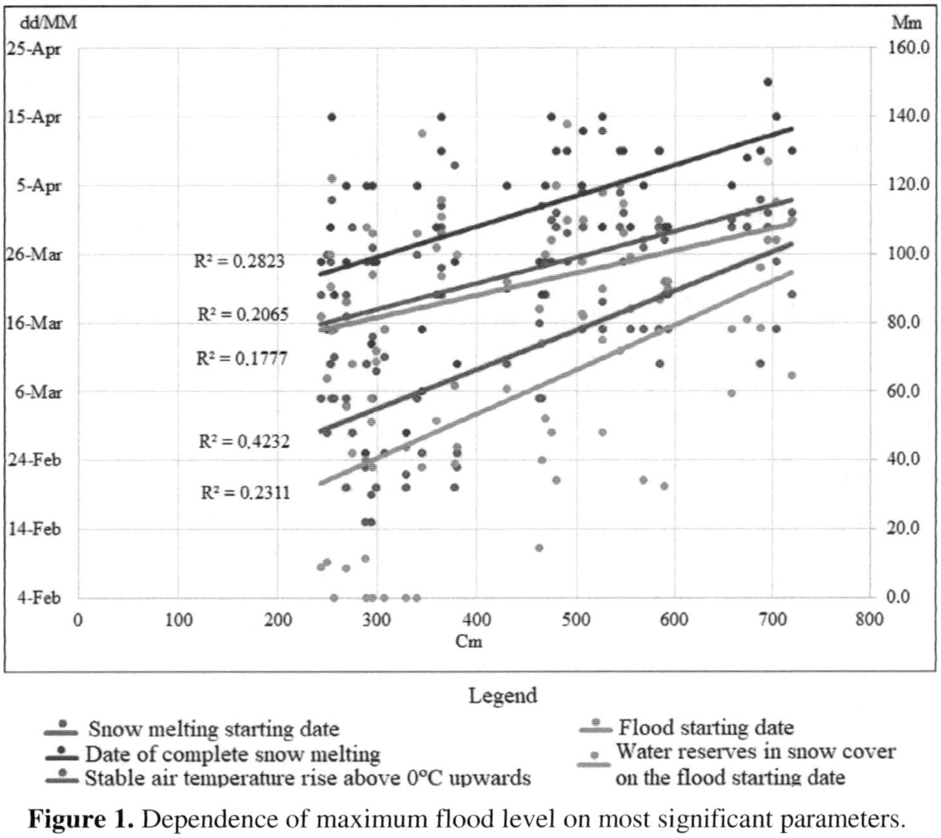

Figure 1. Dependence of maximum flood level on most significant parameters.

In Figure 1 we can see that the maximum flood level of the above-listed climatic parameters is most correlated with the snow melting starting date, the date of stable transition of the average daily air temperature above °0 C upwards, the water reserves in the snow cover on the flood starting date, and the date of complete snow melting.

The snow melting starting date (r=0.65) and the date of complete of snow melting (r=0.53) have an average correlation. For other 3 parameters the correlation coefficient ranges from 0.42 to 0.49.

Then we sorted all parameters by the most significant one, the snow melting starting date. The values of all parameters relating to years with the same snow melting starting dates were combined by the following intervals: February 15-28 (29); March 1-15; and March 16-31.

Also, according to the State Hydrological Institute guidelines, we ranked the maximum flood levels as low ($P<33.3\%$), average ($33.3\%\leq P\leq 66.7\%$), and high ($P>66.7\%$) [31].

As a result of the analysis, we determined that when snow starts melting before the 28th (29th) of February, floods are low and sometimes middle. If the snow melting starting date is in the range from the 1st to the 15th of March, we can observe low, middle, and high floods equally. But if the snow melting starts from the 16th to 31st, the floods are average and high. Such a distribution of the maximum flood level depending on the start of snow melting allows us to make preliminary qualitative assessments of the upcoming flood based on this date.

The next stage was analysis of the links between the maximum flood level and the different components of its formation: the sum of average daily air temperatures above 0 °C, the sum of precipitation, the soil freezing depth, and the water reserves in the snow cover depending on the snow melting starting date. We used regression analysis with a reliability level of 95%.

Table 1. Assessment of the contribution of climatic factors to flood formation

	Coefficients	t-statistic	P-value
Snow melting starting date	13.017	4.563	0.001
Date of stable transition of the average daily air temperature above 0°C upwards	2.837	0.788	0.436
Date of complete snow melting	-0.472	-0.131	0.897
Flood starting date	-7.808	-1.852	0.074
Water reserves in snow on the snow melting starting date	-0.596	-0.694	0.493
Water reserves in snow on the date of transition above 0°C	-1.598	-2.086	0.045
Water reserves in snow on the flood starting date	2.209	2.463	0.020
Sums of positive temperatures from the snow melting starting date to flood peak	2.809	1.101	0.279
Sums of positive temperatures from the flood starting date to flood peak	-7.857	-2.196	0.036
Sums of positive temperatures from the date of transition above 0°C to flood peak	4.443	1.659	0.107
Sums of precipitation from the date of transition above 0°C to flood peak	3.919	1.525	0.137
Sums of precipitation from the snow melting starting date to flood peak	2.018	0.847	0.404
Sums of precipitation from the flood starting date to peak	-0.110	-0.044	0.965
Soil freezing depth on the snow melting starting date	2.193	1.170	0.251
Soil freezing depth on the date of transition above 0°C	0.917	0.447	0.658
Soil freezing depth on the flood starting date	-3.104	-1.218	0.232

The multiple correlation coefficient is 0.85, and the standard error is 94.

According to Table 1, we determined several most statistically significant parameters which affect the flood. These are the snow melting starting date, the water reserves in snow cover on the flood starting date, and the sum of average daily air temperatures above 0°C from the flood starting date to the maximum flood level date.

4. Conclusions

According to the results of our research, the flood height has a strong dependence on the snow melting starting date. This parameter can be used for qualitative assessment of the height of an upcoming flood.

The probability of a high flood is one third when the snow melting starting date is from the 15th of February to the 28th (29th) of March. And if it the melting starts after the 16th of March, the probability of a high flood is 50%.

In our research we have analyzed the relationships between 16 climatic parameters and maximum flood water levels. The most significant factors which determine a flood peak are the snow melting starting date, the water reserve in the snow cover at the flood starting time, and the sum of average daily air temperatures above 0°C from the flood starting time to the maximum flood level time.

References

[1] Dudnik S N, Bukovskiy M E, Kolkova K S, Surovikina I V and Chernova M A 2014 Dynamics of ice phenomenons on the rivers of Don basin within the territory of Tambov region *Geopolitics and ecogeodynamics of regions* **10** (1) 393-98

[2] Arzhanykh Yu P, Dolzhenkov V V and Zvyagintseva A V 2014 Forecasting of the hydrological situation (in the period of the high water) on water objects of the Voronezh region with use of geographical information systems *Heliogeophysical researches* **9** (9) 89-98

[3] Malnes E, Solbø S, Lauknes I, Evertsen G J, Tøllefsen T A, Solheim I and Indregard M 2005 FLOODMAN-Global near-real time flood monitoring for hydrological users *ACTIF Proc. Int. Conf. on Innovation Advances and Implementation of Flood Forecasting Technology (Bergen-Tromsø)* (Electronic Materials) p 1-9

[4] Nester T, Komma J and Blöschl G 2016 Real time flood forecasting in the Upper Danube basin *J. Hydrol. Hydromech.* **64** (4) 404-14 doi: 10.1515/johh-2016-0033

[5] Golubev A B and Zemtsov V A 2013 Danger and flood risk assessment in Barnaul (Zaton village) *Tomsk University Bulletin* **373** 183-8

[6] Belyaev B M and Varentsova N A 2018 Development of long-term hydrological spring high-water period forecasts based on modern hydrometeorological information and methods of its processing *3rd Vinogradov Readings «The Edge of Hydrology»* (St. Petersburg: High Technology p 698–703

[7] Lavrov S A and Kalyuzhniy I L 2016 Climatic Changes Impact upon the Spring High-water Runoff and its Formation Factors in the Volga *Basin Russia's water industry: problems, technologies, management* **6** 42-60

[8] Zhao M and Hendon H H 2009 Representation and prediction of the Indian Ocean dipole in the POAMA seasonal forecast model *Q. J. R. Meteorol. Soc.* **135** 337-52 doi: 10.1002/qj.370

[9] Costabile P, Costanzo C and Macchione F 2012 Comparative analysis of overland flow models using finite volume schemes *J. Hydroinform.* **14** 122-35 doi: 10.2166/hydro.2011.077

[10] Xia X, Liang Q, Ming X and Hou, J 2017 An efficient and stable hydrodynamic model with novel source term discretization schemes for overland flow and flood simulations *Water Resour. Res.* **53** 3730-59 doi: 10.1002/2016wr020055

[11] Costabile P and Macchione F 2015 Enhancing river model set-up for 2-D dynamic flood modelling *Environ. Model. Softw.* **67** 89-107 doi: 10.1016/j.envsoft.2015.01.009

[12] Nayak P, Sudheer K, Rangan D and Ramasastri K 2005 Short-term flood forecasting with a neurofuzzy model *Water Resour. Res.* **41** doi: 10.1029/2004wr003562

[13] Kim B, Sanders B F, Famiglietti J S, Guinot V 2015 Urban flood modeling with porous shallow-water equations: A case study of model errors in the presence of anisotropic porosity *J. Hydrol.* **523** 680-92 doi: 10.1016/j.jhydrol.2015.01.059

[14] Burakov D A, Gordeev I N and Kosmakova V F 2009 Applying of regression dependencies in long-term forecasts of maximum water levels on the example of the Yenisei River near Kyzyl *Questions of geography of Siberia* (Tomsk: Tomsk State University) pp 5-6

[15] Burakov D A, Kosmakova V F and Gordeev I N 2012 The results of testing a physical and statistical method for a long-term forecast of the maximum water levels of the spring flood of the Yenisei river near Kyzyl *Results of testing new and improved technologies, models and methods of hydrological forecasts* **39** 121-6

[16] Burakov D A, Kosmakova V F and Gordeev I N 2015 About the results of operational tests of hydrological forecasting methods on the rivers of Eastern Siberia during the spring flood *Results of testing new and improved technologies, models and methods of hydrological forecasts* **42** 50-9

[17] Valipour M, Banihabib M E and Behbahani S M R 2012 Parameters estimate of Autoregressive Moving Average and Autoregressive Integrated Moving Average Models and compare their ability for inflow forecasting *J. Math. Stat.* **8** 330-8 doi: 10.3844/jmssp.2012.330.338

[18] Adamowski J, Fung Chan H, Prasher S O, Ozga-Zielinski B and Sliusarieva A 2012 Comparison of multiple linear and nonlinear regression, autoregressive integrated moving average, artificial neural network, and wavelet artificial neural network methods for urban water demand forecasting in Montreal, Canada *Water Resour. Res.* **48** doi: 10.1029/2010wr009945

[19] Valipour M, Banihabib M E and Behbahani S M R 2013 Comparison of the ARMA, ARIMA, and the autoregressive artificial neural network models in forecasting the monthly inflow of Dez dam reservoir *J. Hydrol.* **476** 433-41 doi: 10.1016/j.jhydrol.2012.11.017

[20] Chow V T, Maidment D R and Mays L W 1988 *Applied hydrology* (New York: MacGraw-Hill) p 149

[21] Aziz K, Rahman A, Fang G and Shrestha S 2014 Application of artificial neural networks in regional flood frequency analysis: A case study for Australia *Stoch. Environ. Res. Risk Assess.* **28** 541-54 doi: 10.1007/s00477-013-0771-5

[22] Arzhakov A N and Kusatov K I 2009 The results of testing the method of forecasting the maximum water levels of the spring flood on the rivers of Yakutia for 2004-2006 *Results of testing new and improved technologies, models and methods of hydrological forecasts* **36** 86-97

[23] Mekanik F, Imteaz M, S Gato-Trinidad and Elmahdi A 2013 Multiple regression and Artificial Neural Network for long-term rainfall forecasting using large scale climate modes *J. Hydrol.* **503** 11-21 doi: 10.1016/j.jhydrol.2013.08.035

[24] Mosavi A, Rabczuk T and Varkonyi-Koczy A R 2017 Reviewing the novel machine learning tools for materials design *Advances in Intelligent Systems and Computing* **660** 50-8 doi: 10.1007/978-3-319-67459-9_7

[25] Kumani M V, Sysemko V I and Apuhtin A V 2012 Multi-level system for monitoring and forecasting the maximum flow rates and water levels of the spring flood *Scientific notes: Electronic scientific journal of Kursk State University* **4-2** (24) 213-7

[26] Kosheleva E D and Kudishin A V 2018 Short-term forecast of Ob river water level near Barnaul city during flood 2018 *News of the Altai Branch of the Russian Geographical Society* **3** (50) 27-37

[27] Lafdani E K, Nia A M, Pahlavanravi A, Ahmadi A and Jajarmizadeh M 2013 Daily Rainfall-Runoff Prediction and Simulation Using ANN, ANFIS and Conceptual Hydrological MIKE11/NAM Models *Int. J. Eng. Technol.* **1** 32-50

[28] Seo D-J and Breidenbach J 2002 Real-time correction of spatially nonuniform bias in radar rainfall data using rain gauge measurements *J. Hydrometeorol.* **3** 93–111 doi: 10.1175/1525-7541(2002)003<0093:RTCOSN>2.0.CO;2

[29] Shustov A M 2004 *Issues of engineering hydrology of the rivers of the Republic of Mordovia* (Saransk: Mordovian University) p 80

[30] Vladimirov A M 2009 The factors determining occurrence of the high flow and highest water level during a flood *Scientific notes of the Russian State Hydrometeorological University* **9** 22-39

[31] Ishkhanyan M V and Karpenko N V 2016 *Econometrics* vol 1 (Moscow: MIIT) p 117
[32] *Guidelines for determining the calculated hydrological characteristics in the absence of data of hydrometric observations* 2009 (Saint Petersburg: Nestor-History) p 193

An experimental study of atmospheric turbulence characteristics in an urban canyon

A D Pashkin[1,2], I A Repina[1,2], V M Stepanenko[2], V Y Bogomolov[3,2], S V Smirnov[3] and A E Telminov[3]

[1]A.M.Obukhov Institute of Atmospheric Physics RAS, Moscow, Russia
[2]Lomonosov Moscow State University, Moscow, Russia
[3]Institute of Monitoring of Climatic and Ecological Systems SB RAS, Tomsk, Russia

E-mail: artem.ifa64@gmail.com

Abstract. This article presents the first results of an experimental study of turbulent heat exchange between a surface surrounded by buildings and the atmospheric boundary layer. Heat and momentum fluxes are measured at three levels in the first part of the experiment and three points in the second part of the experiment by an eddy covariance (EC) technique. The presence of wind-shear effects is supported in our measurements by the fact that the momentum flux increases with height from the surface. The sensible heat flux increases with height in the daytime. The distributions of the dimensionless proportionality coefficients between the third and second moments thus obtained indicate the presence of coherent structures in the surface layer.

1.Introduction

For mathematical modeling of climate and numerical weather prediction, information on turbulent heat and momentum fluxes between the atmosphere and the underlying surface is needed. Moreover, the traditionally applied parametrizations are based on the conclusions of the Monin–Obukhov similarity theory (MOST), which assumes an underlying surface with a uniform distribution of aerodynamic and temperature roughness, as well as stationary turbulent flows horizontally and in the surface layer of the atmosphere [1]. Over inhomogeneous landscapes, the condition of horizontal statistical flow homogeneity is violated, which requires specialized experiments to establish the limits of applicability and accuracy of MOST, as well as to study both the vertical and horizontal turbulence patterns in the surface layer of the atmosphere [2, 3,4]. Also, the presence of spatial inhomogeneities can cause non-closure of the heat balance, an effect revealed in many field experiments [5].

Conventionally, the surface heterogeneities can be divided into three groups that can act in combination: a step of different heights and lengths, which causes the separation of the wind flow (forest edge, seashore, cliff), areas with sharply different roughness (forest glades surrounded by forest lakes, urban squares), and long tunnels (river beds, mountain gorges, city canyons). Full-scale measurements of atmospheric turbulence characteristics under such terrain conditions are few, and estimates of the spatial distribution of the turbulence characteristics in this case are mainly based on the results of LES modeling and laboratory experiments [6, 7].

It is well-known that one of the most important tasks of statistical hydromechanics is to find the relationships between the statistical moments of thermohydrodynamic quantities. The Monin-Obukhov similarity theory can be interpreted as one of the most famous examples in geophysics of the diagnostic relationship between the first and second moments. At the same time, in a heterogeneous landscape the so-called coherent structures have a significant role in the formation of the turbulent regime, in the presence of which a closer connection can be demonstrated by the second and third moments [4, 8]. The authors suggest that in one-type heterogeneous landscapes (for example, in canyons) there are universal forms of dependences between statistical moments in which dimensionless constants can be associated with the geometrical parameters of a particular locality.

Revealing of such dependences will contribute to the development of new methods for parameterizing the exchange between the inhomogeneous surface and the atmosphere by the momentum and scalar characteristics.

The study of the turbulent structure of the atmospheric boundary layer (ABL) within urban development is a particularly important task. Assessment and prediction of the microclimate and air quality in cities and their environs comes down to the task of determining the statistical characteristics of stratified turbulent currents with spatial detail down to scales comparable to individual buildings and streets. It is well-known that the accumulation of pollutants in an urban environment is determined not only by the location of pollution sources, but also by the development structure [9]. Urban canyons significantly change the wind regime in the surface layer of the atmosphere [10], creating inhomogeneities in the deposition of pollution and wind tunnels. This factor leads to significant heterogeneity of the pollution field of the urban environment. Also, the study of the dynamics of the wind flow inside an urban canyon is important for modeling the effects of the urban "heat island", especially in winter and at night under conditions of stable stratification [11].

2. Relationships of statistical moments in the boundary layer in the presence of coherent structures

For the case when a large part of the dispersion of thermohydrodynamic fields falls on large (comparable to the boundary layer thickness) structures in which ordered ascending and descending jets are distinguished, the authors of [9, 12] proposed a so-called bimodal model. In the bimodal model, the temperature dispersion flux $\overline{w'\theta'\theta'}$ is associated with the potential temperature flux $\overline{w'\theta'}$ and, accordingly, with the heat flux through the relation:

$$\overline{w'\theta'\theta'} = C_\theta S_\theta \left(\overline{\theta'^2}\right)^{\frac{1}{2}} \overline{w'\theta'}, \qquad (1)$$

where $S_\theta = \frac{\overline{\theta'^3}}{\overline{\theta'^2}^{\frac{3}{2}}}$ is the asymmetry coefficient of the temperature distribution, and C_θ is the dimensionless constant. The potential temperature flux is similarly related to the heat flux

$$\overline{w'w'\theta'} = C_w S_w \left(\overline{w'^2}\right)^{\frac{1}{2}} \overline{w'\theta'} \ . \qquad (2)$$

If the values of C_θ and C_w from the data of pulsation measurements turn out to be of the order of 1, this indicates the presence of coherent structures (large eddies) and their determining contribution to the vertical turbulent flows. Thus, using equations (1) and (2) in [4] it was shown that turbulent flows over a small lake surrounded by a forest, with the transverse direction of the wind, are formed due to large structures not described by the Monin–Obukhov similarity theory.

3. Experiment setup

This paper presents an experiment in which all-weather monitoring of the temporal variability and spatial structure of atmospheric turbulence is carried out under conditions close to the conditions of an urban canyon. The measurements are carried out at the geophysical observatory of the Institute of Monitoring of Climatic and Ecological Systems (IMCES) SB RAS, Tomsk. The observation platform is located at an altitude of 167 m above sea level inside a multi-story building with building heights from 11 to 21 m (Figure 1). In the southeast of the site there is a forest with a tree height of at least 30 m. On average, southern and southwestern winds prevail over the year, that is, from the open side. The measuring scheme includes seven AMK-03 acoustic anemometers, which allow measuring pulsations of the three components of wind speed and temperature with a frequency of up to 80 Hz (http://meteosap.ru/catalog/amk-03/), a joint production of IMCES SB RAS and LLC "Sibanalitpribor "(Russia). Five anemometers are located on the ground: three AMKs at a height of 2 m, one at a height of 10 m, and another one at the edge of the forest at a height of 28 m. Two anemometers are located on the buildings surrounding the site. The general arrangement of the devices is shown in Figure 1. The data collected by the geophysical observatory are used to analyze the state of the atmosphere and the underlying surface: standard weather data, snow depth, surface and soil temperatures, and components of the radiation balance.

The location scheme of the sensors of the ground site was previously tested when measuring the structure of turbulence over a lake surrounded by forest [2]. It allows one to estimate the terms of the balance equations of statistical moments and, accordingly, the contribution of horizontal and vertical transport to the formation of turbulent flows.

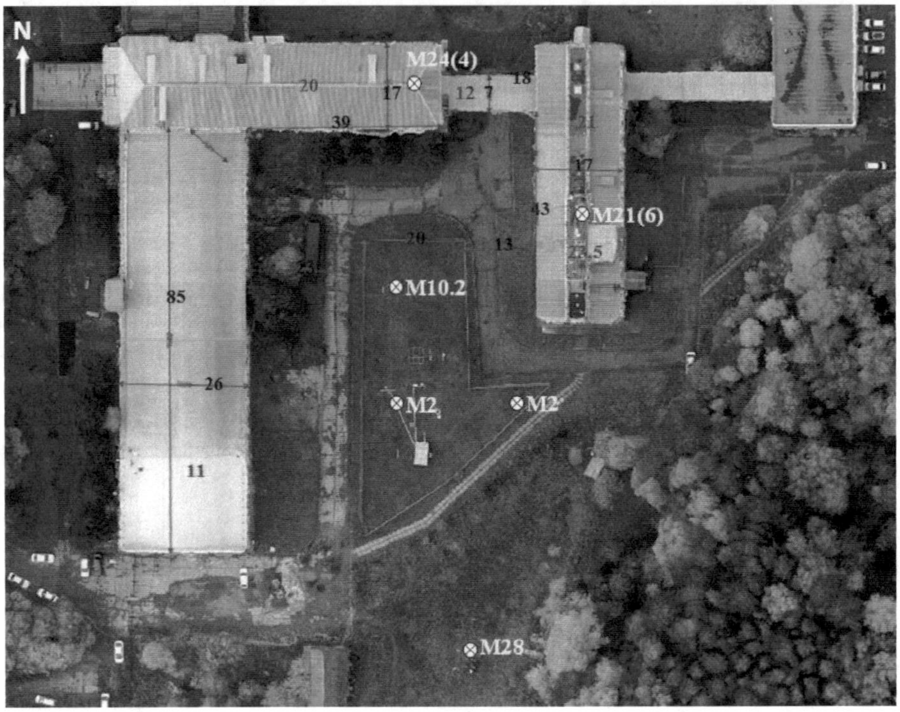

Figure 1. Layout of the masts of the geophysical observatory of IMCES SB RAS. The heights of buildings are shown by red color and horizontal dimensions, by blue one. Masts are denoted by blue circles with crosshairs on a white background with a yellow "M" signature with numbers that indicate the heights of acoustic hot-wire anemometers. For appliances located on the roof, the height above the roof is in brackets. The scheme is oriented to the north.

4. Measurement results

For the consideration, the measurement data of three acoustic hot-wire anemometers located at heights of 2 m and 10 m (mast M10.2) and 28 m (M28), from March 9 to 12, 2019, were selected. Figures 2 and 3 show values with 20-minute averaging; local time (UTC + 7). The temperature with a daily amplitude of about 7°C during the day and in night hours was below zero. The temperature difference between the devices does not exceed 0.4°C, i.e., stratification is close to neutral. Over the entire period the cloud cover was 10 points, and the snow depth was 77 cm. The horizontal wind speed increases with height: the characteristic values at a height of 2 m are in the range of 1-2 m/s, at 10 m in the range of 2-3 m/s, and at 28 m in the range of 3-5 m/s. A pronounced diurnal course of the speed module was not noted. In the wind direction, two modes can be distinguished:

1) wind from the side of the free southern corridor: direction 190°;
2) through the western wing of the building: direction 225°.

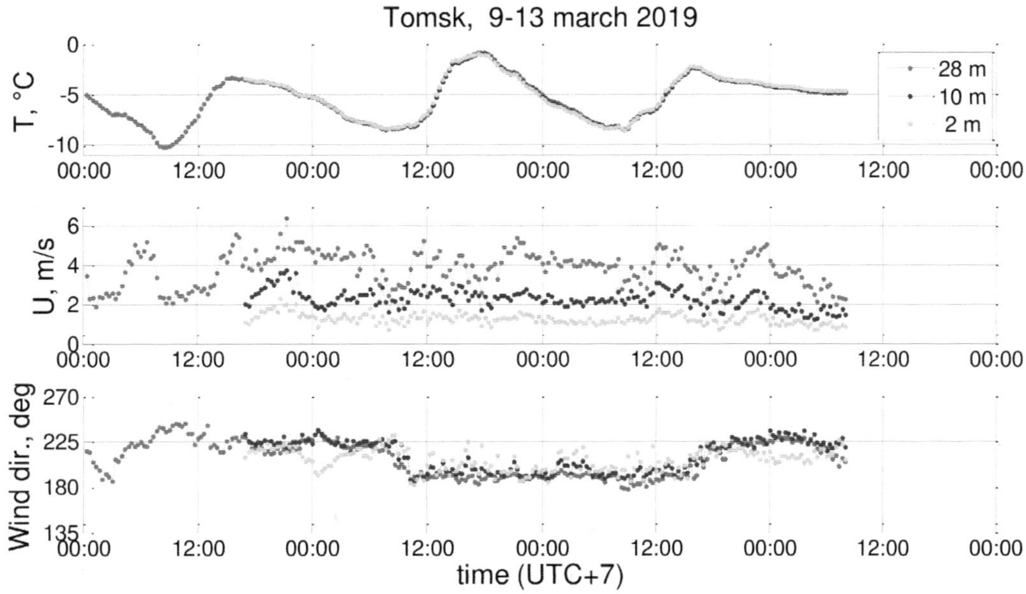

Figure 2. Flow parameters with 20-minute averaging: air temperature, horizontal wind speed, wind direction.

Figure 3 shows the time course of the turbulent flow of sensible heat. The diurnal flow course at heights of 10 m and 28 m stands out clearly. During the day, the positive flow from the surface to the atmosphere at a height of 28 m reaches 200 W/m^2, at a height of 10 m it reaches 80 W/m^2, and at a height of 2 m the flow is close to 0. At night at all levels the flow is close to 0, but at a height of 2 m the flow is directed downward. At night, when the wind blows from the side of the free corridor, the flow at a height of 2 m is directed down to -30 W/m^2, and when the wind blows through the building it is close to 0.

In the daytime, there is no layer of constant flows under both regimes distinguished in the direction of the wind. The heat flow increases with height. This is especially pronounced in the daytime. This can be explained by the fact that the influence area ("footprint") for the device located at a height of 2 m covers an adjacent relatively cold surface covered with snow, for a height of 10 m the part of this area falls on the walls of buildings heated during the day. At a height of 28 m, the presence of nearby mixed forest, which also warms up during the day, can affect the nearby snow. The average value of the pulse flux in the presented segment of recording of the time course of three acoustic hot-wire anemometers also grows with height: at 2 m – 0.27 H/m^2, at 10 m – 0.43 H/m^2, and at 28 m – 0.46 H/m^2. An increase in the momentum flux, as well as the modulus of the heat flux with height during the flow behind the "step", is noted in measurements in natural landscapes [3, 4, 13]. An increase in the pulse flux with height was obtained in similar flows in laboratory experiments [14] and in eddy-resolving modeling [7].

Consider the dimensionless coefficients C_θ and C_w presented in Figure 4 and calculated by equations (1) and (2). On the probability distribution C_w, a shift of the distribution maximum to the right from zero is noticeable: for heights of 2 m and 28 m the maximum frequency falls in the interval from 0 to 2, and for a height of 10 m the maximum is less pronounced and falls in the interval from -1 to 3. The maximum of C_θ for distribution for all three heights falls in the interval from 0 to 2. Note that in [12] $C_\theta = C_w = 1$, while with a less rigorous approach having a wider range of applicability [9] this equality is true in the order of magnitude. Thus, in the city canyon under consideration, one can assume the presence of large vortices and their decisive role in the formation of vertical turbulent flows. At the same time, a significant fraction of the values of C_θ and C_w >>1 means the need for further study of the conditions favorable for the development of coherent structures.

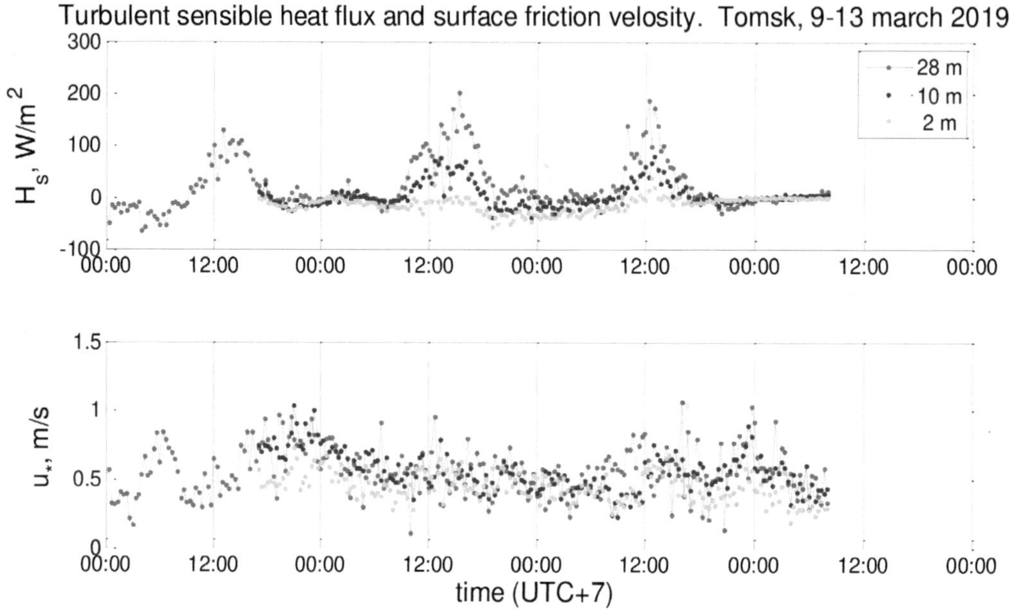

Figure 3. Turbulent sensible heat flux and surface friction velocity.

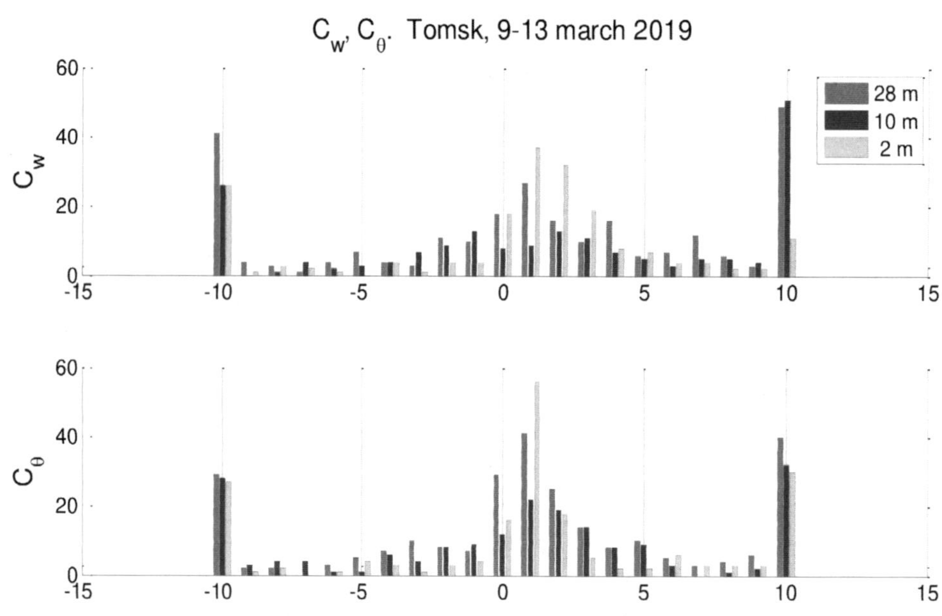

Figure. 4. Diagram of the probabilistic distribution of the coefficients C_θ and C_w (in the extreme columns all values that are beyond the boundaries shown in the figure are accumulated).

5. Conclusions

The measurements and calculations with the turbulent pulsation method in the conditions of an urban canyon have not shown a layer of constant fluxes at heights from 2 m to 28 m in the daytime. The above-obtained distributions of C_θ and C_w (the dimensionless proportionality coefficients between the third $\overline{w'\theta'\theta'}$, $\overline{w'w'\theta'}$ and second $\overline{w'\theta'}$ moments) do indicate the presence of coherent structures in the surface layer.

It is planned to collect observational data in the different seasons and find dependences of the turbulence characteristics on wind direction under various atmospheric stratifications.

Acknowledgments
The data processing and analysis was supported by RSF under grant 17-17-01210, and the experimental work was supported by the Russian Foundation for Basic Research under grant 18-05-60126.

References
[1] Monin A S and Obukhov A M 1954 *Trudy GEOFIAN* **24** (151) 163-87
[2] Stepanenko V M, Repina I A, Barskov K V, Artamonov A Yu, Bogomolov V Yu, Varentsov M I, Gavrikov A V and Pashkin A D 2018 *Turbulence, dynamics of the atmosphere and climate* (Moscow: Fizmatkniga) p 81-93
[3] Barskov K V, Chernyshev R V, Stepanenko V M, Repina I A, Artamonov A Y, Guseva S P and Gavrikov A V 2017 *IOP Conf. Series: Earth and Environmental Science* **96** 12003
[4] Barskov K V, Stepanenko V M, Repina I A, Artamonov A Yu and Gavrikov A A 2019 *Boundary-Layer Meteorology* https://doi.org/10.1007/s10546-019-00469-w
[5] Leuning R, van Gorsel E, Massman W J and Isaac P R 2012 *Agric. Forest Meteorol.* **156** 65-74
[6] Glazunov A V 2014 *Izvestia RAS. Atmospheric and Oceanic Physics* **50** 4 406-19
[7] Glazunov A V and Stepanenko V M 2015 *Izvestia RAS. Atmospheric and Oceanic Physics* **51** (4) 403-15
[8] Zilitinkevich S S, Gryanik V M, Lykossov V N, Mironov D V 1999 *Journal of the Atmospheric Sciences* **56** 3463–77
[9] Baklanov A Mestayer P Clappier A Zilitinkevich S Joffre S Mahura A and Nielsen N W 2008 *Atmospheric Chemistry and Physics* **8** 523-43
[10] Starchenko A V Nuterman R B Danilkin E A 2015 *Numerical modeling of turbulent flows and impurity transport in street canyons* (Tomsk: Tomsk University) p 252
[11] Oke T R Mills G Christen A Voogt J A 2017 *Urban climates* (Cambridge University Press)
[12] Abdella K Mcfarlane N 1997 *Journal of the Atmospheric Sciences* **54** (14) 1850-67
[13] Queck R Bernhofer C Bienert A Schlegel F 2016 *Boundary-Layer Meteorol.* **160** (3) 397-423
[14] Markfort C D Porté-Agel F Stefan H G *Environ.* 2014 *Fluid Mech.* **14** (3) 663–97

Inverse problems for the study of climatic and ecological processes under anthropogenic influences

V V Penenko[1,2] and E A Tsvetova[1]

[1] Institute of Computational Mathematics and Mathematical Geophysics SB RAS, Novosibirsk, Russia
[2] Novosibirsk State University, Novosibirsk, Russia

e-mail penenko@sscc.ru

Abstract. This paper presents a description of algorithms for solving direct and inverse climatic and environmental problems based on a variational principle with weak constraints. The initial first-order system is supplemented by a system of equations in variations. It provides a calculation of all necessary components of the modeling system by combining the model and observation data and including first- and second-order functionals of sensitivity to variations in the model parameters, input data, and observation results.

1. Introduction: variety of inverse problems

To solve environmental problems, we are developing a variational approach that provides great opportunities in the formulation, construction, and implementation of solutions to direct and inverse problems [1-4].

Examples of direct problems are the well-known problems of forecasting the weather and quality of the atmosphere, the problem of climate research, etc. In mathematical sense, they are initial boundary value problems in which the result is obtained by successive propagation in time of the solution of the system of equations of the corresponding process models starting with initial data.

With this approach, it is believed that all necessary parameters of the models, including the initial data, as well as the sources of influences, are known in advance. However, with respect to the parameters and sources, it is not always obvious that they are set reliably. The fact is that usually the parameters are selected from a certain range of values, based on the previous knowledge and experience, which leaves an imprint on the quality of the solution. It is especially important to specify the initial data in cases when their impact on the result is significant, as, for example, in the weather and air quality forecasts.

Therefore, to improve the reliability of the result, researchers often refer to the methods of data assimilation, which are included in the modeling technology. In the history of weather forecasting for many years, several approaches to the methods of accounting for observational data have been developed. They are, for example, methods of subjective and objective analysis, nudging procedures, Kalman filtering methods, variational assimilation of data, etc. For contemporary reviews of works on this subject, see [5, 6] and the bibliography available there.

In fact, the construction of special procedures to obtain initial data for the state functions of the models are procedures for solving inverse problems. For example, we consider the process of data

IOP Conf. Series: Earth and Environmental Science **386** (2019) 012036 doi:10.1088/1755-1315/386/1/012036

assimilation as a sequential solution of interconnected inverse problems. Nowadays, the data assimilation methods make up the largest class of inverse problems used in climate and ecological studies. This is an illustrative example when inverse problems become an urgent need for successful solution of direct problems, justifying the expediency of their coordinated solution.

The methods of data assimilation do not exhaust the entire list of inverse problems used in such studies. Much attention has recently been given to the methods of identification of model parameters and, consequently, the parameters of the simulated processes. Identification of the sources of influence, reconstruction of the coefficients in the equations of models are among such problems. There are also performances of inverse problems which allow one to formulate the so-called continuation problems, which consist in the continuation of the solution from the inside of the region on its border and beyond.

In addition to these tasks, increased attention is paid to the problems of assessing the sensitivity of the models and simulated processes to changes in the parameters. For these purposes, a sensitivity theory is being developed, which makes it possible to indicate the range of changes in the results depending on the range of changes in the parameters. An essential point here is the adoption of the concept of model "flexibility", that is, the abandonment of the "rigid" formulation by including new sought-for functions, named uncertainty functions, in the process model and in the observation model. The benefit of introducing additional unknowns is also that they allow one to construct methods of real time data assimilation [2].

Among the methods gaining popularity in recent years, there are methods for exploring the sensitivity of the second order with special models in variations. Statements of problems of this type are discussed in [7, 8]. They present methods for estimating the sensitivity of the final state of the system (forecast) depending on perturbations in its initial state. For these purposes, methods of solving the spectral problem for the linearized operator in the first-order approximation are used. Accordingly, for approximation of the second order, the spectral methods are applied to matrices of the Hessian of the linearized operator in variations of the second order. In [9], using second-order methods, the sensitivity of the optimal solution and its functionals to errors in the observed data was studied. Despite the fact that methods of the second order are much more time-consuming than those of the first order, they serve to improve the accuracy of calculations. As shown in [8], they extend the predictability intervals in prognostic studies.

This paper presents a description of algorithms for solving direct and inverse climatic and environmental problems based on a variational principle with weak constraints developed at the Institute of Computational Mathematics and Mathematical Geophysics of the Siberian Branch of the Russian Academy of Sciences. The proposed approach provides the calculation of all necessary components of the modeling system by combining the model and observation data, including the study of the first- and second-order sensitivity of the functionals.

2. Key points

It is well-known that in problems of this class it is necessary to work in the conditions of uncertainty in the models of processes, in internal and external sources of influences, in estimates of the initial and current state of the environment, and in the results of observations. To combine all these objects in a mathematical modeling technology, we use the variational principles with weak constraints.

We define the main objects of the modeling technology in operator form [1, 2]:

$$L(\boldsymbol{\varphi}, \mathbf{Y}, \mathbf{f}, \mathbf{r}) \equiv \frac{\partial \boldsymbol{\varphi}}{\partial t} + G(\boldsymbol{\varphi}, \mathbf{Y}) - \mathbf{f} - \mathbf{r} = 0, \quad \boldsymbol{\varphi}(\mathbf{x}, t) \in Q(D_t); \tag{1}$$

$$\boldsymbol{\varphi}^0 = \boldsymbol{\varphi}_a^0 + \boldsymbol{\xi}, \quad t = 0; \tag{2}$$

$$\boldsymbol{\psi} = H(\boldsymbol{\varphi}) + \boldsymbol{\eta}, \tag{3}$$

$$J(\boldsymbol{\varphi}) = \int_{D_t} F(\boldsymbol{\varphi}) \chi(\mathbf{x}, t) dD dt\,. \tag{4}$$

Here $L(\varphi, \mathbf{Y}, \mathbf{f}, \mathbf{r})$ is the model of processes, $Q(D_t) \in L_2(D_t)$ is the real space of the state functions $\varphi(\mathbf{x},t) = \{\varphi_i(\mathbf{x},t), i = \overline{1,n}\}$; $G(\varphi, \mathbf{Y})$ is the nonlinear differential operator of the process model; $\mathbf{f}$ is the function of sources; $\mathbf{Y}$ is the vector of model parameters varying within acceptable values, φ^0 is the initial state, φ_a^0 is its a priori estimate; $D_t = D \times [0, \bar{t}]$, D is the area of change of the spatial coordinates $\mathbf{x} = (x_i, i = \overline{1,3})$, $t \in [0, \bar{t}]$ is the time interval; $J(\boldsymbol{\varphi})$ is the target research (forecasting) functional; $F(\boldsymbol{\varphi})$ is the estimated function; $\chi(\mathbf{x},t)$ is a non-negative weight function defined in D_t; $\boldsymbol{\psi}$, $\mathbf{H}(\boldsymbol{\varphi})$ are the observed data and the observation model. The functions $\mathbf{r}, \xi, \boldsymbol{\eta}$ in (1) - (4) describe the uncertainties and errors in the corresponding objects. The main components in the structure of the model (1) for the vector with components φ are represented by a system of equations describing the processes taking place in the region D_t. The target functional $J(\boldsymbol{\varphi})$ is responsible for the research objectives, for example, it can be the goals of weather forecasting taking into account the observational data, environmental safety assessments, etc.

We define the extended functional of the variational principle for combining all objects in equations (1) - (4) in the form

$$\Phi(\mathbf{Z}) = \left(L(\mathbf{X}), \boldsymbol{\varphi}^*\right) + \alpha_0 J(\boldsymbol{\varphi}) + 0.5\left(\alpha_1 (M_1 \boldsymbol{\eta}, \boldsymbol{\eta}) + \alpha_2 (M_2 \mathbf{r}, \mathbf{r}) + \alpha_3 (M_3 \xi, \xi)\right). \tag{5}$$

Here $\mathbf{Z}(\mathbf{x},t) \equiv (\varphi, \varphi^*, \mathbf{Y}, \mathbf{f}, \mathbf{r}, \xi, \boldsymbol{\eta}, \boldsymbol{\psi})$ is the vector of functional arguments of the entire system, which, in addition to the vector $\mathbf{X} \equiv (\varphi, \mathbf{Y}, \mathbf{f}, \mathbf{r})$ from the system (1), includes components related to the results of observations (3) and functionals of possible types, for example, target, observations, control, etc.; $\varphi^*(\mathbf{x},t) \in Q^*(D_t)$, is the adjoint function representing the vectors of Lagrange multipliers, α_i are real parameters, $M_i, i = \overline{1,3}$ are the weight matrices in the definitions of functionals for estimating the uncertainty functions in equations (1) - (3). The first functional in the right-hand side of equation (5) is an integral identity of the energy type for the process models (1) - (2). Its structure is chosen so that when $\varphi^* = \varphi$ it determines the equation of energy balance of the system (1) - (2).

We construct numerical methods for the so-called "seamless" modeling technology for solving the problems (1) - (4) using the Gateaux definitions of variations of functionals and operators

$$\delta^p \Phi(\mathbf{Z}) = \frac{d}{d\lambda^p} \{\Phi(\mathbf{Z} + \lambda \delta \mathbf{Z})\}_{\lambda=0}, \tag{6}$$

where p is the order of variations and, based on them, the Taylor's formulas

$$\Phi(\mathbf{Z} + \lambda \delta \mathbf{Z}) = \Phi(\mathbf{Z}) + \lambda \delta \Phi(\mathbf{Z}) + \frac{\lambda^2}{2} \delta^2 \Phi(\mathbf{Z}) + \cdots + \frac{\lambda^p}{p!} \delta^p \Phi(\mathbf{Z}) + o(\lambda^p), \ \lambda \to 0. \tag{7}$$

To construct approximations of the numerical models, we use functional decomposition methods, operator splitting methods, finite element/volume methods, and adjoint integrating factor methods [1]-[4]. As a result, approximations are obtained that are generally agreed in the sense of the functional (5). The combination of applied splitting schemes allows us to efficiently use locally one-dimensional problems in the spatial coordinates and time.

Based on the first-order variations of the extended functional (5), we obtain systems of basic and adjoint equations for finding the functions $\varphi, \varphi^*, \mathbf{r}, \xi$ with assimilation of observational data and algorithms for calculating the set of spaces of sensitivity functions of the functionals to variations of the parameters and input data of the process models and to variations of the observation data. From the first three terms in (7), methods of sensitivity theory and methods of second-order forecasting are obtained, respectively.

2.1. Forward / inverse modeling algorithm

We write the forward/inverse modeling algorithm for a discrete analog of the process model operators [1-4], which is obtained from the stationary conditions of the functional (5) to the variations of the functions $\boldsymbol{\varphi}, \boldsymbol{\varphi}^*, \mathbf{r}, \boldsymbol{\xi}$ in accordance with the definition (6) for $p=1$:

$$\frac{\partial \boldsymbol{\varphi}}{\partial t} + G^h(\boldsymbol{\varphi}, \mathbf{Y}) - \mathbf{f} - \mathbf{r} = 0, \quad (8)$$

$$\boldsymbol{\varphi}(\mathbf{x}, 0) = \boldsymbol{\varphi}_a(\mathbf{x}) + \boldsymbol{\xi}(\mathbf{x}), \ \ \mathbf{x} \in D, \ \ t = 0; \quad (9)$$

$$-\frac{\partial \boldsymbol{\varphi}^*}{\partial t} + A^T(\boldsymbol{\varphi}, \mathbf{Y})\boldsymbol{\varphi}^* + \mathbf{d} = 0, \quad (10)$$

$$d = \frac{\partial}{\partial \boldsymbol{\varphi}}\left(\alpha_0 J^h(\boldsymbol{\varphi}) + 0{,}5\alpha_1 \left(\boldsymbol{\eta}^T M_1 \boldsymbol{\eta}^*\right)^h\right) = \alpha_0 \frac{\partial J^h(\boldsymbol{\varphi})}{\partial \boldsymbol{\varphi}} + \left[\frac{\partial H^h(\boldsymbol{\varphi})}{\partial \boldsymbol{\varphi}}\right]^T \alpha_1 M_1 \boldsymbol{\eta}^h, \quad (11)$$

$$\varphi^*(\mathbf{x})\big|_{t=\bar{t}} = 0; \quad (12)$$

$$\mathbf{r}(\mathbf{x}, t) = \alpha_2^{-1} M_2^{-1} \boldsymbol{\varphi}^*(\mathbf{x}, t); \quad (13)$$

$$\boldsymbol{\xi}(\mathbf{x}) = \alpha_3^{-1} M_3^{-1} \boldsymbol{\varphi}^*(\mathbf{x}, 0); \quad (14)$$

$$A(\boldsymbol{\varphi}, \mathbf{Y})\delta\boldsymbol{\varphi} \equiv \frac{\partial}{\partial \alpha}\left[G^h(\boldsymbol{\varphi} + \alpha\delta\boldsymbol{\varphi}, \mathbf{Y})\right]_{\alpha=0}. \quad (15)$$

The superscript h indicates discrete analogs of the corresponding objects. The system (8)-(15) is implemented sequentially in the time intervals $\left\{\left[t_{j-1}, t_j\right], j = \overline{1, j_k}\right\} \in [0, \bar{t}]$.

The system (8) - (15) allows us to find solutions to the direct and adjoint problems, $\boldsymbol{\varphi}$ and $\boldsymbol{\varphi}^*$, to calculate the uncertainty functions $\mathbf{r}, \boldsymbol{\xi}$. After the functions $\{\boldsymbol{\varphi}, \boldsymbol{\varphi}^*, \mathbf{r}, \boldsymbol{\xi}\}$ are calculated, they are used to obtain sensitivity functions of the target functionals of the system (1) - (4) to variations of the model parameters $\mathbf{Y} \in R(D_t)$ and to variations of the observation results $\boldsymbol{\psi}_m \in \Psi_m(D_t^m)$ from (3).

The sensitivity functions are derivatives of the target functionals (5) with respect to the variable parameters. These functions are necessary for organizing algorithms for solving inverse problems for identifying the parameters $\mathbf{Y}$ using the data (3), for assessing trends in the influence of uncertainties in observational data on forecast results, etc.

2.2. Second order algorithms

To construct methods of forecasting and investigating the sensitivity of the second order, in the sense of the definition (6),(7), we use the idea of the method of neighboring extremals [10] and Newton's method for solving nonlinear problems [11]. For this purpose, the objects in the definitions of the second variation are calculated from the following system in perturbations (variations) in the vicinity of the trajectories of the solution of the system of equations (8)-(15):

$$\frac{\partial \delta\boldsymbol{\varphi}}{\partial t} + A(\boldsymbol{\varphi}, \mathbf{Y})\delta\boldsymbol{\varphi} - \delta\mathbf{f} - \delta\mathbf{r} = 0, \quad (16)$$

$$\delta\boldsymbol{\varphi}(\mathbf{x}, 0) = \delta\boldsymbol{\varphi}_a(\mathbf{x}) + \delta\boldsymbol{\xi}(\mathbf{x}), \ \ \mathbf{x} \in D, \ \ t = 0; \quad (17)$$

$$-\frac{\partial \delta\boldsymbol{\varphi}^*}{\partial t} + A^T(\boldsymbol{\varphi}, \mathbf{Y})\delta\boldsymbol{\varphi}^* + \delta\mathbf{d} = 0, \quad (18)$$

$$\boldsymbol{\delta} d = \left[\alpha_0 \frac{\partial J^h(\boldsymbol{\varphi})}{\partial \boldsymbol{\varphi}} + \left[\frac{\partial H^h(\boldsymbol{\varphi})}{\partial \boldsymbol{\varphi}}\right]^T \alpha_1 M_1 \boldsymbol{\eta}^h\right]\boldsymbol{\delta\varphi}, \quad (19)$$

$$\delta\varphi^*(\mathbf{x})\big|_{t=\bar{t}} = 0; \quad (20)$$

$$\boldsymbol{\delta}\mathbf{r}(\mathbf{x}, t) = \alpha_2^{-1} M_2^{-1} \boldsymbol{\delta\varphi}^*(\mathbf{x}, t); \quad (21)$$

$$\delta\boldsymbol{\xi}(\mathbf{x}) = \alpha_3^{-1} M_3^{-1} \delta\boldsymbol{\varphi}^*(\mathbf{x}, 0). \tag{22}$$

In this system, $\delta\boldsymbol{\varphi}, \delta\boldsymbol{\varphi}^*, \delta\mathbf{r}, \delta\boldsymbol{\xi}$ denote the variations of the corresponding objects. The system (16) - (22) has the structure and implementation schemes similar to those in (8) - (15).

Further, analyzing the solutions of the system for variations (16) - (22), we obtain an estimate of the solution of the original problem (1) - (4) with the second order of accuracy and estimates of other objects in the definition of variations (6) of the second order, for example, Hessian operators.

As a result of the implementation of our modeling technology (6) - (22) based on the joint use of models and monitoring data (1) - (5), we obtain a set of information spaces of functions $\{\boldsymbol{\varphi}, \boldsymbol{\varphi}^*, \mathbf{r}, \xi, \delta\boldsymbol{\varphi}, \delta\boldsymbol{\varphi}^*, \delta\mathbf{r}, \delta\boldsymbol{\xi}\}$. Using these spaces, the set of functions of sensitivity of functionals and model operators in the system (1) - (5) to variations of its functional arguments and observation data $\boldsymbol{\Psi}$ taken into account in equation (3) is calculated. These calculations are performed according to the algorithm of formula (6). Denote them by $\partial\boldsymbol{\Phi} / \partial\mathbf{Y}$ and $\partial\boldsymbol{\Phi} / \partial\boldsymbol{\Psi}$.

Such multivariate and multi-scale information on the evolution of the studied processes is required for analyzing the predictability of models (1-4), identifying critical situations and centers of action in the system of studied objects, assessing the quality of models, assessing the influence of disturbances in various components of the modeling system, in the model parameters, in the results of observations, and in organizing targeted monitoring strategies, etc. For these purposes, we use the spectral decomposition methods of the above multidimensional spaces according to the characteristic scales of the processes. A detailed description of the implementation algorithms for such methods for the analysis of multidimensional spaces and some examples of their application to solving environmental forecasting and design tasks are given in [12].

Conclusions

Thus, this paper presented a general scheme for creating a modeling system for the study of climatic and environmental processes, environmental forecasting and design. The initial first-order system was supplemented by a system of equations in variations. The above technology is self-consistent in a variational sense and allows us to solve a wide range of direct and inverse problems.

Acknowledgements

This work was performed under ICM&MG SB RAS project no. 0315-2019-0004 with partial support of the Russian Foundation for Basic Research under project no. 17-01-00137.

References

[1] Penenko V V, Tsvetova E A and Penenko A V 2015 *Izvestiya. Atmospheric and Oceanic Physics* **51** 311–19
[2] Penenko V V, Tsvetova E A and Penenko A V 2015 *Russian Meteorology and Hydrology* **40** 365-73
[3] Penenko V V, Tsvetova E A and Penenko A V 2014 *Comp. & Math. with Appl* **67** 2240–56
[4] Penenko V V, Penenko A V, Tsvetova E A 2019 *Journal of Applied Mechanics and Technical Physics* **60**(2) 392-9
[5] *Seamless Prediction of the Earth System: from Minutes to Months* 2015 WMO-No.1156
[6] Bocquet M *et al* 2015 *Atmos. Chem. Phys.* **15** 5325-58
[7] LeDimet F-X, Navon I M and Daescu D N 2002 *Monthly Weather Rev.* **130** 3 629-48
[8] Godinez H C and Daescu D N 2011 *Computational Geosciences* **15** 3 477-88
[9] Shutyaev V P and Parmuzin E I 2019 *Numerical analysis and applications* **12** 2 191-201
[10] Bryson E R , Ho Yu-Chi 1969 *Applied Optimal Control: Optimization, Estimation and Control* (Waltham: Blaisdell Publishing Company) p 544
[11] Bakhvalov N S, Zhidkov G M and Kobelkov G M 2000 *Numerical methods* (Moscow: Phyzmatlit) p 622

[12] Penenko V and Tsvetova E 2008 *Ecol. Modell.* **217** 279–91

Elements of software package for "cloud" analysis of the environment and climate change

I G Okladnikov[1,2], E P Gordov[1,2], A A Ryazanova[1] and A G Titov[1,2]

[1]Institute of Monitoring of Climatic and Ecological Systems SB RAS, Tomsk, Russia
[2]Institute of Computational Technologies SB RAS, Tomsk branch, Tomsk, Russia

Abstract. The architecture and new basic components of an information and computing software package are presented. It is based on a software framework dedicated to carrying out scientific research related to statistical processing and analysis of spatial geophysical data archives obtained from observations and modelling. To create the package, experience in the development of some information-computational web GISs for processing of large amounts of spatial data has been used. The basic components of this complex are represented by several procedures for searching, sampling, and processing of spatial data arrays, as well as by elements of a graphical user interface. The flexible structure of the computing unit of this software complex provides a possibility of operative expansion of its functionality by using additional procedures for mathematical and statistical analysis, processing, and graphical representation of the results in the form of graphs, diagrams, and plots on maps of the territory of interest.

1. Introduction

The growth of environmental data to the petabyte level and the need to store, search, retrieve, process, visualize, and analyze such geospatial datasets cast doubt on the traditional approaches. This required the creation of new technologies for distributed access and data processing, and turned this area of information technologies into a new scientific field that has recently appeared: a data-intensive application domain, in which appropriate approaches and tools are developed [1-5]. Such tools should provide the functionality of searching for geographic information resources, data sampling and processing, as well as cartographic visualization services along with corresponding client applications [6]. The combination of these tools forms a specialized data processing infrastructure, which has many names nowadays: "virtual research environment", "virtual research laboratory", "scientific gateway", "e-Science", etc. [7, 8], and allows multidisciplinary distributed research teams to perform collaborative research [9].

This infrastructure is commonly based on the modern web and GIS technologies [10-12], which provides functionality for data browsing, navigation, scrolling, scaling and graphic overlay, as well as displaying map legends and corresponding meta-information in a window of a web browser. The advantages of web-oriented GIS technologies are obvious: theoretical independence from web browsers and operating systems, automatic software location, updating, combined use of geographically distributed data sources, and sharing of centralized data warehouses. Thus, the use of web-GIS technologies provides environmental scientists and decision-makers with reliable and understandable tools to study the effects of climate change using a familiar graphical user interface.

There are several computational and/or information web systems working with geophysical data linked to environmental science. Some of them deal with the analysis of real-time meteorological observational data [13]. Some of them, such as NASA's GES-DISC Interactive Online Visualization ANd aNalysis Infrastructure (GIOVANNI), (http://daac.gsfc.nasa.gov/techlab/giovanni/) [14] are designed for analysis and visualization of satellite data. One of the most flexible web-GIS is a Regional Integrated Hydrological Monitoring System (RIMS) created at the University of New Hampshire, USA [15, 16]. As a fully integrated advanced system for online analysis of heterogeneous climate, hydrology, and remote sensing data, the RIMS was a successful attempt to realize a multi-functional GIS as a web application using MapServer (http://mapserver.org/). The RIMS is widely used for various research projects [15, 17] including NEESPI (Northern Eurasian Earth Science Partnership Initiative, http://NEESPI.sr.unh.edu/). The IRI/LDEO Climate Data Library is a powerful and freely accessible online data repository and analysis tool that allows a user to view, analyze, and download climate-related data through a standard web browser (https://iridl.ldeo.columbia.edu/). The EVER-EST (European Virtual Environment for Research – Earth Science Themes: a solution) is a novel project aimed at creating a virtual research environment (VRE) focused on the requirements of the Earth science community allowing one to collaborate, share, analyze, and visualize data over the internet (https://ever-est.eu/).

This paper presents some of the new results of developing a software package "Climate" for distributed "cloud" processing and advanced statistical analysis of large environmental datasets [18, 19]. We describe some of the created new elements of the developed software package for the analysis of environmental and climate changes implemented in the previously proposed specialized environment for the development of thematic information and computing web GIS aimed at the processing of large amounts of spatial data [20].

2. Methods and results

The software package being considered is used to answer three kinds of user queries: plain, specific, and in-depth. Most user requests for climate and environmental data services are simple. The users are interested in the spatio-temporal dynamics and basic statistical characteristics of common climate variables (air temperature and pressure, total precipitation, snow cover, etc.). However, a special analysis (such as "what if") may be requested also. In this case, scientists prefer to perform their own research and use raw (or pre-processed) data accompanied by corresponding metadata. In-depth analysis is required only in rare cases. The first two kinds of queries can be processed online, and the third one is usually considered independently using dedicated software and on-demand.

An architecture that meets these requirements and outlines the main hardware and software components of the software package has been created. It includes geoportal, web client and web services that provide access to cloud computing resources. Such a software package is designed for distributed processing of large sets of climatic spatial data, and can be deployed on several interlinked (LAN, Internet) computing units. Each software complex consists of four typical blocks (Figure 1) [21]:

- structured archives of spatial datasets accompanied by metadata;
- server-side modular computing backend for data searching, sampling, processing, and visualization;
- geoportal providing communication with cartographic web services and running the computing core;
- client-side web-GIS client providing an interactive graphical interface to analysis and visualization tools.

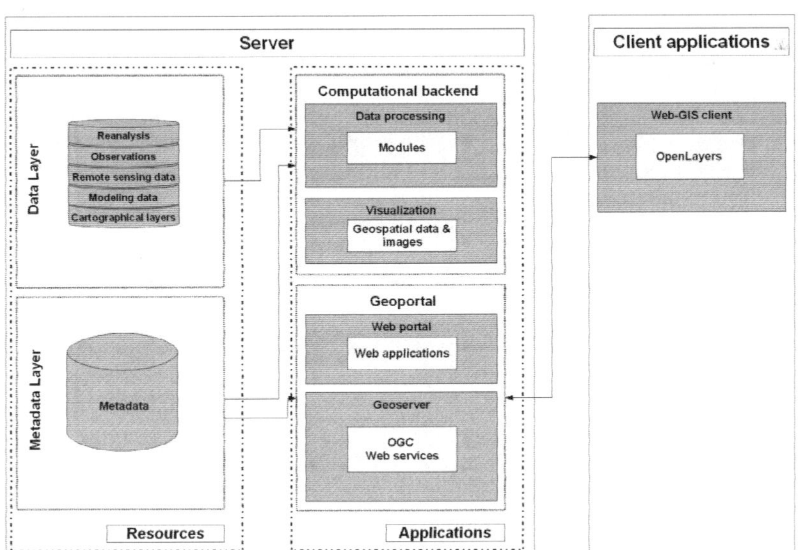

Figure 1. General architecture of the developed software complex.

The structured archives of spatial datasets are represented by heterogeneous modeling (reanalysis, meteorological models) and observation (weather stations, remote sensing) data for various spatio-temporal domains in different file formats. After pre-processing they are stored in high-performance storage systems in accordance with the dedicated storage and processing model developed earlier [22] and accompanied by corresponding metadata. As the main file format for gridded spatial data, the Network Common Data Form (netCDF) was selected. The set of data available for processing can be easily expanded thanks to a dedicated metadata database.

According to this architecture, some key components of the software package were implemented. They provide searching, sampling, processing, and visualization of data stored in the data archives. Representative fragments of some spatial data archives were collected and pre-processed, and some methods for their basic processing were selected and implemented (for the purpose of demonstrating the functionality). Together, these modules form a server-side computing backend running on high-performance computational hardware and orchestrated by the geoportal. After that processing results are written by the core into files in Encapsulated PostScript (EPS), Geo Tagged Image File (GeoTIFF), ESRI Shapefile and NetCDF formats, and are available for download.

A specialized geoportal, also functioning on the server-side, serves as a link between the elements of the developed system, as well as between the system and the end-user. It implements the necessary basic functionality, such as user authentication, connection to databases, use of HTML templates, language localization, the content management system (CMS), and several other features. GeoServer software (http://geoserver.org/) conforming to OpenGIS standards (http://www.opengeospatial.org/standards) is proposed as a technological basis for presenting cartographic information on the Internet.

Access to the resources and interactive tools of the software package is carried out through a specialized web client (Figure 1). It works in a graphical web browser window, and provides a user familiar with geographic information systems (GIS) with a user-friendly and intuitive graphical interface. The web client interface allows you to select geophysical characteristics for analysis, specify a space-time domain, and also set the desired data processing method. After data processing the results are displayed in the graphical interface on the interactive map of the selected region in the form of cartographic layers. The user can additionally download the same results in NetCDF format, as well as through WMS (Web Mapping Service) and WFS (Web Feature Service) interfaces.

The interactive software tools implement both the basic (calculation of averages, WMO climate indices, trends) and advanced (joint analysis of two different georeferenced characteristics of climatic processes) functionality of spatial data analysis. The interface of the application performing analysis of two climatic characteristics is shown in Figure 2.

In the course of expanding the software package functionality, new interactive software tools for statistical analysis of spatial climatological data time series were also developed [23]. These tools are based on reliable algorithms from the packages "extRemes" [24], "quantreg" (https://cran.r-project.org/web/packages/quantreg/), and "copula" [25], which implement time-dependent extremum statistics, quantile regression and copula approach. These packages were written in the R language (https://www.r-project.org/) and provide a flexible API for integration with third-party software.

The results obtained show that the developed software complex and tools will be useful for decision-makers and specialists working in affiliated sciences, with focus on socio-economic and ecological impact assessment, adaptation strategies, science policy administration, and other climate-related activities. On this basis, they will get reliable climate-related characteristics required for studies of economic, political, and social consequences of global climate change at the regional level.

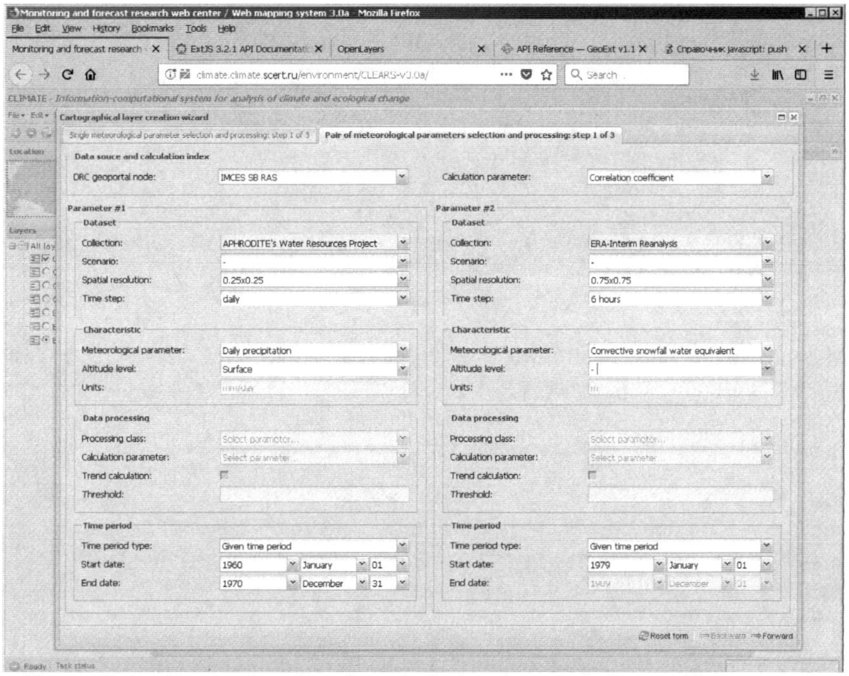

Figure 2. Application interface for calculating the correlation coefficient.

3. Conclusions

This work was aimed at developing a thematic software package for analyzing climatic changes by integrating interdisciplinary (geographic, climatic, meteorological) archives of observations, modeling, and remote sensing in the form of a thematic distributed information and computing system with GIS functionality and "cloud" processing capability. Such a computing complex is the next step in the development of applied information and telecommunication systems. It could provide the specialists in various fields of science with the unique functionalities for performing reliable analysis of heterogeneous geophysical data. Since only proven computational algorithms have been used, the results obtained are reliable. The system is accessible through the Internet and has an ability to work with data without special knowledge in programming and, thus, a wide range of researchers and decision-makers can concentrate on solving their specific problems.

The above-presented software package is based on a modular architecture and can be easily adapted to various sets of geophysical data and tasks from various areas of the Earth sciences. The open-source software used in its development makes its possible to deploy and non-commercially use the system on various software and hardware platforms running under the Linux family OS.

Acknowledgments

This work was performed under state budget theme no. AAAAA-A17-117013050037-0. The integration of new software tools for statistical analysis of time series of spatial climatological datasets was performed under ICT SB RAS state target program no. 0316-2018-0002.

References

[1] Big Data Definition *MIKE 2.0* http://mike2.openmethodology.org/wiki/Big_Data_Definition
[2] Kusnetzky D 16 February 2010 What is "Big Data?" *ZDNet* http://www.zdnet.com/article/what-is-big-data/
[3] Vance A 22 April 2010 Start-Up Goes After Big Data With Hadoop Helper *New York Times Blog* http://bits.blogs.nytimes.com/2010/04/22/start-up-goes-after-big-data-with-hadoop-helper/?dbk
[4] Kalinichenko L A *et al* 2016 Problems of access to data in Data Intensive Domain in Russia *Informatika i ee primeneniya* **10** 3-23
[5] Hilbert M 2016 Big Data for Development: A Review of Promises and Challenges *Development Policy Review* **34** (1) 135-74 doi: http://doi.org/10.1111/dpr.12142
[6] Koshkarev A V 2008 Geoportal as a tool to control spatial data and services *Spatial data* **2** 6-14

[7] Candela L, Castelli D and Pagano P 2013 Virtual Research Environments: An Overview and a Research Agenda *Data Science Journal* **12** GRDI75–GRDI81 DOI:10.2481/dsj.GRDI-013

[8] Bart A, Fazliev A, Gordov E, Okladnikov I, Privezentsev A and Titov A 2018 Virtual Research Environment for Regional Climatic Processes Analysis: Ontological Approach to Spatial Data Systematization *Data Science Journal* **17** 14 DOI: 10.5334/dsj-2018-014

[9] Allan R J 2009 *Virtual Research Environments: from Portals to Science Gateways* (Chandos: Oxford Publishing)

[10] Yakubailik O E 2007 Geoformation geoportal *J. Computational Technologies (Special Issue 3)* **12** 116-25

[11] Dragicevic S, Balram S and Lewis J 2000 The role of Web GIS tools in the environmental modeling and decision-making process *4th Int. Conf. on Integrating GIS and Environmental Modeling (GIS/EM4): Problems, Prospects and Research Needs* (September 2-8, 2000, Banff, Alberta, Canada) http://www.srcosmos.gr/srcosmos/showpub.aspx?aa=5823

[12] Frans J M van der Wel 2005 Spatial data infrastructure for meteorological and climatic data *Meteorol. Appl.* **12** 7-8

[13] Samet R and Tural S 2010 Web based real-time meteorological data analysis and mapping information system *WSEAS Transactions on Information Science and Applications* **7**(9) 1115-25

[14] Berrick S W, Leptoukh G, Farley J D and Rui H 2009 GIOVANNI: A Web service workflow-based data visualization and analysis system *IEEE Trans. Geosci. Remote Sens.* **47**(1) 106-13

[15] Shiklomanov A I 2012 Regional Integrated Monitoring System for the Pan-Arctic (RIMS) *Proc. of the Northern Eurasian Earth Science Partnership Initiative (NEESPI) Regional Science Team Meeting devoted to the High Latitudes (*June 2-6, 2012, Helsinki, Finland) pp 128-30

[16] Shiklomanov A, Prusevich A, Gordov E, Okladnikov I and Titov A 2016 Environmental science applications with Rapid Integrated Mapping and analysis System (RIMS) *IOP Conf. Series: Earth and Environmental Science* **48** 012034 DOI: 10.1088/1755-1315/48/1/012034

[17] Welp L R, Randerson J T, Finlay J C, Davydov S P, Zimova G M, Davydova A I and Zimov S A 2005 A high-resolution time series of oxygen isotopes from the Kolyma River: Implications for the seasonal dynamics of discharge and basin-scale water use *Geophys. Res. Lett.* **32** L14401 DOI: 10.1029/2005GL022857

[18] Gordov E P, Okladnikov I G, Riazanova A A, Titov A G and Gordova Yu E 2018 A virtual thematic research environment for analysis of climate change and its consequences for Northern Eurasia: the current state of the art and perspectives *IOP Conf. Series: Earth and Environmental Science* **211** 012079 DOI: 10.1088/1755-1315/211/1/012079

[19] Gordov E P, Okladnikov I G, Titov A G, Voropay N N, Ryazanova A A and Lykosov V N 2018 Development of Information-computational Infrastructure for Modern Climatology *Russian Meteorology and Hydrology* **43**(11) 722–8 DOI: 10.3103/S106837391811002X

[20] Titov A G and Okladnikov I G 2014 Architecture of web-GIS for climate monitoring based on geospatial data services *Vestnik NGU. Series: Information technologies* **12**(1) 79-88

[21] Okladnikov I G, Gordov E P, Ryazanova A A and Titov A G 2018 Development of a "cloud" thematic software complex for analysis of climatic and environmental changes *IOP Conf. Series: Earth and Environmental Science* **211** 012081 DOI: 10.1088/1755-1315/211/1/012081

[22] Okladnikov I G, Gordov E P and Titov A G 2016 Development of climate data storage and processing model *IOP Conf. Series: Earth and Environmental Science* **48** 012033 DOI:10.1088/1755-1315/48/1/012033

[23] Ryazanova A A, Okladnikov I G and Gordov E P 2017 Integration of modern statistical tools for the analysis of climate extremes into the web-GIS "CLIMATE" *IOP Conf. Series: Earth and Environmental Science* **96** 012014 DOI: 10.1088/1755-1315/96/1/012014

[24] Gilleland E and Katz R W 2016 extRemes 2.0: An Extreme Value Analysis Package in R *Journal of Statistical Software* **72**(8) 39 DOI: 10.18637/jss.v072.i08

[25] Yan J 2007 Enjoy the Joy of Copulas: With a Package copula *Journal of Statistical Software* **21**(4) 21 DOI: 10.18637/jss.v021.i04

Bayesian approach to data assimilation based on ensembles of forecasts and observations

E G Klimova
Institute of Computational Technologies of the Siberian Branch of the Russian Academy of Sciences, Novosibirsk, Prosp. Akad. Lavrentyeva 6, Russia, 630090, klimova@ict.nsc.ru

Abstract. Optimal assessment of geophysical fields with observational data and a mathematical model is a data assimilation problem. To solve it, a Bayesian approach is most often used. In the ensemble algorithms, ensembles of forecasts and observations are used to approximate the covariance matrices considered in the algorithm. If all probability densities are Gaussian, the problem is reduced to that of the ensemble Kalman filter. In the strongly non-Gaussian case, a particle method is used, which is based on a Bayesian approach. Ensembles are also used in this method. In the research devoted to the ensemble Kalman filter much attention is paid to deviation of ensemble elements from an average value – ensemble spread. In this paper, a comparative analysis of spread behavior over time when using different approaches to the improvement of the convergence of the algorithms is performed. The results of numerical experiments with a 1-dimensional test model are discussed. These results show that in stochastic filters the behavior over time of ensemble spread is close to the theoretical error estimate. Some of the approaches that improve convergence in the ensemble filter, such as additive inflation and multiplicative inflation, change the general formula for ensemble spread.

1. Introduction

At present, optimal assessment of geophysical fields with observational data and a mathematical model is one of the most important problems. A Bayesian approach is often used to solve it. If the considered random variables are Gaussian and the forecast and observation models are linear, the problem is equivalent to a variational data assimilation problem (4DVAR) [1].

For approximation of the covariance matrices considered in the algorithm, an ensemble of forecasts and observations can be used [2]. If the optimal estimate is sought at the end of a given time interval, the problem is reduced to the ensemble Kalman filter. If an optimal estimate is considered in a given time interval, this task is called an ensemble smoothing problem [2].

Ensemble filtering and smoothing algorithms can be divided into deterministic and stochastic. In the stochastic Kalman filter, observational data with 'perturbations' are considered. This allows you to make the ensemble of deviations from the mean value the same as the ensemble of estimation errors [3]. The ensemble Kalman filter and ensemble Kalman smoothing, proposed in [2], are stochastic algorithms. Along with these algorithms, there are a large number of 'square root' filters in which the analysis is performed for the mean over the ensemble, and the ensemble of analysis errors is obtained from the ensemble of forecast errors using the transformation matrix. Such algorithms are called deterministic. Comparative analysis of deterministic and stochastic filters is carried out in many papers [3].

Review [3] lists the main approaches to improving the convergence of the ensemble Kalman filter. One of the popular ways to control the convergence of the ensemble Kalman filter is to use the so-called 'inflation factor'. In [3], some ways of improving of the convergence of the ensemble Kalman filter are noted, such as a multiplicative inflation and an additive inflation.

In the nonlinear case, iterative smoothing algorithms are currently used, which are a variational problem using ensembles to calculate the covariances and linearize the prediction and observation operators [1]. In the strictly nonlinear non-Gaussian case, the particle method is used, which is based on the Bayesian approach. Ensembles are also used in this method. These approaches also consider methods for controlling the convergence of the algorithms.

In papers on the ensemble Kalman filter, much attention is given to the behavior of deviations of an ensemble member from the mean (ensemble spread). It is shown in [4] that the spread of an ensemble distribution can in principle give a priori indication of forecast skill and also that the spread correlates with the forecast error. Therefore, it is important for the behavior of the spread in time to correspond to the behavior of the estimation error.

In an ensemble Kalman filter, numerous factors, such as the limited number of ensemble members (sample error), the inaccurately specified random observation errors, model noise, nonlinearity of the model, etc., affect the process of convergence and the relationship between spread and skill.

In the present paper, an approach to analyzing the behavior over time of ensemble spread and its relationship with the behavior of the estimation error based on a general formula for the deviation of an ensemble member from the ensemble mean value are considered. A comparison of the behavior over time of these quantities for deterministic and stochastic filters is made. The effects of major means for improving convergence in the ensemble Kalman filter (additive inflation and multiplicative inflation) on the behavior over time of ensemble spread are investigated.

2. Bayesian approach to data assimilation

Assume that the time change of the estimated quantity $\mathbf{x}^k$ is described by the model

$$\mathbf{x}^{k+1} = f_{k+1,k}(\mathbf{x}^k) + \boldsymbol{\eta}^k,$$

where k is the time step number. In addition, observational data $\mathbf{y}^k$ are known:

$$\mathbf{y}^k = h_k(\mathbf{x}^k) + \boldsymbol{\varepsilon}^k.$$

In these formulas $\boldsymbol{\eta}^k$, $\boldsymbol{\varepsilon}^k$ are random errors of forecast and observations, respectively, $f_{k+1,k}$ is the model operator, and h_k is the operator of transformation of the predicted variable into the observed one.

The Bayesian approach consists of applying the Bayes theorem to obtain an optimal estimate from observational data and a forecast:

$$p(\mathbf{x}\,|\,\mathbf{y}) = \frac{p(\mathbf{y}\,|\,\mathbf{x})\,p(\mathbf{x})}{p(\mathbf{y})}.$$

There are various options for assessing the state of data and forecast: $p(\mathbf{x}_l\,|\,\mathbf{y}_{k,1}), k > l$ - forecast, $p(\mathbf{x}_k\,|\,\mathbf{y}_{k,1})$ - filtering, $p(\mathbf{x}_{k,0}\,|\,\mathbf{y}_{k,1})$ - smoothing, where $\mathbf{x}_{k,0} = \{\mathbf{x}_k, \mathbf{x}_{k-1}, \cdots, \mathbf{x}_0\}$, $\mathbf{y}_{k,1} = \{\mathbf{y}_k, \cdots, \mathbf{y}_1\}$. The notation and definitions are taken from review [1].

In the linear Gaussian case, the solution of the filtering problem is the Kalman filter, and the solution of the smoothing problem is the Kalman smoothing [5]. In [2], it was proposed to use the Monte Carlo method for solving filtering and smoothing problems, the so-called ensemble filtering and smoothing algorithms. In the ensemble Kalman filter in the nonlinear case, the covariance matrix is predicted using a nonlinear model, and the Gaussian condition is violated. Also in this case, the estimate at the analysis step is an approximation of the minimum variance estimate [2].

To solve the smoothing problem in the nonlinear case, iterative smoothing algorithms using ensembles are currently being considered [1]. In the nonlinear non-Gaussian case, the particle method is used, which is also based on the Bayesian approach. To implement a particle filter, algorithms using

IOP Conf. Series: Earth and Environmental Science **386** (2019) 012038 doi:10.1088/1755-1315/386/1/012038

the representation of the probability distribution density as a sum with the coefficients of Gaussian probability distribution densities may be considered. In this case, the algorithm is a generalization of the ensemble Kalman filter [5].

3. The ensemble Kalman filter

Consider a nonlinear dynamic system as a process equation

$$\mathbf{x}_k^t = f(\mathbf{x}_{k-1}^t) + \boldsymbol{\eta}_{k-1}^t \tag{1}$$

and an observation equation

$$\mathbf{y}_k = h(\mathbf{x}_k^t) + \boldsymbol{\varepsilon}_k^t,$$

where $\boldsymbol{\eta}_{k-1}^t$ is a 'model noise' vector, $\boldsymbol{\varepsilon}_k^t$ is an observation error vector, $\mathbf{x}_k^t$ is a vector of the variables being estimated at time t_k, and $\boldsymbol{\varepsilon}_k^t$ and $\boldsymbol{\eta}_{k-1}^t$ are Gaussian random variables: $E(\boldsymbol{\varepsilon}_k^t) = 0,\ E(\boldsymbol{\eta}_{k-1}^t) = 0$, $E[\boldsymbol{\varepsilon}_k^t(\boldsymbol{\varepsilon}_k^t)^T] = \mathbf{R}_k^t,\ E[\boldsymbol{\eta}_{k-1}^t(\boldsymbol{\eta}_{k-1}^t)^T] = \mathbf{Q}_{k-1}^t$. Here $\mathbf{x}_k^t$ is assumed to be the 'true' value.

Consider an ensemble approach in which probability mean is replaced by sample mean. In this case, it is typically assumed that the 'true state' can be initially represented by a sample $\{\mathbf{x}_0 + \boldsymbol{\delta}\mathbf{x}_n, n = 1,\cdots,N\}$, where $\mathbf{x}_0$ is a known estimate of the sought-for quantity, and the sample $\{\boldsymbol{\delta}\mathbf{x}_n, n = 1,\cdots,N\}$ simulates random errors of this estimate. Assume that the sample mean of this error is zero. The time variation of the 'true state' is simulated by the process equation. Then the sample mean will be the optimal estimate, and the deviation from the mean will be the estimation error.

The ensemble Kalman filter consists of an ensemble of forecasts $\{\mathbf{x}_k^{f,n}, n = 1,\cdots,N\}$

$$\mathbf{x}_k^{f,n} = f(\mathbf{x}_{k-1}^{a,n}) + \boldsymbol{\eta}_{k-1}^n \tag{2}$$

and an ensemble of analyses $\{\mathbf{x}_k^{a,n}, n = 1,\cdots,N\}$

$$\mathbf{x}_k^{a,n} = \mathbf{x}_k^{f,n} + \mathbf{K}_k(\mathbf{y}_k^n + \boldsymbol{\varepsilon}_k^n - h(\mathbf{x}_k^{f,n})). \tag{3}$$

The ensembles (2) and (3) provide a sample of 'true' values. Here the sample mean will be the optimal estimate, and the deviations from the mean will be ensembles of analysis and forecast errors, respectively. To implement the ensemble version of the Kalman filter algorithm, an ensemble of observation errors $\{\boldsymbol{\varepsilon}_k^n, n = 1,\cdots,N\}$, an ensemble of forecast errors $\{\mathbf{dx}_k^{f,n} = \mathbf{x}_k^{f,n} - \overline{\mathbf{x}_k^{f,n}}, n = 1,\cdots,N\}$, where $\overline{\mathbf{x}_k^{f,n}} \approx \frac{1}{N}\sum_{n=1}^{N}\mathbf{x}_k^{f,n}$, and an ensemble of model noise $\{\boldsymbol{\eta}_{k-1}^n, n = 1,\cdots,N\}: E[\boldsymbol{\eta}_{k-1}^n(\boldsymbol{\eta}_{k-1}^n)^T] = \mathbf{Q}_{k-1}$ are specified. $\mathbf{K}_k$ is a matrix of the form

$$\mathbf{K}_k = \mathbf{P}_k^f\mathbf{H}_k^T(\mathbf{H}_k\mathbf{P}_k^f\mathbf{H}_k^T + \mathbf{R}_k)^{-1},$$

where $\mathbf{P}_k^f$ and $\mathbf{R}_k$ are matrices that are estimated over the ensemble as

$$\mathbf{P}_k^f = \frac{1}{N-1}\sum_{n=1}^{N}\mathbf{dx}_k^{f,n}(\mathbf{dx}_k^{f,n})^T,\ \mathbf{R}_k = \frac{1}{N-1}\sum_{n=1}^{N}\boldsymbol{\varepsilon}_k^n(\boldsymbol{\varepsilon}_k^n)^T,$$

and $\mathbf{H}_k$ is the linearized operator of $h(\mathbf{x}_k^{f,n})$ with respect to $\overline{\mathbf{x}_k^{f,n}}$:

$$h(\mathbf{x}_k) \approx h(\overline{\mathbf{x}_k^{f,n}}) + \mathbf{H}_k\boldsymbol{\varepsilon}_k^f\ .$$

With a given analysis error $\{\mathbf{dx}_k^{a,n} = \mathbf{x}_k^{a,n} - \overline{\mathbf{x}_k^{a,n}}, n = 1,\cdots,N\}$, where $\overline{\mathbf{x}_k^{a,n}} = \frac{1}{N}\sum_{n=1}^{N}\mathbf{x}_k^{a,n}$, the analysis error covariance matrix can be calculated by the formula

$$\mathbf{P}_k^a = \frac{1}{N-1}\sum_{n=1}^{N}\mathbf{dx}_k^{a,n}(\mathbf{dx}_k^{a,n})^T.$$

Formulas (2) and (3) represent a stochastic ensemble Kalman filter [2]. These formulas can be written in the following form:

$$\mathbf{x}_k^n = (\mathbf{I} - \mathbf{K}_k\mathbf{H}_k)\left[f(\mathbf{x}_{k-1}^n) + \boldsymbol{\eta}_{k-1}^n\right] + \mathbf{K}_k(\mathbf{y}_k^n + \boldsymbol{\varepsilon}_k^n),$$

where $\mathbf{x}_k^n$ is the ensemble of variables being estimated. As noted above, the optimal estimate in the ensemble Kalman filter is the ensemble mean value $\overline{\mathbf{x}_k^n}$. Deviation from the mean (spread) simulates the estimate error $\mathbf{dx}_k^n = \mathbf{x}_k^n - \overline{\mathbf{x}_k^n}$ with the equation

$$\mathbf{dx}_k^n = (\mathbf{I} - \mathbf{K}_k\mathbf{H}_k)(f(\mathbf{x}_{k-1}^n) - \overline{f(\mathbf{x}_{k-1}^n)} + \boldsymbol{\eta}_{k-1}^n) + \mathbf{K}_k\boldsymbol{\varepsilon}_k^n. \tag{4}$$

A 'theoretical' estimation error, $\mathbf{dx}_k^t \triangleq \mathbf{x}_k^t - \overline{\mathbf{x}_k^n}$, satisfies the equation

$$\mathbf{dx}_k^t = (\mathbf{I} - \mathbf{K}_k\mathbf{H}_k)(f(\mathbf{x}_{k-1}^t) - \overline{f(\mathbf{x}_{k-1}^n)} + \boldsymbol{\eta}_{k-1}^t) + \mathbf{K}_k\boldsymbol{\varepsilon}_k^t. \tag{5}$$

The deterministic ensemble Kalman filter (analysis step) consists of the following equation for the mean:

$$\overline{\mathbf{x}_k^{a,n}} = \overline{\mathbf{x}_k^{f,n}} + \mathbf{K}_k(\mathbf{y}_k^n - \overline{h(\mathbf{x}_k^{f,n})})$$

and an estimate of the ensemble of analysis errors such that the corresponding covariance matrix satisfies the Kalman filter equation $\mathbf{P}_k^a = (\mathbf{I} - \mathbf{K}_k\mathbf{H}_k)\mathbf{P}_k^f$.

As noted in some papers, the transformation of an ensemble of forecast errors into analysis errors in a deterministic filter can be represented in the form of left multiplication [6]:

$$\mathbf{dx}_k^{a,n} = \mathbf{A}_k\mathbf{dx}_k^{f,n},$$

where $\mathbf{A}_k$ is an operator that satisfies the equation $\mathbf{A}_k\mathbf{P}_k^f\mathbf{A}_k^T = \mathbf{P}_k^a$, and the operator $\mathbf{A}_k$ may be not unique. The ensemble of analyses of the deterministic filter has the form

$$\mathbf{x}_k^{a,n} = \overline{\mathbf{x}_k^{a,n}} + \mathbf{A}_k(\mathbf{x}_k^{f,n} - \overline{\mathbf{x}_k^{f,n}}).$$

There are many analysis ensembles corresponding to a covariance matrix $\mathbf{P}_k^a$ and a mean value $\overline{\mathbf{x}_k^{a,n}}$. Specifically, deterministic filters calculate an ensemble of fields with a corresponding mean value and covariance matrix [3]. However, the analysis ensemble specified by formula (3) has the following important property: the equation for deviation from the mean ensemble value (4) will have the same form as the equation for the theoretical estimation error (5).

It is shown in [6] that the ensemble of analyses being estimated has different distribution functions in deterministic and stochastic filters. It is evident that the same is true for the ensemble spread.

4. Analysis of ensemble spread behavior over time in the ensemble Kalman filter

The equation for ensemble spread in the stochastic Kalman filter may be written in the following form:

$$\mathbf{dx}_k^n = (\mathbf{I} - \mathbf{K}_k\mathbf{H}_k)(\mathbf{F}_k\mathbf{dx}_{k-1}^n + \boldsymbol{\eta}_{k-1}^n) + \mathbf{K}_k\boldsymbol{\varepsilon}_k^n. \tag{6}$$

Here, instead of the nonlinear model operator *f*, we take the linearized operator $\mathbf{F}_k$. This equation can be rewritten as

$$\mathbf{dx}_k^n = \boldsymbol{\Psi}(k,0)\mathbf{dx}_0^n + \sum_{i=1}^{k}\boldsymbol{\Psi}(k,i)\mathbf{K}_i\boldsymbol{\varepsilon}_i^n + \sum_{i=1}^{k}\boldsymbol{\Psi}(k,i)(\mathbf{I} - \mathbf{K}_i\mathbf{H}_i)\boldsymbol{\eta}_{i-1}^n, \tag{7}$$

where $\boldsymbol{\Psi}(k,i) = \prod_{j=i+1}^{k}(\mathbf{I} - \mathbf{K}_j\mathbf{H}_j)\mathbf{F}_j$, if $k > i$, $\boldsymbol{\Psi}(k,k) \equiv \mathbf{I}$. Note that a similar equation for the classical Kalman filter was obtained in [7, 8].

The estimation error (i.e., the deviation of the mean from the 'true' value) satisfies the following equation (which is similar to (6)):

$$\mathbf{dx}_k^t = (\mathbf{I} - \mathbf{K}_k\mathbf{H}_k)(\mathbf{F}_k\mathbf{dx}_{k-1}^t + \boldsymbol{\eta}_{k-1}^t) + \mathbf{K}_k\boldsymbol{\varepsilon}_k^t. \tag{8}$$

It follows from equations (6) for $\mathbf{dx}_k^n$ and (8) for $\mathbf{dx}_k^t$ that the simulated estimation error tends to the theoretical error if the random vectors of observational errors and model noise being simulated have the same covariance matrices as the true ones. The random vectors of observational errors and model noise being simulated are considered here in the context of the theory of Kalman filtering, that is, they are assumed to be Gaussian white noise with a given covariance matrix [7].

In deterministic filters the analysis is made for the mean. Then some perturbations $\mathbf{dx}_k^{a,n}$ corresponding to the covariance matrix of analysis errors are generated. Writing the formula for the analysis perturbations in terms of 'left multiplication', we obtain the following equation for ensemble spread in the deterministic Kalman filter:

$$\mathbf{dx}_k^n = \mathbf{A}_k(\mathbf{F}_k\mathbf{dx}_{k-1}^n + \boldsymbol{\eta}_{k-1}^n)$$

or

$$\mathbf{dx}_k^n = \boldsymbol{\Psi}_{\det}(k,0)\mathbf{dx}_0^n + \sum_{i=1}^{k}\boldsymbol{\Psi}_{\det}(k,i)\mathbf{A}_i\boldsymbol{\eta}_{i-1}^n,$$

where $\boldsymbol{\Psi}_{\det}(k,i) = \prod_{j=i+1}^{k}\mathbf{A}_j\mathbf{F}_j$. It is evident that the behavior over time of ensemble spread in the deterministic Kalman filter differs from that in the stochastic filter, as well as from the behavior over time of the estimation error. Specifically, the formula for the deterministic filter lacks the term with $\mathbf{K}_i\boldsymbol{\varepsilon}_i^n$, which simulates ensemble spread as a function of observational data distribution and observational error covariances. Note that in the linear case the covariance matrices of the stochastic and deterministic filters differ only by 'sampling error'.

5. Methods of improving convergence in the ensemble Kalman filter

Let us consider the ensemble spread modification in general form. In the first case when the modification is used in the analysis step it has the form

$$\overset{n}{\mathbf{dx}_k} = \alpha_k\mathbf{dx}_k^n + \boldsymbol{\beta}_k^n,$$

where $\boldsymbol{\beta}_k^n$ is a random vector with a specified covariance matrix, and in the second case when the modification is used in the forecast step, the form

$$\overset{n}{\mathbf{dx}_k} = (\mathbf{I} - \mathbf{K}_k\mathbf{H}_k)(\alpha_k\mathbf{F}_{k-1}\mathbf{dx}_{k-1}^n + \alpha_k\boldsymbol{\eta}_{k-1}^n + \boldsymbol{\beta}_k^n) + \mathbf{K}_k\boldsymbol{\varepsilon}_k^n.$$

Then in the first case the formula for a stochastic filter is

$$\overset{n}{\mathbf{dx}_k} = \tilde{\boldsymbol{\Psi}}(k,0)\mathbf{dx}_0^n + \sum_{i=1}^{k}\tilde{\boldsymbol{\Psi}}(k,i)\alpha_i\tilde{\mathbf{K}}_i\boldsymbol{\varepsilon}_i^n + \sum_{i=1}^{k}\tilde{\boldsymbol{\Psi}}(k,i)\alpha_i(\mathbf{I} - \tilde{\mathbf{K}}_i\mathbf{H}_i)\boldsymbol{\eta}_{i-1}^n + \sum_{i=1}^{k}\tilde{\boldsymbol{\Psi}}(k,i)\boldsymbol{\beta}_{i-1}^n,$$

and in the second case,

$$\overset{n}{\mathbf{dx}_k} = \tilde{\boldsymbol{\Psi}}(k,0)\mathbf{dx}_0^n + \sum_{i=1}^{k}\tilde{\boldsymbol{\Psi}}(k,i)\tilde{\mathbf{K}}_i\boldsymbol{\varepsilon}_i^n + \sum_{i=1}^{k}\tilde{\boldsymbol{\Psi}}(k,i)(\mathbf{I} - \tilde{\mathbf{K}}_i\mathbf{H}_i)\alpha_i\boldsymbol{\eta}_{i-1}^n + \sum_{i=1}^{k}\tilde{\boldsymbol{\Psi}}(k,i)(\mathbf{I} - \tilde{\mathbf{K}}_i\mathbf{H}_i)\boldsymbol{\beta}_{i-1}^n,$$

where $\tilde{\boldsymbol{\Psi}}(k,i) = \prod_{j=i+1}^{k}\alpha_j(\mathbf{I} - \tilde{\mathbf{K}}_j\mathbf{H}_j)\mathbf{F}_j$, $\tilde{\mathbf{K}}_j$ is calculated using the modified covariance matrices.

Consider the case of a deterministic filter. Let us denote $\tilde{\boldsymbol{\Psi}}_{\det}(k,i) = \prod_{j=i+1}^{k}\alpha_j\mathbf{A}_j\mathbf{F}_j$. We have for a deterministic filter in the first case (analysis step)

$$\overset{n}{\mathbf{dx}_k} = \tilde{\boldsymbol{\Psi}}_{\det}(k,0)\mathbf{dx}_0^n + \sum_{i=1}^{k}\tilde{\boldsymbol{\Psi}}_{\det}(k,i)\alpha_i\mathbf{A}_i\boldsymbol{\eta}_{i-1}^n + \sum_{i=1}^{k}\tilde{\boldsymbol{\Psi}}_{\det}(k,i)\boldsymbol{\beta}_i^n,$$

and in the second case (forecast step)

$$\overset{n}{\mathbf{dx}_k} = \tilde{\boldsymbol{\Psi}}_{\det}(k,0)\mathbf{dx}_0^n + \sum_{i=1}^{k}\tilde{\boldsymbol{\Psi}}_{\det}(k,i)\alpha_i\mathbf{A}_i\boldsymbol{\eta}_{i-1}^n + \sum_{i=1}^{k}\tilde{\boldsymbol{\Psi}}_{\det}(k,i)\mathbf{A}_i\boldsymbol{\beta}_i^n.$$

The equation for the error when modifying the ensemble spread takes the form

$$\mathbf{dx}_k^t = \tilde{\mathbf{\Psi}}_t(k,0)\mathbf{\Delta}_0^n + \sum_{i=1}^{k}\tilde{\mathbf{\Psi}}_t(k,i)\tilde{\mathbf{K}}_i\mathbf{\varepsilon}_i^t + \sum_{i=1}^{k}\tilde{\mathbf{\Psi}}_t(k,i)(\mathbf{I}-\tilde{\mathbf{K}}_i\mathbf{H}_i)\mathbf{\eta}_{i-1}^t,$$

where $\tilde{\mathbf{\Psi}}_t(k,i) = \prod_{j=i+1}^{k}(\mathbf{I}-\tilde{\mathbf{K}}_j\mathbf{H}_j)\mathbf{F}_j$.

One can see from the above formulas that the perturbation ensembles of deterministic and stochastic filters with the thus modified ensemble spread do not correspond to the error ensemble. For $\mathbf{\beta}_k^n = \mathbf{0}$ we obtain a version of multiplicative inflation. For $\alpha_k = 1$ we obtain a version of additive inflation. In this case, additive inflation can be specified so that the covariance matrix after correction coincides with the matrix obtained when using multiplicative inflation: $\mathbf{\beta}_k^n = \mathbf{\delta Q}_k \xi^n$, where

$$\mathbf{\delta Q}_k = (\alpha_k^2 - 1)\mathbf{P}_k . \tag{9}$$

Here $\mathbf{P}_k$ is the covariance matrix of analysis or forecast errors corresponding to the above versions of modification of ensemble spread, ξ^n is a Gaussian random vector $N(0,\mathbf{I})$, and $\mathbf{I}$ is a unit matrix. Here $N(a,\boldsymbol{B})$ is a random vector distributed normally with mean a and covariance $\mathbf{B}$. In this case we obtain a version of additive inflation in which the covariance matrix is the same as the matrix obtained when using multiplicative inflation, but the ensemble spreads are different.

It follows from the formulas presented in this section that in the case of additive inflation the behavior of stochastic filter spread over time is closer to that of the error than in the case of multiplicative inflation (the matrices $\tilde{\mathbf{\Psi}}$ and $\tilde{\mathbf{\Psi}}_t$ are closer). In the case of additive inflation the formulas after the forecast step and after the analysis step differ only in the multiplier of $\mathbf{\beta}_k^n$. However, when using additive inflation after the analysis step the 'theoretical' relation between the covariance matrices, $\mathbf{P}_k^a = (\mathbf{I}-\mathbf{K}_k\mathbf{H}_k)\mathbf{P}_k^f$ [7], is violated.

It should be noted that in the practical implementation of the particle filter, similar methods are used to improve the convergence of the algorithm over time.

6. Numerical experiments with the Lorenz-96 model

In the present paper, numerical experiments on data assimilation are performed for an ensemble Kalman filter with perturbed observations (a stochastic ensemble Kalman filter) and a version of a deterministic filter (LETKF algorithm [9]).

6.1. Ensemble π-algorithm

A version of a stochastic ensemble Kalman filter in square root form (ensemble π-algorithm) proposed in [10, 11] is used. This algorithm has a property of locality: the analysis error ensemble is calculated at each grid node independently of the others. The algorithm is efficient: the arithmetic operations are performed with matrices whose dimensions are equal to those of the ensemble.

The ensemble π-algorithm is a stochastic filter in which the analysis step is performed only for the ensemble mean. In the ensemble of analysis errors, $\mathbf{D}$ is an ($L \times N$) matrix whose columns are vectors $\{\mathbf{dx}_k^n, n=1,\ldots,N\}$, $\{k=1,\cdots,L\}$ is a grid node number obtained by transforming the ensemble of forecast errors, $\mathbf{B}$ is a matrix with columns $\{\mathbf{b}_k^n, n=1,\ldots,N\}$: $\mathbf{b}_k^n = \mathbf{x}_k^{f,n} - \overline{\mathbf{x}_k^{f,n}}$. $\mathbf{D}^{\mathrm{T}} = (\mathbf{I}+\mathbf{\Pi}^{\mathrm{T}})^{-1}\mathbf{B}^{\mathrm{T}}$, and

$$\mathbf{\Pi}^{\mathrm{T}} = (\mathbf{C}+0.25\mathbf{I})^{\frac{1}{2}} - 0.5\mathbf{I},$$

$$\mathbf{C} = \frac{1}{N-1}\mathbf{F}^{\mathrm{T}}\mathbf{H}^{\mathrm{T}}\mathbf{R}^{-1}(\mathbf{HB}+\mathbf{E}) = \mathbf{C}_1 + \mathbf{C}_2.$$

$\mathbf{E}$ is a matrix whose columns are equal to a vector $\mathbf{\varepsilon}_k^n$, the ensemble of observation errors, and $\mathbf{I}$ is a unit matrix. All details are presented in [10] and [11].

The elements of the matrix $\mathbf{\Pi}$ are calculated for the matrices $\mathbf{H}$ and $\mathbf{R}$ from the ensemble of values $\{\mathbf{dx}_k^n, n=1,\ldots,N\}$ independently of the grid nodes. Therefore, the algorithm can be used locally, and the operations can be performed with matrices the size of the ensemble. Notice that the above properties are valid for the highly popular LETKF algorithm [9], which is a deterministic filter.

6.2. Lorenz-96 model

The Lorenz-96 model [12] is used for the numerical experiments. The equations of the model are

$$\frac{dx_j}{dt} = (x_{j+1} - x_{j-2})x_{j-1} - x_j + F_0,\ j=1,\cdots,J$$

$$x_{-1} = x_{J-1},\ x_{J+1} = x_1,$$

where $x_1,\cdots,x_J$ (J=40) are the variables being forecasted. This is one of the simplest systems that simulate the properties of many atmospheric models. To solve the equations, a fourth-order finite-difference Runge-Kutta scheme is used. The time step $\Delta t = 0.05$, which corresponds to 6 hours (t=1 is taken for five days), F_0=8. To simulate 'true values' in the numerical data assimilation experiments, $\mathbf{x}_0^t \cong N(F_0/4;F_0/2)$ is taken to be an initial value, as in [12], and N_t time steps are made. In the numerical experiments, the results of a forecast for N_t=1000 time steps are taken as initial data for the 'truth'. To obtain initial data for forecasting $\hat{\mathbf{x}}_0$ by using the model, a perturbation is added to the 'true' initial data $\mathbf{x}_d(0) = \mathbf{x}_t(0) + \boldsymbol{\delta}$, $\boldsymbol{\delta} \cong N(0,s_0)$.

6.3. Numerical experiments

The following parameters are specified for the numerical experiments: an ensemble of initial fields $\mathbf{x}^n(0) = \mathbf{x}_d(0) + \boldsymbol{\delta}^n, \boldsymbol{\delta}^n \cong N(0,s_0), n=1,\cdots,N$; observations $\mathbf{y}_0 = \mathbf{x}_t(0) + \boldsymbol{\delta}_0, \boldsymbol{\delta}_0 \cong N(0,\varepsilon_0)$; and an ensemble of observations with perturbations $\mathbf{y}_0^n = \mathbf{y}_0 + \boldsymbol{\delta}_0^n, \boldsymbol{\delta}_0^n \cong N(0,\varepsilon_0), n=1,\cdots,N$. In the experiments it is assumed that the model noise $\boldsymbol{\eta}^n = 0$, and in simulating the 'truth' $\boldsymbol{\eta}^t = 0.01$. The observations are available at each of the J=40 model grid points. The experimental period has length N_t=3000 time steps, with assimilation being done at each time step or at every four time steps. The following values were used in the experiments: $s_0 = \varepsilon_0 = 1$, N =20.

In all of the numerical experiments $\mathbf{R} = \varepsilon_0^2\mathbf{I}$. The observational data from the interval $(l-id, l+id)$ are taken for the analysis at a grid node l. In this case, the matrix $\tilde{\mathbf{R}} = \mathbf{R} \circ e^{0.5(\rho_{il}/bc)^2}$ is used instead of $\mathbf{R}$, where ρ_{il} is the distance between the grid node and the observation, and "$\circ$" denotes element-by-element multiplication. In the experiments we took $id = 5, bc = 5\Delta x$ (Δx is the grid spacing).

The numerical experiments were performed for ten versions of the 'truth', and all estimates were calculated as average values over these ten versions. The following estimates were considered:

$$rms = \frac{1}{K}\sum_{m=1}^{K}\left\{\frac{1}{L}\sum_{i=1}^{L}\overline{(x_{m,i}} - x_{m,i}^t)^2\right\}^{1/2},$$

the root-mean-square error averaged over K=10 versions of calculations (m is the number of a version and i is the number of a grid node), with random values of the initial data errors, observational data errors, and the ensemble of perturbations specified at the initial time in each version. If the 'truth' is considered to be a random quantity, the rms corresponds to the trace of the error covariance matrix (the covariance is estimated from a sample of ten values).

$$sp = \frac{1}{K}\sum_{m=1}^{K}\left\{\frac{1}{L(N-1)}\sum_{i=1}^{L}\sum_{n=1}^{N}(\overline{x_{m,i}^n} - x_{m,i}^n)^2\right\}^{1/2}.$$

IOP Conf. Series: Earth and Environmental Science **386** (2019) 012038 doi:10.1088/1755-1315/386/1/012038

This is the mean value of the covariance matrix trace calculated over K=10 versions of calculations (n is the number of an ensemble member).

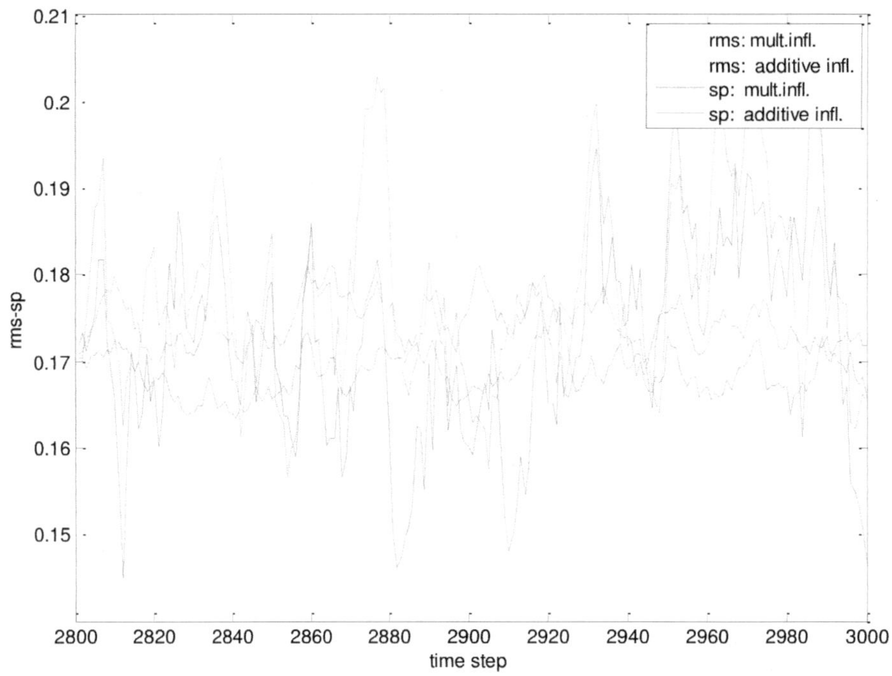

Figure 1. Root-mean-square (*rms*) and covariance matrix trace (*sp*) calculated in experiments 1 – 2 in the first series of experiments for the ensemble π-algorithm.

The numerical experiments with the ensemble π-algorithm described above and the LETKF algorithm were performed under the same conditions. The ensemble spread was modified after the forecast step. Two series of experiments were performed. In the first series observational data were simulated in one time step interval, and in the second series, in four time step intervals. The following experiments were performed in each of the series:

Experiment 1. In this experiment multiplicative inflation α was used (α =1.1 in the first series and α =1.2 in the second series).

Experiment 2. In this experiment additive inflation was used so that the change in the covariance matrix coincided with the change made in experiment 1. In this case formula (9) from Section 4 was used.

Figures 1-3 show the time behavior of *rms* and *sp* obtained in the ensemble π-algorithm (Figures 1 and 3) and the LETKF algorithm (Figure 2).

Figures 1-2 present the results of the first series of experiments for the ensemble π-algorithm (Figure 1) and the LETKF algorithm (Figure 2).

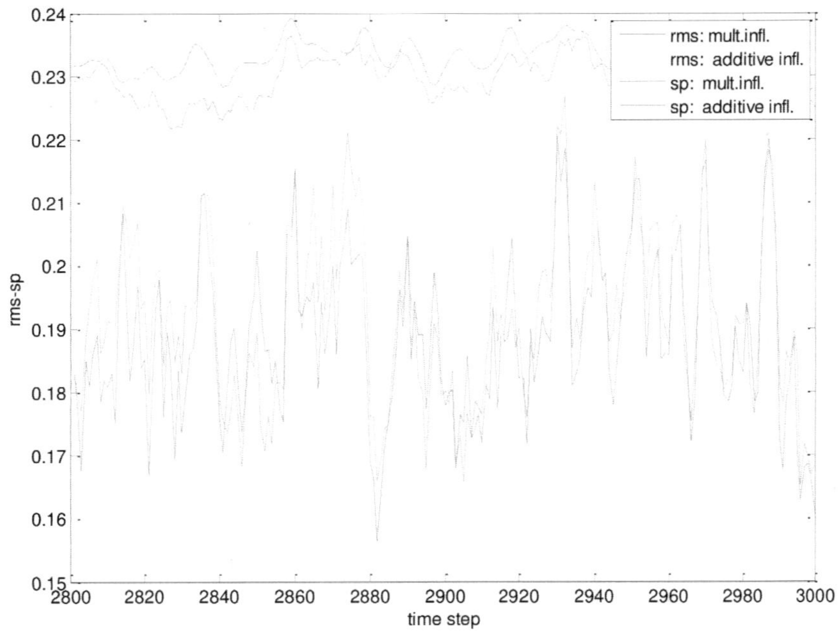

Figure 2. Root-mean-square error (*rms*) and covariance matrix trace (*sp*) calculated in experiments 1 - 2 in the first series of experiments for the LETKF algorithm.

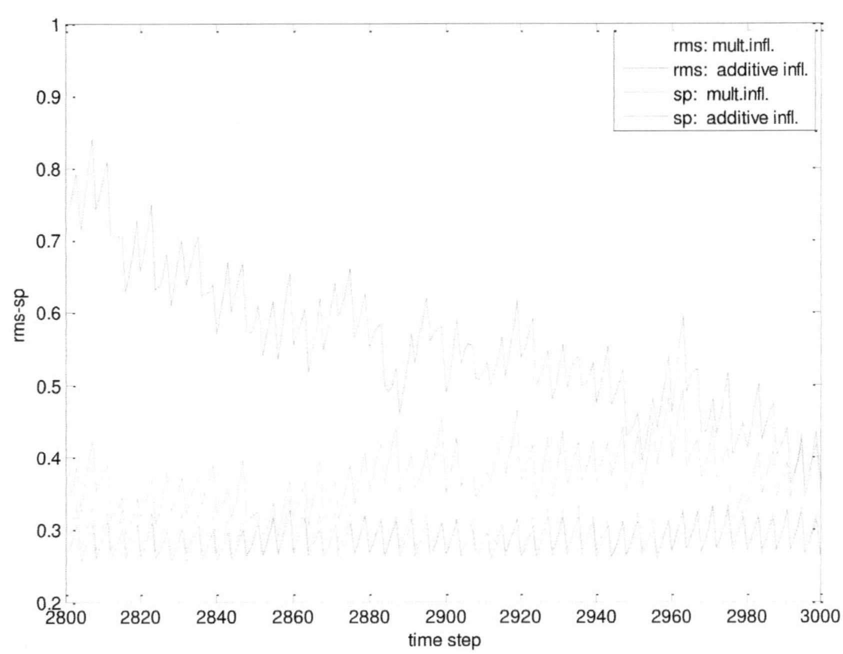

Figure 3. Root-mean-square error (*rms*) and covariance matrix trace (*sp*) calculated in experiments 1 – 2 in the second series of experiments for the ensemble π-algorithm.

Figure 3 presents the results of the second series of experiments for the ensemble π-algorithm.

The goal of the experiments was to make a qualitative comparison of the *rms* and *sp* time behavior depending on the method being used (deterministic or stochastic ensemble filters) and specifications of the inflation parameter and assimilation frequency. One can see from Figures 1-2 that in the ensemble π-algorithm the *rms* is closer to the *sp* than in the LETKF algorithm. In the second series of experiments (Figure 3) the additive inflation in the stochastic filter (ensemble π-algorithm) improves *rms* as compared to multiplicative inflation.

7. Conclusions

A Bayesian approach to data assimilation based on ensembles was discussed. A brief overview of the existing algorithms was presented.

The major factors affecting the behavior of ensemble spread in the ensemble Kalman filter were considered. A general formula for ensemble spread in stochastic and deterministic ensemble Kalman filters has been derived. The impact of some methods that improve convergence on the behavior of the ensemble Kalman filter over time was investigated. The results of this study show that ensemble spread in stochastic filters rather than in deterministic filters is closer to the theoretical estimation error. Multiplicative inflation and additive inflation change the general formula for ensemble spread. The formula for ensemble spread in the stochastic filter with additive inflation rather than with multiplicative inflation is closer to the estimation error.

References

[1] Carrassi A, Bocquet M, Bertino L, Evensen G Data assimilation in the geosciences: An overview of methods, issuers and perspectives 2018 *Wiley interdisciplinary reviews: Climate Change* **131** 5 e535

[2] Evensen G 2009 *Data assimilation. The ensemble Kalman filter* (Berlin Heideberg: Spriger-Verlag)

[3] Houtekamer H L and Zhang F 2016 Review of the ensemble Kalman filter for atmospheric data assimilation *Mon. Wea. Rev.* **144** 4489-532

[4] Murphy J M 1988 The impact of ensemble forecasts on predictability *Quarterly Journal of the Royal Meteorological Society* **114** 463-93

[5] Nakamura G, Potthast R 2015 *Inverse Modeling* (IOP Publishing)

[6] Lei J, Bickel P, Snyder C 2010 Comparison of ensemble Kalman filters under non-Gaussianity. *Monthly Weather Review* **138** 1293-306

[7] Jazwinski A H 1970 *Stochastic Processes and Filtering Theory* (New York: Academic Press)

[8] Cohn S E, Dee D P 1988 Observability of discretized partial differential equations *SIAM J. Numer. Anal.* **25** 3 586-617

[9] Hunt B R, Kostelich E J, Szunyogh I 2007 Efficient data assimilation for spatiotemporal chaos: A local ensemble transform Kalman filter *Physica D* **230** 112-26

[10] Klimova E 2012 A suboptimal data assimilation algorithm based on the ensemble Kalman filter *Quarterly Journal of the Royal Meteorological Society* **138** 2079-85

[11] Klimova E G 2019 The Kalman stochastic ensemble filter with transformation of perturbation ensemble *Siberian. J. Num. Math.* **22** 1 27-40

12] Lorenz E N, Emanuel K A 1998 Optimal sites for supplementary weather observations: simulation with a small model *Journal of the Atmospheric Sciences* **55** 399-414

Creation of a long-term high-resolution hydrometeorological archive for the Russian Arctic: methodology and first results

Vladimir Platonov and Mikhail Varentsov

Lomonosov Moscow State University, Faculty of Geography, Department of Meteorology and Climatology. 119991, Moscow, Leninskie Gory MGU, 1, GSP-1, Phone 8(495)939-29-42
Lomonosov Moscow State University, Research Computing Center, Laboratory of Supercomputer Modelling of Climate System Processes. Moscow, Leninskie Gory MGU, building 4, 119234, Phone 8(495)939-23-53

vplatonov86@gmail.com, mvar91@gmail.com

Abstract. Taking into account that dangerous phenomena on the Arctic coast are increasing in number, providing the region with detailed hydrometeorological and climatic information with a horizontal resolution of at least several kilometers becomes particularly important. In this work, we obtain for the first time a detailed archive of many hydrometeorological parameters with a spatial resolution of less than 5 km based on long-term simulation experiments. Detailed hydrometeorological fields in the Arctic basin over a long period (1980 – 2016) are derived by a two-step downscaling technology with domains of horizontal resolutions of~13 km and ~4 km that cover most of the Russian Artic, by using a regional non-hydrostatic model, COSMO-CLM. First results of verification with hundreds of Russian Arctic stations allow us to select the best configuration of the model and domains, including many turbulent scheme options and starting time of the experiments. The model results for the coast have shown good agreement with observations (with mean errors of 1 - 2 C). The larger temperature biases over the Eastern Siberia inland have been partially reduced by using selected turbulence scheme options (with mean errors of 5 - 6 C to 2 – 3 C). The differences between the ERA-Interim and ERA5 driving conditions are not great. Therefore, the former are chosen as basic reanalysis, taking into account the data volume and limitations on computational resources. A possible future regional reanalysis output would be useful in many applications, e.g. modelling ocean's characteristics, coastal ecosystems, detailed investigation of individual extreme phenomena in nested domains, analysis of trends in the frequency of occurrence of extreme events and features of their spatial distribution, study of the hydrometeorological regime of coastal areas, the climatology and tracking of polar mesocyclones, etc.

1. Introduction

The Arctic is the region that is most sensitive to climate change on the globe; it concerns both physical feedbacks in the climate system and ecological systems [1]. In particular, abrupt sea ice area depletion in the Arctic Ocean starts an albedo positive feedback mechanism, increasing the temperature growth rate [2]. However, the spatial distribution of the climate changes in the region is ambiguous. There are various estimates of the trends of certain meteorological parameters in different water areas and coastal zones of the Russian and foreign Arctic regions [3, 4].

Taking into account the increasing number of dangerous phenomena and outlooks for the Arctic coast and the Northern Sea Route development, the task of providing the region with detailed hydrometeorological and climatic information with a horizontal resolution of at least several kilometers becomes particularly important. As a tool for solving this task, it seems efficient to use regional climate modeling. Long-term detailing of many hydrometeorological fields with a spatial resolution of less than 5 km allows one to obtain more thorough and justified estimates of current regional and mesoscale Arctic climate changes and extreme weather events.

It is necessary to take into account the mesoscale processes to specify and detail the hydrometeorological information in contrast to the current scales resolved in global climate models and reanalyses, including non-hydrostatic effects. This lack could be eliminated by applying mesoscale climate models using global modeling data as input forcing.

2. Data and methods

2.1. Model description
The COSMO-CLM model (ver. 5.0/5.05) is used as the main tool for the creation of this long-term meteorological archive. The COSMO-CLM is a climate version of the well-known mesoscale COSMO model, including some modifications and extensions adapted to the long-term numerical experiments. It was developed by the German Weather Service (DWD) and the CLM-Community (see CLM Community site http://www.clm-community.eu/, [5, 6]). The COSMO-CLM model is based on the primitive Navier-Stokes equations describing the dynamics of compressible fluid in the moist atmosphere. The model equations are solved on the rotational grid 'latitude-longitude' (λ, φ) with a pole tilt. It helps to minimize the problem of convergence of the meridians over the pole. The numerical scheme is realized on an Arakawa C-grid [7], and the vertical coordinate is a hybrid Gal-Chen coordinate [8, 9]. The height-based hybrid Gal-Chen coordinate [8] is given as an analog of the σ-coordinate from the surface to Z_0 level and as the Z-coordinate above the Z_0 level. This representation allows avoiding problems associated with the surface heterogeneity.

The standard configuration of the COSMO-CLM model was applied with the Runge-Kutta integration scheme with the 5th advection order. The Ritter and Geleyn radiation scheme [10] is based on the δ two-stream version of the radiation transfer equation. The precipitation formation is described by a bulk microphysics parameterization, and Tiedtke mass-flux schemes with equilibrium closure based on moisture convergence are used for moist and shallow convection [11]. Turbulence is described by a prognostic TKE-based scheme, with 2.5 order closure [12]; Smagorinsky diffusion is included. There is an option of applying the spectral nudging technique [13]. A full description of the COSMO Model physics, dynamics, and parameterizations is available on (http://www.cosmo-model.org/content/model/documentation/core/default.htm).

2.2. Description of the experiments and downscaling technology
The general scheme proposed is long-term and, therefore, the test experiments are as follows. Reanalyses from ECMWF (ERA-Interim [14], ERA5 [15]), which are widely known and used in international research, will be used as driving conditions for COSMO-CLM regional model runs over 1980 – 2016. Information about the external parameters (e.g., land-sea mask, roughness length, grid sea-ice cover, sea surface temperature, etc.) is obtained from the program utility EXTPAR v.5.1.1 (https://tools.clm-community.eu/web_pep/gui/web_pep.php). Further, it is interpolated on the model grid of the chosen domain as 'ready-to-use'. The 'dynamical downscaling' ('nesting domains') technique will be used to obtain detailed meteorological fields for the last 35 years with an hourly time increment of data. The first model run will be performed over the base raw-resolution domain (grid step: ~13 km) covering most of the area of the Russian Arctic with global reanalysis as forcing. After

that the output data of these experiments will be used as initial and driving conditions for interpolation and model runs over fine-resolution small domains (less than 5 km) for three key regions of the Russian Arctic: the Barents, Kara, and Laptev Seas. A scheme of the model domains is shown in Figure 1. About a hundred of different hydrometeorological parameters would be presented in the output files, including the surface quantities, three-dimensional quantities within the atmosphere and soil. Test and future long-term model runs were made on the MSU Supercomputer Complex "Lomonosov-2" [16]. An example of successful previous implementations by the authors is given in [17].

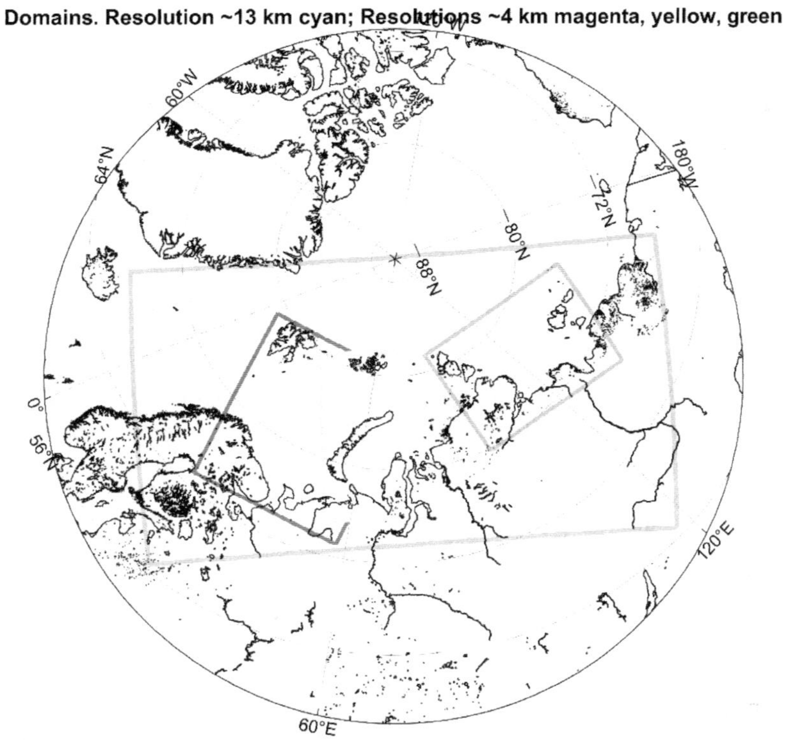

Figure 1. Scheme of the model domains. The cyan rectangle denotes the basic domain with a grid step of ~13 km; and magenta, yellow, and green ones denote nesting modeling domains with grid steps of ~3.3 km for the Barents, Kara, and Laptev Seas, respectively.

The main goal of the test numerical experiments was to choose the best configuration using many parameterizations, options and taking into account the specificity of the Arctic climate system. An optimal configuration is defined based on the verification procedure, i.e. a comparison of model runs with real observations. The test experiments were performed for the base domain with a ~13-km grid step for August-September 2015 and December-January 2012-13. It is important that the experiments were made for different seasons to examine how the model could reproduce the contrasting atmospheric circulation conditions and surface-atmosphere interactions. For instance, the sea-ice conditions are changing significantly over this region during the summer and winter periods and, therefore, the circulation, heat exchange regimes, and stratification are changing as well on the synoptic scale, as on the mesoscale. It was important to estimate the parameters of model experiments which are capable of reproducing the above differences in the contrast conditions.

The main options changed in the experiments are listed below, and their properties and differences are briefly explained. The full list of experiments, their properties and acronyms are given in Table 2 and Appendix.

IOP Conf. Series: Earth and Environmental Science **386** (2019) 012039 doi:10.1088/1755-1315/386/1/012039

1) Different initial and driving conditions: ERA-Interim reanalysis and new reanalysis ERA5.
2) Switching on/off the 'spectral nudging' technique. It means assimilation of the large-scale atmospheric circulation components from reanalysis data not only on the boundaries, but also inside the model domain. It ensures that the mesoscale weather and climate features are simulated in realistic atmospheric circulation conditions.
3) Changing of constants affects the vertical turbulence diffusion parameterization: 4 times reduction of the minimal value of the turbulence drag coefficient (from 0.4 to 0.1) and 5 times reduction of the subgrid temperature heterogeneities scale (from 500 m to 100 m) influencing the TKE generation. These parameters affect the simulation results in the boundary layer stable stratification conditions [18].
4) Two different model versions are used: the 5.0 version as a current stable and the official COSMO-CLM version utilized by the CLM-Community; and the last COSMO 5.05 version. Parameterizations of many physical processes are affected in the latter version, unified with the atmospheric model ICON [19, 20].
5) Different start time of experiments – with or without 'cold start'. 'Cold start' was defined as one month.

3. Results and discussion

Data from 466 meteorological stations over Russia, Norway, Finland, and Sweden covered by the base domain and located north of 60 N were used for the verification of the tests experiments. The data were accessed from the www.rp5.ru site for 2005 – 2017. Each station archive was verified for less than 25 % lacks for 2 m temperature, 10 m wind speed, and sea level pressure separately. Model grids for comparison were defined for each station based on the least RMSE among the 4 nearest grid points. This technique takes into account that mesoscale models tend to reproduce many atmospheric features in the vicinity of the point, but not at the point exactly. The main statistical metrics of verification for each variable were the mean error (ME), the root mean square error (RMSE), and the correlation coefficient. These estimates were made for all stations and for coastal and inland stations separately. The coastal stations were defined by the distance of 20 km or less from the nearest 'sea' grid point. The main results of the verification are listed in Tables 1a-b and Figures 2 – 4. its detailed description is given below.

1) The model reproduces synoptic-scale dynamics almost perfectly within the base domain, which is confirmed by sea level pressure correlation coefficients above 0.98 – 0.99.
2) 'Spectral nudging' ('sn' experiments) significantly reduces the model biases for the temperature and for the wind speed, and increases the correlation between the observations and modeling data compared with experiments without 'spectral nudging'. This result is first obtained for the pan-Arctic spatial scale, in contrast to previous case studies over the Arctic region [21].
3) Correction of the vertical turbulence diffusion parameters ('_turb_sn' experiments) decreases systematic temperature model biases for inland areas, especially for Eastern Siberia and Scandinavia (Figures 2 and 3). The standard configuration of the model significantly overestimates the temperature (up to 5 – 10 C) during calm, clear winter weather conditions. It is a well-known problem of reproduction of stable boundary layers in mesoscale models. However, the sensitivity of the model to the parameters of the turbulence scheme is small for the coastal stations, both for the summer period and for wind speed estimates.
4) Using the new model version ('_v505' experiments) has almost the same effect as the turbulence scheme correction in the old model version. It is associated with the implementation of new physical parameterizations from the ICON model, especially affecting the reproduction of intense mixing in stable stratification conditions. Furthermore, the new model version decreases biases not only for the winter season, but also for the summer season.
5) No significant differences were revealed in biases between the driving conditions (ERA-Interim reanalysis - '_interim' experiments, and ERA5 reanalysis - '_era5' experiments). This allows

one to utilize ERA-Interim reanalysis for the initialization of long-term experiments, because it requires less memory resources and preparation time with almost the same verification results.

6) Experiments with 'cold start' ('_long') have shown some small and multidirected differences. For the wintertime 'cold start' increases model biases for the temperature, and for the summertime the situation is opposite. It is probably associated with differences in the behavior of the inertial components of the model system, the active soil layer in the summertime, and the snow cover in wintertime. Therefore, there are not a certain choice for an optimal model configuration.

Table 1a. RMSE and correlation, for January 2013, all experiments.

Experiment name	T_2M (all)		T_2M (coast)		T_2M (inland)		VEL_10M (all)		VEL_10M (coast)		VEL_10M (inland)		PMSL (all)	
interim	4.22	0.76	3.12	0.78	4.80	0.75	2.30	0.55	3.12	0.59	1.83	0.53	2.98	0.96
_era5	4.19	0.76	3.07	0.79	4.77	0.74	2.30	0.57	3.11	0.62	1.84	0.55	2.77	0.97
interim_sn	3.69	0.83	2.61	0.85	4.25	0.82	2.12	0.65	2.78	0.70	1.74	0.62	2.01	0.99
era5_sn	3.70	0.83	2.62	0.86	4.27	0.81	2.10	0.66	2.76	0.71	1.73	0.63	2.13	0.99
interim_turb	4.10	0.76	3.29	0.78	4.53	0.76	2.30	0.55	3.12	0.59	1.84	0.53	3.24	0.96
era5_turb	4.16	0.76	3.32	0.78	4.60	0.75	2.32	0.57	3.17	0.61	1.84	0.54	3.25	0.96
interim_turb_sn	3.38	0.84	2.65	0.85	3.77	0.83	2.12	0.65	2.79	0.69	1.74	0.63	2.08	0.99
interim_turb_sn_long	3.57	0.83	2.74	0.85	4.00	0.83	2.11	0.65	2.79	0.70	1.72	0.62	2.14	0.99
era5_turb_sn	3.37	0.85	2.67	0.86	3.74	0.84	2.09	0.66	2.77	0.71	1.71	0.64	2.18	0.99
era5_sn_v505	3.34	0.85	2.37	0.88	3.85	0.84	2.22	0.67	2.90	0.71	1.84	0.64	1.69	0.99
interim_v505_long	3.94	0.78	2.99	0.79	4.44	0.78	2.47	0.54	3.30	0.57	2.00	0.53	2.83	0.96
interim_sn_v505_long	3.33	0.85	2.45	0.87	3.80	0.84	2.24	0.65	2.92	0.70	1.86	0.63	1.63	0.99
ERA-Interim	3.10	0.88	2.52	0.89	3.41	0.88	2.31	0.66	2.65	0.72	2.11	0.64	1.10	0.99

Table 1b. RMSE and correlation, for September 2015, all experiments.

Experiment name	T_2M (all)		T_2M (coast)		T_2M (inland)		VEL_10M (all)		VEL_10M (coast)		VEL_10M (inland)		PMSL (all)	
COSMO_interim	2.38	0.77	1.83	0.75	2.69	0.78	2.02	0.65	2.63	0.67	1.65	0.64	1.87	0.99
COSMO_interim_long	2.41	0.76	1.85	0.75	2.72	0.77	2.06	0.64	2.68	0.66	1.68	0.63	1.96	0.99
COSMO_era5	2.34	0.79	1.78	0.78	2.65	0.80	2.00	0.67	2.59	0.69	1.65	0.65	1.70	0.99
COSMO_interim_sn	2.29	0.79	1.72	0.78	2.60	0.79	1.89	0.70	2.41	0.73	1.58	0.68	1.53	1.00

COSMO_era5_sn	2.29	0.81	1.71	0.80	2.61	0.81	1.87	0.71	2.37	0.74	1.58	0.68	1.42	1.00
COSMO_interim_turb	2.43	0.77	1.87	0.76	2.74	0.78	2.03	0.64	2.67	0.66	1.65	0.64	1.91	0.99
COSMO_era5_turb	2.41	0.80	1.82	0.79	2.73	0.80	1.98	0.67	2.55	0.70	1.64	0.66	1.65	0.99
COSMO_interim_turb_sn	2.35	0.79	1.78	0.79	2.67	0.80	1.89	0.69	2.42	0.73	1.58	0.67	1.57	1.00
COSMO_interim_turb_sn_long	2.25	0.80	1.71	0.80	2.54	0.80	1.90	0.70	2.43	0.73	1.58	0.68	1.51	1.00
COSMO_era5_turb_sn	2.35	0.81	1.76	0.81	2.67	0.81	1.88	0.70	2.38	0.74	1.58	0.68	1.45	1.00
COSMO_era5_sn_v505	2.16	0.82	1.60	0.81	2.47	0.83	1.97	0.70	2.47	0.73	1.67	0.68	1.34	1.00
COSMO_interim_v505_long	2.25	0.78	1.75	0.76	2.53	0.80	2.15	0.64	2.78	0.65	1.77	0.63	1.83	0.98
COSMO_interim_sn_v505_long	2.10	0.81	1.58	0.80	2.39	0.82	1.97	0.70	2.48	0.72	1.67	0.68	1.40	1.00
ERA-Interim	1.83	0.87	1.60	0.84	1.96	0.89	1.95	0.72	2.22	0.76	1.78	0.69	0.72	1.00

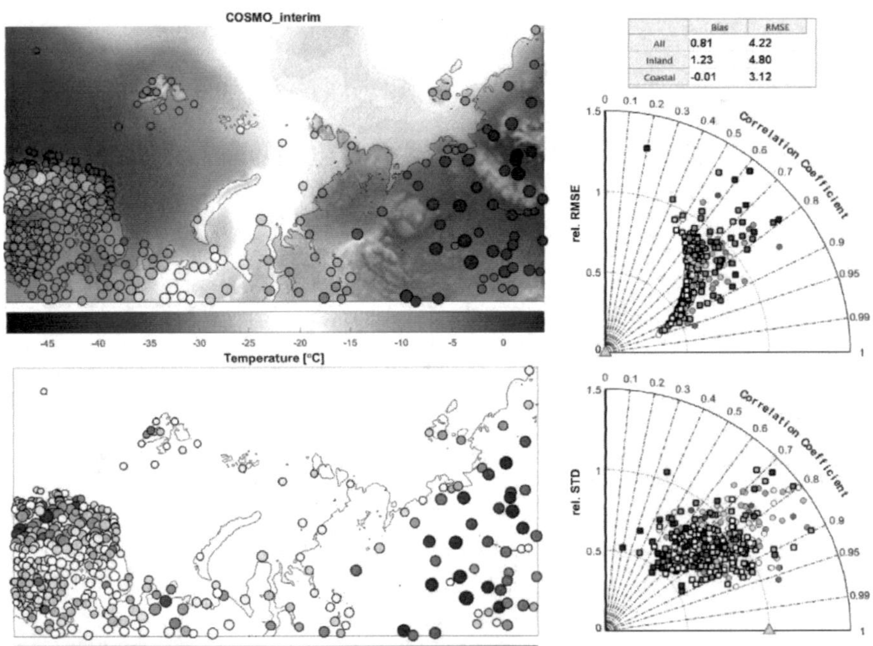

Figure 2. Verification plots for 2 m temperature, January 2013, 'COSMO_interim' experiment. Map of modelled (color background) and observed (round markers) mean monthly temperature (top left). The marker size is proportional to the RMSE value for the given station. Map of mean errors (bottom left). Modified Taylor diagrams show the correlation coefficient and RMSE values normalized on the observed standard deviation for each station (top right); correlation coefficients and ratio between the modelled and observed standard deviations for each station (bottom right). Green triangles show an ideal accordance case. The table (top right) shows mean statistics for the given experiment over the domain: the bias and RMSE for all, inland, and coastal stations.

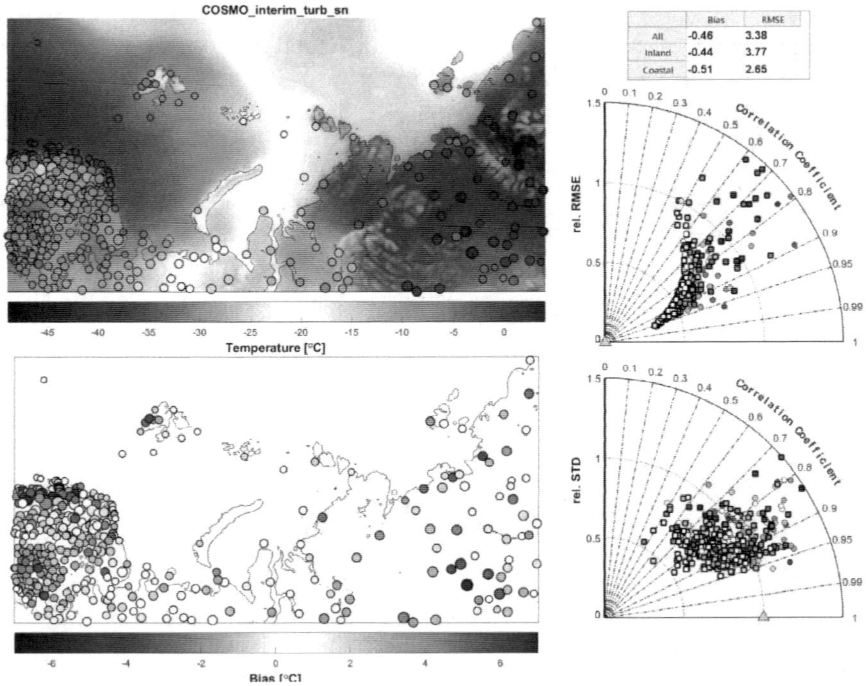

Figure 3. The same as Figure 2, but for 2-m temperature, January 2013, 'COSMO_interim_turb_sn' experiment.

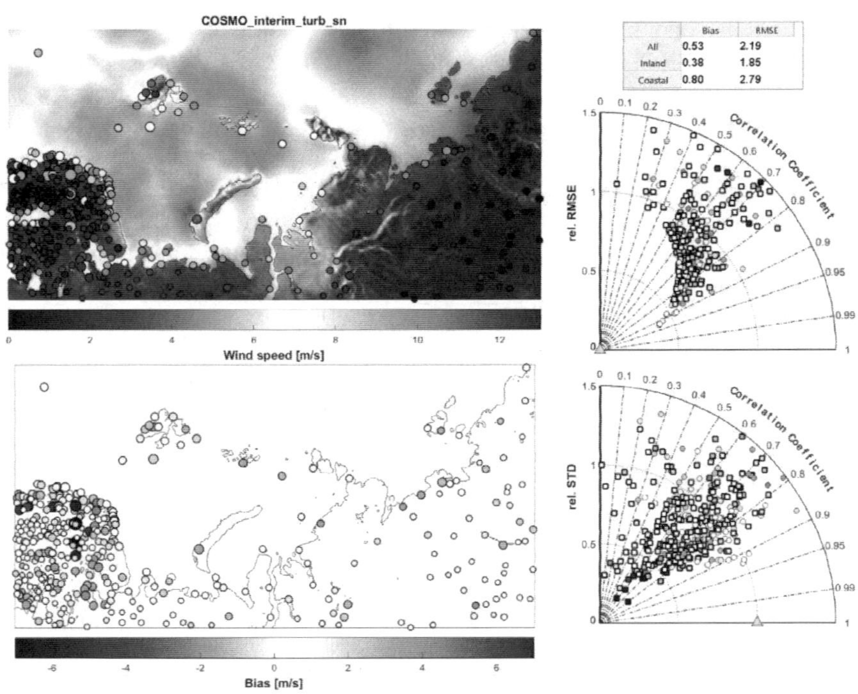

Figure 4. The same as Figure 2, but for 10-m wind speed, January 2013, 'COSMO_interim_turb_sn' experiment.

Generally, the first verification over hundreds of stations of the Russian Arctic has shown the best configuration of the model and domains, including different model versions, spectral nudging, turbulent scheme options, and starting times of the experiments. Many test experiments for summer and winter months were conducted. The default version without spectral nudging has shown the worst results: temperature biases are about 1C, RMSE 3.5 – 4 C. There are specific large errors over the inland Eastern Siberia stations, because the model has strongly underestimated winter boundary layer freezing. Switch on the spectral nudging reduced the RMSE slightly, while many corrections in the turbulent scheme options decreased the RMSE significantly (up to 2.5 C for the coastal stations and 3 C for inlands), and slightly negative biases (approx. -0.5 C) were obtained. The model version 5.05 has also shown a reduction of biases. The same estimates are for the wind speed but, in general, the biases are smaller.

According to the results of the comparison, it could be concluded that an optimal model configuration has to use the 'spectral nudging' technique and correction of the turbulence scheme parameters for the base model version or rely on the new model version. It is also reasonable to use ERA-Interim reanalysis as driving conditions for the main long-term experiments. It is worth noting that this investigation of the sensitivity of the model parameters for such a large Arctic region was carried out at the first time; particularly, a comparison of ERA-Interim and ERA5 driving conditions. A good agreement of these results with many previous case studies confirms relevance and significance of the identified patterns.

4. Conclusions and perspectives

The detailed hydrometeorological fields in the Arctic obtained over a long period (1980 – 2016) can give new, more thorough and justified estimates of current regional climate changes, as well as extreme weather events. This can provide broad opportunities for detailed climate assessments of the Russian Arctic region, research analysis of a large number of climate parameters during the recent decades. In particular, the data can be used in studies on modern environmental changes, e.g., for the construction of various buildings, infrastructure facilities on the coast and the sea shelf, long-term planning of maritime operations, safe navigation along the Northern Sea Route, and many others.

The scientific significance of this regional reanalysis is a possibility of using this information as input data in modeling ocean's characteristics (wind waves and dynamics), coastal ecosystems (turbulent heat and moisture fluxes, greenhouse gases) for a more detailed research of individual phenomena in nested domains (extreme situations, hazardous weather events, etc.), analysis of trends in the frequency of occurrence of extreme events and features of their spatial distributions, hydrometeorological regime in coastal areas, climatology and tracking of polar mesocyclones, etc.

The scientific and applied importance of the above methodology is also in the fact that it can be used for downscaling of the global CMIP5 (or CMIP6) project climate projections and for detailed assessments of climate and environmental changes on a regional scale in the Arctic in the 21st century.

Acknowledgements

This research was carried out using the equipment of the shared research facilities of HPC computing resources at Lomonosov Moscow State University.

This research was supported by the Russian Foundation for Basic Research (RFBR grant no. 18-35-00604) and state project no. AAAA-A16-116032810086-4.

Appendix

Table 2. List of all test experiments with its properties and acronyms.

Experiment's acronym	model version	reanalysis	'cold start'	'spectral nudging'	turbulence scheme correction
COSMO_interim	5.0	ERA-Interim	No	No	Standard

COSMO_interim_long	5.0	ERA-Interim	Yes	No	Standard
COSMO_era5	5.0	ERA5	No	Yes	Standard
COSMO_interim_sn	5.0	ERA-Interim	No	Yes	Standard
COSMO_era5_sn	5.0	ERA5	No	Yes	Standard
COSMO_interim_turb	5.0	ERA-Interim	No	No	Corrected
COSMO_era5_turb	5.0	ERA5	No	Yes	Corrected
COSMO_interim_turb_sn	5.0	ERA-Interim	No	Yes	Corrected
COSMO_interim_turb_sn_long	5.0	ERA-Interim	No	Yes	Corrected
COSMO_era5_turb_sn	5.05	ERA5	No	Yes	Corrected
COSMO_era5_sn_v505	5.05	ERA5	No	Yes	Standard
COSMO_interim_v505_long	5.05	ERA-Interim	Yes	No	Standard
COSMO_interim_sn_v505_long	5.05	ERA-Interim	No	No	Standard

References

[1] IPCC, 2013: *Climate Change 2013: The Physical Science Basis. Contribution of Working Group I to the Fifth Assessment Report of the Intergovernmental Panel on Climate Change*, ed T F Stocker, D Qin, G-K Plattner, M Tignor, S K Allen, J Boschung, A Nauels, Y Xia, V Bex and P M Midgley (Cambridge, United Kingdom and New York, NY, USA: Cambridge University Press) p 1535

[2] Johannessen O M, Kuzmina S, Bobylev L P and Miles M W 2016 Surface air temperature variability and trends in the Arctic: New amplification assessment and regionalization *Tellus* **68**A 28234

[3] Eliseev A V, Semenov V A 2016 Arctic climate changes in XXI century: ensemble model estimates taking into account a realism of reproduction of modern climate *Doklady Akademii Nauk* **471** 2 214 – 18 (In Russian)

[4] Cohen J, Screen J A, Furtado J C, Barlow M, Whittleston D and co-authors Recent Arctic amplification and extreme mid-latitude weather. 2014 *Nature Geosci.* **7** 627 – 37

[5] Rockel B, Will A, Hense A 2008 The regional climate model COSMO-CLM (CCLM) *Meteorologische Zeitschrift* **17** 4 347 – 8

[6] Böhm U, Kücken M, Ahrens W, Block A, Hauffe D, Keuler K, Rockel B, Will A 2006 CLM – The Climate Version of LM: Brief Description and Long-Term Applications *COSMO Newsletters* **6** 225 – 35

[7] Arakawa A, Lamb V R 1977 Computational design of the basic dynamical processes of the UCLA general circulation model *Methods in computational physics* **17** 173 – 265

[8] Gal-Chen T, Somerville R C J 1975 On the use of a coordinate transformation for the solution of the Navier-Stokes equations *Journal of Computational Physics* **17** 2 209 – 28

[9] Schär C, Leuenberger D, Fuhrer O, Lüthi D, Girard C 2002 A new terrain-following vertical coordinate formulation for atmospheric prediction models *Monthly Weather Review* **130** 10 2459 – 80

[10] Ritter B, Geleyn J F 1992 A comprehensive radiation scheme for numerical weather prediction models with potential applications in climate simulations *Monthly Weather Review* **120** 2 303 – 25

[11] Tiedtke M 1989 A comprehensive mass flux scheme for cumulus parameterization in large-scale models *Monthly Weather Review* **117** 8 1779 – 800

[12] Herzog H-J, Vogel G, Schubert U 2002 LLM – a nonhydrostatic model applied to high-resolving simulations of turbulent fluxes over heterogeneous terrain *Theor. Appl. Clim.* **73** 67 – 86

[13] Von Storch H, Langenberg H, Feser F 2000 A spectral nudging technique for dynamical downscaling purposes *Monthly weather review* **128** 10 3664 – 73

[14] Dee D P *et al* 2011 The ERA-Interim reanalysis: Configuration and performance of the data assimilation system *Quarterly Journal of the Royal Meteorological Society* **137** 656 553 – 97

[15] Copernicus Climate Change Service (C3S) ERA5: Fifth generation of ECMWF atmospheric reanalyses of the global climate 2017 Copernicus Climate Change Service Climate Data Store (CDS) *11.08.2019*

[16] Voevodin Vl, Antonov A, Nikitenko D, Shvets P, Sobolev S, Sidorov I, Stefanov K, Voevodin Vad, Zhumatiy S 2019 Supercomputer Lomonosov-2: Large Scale, Deep Monitoring and Fine Analytics for the User Community *Supercomputing Frontiers and Innovations* **6** 2 4 – 11

[17] Platonov V S, Varentsov M I 2018 Supercomputer technologies as a tool for high-resolution atmospheric modelling towards the climatological timescales *Supercomputing Frontiers and Innovations* **5** 3 107 – 10

[18] Cerenzia I, Tampieri F, Tesini M S 2014 Diagnosis of Turbulence Schema in Stable Atmospheric Conditions and Sensitivity Tests *COSMO Newsl.* **14** 28 – 36

[19] Zängl G, Reinert D, Rípodas P, Baldauf M 2015 The ICON (ICOsahedral Non-hydrostatic) modelling framework of DWD and MPI-M: Description of the non-hydrostatic dynamical core *Q. J. R. Meteorol. Soc.* **141** 687 563 – 79

[20] Giorgetta M A *et al* 2018 ICON-A, the Atmosphere Component of the ICON Earth System Model: I. Model Description *J. Adv. Model. Earth Syst.* **10** 7 1613 – 37

[21] Varentsov M I, Verezemskaya P S, Zabolotskikh E V, Repina I A 2016 Quality estimation of polar lows reproduction based on reanalysis data and regional climate modeling *Sovremennye problemy distantsionnogo zondirovaniya Zemli iz kosmosa* **13** 4 168 – 91 (In Russian)

Application of a neural network model to forecasting of El Niño and La Niña

A S Lubkov, E N Voskresenskaya, and O V Marchukova

Institute of Natural and Technical Systems, Lenin str., 28, Sevastopol, 299011, Russia

E-mail: andrey-ls2015@yandex.ru

Abstract. In this paper, a possibility to forecast El Niño and La Niña by using an artificial intelligence model based on neural networks is studied. The quality of such a long-term climate forecast is assessed too. A set of global climatic indices of atmosphere-ocean system oscillations in 1950-2019 is used as input parameters of the model. The Nino3.4 index is calculated by using monthly average 500mb geopotential height and sea surface temperature fields from NCEP/NCAR reanalysis data sets. A verification of the model is carried out by using a control sample of 1950–1957. Additionally, the same indices of 1872-1947 are calculated by using 20th Century Reanalysis (20CR) data sets to test the model. A possibility to predict the Nino3.4 index for 2 to 7 months is shown. However, in spite of a high-level reconstruction of the index dynamics, increasing the time of forecasting is accompanied by decreasing its quality. With 20CR data it is shown that the model is able to successfully predict the beginning of 75% of El Niño and La Niña for 3 months, 66% of the events for 5 months, and only 52% for 7 months in advance.

1. Introduction

The El Niño-Southern Oscillation (ENSO) phenomenon is the strongest global interannual climatic signal of the ocean-atmosphere system of the Pacific equatorial zone. It is characterized by extreme quasi-periodical 2-7 years change of the sea surface temperatures (SSTs) and sea level pressures. The ENSO includes two episodes opposite in sign: warm (El Niño) and cold (La Niña).

In 20th century the scientists identified several typical features of the ENSO phases. It was considered that El Niño is characterized by the eastward transfer of warm equatorial waters, the alignment of the thermocline near the South American coast, weakening of the Walker cell intensity, and a shift of the atmospheric action centers at the equator [1, 2]. La Niña is characterized by the westward transfer of warm equatorial pool, the thermocline levelling up near the South American coast, and intensification of the Walker cell [3]. The Nino3.4 index is usually applied for definitions of El Niño and La Niña events. It is calculated as an SST anomaly in the Nino3.4 region located in the equatorial Pacific within the coordinates (5°S – 5°N, 170°W –120°W).

Manifestations of the ENSO are noted in weather and climate anomalies not only over the Tropical Pacific, but in extra-tropical regions too, through the "atmospheric bridge" [3, 4, 5]. The consequences of these events often lead to serious casualties and economic losses [6, 7]. However, despite the improvement of the modeling approaches and the development of technologies for El Niño and La Niña monitoring, the forecasting of these events remains imperfect [8]. The fact is that, despite many studies [2, 3, 4, 6, 7] and available examples of successful El Niño modeling [9], the mechanism of El Niño and La Niña formation is still one of the most important scientific problems.

In the present paper, we try to use a method based on artificial neural networks (NNs) to forecast the beginning of extreme ENSO phases. NN is a subsection of artificial intelligence in which signals are processed using phenomena similar to those in neurons of living organisms. The first ideas to create the neural networks were proposed at the end of the 19th century by Warren McCulloch and Walter Pitts, who proposed the first neuron binary model [10]. As a result of development of the neural networks theory, in 1975 Paul John Werbos proposed a new method of machine learning – backpropagation [11]. The backpropagation algorithm is currently considered as one of the most efficient machine learning algorithms for multilayer NNs.

Previously we attempted to use the NN to forecast. Methods based on NNs were used to predict the Southern Oscillation Index [12]. However, at that time the atmospheric oscillation indices of the subequatorial and equatorial Pacific were not used between input parameters in the NN model, and the set of indices of the Southern Hemispheric baric systems was not complete. In this paper, the quality assessment of the Nino3.4 index forecasting using a NN model is made.

2. Data and methods

Monthly SST data sets from the Met Office Hadley Centre (HadISST) were used to calculate the *Nino 3.4* index. It was calculated as the difference between the observed values and the 30-year moving average. Atmospheric patterns were selected based on the results of the space-time correlation analysis between the Nino 3.4 index and baric atmospheric fields. The monthly indices of selected atmospheric patterns were calculated and used in the proposed model. Space-time correlation analysis and calculation of the atmospheric indices were carried out using monthly 500mb geopotential heights and sea level pressures of NCEP/NCAR reanalysis data sets for the available period of observations (1948-2019). The study also includes a retrospective forecast. In this case the atmospheric indices were calculated using 20th Century Reanalysis V2c (20CR) data in 1870-1947. Thus, the following ocean-atmosphere system indices were used for modeling: the North Atlantic Oscillation (NAO), the Arctic Oscillation (AO), the East Atlantic Oscillation (EA), the Scandinavian Oscillation (SCAND), the Polar Oscillation (POL), the West Pacific Oscillation (WP), oscillations between the East Atlantic and West Eurasia (EA/WR), between the eastern and northern parts of the Pacific (EP-NP), between the Pacific and North America (PNA), oscillations in the middle latitudes of the Northern Hemisphere (MNH) and Southern Hemisphere (MSH), the South Atlantic Oscillation (SAO), the South Pacific Oscillation (SPO), oscillations between the South Atlantic maximum and the South Ocean minimum located south of the coast of Africa (SA/S), the South Indian Ocean Zonal Oscillation (SIZO), the South Indian Ocean Meridional Oscillation (SIMO), the Equatorial Oscillation between the Pacific Ocean and Indonesia (EP/EI), Equatorial Oscillation between the Atlantic and Pacific Oceans (EA/EP), the sum of the geopotential heights of middle latitudes over the Pacific Ocean (MNP/MSP) and the average pressure in equatorial latitudes (EqO).

Unidirectional hetero-associative neural network (one hidden layer) with a teacher was used for Nino 3.4 forecasting. The neurons of input, hidden, and output layers are represented by a sigmoidal bipolar function: $f(x) = \tanh(\beta x)$ [10,13]. The set of the above-mentioned atmospheric indices was used as input parameters. The model was trained using the backpropagation algorithm.

From paper [14] it is known that the quality of the NCEP/NCAR reanalysis data is better than that of the 20CR reanalysis data, which can be explained by differences in the algorithm of assimilating meteorological data by reanalyses. Therefore, in this paper we will use the NCEP/NCAR reanalysis data for modeling. The series of simulated parameter values and input atmospheric indices were divided into training, test, and control samples. In an earlier paper of the authors [15] it was shown that 30-year periods must be used for the testing and training to improve the quality of model training. In the present study, the sample for 1958-1987 was used for training and the sample for 1988-2019, for testing. The control sample did not participate in the modeling process. The eight-year period of 1950-1957 was considered as a control sample. It included two La Niña events (1949-1951 and 1954-

1956) and two El Nino events (1951-1952 and 1957-1958). Additionally, to estimate the forecasting in advance, a retrospective forecast was carried out on the basis of an efficient model for 1872-1947.

The process of adaptation of the models included three stages: pre-processing, modeling, and final processing of the results [15].

The *first stage* can be divided into 2 parts. First the series of each of the input indices was converted into four series: the original index, the same index with 30-year variability removed, and the same two series using a three-month average sliding filter. Averaging windows are the empirical choice of the authors. Thus, groups of series were formed, and each of them corresponded to one atmospheric index. After that the correlation between the atmospheric indices and the predicted parameter was calculated. During the filtering process, the selected indices were ranked by the correlation coefficient. Repetitions were excluded for the group of series of each index. As a result, 22-24 indexes were fed into the model input.

Modeling. The practical modeling showed that the behavior of the NN is not always predictable. The model not always had satisfactory results when run on the basis of indices selected at the preliminary stage, probably because it used indices not related directly to the modeled parameter or some input indices were not orthogonal. Therefore, the model was run many times with different combinations of input parameters selected at the preliminary stage. For each NN combination at the time of maximum training the correlation was calculated between the actual and estimated values in the test sample. The maximum training was determined on the basis of a comprehensive assessment of the Euclidean distance and correlation coefficients of the training and test samples [15]. The NN combinations, the NN parameters at the time of maximum training, and the correlation estimates were recorded in a log file.

At the *final stage*, on the basis of the obtained correlation coefficients of the model data with the test sample, selection of the 20 best NN combinations was performed. Since the test sample is involved in the ranking, it is probable that the selected NN combinations are mistakenly trained or retrained. Therefore, additional independent assessment of the trained model results was carried out using the control sample. If one or more NN combinations were significantly different from the rest, these combinations were excluded. Further, an ensemble was calculated from the remaining NN combinations, and its compliance with the control sample was checked.

For this, the correlation coefficient was calculated:

$$r = \frac{cov(x_i\, y_i)}{\sigma_x \sigma_y}, \tag{1}$$

where σ_x and σ_y are the standard deviations (SD or σ) of the samples x and y, which represent the simulation result and the observed Nino 3.4 index; the root mean square error (RMSE or E):

$$RMSE = \sqrt{\frac{\sum (x_i - y_i)^2}{n}}, \tag{2}$$

and the ratio of the RMSE to the standard deviations.

3. Results and discussion

The ability to predict the Nino 3.4 index was studied for each month taking into account different forecasting in advance from 2 to 9 months.

First we will evaluate the correlation coefficients between the model calculations for the forecast for 2, 3, 4, 5, 7, and 9 months and the observed Nino 3.4 index for each month within the test sample (1988-2018). The differences of intra-annual values are demonstrated in Figure 1. It shows that, on average, the correlation coefficient is smaller in June, July, August, and September. This may be due to the fact that the El Niño and La Niña events begin mainly from May to July. The correlation coefficients decrease with an increase in the advance forecasting for the months from July to January, and they are approximately equal for February, March, April, and June. The lowest correlations are typical for the 9-month advance forecast and the highest correlations, for the 2-month advance forecast.

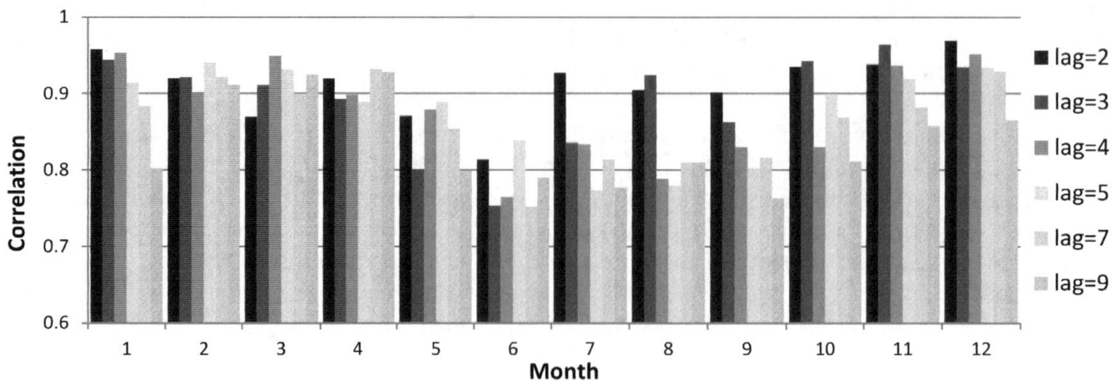

Figure 1. Estimation of correlation coefficients for different forecasting in advance: from 2 to 9 months.

Next, we evaluate the quality of modeling in the control sample. Table 1 shows the statistical characteristics (r, E, E/σ) calculated by formulas (1) and (2), and the SD of the NN combinations ensemble (σ_m). It can be noted that the correlation of the control sample (all months for 1950-1957), as well as in the case of the test sample, decreases with increasing time of forecasting in advance and the RMSE and SD of the NN combinations ensemble increase. All these indicators lead to the conclusion that the quality of modeling worsens with increasing time of forecasting in advance.

Table. 1. Assessment of the model quality for each month and the entire period. Adv is the forecasting in advance of 2 to 9 months, r is the Pearson correlation coefficient, E or RMSE is the root mean square error, E/σ is the the ratio of RMSE to standard deviations of the Nino3.4 index, and σ_m is the ensemble standard deviation.

	Adv	2	3	4	5	7	9		Adv	2	3	4	5	7	9
Jan	r	0.95	0.84	0.89	0.86	0.73	0.78	Jul	r	0.97	0.88	0.96	0.92	0.90	0.90
	E	0.45	0.49	0.55	0.40	0.53	0.64		E	0.27	0.39	0.30	0.28	0.40	0.33
	E/σ	0.65	0.70	0.79	0.57	0.77	0.92		E/σ	0.35	0.52	0.40	0.37	0.52	0.44
	σ_m	0.35	0.37	0.40	0.46	0.47	0.61		σ_m	0.17	0.27	0.43	0.36	0.31	0.37
Feb	r	0.91	0.83	0.83	0.80	0.70	0.75	Aug	r	0.95	0.98	0.86	0.80	0.93	0.93
	E	0.42	0.43	0.34	0.44	0.50	0.41		E	0.22	0.17	0.35	0.42	0.24	0.24
	E/σ	0.66	0.68	0.53	0.68	0.79	0.65		E/σ	0.32	0.25	0.50	0.60	0.34	0.34
	σ_m	0.42	0.29	0.39	0.31	0.27	0.35		σ_m	0.21	0.25	0.40	0.33	0.39	0.39
Mar	r	0.89	0.87	0.77	0.81	0.73	0.84	Sep	r	0.95	0.93	0.96	0.90	0.88	0.87
	E	0.32	0.34	0.38	0.31	0.38	0.32		E	0.28	0.32	0.26	0.48	0.45	0.52
	E/σ	0.60	0.63	0.73	0.59	0.72	0.59		E/σ	0.35	0.40	0.32	0.61	0.57	0.65
	σ_m	0.20	0.23	0.18	0.28	0.26	0.27		σ_m	0.33	0.37	0.29	0.35	0.43	0.37
Apr	r	0.96	0.82	0.81	0.85	0.67	0.87	Oct	r	0.95	0.91	0.93	0.92	0.94	0.84
	E	0.19	0.35	0.36	0.33	0.45	0.32		E	0.30	0.36	0.35	0.35	0.39	0.61
	E/σ	0.30	0.55	0.57	0.52	0.72	0.51		E/σ	0.38	0.46	0.45	0.44	0.50	0.77
	σ_m	0.14	0.23	0.22	0.23	0.28	0.27		σ_m	0.39	0.34	0.36	0.33	0.41	0.48
May	r	0.95	0.75	0.82	0.77	0.60	0.75	Nov	r	0.97	0.83	0.81	0.83	0.95	0.93
	E	0.25	0.45	0.34	0.37	0.47	0.40		E	0.32	0.63	0.63	0.67	0.48	0.58
	E/σ	0.41	0.74	0.56	0.60	0.77	0.66		E/σ	0.31	0.62	0.62	0.66	0.47	0.56
	σ_m	0.14	0.27	0.34	0.31	0.31	0.32		σ_m	0.38	0.24	0.37	0.31	0.46	0.46
Jun	r	0.90	0.82	0.78	0.82	0.87	0.91	Dec	r	0.92	0.93	0.92	0.91	0.98	0.97
	E	0.30	0.33	0.38	0.35	0.29	0.28		E	0.43	0.39	0.51	0.51	0.36	0.52
	E/σ	0.50	0.54	0.63	0.59	0.49	0.47		E/σ	0.51	0.46	0.60	0.59	0.42	0.60
	σ_m	0.31	0.22	0.25	0.34	0.27	0.41		σ_m	0.33	0.39	0.47	0.45	0.46	0.48
								All time	r	0.92	0.85	0.84	0.84	0.84	0.79
									E	0.32	0.40	0.41	0.42	0.42	0.51
									E/σ	0.41	0.53	0.55	0.55	0.55	0.63
									σ_m	0.28	0.29	0.34	0.34	0.36	0.39

Now we compare the modeling results for the 3- and 9-month advance forecasts with a real Nino3.4 index during the control sample period. Figure 2 shows the time course of the observed and calculated Nino3.4 index smoothed by a 3-month sliding filter. In the time of two-year La Niña (1949-1950) it can be seen that the presence of a negative SST anomaly is during the first mature phase of this phenomenon in the model calculations, despite a slight weakening of the index. As for the second minimum of this La Nina event in 1950, the model of forecasting in advance of 3 months shows it well, but the model of forecasting in advance of 9 months has rather high index values, in connection with which one can even make an erroneous conclusion on premature completion of the La Niña. In this case the erroneous conclusion about early completion of the La Niña may be made. Following the considered La Niña, the 1951 El Nino event occurred. Both models predict it well. The dynamics of this modelled El Nino evolution is somewhat overestimated as compared to the real Nino3.4 index. The graph shows that extreme ENSO events were absent from 1952 to 1954, but the model of forecasting in advance of 9 months erroneously indicated the presence of a weak El Niño in 1953. The beginning of the two-year La Nina (1954-1956) was determined by both models with a delay of 4 months, but the dynamics were modeled well. The start of the 1957 El Nino was predicted by the models a month later, and the dynamics were generally overestimated.

Next, we evaluated the results of a retrospective forecast of the Nino3.4 index. For this forecast, the 20CR reanalysis was used. Since the algorithms of assimilating the data in 20CR and NCEP/NCAR reanalyses differ, there are some differences in the geopotential and baric fields [14]. Therefore, when analyzing the results of retrospective modeling, emphasis was placed on qualitative assessment. First some statistical characteristics were estimated. Fig. 3 shows that the best similarity with the real index is observed in the 3-month advance forecast (Fig. 3a). The correlation of this model is 0.75, and the RMSE is 0.56 °C (the presence of fluctuations in the model index during the El Niño and La Niña can be explained by a high RMSE). Models with the ability of forecasting 5 and 7 months in advance noticeably do worse modeling of the Nino3.4 index. The correlation of these models with the Nino3.4 index is 0.61 and 0.46, and the RMSE is 0.67 °C and 0.77 °C, respectively. Figure 4 shows the time course of the Nino 3.4 index and modeling 3, 5, and 7 months in advance. All graphs are smoothed out by a 3-month moving average filter. It can be seen that the model with the forecast of the 3-month period describes the Nino3.4 index better.

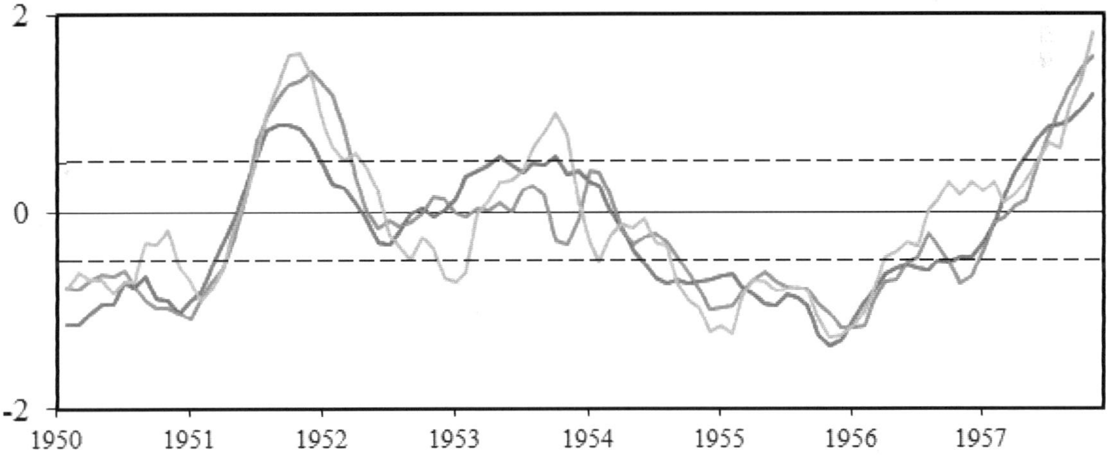

Figure 2. Three-month mean average Niño 3.4 (red line), forecasting Niño 3.4 3 month in advance (green line) and 9 month in advance (blue line) for 1950-1957. The threshold for El Niño and La Niña selections is shown by a dotted line.

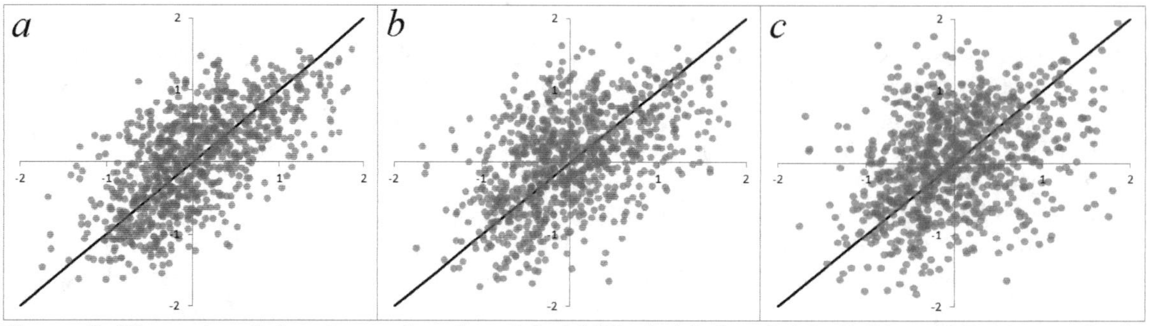

Figure 3. The ratio of the observed and modeled Niño 3.4 index 3 (c), 5 (b), and 7 (c) months in advance for 1872-1947.

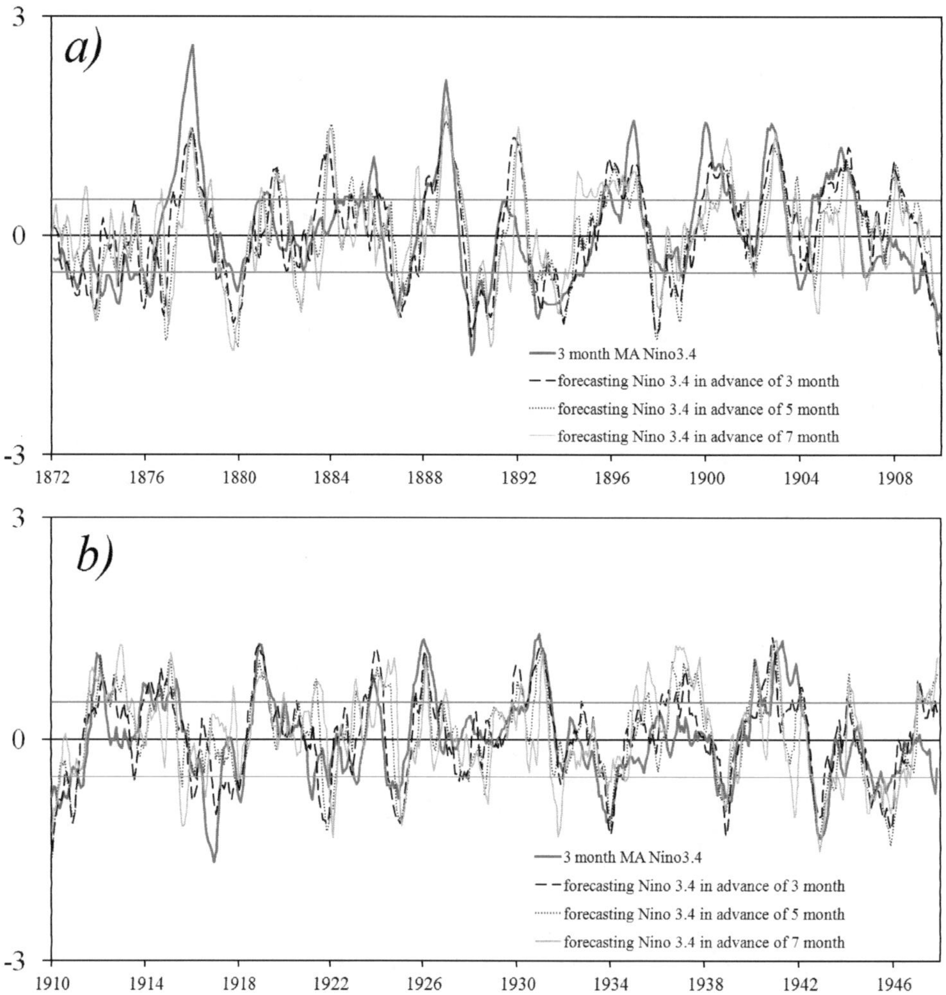

Figure 4. Three-month mean average Niño 3.4 (red line), forecasting Niño 3.4 3 months in advance (dotted line), 5 month in advance (pointed line), and 7 month in advance (blue line) for 1872-1909 (a) and 1910-1947. The threshold for El Niño and La Niña selections is shown by a green line.

IOP Conf. Series: Earth and Environmental Science **386** (2019) 012040 doi:10.1088/1755-1315/386/1/012040

Table. 2. Assessment of the possibility of predicting the start and maximum phase of El Niño (EN) and La Niña (LN) events in advance. Adv is forecasting 3, 5, 7, and 9 months in advance, r is the Pearson correlation coefficient. If the modeled index is delayed, the value is negative.

ENSO	Years	max(min) nino3.4 / duration	Adv=3 r=.75		Adv=5 r=.61		Adv=7 r=.46		Adv=9 r=.40	
			start	max. phase	start	max. phase	start	max. phase	start	max. phase
LN	1872	-0.97/8	-3	0	—	—	—	—	—	—
LN	1873	-1.06/15	—	—	—	—	—	—	—	—
LN	1875	-0.89/14	—	—	—	—	—	—	—	—
EN	1877	2.71/17	0	-1	-5	-1	—	-1	-8	0
EN	1885	1.03/5	—	—	—	—	—	—	—	—
LN	1886	-1.16/13	-3	0	—	-2	-7	0	-7	-2
EN	1888	2.37/16	0	0	-5	+2	—	+1	-7	-1
LN	1889	-2.12/16	-3	-1	-5	-1	-6	-2	—	—
LN	1892	-1.30/22	+3	-1	+1	-3	—	—	—	—
EN	1895	0.75/5	-2	-2	-3	-5	0	-1	—	—
EN	1896	1.64/9	-1	-1	+1	+4	+1	-2	—	—
EN	1899	1.59/15	-3	-2	—	—	—	—	—	—
EN	1902	1.61/12	0	-1	-2	-1	-7	-1	-6	-1
LN	1903	-0.94/5	—	—	—	—	—	—	—	—
EN	1904	1.46/17	0	-4	—	-2	—	-4	—	—
LN	1909	-1.32/17	0	-1	-2	-1	-3	-1	-5	-1
EN	1911	1.51/6	-3	-2	-2	-1	+1	-2	—	—
LN	1916	-1.69/9	—	—	—	—	—	—	—	—
EN	1918	1.36/10	0	+2	0	+2	0	+2	-2	0
LN	1924	-0.98/9	-1	+1	-5	+1	—	—	—	—
EN	1925	1.50/13	—	-1	-5	-3	—	—	—	—
EN	1931	1.75/12	0	-1	-3	-2	-6	-2	-6	-2
LN	1933	-1.22/10	-2	-1	-5	0	-5	0	—	—
LN	1938	-1.00/10	—	+3	-1	+2	-1	+3	—	—
EN	1940	1.15/8	-3	-6	0	-1	0	0	0	0
EN	1941	1.39/14	0	+5	+3	+4	+1	+4	—	—
LN	1942	-1.44/9	0	0	0	+1	-1	+1	-1	0

Let us analyze the possibility of predicting upcoming of El Niño and La Niña events in advance. Table 2 shows all recorded extreme ENSO phases from 1872 to 1947. As a criterion, it was assumed that the Nino3.4 index should be greater/less than 0.5/ -0.5 °C and that the duration of the anomaly should exceed 5 months. If the beginning of the phenomenon was determined by the model with a delay greater than the time of forecasting in advance, then it was believed that the model does not reproduce its beginning. Thus, the model with a 3-month advance forecast predicts 20 from 27 extreme events. It is slightly more than 74%. Note that 7 event starts could not be modeled (5 events from 7 are La Nina). The model with 5-month advance forecast predicts 18 events from 27. This is slightly more than 66%. If the forecasting in advance is 7 months, 14 from 27 extreme events were predicted, while in the case of forecast of 9 months only 9 from 27 events were identified in advance.

It is obvious that the quality of the retrospective forecast is much worse in comparison with the forecast with the control sample. There may be several reasons for this: the use of other data in the retrospective forecast, low availability of input data during assimilation by the 20CR reanalysis for 1872-1947, and the "memory" of the model, which was trained and tested in one climate but verified in another climate. It follows that before each forecast the neural network model must be retrained based on new data.

4. Conclusions

With a neural network model, a possibility of a Nino 3.4 index long-term 2-9 month advanced forecast was studied in this paper. The quality of the model was assessed by using NCEP/NCAR reanalysis and 20CR data.

On a test sample, the quality of the modeling worsens with increasing time of forecasting. The correlation coefficients between the observed and modelled indices for 2-3 months in advance are 0.95–0.85, while in the case of a 7-9 month advance forecast they are reduced to 0.91–0.76.

On a control sample, the quality of the NN model does not change fundamentally when the series is evaluated as a whole. The model calculations for a 3 month forecast showed a correlation coefficient of 0.92, while for a 9 month one it is 0.79 with the observed ENSO index. The RMSE and ensemble standard deviation increased with increasing time of forecasting. A qualitative comparison of the El Niño and La Niña forecasting on the control sample showed no significant disagreement with the observed indices when modeling for 3-9 months in advance. At the same time, in forecasting for 9 month in advance, erroneous ENSO events are sometimes modelled.

A retrospective forecast, carried out for an additional verification by using reanalysis data of 20CR in 1872-1947, showed a possibility of successful forecasting of up to 75% of the events of El Niño and La Nina for 3 months, up to 66% in forecasting for 5 months, and only up to 52% in forecasting for 7 months.

References

[1] Larkin N K and Harrison D E 2005 *Geophys Res Lett.* **32** L13705
[2] Rasmusson E M and Carpenter T H 1982 *Monthly Weather Review* **110** 354–84
[3] Philander S G 1990 *El Nino, La Nina and the Southern Oscillation* (CA: Academic Press, San Diego) 304
[4] McPhaden M J, Zebiak S E, Glantz M H 2006 *Science* **314** 1740–45
[5] Alexander M A, Bladé I, Newman M, Lanzante J R, Lau N-C, Scott J D 2002 *Journal of Climate* **15(16)** 2205–31
[6] Ashok K, Behera S K, Rao S A, Weng H and Yamagata T 2007 *J. Geophys. Res.* **112** C11007
[7] Yuan Y, Yang H, Li C Y 2012 *Acta Meteorol Sin* **70** 467–78
[8] Glantz M H 2015 *Int. J. Disaster Risk Sci.* **6** 94–103
[9] Hendon H H, Lim E, Wang G, Alves O, Hudson D 2009 *Geophysical research letters* **36** L19713
[10] Osovskii S 2002 *Neural Networks for Data Processing* (Finansy i statistika, Moscow) 344
[11] Werbos P 1990 *Proceedings of the IEEE* **78 (10)** 1550–60
[12] Lubkov A S, Voskresenskaya E N, Marchukova O V 2017 *Vestnik of Saint Petersburg University Earth sciences* **4(62)** 370–88
[13] Haykin S 1994 *Neural networks, a comprehensive foundation* (N.Y.: Macmillan College Publishing Company) 823
[14] Xu J and Powell A M J 2012 *Theor Appl Climatol* **108** 505–18
[15] Lubkov A S, Voskresenskaya E N, Kukushkin A S 2016 *Atmos Ocean Opt* **29** 457–64

IOP Conf. Series: Earth and Environmental Science **386** (2019) 012041 doi:10.1088/1755-1315/386/1/012041

Numerical analysis of an inverse coefficient problem for a chemical transformation model

A V Penenko[1,2], **Zh S Mukatova**[1,2], **A B Salimova**[2]

[1] Institute of Computational Mathematics and Mathematical Geophysics SB RAS, ICM&MG SB RAS, prospekt Akademika Lavrentjeva 6, Novosibirsk, 630090, Russian Federation
[2] Novosibirsk State University, Pirogova st. 1, Novosibirsk, 630090, Russian Federation

a.penenko@yandex.ru

Abstract. Inverse coefficient problems for a non-stationary chemical transformation model are considered. The objective of this work is to test an approach consisting in reducing the inverse problem to a quasi-linear matrix equation based on sensitivity operators constructed from an ensemble of independent solutions of adjoint equations. A Newton-Kantorovich-type algorithm is used to solve the thus obtained matrix equations. This approach is tested on a chemical transformation scheme with 22 species and 20 reactions. The reconstruction results are compared with several sets of unknown reaction rates according to the influence characteristics. The analysis seems to be useful for selecting sets of reaction rates that can be reconstructed by the inverse problem solution.

1. Introduction

Production-destruction type models are used in a wide range of applications, including modeling of the chemical transformation process. The chemical reaction rates can be potentially determined in laboratory conditions. Another way is to identify them implicitly by solving the inverse problems. A possible application of this approach is to indirectly estimate the reaction conditions that affect the reaction rates, e.g., temperature, pressure, or incoming solar radiation fluxes in the case of photochemical reactions.

In this paper, we study an approach based on the idea of G.I. Marchuk [1]. In this approach the inverse problems are reduced to the family of quasi-linear matrix equations based on the ensemble of independent solutions of adjoint equations. A Newton-Kantorovich-type method is used to solve the resulting quasilinear matrix equations. In the general case Newton-type algorithms are locally-convergent, i.e., their performance strongly depends on the choice of the initial guess.

In previous works we considered the inverse problems when the number of unknowns is comparable to the number of measurement data elements [2], [3]. In the case being considered the number of unknowns is less than the number of measurement data elements. The objective of this paper is to study numerically the local convergence properties of the adjoint ensemble-based algorithm, with regard to the different number of reconstructed reaction rates for an atmospheric chemistry transformation model with 22 species and 20 chemical reactions. This work is an extension of [4] by the application to the chemical transformation model.

IOP Conf. Series: Earth and Environmental Science **386** (2019) 012041 doi:10.1088/1755-1315/386/1/012041

2. Inverse Problem

Consider the time interval $[0,T]$ and the Cauchy problem for the system of ordinary differential equations in the form of production-destruction:

$$\frac{\partial \varphi_l}{\partial t} + P_l(t,\boldsymbol{\varphi},\mathbf{y})\varphi_l = \Pi_l(t,\boldsymbol{\varphi},\mathbf{y}) + f_l + r_l,\ t \in [0,T]. \tag{1}$$

$$\varphi_l(0) = \varphi_l^0, l = 1,\ldots,N_c, \tag{2}$$

where N_c is the number of components under consideration, $\varphi_l(t)$ is the value of component l at the time moment t, and $\boldsymbol{\varphi}(t) \in \mathrm{R}^{N_c}$ is a vector function with elements $\varphi_l(t)$. Let Y be a set of admissible coefficients. According to the inverse modeling approach developed in [5], the constant coefficients vector $\mathbf{y} \in Y \subset \mathrm{R}^{N_y}$ is called the uncertainty vector. The vector-functions $P,\Pi:[0,T]\times\mathrm{R}^{N_c}\times Y \to \mathrm{R}^{N_c}$ describe the destruction and production processes, respectively. The set $\Phi \subset L^2(0,T;\mathrm{R}^{N_c})$ denotes the solutions of (1)-(2) corresponding to Y. Suppose that all involved functions are smooth enough for all subsequent calculations to make sense.

The problem of determining $\boldsymbol{\varphi}$ from (1)-(2) by known $\boldsymbol{\varphi}^0 \in \mathrm{R}^{N_c}$ and $\mathbf{y} \in Y$ is called a direct problem. Let us denote its solution by $\boldsymbol{\varphi}[\mathbf{y}] \in \Phi$.

The set of indices L_{meas} denotes the observed state function components. Define the subspace $U_{meas} \subset L^2(0,T;\mathrm{R}^{N_c})$ and the operator that maps the coefficients to the observed time-series at their respective positions and place zero time-series in the rest of the positions:

$$U_{meas} = \left\{ \left\{ \begin{cases} h_l, l \in L_{meas} \\ 0, l \notin L_{meas} \end{cases} \right\}_{l=1}^{N_c} \mid h_l \in L_2(0,T) \right\}, \qquad A: \begin{cases} Y \to U_{meas} \\ \mathbf{y} \mapsto \left\{ \begin{cases} \varphi_l[\mathbf{y}], l \in L_{meas} \\ 0, l \notin L_{meas} \end{cases} \right\}_{l=1}^{N_c}, \end{cases}$$

where $\{a_l\}_{l=1}^{N_c}$ is a vector with elements $a_1,\ldots,a_{N_c}$. Let there be some $\mathbf{y}^{(*)} \in Y$, which we call the "exact solution", and define measurements data vector:

$$\mathbf{I} = A\left(\mathbf{y}^{(*)}\right). \tag{3}$$

The problem of finding $\mathbf{y}^{(*)}$ by $\mathbf{I}$ and (3) is called the inverse coefficient problem.

In practical applications, instead of the continuous concentration function available, there is usually a sequence of concentration measurements at separate time moments. We need the following definitions:

$$.\#.: \begin{cases} \mathrm{R}^{N_c}\times\mathrm{R}^{N_c} \to \mathrm{R} \\ \{a,b\} \mapsto a^T diag(\mathrm{r}) b = \sum_{l=1}^{N_c} a_l b_l \rho_l \end{cases}, \qquad ."\,.: \begin{cases} \mathrm{R}^{N_y}\times\mathrm{R}^{N_y} \to \mathrm{R} \\ \{a,b\} \mapsto a^T diag(\mathrm{c}) b = \sum_{l=1}^{N_y} a_l b_l \chi_l \end{cases},$$

where the points "below" in the notation of the scalar products mean that the corresponding entry contains two arguments, $.^T$ is the transposition operation, $\rho_l, \chi_l > 0$ are some weights, and $\rho = \{\rho_l\}_{l=1}^{N_c}$, $\mathrm{c} = \{\chi_l\}_{l=1}^{N_y}$, $diag(\mathbf{x})$ is the diagonal matrix with elements of the vector $\mathbf{x}$ on the diagonal, $\mathrm{R}_\rho^{N_c}$ is the Euclidean space with the scalar product .# ., and $\mathrm{R}_\chi^{N_y}$ is the Euclidean space with the scalar product ." . Let $S^\#$ denote the adjoint operator for $S:\mathrm{R}_\rho^{N_c} \to \mathrm{R}_\rho^{N_c}$ and $S^"$, the adjoint operator for $S:\mathrm{R}_\rho^{N_c} \to \mathrm{R}_\chi^{N_y}$. The matrices for the operators in the canonical bases are

$$S^\# = \left(diag(\mathrm{r})\right)^{-1} S^T diag(\mathrm{r}), S^" = diag(\mathrm{c})^{-1} S^T diag(\mathrm{r}).$$

Let there be a sequence of time moments $\Theta = \{\theta^m\}_{m=1}^{M} \subset [0,T]$, $\theta^1 > 0$, $\theta^M = T$ and a set of matrices $\{\mathbf{H}^{(\xi)} \in \mathrm{R}^{N_c \times M} \mid \xi = 1,\ldots,\Xi\}$. This type of reduction [3] can correspond to an *in situ* monitoring system that performs measurements regularly. We define a bilinear map corresponding to one composite measurement result and the corresponding operator H_Θ:

$$\langle .,. \rangle_\Theta : \begin{cases} \mathrm{R}^{N_c \times M} \times \Phi \to \mathrm{R} \\ \{\mathbf{h}, \boldsymbol{\varphi}\} \mapsto \sum_{m=1}^{M} \mathbf{h}^m \# \varphi(\theta^m) \end{cases}, \qquad H_\Theta : \begin{cases} \Phi \to \mathrm{R}^{\Xi} \\ \boldsymbol{\varphi} \mapsto \sum_{\xi=1}^{\Xi} \langle \mathbf{H}^{(\xi)}, \boldsymbol{\varphi} \rangle_\Theta \mathbf{e}^{(\xi)} \end{cases},$$

where h^m is the m-th column of the matrix $\mathbf{h}$. For example, the result of the measurements of the l-th component of the state function j at the time moment θ^m can be written as $\varphi_l(\theta^m) = \langle h, \boldsymbol{\varphi} \rangle_\Theta$, where $\mathbf{h} \in \mathrm{R}^{N_c \times M}$ is the matrix with all zero columns except for the m-th one, and the m-th column is the vector whose l-th element is 1 and the others are zeros. Let $H^{(\xi)}$ rows be nonzero only for the indices from L_{meas}.

3. Quasi-linear matrix equation

The solution algorithm of the inverse problem is taken from [4]. It is based on the ensembles of adjoint problem solutions. To define the adjoint problem, we need the definition of the divided difference operators $\bar{\nabla}_\varphi$ and $\bar{\nabla}_y$ that map the vector function $S:[0,T] \times \mathrm{R}^{N_c} \times \mathrm{R}^{N_y} \to \mathrm{R}^{N_c}$ to vector functions $\bar{\nabla}_\varphi S:[0,T] \times \mathrm{R}^{N_c} \times \mathrm{R}^{N_c} \times \mathrm{R}^{N_y} \to \mathrm{R}^{N_c \times N_c}$ and $\bar{\nabla}_y S:[0,T] \times \mathrm{R}^{N_c} \times \mathrm{R}^{N_y} \times \mathrm{R}^{N_y} \to \mathrm{R}^{N_c \times N_y}$ such that for any $t \in (0,T]$ and $\boldsymbol{\varphi}^{(m)} \in \mathrm{R}^{N_c}$, $\mathbf{y}^{(m)} \in \mathrm{R}^{N_y}$, $m = 1,2$,

$$S(t,\boldsymbol{\varphi}^{(2)},\mathbf{y}) - S(t,\boldsymbol{\varphi}^{(1)},\mathbf{y}) = \bar{\nabla}_\varphi S(t,\boldsymbol{\varphi}^{(2)},\boldsymbol{\varphi}^{(1)};\mathbf{y})\left(\boldsymbol{\varphi}^{(2)} - \boldsymbol{\varphi}^{(1)}\right),$$
$$S(t,\boldsymbol{\varphi},\mathbf{y}^{(2)}) - S(t,\boldsymbol{\varphi},\mathbf{y}^{(1)}) = \bar{\nabla}_y S(t,\boldsymbol{\varphi};\mathbf{y}^{(2)},\mathbf{y}^{(1)})\left(\mathbf{y}^{(2)} - \mathbf{y}^{(1)}\right).$$

Let $\boldsymbol{\varphi}^{(m)} = \boldsymbol{\varphi}[\mathbf{y}^{(m)}]$, $m = 1,2$. We define the adjoint problem with the solution Υ for any function $\mathbf{h} \in L^2(0,T;\mathrm{R}^{N_c})$ and matrix $\mathbf{H} \in \mathrm{R}^{N_c \times M}$:

$$-\frac{\partial \Psi}{\partial t} + W(t,\boldsymbol{\varphi}^{(2)},\mathbf{y}^{(2)},\boldsymbol{\varphi}^{(1)},\mathbf{y}^{(1)})^{\#}\, \Psi = 0, \quad t \in \left[\theta^m, \theta^{m+1}\right], m = 0,\ldots,M-1, \tag{4}$$

$$W(t,\boldsymbol{\varphi}^{(2)},\mathbf{y}^{(2)},\boldsymbol{\varphi}^{(1)},\mathbf{y}^{(1)})^{\#} = diag(P(t,\boldsymbol{\varphi}^{(2)},\mathbf{y}^{(2)}))$$
$$+\bar{\nabla}_\varphi P(t,\boldsymbol{\varphi}^{(2)},\boldsymbol{\varphi}^{(1)};\mathbf{y}^{(1)})^{\#}\, diag(\boldsymbol{\varphi}^{(1)}) - \bar{\nabla}_\varphi \Pi(t,\boldsymbol{\varphi}^{(2)},\boldsymbol{\varphi}^{(1)};\mathbf{y}^{(1)})^{\#}, \tag{5}$$

$$\left[\Psi\right]_{t=\theta^m} = -\mathbf{H}^m, m = 1,\ldots,M, \tag{6}$$

$$\Psi(T+) = \mathbf{0}, \tag{7}$$

where $\theta^0 = 0$, the expressions $f(t-)$ and $f(t+)$ denote the limits from below and top, respectively, $\left[f\right]_{t=\theta^m} = f(\theta^m+) - f(\theta^m-)$ is the jump of the function f. Let us denote the solution of the adjoint problem (4)–(7) with $\Psi\left[\mathbf{y}^{(2)},\mathbf{y}^{(1)};\mathbf{H}\right]$ and define the sensitivity function $M[\mathbf{y}^{(2)},\mathbf{y}^{(1)};\mathbf{H}]$ in the following way:

$$M[\mathbf{y}^{(2)},\mathbf{y}^{(1)};\mathbf{H}] = \int_0^T K(t,\boldsymbol{\varphi}^{(2)},\boldsymbol{\varphi}^{(1)},\mathbf{y}^{(2)},\mathbf{y}^{(1)})\Psi\left[\mathbf{y}^{(2)},\mathbf{y}^{(1)};\mathbf{H}\right]dt.$$

$$K(t,\boldsymbol{\varphi}^{(2)},\mathbf{y}^{(2)},\boldsymbol{\varphi}^{(1)},\mathbf{y}^{(1)}) = \bar{\nabla}_y \Pi(t,\boldsymbol{\varphi}^{(2)};\mathbf{y}^{(2)},\mathbf{y}^{(1)})'' - \bar{\nabla}_y P(t,\boldsymbol{\varphi}^{(2)};\mathbf{y}^{(2)},\mathbf{y}^{(1)})''\, diag(\boldsymbol{\varphi}^{(1)}).$$

Hence, the measurement data projection system defines the adjoint problem solution ensemble. We call the aggregate of an ensemble of the sensitivity functions as the sensitivity operator. The elements of the ensemble are evaluated in parallel. For any $\mathbf{y}^{(m)}, \in Y$, $m = 1,2$ the following relation holds:

$$H_\Theta A(\mathbf{y}^{(2)}) - H_\Theta A(\mathbf{y}^{(1)}) = M_\Theta[\mathbf{y}^{(2)},\mathbf{y}^{(1)}]\delta\mathbf{y}, \tag{8}$$

where

$$M_\Theta[\mathbf{y}^{(2)},\mathbf{y}^{(1)}]:\begin{cases} Y \to \mathrm{R}^{\Xi_\Theta} \\ \mathbf{z} \mapsto \sum_{\xi=1}^{\Xi_\Theta} \mathbf{e}^{(\xi)} M[\mathbf{y}^{(2)},\mathbf{y}^{(1)};\mathbf{H}^{(\xi)}]'' \ \mathbf{z} \end{cases},$$

Choosing $\mathbf{y}^{(2)} = \mathbf{y}^{(*)}$ in (9), we come to a parametric family of operator equations, which can be used to solve the inverse problem:

$$H_\Theta(I - A(\mathbf{y})) = M_\Theta[\mathbf{y},\mathbf{y}](\mathbf{y}^{(*)} - \mathbf{y}) + Q[\mathbf{y}^{(*)},\mathbf{y}],$$
$$Q[\mathbf{y}^{(*)},\mathbf{y}] = (M_\Theta[\mathbf{y}^{(*)},\mathbf{y}] - M_\Theta[\mathbf{y},\mathbf{y}])(\mathbf{y}^{(*)} - \mathbf{y}).$$

Thus, any pair of measurement data reduction operator H_Θ and "uncertainty" of the model (in our case it is $\mathbf{y}$) generates the sensitivity operator, which is the characteristic of this inverse problem. To solve the matrix equation we use a Newton-Kantorovich-type algorithm similar to [2]. In the algorithm, the ill-conditioned sensitivity operator matrix inversion is regularized by the truncated SVD with the sequential increase of the singular values being considered.

4. Numerical experiments

Let us consider a numerical experiment on the identification of the reaction rates for the atmospheric chemistry transformation model with 22 species and 20 chemical reactions from [6] augmented with a SO_2 reaction from [7]:

$$\begin{array}{cc}
NO_2 + hv \to NO + O_3P & hv + O_3 \to O_1D + O_2 \\
HCHO + hv \to CO + 2H_2O & HCHO + hv \to CO + H_2 \\
O_2 + O_3P \to O_3 & N_2 + O_1D \to N_2 + O_3P \\
O_1D + O_2 \to O_2 + O_3P & H_2O + O_1D \to 2OH \\
HO_2 + NO \to NO_2 + OH & NO + O_3 \to NO_2 + O_2 \\
NO + RO_2 \to HCHO + HO_2 + NO_2 & CO + OH \to CO_2 + HO_2 \\
HC + OH \to H_2O + RO_2 & HCHO + OH \to CO + H_2O + HO_2 \\
NO_2 + OH \to HNO_3 & 2HO_2 \to H_2O_2 + O_2 \\
H_2O + 2HO_2 \to H_2O + H_2O_2 + O_2 & HO_2 + RO_2 \to O_2 + ROOH \\
2RO_2 \to \mathrm{Pr}od & OH + SO_2 \to HO_2 + Sulf
\end{array}$$

As the «exact» solution $\mathbf{y}^{(*)}$, we consider the reaction rates from [6], [7], and [8] corresponding to noon (Figure 1 a)).

Similarly to [2], for the numerical computations we use a first-order discrete-analytical numerical scheme, which is equivalent to the QSSA scheme [9]. The scheme is explicit and preserves nonnegative concentrations. The numerical schemes for the direct and adjoint problems are constructed to satisfy the discrete analog of the sensitivity relation that links the variation of the state function to the variation of the model coefficients. The grid parameters in the computations are $T = 3600s$, $N_t = 500$, $N_c = 22$, $N_y = 20$, where N_t is the number of the temporal grid points. The point-wise measurements are taken with the uniform time interval $\Delta\theta = 360$. In this case the number of the adjoint ensemble members is $\Xi = |L_{meas}|[T/\Delta\theta]$. Let all concentrations be measured ($L_{meas} = 1,\dots,N_c$).

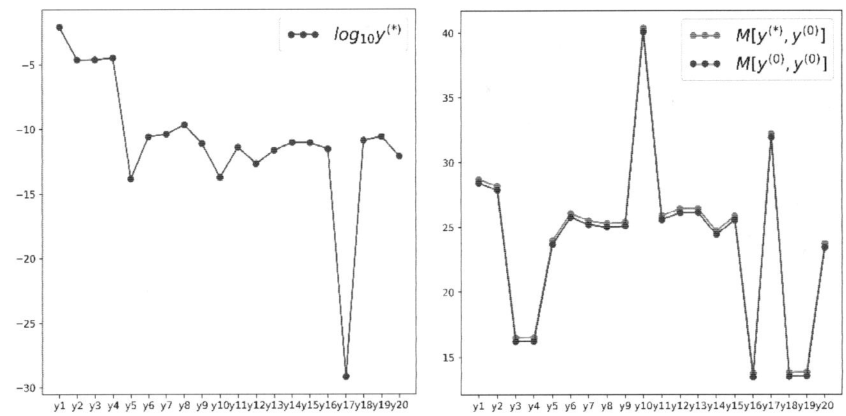

a) b)

Figure 1. «Exact» solution in $\log_{10}$ scale a) and norms of the sensitivity operator matrix columns b).

In the first experiment we calculate the sensitivity operator matrices for $M_{\Theta}[\mathbf{y}^{(*)},\mathbf{y}^{(0)}]$ and $M_{\Theta}[\mathbf{y}^{(0)},\mathbf{y}^{(0)}]$ with $\mathbf{y}^{(0)}=2\mathbf{y}^{(*)}$. The norm of the column of a sensitivity operator matrix describes the sensitivity of measurements to the variation of a particular reaction rate. The norms are presented in Figure 1 b). Let us call these norms the influences of the reaction rates.

Let us evaluate the reconstruction efficiency for the reaction rates corresponding to the reaction rates from different sensitivity groups. In Figure 3 we present examples of the outcomes of reconstruction for different sets of reconstructed rates L_{cons} and for different initial guesses:

$$y_l^{(0)} = y_l^{(*)} + \begin{cases} 0, \; l \notin L_{cons} \\ (g-1)\, y_l^{(*)}, \; l \in L_{cons} \end{cases}.$$

In Figure 2 a) we present a relative reconstruction error for the set of rates $L_{cons}=\{y_1, y_3, y_4\}$. As we see, y_1 has the highest influence among L_{cons} (see Figure 1), but its reconstruction is not the best one. In Figure 2 b) c) and d), $\{y_1, y_3, y_4\}$ is augmented with y_{18}, y_{20}, and y_{10}, respectively: y_{18} has the lowest influence, y_{20} has a medium one, and y_{10} has the highest one. In the experiment with y_{18}, the reconstruction error of $\{y_1, y_3, y_4\}$ was not affected (Figure 2 b). In the experiment with y_{20}, it was slightly modified (Figure 2 c)), and the reconstruction was unsuccessful in the experiment with y_{10}, i.e. the resulting reconstruction error is equal to the initial guess error (Figure 2 d).

5. Discussion

In the numerical experiments, the resulting reconstruction error is dependent on the initial guess, i.e., the algorithm shows local convergence. The efficiency of reconstruction of an individual reaction rate is affected by the complete set of unknowns (i.e., it depends on the context). The reaction rate with a high influence can change the reconstruction efficiency of the other rates from the set. In the case being considered the highest influence was not connected with better reconstruction within the set of reaction rates.

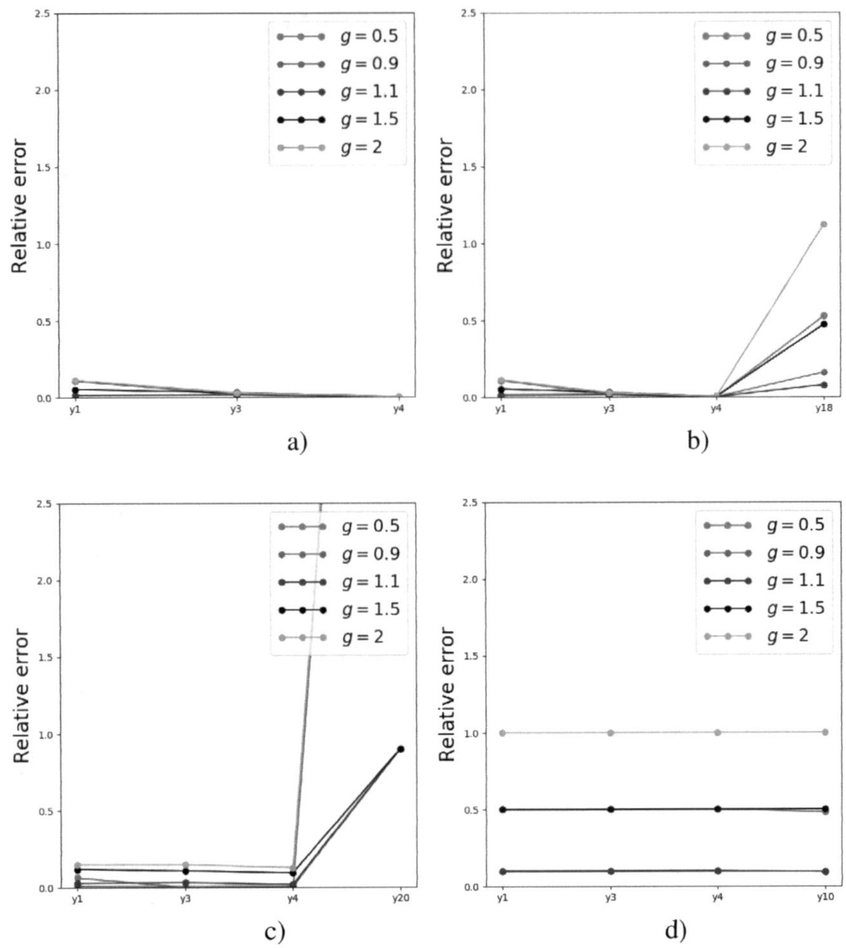

Figure 2. Relative reconstruction error for different sets L_{cons} of the unknown reaction rates (a),b),c),d)) and for different initial guesses (different colors) with all concentrations measured.

6. Conclusions

In the paper, an inverse coefficient problem solution algorithm was tested for a 0-D atmospheric chemistry transformation model. With the help of an adjoint equation solutions ensemble, the inverse problem can be reduced to a parametric quasi-linear matrix equations family. In the above-considered numerical experiment, there were sets of reaction rates that can be reconstructed by the solution of the inverse problem. We considered the reconstruction error together with a characteristic that measures the influence of the reaction rate on the measurement data. In the numerical experiment presented above the reaction rates with high influence characteristics were able to modify the reconstruction efficiency of a set of reaction rates. This analysis seems to be useful for selecting sets of reaction rates that can be reconstructed by the inverse problem solution.

Acknowledgments

The work was supported by the Russian Foundation for Basic Research under grant 19-07-01135 (inverse coefficient problem with point-wise measurements).

IOP Conf. Series: Earth and Environmental Science **386** (2019) 012041 doi:10.1088/1755-1315/386/1/012041

References

[1] Marchuk G 1964 *Reports of the USSR Academy of Sciences Ed. Science* **156** 503–06 (In Russian)
[2] Penenko A V 2019 *Numerical Analysis and Applications* **12** 51-69
[3] Penenko A V 2018 *Numerical Analysis and Applications* **11** 73-88
[4] Penenko A Mukatova Z Salimova A 2019 preprint URL https://www.researchgate.net/publication/334249101_Numerical_study_of_the_coefficient_identification_algorithm_based_on_ensembles_of_adjoint_problem_solutions_for_a_production-destruction_model
[5] Penenko V V 1996 *Bull. Nov. Comp. Center, Series Num. Model. Atmosph.* **4** 31-52
[6] Stockwell W and Goliff W 2002 *Journal of Geophysical Research* **107** 4643–50
[7] Visscher A D 2013 *Air Dispersion Modeling: Foundations and Applications* (JOHN WILEY & SONS INC).
[8] Schaap M, Roemer M, Sauter F, Boersen G, Timmermans R and Builtjes P J H 2005 *LOTOS-EUROS Documentation* (TNO report)
[9] Hesstvedt E, Hov O and Isaksen I 1978 *International Journal of Chemical Kinetics* **10** 971-94

Wind projections for the territory of Russia considering the development of wind power

E V Fedotova

Moscow Power Engineering Institute, Krasnokazarmennaya 14, Moscow

ek.v.fedotova@gmail.com

Abstract. The aim of the present work is to obtain surface wind speed projections which could be used as guidelines for long-term planning of wind power construction in Russia. A classical multi-model ensemble approach is implemented by using CMIP5 simulation results. The reliability of the ensemble estimation is assessed by a comparison of three different ensemble versions, which are validated against reanalysis data for the whole 20th century and have been found to give consistent results since 1950. Agreement between the results of all the assembling approaches has been found to be quite good for the mid-twenty-first century. All ensembles being considered agree that a considerable decrease in wind resources should be expected in the European part of Russia and in the south of Western Siberia towards 2050. Another robust output of the analysis is an increase in annual wind speed in the Southern Russian Far East. The wind change during the considered 40-year period is in the range from -6 to +6%, which means a -18 to +18% change in potential wind generation. The main output of the present work is that climate change by no means can be seen to be an obstacle to the development of renewable power in Russia. However, the climate change associated alteration of wind regime should be necessarily taken into account when establishing long-term plans for wind farm construction in Russia.

1. Introduction

The energy systems are in transition around the world. One of the largest drivers of this transformation process are the renewable electricity generation technologies that seemed to be of a purely academic interest only twenty years ago. Nowadays the development of wind generators is one of the most impressive examples of rapid development of renewable technology. Wind generators have operated since the first decades of the twentieth century, but have played a minor role in the energy systems for many decades. Today wind power all over the world has become one of the leading generation technologies as a part of real industrial power systems. The total wind power output is approaching the nuclear power plants.

The wind generation is directly linked with the wind speed fluctuations, which mean that the wind power economics is greatly influenced by the climatic conditions. Integration of an increasing wind power share into the energy systems is possible only if the wind regime in a particular area is well understood [1]-[3]. Multi-annual wind speed monitoring has become a common step of wind farm procurement.

Wind energy in Russia is currently at the very early stage of development. The overall installed wind power across the country is currently approaching 1 GW. Several new wind farms should be launched across the country during the next decades, multiplying the installed capacity value.

Assessment of the long-term wind speed dynamics across the country is becoming crucially important for the support of this development process.

2. Related research

The long-term wind speed dynamics for different regions of the world is being intensively studied during the last decade. A recent paper [4] has demonstrated that the modern climate models tend to heavily underestimate the decreasing wind trends observed in the Northern Hemisphere during the twentieth century. This makes evident the importance of careful consideration of the models selection used to project the wind speed in a given world area.

2.1. Meteorological observations

A continuously decreasing wind speed trend in the surface has been observed in Russia since the 1970s [5]. A comprehensive meteorological analysis performed by Roshydromet experts has allowed one to conclude that the reduction of the average wind speed is typical for all seasons and occurs almost everywhere [6]. This effect is especially pronounced in the European territory of Russia, where most wind farms to introduce to construction during the next decades should be located. The authors of [6] definitely attribute the observed tendency to the modern climate change.

2.2. Projections

Comprehensive work on climate projections for the Russian territory was performed at the Voeikov Main Geophysical Observatory [7] about five years ago. These calculations utilized the CMIP5 (Coupled Model Intercomparison Project) Phase5. The authors have made validation of the available global climate circulation models to select sixteen models which reproduce seasonal variations of the temperature, precipitation, and pressure in a way that is most consistent with the observations. The authors have stated that the wind speed changes up to +/- 1 m/s are projected but do not provide any robustness assessment of the projected wind speed fields.

The authors of the Roshydromet Assesment Report [5] make reference to the ensemble calculations of the surface wind speed for the rcp 8.5 scenario, but the results of these calculations are not presented. The authors mention that the wind speed change obtained with this assessment is in the range of +/-1m/s, and believe this change to be insignificant not to discuss any further details.

Recently, paper [8] has presented a detailed analysis of the CMIP5 results for the Russian Arctic territories with a careful consideration of the validation issues for the used models. However, the focus of the authors was only on the Arctic seas. In addition, calculations have been made for the catastrophic climatic scenario of rcp 8.5 which, fortunately, should be treated as a very likely one. This does not allow us to utilize the results obtained in [8] even for the most preliminary conclusions about the real wind conditions which will impact the operation of wind parks in Russia during the next decades.

2.3. Technology status

It is difficult to agree with the conclusion about the insignificance of the observed wind speed changes. The wind turbines are quite sensitive to the wind speed variations due to the fact that wind turbine power generation depends on the kinetic energy of the wind flow P_{wind}, which is proportional to the cube of the wind speed w:

$$P_{wind} = \frac{1}{2} w^3, \qquad (1)$$

where ρ is the air density.

Wind speed changes of about 1 m/s mean a change in the wind turbine output by two to three times at a typical surface wind speed of 3...5 m/s, which is usual on the Russian territory. Thus, the climate change seems to inevitably affect the development of the wind energy in Russia. Nowadays, at the very beginning of this development process, there is an urgent need to obtain at least some robust projections of the surface wind speed in Russia under realistic climate scenarios.

3. Motivation

The presented work is intended for obtaining projections of the surface wind speed, which could be used as a guideline for long-term planning of the wind power construction in Russia, particularly when assessing the future energy system options and control strategies. We will use a classical ensemble approach considering different ensembling approaches.

4. Methods

The moderate climate scenario rcp 4.5 was taken for further calculations, as the closest one to the likely real climate changes during the 21st century. The soft scenario rcp 2.6 and the dramatic 8.5 were considered to estimate the robustness of the calculated projections. The historic climate experiments were used in the validation procedure. Our calculations used the results of the CMIP5 project that up to date provides access to the most comprehensive set of the global climate models simulation results. We used the monthly output of 27 CMIP5 models available for the near-surface wind speed [9].

Table 1. List of CMIP5 models used for ensemble estimations.

Model Name	Institute
ACCESS 1.0	Commonwealth Scientific and Industrial Research Organization/Bureau of Meteorology, Australia (CSIRO-BOM)
ACCESS 1.3	
bcc-csm1-1-m	Beijing Climate Center, China (BCC)
bcc-csm1-1	
BNU-ESM	Beijing Normal University, China(BNU)
CanESM2	Canadian Centre for Climate Modelling and Analysis, Canada (CCCma)
CMCC-CESM	Centro Euro-Mediterraneo sui Cambiamenti Climatici, Italy (CMCC)
CMCC-CMS	
CMCC-CM	
CNRM-CM5	Centre National de Recherches Météorologiques, Centre Européen de Recherche et de Formation Avancée en Calcul Scientifique, France (CNRM-CERFACS)
CSIRO-Mk 3.6.0	Commonwealth Scientific and Industrial Research Organization/Queensland Climate Change Centre of Excellence, Australia (CSIRO-QCCCE)
GFDL-CM3	Geophysical Fluid Dynamics Laboratory, USA (NOAA GFDL)
GFDL-ESM2G	

Model Name	Institute
GFDL-ESM2M	
HadGEM2-CC	Met Office Hadley Centre, UK (MOHC)
HadGEM2-ES	
inmcm4	Russian Academy of Sciences, Institute of Numerical Mathematics, Russia (INM)
IPSL-CM5A-LR	Institut Pierre Simon Laplace, France (IPSL)
IPSL-CM5A-MR	
IPSL-CM5B-LR	
MIROC-ESM	Atmosphere and Ocean Research Institute (The University of Tokyo), National Institute for Environmental Studies, and Japan Agency for Marine-Earth Science and Technology, Japan (MIROC)
MIROC-ESM-CHEM	
MIROC5	
MPI-ESM-LR	Max Planck Institute for Meteorology, Germany (MPI-M)
MPI-ESM-MR	
MRI-CGCM3	Meteorological Research Institute, Japan (MRI)
MRI-ESM1	

4.1. Ensemble strategy

The used ensemble strategies accounted for [10] CMIP5 validation results on the high-resolution reanalysis data. The essence of this validation procedure was the reproducibility testing of the daily wind speed distributions in each computational grid cell by each of the CMIP5 models in the European domain. The model quality metric was defined as a cell number where the Kolmogorov-Smirnov test gave a positive result.

In our calculations we have considered three ensembles:

1) the ensemble recommended in [10] consisted of eight models which model the distribution of wind speeds corresponding to the reanalysis data for 75% of the model cells in a considered area;

2) an ensemble of nine models which give 70% of the cells share with a correctly reproduced daily wind speed distribution;

3) an ensemble of all available models.

The geographical restriction of [10] seems not to be critical for the considered problem, since most wind farms in Russia planned for construction are located in the European part of the country. Besides, the global climate models ensemble estimations may be treated for the wind speed only as a quite approximate assessment. Downscaling approaches should be applied to produce wind speed projections suitable for use in the energy systems modeling and planning. What we are focused on is obtaining a general picture of the muldidecadal wind speed trends.

4.2. Validation

A simplified validation procedure was performed to check the reproducibility of the long-term wind speed dynamics across the country by the considered ensembles. The results of the ensemble estimations for the historic CMIP5 experiment were compared with the 20th Century Reanalysis V2c [11] reanalysis data for the whole twentieth century.

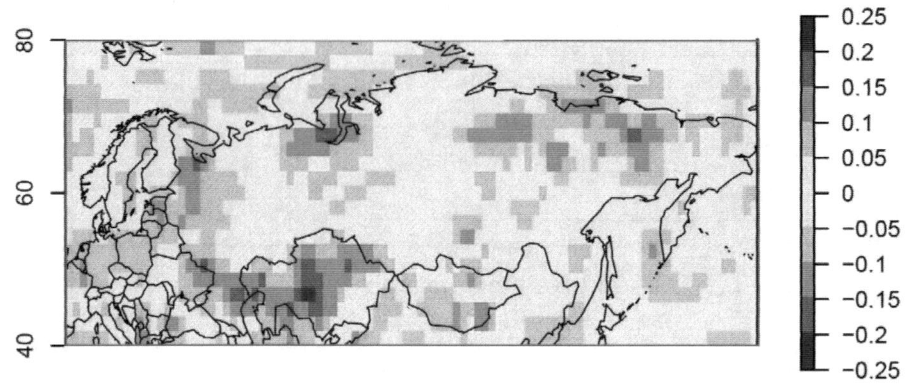

a

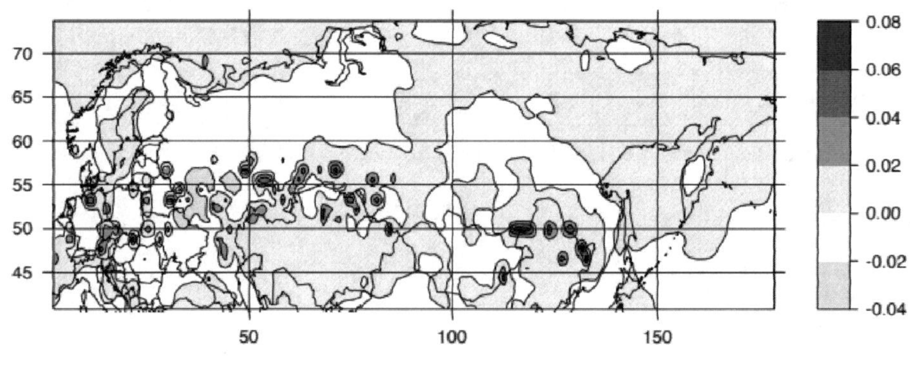

b

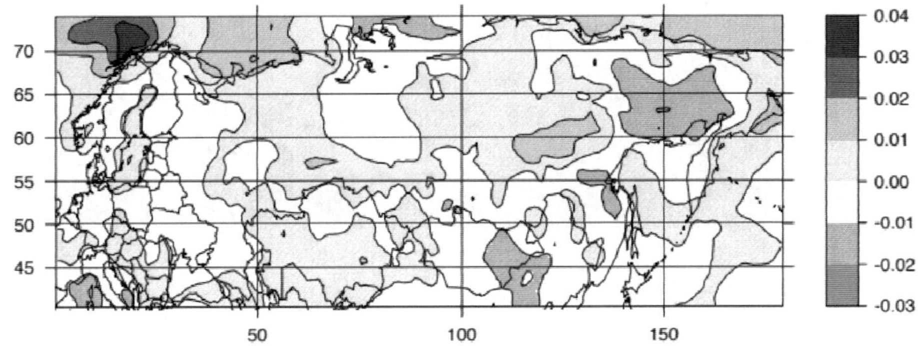

c

Figure 1. Change in the annual surface wind speed in 1995–2004 compared with 1977–1986: a – according to 20Cv2 reanalysis, b – calculated by the eight-models ensemble, c – calculated by the whole set of all available models.

It was found that the eight-model ensemble performs better as compared with the all-models ensemble (Figure 1). The correspondence is more qualitative than quantitative and is satisfactory for the whole Russian area only since 1940. The discrepancy for the earlier dates may obviously result both from the ensemble uncertainties and flaws of the reanalysis data. A more detailed analysis seems to be necessary to clarify the long-term trends of the wind speed during the twentieth century. However, the ensemble estimations seem to be quite consistent with the reanalysis data on fifty-sixty years' time horizons, which is quite sufficient for the purpose of our work.

5. Results and discussion

The three above-considered ensembles give close results for the relative wind speed change across the country during the twenty-first century. However, they differ significantly in details. The range of changes in the territory is the same for all three cases varying from -6...-4% to +5...+6% (Fig. 2). The ensembles also agree that the most pronounced decrease in wind speed is to be expected in the central and southern parts of the European territory of Russia, as well as in the south of Western Siberia, which is consistent with the observation data of [6].

A second conclusion which follows from this comparison is ae likely increase in the average wind speed in Primorye and on the Sakhalin shelf. This effect is apparently linked to the atmospheric circulation shift, in particular, to the weakening of the Siberian High and intensification of the cyclonic activity over the Pacific observed during the last decades and attributed to climate change. This result is remarkably stable for all ensembles considered. Moreover, this output is consistent even with different climate scenarios. A control ensemble calculation run has given us qualitatively the same wind speed increase in the Far East for rcp 2.6 and rcp 8.5 as well.

It should be emphasized again that the present work is intended to obtain robust large-scale projections of long-term forecast changes in the wind potential. The effects of mesoscale atmospheric hydrodynamics combined with local topography have a huge impact on wind farm operation. That is why the application of dynamic or statistical downscaling procedures is a necessary step in using the above-obtained national-scale results in energy system optimization models.

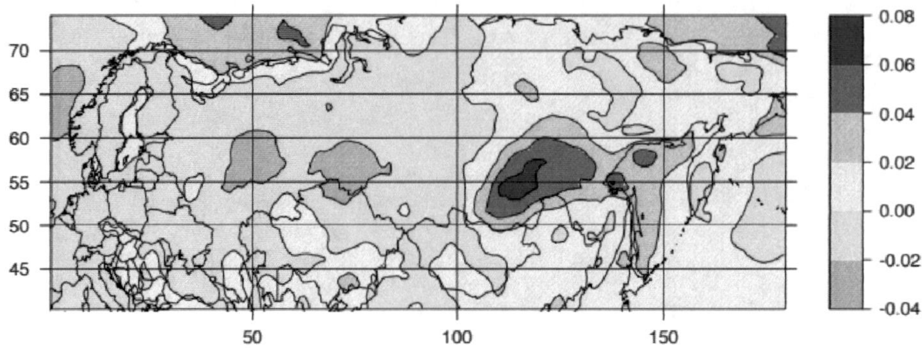

Figure 2. Projection of the annual surface wind speed change in 2065–2074 as compared with 2007–2016 eight-models ensemble.

What may be concluded right now is that the Russian Far East will definitely obtain a chance to renovate the local energy systems on a much cleaner basis as compared to the coal power plants in operation there today. Apart from the anticipated increase in wind potential, the Far East is one of the Russian regions where a significant rise in bio-productivity is very likely during the 21st century [12]. Climate change seems to favor the renewable renovation plans whose realization is gradually starting now in the energy systems of the Far East. However, the projections for the wind velocity changes in Kamchatka and Chukotka areas are far less certain and require a more detailed analysis, since wind power there is of highest interest for the regional remote energy systems.

There are some areas where the decrease in wind speed may lead up to a 12% decrease in wind specific energy. Fortunately, it does not mean any serious risks for the development of national wind power, since the areas of noticeable wind speed decrease are located in regions where no extensive wind power development is under discussion. The south of European Russia, the Baltic Sea coast, and the Kola Peninsula, where wind park constructions are planned, are likely to be subjected to relatively low changes in surface wind speed, in the range of +/-2%. Climate change seems to provide more possibilities than threats for the development of renewable power in Russia. It should be kept in mind, however, that the integration of an intermittent wind generation into the energy systems requires careful consideration of wind speed dynamics and a shift in the control paradigm of energy systems.

Acknowledgments

This work was supported by the Russian Science Foundation (project no. 18-79-10255) in part of developing the methodology and by the Russian Foundation for Basic Research (grant no. 17-08-00134) in part of making the calculations. The author highly acknowledges the participation of the modeling groups of the CMIP5 project listed in Table 1 and the World Data Center for Climate in Hamburg for providing access to the CMIP5 simulation data.

References

[1] Gkarakis K C 2018 Performance analysis of an operating windfarm of 21 MW in Greece for a period of three years *J. of Power Technol.* **98** 5 396-402

[2] Elistratov V V, Diuldin M V, Denisov R S 2018 Justification of project and operation modes of hybrid energy complexes for arctic conditions *IOP Conf. Ser.: Earth and Environ. Science* **180**

[3] Bezrukovs V, Bezrukovs V 2011 Problems in Assessment of Wind Energy Potential and Acoustic Noise Distribution when Designing Wind Power Plants *Environ. and Climate*

Technol. **6** 1 9-16
[4] Tian Q, Huang G, Hu K, Niyogi D 2019 Observed and global climate model based changes in wind power potential over the Northern Hemisphere during 1979-2016 *Energy* **167** 1224-35
[5] *Second Roshydromet Assessment Report on the Climate Change and Its Consequences in Russian Federation* 2014 (Moscow: Roshydromet) [in Russian]
[6] Bulygina O N, Korshunova N N, Razuvaev V N 2013 Change of the wind regime in Russia *GGO Proc.* **586** 156–72 [in Russian]
[7] Shkol'nik I M, Meleshko V P, Stadnik V V, Khlebnikova E I, Akent'eva E M, Genrikhovich E L, Kiselev A A 2014 Climate change impact of the energy production and consumption in Russia *GGO Proc.* **573** 92-222 [in Russian]
[8] Surkova G V and Krylov A V 2018 Changes of mean and extremal wind speed values in Arctic to the end of 21st century *Arctic and Antarctic* **3** 26–36 [in Russian]
[9] Knutti R 2014 *IPCC Working Group I AR5 snapshot: The rcp45 experiment* (World Data Center for Climate)
[10] Carvalho D, Rocha A, Gomez-Gesteira M, Silva Santos C 2017 Potential impacts of climate change on European wind energy resource under the CMIP5 future climate projections *Renewable Energy* **101** 29-40
[11] Compo G P *et al* 2011 The Twentieth Century Reanalysis Project *Quarterly J. of the Royal Meteor. Soc.* **137** 654 1-28
[12] Klimenko V, Tereshin A, Mikushina O 2019 Increase of Energy Potential of Russian Forest resources due to Climate Change and CO_2 Fertilization *E3S Web of Conf.* **103**

IOP Conf. Series: Earth and Environmental Science **386** (2019) 012043 doi:10.1088/1755-1315/386/1/012043

Stochastic quasi-Gaussian models of atmospheric clouds

E G Kablukova[1], V A Ogorodnikov [1,2], S M Prigarin[1,2] and M S Yudin[1]

[1]Institute of Computational Mathematics and Mathematical Geophysics SB RAS, Novosibirsk, Russia
[2]Novosibirsk State University, Novosibirsk, Russia

E-mail: kablukovae@sscc.ru, ova@osmf.sscc.ru, sergeim.prigarin@gmail.com, m.yudin@ommgp.sscc.ru

Abstract. Random field simulation methods that can be used to simulate the stochastic structure of three-dimensional clouds are discussed. Examples of homogeneous random field realizations that reproduce one-dimensional distributions and correlation functions of experimental fields of stratus optical thicknesses are presented.

1. Introduction
Cloudiness is a major factor affecting the radiation balance of the Earth's atmosphere. However, the stochastic structure of clouds brings considerable uncertainty into climatic models. An important problem is to construct numerical models that take into account the random optical properties and geometry of clouds. There are a lot of approaches to solving this problem (see, for example, [1] and the references therein). To simulate random fields of some of the characteristic properties (for instance, extinction coefficient or water content) of stratus clouds, we have analyzed observation data [2] and made some assumptions about the correlations along the vertical axis. We have used nonlinear transformations of Gaussian random fields [3] and simulated various models of Gaussian random fields based on a vector autoregressive scheme and a spectral decomposition. One of the objectives of this paper is to present some new simulation examples and to demonstrate, in addition to the results of some previous investigations, how the numerical methods work. These simulation methods seem to be universal (see [3, 4, 5], where similar approaches are used for various applications).

2. Quasi-Gaussian models
Assume that the stratus extinction coefficient $\sigma(x, y, z)$ is a homogeneous random field and its one-dimensional distribution function F and its autocovariance function $r(x, y, z)$ are defined. A well-known method to numerically simulate such random fields is based on nonlinear transformations of Gaussian homogeneous fields [6, 3]. The simulation formula is

$$\tilde{\sigma}(x,y,z) = F^{-1}\left(\Phi\left(\omega(x,y,z)\right)\right). \quad (1)$$

Here $\Phi(\omega)$ denotes a standard normal distribution function and $\omega(x, y, z)$ is a homogeneous Gaussian random field with zero expectation, $\mathbf{E}\omega(x, y, z)=0$, unit variance, $\mathbf{V}\omega(x, y, z)=1$, and auto-covariance function $\rho(x, y, z) = \mathbf{E}\omega(x, y, z)\omega(0, 0, 0)$. The autocovariance function $r(x, y, z)$ of the field $\sigma(x, y, z)$ has the form

$$r(x,y,z)=\mathbf{E}\sigma(x,y,z)\sigma(0,0,0)=R_F\left(\rho(x,y,z)\right)=\int_{-\infty}^{\infty}\int_{-\infty}^{\infty}F^{-1}\left(\Phi(\xi)\right)F^{-1}\left(\Phi(\eta)\right)\varphi_\rho(\xi,\eta)d\xi d\eta,$$

(2)

and $\varphi_\rho(\xi,\eta)$ is the probability density of a two-dimensional Gaussian random vector with zero mean, unit variance and correlation coefficient ρ between the components

$$\varphi_\rho(\xi,\eta)=\left(2\pi\sqrt{1-\rho^2}\exp\left(\frac{\xi^2+\eta^2-2\rho\xi\eta}{2(1-\rho^2)}\right)\right)^{-1}$$

Thus, the method implies the calculation of the function $\rho(x,y)=R_F^{-1}\left(r(x,y)\right)$, the subsequent simulation of the Gaussian field with this autocovariance function and, then, the transformation (1).

3. Statistical analysis of experimental fields

We have analyzed a random field of the Arctic stratus optical thickness τ (x, y) retrieved from aircraft measurements obtained during f VERDI campaign 2012 [2 , 7]. The estimations of the random field one-dimentional distribution densities $f(\tau)$ have been made. Note that the one-dimensional distribution $f(\tau)$ of τ (x, y) can be approximated by a truncated Gaussian distribution in [0, u_{max}] with the density

$$f_{\mu,\sigma}(u)=\frac{\exp\left(-(u-\mu)^2/2\sigma^2\right)}{\left|\Phi(u_{\max})-\Phi(\mu)\right|\sigma\sqrt{2\pi}},\ \mu>0,$$

or a Gamma distribution with the density

$$g_{\lambda,\nu}(u)=\frac{\lambda^\nu u^{\nu-1}}{\Gamma(\nu)}e^{-\lambda u},\quad \lambda>0,\ \nu>0,\ u>0.$$

Examples of one-dimensional distributions $f(\tau)$ of the fields $\tau(x, y)$ in question and some approximations of $f(\tau)$ are shown in Figure1.

When modeling cloud fields using the inverse function method, it is necessary, at first, to estimate the correlation functions distortion. For the one-dimensional distributions $F(.)$ of the observed optical thickness fields $\boldsymbol{\tau}(x, y)$ St-01 - St-10, the normalized autocorrelation functions $r(x,y)=R_F\left(\rho(x,y)\right)$ of the corresponding quasi - Gaussian processes $\tilde{\tau}(x,y)=F^{-1}\left(\Phi\left(\omega(x,y)\right)\right)$ from (1) are computed according to the relation (2). In Table 1, the maximum values of the correlation distortion for the clouds St-01 -- St-10 are given for some intervals of ρ. These results show that the distortion of the correlations in the nonlinear transformation for the above-considered one-dimensional distributions does not exceed 6 percent, and they can be neglected when constructing a numerical model, i.e. one can use the correlation function $r(x, y)$ of a non-Gaussian field $\boldsymbol{\tau}(x, y)$ for constructing the Gaussian field $\omega(x, y)$. In a statistical analysis of the available random fields τ (x, y), unbiased $\hat{r}(k,l)$ and biased $r(k,l)$ estimations of the correlation function are calculated using the following formulas:

$$\hat{r}(k,l)=\frac{1}{(M-k)(N-l)\mathbf{V}\tau}\sum_{i=1}^{M-1-k}\sum_{j=1}^{N-1-l}\left(\tau\left(x_i+kh_x,y_j+lh_y\right)-\mathbf{E}\tau\right)\left(\tau\left(x_i,y_j\right)-\mathbf{E}\tau\right),$$

$$r(k,l)=\frac{1}{MN\mathbf{V}\tau}\sum_{i=1}^{M-1-k}\sum_{j=1}^{N-1-l}\left(\tau\left(x_i+kh_x,y_j+lh_y\right)-\mathbf{E}\tau\right)\left(\tau\left(x_i,y_j\right)-\mathbf{E}\tau\right),$$

$$\mathbf{E}\tau=\frac{1}{MN}\sum_{i=1}^{N}\sum_{j=1}^{M}\tau\left(x_i,y_j\right),\quad \mathbf{V}\tau=\frac{1}{MN}\sum_{i=1}^{N}\sum_{j=1}^{M}\left(\tau\left(x_i,y_j\right)-\mathbf{E}\tau\right)^2$$

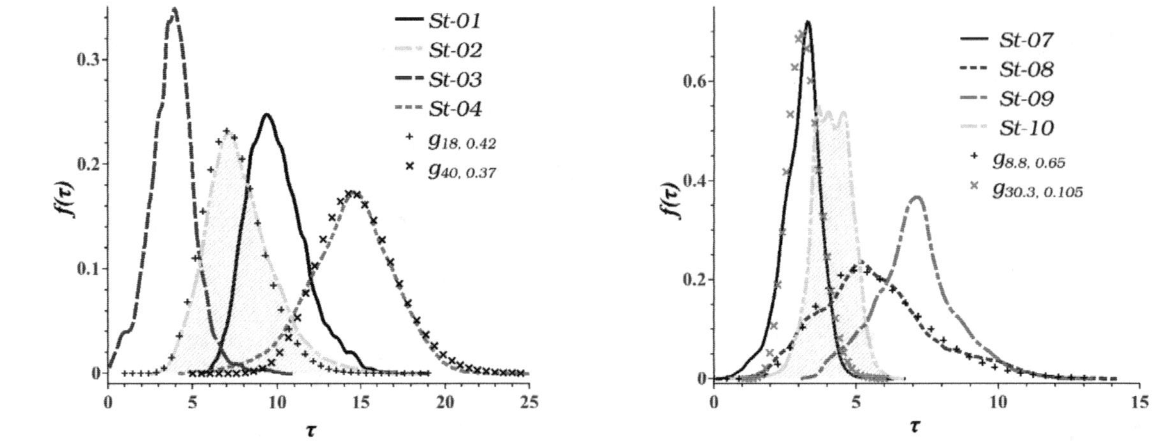

Figure 1. One-dimensional distribution density of cloud optical thickness field and its approximation.

According to the estimations obtained above, the one-dimensional correlation functions can be approximated with satisfactory accuracy by an exponential correlation function, at least, it can be made at distances not exceeding the correlation radius (the distance at which the correlation function decreases e times). Analyzing estimations of the correlation radii of arctic stratus fields [7], note that the mean value of the correlation radius is approximately equal 0.5 km, the minimal correlation radius is 0.09 km, and the largest correlation radius is 1.12 km. At the same time, the average correlation radius in the horizontal direction of the stratus water content of the temperate climatic zone is 1.4–1.5 km [8].

Table 1. Correlation distortion of quasi-Gaussian cloud models St-01 – St-10.

	$\max\lvert r(\rho)-\rho\rvert$ $\rho\in(-1,1)$	$\max\lvert r(\rho)-\rho\rvert$ $\rho\in(-0.5,1)$	$\max\lvert r(\rho)-\rho\rvert$ $\rho\in(0,1)$
St-01	0.04417	0.01672	0.00618
St-02	0.05408	0.02086	0.00782
St-03	0.01385	0.00767	0.00675
St-04	0.01301	0.00525	0.00248
St-05	0.00664	0.00382	0.00269
St-06	0.00386	0.00361	0.00361
St-07	0.03071	0.01253	0.00597
St-08	0.03735	0.01483	0.00638
St-09	0.00885	0.00546	0.00422
St-10	0.00846	0.00408	0.00321

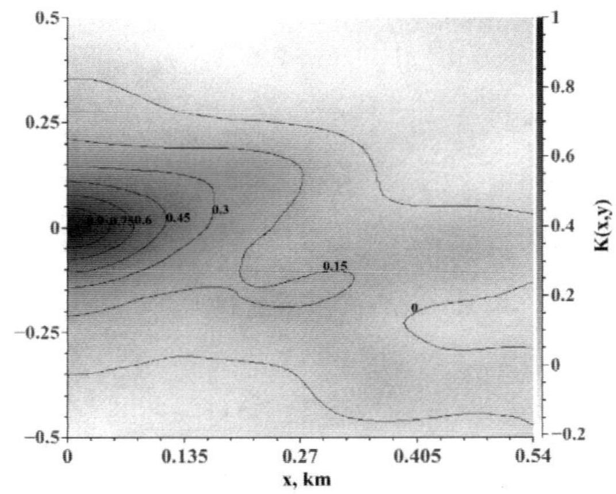

Figure 2. Estimation of the autocorrelation function $r(x, y)$ of the optical thickness field St-01.

4. Numerical models of Gaussian stochastic fields

4.1. Multiplicative representation of the correlation function of a three-dimensional field

There are various ways of constructing three-dimensional random fields with a given correlation structure. Assuming that the correlation function of a homogeneous three-dimensional random field $u(x, y, z_i)$, $z_i = ih_z, i = 1,..., N$ is represented by the product of two-dimensional and one-dimensional correlation functions $R(x, y, i) = \mathbf{E}u(x, y, ih_z)u(0,0,0) = R_{xy}(x, y)R_z(i)$, it is possible to build a combined algorithm for modeling such a random field, by constructing, at the first stage, a given number N of two-dimensional fields $\omega(x, y, z_k) = \omega_k(x, y)$, $k = 1,\dots, N$, for instance, using vector methods of conditional distributions and autoregressive models [3, 13]. This set of two-dimensional random fields can be considered as a set of dependent random vectors $\vec{\omega}_N(x, y) = (\omega_1(x, y),..., \omega_N(x, y))$ with independent components. To obtain a three-dimensional random field with the correlation function R(x, y, z) for each fixed point (x, y), we use a method based on linear transformations, $\vec{u}_N(x, y) = A\vec{\omega}_N(x, y)$, where $A = (a_{ij})$ is a matrix such that $AA^* = B = (b_{ij}), b_{ij} = R_z(i - j)$. The factorization $AA^* = B$ can be obtained, for example, by using Cholesky decomposition [3, 12]. This algorithm is convenient if the two-dimensional fields are simulated on a discrete grid. A vector "conditional distributions" algorithm and a vector autoregression scheme will be briefly described in the next subsection.

To simulate an isotropic continuous Gaussian field with a correlation function $\rho(s) = \rho\left(\sqrt{x^2 + y^2 + z^2}\right)$ one can use a spectral model [10] of a random field $\tilde{\omega}(\vec{r}) = \tilde{\omega}(x, y, z)$.

If the radial correlation function is exponentially decaying, $\rho(s) = \exp(-\lambda s)$, there is an algorithm [11] for simulating random harmonics of the spectral model $\tilde{\omega}(x, y, z)$. It will be described in subsection 3 of this section.

4.2. Vector methods of conditional distributions and autoregression [3, 13]

To simulate the homogeneous Gaussian random field $\omega(x, y)$ on a grid (ih_x, jh_y), $i, j=1,\dots,N$ we used a two-dimensional autoregressive scheme [3, 13]. The main idea of the method is an iterative step by step simulation of the vectors ξ_i, taking into account the correlations both between the random elements of the simulated vector ξ_i and between the elements of this vector and the n previous vectors ξ_{i-j}, $j=1,n-1$ according to a given correlation function.

To simulate the stationary vector-valued autoregressive processes ξ_t with a given covariance structure, the initial vectors $\xi_1, \ldots, \xi_n$ are calculated as

$$\begin{aligned}
&\xi_1 = C_0\varphi_1 \\
&\xi_2 = B_1^T[1]\xi_1 + C_1\varphi_1 \\
&\ldots\ldots\ldots\ldots\ldots\ldots\ldots \\
&\xi_n = B_1^T[n-1]\xi_{n-1} + \ldots + B_{n-1}^T[n-1]\xi_1 + C_{n-1}\varphi_n
\end{aligned} \tag{3}$$

and then the following autoregressive scheme is used:

$$\xi_t = B_1^T[n-1]\xi_{t-1} + \ldots + B_{t-1}^T[n-1]\xi_{t-n+1} + C_{n-1}\varphi_t, t \ge n.$$

To calculate the matrices $B_i[j]$, it is necessary to solve systems of linear equations with a block Toeplitz matrix at each step. Details and various modifications of this method can be found in [3]. A general description of this problem solution is presented in [13]. Examples of simulating two-dimensional random fields of optical thicknesses are presented in [9].

4.3. Spectral model of Gaussian fields

We have simulated approximations of the homogeneous Gaussian fields $\hat{\omega}(x,y,z)$ based on a spectral decomposition [10, 11]. We applied an isotropic stochastic field with a correlation function $\rho(s)$ depending on the distance between spatial points, $s=\sqrt{x^2+y^2+z^2}$. The correlation function of the field $\hat{\omega}(x,y,z)$ can be represented in the form $\rho(s)=\int_0^\infty \frac{\sin(\gamma s)}{\gamma s} h(\gamma)d\gamma$, where $h(\gamma)$ is the radial spectral distribution density on $[0, +\infty)$ of a homogeneous isotropic random field $\hat{\omega}(x,y,z)$:

$$h(\gamma) = \frac{2}{\pi}\int_0^\infty \gamma s \sin(\gamma s)\rho(\gamma s)ds.$$

For the numerical simulation of the Gaussian field we use the following spectral model

$$\hat{\omega}_{NM}(\vec{r}) = \sum_{n=1}^{N} a_n \omega_M^{(\gamma_n)}(\vec{r}), \quad \omega_M^{(\gamma)}(\vec{r}) = \frac{1}{\sqrt{M}}\sum_{m=1}^{M}\left[\xi_m \cos\langle\lambda_m,\vec{r}\rangle + \eta_m \sin\langle\lambda_m,\vec{r}\rangle\right],$$

where ξ_m and η_m are independent standard normal variables, $\langle\lambda_m,\vec{r}\rangle$ is the scalar product of vectors in R^3. For a model without splitting the spectrum, the vectors λ_m have to be uniformly distributed on a hemisphere $S \in \mathbf{R}^3$ of radius γ: $\lambda_m = \gamma\boldsymbol{e}$, where $\boldsymbol{e} = (\boldsymbol{e}_1, \boldsymbol{e}_2, \boldsymbol{e}_3)$ is an isotropic vector, $a_n^2 = 1/N$, γ_n are independent random variables identically distributed on $[0, +\infty)$ with density h (γ). It is convenient to model the isotropic direction in the hemisphere S by the algorithm

$$\mathbf{e}_1 = \alpha_1, \quad \mathbf{e}_2 = \sqrt{1-\mathbf{e}_1^2}\cos(2\pi\alpha_2), \quad \mathbf{e}_3 = \sqrt{1-\mathbf{e}_1^2}\sin(2\pi\alpha_2).$$

Here α_1, α_2 are independent uniformly distributed on (0, 1) random values. For an exponential correlation function $\rho(s)=\exp(-\lambda s)$ of the random field $\hat{\omega}(x,y,z)$, the radial spectral distribution density has a form

$$h(\gamma) = \frac{4\lambda\gamma}{\pi(\lambda^2+\gamma^2)^2}.$$

To simulate the random radius γ, one can use a special algorithm [11] which is based on simulating random variables by using the gamma distribution $\gamma = \lambda\sqrt{\zeta_1^2 - 2\ln\alpha_1}/|\zeta_2|$, where ζ_1, ζ_2 are independent standard normal variables and α_1 is a uniformly distributed variable on (0,1).

5. Specific conversion of Gaussian fields

For obtaining a stochastic field with the marginal gamma distribution

$$f_{\nu,\lambda}(x) = \frac{\lambda^\nu x^{\nu-1}}{\Gamma(\nu)} e^{-\lambda x}, \lambda > 0, \nu > 0, x \geq 0$$

,

where ν is an integer parameter, it is convenient to apply a nonlinear transformation of the Gaussian fields $\omega^{(k)}, k = 1, \ldots 2\nu$:

$$\sigma_V(\vec{r}) = V\left(\omega^{(1)}(\vec{r}), \ldots, \omega^{(2\nu)}(\vec{r})\right) = \delta^2\left(\left(\omega^{(1)}(\vec{r})\right)^2 + \ldots + \left(\omega^{(2\nu)}(\vec{r})\right)^2\right).$$

A similar transformation with ν = 1 is described in [10, 11]. With this transformation, the autocorrelation function of the field $\sigma_V(\vec{r})$ has the form $r(s) = \rho^2(s)$, where $s = \sqrt{x^2 + y^2 + z^2}$, the mathematical expectation of the marginal distribution is $\mathbf{E}\sigma_V = 2\nu\delta^2$, the variance is $\mathbf{V}\sigma_V = 4\nu\delta^4$, abd the scale parameter is $\lambda = 1/2\delta^2$.

6. Numerical results

A three-dimensional random field of optical attenuation coefficients σ (x, y, z) of visible radiation is constructed under the assumption that the one-dimensional distribution function of the two-dimensional optical thicknesses field is a gamma function with parameters λ and ν (see Fig. 1), and the correlation function along the z axis is exponential with correlation radius ρ_z. Let the geometric thickness of the cloud layer be H = 0.5 km, and the geometric length of the field in the horizontal direction $a = b = 1$ km. The number of field partitions along the z axis is $N_z = 5$, along the x and y axes: $N_x = N_y = 100$. For each elementary three-dimensional cell τ_{ijk} , the optical thickness is simulated, $i = 1 \ldots N_x$, $j = 1 \ldots N_y$, $k = 1 \ldots N_z$. The attenuation coefficient in the cell ijk is constant , $\sigma_{ijk} = \tau_{ijk} * Nz / H$. Due to the infinite divisibility of the gamma distribution [12], the one-dimensional distribution of the optical thickness field τ_{ijk} of unit cells is a gamma distribution with parameters λ and ν/N_z. Successive horizontal sections of the attenuation coefficient field $\sigma(x, y, z)$, simulated according to the algorithms in paragraphs 4.1 and 4.2, with initial data obtained from statistical analysis of the field St-01 [2] are presented in Fig. 3. The parameters of this model are as follows: $\rho_z = 0.2$, $\lambda = 0.26$, $\nu = 37$, $\nu / N_z = 7.2$. For such a one-dimensional distribution of the field, the distortion of the horizontal correlation function $R_{x,y}(x, y)$ in the inverse function method does not exceed 6%, therefore, to simulate the two-dimensional Gaussian fields, a correlation function estimated from the experimental data (St-01) was used.

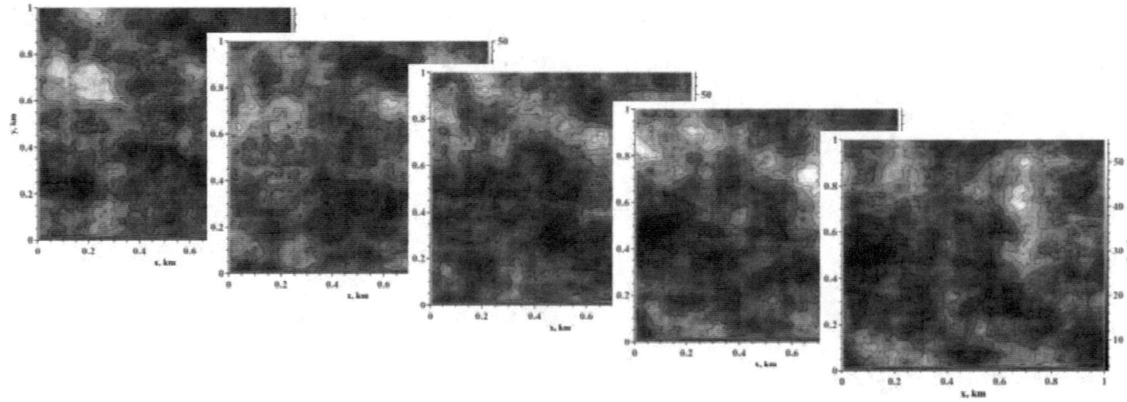

Figure 3. Successive horizontal sections of realizations of the attenuation coefficient field σ (x, y, z) obtained by the algorithms of conditional distributions, inverse functions, and autoregression.

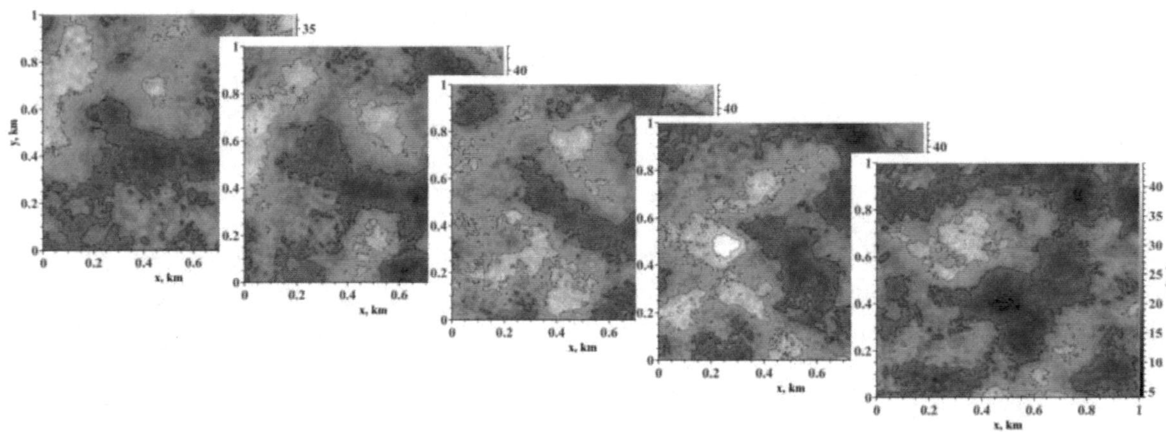

Figure 4. Successive horizontal sections of realizations of an isotropic spectral model of the attenuation coefficient field σ (x, y, z) obtained with a one-dimensional gamma distribution and an exponential correlation function.

Figure 4 shows the attenuation coefficients σ (x, y, z) for given fixed values of the vertical coordinate $z = (k\text{-}0.5)\ h_z$. $k = 1, ..., 5$, $h_z = 0.1$ km, simulated by using an isotropic spectral model (subsection 4.3) and a special nonlinear transformation (section 5). The correlation radius of the exponential function of the field σ (x, y, z) is $\rho(s) = 0.25$ km, the number of partitions of the spectral space $N = 20$, $M = 100$, and the number of terms in the nonlinear transformation $K = 2\nu = 20$, $\delta^2 = 1$.

7. Conclusions

The computation results show that the above-considered numerical methods for the stochastic field simulation (autoregressive schemes, spectral models, nonlinear transformations of Gaussian functions) can be effectively used in reproducing the geometrical and optical properties of atmospheric clouds. It is suggested to incorporate these methods into atmospheric models to study, in particular, the effects of clouds on gravity currents in the atmosphere [14, 15].

Acknowledgements
This work was supported by the Russian Foundation for Basic Research under grants 18-01-00149, 18-01-00609, and 17-01-00137, and ICMMG SB RAS under target program 0315-2019-0004.

References

[1] Prigarin S M and Marshak A L 2009 A simple stochastic model for generating broken cloud optical depth and cloud top height fields *J Atmos Sci.* **66**(1) 92–104

[2] Schafer M, Bierwirth E, Ehrlich A, Jakel E, Werner F and Wendish M 2017 *PANGAEA. Data Publisher for Earth and Environmental Science,* DOI: https://doi.org/10.1594/PANGAEA.874798

[3] Ogorodnikov V A and Prigarin S M 1996 *Numerical modelling of random processes and fields: algorithms and applications* (Utrecht, VSP)

[4] Ogorodnikov V A 1989 Testing of stochastic models for series of dry and rainy days *Proc. ZapSibNIGMI Goskomgidrometa* **86** 74–9

[5] Wilson R J and Nott D J 2001 Review of recent results on excursion set models *Image Analysis and Stereology* **20**(2) 71–8

[6] Piranashvili Z A 1966 Some problems of statistical probabilistic modelling of random processes *Problems of operations research* (Tbilisi: Metsniereba) pp 53–91

[7] Schafer M, Bierwirth E, Ehrlich A, Jakel E, Werner F and Wendish M 2017 Directional horizontal inhomogeneities of cloud optical thickness fields retrieved from ground-based and airborne spectral imaging *Atmos Chem Phys* 17 2359–72

[8] Kosarev A L, Mazin I P, Nevzorov A N and Shugaev V F 1976 Optical density of clouds (experimental data) *Proc. of Central Aerological Observatory* (Moscow: Gidrometeoizdat) issue 124

[9] Ogorodnikov V A, Kablukova E G and Prigarin S M 2018 Stochastic models of atmospheric clouds structure *Statistical Papers* **59**(9) 1521–32 DOI: 10.1007/s00362-018-1036-7

[10] Prigarin S M 2001 *Spectral models of random fields in Monte Carlo methods* (Utrecht, VSP)

[11] Shalygin A S and Palagin Yu I 1986 *Applied methods of statistical simulation* (Leningrad: Mashinostroenie)

[12] Mikhailov G A and Voytushek A V 2006 *Numerical statistical simulation. Monte Carlo methods* (Moscow: Academy)

[13] Marple S L Jr. 1987 *Digital spectral analysis with applications* (Englewood Cliffs, N.J.: Prentice-Hall)

[14] Yudin M S and Wilderotter K 2006 Simulating Atmospheric Flows in the Vicinity of a Water Basin *Computational Technologies* **11** 128-34

[15] Yudin M S 2016 A numerical study of gravity waves in the atmosphere: smooth and steep orography effects *IOP Conference Series: Earth and Environmental Science* **48** DOI http://dx.doi.org/10.1088/1755-1315/48/1/012024

Simplified method for monitoring of cumulus clouds by using global horizontal irradiance

S V Zuev[1] **and N P Krasnenko**[1,2]

[1] Institute of Monitoring of Climatic and Ecological Systems SB RAS, Tomsk, Russia
[2] Tomsk State University of Control Systems and Radioelectronics, Tomsk, Russia

E-mail: zuev@imces.ru

Abstract. A simplified method for monitoring of cumulus clouds which is based on a variational analysis of a time series of measured global horizontal irradiance is proposed. A distinctive feature of the technique is that it does not use an amplitude analysis of the time series and, as a result, there is no need to use any clear-sky model.

1. Introduction

Cumulus clouds have a significant impact on the formation of weather and climate of regions, because they are capable of carrying huge masses of water over considerable distances and heights. Besides, there is the need of continuous monitoring of cumulus clouds which, in turn, are the basis for the formation of cumulonimbus clouds that carry showers, thunderstorms, squalls, and hail. This paper describes an approach for the monitoring of cumulus clouds in the daytime using only a variational analysis of a time series of measured global horizontal irradiance (GHI).

The problem of automating observations of the sky state from the earth's surface is very difficult. One of the ways to replace the human observation of clouds could be analysis of operational actinometric information, in particular, analysis of global [1] or, separately, direct and diffuse irradiance [2]. In the standard approach measured solar irradiance is compared with the model values under clear sky, and then, using variance and amplitude analysis, the state of the sky is determined.

Variational analysis is carried out with the help of the standard deviation or the coefficient of variation (CV) of a time series. It shows how this series is variable in time. As applied to the measurement of solar irradiance, variational analysis gives an estimate of the frequency of changing the degree of coverage of the solar disk with clouds of different densities (at the analysis of direct irradiance) and the speed of movement of clouds in the zenith and near-zenith zones above the observation point (at the analysis of diffuse irradiance). In the process of analyzing the global irradiance, there is a combination of the influence of direct and diffuse irradiance and, accordingly, a combination of assessing the state of the solar disk and the state of the sky in the zenith and near-zenith zones.

According to the genetic classification of clouds, they can be stratiform, waveform, and cumuliform. Stratiform clouds (such as *Cs*, *As*, and *Ns*), stratus clouds *St* and the clear sky state have low variability in time and, accordingly, a small CV (close to zero). Waveform clouds (such as *Cc*, *Ac*, and *Sc*) are more dynamic and have medium values of the CV. Cumuliform clouds (such as *Ac cuf.*, *Sc cuf.*, and *Cu*) are capable of completely blocking direct solar radiation with some frequency, so they are characterized by large values of the CV.

The purpose of the amplitude analysis of a time series is to obtain a measure of the difference in the value of the measured solar radiation from its value under clear sky. This is usually done by calculating their ratio, and the result is a time series normalized in amplitude. The main problem in this case is the development of a clear sky model which should be correlated with the real state of the atmosphere.

2. Theory

If small segments of a time series are used for the analysis, for example, ± 10 minutes near the analysis point, the values of the measured GHI can be considered normalized and amplitude analysis will not be required. Figure 1 shows an example of daily variation of the global horizontal irradiance for clear sky (Ogurtsovo, WMO 29638), and Figure 2 shows its values for the analysis points 8:00, 12:00 and 17:00 GMT+6 (points A, B and C, respectively). It can be seen that these 21-min segments have small changes and, therefore, they can be taken as samples of the time series of the GHI with the same scale.

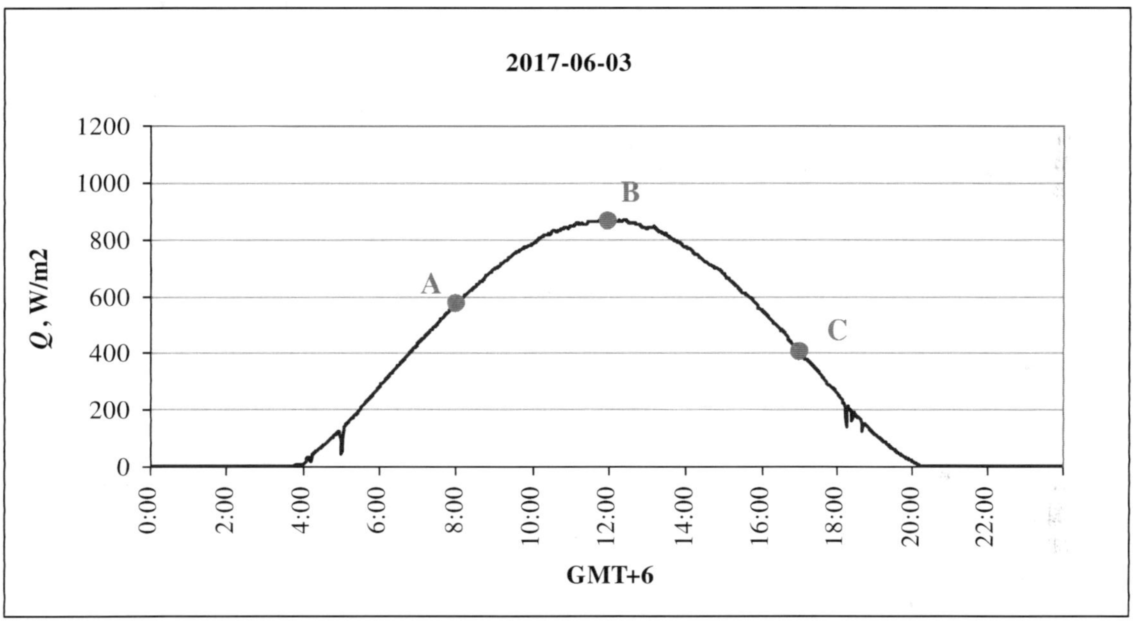

Figure 1. Daily time series of GHI for clear sky.

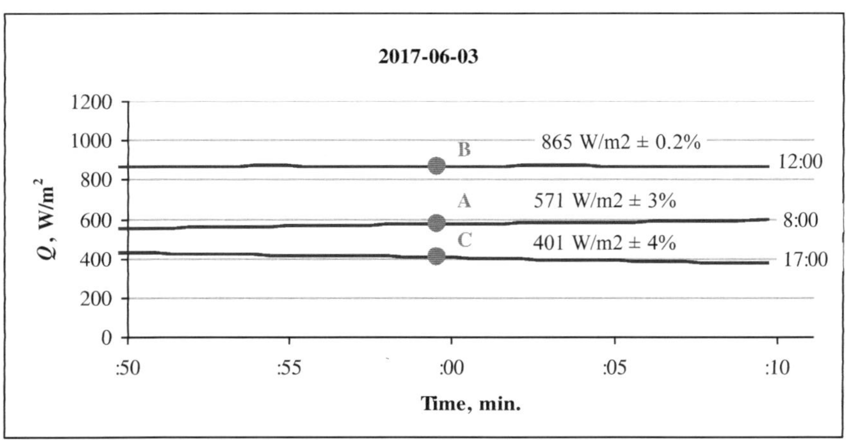

Figure 2. GHI in the range of ± 10 min for analysis points at 8:00 (A), 12:00 (B), and 17:00 (C).

That is, the 21-minute segments of the time series of global horizontal irradiance can be considered normalized, and the coefficient of variation can be used to analyze their homogeneity. Figure 3 shows typical variations of the GHI under clear sky and for some types of clouds and approximate values of the coefficient of variation V_Q for a 21-minute moving window.

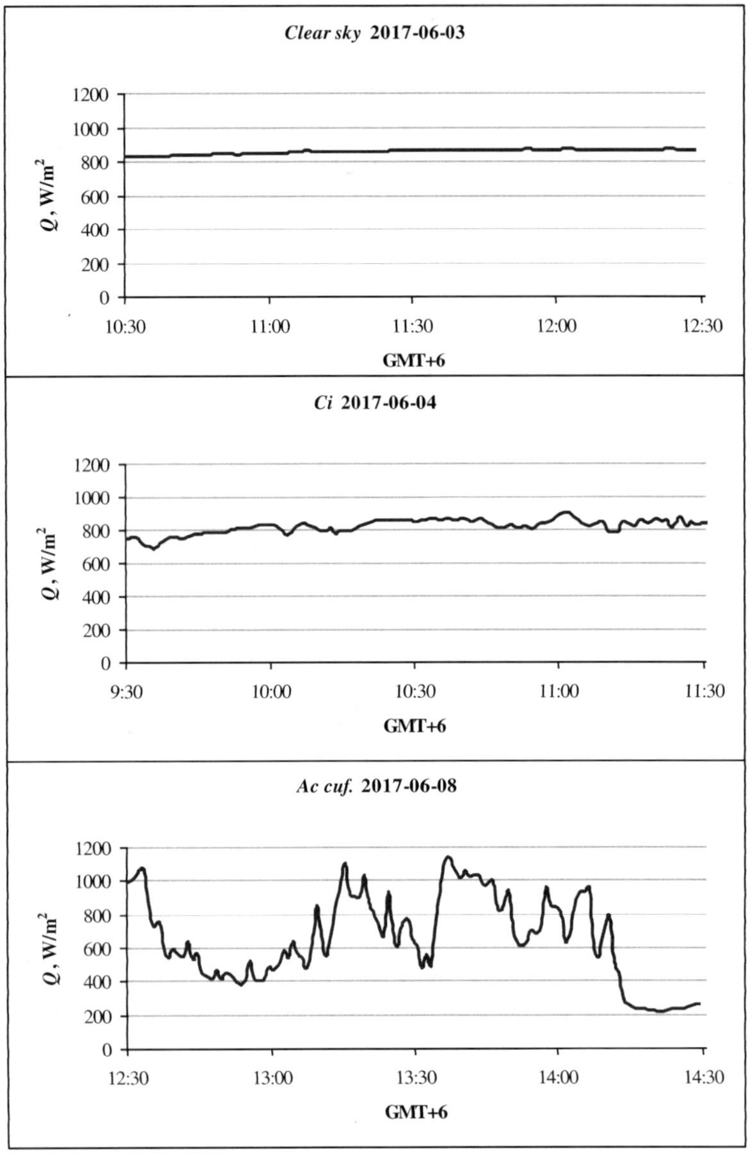

a)

$V_Q = 0.13 \div 0.31$

b)

$V_Q = 0.05 \div 0.18$

c)

$V_Q = 0.13 \div 0.31$

Figure 3. Typical time series of GHI under *a*) clear sky, *b*) cirrus clouds *Ci*, and *c*) altocumulus clouds *Ac cuf.*

Unlike other sky states, cumulus clouds clearly form a very inhomogeneous time series of global horizontal irradiance. Figure 4 shows a typical time series of global irradiance under cumulus clouds and the corresponding series of the CV. Values above the threshold of 0.33 correspond to moments of medium or large cumulus cloud amount. In this case, the task of monitoring of cumulus clouds can be reduced to the detection of such moments when the coefficient of variation for a 21-minute moving window of the measured GHI exceeds a certain threshold $V_Q > 0.33$.

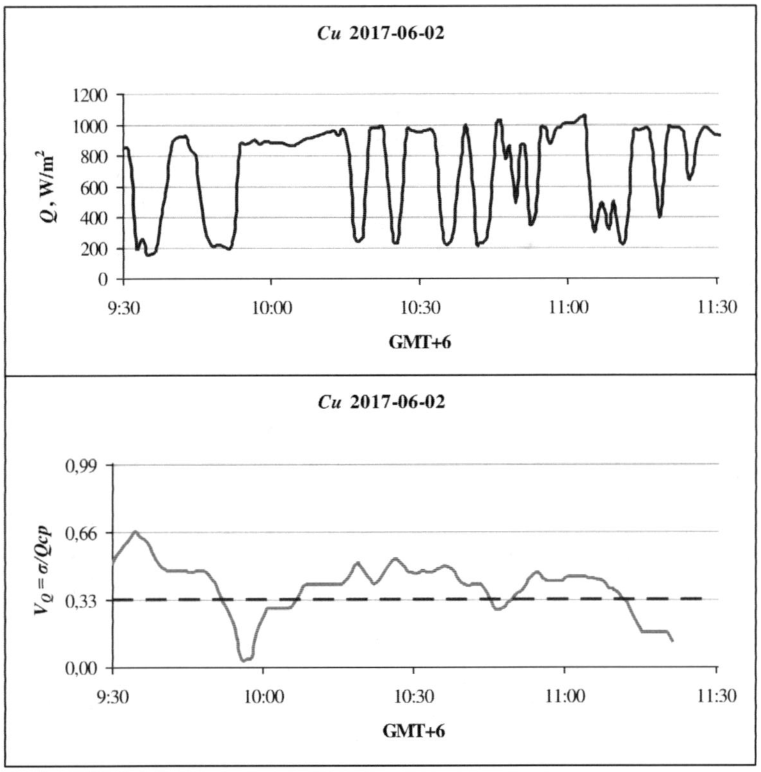

a)

b)

Figure 4. Typical time series of GHI *a*) and CV *b*) under cumulus clouds *Cu*.

3. Practical results

To test the proposed method, we used the measured GHI of an unshaded pyranometer CM11 (Kipp & Zonen) of the meteorological observatory of the IMCES SB RAS for the warm season of 2018 (at the Sun altitude $h > 30^o$). Verification of the results was carried out using a binary classifier by visual comparison of color panoramic images of the sky from the All-Sky camera installed next to the pyranometer. Comparison of *Precision* and *Recall* of the *Cu* classification was performed for $V_Q >$ 0.30, 0.33, and 0.36 by using the *F-measure*. For these values of V_Q, no significant differences were found.

Table 1. Results of the cumulus cloud classification.

Month	Number of analysis points ($h > 30^o$)	Threshold V_Q	Number of *Cu* points (*True positive*)	*Precision*	*Recall*	*F-measure*	Number of *Cu* error points (*False positive*)	Error rate, %	
								Sc	*Ac*
May	14806	0.30	3225	0.82	0.92	0.87	694	51	37
		0.33	2821	0.86	0.90	0.88	445	55	44
		0.36	2387	0.88	0.81	0.85	331	54	46
June	13255	0.30	2182	0.69	0.79	0.73	1002	31	49
		0.33	1885	0.72	0.74	0.73	723	31	55
		0.36	1679	0.77	0.69	0.73	498	33	44
July	16913	0.30	2847	0.82	0.85	0.84	614	37	31
		0.33	2628	0.85	0.82	0.84	452	33	39
		0.36	2370	0.88	0.77	0.82	332	32	31
August	14784	0.30	2970	0.81	0.79	0.80	706	48	38
		0.33	2610	0.83	0.74	0.78	537	49	38
		0.36	2262	0.85	0.68	0.76	395	51	36

As a threshold of V_Q, it is convenient to use a value of 0.33, which is usually used as a boundary to separate homogeneous and inhomogeneous time series. The results of verification of the simplified method for monitoring of cumulus clouds are shown in the table 1.

It should be noted that most cases of incomplete sampling are due to a small *Cu* amount. Errors of the classification accuracy (*false negative* and *false positive*) are mainly caused by the simultaneous presence of more than one type of clouds and errors of visual determination of the type and amount of clouds by All-Sky images. Stratocumulus *Sc* and altocumulus *Ac* clouds are most often mistakenly classified as cumulus clouds. Taking into account the difficulty of separating cumulus and cumulonimbus clouds over the observation point, the classification of *Cb* as *Cu* was taken as a positive result of the method (*true positive*).

4. Conclusions

The above-proposed simplified method allows, in the daytime (at the Sun altitude $h > 30^o$) and in almost real time, the monitoring of cumulus clouds of medium and large size, which are a basis for the formation of cumulonimbus clouds. First of all, it is necessary to prevent dangerous meteorological phenomena (showers, thunderstorms, squalls, hail) whose sources are just cumulonimbus clouds. There is no need to develop a clear sky model. This allows implementing the method by using small computational resources. This is very important for autonomous systems of measuring solar radiation, which are equipped, as a rule, with only one unshaded pyranometer for measuring global horizontal irradiance.

Acknowledgments

This research was carried out with financial support under project of fundamental research of the SB RAS No. IX.138.2.5 and assignment of the Ministry of Science and Higher Education of the Russian Federation No. 5.3279.2017/4.6.

References

[1] Duchon C E and O'Malley M S 1999 Estimating cloud type from pyranometer observations *J. Appl. Meteorol.* **38** 132-41

[2] Zuev S V and Krasnenko N P 2017 Cloud identification using actinometrical data *IOP Conf. Series: Earth and Environmental Science* **96** 012013

High-resolution simulation of particle transport in the urban atmospheric boundary layer

A I Varentsov[1], V M Stepanenko[2,1] and P I Konstantinov[1]

[1] Department of Meteorology and Climatology, Faculty of Geography, Lomonosov Moscow State University, 119991, Leninskie gory, 1, bld.1, Moscow, Russia
[2] Research Computing Center, Lomonosov Moscow State University, 119991, Leninskie gory, 1, bld.4, Moscow, Russia

E-mail: jdeadfox17@gmail.com

Abstract. This paper presents a model simulating the transport of particles of various sizes and masses in complex atmospheric boundary layer domains over urban areas by using a Lagrangian approach. The model takes input fields of air velocity and turbulence characteristics from various external RANS, LES, or DNS models (in this study, ENVI-met model is used), and includes a parameterization of subgrid stochastic velocity fluctuations. It allows us to estimate the movement of particles, their sedimentation, accumulation on the surface, and the variability of these processes depending on meteorological conditions. The results of test numerical experiments for aerosol transport over an idealized urban canopy are presented. The model has shown a good degree of qualitative conformity with the ENVI-met model.

1. Introduction

Air is a crucial resource of human life with a complex gas composition. It contains a multitude of different suspended particles, whose influence on people and nature is of great importance. In pristine nature, the phenomenon of particle transport rarely poses a serious danger to humans or ecosystems. In contrast, urban environment is a significant source of technogenic emissions, including various solid and liquid particles that are involved in the airflow and often have a stronger (compared to natural particles) negative impact on human, animal, and plant health.

Spatial heterogeneity of emission sources and complex geometry of the urban surface greatly complicate the process of particle transport. Thus, it can lead to overall high concentrations and the accumulation of particles in limited areas [1]. In the urban environment, both anthropogenic and natural aerosols are present and, therefore, the atmospheric transport becomes intricate for natural particles (plant pollen, snow, water drops, dust, and others) as well. Also, burning of various substances and the release of harmful combustion products into the atmosphere are typical for industrial areas and landfills. Under high wind speed conditions, the emission rate of heavy particles, such as sand, soil particles, and road dust, can significantly increase. The capabilities of modern measurement techniques to provide the spatiotemporal pattern of aerosols in an urban atmospheric boundary layer are limited because of high variability of particle concentrations on all scales. Therefore, the direct influence of the particle transfer in large cities on human health and the environment motivates the development of computational tools for modeling and, ultimately, forecasting the air quality determined by the aerosol content.

Due to the present-day fast increase of computer performance, the spatial and temporal resolution of hydrodynamic models of atmosphere steadily improves. However, the available microscale models of particle transport do not always allow one to adequately reproduce the transfer in the atmosphere of the entire diversity of particles characterized by different shapes, sizes, masses, and chemical compositions, since only a limited number of aerosol parameters are typically taken into account. It limits the possibilities of these models for studying the transport and deposition of particles. Also, computational modules implementing the transport of various pollutants which are integrated in complex microclimatic or meteorological models are linked to a specific software product and, hence, inherit possible limitations of the mathematical formulation and programming technologies chosen by the code developers. Most of such modules follow the Eulerian approach – disabling the possibility of tracking trajectories of individual particles, whereas the latter provides useful information in many studies. Taking into account the above, this work is aimed at developing a new Lagrangian model for the motion of particles in geophysical media, especially in the urban atmospheric boundary layer, with different options to parameterize turbulence, microscale physical and chemical processes. This paper presents first results of the model development.

2. Model description

2.1. *Governing equations*

In this model, the movement of particles and their interaction with the surrounding medium is described on the basis of known axioms and laws of physics using semi-empirical parameterizations.

The simplest concept to represent a particle suspended in the air is a material point. The mathematical formulation of how a material point or body moves in the three-dimensional space is the equation of motion, which describes the dependence of the spatial coordinates of a point on time:

$$\mathbf{r} = \mathbf{r}(t) = x(t)\mathbf{e_x} + y(t)\mathbf{e_y} + z(t)\mathbf{e_z}\,, \tag{1}$$

where $\mathbf{r}$ is the material point position, t is the time, x, y, z are the Cartesian coordinates, and $\mathbf{e_x}$, $\mathbf{e_y}$, $\mathbf{e_z}$ are unit basis vectors.

The equation of motion for particles suspended in the air and affected by the buoyancy and air resistance forces is as follows:

$$\mathrm{d}\mathbf{u_p}/\mathrm{d}t = \mathbf{g}(\rho_\mathrm{p} - \rho)/\rho_\mathrm{p} + F_\mathrm{D}(\mathbf{u} - \mathbf{u_p})\,, \tag{2}$$

where $\mathbf{u_p} = \mathrm{d}\mathbf{r}/\mathrm{d}t$ is the particle velocity, $\mathbf{g}$ is the gravitational acceleration, $\rho_\mathbf{p}$ is the particle density, ρ is the density of air (or other medium), F_D is the drag coefficient, and $\mathbf{u}$ is the ambient flow velocity.

The drag coefficient F_D is given by the following expression [2]:

$$F_\mathrm{D} = \frac{3\mu C_\mathrm{D} Re}{4\rho_\mathrm{p} d_\mathrm{p}^2}\,, \tag{3}$$

where μ is the dynamic air viscosity, C_D is the drag coefficient, Re is the Reynolds number for a particle, and d_p is the particle diameter.

Usually the Stokes law is used instead of this formula, but it does not take into account the flow turbulence. In the lower troposphere at the microscale level the influence of turbulence becomes very important. Hence, in equation (3) the Reynolds number accounts for the turbulence created by the particle motion relative to the ambient flow:

$$Re \equiv \frac{\rho d_p |\mathbf{u_p} - \mathbf{u}|}{\mu}. \tag{4}$$

Tor calculate the drag coefficient C_D, different methods are used depending on the shape of particles. In our model, particles are assumed to be spherical, and formulas from [3] are used:

$$C_\mathrm{D} = a_1 + \frac{a_2}{Re} + \frac{a_3}{Re^2}\,, \tag{5}$$

where a_1, a_2, a_3 are the empirical coefficients depending on the range of Reynolds numbers.

Various explicit and implicit finite-difference schemes are available for approximate solution of differential equations. In this model, to solve the equation of particle motion (2), an explicit 4-order Runge-Kutta method is used, which is chosen for high accuracy and ease of implementation.

IOP Conf. Series: Earth and Environmental Science **386** (2019) 012045 doi:10.1088/1755-1315/386/1/012045

2.2. *Model structure and operation scheme*

The model is implemented in a limited three-dimensional area with a given boundary geometry and continuous medium characteristics. The model computes the motion of an arbitrary number of particles, tracking the trajectory of each particle, diagnosing the particle concentrations and the surface deposition rate. The model code is implemented in C++.

The dynamic and thermodynamic properties of the air are assumed to be stationary and are read from the input file. The influence of the transported particles on the characteristics of the medium is taken negligibly small in comparison with the influence of external factors, the initial and boundary conditions.

The model algorithm is a sequence of iterations limited by a given modeling period. At each iteration, a cycle of logical blocks takes place, and each of them is responsible for a certain function. The first step in the algorithm chain is the reading of the input data and the initialization of the computational domain. Then the particle characteristics are calculated during the simulation period in a time-stepping cycle and, in the end, the results are written in files for subsequent analysis and visualization.

Input data (stationary air velocity field, turbulence properties) are assigned to the computational grid nodes. The last one is attached to the Cartesian coordinate system. Trilinear interpolation is used to calculate medium's parameters at the current location of each particle.

The domain boundary conditions for the particles in the model are realized as follows. When there is no solid surface at the boundary, the periodic conditions or transparent boundary conditions are applied. The interactions of particles with solid surfaces are divided into two types: deposition and elastic rebound (the velocity component normal to the boundary reverses the sign after the interaction) – the type of interaction can be set for each solid surface.

2.3. *Parameterization of turbulent velocity fluctuations*

In turbulent flows of fluid or gas the velocity may be represented as the sum of the averaged and fluctuation components:

$$\mathbf{u} = \bar{\mathbf{u}} + \mathbf{u}' , \tag{6}$$

where $\mathbf{u}$ is the flow velocity, $\bar{\mathbf{u}}$ is the averaged flow velocity, and $\mathbf{u}'$ is the fluctuation component of the flow velocity.

The averaged velocity in the model depends only on the spatial coordinates and is taken from the input data. The velocity fluctuation characterizes the effects of subgrid turbulent vortices and depends on both the coordinates and time. Since these vortices are not resolved explicitly, various parameterizations can be used to take into account their influence.

Two different parameterizations are realized in the model. In the first one (hereafter referred to as "simple parameterization"), the fluctuation component is a Gaussian random variable generated at each time step along a particle trajectory. Its mean value is zero, and the standard deviation is calculated as follows:

$$\sigma = \sqrt{2k/\epsilon} , \tag{7}$$

where σ is the standard deviation, k is the turbulent kinetic energy, and ϵ is the turbulent kinetic energy dissipation rate.

The second option to parameterize turbulence implemented in this work is a discrete random walk (DRW) model [4]. In this method, the fluctuation component is the Gaussian random variable, too; it has zero mean, and equation (7) is employed for the standard deviation, but the new value of this variable is not generated at each time step. The time of interaction of a particle with a turbulent eddy is introduced, during which the fluctuation component acting on the particle is considered constant. This time is calculated by the following equation:

$$t_{\mathrm{int}} = \min\left(t_{\mathrm{e}}, t_{\mathrm{R}}\right) , \tag{8}$$

where t_{int} is the time of interaction of a particle with an eddy, t_{e} is the eddy lifetime, and t_{R} is the time for the particle to cross the eddy (transit time scale).

The eddy lifetime t_{e} is estimated as follows:

$$t_e = l_e/|\mathbf{u}'|\,, \tag{9}$$

where l_e is the characteristic size of a randomly sampled vortex, which is assumed as the dissipation length scale given in [4].

The transit time scale t_R is estimated by the following expression:

$$t_R = -\tau \ln\left[1 - l_e/\left(\tau|\mathbf{u} - \mathbf{u_p}|\right)\right], \tag{10}$$

where τ is the droplet "relaxation" time defined in [4].

The parameterizations described above require accurate turbulent flow characteristics. Recent studies show that sufficient accuracy of such data for the particle transport problem is provided only by large eddy simulation (LES) models and direct numerical simulation (DNS) [5]. Therefore, careful analysis of the results is required if using this model with the input data obtained from Reynolds-averaged Navier-Stokes (RANS) models.

3. Results

The model has been tested in a number of idealized experiments, including a circular flow with a known analytical solution for trajectories, with the presence and absence of turbulence. In what follows we present the results of a particle transport simulation in a two-dimensional urban canyon.

To provide the Lagrangian model with external data on wind velocity, turbulent kinetic energy and dissipation rate, ENVI-met, a holistic microclimate modelling system [6] was used. *Inter alia*, this system includes a block for particle concentration transport, which uses the Eulerian approach [7].

The developed Lagrangian model showed qualitative agreement with the ENVI_MET Eulerian model in terms of concentrations and sedimentation rates in experiments with a flat surface, a single point source of particles, and a steady unidirectional wind.

For 2D modeling of the particle transport in an urban canyon, the inflow wind speed was set in the ENVI-met by a logarithmic profile with 5 m/s at 10 m and directed perpendicular to a set of parallel canyons. The atmosphere stratification was set neutral. Six buildings 30 meters high and 18 meters wide were placed in the domain. All buildings were separated by intervals of 30 meters, which formed five consecutive identical city canyons with an aspect ratio of 1. A point source was placed inside the second downwind canyon to simulate the automobile emissions and study the quality of ventilation of the canyon with this aspect ratio and perpendicular flow velocity. All particles had a diameter of 2.5 μm and a density of 1000 kg/m^3. The simulation period was set to 600 seconds.

The results of the simulations show (Figure 1.a and 1.b) that most of the particles emitted into the air remain in the source canyon, and maximum concentrations are observed there. The model also shows that a small part of the particles rises above the canyon and spreads further, flowing and settling in the downwind canyons. The ability of the particles to enter downwind canyons is also found when using more accurate large eddy simulation models [8]. The developed Lagrangian model with simple parameterization of turbulence demonstrates (Figure 1.a) the capture of particles by a large vortex inside the canyon and their uneven rise to the roof level. The more advanced DRW method shows (Figure 1.b) an uneven distribution of the concentrations within the canyon. In addition, the plume is wider when using this stochastic parameterization, because of longer influence of each generated speed fluctuation on the particle, which seems more realistic in specified conditions. However, the advanced method is more sensitive to input data on turbulence characteristics and, therefore, it may not be applicable in some cases due to the suppression of other effects by turbulence. This indicates the sensitivity of the model to the turbulence parameterization being used and the need to explore this aspect of Lagrangian modelling more.

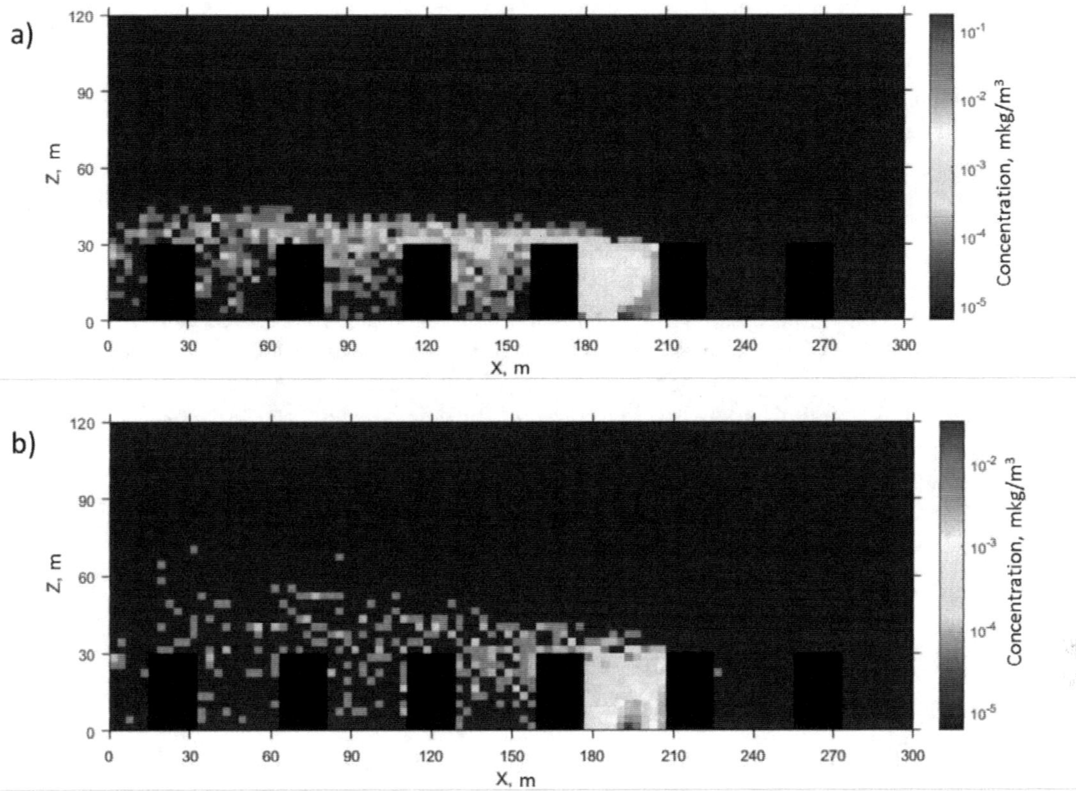

Figure 1. Particle concentrations calculated by Lagrangian model using two turbulence (stochastic) parameterizations: a) simple parameterization; b) discrete random walk model.

4. Conclusions

This work presents first steps in the development of a Lagrangian model of particle advection in the urban atmospheric boundary layer with the possibility of receiving fields of hydrodynamic and thermodynamic variables from various possible sources (RANS, LES, DNS models) and simulating aerosols in a wide range of physical and chemical properties.

The model has been tested in a number of idealized flow cases with input data from the ENVI-met model. The results of modeling and comparison showed a high degree of conformity between the above-developed technology and the ENVI-met model. Two parameterizations of subgrid turbulent (stochastic) velocity fluctuations were introduced, and simulations of aerosol transport in a 2D domain over a sequence of parallel canyons demonstrated the sensitivity of model results to the choice of stochastic parameterization.

Further steps in the development of the Lagrangian model will be in coupling to a RANS urban boundary layer model parallelized for supercomputers in order to run simulations in larger domains and in wider ranges of boundary layer parameters and surface geometries. Another direction of the model improvement is more accurate accounting for the physical properties of specific classes of particles, both natural and anthropogenic, e.g. their shape and size distributions.

Acknowledgments
This research was partially supported by the Russian Foundation for Basic Research (project no. 18-05-60126).

IOP Conf. Series: Earth and Environmental Science **386** (2019) 012045 doi:10.1088/1755-1315/386/1/012045

References

[1] Droste A M, Steeneveld G J and Holtslag A A M 2018 Introducing the urban wind island effect *Environ. Res. Lett.* **13** 094007
[2] *ANSYS Fluent Theory Guide* (Release 15.0 November 2013), ANSYS, Inc., Pennsylvania, p 374
[3] Morsi S A and Alexander A J 1972 An investigation of particle trajectories in two-phase flow systems *J. Fluid Mech.* **55** 193
[4] Gosman A D and loannides E 1983 Aspects of Computer Simulation of Liquid-Fueled Combustors *Journal of Energy* **7** 482–90
[5] Blocken B 2018 LES over RANS in building simulation for outdoor and indoor applications: A foregone conclusion? *Build. Simul.* **11** 821–70
[6] ENVI_MET - Decoding urban nature https://www.envi-met.com/ Accessed 20 Mar 2019
[7] Bruse M 2007 ENVI-met implementation of the gas/ particle dispersion and deposition model PDDM http://envi-met.net/documents/sources.PDF Accessed 12 Apr 2019
[8] Glazunov A V 2018 Numerical simulation of turbulence and transport of fine particulate impurities in street canyons *Vychisl. Metody Programm.* **19** 17–37

Real-time microscale modeling of thermal comfort conditions in Moscow region

A A Perkhurova[1], P I Konstantinov[1], M I Varentsov[2], N I Shartova[1] , T E Samsonov[1] and V N Krainov[1]

[1] Lomonosov Moscow State University, Faculty of Geography, Russia, Moscow, GSP-1, Leninskiye gory, 119991

[2] Lomonosov Moscow State University, Research Computing Center, Russia, Moscow, Leninskiye gory, 119991

E-mail: an.perkh@gmail.com, kostadini@mail.ru

Abstract. Urban climate comfort is an indicator of a set of parameters, such as temperature, humidity, and solar radiation for a person's sensation of being favorable to being outdoors or indoors. In this study, an attempt to develop a technology of real-time prediction of thermal comfort conditions in urban landscape is described (based on the example of Moscow State University campus). For this, the authors used a RayMan model-based algorithm for calculating three most popular worldwide comfort indexes. In the scripting method, predictive data of the Canadian global model meteorological parameters are automatically transferred to the RayMan-model (with an implementation of the unique thermal and radiation properties of the Moscow State University campus landscape) by using an autoclicker software. For the convenience of perception of the information, the results of calculations are visualized on the basis of a free web mapping service. Thus, the main idea of the work is that any user with minimum expenditure of his time resource and without knowledge of the model's work can launch the program and receive an individual forecast of comfort conditions for the next few hours in a visually understandable format. It is suggested that the developed methodology will be used for calculations on projected areas to identify the safest construction option. Such real-time forecasting will continue to be of particular importance for urban infrastructure.

1. Introduction and motivation

The health of the urban population is influenced by many factors, among which there are meteorological parameters. Moreover, the impact is not expressed in individual indicators, but in their totality, which is displayed by thermal comfort indices [1]. Basically, they are identified empirically and are presented in the format of the outdoor temperature for ease of perception of information.

Due to the heterogeneity of urban development and urban climate in the Moscow region [17], the thermal comfort conditions at different points in the same territory will differ noticeably with the same meteorological parameters. Thus, it is necessary to study the parameters of comfort on the microscale. Therefore, within the framework of this study, in order to inform the public about the negative impact of weather, and further to minimize the consequences on the human body, an attempt was made to develop an operational system for predicting dangerous conditions of thermal comfort.

By now, more than 100 indices have been developed in world practice to determine the bioclimatic comfort of human life [2]. Such a huge number of indexes developed up to this point is due to the fact that there is no universal indicator due to very different weather conditions in different parts of the planet. Indices that give a more accurate value of cold stress and are based on experiments in a frosty environment are called cold ones. Accordingly, those that better determine heat stress and are based on experiments in a hot environment are called thermal ones. Also, special models have been developed to take into account the individual characteristics of a person when calculating thermal conditions.

IOP Conf. Series: Earth and Environmental Science **386** (2019) 012046 doi:10.1088/1755-1315/386/1/012046

Heat balance models take into account many aspects of the heat transfer mechanism. In this study, the authors use three most popular comfort indexes: PET, UTCI, and mPET.

The PET (Physiologically Equivalent Temperature) [3] is a parameter calculated using the Munich Energy-balance Models for Individuals (MEMIs) and defined as the air temperature at which the heat balance for normal room conditions for the human body remains unchanged with the temperature values of the internal organs and skin temperature for this situation. In addition to standard meteorological parameters, such as air temperature, wind speed, and relative humidity, the calculation of PET takes into account the processes of human metabolism, activity, and the thermal insulation properties of clothes. Currently, PET is one of the most commonly used indicators of human comfort and its influence on mortality during heat waves [14].

The mPET is a modified PET [4]. This index represents an improved PET with corrected weaknesses, which were: poor ability to predict thermophysiological parameters and poor response to both clothing insulation and wet conditions.

The parameter UTCI (Universal Thermal Climate Index) is also based on the model of human heat budget [5]. It is expressed as the equivalent ambient temperature of the reference environment providing the same physiological reaction of the reference person in a real environment. In contrast to the PET, when calculating the UTCI not only the heat-insulating properties of clothes are taken into account, but also vapor tightness, insulation of the air surface layers that undergo changes with changing wind speed and the movement of the human body and, as a consequence, affect the physiological reactions. Thus, the modified insulating properties of clothing in a real environment are taken into account.

In 2018, a real-time heat stress forecasting service in Greece was developed based on a thermophysiologically based index. This work is considered in the article “Operational Forecasting of Human-Biometeorological Conditions” [6]. The service is based on an autonomous connection between a model of numerical weather prediction operating at the National Observatory in Athens and a human biometeorological model. The main purpose of the service is to provide information about comfort conditions in a form that will be easily understood by the public. The forecast is provided by the service in 2 forms: table format for settlements and map format. The described service is one of the first of its kind. Thus, in this article the authors show that biometric information about a person can and should be transmitted to the population in an effective and easy to understand way.

In this paper, an attempt is made to develop an operational forecast of thermal comfort indices for the campus of Moscow State University. However, unlike the Greek service, it was planned to write an algorithm based on open forecast data of meteorological parameters. Consequently, it will be possible to implement a real-time prediction based on any other forecast models.

2. Materials and methods

To assess the urban climate in a physiologically significant manner, it is required to use methods of modern human-biometeorology dealing with the effects of weather, climate, and air quality on human organism [7]. The model **RayMan** [10] estimates the radiation fluxes and the effects of clouds and solid obstacles on short wave radiation fluxes. The model, which takes into account complex structures, is suitable for utilization and planning purposes on local and regional levels. The final output of this model is the calculated mean radiant temperature, which is required in the energy balance model for humans. Consequently, it is also required for the assessment of urban bioclimate and thermal indices, such as the Predicted Mean Vote (PMV), the Physiologically Equivalent Temperature (PET), and the Standard Effective Temperature (SET*).

The development of the model is based on the German VDI-Guidelines 3789, Part II: Environmental Meteorology, Interactions between Atmosphere and Surfaces; Calculation of the Short- and Long Wave Radiation and VDI-3787: Environmental Meteorology, Methods for the Human-Biometeorological Evaluation of Climate and Air Quality for the Urban and Regional Planning at Regional Level. Part I: Climate [8,9]. For the calculation of thermal indices based on the human

IOP Conf. Series: Earth and Environmental Science **386** (2019) 012046 doi:10.1088/1755-1315/386/1/012046

energy balance, meteorological (air temperature, wind speed, air humidity, and short and long wave radiation fluxes) and thermo physiological (activity and clothing) data are required [10,11,12].

Sensitivity test. To test the RayMan model for accuracy, a traditional sensitivity test was performed. The input information includes several meteorological and physiological parameters: air temperature, wind speed, relative humidity, cloud cover, physical activity, which is measured in watts [13], and others. The indexes were calculated for data series with one parameter changing and the others stable. Then, based on the values obtained, a graph was constructed, and the regression equation was determined. That is, you can trace how much the indices vary when one of the parameters included in the calculation deviates. The data are presented in Table 1.

Table 1. Change in comfort indices (modulo) when changing the parameter by one (for humidity - by 10%, for activity - by 10 W). The parameters are arranged in decreasing order of influence on the PET index.

Parameter	PET	UTCI	mPET
Wind speed (for Δ1 m/s)	-2.50	-1.65	-1.93
Temperature (for Δ1 °C)	1.00	0.94	0.86
Rel.humidity (for Δ10%)	0.10	0.04	0.04
Cloud cover (for Δ1 oct)	0.09	0.05	0.07
Human activity (for Δ10W)	0.00	0.00	0.44

On all graphs of the dependence of the index on the parameter, a linear regression was constructed. In almost all cases, it turned out to be positive, which means that with increasing values of the abscissa axis, the values of the ordinate axis also increase. Only for wind the dependence turned out to be negative (the higher the wind speed, the lower the value of the indicator). The dependences of the UTCI and PET indices on physical activity are not observed, which is apparently due to the fact that a certain activity value is included in the calculation models of these parameters and, therefore, activity is not input information.

Table 1 shows that the calculation models for all 3 indices are most sensitive to changes in wind speed. The sensitivity of the parameters to changes in relative humidity, cloud cover, and activity is weakly expressed. In relation to all parameters (except for human activity), the PET is more dependent than the others. On the contrary, the UTCI is the least sensitive to changes in all parameters (except for temperature).

The sensitivity of the UTCI to wind is less than the sensitivity of the PET due to the fact that when calculating the first index heat loss in clothing deformation during strong winds is better taken into account than the PET.

There is no clearly positively distinguished index among the rest; each has its own advantages and disadvantages. Therefore, in relation to the study at the Moscow State University campus, calculations will be performed for all 3 indices considered above.

3. Results

Development of Real-Time Forecast Technology

Diagnostic models for calculating comfort indices have already been invented in the biometeorology industry. However, to minimize the harm caused to health, a person must be informed about the conditions of conditional comfort, and so there is a need to create a prognostic model. Nevertheless, the calculation program is hidden from the user. In this paper, an attempt is made to solve this problem. Thus, the main idea of this work is that any user with the minimum expenditure of his time

IOP Conf. Series: Earth and Environmental Science **386** (2019) 012046 doi:10.1088/1755-1315/386/1/012046

resource and without knowledge of the model's work could launch the program and receive an individual forecast of comfort conditions for the next hours in a visually understandable format.

When developing the technology, it was necessary to create several sequentially executed blocks. The flowchart is shown in Figure 1. Further, each of the blocks will be considered in more detail.

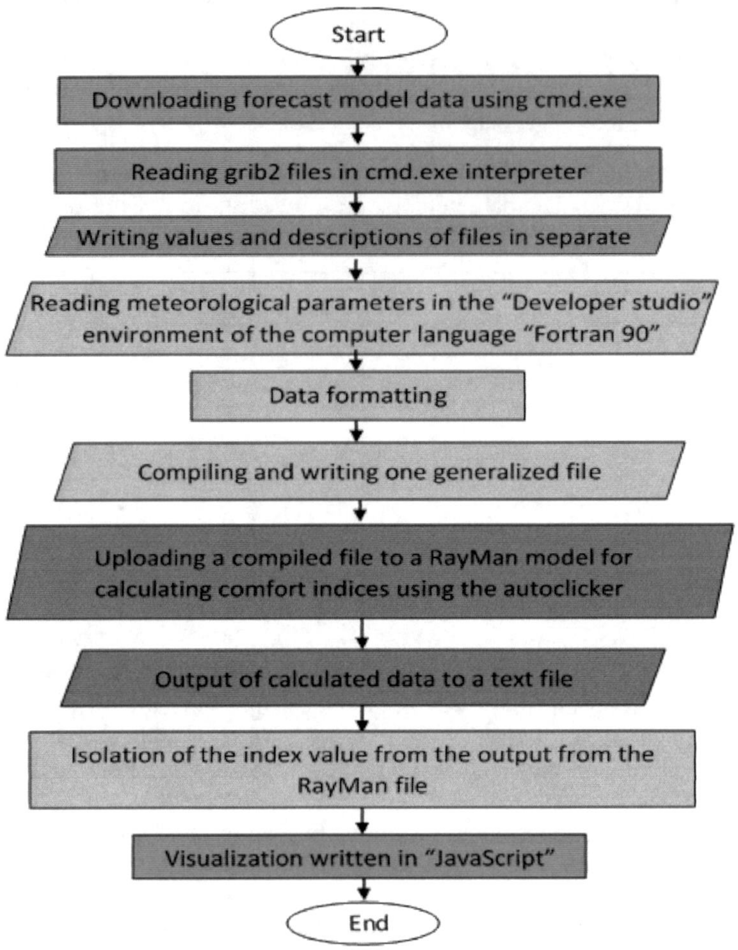

Figure 1. Flowchart of technology for real-time modeling of thermal comfort indices. Blocks operating in the same environment are indicated by one color.

3.1. Download forecast data

For the study, it was decided to use the forecast data of the Canadian GEM global model. The Global Environmental Multiscale Model (GEM) [15] is an integrated forecasting and data assimilation system developed by Environment Canada to meet the operational weather forecasting needs of Canada for the foreseeable future.

These needs include short-range regional forecasting, medium-range global forecasting, and data assimilation; the model is currently operational for the global data assimilation cycle and medium-range forecasting, the regional data assimilation spin-up cycle and short-range forecasting. Overnight mesoscale forecasts are also produced, and are available to operational forecasters. A growing number of meteorological applications now either use or are based on the GEM model.

IOP Conf. Series: Earth and Environmental Science **386** (2019) 012046 doi:10.1088/1755-1315/386/1/012046

It was decided to use the forecast data of the Canadian GEM global model because, first, this information is publicly available, and second, the values of different meteorological parameters are located in separate files for easy pumping and processing.

Downloading is performed using the "cmd.exe" interpreter in the Windows environment, after installing the *wget* utility on the hard disk. For further calculations, it is necessary to download data on surface temperature, surface atmospheric pressure, total cloud cover, wind speed 10 meters above surface level, and specific humidity.

3.2. File conversion

The need for this module was that the GEM global model provides its information in grib2 format, and the RayMan model which calculates comfort indices works with text files.

Conversion is also performed in the cmd.exe interpreter in the Windows environment using the *wgrib2* utility. First you need to set the path to the folder in which this utility is located, and then connect it. The next step is to retrieve the value at the desired point. The resolution of the Canadian model is 0.6 degrees. After preliminary calculating the number of the cell in which the values of the meteorological parameters for Moscow are located, the required information is obtained using the "-ij" option. This operation must be performed 5 times for each of the downloaded files. After extraction, the forecast meteorological variables are written to separate text files. Description information is also recorded.

3.3. Data formatting

The next stage of the program is data formatting and combining all necessary information into one file using the computer language "Fortran 90" in the "Developer studio" environment. Correction of the data is required to conform to the format requested by the RayMan model. In this block, specific humidity is converted to relative one, the cloud units are converted from percent to octants, the wind speed is reduced from 10 m to 2 m, the temperature is converted from degrees Celsius to degrees Kelvin. Then all formatted data are combined into one text file, which will serve as input to the RayMan model. As a result, a file with a program of the .exe format is downloaded to include the module in a common file.

3.4. Calculation of comfort indices

The file obtained at the previous stage is loaded into the model using the "AutoClickExtreme" auto clicker [18]. Autoclickers are a type of software or macro that can be used to automate clicking. They can be triggered to generate input that was recorded earlier or generated from various current settings.

Comfort indices are calculated based on the heat-energy balance model. All formulas are embedded in the RayMan model and are hidden from an ordinary user, which makes it difficult to include the calculation stage in a common script. This was the need to use the autoclicker. It recorded mouse movements: using the file obtained at the previous step, as well as the obstacle file and marking the location of the point on it as input; ticking the used weather data; start of index calculation. After completing all necessary actions, it is possible to deflate the autoclicker record in a separate file, so that later it can be included in the technology.

The choice of the clicker was ensured by its advantages: the ability to run from the command line, tracking and recording computer mouse movements, typing on the keyboard, and playing back actions almost without errors.

To calculate the comfort conditions, 10 points were chosen on the Moscow State University campus. The points are located on all 4 sides of the main building, in the botanical garden, and on the open football field, and so at each moment of time they are in a different exposure due to the shading conditions. This fact gives reason to consider a representative picture made up of selected points to be representative, as it was made for Moscow-2010 heat wave conditions [16].

The comfort indices were calculated for a conventional person with standard physiological parameters (male, height - 180 cm, weight - 75 kg, age - 35 years) using one file with the values of meteorological elements and 3 obstacle files.

3.5. Formatting a file with calculated values
The next step is formatting again. At this point the result file changes, in which there is a large amount of information that is unnecessary for visualization and complicates it. Using the “Fortran 90” language, a program was implemented to extract the desired value from the file.

3.6. Visualization
After receiving the results, there is a need to display them in an easily perceptible format. For visualization, the calculated values were plotted on the campus map of Moscow State University in the form of punsons at the studied points. At this stage, the multifunctional mapping library “Leaflet” was used, working on the basis of the computer language “JavaScript”. The result (Figure 2) is displayed on the website http://urbanreanalysis.ru/PET_forecast.html.

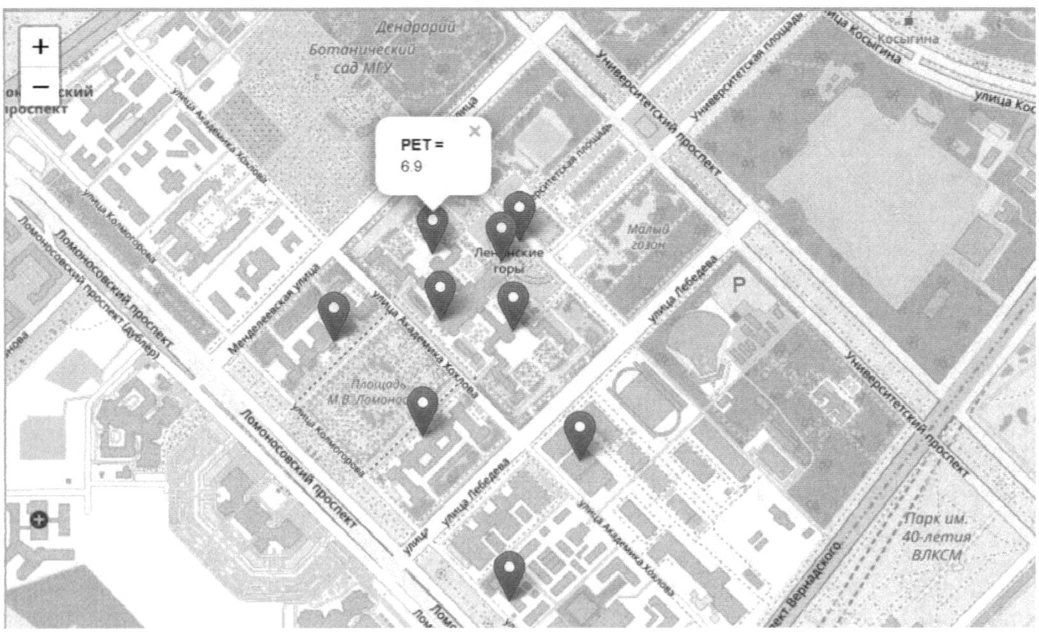

Figure 2. Visualization of calculated PET-values.

The last step in creating the technology was to combine all modules into one script for better usability. This stage was carried out using the “Bash” language in the Windows environment. The result of the work is the ability of the user to launch the finished bat-file with two clicks without a detailed study of the essence of the technology and in a few minutes to receive the spatial distribution of the predicted comfort indices in the area of interest.

4. Technology approbation and discussion

Climatic calculations of thermal comfort indices on the campus of Moscow State University

To track the differences in the comfort conditions on a climatic scale on the university campus, the indices were calculated from 1980 to 1999, provided by the Moscow State University Meteorological Observatory. The selected period was 20 years, since the calculation time is limited in the RayMan model. To search for maximum discrepancies, only the July period was taken from a series of climatological data. The authors also analyzed two points that were different in terms of shading: the first is located on the football field (open), and the second in the botanical garden (covered by tree crowns). The obtained values of the PET index were distributed according to the repeatability of stress levels and are illustrated in pie charts (Figure 3).

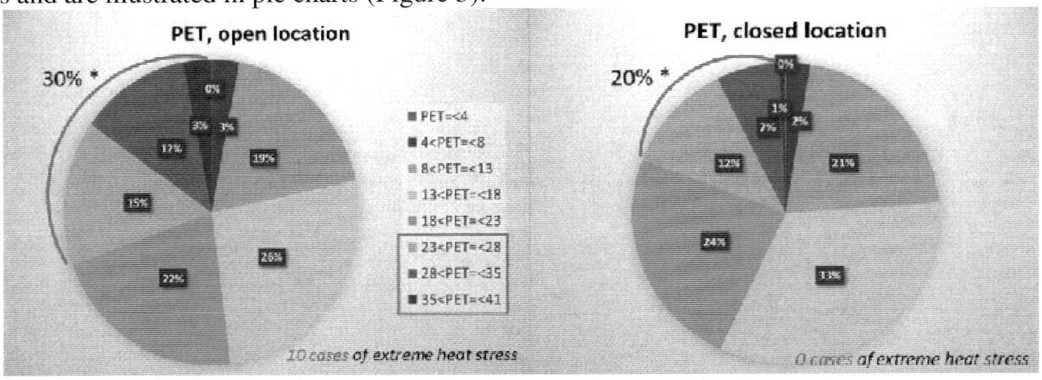

Figure 3. Frequency of PET Thermal Comfort Index values distributed across various comfort levels. *repeatability of cases of mild, moderate, and severe heat stress ($28^{\circ}C < PET < 41^{\circ}C$).

In Figure 3 one can see that on the football field in July the cases of weak, moderate, and severe heat stress ($28^{\circ}C < PET < 41^{\circ}C$) are repeated 10% more often than in the botanical garden. It is also worth noting that at the first point the number of cases of extreme heat stress ($PET > 41^{\circ}C$) is 10, and at the second one it is 0.

This confirms the conclusion that in the summer in Moscow in the open area it is much more dangerous than in the shaded one. Moreover, if we are talking about a football field, a person receives heat not only due to microclimatic features, but also due to increased physical activity. This can lead to health problems, and in the case of abuse of activity in a non-shaded area, also to more serious illnesses (heat strokes, etc.).

5. Conclusions

The main result of this study is the following: for the first time in Russia, a real-time prediction technology of thermal comfort conditions has been developed. The result is illustrated in a form that is visually understandable to the common user. In the future, the technology can be applied not only to the campus of Moscow State University, but also to any other subject with the necessary input data: an obstacle file, a file of forecast meteorological parameters. Also, if desired, the user can set other physiological parameters in order to project the obtained PET-values for a specific person.

Also, a sensitivity test was carried out, which showed a great dependence on the meteorological parameters of the PET index compared to the UTCI and mPET. On the activity of a person using the 3 considered indices, only the latter depends. A strong sensitivity of all 3 indices to changes in wind speed has been obtained.

Calculations based on climatic data for the 2 most different in terms of shade points were made. It has been shown that there are 10% more cases at an open point with conditions of weak, moderate, and severe heat stress than at a point covered with tree crowns.

This work can be continued for other cities, as well as megacities. As part of the technology, in the future it is possible to use the outputs of regional models as initial data, which ultimately may lead to an improvement in the quality of climate services for the population of Moscow and the Moscow region. Real-time forecasting is a new direction in biometeorology and, quite possibly, it will continue to be of particular importance for urban infrastructure.

Acknowledgements

Much of the research (the development of a real-time prediction technology for thermal comfort conditions) was supported by the Russian Foundation of Basic Research and Moscow city government (project no. 19-35-70009 mol_a_mos). The thermal comfort indices assessment and computing at the MSU campus (performed by A.A.Perhurova, N.V.Shartova, and P.I. Konstantinov) were supported by the Russian Science Foundation (project no. 17-77-20070 "Initial Assessment and Projection of Bioclimatic Comfort in Russian Cities in the 21st Century under Climate Change").

References

[1] Emelina S V, Konstantinov P I, Malinina E P and Rubinstein K G 2014 Evaluation of the informativeness of some biometeorological indices for different regions of Russia *J. Russ. Meteorol. Hydrol.* **7** 25-37
[2] Blazejczyk K *et al* 2012 Comparison of UTCI to Selected Thermal Indices *Int. J. Biometeorol.* **56** 515–35
[3] Höppe P 1984 Die Energiebilanz des Menschen *Wiss Mitt Meteorol Inst Univ München* **49**
[4] Lin T P, Yang S R, Chen Y C and Matzarakis A 2019 The potential of a modified physiologically equivalent temperature (mPET) based on local thermal comfort perception in hot and humid regions *Theor. Appl. Climatol.* **135** 873–6
[5] Weihs P *et al* 2012 The Uncertainty of the UTCI Due to the Uncertainties of the Meteorological Data *Int. J. Biometeorol.* **56** 537–55
[6] Giannaros T M, Lagouvardos K, Kotroni V and Matzarakis A 2018 Operational Forecasting of Human-Biometeorological Conditions *Int. J. Biometeorol.* **62** (7) 1339–43
[7] Mayer H, 1993 Urban bioclimatology *Experientia* **49** 957-63
[8] VDI 1994 VDI 3789 Part 2: Environmental Meteorology, Interactions between Atmosphere and Surfaces; Calculation of the short- and long wave radiation (Düsseldorf: VDI/DIN-Handbuch Reinhaltung der Luft, Band 1b)
[9] VDI 1998 VDI 3787 Part I: Environmental meteorology, Methods for the human biometeorological evaluation of climate and air quality for the urban and regional planning at regional level Part I: Climate (Düsseldorf: VDI/DIN-Handbuch Reinhaltung der Luft, Band 1b)
[10] Matzarakis A and Rutz F 2005 Application of RayMan for tourism and climate investigations *Annalen der Meteorologie* **41** 2 631-6
[11] Matzarakis A, Rutz F and Mayer H 2007 Modelling Radiation fluxes in simple and complex environments – Application of the RayMan model *Int. J. Biometeorol.* **51** 323-34
[12] Matzarakis A, Rutz F and Mayer H 2010 Modelling Radiation fluxes in simple and complex environments – Basics of the RayMan model *Int. J. Biometeorol.* **54** 131-9
[13] Fröhlich D and Matzarakis A 2018 RayMan Pro, modelling of Mean Radiant Temperature and Thermal Indices. RayMan manual version 0.1
[14] Shartova Natalia, Shaposhnikov Dmitry, Konstantinov Pavel and Revich Boris 2018 Cardiovascular mortality during heat waves in temperate climate: an association with bioclimatic indices *International Journal of Environmental Health Research* **28**(5) 522–34
[15] Kao-San Yeh, Côté Jean, Gravel Sylvie, Méthot André, Patoine Alaine, Roch Michel, Staniforth Andrew 2002 The CMC–MRB Global Environmental Multiscale (GEM) Model Part III: Nonhydrostatic Formulation *Monthly Weather Review* **130** (2) 339–56
[16] Konstantinov P I, Varentsov M I, Malinina E P 2014 Modeling of thermal comfort conditions inside the urban boundary layer during Moscow's 2010 summer heat wave (case-study) *Urban Climate* **10** (3) 563–72 DOI:10.1016/j.uclim.2014.05.002
[17] Varentsov M, Wouters H, Platonov V and Konstantinov P 2018 Megacity-Induced Mesoclimatic Effects in the Lower Atmosphere: A Modeling Study for Multiple Summers over Moscow, Russia *Atmosphere* **9** 50 DOI:10.3390/atmos9020050
[18] http://www.autoclickextreme.com/ (date of the application 24.02.2019)

Analysis of circulation near the coast of Western Crimea and the region of Sevastopol with assimilation of temperature and salinity observations

N Evstigneeva and S Demyshev

Marine Hydrophysical Institute RAS, Sevastopol, Russian Federation

E-mail: naevstigneeva@yandex.ru

Abstract. Hydrophysical fields which are continuous in time and space are reconstructed in the coastal zone of the Black Sea (near the coast of Western Crimea and the region of Sevastopol) by a three-dimensional nonlinear hydrodynamic model with assimilation of observational data of the 2007 hydrological survey. A sequential optimal interpolation of temperature and salinity observations is used as the assimilation procedure. We use real atmospheric forcing and a high resolution (a horizontal grid of ~1.6×1.6 km and 31 vertical layers from 1 m to 1300 m). Mesoscale features of the currents are obtained, and coastal upwelling in the Kalamitsky Bay is reconstructed and registered in satellite observations.

1. Introduction

The reconstruction of the spatio-temporal structure of the coastal circulation of the Black Sea, close to the observed one, allows one to determine the regions of generation of mesoscale and submesoscale eddies, frontal zones and jets. It is important in solving problems connected with the construction of coastal and port facilities, navigation, mining, forecasting the impact of accidental emissions of pollutants on the marine environment. Overwater and underwater shipping, intensive exploitation of the port in Sevastopol, which is the naval base of Russia, require analysis and forecast of hydrophysical fields throughout the entire sea. It is practically impossible to solve these problems without reconstruction of the circulation on the basis of the synthesis of a numerical model of high spatial resolution and observational data from hydrological surveys.

Some calculations of hydrophysical fields on the basis of the assimilation of observational data in numerical models were carried out earlier. A problem of variational assimilation of satellite observational data on the temperature and the ocean surface level and temperature and salinity data on the basis of a three-dimensional model of hydrothermodynamics of INM RAS was formulated in [1], and the expediency of using the model with the assimilation unit of operational observational data was demonstrated on the example of a numerical experiment for the Indian Ocean. The main steps in improving the methods for calculating the dynamics of the ocean with observations of the temperature and salinity of sea water were listed in [2], and the problem of integrated use of measurements of currents, while assimilating them in a hydrodynamic model using the Kalman filter, was considered for the western Atlantic equatorial region. The Argo-based Model for Investigation of the Global Ocean (AMIGO) was described in [3]. It consists of a block of variational interpolation onto a regular grid of profiling data for drift Argo meters and a block of model hydrodynamic adaptation of variationally interpolated fields, which allowed one to obtain a complete set of oceanographic characteristics from irregularly located measurement data. A technique for modeling the dynamics of the Black Sea for calculating the distribution of pollutants in the coastal zone of the Adler – Sochi region was proposed in [4] on the basis of the hydrodynamic model of INM RAS implemented in two versions (with a uniform spatial resolution of ~ 4 km and a non-uniform one, with a step reduction of up to 50 m). It was shown that the model version with higher spatial resolution reproduced the eddy circulation more adequately. The results of modeling the submesoscale variability and eddy formation of the northeastern shelf of the Black Sea based on a numerical model with high spatial resolution were presented in [5], and a comparison of these results with direct measurements for September 2008, January 2011 and March 2013 was performed. Several hydrological surveys were analyzed in [6, 7] in the region of the western coast of Crimea and on the north-western shelf of the Black Sea, using a hydrodynamic model and an assimilation procedure on the basis of e sequential correction of the model values by the method of optimal interpolation of residuals.

The Marine Hydrophysical Institute (MHI) regularly performs expeditionary studies in the coastal zone of the Black Sea, whose tasks are to obtain experimental data on the vertical and lateral distributions of hydrological characteristics. This study is devoted to an analysis of the results of numerical modeling of the coastal circulation of the Black Sea on the basis of the assimilation of experimental data on temperature and salinity measurements in a hydrodynamic model. A z-coordinate three-dimensional nonlinear model of the Black Sea [8], which was developed at MHI and adapted to the conditions of the coastal zone in [6], was chosen as a research tool. Sequential optimal interpolation of the observational data on temperature and salinity was used as an assimilation procedure.

A hydrological survey of the scientific research vessel "Experiment" in September 2007, in which experimental data on temperature and salinity in the coastal region of the Black Sea were obtained (Figure 1), was chosen for analysis.

The purpose of this work is to reconstruct and analyze three-dimensional fields of currents, temperature and salinity, continuous in time and space, with a horizontal resolution of ~1,6 km on the basis of the assimilation in the hydrodynamic model of observational data obtained from 16 to 22 September, 2007 in the western region of Crimea and near Sevastopol, to analyze the mesoscale features of the coastal circulation, and to show the effectiveness of applying the assimilation procedure in a numerical model while reconstructing hydrophysical fields (on the basis of comparison of the results of two numerical experiments without and with observational data).

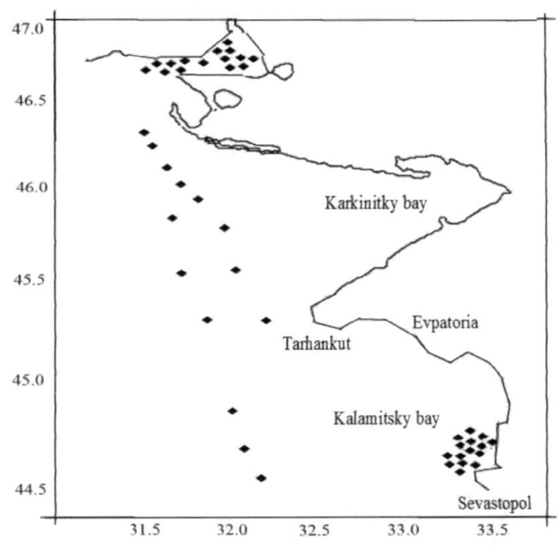

Figure 1. Scheme of stations of the scientific research vessel "Experiment" from 16 to 22 September, 2007.

2. Parameters of the model and description of the numerical experiment

The system of model equations of the dynamics of the Black Sea on the basis of the complete equations of thermohydrodynamics of the ocean in the Boussinesq approximation, hydrostatics and incompressibility of the sea water in the Gromeka-Lamb form, the boundary conditions on the surface, on the bottom, and on the solid lateral walls were described in detail in [8]. We considered a region of the Black Sea located between the meridians of 28.5 and 33.5° E with an open southern boundary, corresponding to the parallel of 44,4°N (we used a detailed presentation of the bottom topography with a resolution of ~1.6 km). The numerical experiment was carried out with a resolution of~1.6 km. The time step was equal to 30 s. The horizontal coefficients of turbulent viscosity and diffusion were equal

$\nu_H = 5 \cdot 10^5 \mathrm{cm}^2/s$, $\kappa_H = 5 \cdot 10^5 \mathrm{cm}^2/s$. The total period of integration of the model equations was 10 days (from 14 to 24 September 2007). Along the vertical, the horizontal components of the current velocity, temperature, and salinity were computed at 31 depths: 1; 3; 5; 7; 10; 13; 16; 20; 25; 30; 36; 42; 48; 55; 65; 80; 95; 120; 150; 200; 300... 1300 m. The vertical component of velocity was calculated for intermediate horizons. The vertical coefficients of turbulent exchange of momentum and diffusion were calculated according to the Philander–Pacanowski approximation [9]. The fields of currents, temperature, and salinity obtained by the model for the entire sea on a 5 × 5 km horizontal grid were used to specify initial and boundary conditions on the open boundary of the domain. The following conditions were selected on the open southern boundary. The components of the current velocity, temperature, and salinity (the Dirichlet conditions) were specified in the boundary regions where the water flowed into the domain ($v > 0$); conditions $\partial u/\partial n = 0$, $\partial v/\partial n = 0$, and radiation conditions for T and S were specified in the boundary regions where the water flowed out of the domain ($v < 0$). We took into account the discharges of the following rivers: Danube, Dnieper, Dniester, and South Bug. The fields of tangential wind stress, heat fluxes, shortwave radiation fluxes, precipitation and evaporation obtained from data of the regional atmospheric model ALADIN [10] with a spatial resolution of 1/4 ° × 1/4 ° provided by the Department of Marine Forecasts [11] of MHI and bilinearly interpolated to the selected grid were specified for each day on the sea surface. It was found by analysis of the wind fields that a south-western wind prevailed with a maximum velocity of 11 m/s from 14 to 18 September, and north and north-eastern winds with a maximum velocity of 16 m/s were acting from 19 to 24 September.

A simplified procedure of four-dimensional analysis [12, 13], based on the method of sequential optimal interpolation, when the covariance functions of the temperature and salinity fields were calculated under the assumption of homogeneity and isotropy, was used to realize the procedure of observational data assimilation.

The maximum depth to which soundings were carried out in the hydrological survey of the scientific research vessel "Experiment" varied from 5 to 300 m, and the number of stations was 44. The coordinates of all the stations where the temperature and salinity were measured were correlated with the calculated model grid, and the data were prepared for assimilation in the hydrodynamic model. A correlation radius of ~20 km was determined from the analysis of the statistical structure of the fields (calculation of the spatial correlation functions of the fields of temperature and salinity). We approximated the covariance functions of the thermohaline fields by a function of the exponential type $\exp\left(-\lambda\left[(x-x')^2+(y-y')^2\right]\right)$ [13], where λ is a dimensional parameter (equal to $0.016 \cdot (\Delta x)^{-2}$), corresponding to a correlation function value of 0.1. The hydrological survey data were grouped for six days, and assimilation was carried out once a day: September 16 – 3, September 17 – 7, September 18 – 8, September 19 – 7, September 20 – 4, and September 22 – 15 stations. Thus, thermohaline fields were calculated using the model equations until the moments of receiving of the data observations. Correction of the temperature and salinity fields was performed at the moments of assimilation taking into account the correlation radius.

The region located between the meridians 32.2 and 33.8 °E and the parallels 44.4 and 45.5°N was analyzed in detail.

3. Fields of currents

The level fields for September 15, 20 and 23, 2007 are presented in Figures 2a, b, c (the contours were drawn every 0.2 cm), on which circulation elements, such as eddies and jets, were clearly observed. An eddy with a radius of about 15 km (most noticeable in Figure 2b, c) and an eddy between 32.2 and 32.4°E associated with the meandering of the Rim Current (RC) (most noticeable in Figure 2b) were generated during the calculation period. These eddies were repeatedly recorded on satellite images of surface temperature from the *NOAA* satellite. Thickening of the isolines near Sevastopol and along the western coast of Crimea associated with the formation of a jet was noted in the fields of level on September 23 (Figure 2c), in addition to the eddy in the Kalamitsky Bay. We observed variability of the level fields with time (Figures 2b and c) associated with the influence of the observational data of September 22 located between 33 and 33.5° E (Figure 1).

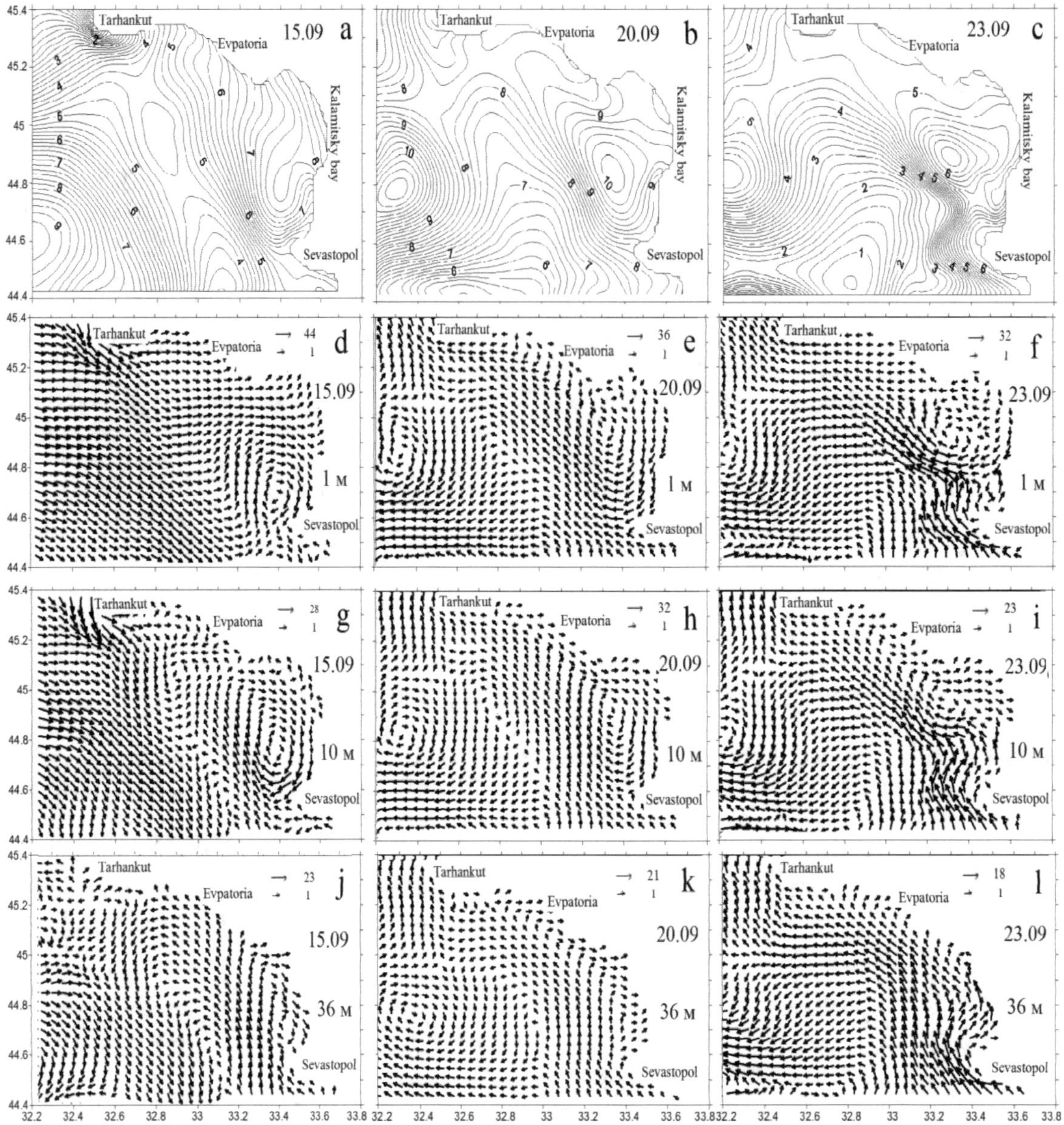

Figure 2. Level fields (cm) a: September 15; b: September 20; c: September 23 and current fields (cm/s) d: September 15 at the depth of 1 m; e: September 20 at the depth of 1 m; f: September 23 at the depth of 1 m; g: September 15 at the depth of 10 m; h: September 20 at the depth of 10 m; i: September 23 at a depth of 10 m; j: September 15 at a depth of 36 m; k: September 20 at a depth of 36 m; l: September 23 at a depth of 36 m.

The fields of currents are presented in Figures 2d, e, f, g, h, i, j, k, l at different horizons for September 15, 20 and 23, 2007 (every third arrow was shown). Due to the shallow water of the region, the wind regime had a great influence on the formation of circulation.

Under the influence of the south-west wind acting from 14 to 18 of September the main direction of the surface currents was eastern. We present the current fields for September 15 (Figures 2d, g, j) as an illustration. An anticyclonic eddy with a radius of about 15 km was obtained in the Kalamitsky Bay in the upper 36-meter layer, which was most noticeable in the 5–20 m layer (Figure 2g) and whose orbital velocity decreased with depth from 24 to 12 cm/s (Figure 2j).

Intensification of the currents was noticed near Cape Tarkhankut to values of 44 cm/s on the upper horizon (Figure 2d), and a cyclonic eddy with a radius of about 10 km near Evpatoria was generated in a layer of 10–20 m (Figure 2g). A possible mechanism for the formation of eddies was the flow of a current over inhomogeneities of the bottom topography and coastal orography.

We noted that no eddy was observed in the Kalamitsky Bay in the upper layer at sufficiently high wind velocities (about 10 m/s); the current velocity reached 54 cm/s. The features of a circulation associated with meandering of the RC (which is located to the south of the calculation area) passing along the depth slope began to form from September 17 below the 10-m horizon.

Under the influence of the north and north-east winds acting from 19 to 24 of September the main direction of the surface currents was western, an element of meander of the RC (an anticyclonic eddy with a radius of about 15 km) between 32.2 and 32.6° E was observed in the entire water layer. We noted that it was also obtained in the results of the model with a step of 5 km. We present the current fields for September 20 (Figures 2e, h, k) as an illustration.

An intense jet formed in the upper 36-meter layer (with a maximum velocity of 32 cm/s on the upper horizon) on September 22 and 23 near Sevastopol and along the western coast of Crimea directed to the north and north-west, an anticyclonic eddy was observed in the Kalamitsky Bay. These features of the coastal circulation in the current fields for September 23 are presented in Figures 2f, i, l.

4. Fields of temperature and salinity

The initial field of surface temperature was characterized by values from 19 to 22°C; the maximum values were obtained in the Kalamitsky Bay, and the minimum ones along the coast to the east of Cape Tarkhankut. An area with increased temperature of about 21°C relatively adjacent waters was observed near Evpatoria. The initial field of surface salinity was characterized by values from 17.7 to 18.3‰; the maximum values were obtained in the western part of the region, and the minimum ones along the coast to the east of Cape Tarkhankut. There was a jump layer at a depth of ~20–25 m in the vertical profiles. Figure 3 shows the surface temperature fields for September 15, 17, 19, 21, 22, 23 and the surface salinity fields for September 15, 19 and 23.

During the calculation period there was some cooling of the surface waters, which was most noticeable along the coast to the east of Cape Tarkhankut (the temperature decreased from 18°C on September 15 (Figure 3a), to 16°C on September 17 (Figure 3b), to 15°C on September 19 (Figure 3c), and to 14°C on September 21 (Figure 3d)). The water temperature decreased to 21°C on September 19 (the fifth day of calculations, Figure 3c) in the Kalamitsky Bay, and near Evpatoria to 17.5°C. The salinity fields (Figures 3g and h) changed insignificantly during the calculating period until September 21.

According to the measurements and images from the NOAA satellite (NOAA-17 and NOAA-18 with a resolution of 1 km, data server http://dvs.net.ru/mp/data/200709bs_sst_ru.shtml), a rise of the cold salt waters was observed in the southern part of the Kalamitsky Bay on September 22 and 23, due to the action of the north and north-eastern winds. Water with a temperature below 14°C and salinity more than 18‰ was observed on the surface. That was confirmed by the results of numerical calculation (Figures 3e, f, i). We noted a distribution of cold water in the model fields (Figures 3d, e, f, i) of temperature from September 21. The maximum values of the vertical velocity (0.01 cm/s) were obtained along the coastline.

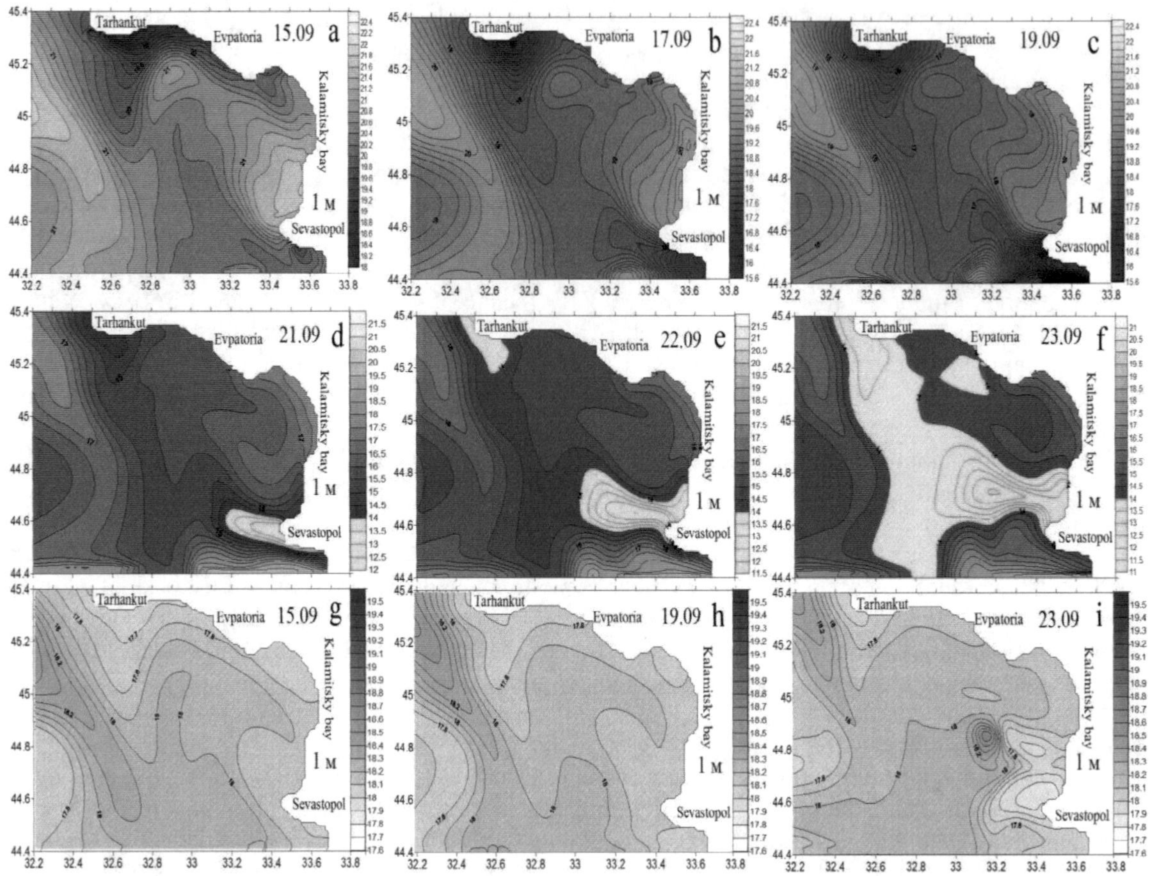

Figure 3. Temperature fields (°C) in the upper water layer a: September 15, b: September 17; c: September 19; d: September 21; e: September 22; f: September 23, and salinity fields (‰) in the upper water layer g: September 15; h: September 19; i: September 23.

5. Analysis of the influence of observational data assimilation on the accuracy of the reconstruction of fields of currents, temperature and salinity

The influence of the assimilation procedure in the numerical model on the formation of fields of currents, temperature, and salinity was analyzed on the basis of comparison of the results of two numerical experiments on the calculation of hydrophysical fields without assimilation and with assimilation of temperature and salinity data measurements.

Taking into account the observational data led to some qualitative and quantitative differences in the structure of the fields of currents (intensification of currents, change in the direction of currents, eddy formations were more noticeable). We present Figures 4a, b, c, and d as an illustration, which shows the current fields at a depth of 5 m for the two experiments. We noted an intensification of the currents in the central part and a change in the direction of the currents at the southwestern boundary of the region for September 17 (Figure 4b) compared with the results presented in Figure 4a. An intense jet along the western coast of Crimea was more noticeable on September 22 (Figure 4d) in comparison with the results presented in Figure 4c, the maximum velocity increased from 27 to 34 cm/s, and an anticyclonic eddy was obtained between 32.2 and 32.4°E (Figure 4d).

Taking into account the observational data led to a more accurate reproduction of the upwelling structure in the fields of temperature and salinity, which was also observed in the calculation without data assimilation. Figures 4e, f, g, and h shows the temperature and salinity fields at a depth of 5 m for the two experiments. A rise of the water with a minimum temperature of 8°C and a maximum salinity of 18.4‰ was observed in the upper layer taking into account the observational data (Figures 4f and h).

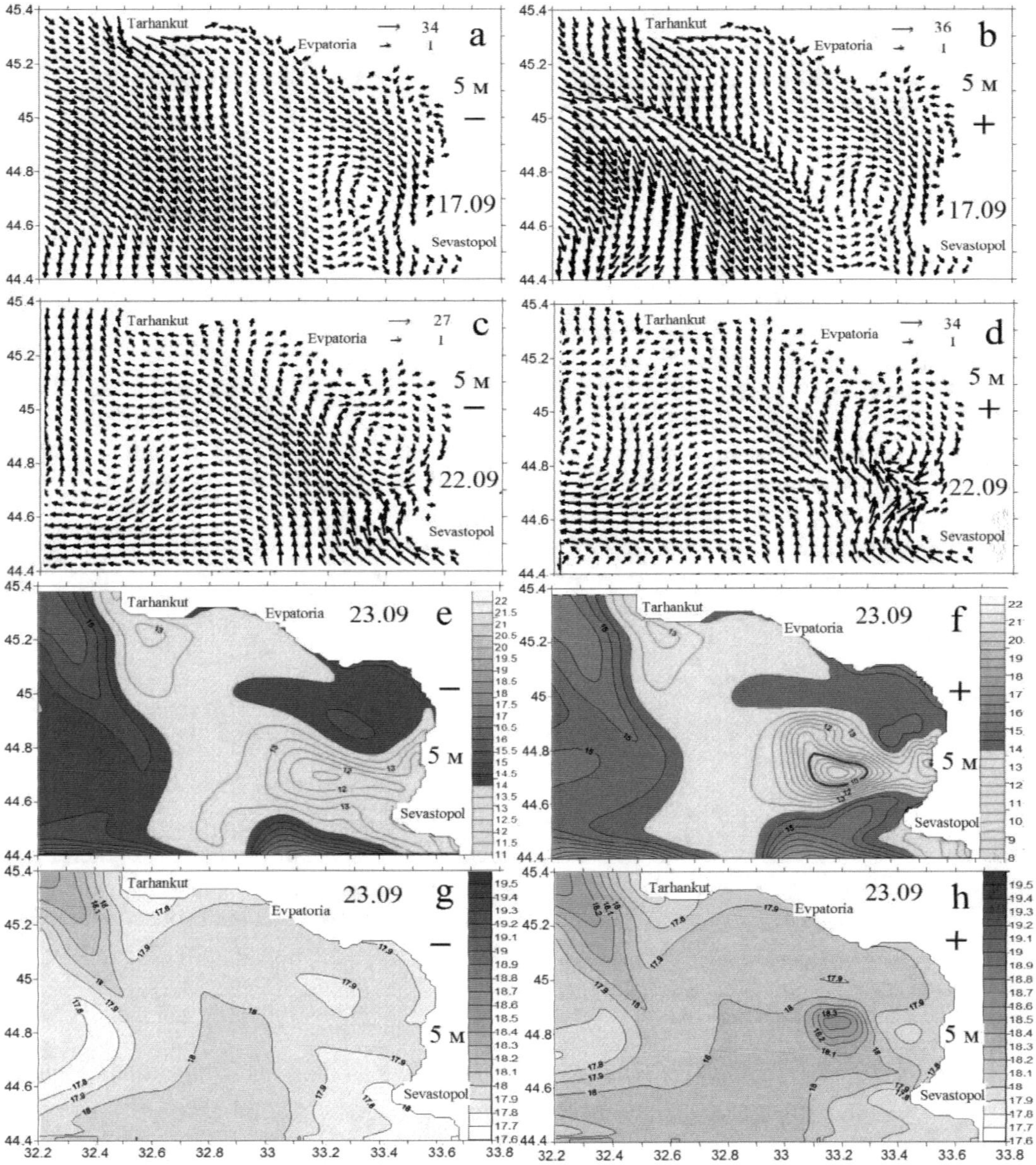

Figure 4. A: Current field (cm/s) at a depth of 5 m on September 17 calculated not taking into account the observational data; b: velocity field (cm/s) at a depth of 5 m on September 17 calculated taking into account the observational data; c: velocity field (cm/s) at a depth of 5 m on September 22 calculated not taking into account the observational data; d: velocity field (cm/s) at a depth of 5 m on September 22 calculated taking into account the observational data; e: temperature field (°C) at a depth of 5 m on September 23 calculated not taking into account the observational data; f: temperature field (°C) at a depth of 5 m on September 23 calculated taking into account the observational data; g: salinity field (‰) at a depth of 5 m on September 23 calculated not taking into account the

observational data; h: salinity field (‰) at a depth of 5 m on September 23 calculated taking into account the observational data.

6. Conclusions

In this paper, fields of sea level, currents, temperature, and salinity were reconstructed in the coastal area of the Black Sea (near the coast of Western Crimea and the region of Sevastopol) with a high spatial resolution (a horizontal grid of~1.6×1.6 km and 31 vertical layers) by a hydrodynamic model and observational data of a hydrological survey performed by a scientific research vessel, "Experiment", in 2007. The areas of generation of mesoscale and submesoscale eddies and jets were determined. An anticyclonic eddy with a radius of about 15 km in the Kalamitsky Bay in the upper water layer and a cyclonic eddy with a radius of about 10 km near Evpatoria in a layer of 10–20 m were formed in a current flow over the coastal orography and inhomogeneities of the bottom relief. The anticyclonic eddy with a radius of about 15 km between 32.2 and 32.6°E in the entire water layer was associated with meandering of the RC. An intensification of the currents near Cape Tarkhankut was revealed. An intensive current near Sevastopol and along the western coast of Crimea, directed to the north and north-west, was observed.

The spatio-temporal structure of temperature and salinity fields was reconstructed along the western coast of Crimea and near the region of Sevastopol. Cooling and desalination of the surface waters, which was most noticeable along the coast to the east of Cape Tarkhankut, was detected. Coastal upwelling in the Kalamitsky Bay was reconstructed on September 21–23, 2007 and registered in data observations, as a result of a rise in the underlying cold waters.

Based on a comparison of the results of two numerical experiments with and without observational data, it has been shown that using the observational data made it possible to more accurately reconstruct the mesoscale features of the coastal circulation near the western and north-western coast of Crimea.

Acknowledgments. This work was carried out with financial support by the Russian Foundation for Basic Research and the city of Sevastopol under scientific project No. 18-45-920019.

References

[1] Agoshkov V, Ipatova V, Zalesny V et al. 2010 Problems of variational assimilation of observational data for ocean general circulation models and methods for their solution *Izvestiya – Atmospheric and Ocean Physics* **46** 6 677–712

[2] Knysh V, Korotaev G, Mizyuk A *et al* 2012 Assimilation of hydrological observation data for calculating currents in seas and oceans *Izvestiya – Atmospheric and Ocean Physics* **48** 1 57–73

[3] Lebedev K An argo-based model for investigation of the Global Ocean (AMIGO) 2016 *Oceanology* **56** 2 172–81

[4] Diansky N, Fomin V, Zhokhova N *et al* 2013 Simulations of currents and pollution transport in the coastal waters of Big Sochi *Izvestiya – Atmospheric and Ocean* Physics **49** 6 611–21

[5] Divinsky B, Kuklev S, Zatsepin A *et al* 2015 Simulation of submesoscale variability of currents in the Black Sea coastal zone Oceanology **55** 6 814–19

[6] Demyshev S, Evstigneeva N 2013 Numerical analysis of hydrophysical fields on the northwestern shelf of the Black Sea *Oceanology* **53** 5 511–24

[7] Demyshev S, Evstigneeva N 2012 Numerical modeling of hydrophysical fields with the assimilation of observational data in the coastal zone of the Southern coast of Crimea in July 2000 *Environmental safety of the coastal and shelf zones and integrated use of shelf resources* **26** 2 27–39 (in Russian)

[8] Demyshev S A numerical model of online forecasting Black Sea currents 2012 *Izvestiya. Atmospheric and Oceanic Physics* **48** 1 120–32

[9] Pacanowski R, Philander S 1981 Parameterization of vertical mixing in numerical models of

tropical oceans *J. Phys. Oceanogr* **11** 11 1443–51
[10] Mesoscale modelling seen as a basic tool for weather forecasting and atmospheric research / members of the ALADIN international team the ALADIN PROJECT 1997 *World Meteorological Organization Bulletin* **46** 4 317–24
[11] Ratner Yu, Martynov M, Bayankina T *et al* 2005 Information flows in the system of operational monitoring of hydrophysical fields of the Black Sea and automation of their processing *Environmental Monitoring Systems* 140–49 (in Russian)
[12] Knysh V, Moiseenko V, Sarkisyan A *et al* 1980 Complex use of measurements at the hydrophysical ocean polygons in four-dimensional analysis *Reports of AS USSR* **252** 4 832–36 (in Russian)
[13] Knysh V 1981 Hydrothermodynamic ocean models in the algorithm of the multielement four-dimensional analysis of hydrophysical fields: A thesis for a scientific degree of doctor of science (in Russian)

IOP Conf. Series: Earth and Environmental Science **386** (2019) 012048 doi:10.1088/1755-1315/386/1/012048

Long Periods of Increased/Decreased Runoffs of Large Russian Rivers

A G Georgiadi, E A Kashutina and I P Milyukova

Institute of Geography, RAS, Moscow, Russia

E-mail: georgiadi@igras.ru

Abstract. Long series of annual and seasonal runoffs of large Russian rivers (the Volga at Volgograd, the Don at Razdorskaya, the Yenisei at Igarka, and the Lena at Kyusyur) since the 1870s for the Russian Plain rivers and since the 1930s for the Siberian rivers have been analyzed. Long periods (phases) of increased and decreased water flow have been identified. The boundaries of contrast phases are determined with cumulative deviation curves in combination with Student's test. The duration of the phases varies from 20–25 years to many decades (for runoff in low-water seasons of the Volga and Don). For the Volga and Don, the phases of decreased runoff are generally longer than increased flow phases (this is especially true for low-water seasons). The identified contrast phases show a statistically significant difference between the annual and seasonal runoffs, which varies from 10 to 65%. In the phases of increased flow, high-water years occur much more often than low-water years, and vice versa. In the period of current climate warming (since 1981), the changes in annual runoff relative to the calculated runoff of the reference period (from the 1930s to 1980) are opposing, i.e., the runoff decreased in the Don and increased in the Volga, Yenisei, and Lena. In this case, the changes are most significant in the Don. In the period of current climate warming, both the winter and snow-melt flood runoff has increased only for the Lena. At the same time, the contributions of climatic and anthropogenic factors to the changes in runoff differ significantly for each of the rivers under study.

1. Introduction

Studies show that the long-term changes of the annual and seasonal river flow are characterized by periods (phases, according to the terminology accepted in Russian hydrology) of decreased and increased values with different duration [1–3]. An important result, in particular for forecasting, is that long contrast phases with a duration of 10–15 years and more can be identified. They are observed against the background of corresponding climate changes. The anthropogenic factors can considerably transform both runoff values and the boundaries between contrast phases.

For several years, the authors have been studying the phases of decreased/increased flow in large rivers in Northern Eurasia [4–7].

This article presents the results of a comparison of the characteristics of the contrast phases of the largest rivers of the Russian Plain (the Volga and the Don) and Siberia (the Yenisei and Lena) over the period of observations (since the 1870s in the Russian Plain rivers and since the 1930s in Siberian rivers).

2. Methods to Analyze the Annual and Seasonal Runoff Data Sets

The approach to the analysis of long phases of multi-year variations of the annual and seasonal runoff caused by climate changes is based on its naturalization (the elimination of changes caused by the anthropogenic impact), the use of cumulative deviation curves, criteria of statistical homogeneity of the average flow values, and analysis of the characteristics of the identified contrast phases.

The cumulative deviation curves reflect the cumulative sum of deviations of a characteristic from its average value calculated for the entire observation period. In some cases, the deviations are normalized by the coefficient of variation in order to compare the temporal changes of different types of characteristics. The cumulative deviation curves allow one to identify long-term periods (phases) during which the values of a characteristic below (or above) its long-term mean are predominant.

The boundaries of hydrological seasons (snow flood, low-water seasons) were determined using runoff hydrographs throughout the observation period. The data on the long-term dates of the beginning and the end of snow flood and winter freeze-up were also used.

The naturalization of the runoff assumes that the long-term hydrological series for the rivers under consideration are not homogeneous in terms of the impact of anthropogenic factors, and consist of two parts. The first part of the series included long-term data for the period before the effect of anthropogenic factors became appreciable. The second part consisted of a long-term series in which the annual runoff changed to a different extent under the effect of anthropogenic factors (primarily, reservoirs).

The flow for that period was reconstructed by two methods. One method [6], which has been used to reconstruct the Volga and Don flow, is based on the regression relationships between the flow of the main river and those of the rivers used as indicators of climate conditions (tributaries and the upper reaches of the main river), which show relatively weak anthropogenic perturbations in their water regime. The study of the phases of the Yenisei and Lena was based on long-term series of mean daily water discharges, naturalized in [8, 9] using a method for the transformation of runoff hydrograph, which is based on the influence function of G.P. Kalinin and P.I. Milyukov [10].

The part of the long-term runoff series reconstructed by one of the methods mentioned above was combined with the other part, in which the flow was not greatly affected by the anthropogenic impact. This is how the common series of conditionally natural runoff were constructed.

The time boundary of the change in the long-term phases of increased/decreased runoff values is determined by identifying the minimal and maximal values of the coordinates of cumulative deviation curves in combination with Student's test. The long-term phases of the increased water flow correspond to a unidirectional tendency in an increase in the coordinates of the cumulative deviation curve, while the phases of the decreased flow correspond to a decrease in its coordinates. Short-period runoff changes can be seen against the background of such long-term trends.

3. Results and discussion

3.1. Long-Term Phases of Multi-Year Variations in the Conditionally Natural Annual and Seasonal Runoff

In the period of observations of the winter runoff (in all rivers under consideration), the summer–autumn runoff (in rivers on the Russian Plain), as well as the snow-flood runoff and the annual runoff of Siberian rivers, two major long-term phases can be identified. The phase of decreased flow is associated with the early observation period, and the following phase of increased flow begins in the 1970s–1980s (which is close to the beginning of the modern global climate warming) and continues up to the 2000s (Fig. 1). The snow-melt flood runoff of the Don also shows two long phases. However, the first phase (since the 1870s) showed an increased flow, followed by a phase of decreased flow (since 1972). In the Volga, a phase of increased snow flood runoff was also observed since the 1870, but as soon as the 1930s it was replaced by a phase of decreased water flow, with a new phase of increased runoff starting since the 1980s. And since the beginning of the 2000s again there is a phase of decreased snow flood flow. The number and sequence of replacement in the contrasting phases of the annual runoff on the Volga and Don is more complex than for the snow flood.

The time boundary of the change in the phases of contrast water flow varies appreciably for both the annual runoff and the runoff of hydrological seasons both for each river and between the rivers (Figure 1).

3.2. The Duration of Phases

The duration of the phases varies from 20–25 to 90 years (for the low water-season runoff in the Volga and Don). As a rule, the duration of the phases of decreased runoff is longer (and often, considerably) than the duration of increased flow phases, especially for the low water-season runoff of the Volga and Don. In the Lena and Yenisei, the duration of the contrast phases for the annual and winter runoff was similar. However, the phase of increased snow flood flow in these rivers was much shorter than the phase of its decreased flow.

IOP Conf. Series: Earth and Environmental Science **386** (2019) 012048 doi:10.1088/1755-1315/386/1/012048

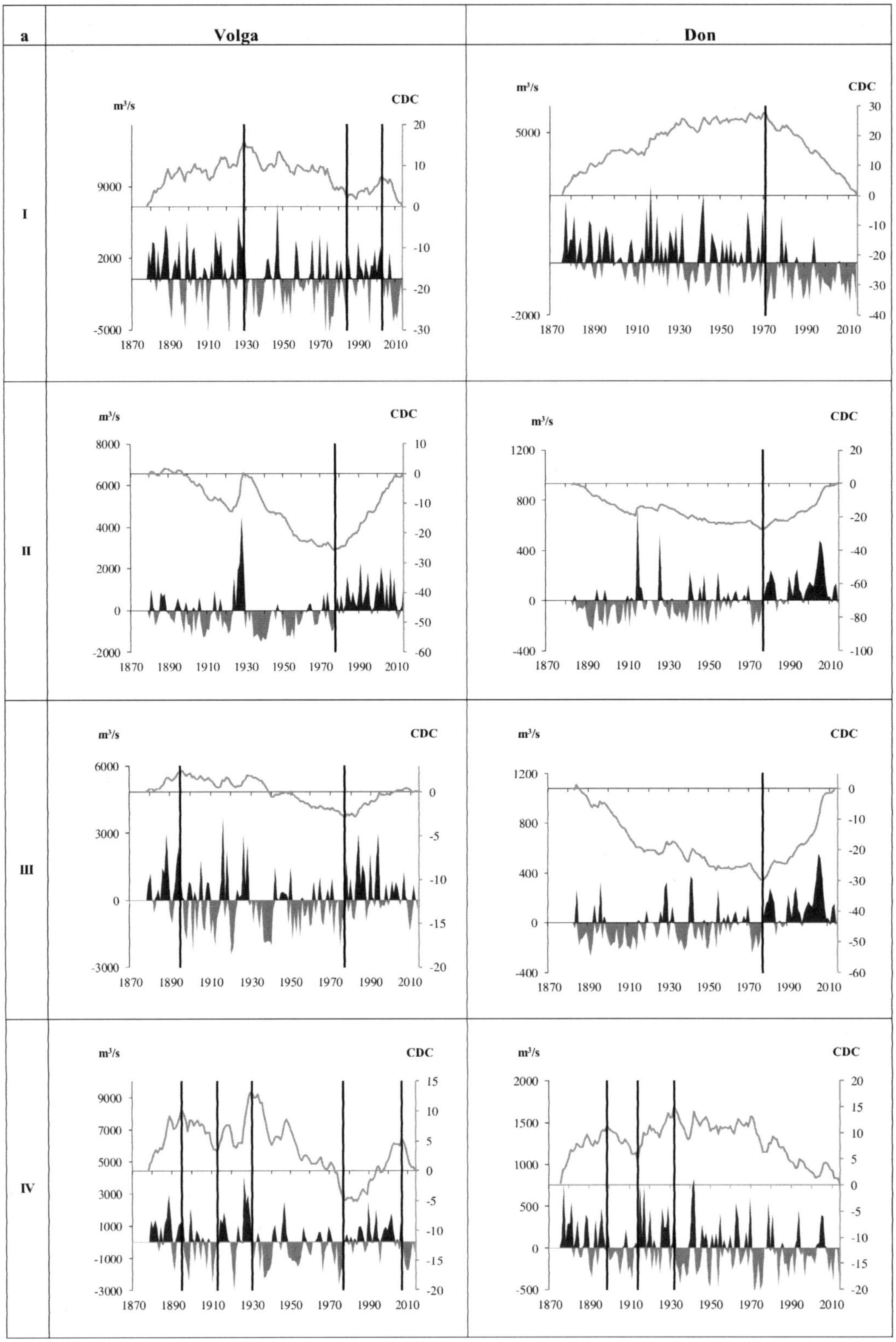

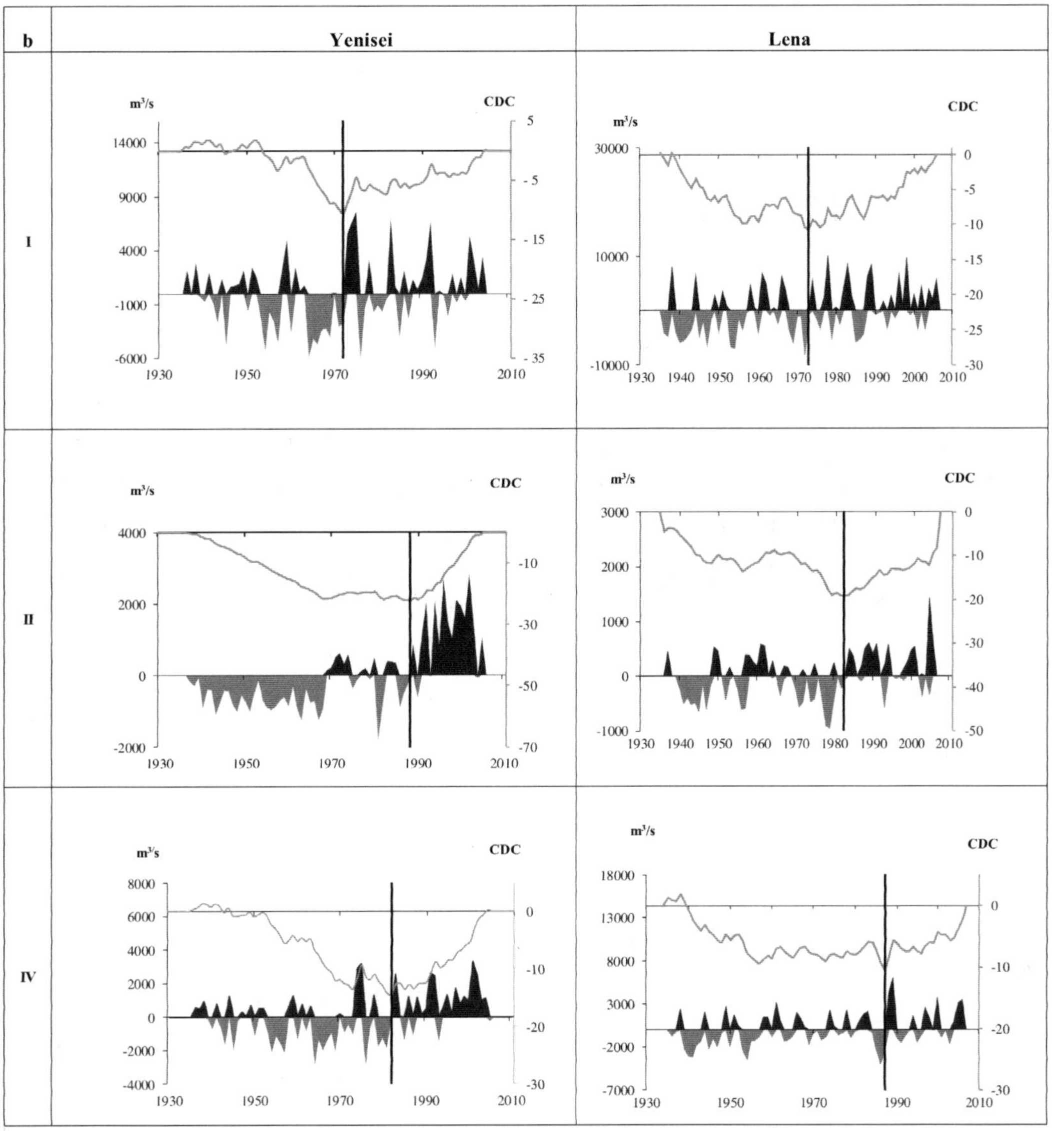

Figure 1. Long-term changes in naturalized (I) snow flood flow, (II) winter water flow, (III) summer-autumn water flow, and (IV) mean annual water flow in: (a) the Volga at Volgograd and the Don at Razdorskaya, (b) the Yenisei River at Igarka and the Lena River at Kyusyur (blue and red) as deviations from their long-term mean values: (green) in the form of normalized cumulative deviation curves (CDCs). The vertical black lines show phase boundaries (shift points) of increased/decreased values of the naturalized water flow.

3.3. Differences in the Runoff of Contrast Phases

The difference between the average runoff values in the phases of higher and lower flow is more than 12-18% for the annual runoff for both rivers; the same differences are 12 and 65% for the spring flood runoff, about 35% and 53% for the winter runoff, and 21 and 52% for the summer-autumn runoff in the Volga and Don, respectively.

In the Siberian rivers, the difference between the runoff in the phases of increased and decreased flow in winter is more than 40% in the Yenisei and about 20% in the Ob and Lena. The difference in the annual and spring flood runoff in these phases is smaller, varying within 10%.

3.4. Runoff Frequency Structure in the Phases of Increased and Decreased Runoff

The numbers of high-water and low-water years in the long-term phases of the contrast water flow differ considerably. Thus, for the Volga and Don significant differences were found to exist between the ratios of the numbers of high-water and low-water years for the long-term phases of increased (PIR) and decreased (PDR) annual and seasonal water flow. A year was classified as high-water one if its annual and seasonal runoff values were not less than the runoff of 25% exceedance probability and as low-water if both the annual and seasonal runoffs were not greater than its value with 75% exceedance probability. In the phases of increased annual runoff, snow flood runoff, and the runoff of the winter and summer–autumn low-water seasons (Figure 2), the numbers of high-water years were 30, 26, 53, and 42% of the total number of years in the PIR, respectively; the respective numbers for the Don were 44, 35, 63, and 59%. The proportion of low-water years during the PIR was much lesser. In the Volga, it was 12, 19, 13, and 18%, and in the Don, 18, 7, 9, 10%, respectively. The situation at the PDR was inverse.

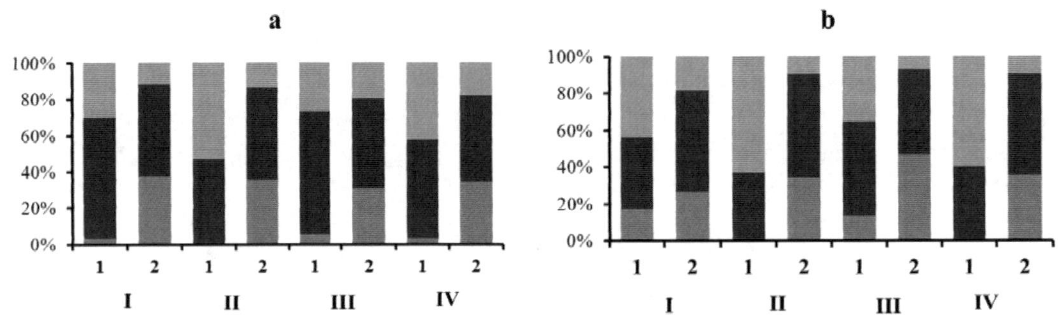

Figure 2. Frequency structure of (I) annual, (II) snow flood, (III) winter, and (IV) summer-autumn runoff in the long-term phases of: (1) increased and (2) decreased runoff (a) Volga at Volgograd, (b) Don at Razdorskaya Proportion of years with runoff: (red) equal to or less than that corresponding to 75% exceedance probability; (violet) greater than that corresponding to 75% and less than that corresponding to 25% exceedance probability; (blue) equal to or greater than that corresponding to 25% exceedance probability.

At the same time, the total proportion of low- and high-water years is relatively stable, varying mostly within the range of 40–60%. Accordingly, the proportion of the years with water discharges lying within the range of 25 and 75% exceedance probabilities is also relatively stable, varying within about the same limits (40–60%). For the annual runoff and the snow flood runoff, the proportion of such years is greater in the PIR than in the PDR (except for the annual runoff of the Don). Conversely, the proportion of the years falling within this range for the winter runoff and the runoff of the summer–autumn low-water season is greater in the PLR than in the PHR (except for the summer–autumn runoff of the Volga).

3.5. Runoff Changes in the Epoch of Modern Climate Warming

3.5.1. Overall changes in runoff caused by climate and anthropogenic factors. As one can see from Table 1, changes in the annual runoff in the rivers under consideration had different signs: in the period of current climate warming the runoff was decreasing in the Don and increasing in the Volga, Yenisei, and Lena. These changes were most significant in the annual Don runoff. Its decrease in that period was equal to more than five annual volumes of the runoff of the basic period, a fact that had a very adverse effect on the water-management and hydroecological situation in its basin and on the state of the Sea of Azov. At the same time, the total increase in the runoff of the Volga at Volgograd, the Yenisei at Igarka, and the Lena at Kyusyur over the period of current climate changes was approximately equal to the annual runoff volume of the basic period for the respective rivers.

Both the winter runoff and the runoff of the snow flood have increased over the period of modern climate warming only in the Lena River (Table 2). In the other rivers the changes in the runoff of the winter low-water season and the snow flood in this period had different signs. The relative increase in the winter runoff was greater than the decrease in the spring flood runoff.

Table 1. Changes in the average annual runoff under the effect of climate and anthropogenic factors in the period of modern climate warming relative to its naturalized values for the basic period (from the 1930s to 1980).

River–gage	Average annual runoff of the basic period		Period of modern climate warming				
			Average annual runoff		Changes in the average annual runoff		
	years	runoff volume, km^3/year	years	runoff volume, km^3/year	on the average for a year, km^3	on the average for a year, %	total over the period, km^3
Volga–Volgograd	1930-1980	246.7	1981-2014	255.1	8.4	3.4	285.6
Don–Razdorskaya	1930-1980	25.4	1981-2014	20.7	-4.7	-18.5	-159.8
Yenisei–Igarka	1936-1980	577.6	1981-2004	608	30.4	5.3	729.6
Lena–Kyusyur	1936-1980	520.5	1981-2007	545.4	24.9	4.8	672.3

Table 2. Changes in the average winter runoff and the runoff of the snow flood under the joint effect of climatic and anthropogenic factors in the period of current climate warming relative to their naturalized values for the basic period (from the 1930s to 1980).

River–gage	Average annual runoff of the reference period, km^3		The period of modern climate warming					
			Average annual runoff, km^3		Changes in the average annual runoff			
	winter	snow flood	winter	snow flood	winter, km^3	winter, %	snow flood, km^3	snow flood, %
Volga–Volgograd	27.2	166.3	66.9	125.3	39.7	146.0	-41	-24.7
Don–Razdorskaya	2.5	16.6	4.7	6.7	2.2	88.0	-9.9	-59.6
Yenisei–Igarka	74.8	418.5	137.3	389.6	62.5	83.6	-28.9	-6.9
Lena–Kyusyur	32.4	387.8	48.4	397.1	16	49.4	9.3	2.4

3.5.2. Contribution of climate warming and anthropogenic impact to the total runoff changes. The contributions of climate changes and anthropogenic impact to the total runoff changes over the period of modern climate warming were evaluated by comparing the naturalized average annual and seasonal runoff for the basic period (1930–1980) and the naturalized and actual runoff for the period since 1981. The boundaries of the periods in the rivers differ (see Table 1). This is a result of later start of observations at Siberian rivers and the lack of data on their naturalized runoff.

The difference between the naturalized runoff for the basic period and the following period characterizes the effect of climate changes (under the assumption that they are not caused by human activity). The runoff changes that are due to the anthropogenic impact were calculated by the difference between the estimates of climate-induced changes in the runoff and its changes caused by a joint effect of climatic and anthropogenic factors in the period of modern climate warming (see Tables 1 and 2).

The anthropogenic and climate-induced changes in the annual runoff were both directed toward its decrease only in the Don (Table 3). In this case, a much larger contribution to the decrease in the Don runoff was due to the anthropogenic factors, the most important among them being the construction of the Tsimlyansk Reservoir and consumptive water use for irrigation (Figure 3).

Table 3. Changes in the average annual runoff over the period of modern climate warming relative to their naturalized values for the basic period (from the 1930s to the 1980s).

River–gage	Anthropogenic changes		Climate changes over	
	total over the period, km^3	average over year, km^3	total over the period, km^3	average over year, km^3
Volga–Volgograd	-370.6	-10.9	656.2	19.3

Don–Razdorskaya	-119	-3.5	-40.8	-1.2
Yenisei–Igarka	-31.2	-1.3	760.8	31.7
Lena–Kyusyur	*	*	669.6	24.8

*anthropogenic changes are nearly zero

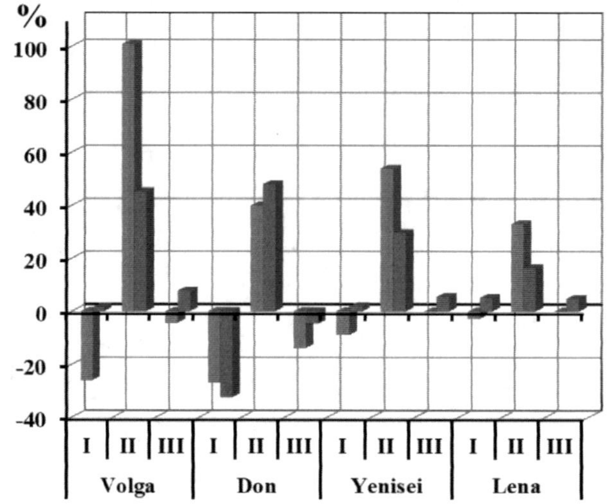

Figure 3. Changes in (I) snow flood, (II) winter and (III) annual runoff under the effect of (red) anthropogenic and (blue) climatic factors in percent of the naturalized annual and seasonal runoff of the basic period (the 1930s–1980).

In the Lena, the changes in the annual runoff were due to climate changes, which caused its increase. The anthropogenic impact has not caused any appreciable changes in the Lena annual runoff. In the Volga and Yenisei, the effects of these factors were differently directed. Note that the climate-induced changes in the annual runoff in the Yenisei were larger, and the Volga runoff changes under the dominating effect of climate, the anthropogenic changes in the runoff were also significant.

The contributions of the climate-induced and anthropogenic factors to the runoff changes in the seasons with contrast hydrological seasons (snow flood and winter low-water season) in the modern climate change period differed considerably. The changes in the winter runoff under anthropogenic and climatic factors were in one direction—toward its increase (Table 4, Figure 3). Their contributions to the changes in the Don winter runoff were nearly the same, while in the other rivers the contribution of climate changes to the increase in the winter runoff was twice as large as that of anthropogenic factors.

The effects of these factors on the snow flood runoff were cardinally different. The anthropogenic factors caused a decrease in the spring flood runoff in these rivers (the largest one in the relative characteristics in the Volga and Don) (Figure 3). Conversely, the climate changes contributed to an increase in the snow flood (except for the Don). Their effect in the Lena was much more considerable than that of the anthropogenic factors. In the Volga and Yenisei, conversely, the contribution of anthropogenic factors to the decrease in the spring flood runoff was much larger, resulting in its decrease in the period of current climate warming.

Table 4. Anthropogenic and climate-induced changes in the average annual runoff over winter season and snow flood on average over the season over the period of modern climate warming (km^3) relative to their naturalized values for the basic period (from the 1930s to the 1980s).

River–gage	Winter		Snow flood	
	anthropogenic changes	climate-induced changes	anthropogenic changes	climate-induced changes
Volga–Volgograd	27.4	12.3	-43.4	2.4
Don–Razdorskaya	1	1.2	-4.5	-5.4
Yenisei–Igarka	40.3	22.2	-37.2	8.3
Lena–Kyusyur	10.7	5.3	-10.6	19.9

4. Conclusions

During the periods of observations of winter runoff (for all above-considered rivers), summer–autumn runoff (for the rivers of the Russian Plain), spring flood and annual runoff for the Siberian rivers, two major long-term phases can be identified: the phase of decreased runoff at the beginning of the observation period, and the following phase of increased runoff that begins in the 1970–1980s (which is close to the onset of current global climate warming).

The snow-melt flood runoff of the Don also shows two long phases. The first phase (since the 1870s) showed higher runoff, and the following phase (since 1972) was a phase of lower runoff. In the Volga, first (since the 1870s) a phase of higher snow flood runoff was also observed; however, as soon as the 1930s it was replaced by a phase of lower runoff, which was followed again, since the 1980s, by a phase of higher runoff. Since the mid-1990s–early 2000s there was a tendency towards the passage to a phase of lower, spring flood, and summer–autumn runoff.

The time boundary between the higher and lower runoff phases varies appreciably for the annual runoff and the runoff of hydrological seasons, both for each single river and between the rivers. The phase duration is also very volatile and lies mostly within the range from 20–25 years to many decades (for the winter and summer–autumn runoffs of the Volga and Don). The average runoffs of contrast phases differ by 10–65%.

For the Volga and Don, the number of high-water years during increased runoff phases is much larger than the number of low-water years, and vice versa.

In the period of current climate warming (since 1981), compared with the calculated runoff of the basic period (from the 1930s to 1980), the changes in the annual runoff of the rivers under consideration were opposing: in the period of current climate warming the Don runoff was decreasing, and that of the Volga, Yenisei, and Lena is increasing. The changes were the largest in the annual runoff of the Don. Both winter and snow flood runoff increased in the period of current climate warming only for the Lena. In the other rivers in this period, the runoff of the winter low-water season and that of the snow flood changed in different directions. The relative increase in the winter runoff was more considerable than the decrease in the spring flood runoff. The anthropogenic and climate-induced changes in the annual runoff had the same direction (towards a decrease) only for the Don. The contribution of anthropogenic factors to the decrease in the Don runoff was much greater. In the Lena, the decrease in the annual runoff was caused by climate change. In the Volga and Yenisei, these factors acted in different directions. In the Yenisei, the climate-induced changes in the annual runoff were larger, and in the Volga, at the predominant effect of climate, its anthropogenic changes were considerable.

The contributions of climatic and anthropogenic factors to the changes in the runoff of contrast hydrological seasons (snow flood versus winter low-water season) in the period of current climate warming differed considerably. The winter runoff was steadily increasing under the effect of anthropogenic and climatic factors, and the contributions of these factors to the changes in the Don winter runoff were nearly the same, while the contribution of climate changes to the increase in the winter runoff of the other rivers was twice as much as that of the anthropogenic factors.

The effects of these factors on the snow flood runoff were cardinally different. The anthropogenic factors caused a decrease in the spring flood runoff of these rivers (which was the largest in relative values for the Volga and Don), while climate changes contributed to an increase in the snow flood runoff in all rivers, except for the Don. Their effect on the Lena runoff was much more considerable than the effect of anthropogenic factors, while for the Volga and Yenisei, on the contrary, the contribution of anthropogenic factors to the decrease in spring flood runoff was much larger, resulting in its decrease in the period of current climate warming.

Acknowledgments

The study was carried out under state order 0148-2019-0007 and was supported by the Russian Foundation for Basic Research under projects 18-05-60240 and 17-05-00948.

References

[1] *Water Resources of Russia and Their Use* 2008 ed I A Shiklomanov (St. Petersburg: State Hydrological Institute) p 600
[2] Yeh C F, Wang J, Yeh H F and Lee C H 2015 Spatial and Temporal Streamflow Trends in Northern Taiwan *Water* 7 (2) pp 634-651 doi:10.3390/w7020634

[3] Sharma S and Singh P K 2017 Long Term Spatiotemporal Variability in Rainfall Trends over the State of Jharkhand, India. *Climate* 5 pp 2-18 doi:10.3390/cli5010018

[4] Georgiadi AG and Kashutina E A 2011 Regional features of temporal changes in freshwater inflow into the Arctic Ocean ed V M Kotlyakov *Polar Cryosphere and Continental Waters. Russia's Contribution to IPY 2007/08* (Moscow – St. Petersburg: Paulsen) pp 252-64 (in Russian)

[5] Georgiadi A G and Kashutina E A 2016 Long-Term Runoff Changes of the Largest Siberian Rivers *Izvestiya Rossiiskoi Akademii Nauk, Seriya Geograficheskaya* **5** pp 70-81 (in Russian)

[6] Georgiadi A G, Koronkevich N I, Milyukova I P, Kashutina E A and Barabanova E A 2014 *Present-day and Scenario Changes in River Runoff in the Basins of Largest Russian Rivers Part 2 Basins of the Volga and Don* (Moscow: MAKS Press) p 214

[7] Georgiadi A G, Kashutina E A and Milyukova I P 2018 Long-term Changes of Water Flow, Water Temperature and Heat Flux of the Largest Siberian Rivers *Polarforschung* **87** (2) 167–76

[8] Shiklomanov A I., Golovanov O, Lammers R B, Tretjyakov M and Yang D 2011 Dam/ Reservoir-Induced Hydrological Changes in Large Siberian Rivers *American Geophysical Union, Fall Meeting 2011. Abstract* #C31B-06

[9] Shiklomanov A I, Lammers R B, Lettenmaier D R, Polischuk Yu *et al* 2013 Hydrological Changes: Historical Analysis, Contemporary Status, and Future Projections *Regional Environmental Changes in Siberia and Their Global Consequences* (Dordrecht:Springer Science+Business Media) Chapter 4 pp 111–54

[10] Kalinin G P, Milyukov P I 1958 Approximate calculation of transient flow of water masses *Trydy TsIP* **66** (Leningrad: Gidrometeoizdat) p 72

Anthropogenic heat fluxes in urban agglomerations and their impact on meteorological processes

A S Ginzburg[1,2] and S A Dokukin[1,3]
[1] A.M. Obukhov Institute of Atmospheric Physics, Russian Academy of Sciences, Pyzhevsky 3, Moscow 119017, Russian Federation
[2] Moscow Technological Institute, Moscow, Russia
[3] Moscow State University, Moscow, Russia

[*] Corresponding author: gin@ifaran.ru

Abstract. One of the main factors of the impact of urbanization on mesoscale atmospheric and climatic processes are the anthropogenic heat fluxes (AHFs) caused by all types of heat sources in urban areas – from industry to metabolism in residents. A calculation of the influence of energy consumption in urban weather and climate made by the COSMO-CLM model with the TERRA-URB scheme shows that anthropogenic heat fluxes have a noticeable effect on urban temperature and wind regime. In Moscow's agglomeration, the AHF contribution results in an increase of the mean annual temperature by 2°C and of the mean annual wind speed by more than 1 m/s, while the prevailing wind direction changes only slightly.

1. Background

Since the beginning of the 21st century, more and more attention has been paid to studying the influence of world energy not only on greenhouse gases emissions, but also on thermal pollution of the atmosphere of urban agglomerations.

The scales of the urban atmospheric processes are in the middle of the time and horizontal length scales of the meteorological processes [1, 2] and for understanding and modeling of the anthropogenic impact on meteorological and climatic characteristics within large urban agglomerations it is necessary to consider a wide range of atmospheric processes.

Anthropogenic heat fluxes (AHFs) arising from the energy consumption in urban areas strongly affect the mesoscale atmospheric processes in densely populated regions of the world. AHFs are one of the main factors in the formation of the urban heat island and have a significant impact on the city's wind regime, humidity, and cloud cover.

Estimates of anthropogenic heat fluxes in different countries and specific urban agglomerations vary greatly due to different methods for their determination and significant uncertainty and initial statistical data. The AHFs are responsible for a significant part of the urban heat island intensity [3, 4]. Regional atmospheric advection also significantly affects the intensity and shape of the urban heat island. To describe the features of the urban heat island dynamics and assess the energy needs of urban economy, it is necessary to consider the feedbacks between the temperature regime and the urban energy consumption.

The anthropogenic heat fluxes within large urban agglomerations could be estimated by empirical assessment based on reliable data on the population and energy consumption of the urban economy. Another approach to the AHF estimation is inventory and summation of all consumers of the heat and electric energy in the city.

These two main approaches to the determination of anthropogenic heat fluxes and methods for their assessment are described in detail in [5, 6].

A series of studies on the empirical assessment of anthropogenic heat fluxes in the largest urban agglomerations of the world was performed at the Laboratory of Mathematical Ecology of the A.M. Obukhov Institute of Atmospheric Physics (IAPh) RAS. The method of this empirical assessment is based on the most reliable data on the population and energy consumption of the urban economy of megalopolises. Calculation of the influence of energy consumption in urban areas on mesoscale atmospheric processes was carried out using the COSMO-CLM [7] model with a TERRA-URB scheme [8-12].

2. Methods

It is well known that large cities create their own climate and anthropogenic heat fluxes play a crucial role in the formation and dynamics of the urban climate. However, obtaining reliable data on the magnitude and spatiotemporal distribution of anthropogenic heat fluxes in megacities of various countries of the world is associated with many difficulties.

One of them is the inclusion of different anthropogenic heat sources which exist on urban territories: electric and heat production energy consumption, transport, boilers, and even human metabolism. Electric energy is usually generated outside urban territories, and this AHF source is not included in city statistical data. Another difficulty is that the administrative boundaries of a city often include not only downtowns, business and residential blocks, but also industrial and suburban territories.

The main useful method for the estimation of the AHFs is the inventory and summation of all consumers of heat and electric energy in the city taking into account the population size and the influence of transport, the length of roads, and the engineering communications (see, for example, [13]):

$$Q_a = Q_V + Q_b + Q_m, \quad (1)$$

where Q_a is the total AHF, Q_V is the heat generated by vehicles, Q_b is the heat from buildings, and Q_m is the human metabolism. The COSMO consortium team is using a similar approach.

An alternative method of the AHF estimation was proposed in **[5, 6].** It is based on the most reliable data on the population and energy consumption of the urban economy. In this case

$$Q_a = k \times PD \times EC, \quad (2)$$

PD is the population density within the urban administrative boundaries, EC is the energy consumption per capita for the country where this urban agglomeration is located. If Q_a is in W/m^2, PD in people per sq. km, and EC in kg o.e., the coefficient will be k = 1.325.

3. Results

In [5, 6] the available data provide estimates of the mean annual AHF for several world megacities and capitals, such as New York, London, Beijing, Tokyo, Seoul, Jakarta, as well as Moscow within the 2011 borders and after the city borders expansion in 2012.

A difficulty of such estimates, apart from the uncertainty (and not always reliability) of energy consumption data, lies in the fact that in various cases sparsely populated (essentially, suburban) areas are included or not included in the administrative boundaries of cities. For example, until 2012 the administrative area of St. Petersburg was almost twice as large as Moscow, while the population of St. Petersburg was less than twice as large as that of Moscow. Similar situations are in New York, London, and Beijing, where there is a “city core” comparable to Seoul or Moscow (until 2012), and the suburbs administratively subordinate to the city authorities, comparable to new Moscow (territories annexed in 2012).

Table 1. Administrative area, population, energy consumption, and mean AHF for several world cities.

City	Administrative area, km^2	Population, millions	Mean energy consumption in the country, kg o.e. per capita	Mean anthropogenic heat flux, W/m^2
New York	1215	8.4	6949	63.7
Moscow (June 2012)	1081	11.3	4943	68.5
Moscow (July 2012)	2510	11.6	4943	30.3
London	1572	8.9	2777	20.8
Beijing	1288	11.1	2237	25.5
St. Petersburg	1439	5.4	4943	24.6
Novosibirsk	503	1.6	4943	20.8

In Table 1, the data are taken from Wikipedia [14-19] for the city area and population and from the UN Little Green Data Book [20] for the national mean energy consumption per capita. The AHF values are calculated by equation (2).

In numerical climate models, the databases created at the beginning of the 21st century and most fully presented in [21] are usually used to describe the AHF field. This work uses statistics from 2005. It should be noted that over the past decade data on the energy consumption per capita in many countries have changed significantly. Thus, in the UK this value has decreased by a quarter and in China, by three quarters. In Russia and the USA this indicator has not changed so much over the past decade.

To study the influence of the AHF on mesoscale atmospheric processes, we use the regional climate model COSMO-CLM [7], where the initial average annual AHF fields are set based on data [21] with additions and corrections based on NOOA and EEA datasets. Figure 1 shows the average annual AHF fields used in the COSMO-CLM for New York, London, and Beijing, as well as for the three largest cities in Russia - Moscow, St. Petersburg, and Novosibirsk. A comparison of the AHF values given in Table 1 and presented in Figure 1 shows good agreement between the estimates obtained by various methods.

We performed a numerical simulation of the anthropogenic heat flux impact on the air temperature and wind within Moscow region using the COSMO-CLM model coupled with a TERRA-URB surface energy balance scheme. The model and the energy balance scheme were adapted to the capabilities of the IAPh cluster (see [22, 23]). The calculations were carried out on two nested domains in the European part of Russia. Each of these domains contains 10,000 cells of a horizontal grid corresponding to a size of 17x17 km and 5x5 km, 40 layers in the atmosphere and 9 layers in the underlying layer. Calculating a day for these domains using two quad-core processors of the IAPh cluster takes approximately 10 and 30 minutes of computational time, respectively.

Figure 2 shows some results of temperature and wind fields for several days at UTC 0 from simulations during 2016 and 2017. The air temperature over Moscow is higher when considering the AHF. It means that the formation of an urban heat island is observed. We can see that the wind speed is higher in the figures with the consideration of the AHF (especially on January 5, 2016 and July 6, 2017).

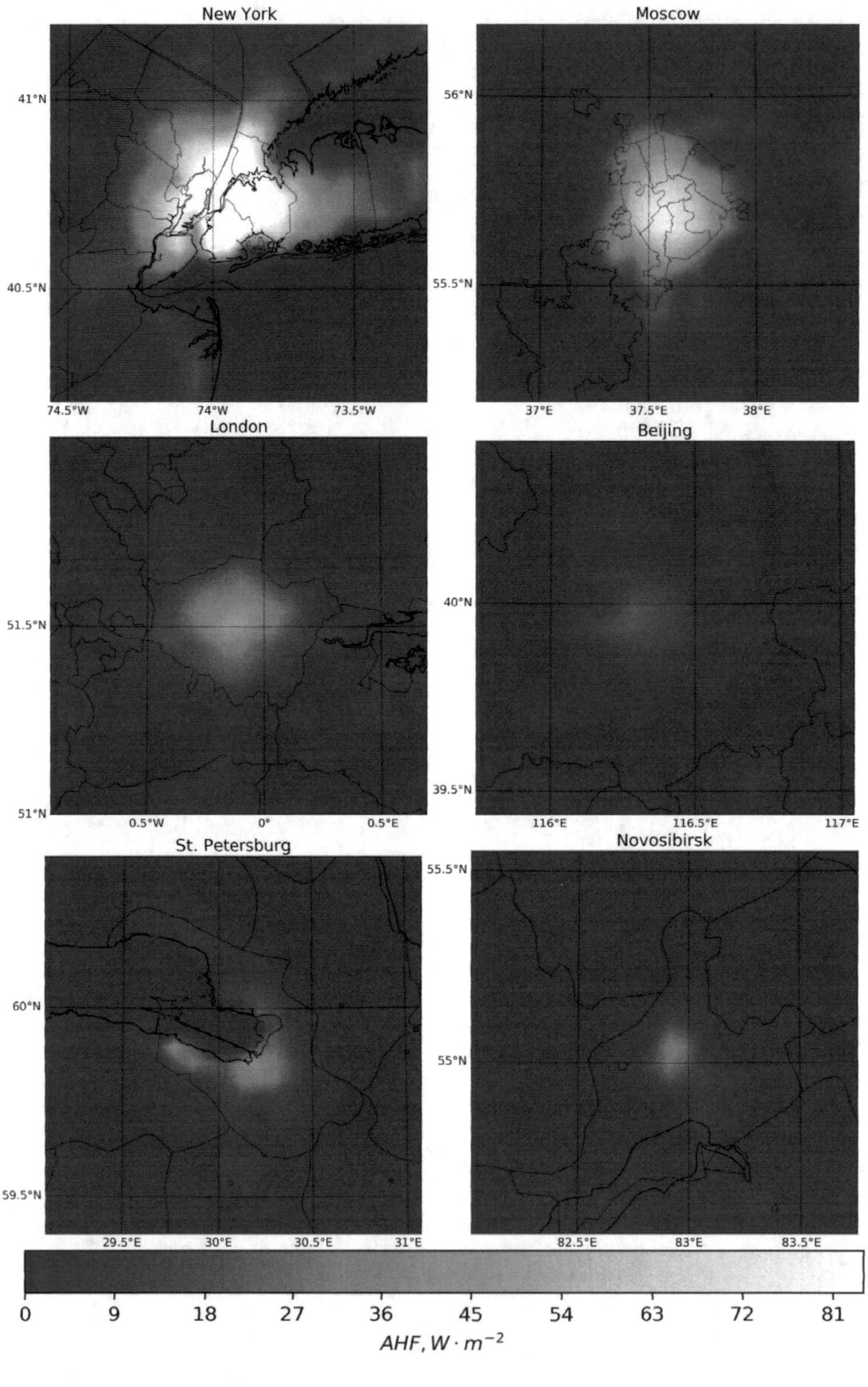

Figure 1. Mean annual anthropogenic heat flux spatial distribution for selected large world and Russian cities (data from [21] accessed using WebPEP tool of CLM-Community).

IOP Conf. Series: Earth and Environmental Science **386** (2019) 012049 doi:10.1088/1755-1315/386/1/012049

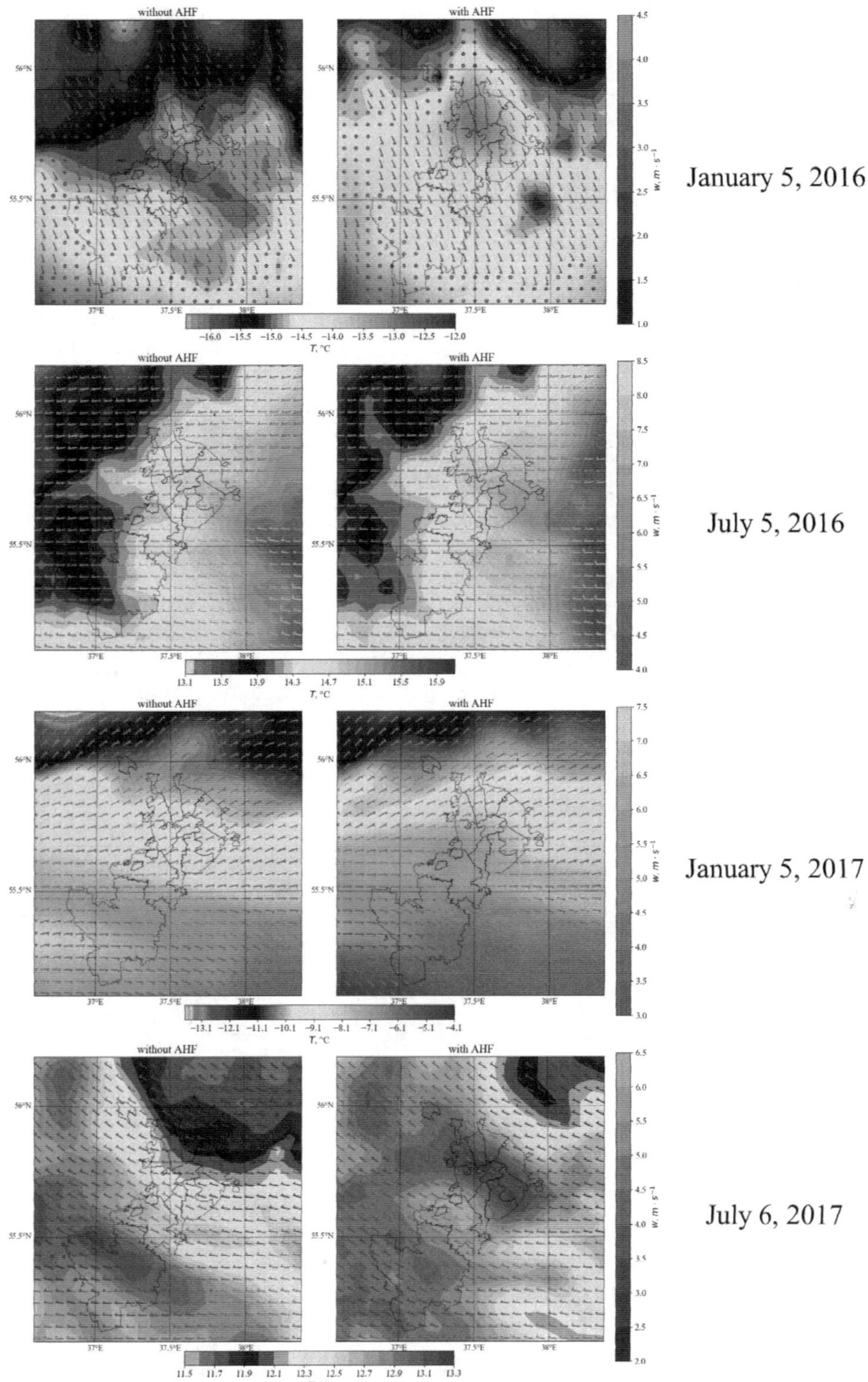

Figure 2. Temperature and wind fields in Moscow region from COSMO-CLM simulations (left figures – without AHF, right figures – with AHF).

Table 2. Mean annual air temperature and wind velocity in Moscow region during 2017 calculated with COSMO-CLM model.

Parameter	Without AHF	With AHF
Air temperature, °C	1.3	3.1
Wind velocity, m/s	2.9	3.9

Table 2 presents the average annual air temperature and wind velocity in Moscow region in 2017. The air temperature in the model with the consideration of the AHF is higher than in the model without the consideration of the AHF because, according to the Fourier law, heat fluxes increase the temperature. The average wind speed is also higher in the model with the consideration of the AHF because, as shown previously, the consideration of the AHF results in strengthening of the urban breeze circulation [24-25].

4. Discussion and conclusions

Anthropogenic heat fluxes are major factors in the formation of an urban heat island. AHFs also significantly affect the wind regime of a megalopolis. In Moscow's agglomeration, the average wind speed increases by more than 1 m/s, while the prevailing wind direction changes slightly. A similar effect is described in [24] and some other publications.

In turn, regional atmospheric advection significantly affects the intensity of an urban heat island, by strengthening or weakening the feedback between the temperature regime and the energy consumption of urbanized territories.

The feedback is negative during the heating season and positive during the indoor air conditioning period [26, 27]. The significance of such feedbacks substantially depends on the climate and relief of the region in question, the heat and power supply systems of the city, and other factors.

Anthropogenic heat fluxes are very strongly interconnected with the characteristics of the heating season [28] and with the carbon footprint of large cities [29].

Acknowledgments. The investigation was supported by the Russian Science Foundation (project no. 16-17-00114).

References

[1] Orlanski I 1975 A rational subdivision of scales for atmospheric processes *Bulletin of the American Meteorological Society* **56** (5) 527–30
[2] Markowski P, Richardson Y 2010 Mesoscale Meteorology in Midlatitudes *Wiley-Blackwell* **414**
[3] Landsberg H E 1981 *The Urban Climate* (New-York: Academic press) p 275
[4] Oke T R, Mills G, Christen A, Voogt J A *Urban climates* (Cambridge: Cambridge University Press) p 509
[5] Ginzburg A S, Belova I N, Raspletina N V 2011 Anthropogenic heat fluxes in urban
[6] agglomerations *Dokl. Earth Sci.* **439** 1006–09
[7] Aleksandrov G G, Belova I N, Ginzburg A S 2014 Anthropogenic heat flows in the capital agglomerations of Russia and China *Dokl. Earth Sci.* **457** 850-54
[8] Rockel B, Will A, Hense A 2008 The regional climate model COSMO-CLM (CCLM) *Meteorologische Zeitschrift* **17** 347–48 DOI: 10.1127/0941-2948/2008/0309
[9] http://www.cosmo-model.org/content/tasks/workGroups/wg3b/docs/terra_urb_user.pdf
[10] Wouters H, Demuzere M, Blahak U, Fortuniak K *et. al* 2016 The efficient urban canopy dependency parametrization (SURY) v1.0 for atmospheric modelling: Description and application with the COSMO-CLM model for a Belgian summer *Geoscientific Model Development* **9** 3027–54 DOI: 10.5194/gmd-9-3027-2016

[11] Varentsov M I, Konstantinov P I, Samsonov T E 2017 Mesoscale modelling of the summer climate response of Moscow metropolitan area to urban expansion *IOP Conference Series: Earth and Environmental Science* **96** 12009 DOI: 10.1088/1755-1315/96/1/012009
[12] Varentsov M I, Samsonov T E, Kislov A V, Konstantinov P I 2017 Simulations of Moscow agglomeration heat island within framework of regional climate model COSMO-CLM [in Russian] *Moscow University Vestnik. Series 5. Geography* **6** 25-37
[13] Varentsov M, Wouters H, Platonov V, Konstantinov P 2018 Megacity-Induced Mesoclimatic Effects in the Lower Atmosphere: A Modeling Study for Multiple Summers over Moscow, Russia. *Atmosphere* **9** 50 DOI: 10.3390/atmos9020050
[14] Allen L, Lindberg F, Grimmond C S B 2011 Global city scale urban anthropogenic heat flux: model and variability *Intern. J. Climatol* **31** 1990–2005
[15] https://en.wikipedia.org/wiki/New_York_City
[16] https://en.wikipedia.org/wiki/Moscow
[17] https://en.wikipedia.org/wiki/London
[18] https://en.wikipedia.org/wiki/Beijing
[19] https://en.wikipedia.org/wiki/Saint_Petersburg
[20] https://en.wikipedia.org/wiki/Novosibirsk
[21] *The Little Green Data Book 2017* (Washington, DC: World Bank) doi:10.1596/978-1-4648-1034-3
[22] Flanner M G 2009 Integrating anthropogenic heat flux with global climate models *Geophysical Research Letters **36** L02801* doi:10.1029/2008GL036465
[23] Alexandrov G G, Belova I N, Varentsov M I, Dokukin S A 2017 Impacts of climate change on energy consumption of Russian cities in the winter period *Proc. SPIE 10466, 23rd International Symposium on Atmospheric and Ocean Optics: Atmospheric Physics 104666K (30 November 2017)* doi: 10.1117/12.2287766
[24] Ginzburg A S, Dokukin S A 2018 Numerical modeling of anthropogenic heat flux impact on air temperature in Moscow in wintertime *IOP Conference Series: Earth and Environmental Science* **211** 012019 doi:10.1088/1755-1315/211/1/012019
[25] Lemonsu A, Masson V 2002 Simulation of a summer urban breeze over Paris *Boundary-Layer Meteorology* **104** 463–90 DOI: 10.1023/A:1016509614936
[26] Chen Y, Jiang W M, Zhang N *et al* 2009 Numerical simulation of the anthropogenic heat effect on urban boundary layer structure *Theor. Appl. Climatol.* **97** 123-34 *https://doi.org/10.1007/s00704-008-0054-0*
[27] Ginzburg A S, Demchenko P F 2017 Air temperature and energy consumption feedbacks within urbanized areas *Izv. Atmos. Ocean. Phys.* **53** 487–94
[28] Demchenko P F, Ginzburg A S 2018 Influence of feedbacks in the climate–energetics system on the intensity of an urban heat island *Izv. Atmos. Ocean. Phys.* **54** 313-21. DOI: 10.1134/S0001433818040217
[29] Belova I N, Ginzburg A S, Krivenok L A 2018 Heating seasons length and degree days trends in Russian cities during last half century *Energy Procedia* **149** 373-79 DOI: 10.1016/j.egypro.2018.08.201
[30] Alexandrov G G, Ginzburg A S 2018 Anthropogenic impact of Moscow district heating system on urban environment *Energy Procedia* **149** 161-69 DOI: 10.1016/j.egypro.2018.08.180

IOP Conf. Series: Earth and Environmental Science **386** (2019) 012050 doi:10.1088/1755-1315/386/1/012050

IMCES Geophysical Observatory for studies of surface-atmosphere interactions

E P Gordov, V Yu Bogomolov, E A Dyukarev, I G Okladnikov and S V Smirnov

Institute of Monitoring of Climatic and Ecological Systems SB RAS, Tomsk, Russia
E-mail: gordov@scert.ru

Abstarct. This paper describes a geophysical observatory in use at the Institute of Monitoring of Climatic and Ecological Systems SB RAS and the data it provides. These data can be used for the development of new parameterizations in an active layer model. This model will allow us to reproduce the physical processes in the soil and on its surface and make a detailed description of the exchanges between the surface and the atmosphere. This, in turn, could significantly increase the reliability of regional weather forecasts, especially of extreme weather events. The anticipated improvement in regional meteorological forecasting, as well as in a variety of characteristics measured at the geophysical observatory and the supporting information infrastructure open up possibilities for novel practical applications. In particular, we plan to use the ongoing measurements and data obtained for agrometeorological applications in Western Siberia.

1.Introduction

A complete cycle of obtaining new knowledge in meteorology and climatology, as well as new opportunities for its applications, starts from measurements of meteorological characteristics. At the next step, researchers are trying to study relevant processes and create phenomenological models, which eventually lead to physical models providing mathematical description. The use of the models created allows one to determine the meteorological characteristics for some space domain and forecast their evolution in time. Joint use of observations and models forms a basis for monitoring and prognosis of regional meteorological characteristics and processes [1].

The development of novel instrumentation automatically measuring relevant characteristics and sending data obtained to a dedicated server, as well as the mature state of meteorological models and information-computational technologies, provide us with opportunities to add new features to the cycle mentioned above and significantly improve both models and relevant prognoses, thus leading to novel applications of the knowledge. Timely employment of local meteorological characteristics can be used to significantly enhance both the knowledge of the space distribution of the measured local values and their time evolution. Two types of approaches can be employed to reach these objectives. These are the data assimilation techniques widely used now in meteorological forecasting and fast bias correction procedures to correct biased regional climate models outputs [2]. As a result, systematic local measurements might lead to reliable calculations of high resolution spatial and temporal behavior of meteorological conditions, thus opening a way to novel applications in experimental areas like agriculture.

In this paper, we first describe the state of the art of a Geophysical Observatory (GO) created at the Institute of Monitoring of Climatic and Ecological Systems (IMCES) SB RAS near the city of Tomsk and equipped with well-established standard meteorological instruments, as well as with modern measuring devices developed at IMCES. Then some results of a research of atmospheric and surface processes performed by using the ongoing GO observations will be presented. Finally, we discuss the use of knowledge gained in the development of new parameterizations for meteorological and climatic models, as well as approaches to enhancing the reliability of the current and predicted meteorological characteristics calculated with a high spatial resolution for West Siberia region.

2. The IMCES Geophysical Observatory

2.1. General Description

The geophysical Observatory (GO) is located at the premises of IMCES (coordinates: 56°28′32″ N, 85°03′17″ E, the altitude elevation is 167 m, the zero point elevation of the barometer is 188.4 m). The GO infrastructure is located in the territory and in the buildings of IMCES SB RAS in Akademgorodok, in the south-eastern suburban area of Tomsk. The observatory has two main observation sites, several laboratory rooms with measuring and recording equipment and instrumentation for office processing of field research data. It has been working since 2007.

Currently, instrumental and visual meteorological, aerological, spectrophotometric, atmospheric electric, aerosol spectrometric, and gas analytical observations are being performed at the observatory. The observation modes are either continuous or periodic, depending on their type and tasks being solved.

The main ground-based meteorological observations, including atmospheric-soil, actinometric, and atmospheric-electric ones, are carried out using an automated meteorological information-measuring system (AMIMS) with high temporal resolution (continuous measurements with half or minute averaging). Visual hourly meteorological observations provide information on the cloud cover, the state of the atmosphere, and the underlying surface and weather patterns.

Using an automated meteorological information-measuring complex (AMIMC) based on spatially separated ultrasonic automatic meteorological stations (AMSs), gradient high-frequency (up to 80 Hz) measurements are carried out at heights of up to 30 m (it is planned to increase them to 40 m), providing information on both the basic meteorological values and the characteristics of atmospheric turbulence in the surface layer, including heat, moisture, and momentum fluxes with a high temporal resolution.

Remote sensing of the vertical distribution (VD) of the wind speed and direction, as well as atmospheric turbulence characteristics in the atmospheric boundary layer (ABL) is carried out automatically with preset time intervals using a Doppler sodar (a meteorological wind profiler). The information is provided with high spatial resolution (up to 10 m) at heights of up to 1 km.

Spectrophotometric observations provide data on the total ozone content (TOC), the characteristics of ultraviolet radiation (UV), cloud cover, spectral transparency of the atmosphere (spectral aerosol optical depth, AOD), and the atmospheric moisture content. UV and TOC measurements are performed using multichannel medium resolution filter radiometers in manual (hourly) and automatic (continuous with minute averaging) modes. The AOD and moisture content are measured automatically every minute using a multi-channel solar photometer under direct solar radiation. Registration of the sky state is made automatically with a preset time interval using a panoramic sky scanner based on a high-resolution digital camera.

Spectral measurements of the surface aerosol particulate number and light scattering coefficients are performed using a set of aerosol spectrometers and a nephelometer in the periodic mode with a specified time interval.

Gas analysis of radon, carbon dioxide, and water vapor surface concentrations is carried out by automatic gas analyzers by air sampling at ground levels, 2 and 10 m, or from accumulation chambers and wells (for estimating gas emissions from the soil) at specified time intervals. Physicochemical studies of air and water samples collected during field expeditions are carried out in the observatory using chromatographs.

All-year chamber measurements of carbon dioxide fluxes using an automated soil CO_2 flux system LI-8100a allow one to estimate the carbon balance of the urban grassland. A model of net ecosystem exchange (NEE) was suggested to study the influence of different environmental factors. The model uses air temperature, incoming photosynthetically active radiation, vapor pressure deficit, and leaf area index as the explanatory factors for gross primary production, heterotrophic and autotrophic respiration. Observation results obtained at two sites (bare and vegetated soil) allow one to calculate the temperature sensitivity for soil and plant respiration. The total annual NEE is 163.5 g C

m^{-2}. Growing plants accumulate 522.7 g C m^{-2} in total, but the net annual release of CO_2 is higher (686.2 g C m^{-2}). The ecosystem being studied is a source of carbon according to modeling and observation results [3].

The observational data acquisition and storage in the GO is hierarchical. Raw data are recorded in the measuring instruments, databases of information-measuring systems and complexes, and an electronic journal of weather observers. Then the data are transferred to the database of the integrated information-computational system of IMCES SB RAS, where they undergo quality assessment, archiving, visualization in graphical and digital forms for the users.

Currently the GO regularly transmits measurement data on total ozone concentration and UV radiation to the State Geophysical Observatory, the Central Aerological Observatory, and o the PEEX *In-Situ* Atmospheric-Ecosystem Collaborating Stations Network in the Russian Federation. It should be added that the GO is also used for education and research works of pupils, students, and post-graduates.

2.2. Radiation measurements with a high resolution

The radiation measurements are performed by a multifilter radiometer NILU-UV-6T [4] and a pyranometer Keep & Zonen CM11 [5] on a roof observation platform with 1-min resolution. The radiometer measures UV radiation at 305, 312, 320, 340, and 380 nm (FWHM=10 nm) and PAR. The measurement accuracy is less than 10%. The data processing software calculates the daily average and maximum dose rates and integrated daily doses of UV-A, UV-B, PAR, CIE-, and CLW-weighted irradiance [6–9]. The software also provides cloud cover assessment (spectral transmittance of cloudiness) and total ozone column.

The pyranometer CM11 measures global radiation in the range of 305–2800 nm. The measurement accuracy is 2–3 %.

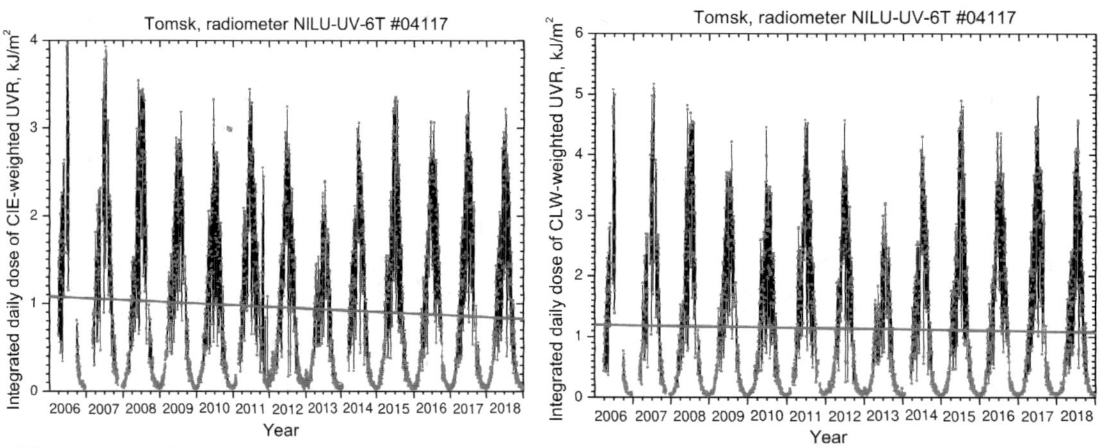

Figure 1. Long-term observations of CIE-weighted irradiance, long-term observations of CLW-weighted irradiance.

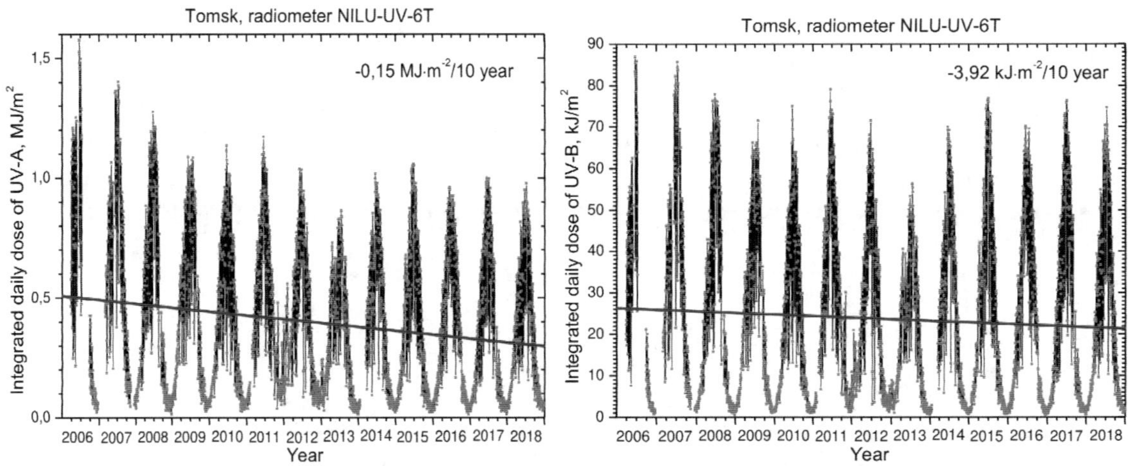

Figure 2. Long-term observations of UV-A irradiance, long-term observations of UV-B irradiance.

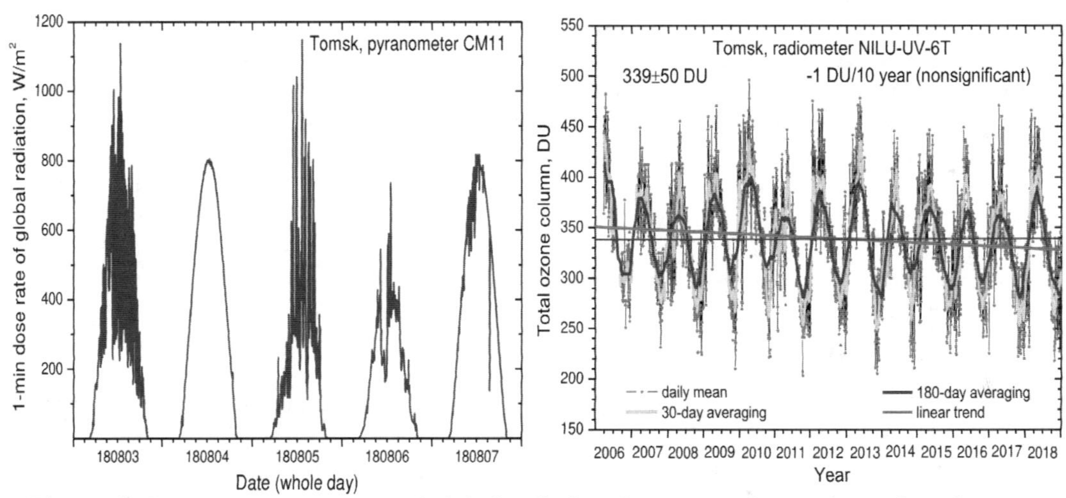

Figure 3. Current measurements of global radiation, long-term observations of total ozone.

Long-term (2006–2018) observations of incoming short-wave solar radiation carried out with a high temporal resolution, total ozone and cloudiness in Tomsk (Western Siberia) led to the following results:

- time series of UV-A/B irradiance have significant negative linear trends and the slope of UV-A trend is the greatest one;
- trend of the linear PAR is insignificant;
- diminishing of the total ozone concentration over Tomsk region stopped and, thus, observed negative linear UV trends can be associated with the total ozone stabilization;
- the influence of cloudiness on radiation fluxes is stronger in the UVI range than in the entire UV range;
- variations of the global radiation and PAR induced by clouds are up to 30 %, for UV radiation they are up to 10 %, and Ci, Ac, and Cu clouds have a maximum impact;
- dense cloudiness (As, Ns, Sc, Cb) and overcast completely attenuate radiation fluxes in the UVI range.

IOP Conf. Series: Earth and Environmental Science **386** (2019) 012050 doi:10.1088/1755-1315/386/1/012050

2.3. Measurements of surface and soil characteristics

The measurement subsystem includes (1) a programmable microcontroller ATMEGA1280 that collects measurement data from sensors, sequentially polls through a 1-Wire digital interface, stores them in non-volatile Flash memory (AT45DB321, 4 Mb or 90,000 measurements), and transfers them to the server according to a schedule via a Wi-Fi network or FT232RL USB interface, (2) a power supply containing a network converter and a battery, and (3) digital and analog measuring converters (sensors):

- sensors of air temperature DS18B20 and air relative humidity HIH5031 (in a radiation shield at 2-m height above the surface);
- liquid and solid precipitation sensor Davis 7852M (at 2-m height);
- soil temperature probe with DS18B20 temperature sensors (depth levels: 0, 2, 5, 10, 15, 20, 30, 40, 60, 80, 100, 120, 160, 200, 240, 280, 320, 400, 450, and 500 cm);
- single soil temperature sensors DS18B20 in the soil section (depth levels 0, 2, 5, 10, 15, 20, 30, 40, 60, 80 and 100 cm) and in soil cores (depth levels: 20, 40, 60, 120, 160, 240, and 320 cm);
- Vaisala soil temperature sensors DTS12G1 (depth levels: 0, 10, and 20 cm);
- snow temperature probe with DS18B20 temperature sensors (levels from 0 to 120 cm in increments of 5 cm; in-house development);
- ultrasonic snow depth sensor MB7389;
- sensor of electrical conductivity of snow cover (in-house development);
- atmospheric pressure sensor MPL3115A2.

The technical characteristics of the main sensors are given in Table 1.

The controller is located in a protecting box (Ingress Protection Code IP65), which has thermal insulation, which allows the controller to operate without additional heating in the temperature range from –50 to + 50 °C. Measurements can be performed at intervals of 2 s or more depending on the task. The measurement period is currently 10 minutes. Solid precipitation in the precipitation meter is pre-melted. The software allows one to remotely operate the controller, set the order (periodicity) of reading measurement data, plot measurement results, and export data into “csv” text format.

The number of soil and snow temperature sensors and their installation depths are stipulated by the need to obtain high-quality information on the distribution and variability of temperature in the soil layer and snow cover with high spatial and temporal resolution, which is required for modeling heat-mass transfer processes in the soil, heat transfer between the underlying surface and the atmosphere. In particular, the placement of several sensors at the same depth is caused by the need to ensure representativeness of data on soil and snow temperature obtained by temperature probes of a specific design, which are used in field studies of the soil and peat temperatures in different landscapes.

Table 1. Technical characteristics of the temperature, humidity, and precipitation sensors.

Parameter	Value
Range and accuracy of air/soil/snow temperature sensor	–55…+50°C, ±0.1°C
Range and accuracy of air humidity sensor	0…100 %, ±3.5 %
Range and accuracy of liquid precipitation sensor	0…1000 mm/hr, ±5 %

Measurements of meteorological values are carried out in an automatic mode, independent of external power, with a high time resolution. The acquisition, operational storage, and transmission of measurement data is carried out both via wired and wireless communication channels to the observatory database.

As an example, Figure 4 shows some measurement results obtained using a modernized AMIIS. The temporal variations of air temperature (Ta), precipitation amount (P) for 10-minute intervals, and soil temperature (Ts) on the surface (Ts0) and depths from 5 to 500 cm for the period from March to August 2019 are shown. Despite significant fluctuations of the air temperature, the temperature of the surface and upper soil layers under the snow cover (until April 13, 2019) changed slightly. The snow depth in March 2019 reached 90 cm, and the soil froze to a depth of 15 cm only. After quick snow melting and short but intense thaw, the soil thawed completely. In summer, the temperature fluctuations on the surface and in the upper layers of the soil actually repeat the air temperature fluctuations. With depth, the amplitude of diurnal temperature fluctuations decreases significantly and deeper than 40 cm the diurnal course of the soil temperature does not exist. The annual course of soil temperature can be traced at all depths. The maximum values of soil temperature at a depth of 80 cm were observed in August, at a depth of 160 cm in September, and deeper the temperature will increase until December.

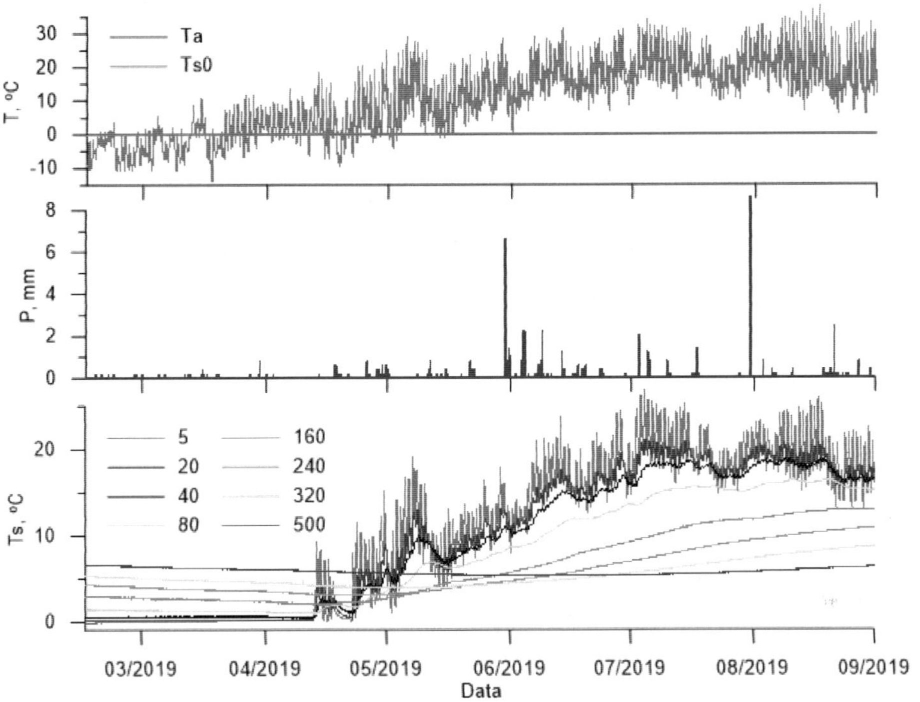

Figure 4. Variations of air temperature, precipitation amount, and soil temperature at the surface and depths for March - August 2019 according to measurements on the ground observation site of GO.

3. Use of measured data for the development of new parameterizations

A further step in understanding the physical processes is to perform mathematical and numerical simulations using the available measurement data for the particular area. At present, the data of measurements performed at the GO are used in the development of a new block of the active land layer for the climatic model of INM RAS [10]. The active layer model includes a multilayer (23 levels) soil model that describes the transfer of heat, liquid moisture taking into account its freezing / thawing, and water vapor [11], as well as a multilayer heat transfer scheme in snow cover. Calculation of the surface and subsurface runoff, as well as the influence of vegetation on the soil water balance, is carried out according to the parameterizations used in the ECHAM3 model (The ECHAM3 atmospheric general circulation model, 1992). The current version of the active layer model of INM RAS-MSU considers that a land surface cell may contain the following surface types: vegetation, open soil, snow, and inland waters. The proportion of snowless surface occupied by vegetation, inland waters, and open soil is prescribed according to [12].

IOP Conf. Series: Earth and Environmental Science **386** (2019) 012050 doi:10.1088/1755-1315/386/1/012050

Detailed measurements of the thermal and radiation balance on the surface, as well as the thermal regime in the soil carried out at the GO, will be used to validate and verify the model of the active land layer for specific soil types. In particular, this will make it possible to determine the contribution of specific parameterizations included in the model of the active land layer, which significantly affect the calculated moisture and soil temperature. These parameterizations, which are some approximations obtained from physical considerations and empirical data, are then used to calculate important soil characteristics, such as thermal conductivity (Johansen, O. et. All). In this case, the calculated values can give results significantly different from the measurement data (Figure 5).

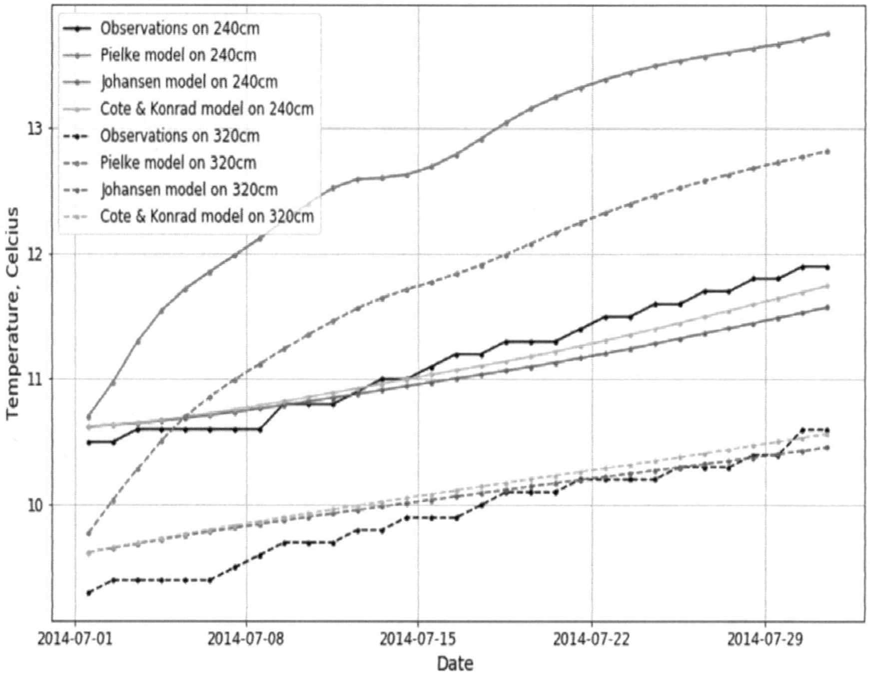

Figure 5. Soil temperature at different depths according to observations at Meteorological Observatory and model calculations.

The results of detailed measurements will allow us to find limits of applicability of the parameterizations and determine the possibility of using them for various types of mineral soils. Modeling of processes in the soil should make it possible to determine the coefficients through quantities that depend on a number of moisture, density, and temperature factors. Knowledge of these dependences will allow one not only to reproduce all physical processes occurring in the soil and on its surface, but will allow a more detailed description of exchanges between the surface and the atmosphere. Of course, the latter is significant for climate modeling, but these processes are even more important for mesoscale meteorological modeling. A correct description of exchanges between the surface and the atmosphere will significantly increase the reliability of regional weather forecasts, especially forecasting of extreme weather events.

4. Planned use of GO data for practical applications

The anticipated improvement of regional meteorological prognoses, as well as a variety of characteristics measured at the GO and a mature level of the supporting information infrastructure, opens up opportunities for novel practical applications. In particular, we plan to use the GO ongoing measurements and data obtained for agrometeorological applications in West Siberia.

Indeed, the climatic and meteorological conditions, especially extreme hydrothermal phenomena, significantly affect the productivity of the agricultural sector. The traditional calculation of agrometeorological characteristics on the basis of data measured at Roshydromet weather stations does not provide the spatial resolution necessary for agricultural producers and does not provide them with predictive characteristics. Plain expansion of the measurement network, rapid transfer of data, and their analysis also cannot provide a consumer with the required information. To solve this problem, we propose using the Web GIS Climate previously developed for climate applications (http://climate.scert.ru) [13, 14, 15]. To do this, its functionality will be expanded with new algorithms for calculating agrometeorological characteristics, and the transition to quasi-operational data analysis will be completed.

It is planned that a high-speed data stream (selected meteorological characteristics) from the GO and several autonomous weather stations deployed by an agricultural producer will be obtained in the real-time mode by a dedicated server, pre-processed, and stored in an intermediate data storage. At the same time, every 24 hours the results of predictive mesoscale meteorological modeling of high resolution will be transferred to the same dedicated server. They will undergo a quick bias correction procedure and then stored in a high-performance data archive of the web-GIS Climate. Later, these data will be used by the web-GIS Climate for quick calculation of corresponding agrometeorological characteristics and production of digital maps of agrometeorological risks for a region of interest. As a result, such an approach will provide, in a quasi-operational mode, accurate agrometeorological characteristics and their forecasts with the spatial resolution required by stakeholders. It might also lead to the development of digital intelligent farming and land management systems.

5. Conclusions

The above description was given to illustrate the use of data measuring at the GO at the different steps of knowledge and applications. The use of measured data allows one to understand the physical processes of surface-atmosphere interaction and eventually could lead to more reliable results of climatic and meteorological modeling. The GO ongoing measurements and data obtained, as well as the anticipated information, open up possibilities for novel applications, like the above-described agrometeorological applications in Western Siberia.

It should be added that the variety of characteristics measured at the GO and the present mature level of the supporting information infrastructure of the GO provide a potential that can be widely used in national and international research cooperation.

Acknowledgments

This study was supported by project AAAA-A17-117013050037-0 of the RAS Basic Research Program.

References

[1] Kabanov M V and Lykosov V N 2007 Monitoring and modeling of climatic changes in Siberia *Atmos. Oceanic Opt.* **19** (9) 675-85

[2] Teutschbein C and Seibert J 2012 Bias correction of regional climate model simulations for hydrological climate-change impact studies: Review and evaluation of different methods *J. Hydrol.* **456** 12–29

[3] Dyukarev E A 2017 *Agric. For. Meteorol.* 239 236–48 doi:10.1016/j.agrformet.2017.03.011

[4] Dahlback A 1996 Measurements of biologically effective UV doses, total ozone abundance and cloud effects with multi-channel moderate bandwidth filter instruments *Appl. Opt.* **35** 6514–21

[5] *Http*://www.biospherical.com/nsf/login/GUV/Caldwell.txt

[6] McKinlay A F and Diffey B L 1987 A reference action spectrum for ultraviolet induced erythema in human skin *CIE Journal* **6** 17–22

[7] Caldwell M M 1971 Solar UV irradiation and the growth and development of higher plants *Photophysiology* **6** (4) 131–77

[8] Green A E S, Sawada T and Shettle E P 1974 The middle ultraviolet reaching the ground *Photochemistry and Photobiology* **19** 251–59

[9] Johansen O 1975 Thermal conductivity of soils (Trondheim: University of Trondheim) p 637

[10] Bogomolov V Yu and Stepanenko V M 2016 Development of lake parametrization in the INMCM climate model *IOP Conference Series: Earth and Environmental Science* **48**(1) doi:10.1088/1755-1315/48/1/012005

[11] Volodin E M and Lykosov V N 1998 Parametrization of heat and moisture transfer in the soil-vegetation system for use in atmospheric general circulation models: 1. Formulation and simulations based on local observational data *Izvestiya. Atmospheric and Oceanic Physics* **34** (4) 405-16

[12] WilsonM G F and Henderson-Sellers A 1985 A global archive of land cover and soils data for use in general circulation climate models *Journal of Climatology* **5** (2) 119-43

[13] Okladnikov I G, Gordov E P and Titov A G 2016 Development of climate data storage and processing model *IOP Conf. Series: Earth and Environmental Science* **48** 012033 DOI:10.1088/1755-1315/48/1/012033

[14] Ryazanova A A, Okladnikov I G and Gordov E P 2017 Integration of modern statistical tools for the analysis of climate extremes into the web-GIS “CLIMATE” *IOP Conf. Series: Earth and Environmental Science* **96** 012014 DOI: 10.1088/1755-1315/96/1/012014

[15] Gordov E P, Okladnikov I G, Riazanova A A, Titov A G and Gordova Yu E 2018 A virtual thematic research environment for analysis of climate change and its consequences for Northern Eurasia: the current state of the art and perspectives *IOP Conf. Series: Earth and Environmental Science* **211** 012079 DOI: 10.1088/1755-1315/211/1/012079

Role of sea surface warming in convective activity over Europe and Northern Eurasia: estimates with sensitivity model experiments

Yu I Yarinich[1,2], A V Chernokulsky[1], V A Semenov[1,3], M Latif[4]

[1]A.M. Obukhov Institute of Atmospheric Physics, Russian Academy of Sciences, Pyzhevsky 3, Moscow, 119017, Russia
[2] Lomonosov Moscow State University, Leninskie Gory 1, Moscow, 119991, Russia
[3] Institute of Geography, Russian Academy of Sciences, Staromonetny 29, Moscow, 119017, Russia
[4] GEOMAR Helmholtz Centre for Ocean Research Kiel, 24105 Kiel, Germany

E-mail: a.chernokulsky@ifaran.ru

Abstract. The influence of sea surface warming on convective activity over Europe and Northern Eurasia is estimated from sensitivity model experiments by an atmospheric general circulation model, ECHAM5, with prescribed boundary conditions ("warm" and "cold" sea surface). Convective activity is analysed by using various indices (thermodynamic, dynamic, and composite). It is shown that warmer sea surface leads to a general increase in the thermodynamic indices that is broadly consistent with observations. Particularly, the observed increase in CAPE over the eastern part of the Mediterranean Sea, the Black Sea, and Eastern Europe is well reproduced in the sensitivity experiments. At the same time, the shear and helicity instability indices depend little on sea surface warming. The experiment with only Mediterranean and Black Seas warming tends to overestimate the increase in the thermodynamic indices near these seas and underestimate the increase in the other regions. There are several regions (the Iberian Peninsula, Mongolia, and Northern China) where the observations show a decrease in the convective indices. These negative changes are not reproduced in the model experiments, because their nature is, apparently, not related to sea surface warming.

1. Introduction

The convective processes over Europe and Northern Eurasia become more prominent, which is manifested in an increase of the convective cloudiness [1-4], convective precipitation [5-8], the occurrence of thunderstorms [9], and severe convection environments [10-12]. In particular, Figure 1 shows the difference in CAPE (convective available potential energy) between 2000–2012 and 1979–1999 in the summer season based on ERA-Interim data [13]. With some regional exceptions, both the mean and frequency of exceeding the threshold tend to increase in the second period.

These changes may depend on different factors including thermodynamics (e.g., increase in the surface temperature and humidity, changes in the lapse rate [14,15]), dynamics (e.g., changes in cyclones and activity of atmospheric fronts [16-18], moisture convergence [19,20]), and microphysics (e.g., influence of aerosols [21]). In turn, both local and global warming and moistening of the lower atmosphere can be important. In particular, an important role of warm seas (the Mediterranean, Black, and Caspian seas) in the formation of strong tornadoes [22-24] and heavy convective precipitation events [25, 26] over Europe and Northern Eurasia was highlighted. It should be noted that the largest increase of the CAPE (Fig.1) is observed above the Eastern Mediterranean, Black, and Caspian Seas. The role of different factors in changes of the convective activity in the Eurasian midlatitudes should be quantitatively estimated for increasing our ability to predict severe convective events and project their changes in the 21st century.

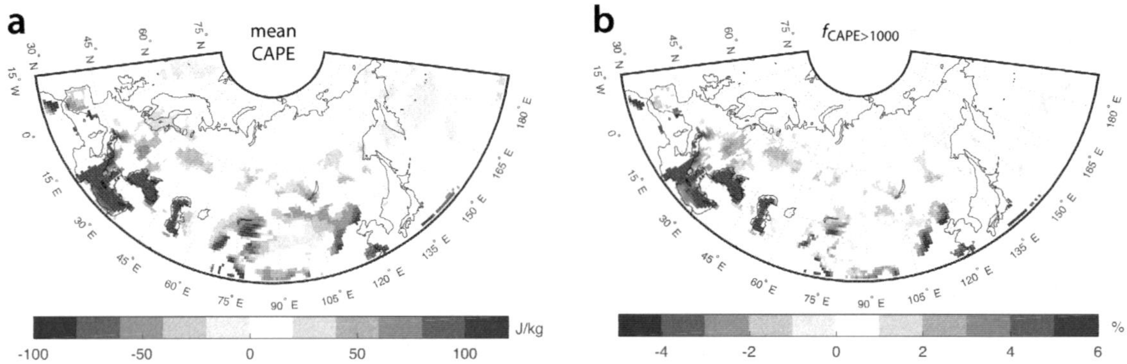

Figure 1. Difference between 2000–2012 and 1979–1999 of (a) summer-mean CAPE and (b) frequency of CAPE>1000 J kg^{-1} in summer based on ERA-Interim reanalysis data. Only significant changes are shown (see the definition of *significant changes* in Section 2).

One of the approaches to discriminate the role of different factors is to analyze specially designed sensitivity model experiments [25-27]. Particularly, Meredith and co-authors [25] used model simulations with warm and cold SST and highlighted the enhancement of local lower tropospheric instability due to the current warmer Black Sea. This enhancement triggers deep convection and increases precipitation by more than 300% between warm and cold SSTs during single events, as in July 2012 near the town of Krymsk [25]. A similar approach was used to estimate the influence of Mediterranean Sea warming on extreme precipitation formation in Europe [26].

In this study, the impact of sea-surface warming (global ocean warming and the Mediterranean and Black Seas only warming) on the convective activity changes over Northern Eurasia and Europe is estimated based on sensitivity experiments with the atmospheric general circulation model (AGCM). Convective activity in the atmosphere is estimated based on analysis of various convective indices. The approach for analysis of convective indices using the results of global climate model simulation was used previously for the USA [28,29], Australia [30,31], and Europe [32].

2. Data and method

We analyzed three numerical experiments with the AGCM ECHAM5 that are identical except for sea-surface temperature (SST) and sea ice concentration (SIC). The control experiment was forced globally with monthly climatological fields of 1970–1999 SST and SIC [26] (so-called 'cold ocean experiment', CO). In the warm Mediterranean experiment (WM), the warmer 2000–2012 SST climatology was employed only in the Mediterranean and Black Seas. In the global warm experiment ('warm ocean', WO), the ECHAM5 was forced globally with monthly climatological fields of 2000–2012 SST and SIC. For each experiment, a 40-member ensemble of one year was computed. The model was integrated at a relatively high horizontal resolution for a global atmosphere model of T159 (approximately 75 km) and with 31 vertical levels. Forcing conditions for all experiments were taken from the Hadley Centre Sea Ice and Sea Surface Temperature dataset [33]. Radiative and greenhouse gas forcing was fixed to the present-day levels [26].

Based on 6-hour model data, various convective instability indices were calculated including thermodynamic (CAPE, CIN) [34-36] and dynamic indices (SRH, DLS) [37,38], as well as composite parameters (WMAXSHEAR, EHI) [35,39] and a simplified index 3D [40,41] (Table 1).

Table 1. Analysed convective indices.

Index (acronym)	Index (full name)	Formula	Threshold
CAPE*, J Kg^{-1}	Convective available potential energy	$\mathrm{CAPE} = \int_{LFC}^{EL} g\left(\frac{T_{v,parcel} - T_{v,env}}{T_{v,env}}\right)dz$	0, 100, 1000, 2500
CIN*, J Kg^{-1}	Convective inhibition	$\mathrm{CIN} = \int_{surf}^{LFC} g\left(\frac{T_{v,parcel} - T_{v,env}}{T_{v,env}}\right)dz$	10, 30, 50
SRH, m^2 s^{-2}	Storm relative helicity	$\mathrm{SRH} = \int_{surf}^{3km} (\vec{v}_{hor} - \vec{c})\vec{\omega}_{hor} dz$	160, 300, 400
DLS, m s^{-1}	Deep layer shear	$\mathrm{DLS} = \Delta V_{500-950}$	20, 30, 40
WMAXSHEAR, m^2 s^{-2}	WMAXSHEAR	$\mathrm{WMAXSHEAR} = \mathrm{DLS} \times (2\mathrm{CAPE})^{-1/2}$	300, 450, 600
EHI	Energy-helicity index	$\mathrm{EHI} = \mathrm{CAPE}\times\mathrm{SRH} / 160\,000$	1, 2, 3
3D, °C	Dewpoint, depression of dewpoint index	$3\mathrm{D} = D_{surf} - DD_{surf}$	15.5, 17, 19

The following notation is used: LFC – height of level of free convection, EL – height of equilibrium level, $T_{v,parcel}$ – virtual temperature of a specific parcel, $T_{v,env}$ – virtual temperature of the environment, D – dew point temperature, DD – depression of dew point temperature, V – wind speed. Subscripts denote vertical levels (surf – near surface; 950 and 500 – height of corresponding geopotential height).

* Mean-level (ML) (1 km) and most-unstable (MU) CAPE and CIN were calculated.

We estimated the difference in the index values (mean, maximum, frequency of exceeding thresholds) between a pair of experiments (WO–CO, WM–CO, WO–WM) calculated for 40 modeled summer months (June, July, August). The significance of the difference was estimated based on the non-parametric Mann-Whitney U-test (the difference was called *significant* for p-values ≤0.05) [42]. In addition to the experiments with the ECHAM5 model, data from ERA-Interim reanalysis (for 1979–2012) were used as well.

3. Results

The warmer Global Ocean favors warmer and more humid near-surface air, which results in a significant increase of the CAPE (ML and MU), WMAXSHEAR, and 3D indices (both the means and frequencies of exceeding the threshold (Figures 2–4)). Exceeding of higher thresholds (e.g., 2500J kg^{-1} for CAPE and 600 m^2 s^{-2} for WMAXSHEAR) also occurs much more often in the WO experiment (not shown).

The increase in the thermodynamic convective indices is especially pronounced over marginal seas of the Atlantic and Pacific Oceans (the Mediterranean, Black, Baltic, Japan, and Okhotsk Seas), which broadly corresponds to the observed changes of CAPE (Figure 1). The greatest increase of CAPE is found in the Black Sea and Eastern Mediterranean in both the reanalysis and sensitivity model simulations. The observed increase of CAPE over Europe (Figure 1) can be attributed to the influence of sea surface warming. Specifically, both the ML CAPE and MU CAPE (their means and frequencies of exceeding the threshold) are significantly larger in this region in the WO experiment compared to the CO. On the contrary, no positive statistically significant difference of the CAPE between the WO and CO over eastern China is revealed in the observations. No negative WO–CO difference of the CAPE and 3D was found, which is in contrast to the reanalysis data that show a decrease of the CAPE over the Iberian Peninsula, Mongolia, and Northern China (Figure1).

IOP Conf. Series: Earth and Environmental Science **386** (2019) 012051 doi:10.1088/1755-1315/386/1/012051

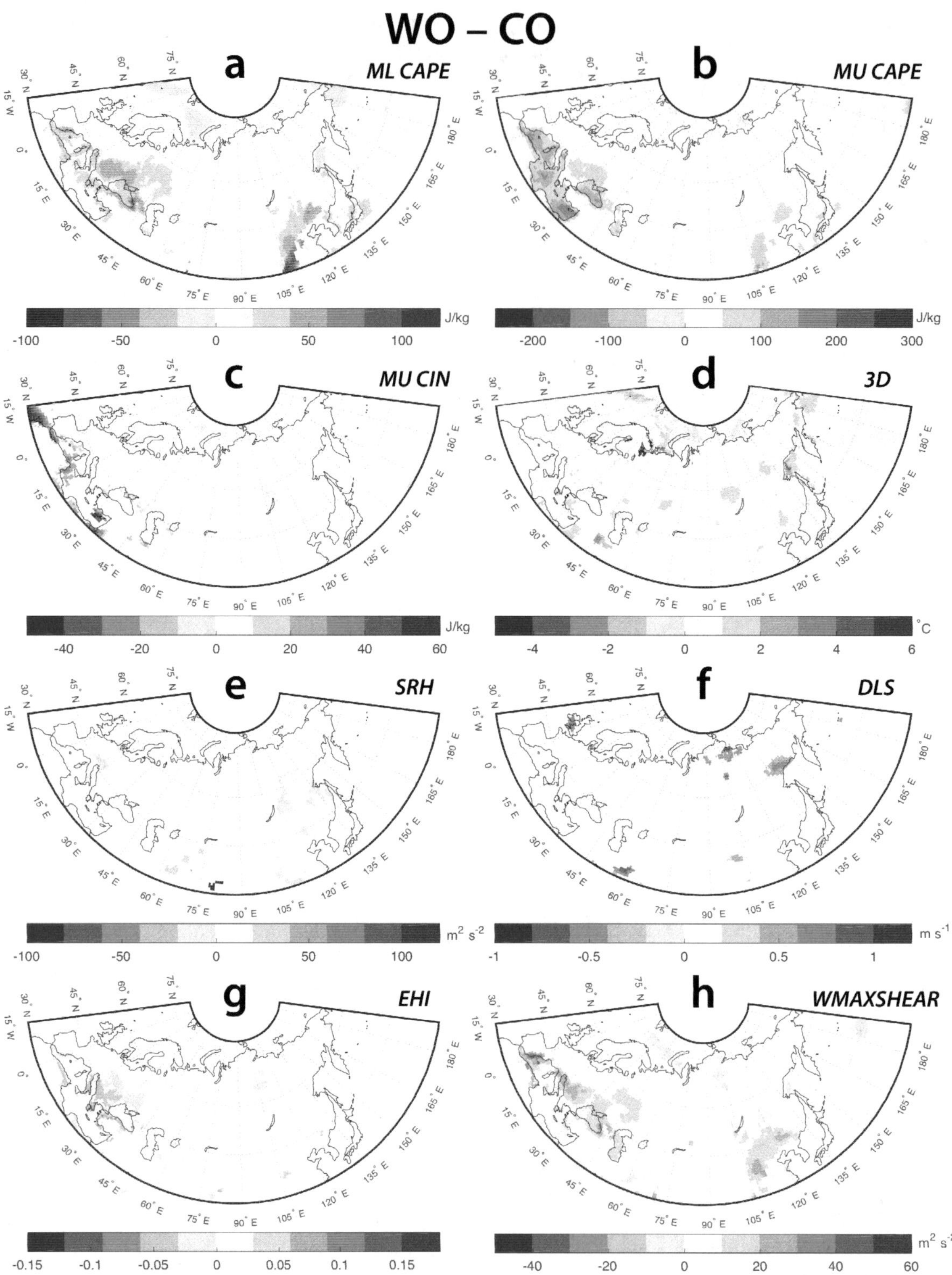

Figure 2. Difference between WO and CO for summer means of various indices: (a) ML CAPE, (b) MU CAPE, (c) MU CIN, (d) 3D, (e) SRH, (f) DLS, (g) ML EHI, (h) ML WMAXSHEAR. Only significant difference is shown.

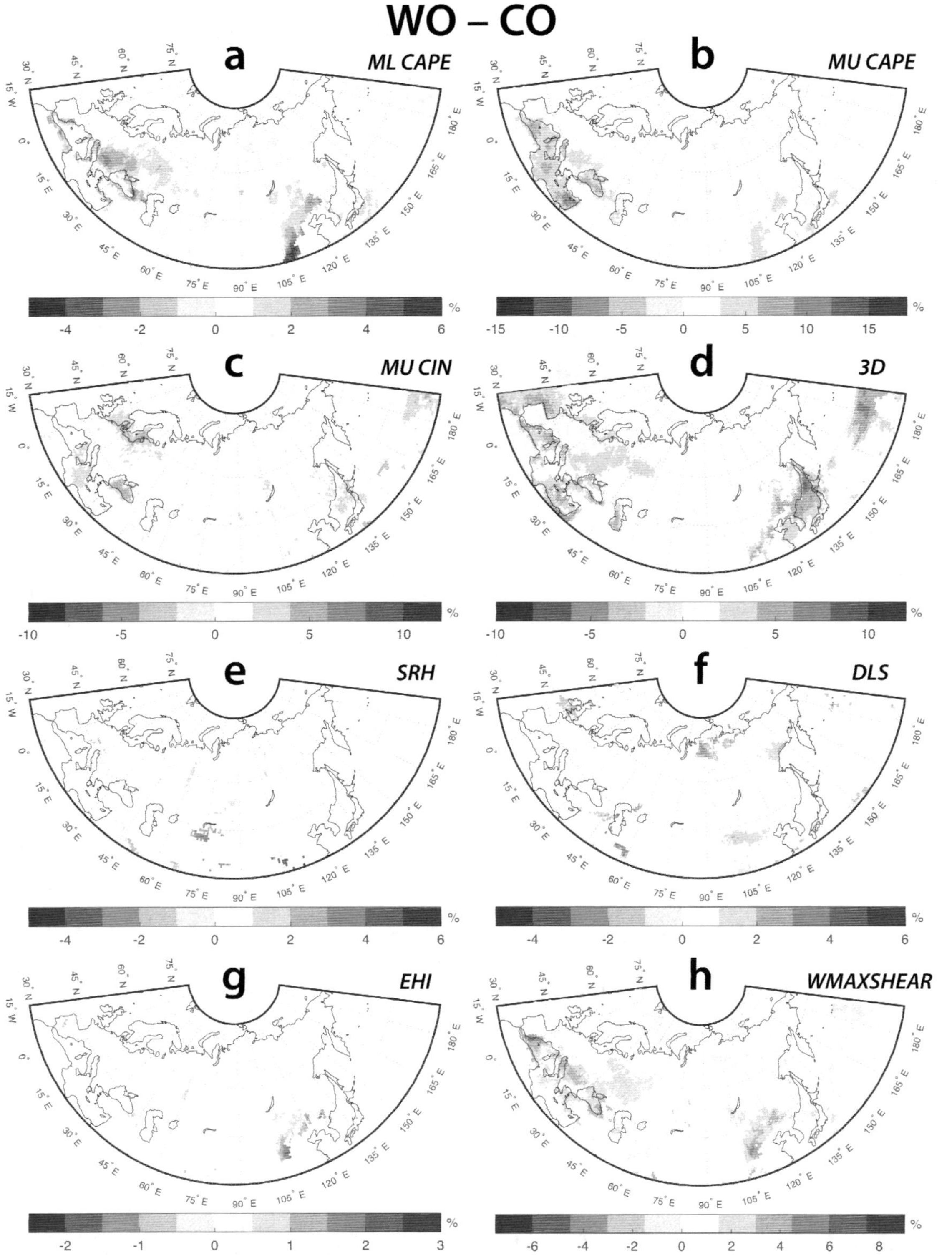

Figure 3. Difference between WO and CO for the frequency of exceeding the threshold in summer for different indices: (a) ML CAPE>1000 J kg^{-1}, (b) MU CAPE > 1000 J kg^{-1}, (c) MU CIN < 30 J kg^{-1}, (d) 3D > 15.5 °C, (e) SRH > 160 m^2 s^{-2}, (f) DLS > 20 m s^{-1}, (g) ML EHI > 1, (h) ML WMAXSHEAR > 300 m^2 s^{-2}. Only significant changes are shown.

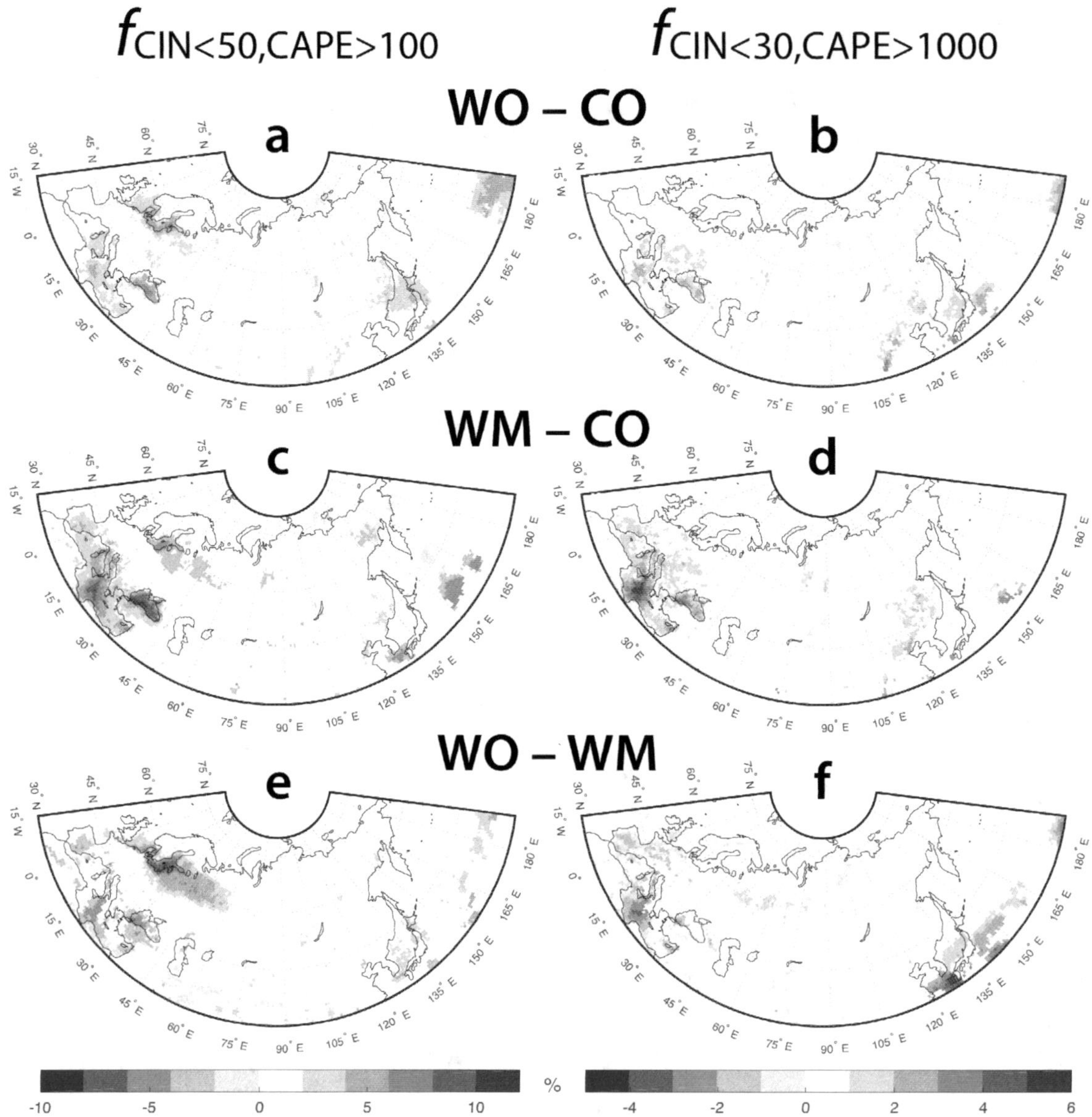

Figure 4. Differences between WO and CO (a, b), WM and CO (c, d), WO and WM (e, f) for the frequency of exceeding the threshold in summer for different indices: (a, c, e) MU CAPE>100 J kg^{-1} and MU CIN < 50 J kg^{-1}, (b, d, f) MU CAPE>1000 J kg^{-1} and MU CIN < 30 J kg^{-1}. Only significant changes are shown.

The wind shear and helicity indices (their summer means and frequencies of exceeding the threshold) do not exhibit significant changes between the CO to the WO (Figures 2 and 3).

Sea surface warming results in changes in convective inhibition. Large values of CIN (i.e., more than 100 J kg^{-1}) prevent the development of convection. Although the mean CIN is generally larger in the WO than in the CO (except for the eastern part of the Mediterranean Sea) (Figure 2), the frequency of a very low CIN is also larger in the WO (Figure 3). If these low-CIN environments co-occur with a high-CAPE, this likely initiates a severe storm [28]. The frequency of such environments is significantly larger in the WO experiment as compared to the CO (Fig. 4a, b) in many regions (primarily over the marginal seas of the Atlantic and Pacific Oceans and Eastern Europe).

Only the Mediterranean and Black Sea surface warming (the WM experiment) results in a larger (as compared to the WO) increase of such indices as CAPE, 3D, and WMAXSHEAR (their means and frequency of exceeding the threshold) and the frequency of low-CIN/high-CAPE environments in these regions, and a lower increase (or even decrease) in other regions (Figures 4c-f).

4. Conclusions

In this paper, sensitivity experiments were performed to estimate the role of sea surface warming in convective activity over Europe and Northern Eurasia. The convective instability indices were calculated with 6-hour data from three model experiments with different boundary conditions (warm global ocean, cold global ocean, and cold ocean – warm Mediterranean and Black Seas).

A comparison of the results of the sensitivity experiments has shown that warmer sea surface leads to a general increase in the thermodynamic convective instability indices. This increase is broadly consistent with observations. Particularly, the sensitivity experiments reproduce well the observed increase in CAPE over the eastern part of the Mediterranean Sea, the Black Sea, and Eastern Europe. The 3D and WMAXSHEAR indices grow as well in the WO experiment compared to the CO. The shear and helicity instability indices depend little on sea surface warming. The WM experiment shows a greater (compared to the WO) increase in the thermodynamic indices in these very regions and a smaller increase (or even decrease) in the indices in the other regions. In particular, the observed increase in CAPE in Eastern Europe cannot be explained by the warming of the Mediterranean and Black Seas alone. Note that the above-presented results are preliminary, and further analysis is required.

The reanalysis data show a decrease in the convective indices in several regions (the Iberian Peninsula, Mongolia, and Northern China). These negative changes are not reproduced in the model experiments, because their nature is, apparently, not related to sea surface warming. For instance, a decrease in the lapse rate may prevail here over an increase in the surface temperature. Further research is needed to establish the causes of these changes. For instance, sensitivity experiments with interactive greenhouse gases should be carried out to clarify the role of lapse rate changes in the convective activity variability. Given the scale of convective phenomena, it is desirable to use regional climate models [43] or dynamic downscaling of the AGCM [44].

Acknowledgements

This study was supported by the Russian Science Foundation under project no. 18-77-10076. The model simulations were performed under Russian Foundation for Basic Research grant 17-05-00561 and DAAD project 57314018.

References

[1] Sun B, Groisman P Y and Mokhov I I 2001 Recent Changes in Cloud-Type Frequency and Inferred Increases in Convection over the United States and the Former USSR *Journal of Climate* **14** 1864–80

[2] Chernokulsky A V, Bulygina O N and Mokhov I I 2011 Recent variations of cloudiness over Russia from surface daytime observations **6** 035202

[3] Chernokulsky A V, Esau I, Bulygina O N, Davy R, Mokhov I I, Outten S and Semenov V A 2017 Climatology and Interannual Variability of Cloudiness in the Atlantic Arctic from Surface Observations since the Late Nineteenth Century *Journal of Climate* **30** 2103–20

[4] Chernokulsky A and Esau I 2019 Cloud cover and cloud types in the Eurasian Arctic in 1936–2012 *Int. J. Climatol.* **16** 14343–20

[5] Rulfová Z and Kyselý J 2014 Trends of Convective and Stratiform Precipitation in the Czech Republic, 1982–2010 *Adv. Meteorol.* **2014** 1–11

IOP Conf. Series: Earth and Environmental Science **386** (2019) 012051 doi:10.1088/1755-1315/386/1/012051

[6] Han X, Xue H, Zhao C and Lu D 2016 The roles of convective and stratiform precipitation in the observed precipitation trends in Northwest China during 1961–2000 *Atmospheric Research* **169** 139–46

[7] Ye H, Fetzer E J, Wong S and Lambrigtsen B H 2017 Rapid decadal convective precipitation increase over Eurasia during the last three decades of the 20th century *Science Advances* **3** e1600944

[8] Chernokulsky A, Kozlov F, Zolina O, Bulygina O, Mokhov I I and Semenov V A 2019 Observed changes in convective and stratiform precipitation in Northern Eurasia over the last five decades *Environmental Research Letters* **14** 045001–17

[9] Taszarek M, Allen J, Púčik T, Groenemeijer P, Czernecki B, Kolendowicz L, Lagouvardos K, Kotroni V and Schulz W 2019 A Climatology of Thunderstorms across Europe from a Synthesis of Multiple Data Sources *Journal of Climate* **32** 1813–37

[10] Riemann-Campe K, Fraedrich K and Lunkeit F 2009 Global climatology of convective available potential energy (CAPE) and convective inhibition (CIN) in ERA-40 reanalysis *Atmospheric Research* **93** 534–45

[11] Chernokulsky A V, Kurgansky M V and Mokhov I I 2017 Analysis of changes in tornadogenesis conditions over Northern Eurasia based on a simple index of atmospheric convective instability *Doklady Earth Sci.* **477** 1504–9

[12] Pistotnik G, Groenemeijer P and Sausen R 2017 Validation of Convective Parameters in MPI-ESM Decadal Hindcasts (1971–2012) against ERA-Interim Reanalyses *Meteorol. Z.* **25** 753–66

[13] Dee D P *et al* 2011 The ERA-Interim reanalysis: configuration and performance of the data assimilation system *Quarterly Journal of the Royal Meteorological Society* **137** 553–97

[14] Berg P, Moseley C and Haerter J O 2013 Strong increase in convective precipitation in response to higher temperatures *Nature Geosci.* **6** 181–5

[15] Berg P and Haerter J O 2013 Atmospheric Research *Atmospheric Research* **119** 56–61

[16] Mokhov I I, Chernokul'skii A V, Akperov M G, Dufresne J-L and Le Treut H 2009 Variations in the Characteristics of Cyclonic Activity and Cloudiness in the Atmosphere of Extratropical Latitudes of the Northern Hemisphere Based from Model Calculations Compared with the Data of the Reanalysis and Satellite Data *Doklady Earth Sci.* **424** 147–50

[17] Berry G, Jakob C and Reeder M 2011 Recent global trends in atmospheric fronts *Geophysical Research Letters* **38** L21812

[18] Schemm S, Sprenger M, Martius O, Wernli H and Zimmer M 2017 Increase in the number of extremely strong fronts over Europe? A study based on ERA-Interim reanalysis (1979-2014) *Geophysical Research Letters* **44** 553–61

[19] Loriaux J M, Lenderink G and Siebesma A P 2017 Large-Scale Controls on Extreme Precipitation *Journal of Climate* **30** 955–68

[20] Lenderink G, Barbero R and Loriaux J M 2017 Super Clausius-Clapeyron scaling of extreme hourly convective precipitation and its relation to large-scale atmospheric conditions *Journal of Climate* **30** 6037–52

[21] Tao W-K, Chen J-P, Li Z, Wang C and Zhang C 2012 Impact of aerosols on convective clouds and precipitation *Reviews of Geophysics* **50** RG2001

[22] Finch J and Bikos D 2012 Russian Tornado Outbreak of 9 June 1984 *E-journal of Severe Storms Meteorology* **7** 1–28

[23] Miglietta M M, Mazon J, Motola V and Pasini A 2017 Effect of a positive Sea SurfaceTemperature anomaly on aMediterranean tornadic supercell *Sci Rep* 1–8

[24] Shikhov A and Chernokulsky A 2018 A satellite-derived climatology of unreported tornadoes in forested regions of northeast Europe *Rem. Sens. Environ.* **204** 553–67

[25] Meredith E P, Semenov V A, Maraun D, Park W and Chernokulsky A V 2015 Crucial role of Black Sea warming in amplifying the 2012 Krymsk precipitation extreme *Nature Geosci.* **8** 615–9

[26] Volosciuk C, Maraun D, Semenov V A, Tilinina N, Gulev S K and Latif M 2016 Rising Mediterranean Sea Surface Temperatures Amplify Extreme Summer Precipitation in Central Europe *Sci Rep* **6** 1–7
[27] Semenov V A and Latif M 2015 Nonlinear winter atmospheric circulation response to Arctic sea ice concentration anomalies for different periods during 1966–2012 **10** 1–8
[28] Diffenbaugh N S, Scherer M and Trapp R J 2013 Robust increases in severe thunderstorm environments in response to greenhouse forcing *Proceedings of the National Academy of Sciences* **110** 16361–6
[29] Seeley J T and Romps D M 2015 The Effect of Global Warming on Severe Thunderstorms in the United States *Journal of Climate* **28** 2443–58
[30] Allen J T, Karoly D J and Walsh K J 2014 Future Australian Severe Thunderstorm Environments. Part I: A Novel Evaluation and Climatology of Convective Parameters from Two Climate Models for the Late Twentieth Century *Journal of Climate* **27** 3827–47
[31] Allen J T, Karoly D J and Walsh K J 2014 Future Australian Severe Thunderstorm Environments. Part II: The Influence of a Strongly Warming Climate on Convective Environments *Journal of Climate* **27** 3848–68
[32] Marsh P T, Brooks H E and Karoly D J 2009 Atmospheric Research *Atmospheric Research* **93** 607–18
[33] Rayner N A, Parker D E, Horton E B, Folland C K, Alexander L V, Rowell D P, Kent E C and Kaplan A 2003 Global analyses of sea surface temperature, sea ice, and night marine air temperature since the late nineteenth century *Journal of Geophysical Research* **108** 14
[34] Moncrieff M W and Miller M J 1976 The dynamics and simulation of tropical cumulonimbus and squall lines *Quarterly Journal of the Royal Meteorological Society* **102** 373–94
[35] Doswell C A III and Schultz D M 2006 On the use of indices and parameters in forecasting severe storms *E-journal of Severe Storms Meteorology* **1** 1–24
[36] Kurgansky M V, Chernokulsky A V and Mokhov I I 2013 The tornado over Khanty-Mansiysk: An exception or a symptom? *Russian Meteorology and Hydrology* **38** 539–46
[37] Rasmussen E N and Blanchard D O 1998 A Baseline Climatology of Sounding-Derived Supercell and Tornado Forecast Parameters *Weather and Forecasting* **13** 1148–64
[38] Davies-Jones R, Burgess D and Foster M 1990 Test of helicity as a tornado forecast parameter. 16th Conf. on Severe Local Storms (Kananaskis Park, AB, Canada) pp 588–92
[39] Taszarek M, Brooks H E, Czernecki B, Szuster P and Fortuniak K 2018 Climatological Aspects of Convective Parameters over Europe: A Comparison of ERA-Interim and Sounding Data *Journal of Climate* **31** 4281–308
[40] Livingston R 1984 *The subsynoptic pre-tornado Environment* (Columbia: University of Missouri)
[41] Chernokulsky A V, Kurgansky M V, Zakharchenko D I and Mokhov I I 2015 Genesis environments and characteristics of the severe tornado in the South Urals on August 29, 2014 *Russian Meteorology and Hydrology* **40** 794–9
[42] Wilks D 2011 *Statistical Methods in the Atmospheric Sciences* vol 100 (Academic Press)
[43] Púčik T, Groenemeijer P, Rädler A T, Tijssen L, Nikulin G, Prein A F, van Meijgaard E, Fealy R, Jacob D and Teichmann C 2017 Future Changes in European Severe Convection Environments in a Regional Climate Model Ensemble *Journal of Climate* **30** 6771–94
[44] Gensini V A and Mote T L 2014 Estimations of Hazardous Convective Weather in the United States Using Dynamical Downscaling *Journal of Climate* **27** 6581–9

IOP Conf. Series: Earth and Environmental Science **386** (2019) 012052 doi:10.1088/1755-1315/386/1/012052

Spatial distribution of extreme wind speeds over Sakhalin Island based on observations and high-resolution modelling data

Vladimir Platonov, Alexander Kislov

Lomonosov Moscow State University, Faculty of Geography, Department of Meteorology and Climatology. 119991, Moscow, Leninskie Gory MGU, 1, GSP-1, Phone 8(495)939-29-42

vplatonov86@gmail.com, avkislov@mail.ru

Abstract. An analysis of extreme wind speeds over Sakhalin region shows that a set of wind speed extremes obtained from observations is a mixture of two different subsets, each of them neatly described by a Weibull distribution. The empirical tail of the pdf diverges from a linearized Weibull model, indicating that another model could fit the most extreme wind speed data better. Each of these subsets is characterized by parameters k and A, regarded as coefficient and free terms of the linear Weibull model, these samples are labelled BSs and Ds. The Ds are responsible for the strongest extremes. Mesoscale modelling is used to investigate a possibility of the models to reproduce these statistical features. A detailed hydrodynamic simulation of major meteorological parameters (1985 – 2014) has been performed for the Sea of Okhotsk and Sakhalin Island with horizontal resolutions of ~13.2, ~6.6, and ~2.2 km by using a regional climate model, COSMO-CLM. This dataset is utilized for an investigation of the statistical structure of extreme wind speeds. First, it is shown that a model with a detailed spatial resolution is able to reproduce the statistical structure in general, and a special mechanism is responsible for the generation of the largest of wind extremes. However, a model with a resolution of ~13,2 km could not reproduce some essential parts of the wind speed maximum' statistical properties, underestimating the parameters k and A of Ds Weibull distribution significantly. This gap could be covered by using a higher resolution, as well as by areal estimation techniques and many others.

1. Introduction

In this paper, we focus on statistical analysis of extreme wind events. Statistical methods of the extreme value analysis of wind speeds are important not only due to their practical value, but also because they allow us to detect their origin, since the same statistical distribution very often suggests a common originating mechanism. From this point of view, it is substantial to understand an extension of the ability of reproduction of wind speed extremes by atmospheric models. Establishment of the correspondence between wind simulation products and observations could perform the assessment of this reproduction and ability. The use of station data will make it possible to evaluate the consistency, in terms of reproduction of the statistical behavior, between the model simulation products and near-surface observations. Studying the statistical properties of extrema, determining their genesis and mechanisms are the basis for constructing a model of complex extreme hydrometeorological phenomena. Wind speed extremes in the sub-Arctic realm of the North-East Pacific region were investigated by extreme value analysis of wind speed obtained from atmospheric simulations of the COSMO-CLM mesoscale model, as well as using observational data.

In our previous paper we analysed datasets of wind observations over the Eurasian part of the Arctic region [1]. Extreme value analysis showed that the observed samples of wind speeds are strictly divided into two sets of variables. Moreover, this feature is marked out at many stations both for the summer and winter seasons [2]. Events exceeding the threshold value are much more pronounced than the members of the other group (located below the threshold value) predicted by the extrapolation of law distributions to their tail. The situation is the same in different areas of science where the data referring to the same nomenclature adhere to different statistical models. They were termed "Black Swans" (BSs) [3] and "Dragons" (Ds, or "kings") [4]. This result motivates our interest in the ability to detect, analyze, and understand such different extremes and their nature.

In this paper we extend the analysis of extreme wind speed events in the Sub-Arctic region. The next step of our analysis is an investigation of the above-mentioned peculiarities of wind extremes simulated with a fine spatial resolution by a mesoscale atmospheric model. We used the COSMO-CLM regional climate model developed within the framework of the Consortium for Small-scale Modeling (http://www.cosmo-model.org) and CLM-Community (http://clm-community.eu).

2. Data and methods

2.1. Weibull distribution approach

Research in the statistical analysis of extreme values has flourished over the past several decades: new probability models, inference and data analysis techniques have been introduced; and new application areas have been explored [5 – 7]. The statistical method of extreme value analysis of wind speeds is important, because it allows us to detect their statistical model. Note that the same statistical distribution suggests a common originating mechanism. We plan to use this idea to interpret extreme wind records. In the extreme value theory, it was noted [8] that for sufficiently long sequences of independent and identically distributed random variables the maxima of wind speed samples can fit one of three limiting distributions. One representative of these three limiting distributions is the Weibull distribution, which has traditionally been used for the statistical modelling of wind extremes [9]. In this study we model empirical extremes by the Weibull cdf using the following form:

$$\frac{n}{N} \approx F(U) = 1 - e^{-AU^k}. \quad (1)$$

This expression can be replaced by

$$ln\left[-ln\frac{N-n}{N}\right] = klnU + lnA. \quad (2)$$

Thus, this expression allows one to represent the empirical function in the linear Cartesian coordinates of the Weibull distribution ('Weibull plots'). The model parameters (A and k) can be estimated by using the maximum likelihood approach. The coefficient of determination (R^2) was calculated to estimate the success of linear approximation.

It is also an important condition of extreme value analysis that the extremes selected have to be statistically independent. However, 3-hourly observational data could have several clustered maximum speeds from a single storm and, therefore, such events are unlikely to be statistically independent. A simple method to remove the dependence is to require a minimum time separation or "deadtime" between the selected events, which is equivalent to e-folding time of the autocorrelation function. It was shown earlier that this time is equal to 48 – 72 hours and, therefore, we used a "deadtime" of 72 hours. Moreover, the principal differences between the atmospheric circulation regimes were taken into account. Thus, the summertime and wintertime extremes were analyzed separately and statistically independent. The summer season was defined as June–September and the winter one, as November–March considering monsoonal circulation conditions over the Okhotsk Sea and Sakhalin Island regions.

Many interesting features were revealed in our previous work, concerning the cdf form of wind speed extremes [1, 2, 10]. We found that the empirical cdfs of wind speed extremes at all Russian Arctic stations consistently deviate from the theoretical line starting with certain large threshold values (U_{th}). This deviation means that the standard Weibull model does not describe the empirical cdf starting from a certain threshold $U > U_{th}$. In this way, the greatest extremes, namely, which are most important, should be described by some other law, but not by the common Weibull distribution. The shape of the curve suggests that the sample is composed of two sets of variables, each of them described by its own Weibull function. Very large R-squared values in all cases show great success of this approximation. The coefficients of regression equations allow us to estimate the parameters k and A in each case. A more detailed description of the Weibull distribution splitting and its application to wind speed extremes data over the Sakhalin island is given in [1, 11]. Finally, the first sample obeying the common Weibull distribution was named “Black Swans” (BSs) according to the metaphorical terminology from [3]; most extremes not obeying the common Weibull distribution were called “Dragons” (Ds) according to [4]. It is worth noting that the coefficients k and A of the Weibull distribution differ significantly in these splitting samples and should be investigated further in observational and modelling data.

2.2. Description of the atmospheric model

We used a wind simulation dataset of the COSMO-CLM regional model for 1985 – 2014 [12, 13] as a source of model data. The COSMO-CLM model was applied as a detailing tool of the reanalysis data. The regional COSMO-CLM model was developed by the CLM-Community (http://clm-community.eu) [14]. Dynamic downscaling during the COSMO-CLM simulation was performed through three domains in a nesting scheme (Figure 1). The outer, ‘basic’ domain (grid spacing: ~13.2 km) covers the Sakhalin Island, the Sea of Okhotsk, the Kamchatka Peninsula, the surrounding regions of the Pacific Ocean, and the Asian continent. The model output on the outer domain provides meteorological fields at the lateral boundaries of the intermediate domain (grid spacing: 6.6 km), and the latter, in turn, provides output at the lateral boundaries of the inner domain (grid spacing: 2.2 km). 40 model levels were applied in the vertical direction for the basic and intermediate domains and 50 model levels, for the inner domain. In this paper we will consider model data and statistical analysis over the basic domain only. The experience of the authors in the creation of such a long-term atmospheric model archive based on supercomputer technologies is described in [15].

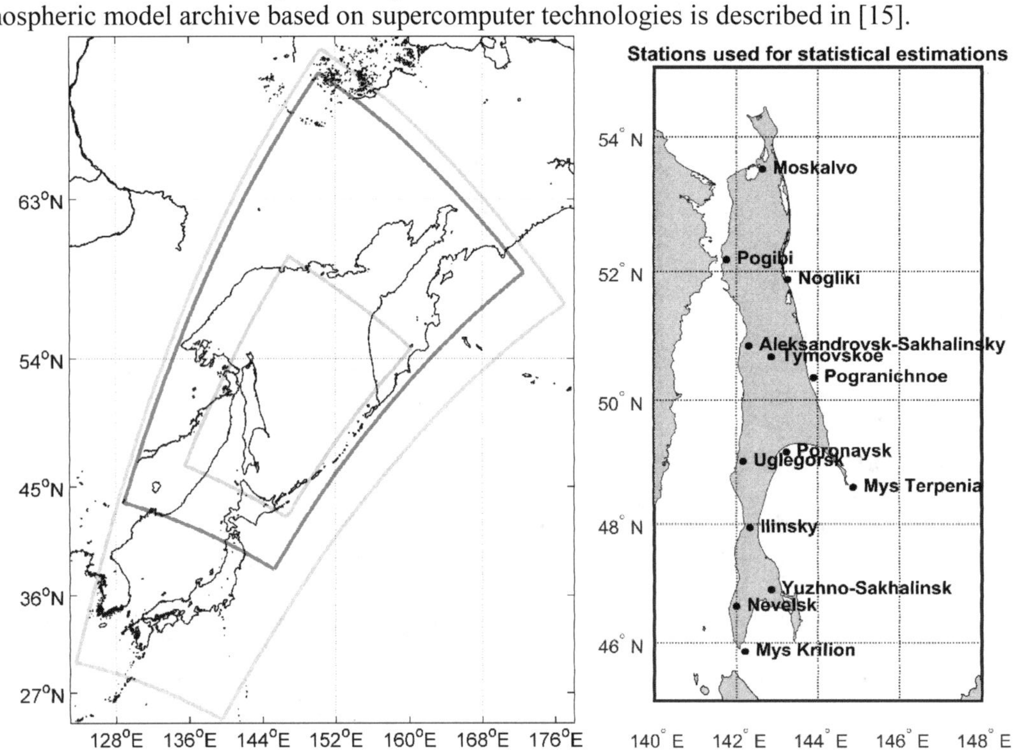

Figure 1. Left: map of the boundaries of model domains with a grid spacing of ~13.2 (cyan), ~6.6 (magenta), and ~2.2 km (green). Right: map of meteorological stations used for statistical analysis.

The COSMO-CLM model based on fully compressible fluid non-hydrostatic equations (Reynolds equations) was realized on a staggered Arakawa grid-C [16]. The COSMO-CLM uses a hybrid Gal-

Chen vertical coordinate system represented as the sigma-coordinate from the surface to the level (Z_F), and as the Z-coordinate above Z_F [17, 18]. The physical parameterizations chosen for the COSMO-CLM are based on extensive development and testing of the COSMO-CLM over a wide-range of Arctic environments [19 – 22]. We used the Tiedtke scheme [23] for the parameterization of convection, the two-stream Ritter-Geleyn [24] radiative transfer model, and the Mellor–Yamada 2.5–level planetary boundary layer and complementary surface layer schemes [25]. More detailed documentation can be found on [http://www.cosmo-model.org/content/model/documentation/core/default.htm].

The initial and lateral boundary conditions for the outer domain in the COSMO-CLM are provided by ERA-Interim reanalysis [26] for each 6 hours during 1985 – 2014. To avoid the model drift in the atmospheric circulation, the spectral nudging technique [27, 28] was implemented for temperature, geopotential height, and wind components above 850 hPa in the outer domain. We used the wavenumber 11 to take into account the large-scale synoptic processes (wavelengths: > 1000 km) only. The time output in the entire experiment (1985 – 2014) was 1 hour for all variables. More detailed information about this long-term detailed archive and its verification on observational data can be found out in [12].

3. Results and discussion

3.1. Statistical features in observational data

Figure 2 gives an example of empirical cdfs based on station measurements plotted in the 'Weibull' linear coordinates. We found that the empirical cdfs deviate consistently from the theoretical line starting with certain large threshold values (U_{th}); it is typical for all sites. This means that the empirical tail diverges from the Weibull model, indicating that a different model might describe the data well. This is exactly the situation discussed above in Section 2.1. Therefore, we can approximate the common cdf as two different samples, describing each of them by their own Weibull distribution parameters (Figure 3). There are very large R-squared values at all sites ($R^2 > 0.9$ in almost all cases), and the linear terms of the approximation equation give the Weibull parameters *k* and *A* for each station and season.

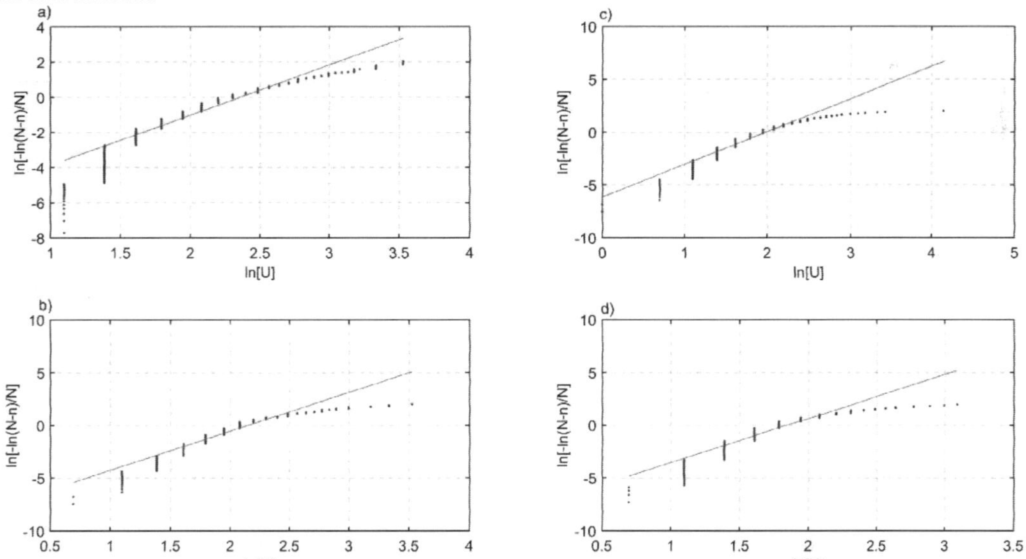

Figure 2. Cumulative distribution functions of wind speed maxima (station observations) in the linear coordinate axis of the Weibull distribution, and linear regression line corresponding to the Weibull function. (a), (b): Aleksandrovsk-Sakhalinsky (winter and summer seasons); (c), (d): Yuzhno-Sakhalinsk (winter and summer seasons).

The Kolmogorov-Smirnov (K-S) test was used to determine whether or not these separated samples referred to the Weibull distribution [1]. It was concluded that in both cases ($U > U_{th}$) and ($U < U_{th}$) the Weibull distribution fits well (but with different parameters *A* and *k*) for all stations and seasons.

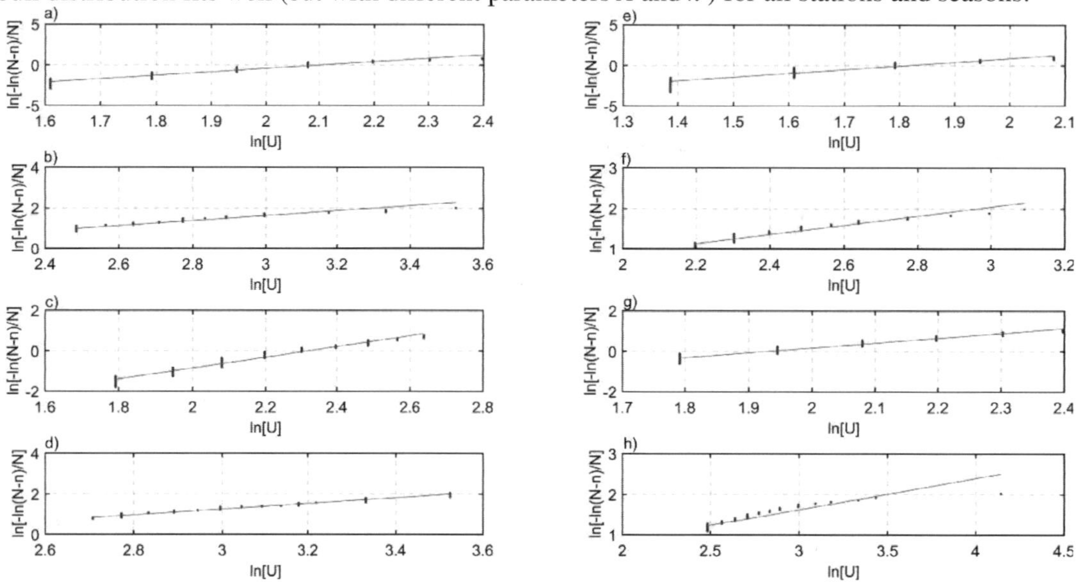

Figure 3. Cumulative distribution functions of wind speed maxima (station observations) in the linear coordinate axis of the Weibull distribution, and linear regression line corresponding to the Weibull function. (a), (b): Aleksandrovsk-Sakhalinsky (summer season); (c), (d) Aleksandrovsk-Sakhalinsky (winter season); (e), (f): Yuzhno-Sakhalinsk (summer season); (g), (h): Yuzhno-Sakhalinsk (winter season). (a), (c), (e), (g) denotes the Weibull distribution for $U \leq U_{th}$ (BSs); (b), (d), (f), (h) denotes the Weibull distribution for $U > U_{th}$ (Ds). $R^2 > 0.96$ in all cases.

A very important feature is that events adhering to the BSs or Ds could be diagnosed easily, and the Ds can be detected evidently based on obvious breaks in the tail of the wind distributions. The estimated parameters of each distribution allow us to calculate the quantiles for both Weibull distributions and use them as metrics of extremes frequency within the BSs and Ds separately, and their ratio for each station and season. In this paper, we have calculated *U(0.99)* quantile values for all stations and both seasons. An example of these values is given in Table 1 together with the parameters *k* and *A*. The ratio between the Ds and BSs can reach up to 30%. The most pronounced feature of the seasonal distribution of the quantile wind speed values is that the maxima (both of the BSs and Ds) are observed in the winter period. Such "winter acceleration" of the wind speed is determined by storms that are typically much more active during the cold season.

3.2. Statistical features in model archive

The next step of the analysis is to investigate how great is the ability of the regional mesoscale atmospheric model to reproduce the peculiarities of wind extremes based on the long-term data archive described in Section 2.2. We used specific model grid points for comparison with each station, defined in [12] as one of the four closest grid points having the least wind speed RMSE.

In Figure 4, several cdfs based on the model archive wind speed data displaying clear deviation of the most extremes from the common linear Weibull law are plotted. Again, as noted earlier in the analysis of the observation data, we propose that these curves show approximations of the cdfs of wind velocity extremes by two different Weibull laws. These features are denoted almost at all grid points and in both seasons. The result of splitting of the original sample is shown in Figure 5. Using our terminology, we can conclude that the COSMO-CLM model wind speed reproduces extremes are both the DSs and the Ds. It refers to the ability of the model with a detailed spatial resolution to reproduce the statistical structure and the special mechanism responsible for the generation of the largest wind extremes. However, the observed extremes characterized by *U(0.99)* are almost half times greater than the modeled values. Apart from this, the ratio between the modelled Ds and BSs can reach only up to 10%. It is significantly less than what was established for the observations (up to 30%, see above). However, the most extreme wind speed parameters referred to the same stations in the observations and model data, and the model reproduced well the fact of increased winter extremes compared to the summer ones.

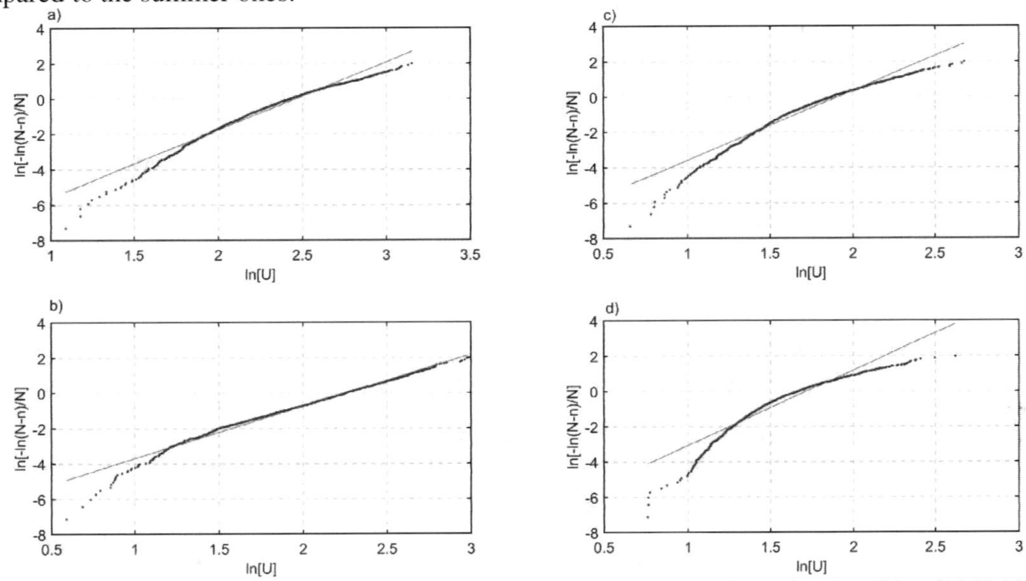

Figure 4. Cumulative distribution functions of wind speed maxima simulated by COSMO-CLM at grid points corresponding to Aleksandrovsk-Sakhalinky (a), (b); Yuzhno-Sakhalinsk (c), (d); ((a), (c) – winter season, (b), (d) – summer season) in the linear coordinate axis of the Weibull distribution, and the linear regression line corresponding to the Weibull function. In all cases $R^2 > 0.94$.

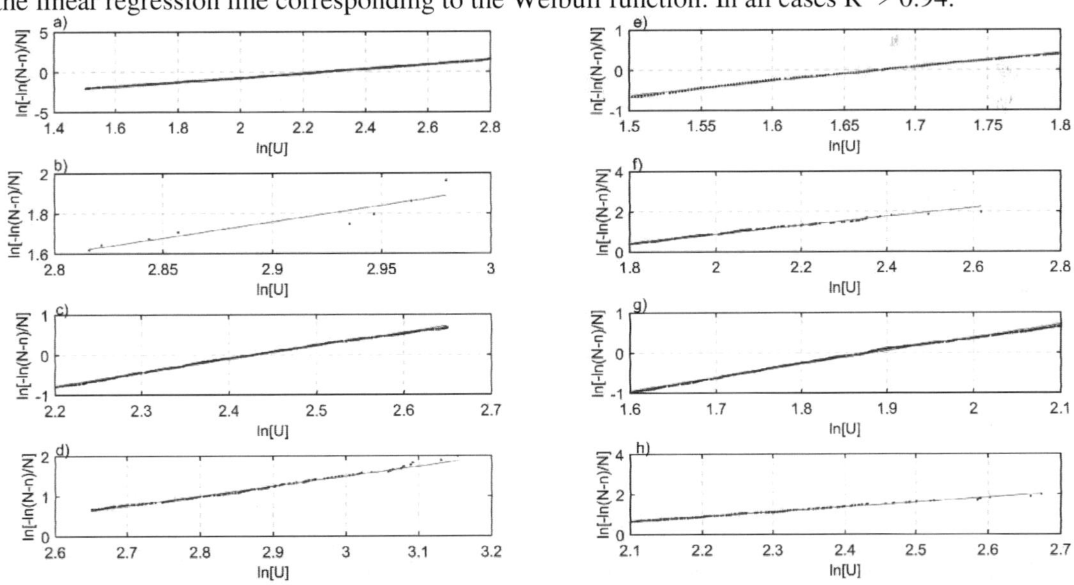

Figure 5. Cumulative distribution functions of wind speed maxima simulated by COSMO-CLM at grid points corresponding to Aleksandrovsk-Sakhalinky ((a) and (b) – summer season), Yuzhno-Sakhalinsk ((c) and (d) – summer season), Aleksandrovsk-Sakhalinky ((e) and (f) – winter season);

Yuzhno-Sakhalinsk ((g) and (h) - winter season) in the linear coordinate axis of the Weibull distribution, and the linear regression line corresponding to the Weibull function. (a), (c), (e), (g) the Weibull distribution for $U \leq U_{th}$ (BSs); (b), (d), (f), (h) the Weibull distribution for $U > U_{th}$ (Ds). In all cases $R^2 > 0.96$.

Further analysis was made to compare the coefficients k and A and *U(0.99)* quantiles between the model and observational data according to BSs and Ds samples separately (Tables 1 and 2). A good agreement was shown for BSs, and it is truer for *k*-values, which are mostly belonging to the interval 3 – 5. At the same time, the coefficients *A* are overall overestimated by the model. This means that the model slightly marks up the extremity of the BSs wind speed sample. However, the Weibull distribution parameters related to the Ds samples obtained by the model are significantly different. For instance, the *k*-values, which are responsible for the slope of the approximation line, are overestimated (approx. 1 – 2 more). In this way, the Ds' slope is closer to the BSs' slope, and the difference between the BSs and Ds is getting smaller than in reality. Another important feature is overestimation of the coefficients *A* by almost an order of magnitude, which corresponds to significant underestimation of the extremity of the Ds wind speed samples by the mesoscale model.

The value of *U(0.99)* for summer characterizes the maximum wind speed, which is exceeded, on average, once every five warm seasons. Similarly, the value of *U(0.99)* for the cold period of the year characterizes the maximum wind speed, which is exceeded, on average, two times every three cold seasons. The *U(0.99)* quantiles for the Ds sample reproduced by the model are lower (12 – 22 m/s at most), i.e. the extremity of the wind speed maximum is underestimated. The quality of reproduction of the statistical parameters by the model also depends on the characteristics of the underlying surface and the sea-land spatial distribution. The spread commonly decreased on the flat seashore stations, and increased over inlands and highly indented coastlines [11].

Table 1. Weibull cdf linear coefficients *k* and *A* and *U(0.99)* quantiles for BSs and Ds samples for each station for summer seasons samples.

Station name	k, BSs	A, BSs	U(0.99), BSs	k, Ds	A, Ds	U(0.99), Ds
Aleks-Sakhalinskiy	4.1865	0.0001	11.8	1.2366	0.1244	18.6
Ilinsky	3.9613	0.0001	14.2	1.2720	0.1023	19.9
Mis Krilion	3.6026	0.0001	20.4	1.2659	0.0611	30.4
Mys Terpenia	3.7125	0.0001	16.2	1.5952	0.0331	22.1
Moskalvo	4.1973	0.0001	13.5	0.7612	0.4237	23.0
Nevelsk	2.0733	0.0077	21.9	0.5785	0.7964	20.8
Nogliki	5.0588	0.0000	9.9	0.9384	0.3619	15.0
Pogibi	3.1347	0.0006	17.4	0.6615	0.6039	21.6
Pogranichnoe	3.1523	0.0034	9.9	0.8091	0.5757	13.1
Poronaysk	4.2868	0.0002	11.0	0.9296	0.3872	14.3
Timovskoe	4.7739	0.0001	10.0	1.6805	0.0669	12.4
Uglegorsk	3.9439	0.0002	12.1	1.3076	0.0970	19.1
Yuzhno-Sakhalinskiy	4.5687	0.0003	8.6	1.1295	0.2591	12.8

Table 2. Weibull cdf linear coefficients *k* and *A* and *U(0.99)* quantiles for BSs and Ds samples for model grid points corresponding to each station for summer seasons samples.

Station name	k, BSs	A, BSs	U(0.99), BSs	k, Ds	A, Ds	U(0.99), Ds
Aleks-Sakhalinskiy	2.7143	0.0021	16.9	1.6214	0.0527	15.6

Ilinsky	3.5517	0.0016	9.4	2.4723	0.0135	10.6
Mis Krilion	3.8268	0.0004	11.2	2.2385	0.0135	13.5
Mys Terpenia	3.0395	0.0008	17.6	2.1199	0.0081	19.9
Moskalvo	3.0096	0.0019	13.2	1.8035	0.0313	15.9
Nevelsk	2.8194	0.0017	16.6	2.1632	0.0092	17.7
Nogliki	4.6550	0.0001	9.7	2.2289	0.0177	12.1
Pogibi	4.1251	0.0001	14.0	2.7077	0.0029	15.3
Pogranichnoe	2.5881	0.0064	12.7	2.0431	0.0210	14.0
Poronaysk	4.6764	0.0002	8.8	2.1930	0.0233	11.1
Timovskoe	3.7620	0.0020	7.8	2.7027	0.0137	8.6
Uglegorsk	3.0589	0.0039	10.1	2.1907	0.0228	11.3
Yuzhno-Sakhalinskiy	3.5080	0.0028	8.2	2.2061	0.0291	9.9
Val	3.8642	0.0002	11.1	2.0280	0.0168	13.3
Komrvo	3.1297	0.0017	12.6	2.1544	0.0148	14.3

4. Conclusions and perspectives

An extreme value analysis has been implemented to estimate the statistical properties of extreme wind speeds over the Sakhalin Island region based on the long-term COSMO-CLM model dataset and observations.

It was shown for all stations that the observational samples of extreme wind speed are composed of two sets of variables. All samples of each population have the same statistical properties, but one population is sharply different from another. Using metaphoric terminology, we marked these samples BSs and Ds. Within the framework of a 30–year high-resolution regional data set of the sub-Arctic region, generated by the COSMO-CLM model, we reproduced several important aspects of wind speed extremes. First of all, we found among them samples adhering to various populations which can be interpreted as BSs and Ds. The analysis showed the ability of the mesoscale atmospheric model to capture the specific physical mechanism that generates wind speed extremes. However, the model with the given resolution was not able to reproduce some essential parts of the statistical properties of wind speed maximum: there are large discrepancies concerning the absolute values of the modelled extremes, which are almost two times smaller than the observed values. Although the main statistical characteristics of the BSs samples were reproduced satisfactorily, the time parameters of the Ds samples were significantly underestimated by the mesoscale atmospheric model. These gaps could be covered by using a higher resolution, as well as by other areal estimation techniques, and many others. Future investigations will be dedicated to specific physical mechanisms of Ds formation, scaling problems, and resolution dependencies.

Acknowledgements

The authors express their appreciation to director of the “Hydrometcenter of Russia” Roman M. Vilfand and director of the “Main Computational Center of Roshydromet” Sergey V. Lubov for their continuing attention to this work and assistance in its implementation.

Support for this work was provided by the Russian Foundation for Basic Research (grant 18-05-60147) and Lomonosov Moscow State University (grant AAAA-A16-116032810086-4).

References

[1] Kislov A, Matveeva T 2016 An extreme value analysis of wind speed over the European and Siberian parts of Arctic region. *Atm. and Clim. Sci.* **6** 205 – 23

[2] Kislov A V, Matveeva T A, Platonov V S 2015 Wind speed extremes in Arctic *Fund. and Appl. Climat.* **2** 63 – 80 (In Russian)

[3] Taleb N *The black swan: The impact of the highly improbable fragility.* 2010 (New York, Random House) p 300

[4] Sornette D Dragon-Kings 2009 Black Swans and the Prediction of Crises *International Journal of Terraspace Science and Engineering* **2** 1 – 18

[5] Beirlant J, Goegebeur Y, Segers J, Teugels J, De Waal D, Ferro C *Statistics of extremes: theory and applications* 2004 (Wiley Series in Probability and Statistics, John Wiley & Sons Ltd., Chichester) p 497

[6] Coles S, Bawa J, Trenner L, Dorazio P *An introduction to statistical modeling of extreme values* 2001 (Springer Series in Statistics, Springer-Verlag) p 208

[7] de Haan L and Ferreira A *Extreme value theory: an introduction* 2006 (Springer Series in Operations Research and Financial Engineering, Springer) p 417

[8] Fisher R and Tippett L 1928 Limiting forms of the frequency distribution of the largest or smallest members of a sample *Proceedings of the Cambridge Philosophical Society* **24** 180-90

[9] Palutikof J P, Brabson B B, Lister D H and Adcock S T. A 1999 Review of Methods to Calculate Extreme Wind Speeds *Meteorological Application* **6** 119 – 32

[10] Platonov V S and Kislov A V 2016 Mesoscale high-resoluion modeling of extreme wind speeds over western water areas of the Russian Arctic *IOP Conference Series: Earth and Environmental Science* **48**

[11] Kislov A, Platonov V 2019 Analysis of observed and modelled near-surface wind extremes over the sub-arctic northeast pacific *Atmospheric and Climate Sciences* **9** 1 146 – 58

[12] Kislov A V, Rivin G S, Platonov V S, Varentsov M I, Rozinkina I A, Nikitin M A, and Chumakov M M 2018 Mesoscale atmospheric modelling of extreme velocities over the Sea of Okhotsk and Sakhalin Island *Izvestiya, Atmospheric and Oceanic Physics* **54** 322 – 6

[13] Platonov V, Kislov A, Rivin G, Varentsov M, Rozinkina I, Nikitin M, and Chumakov M 2017 Mesoscale atmospheric modelling technology as a tool for creating a long-term meteorological dataset *IOP Conference Series: Earth and Environmental Science* **96**

[14] Rockel B, Will A, Hense A 2008 The regional climate model COSMO-CLM (CCLM) *Meteorologische Zeitschrift* **17** 4 347 – 8

[15] Platonov V S, Varentsov M I 2018 Supercomputer technologies as a tool for high-resolution atmospheric modelling towards the climatological timescales *Supercomputing Frontiers and Innovations* **5** 3 107 – 10

[16] Arakawa A, Lamb V R 1977 Computational design of the basic dynamical processes of the UCLA general circulation model *Methods in computational physics* **17** 173 – 265

[17] Gal-Chen T, Somerville R C J 1975 On the use of a coordinate transformation for the solution of the Navier-Stokes equations *Journal of Computational Physics* **17** 209 – 28

[18] Schär C, Leuenberger D, Fuhrer O, Lüthi D, and Girard C 2002 A new terrain-following vertical coordinate formulation for atmospheric prediction models *Mon. Wea. Rev.* **130** 2459 – 80

[19] Rockel B, Geyer B 2008 The performance of the regional climate model CLM in different climate regions, based on the example of precipitation *Meteorologische Zeitschrift* **17** 4 487 – 98

[20] Van Pham T, Brauch J, Dieterich C, Früh B, Ahrens B 2014 New coupled atmosphere-ocean-ice system COSMO-CLM/NEMO: assessing air temperature sensitivity over the North and Baltic Seas *Oceanologia* **56** 2 167 – 89

[21] Haas R and Pinto J G 2012 A combined statistical and dynamical approach for downscaling large-scale footprints of European windstorms *Geoph. Res. Let.* **39** L23804 1 – 6

[22] Varentsov M I, Verezemskaya P S, Zabolotskikh E V, Repina I A 2016 Quality estimation of polar lows reproduction based on reanalysis data and regional climate modeling

Sovremennye problemy distantsionnogo zondirovaniya Zemli iz kosmosa **13** 4 168 – 91 (In Russian)
[23] Tiedtke M 1989 A comprehensive mass flux scheme for cumulus parameterization in large-scale models *Mon. Wea. Rev.* **117** 1779 – 800
[24] Ritter B, Geleyn J F 1992 A comprehensive radiation scheme for numerical weather prediction models with potential applications in climate simulations *Mon. Wea. Rev.* **120** 303 – 25
[25] Herzog H-J, Vogel G, Schubert U 2002 LLM – a nonhydrostatic model applied to high-resolving simulations of turbulent fluxes over heterogeneous terrain *Theor. Appl. Clim.* **73** 67 – 86
[26] Dee D P *et al* 2011 The ERA-Interim reanalysis: Configuration and performance of the data assimilation system *Quart. J. R. Met. Soc.* **137** 553 – 97
[27] Von Storch H, Langenberg H, Feser F 2000 A spectral nudging technique for dynamical downscaling purposes *Monthly weather review* **128** 10 3664 – 73
[28] Feser F, Barcikowska M 2012 The influence of spectral nudging on typhoon formation in regional climate models *Environmental Research Letters* **7** 1 1 – 10

IOP Conf. Series: Earth and Environmental Science **386** (2019) 012053 doi:10.1088/1755-1315/386/1/012053

Using information and computing system "Climate" to raise awareness among the population and decision makers about regional effects of climate change

Yu E Gordova, A A Riazanova, A G Titov and E P Gordov

Institute of Monitoring of Climatic and Ecological System SB RAS, Tomsk, Russia

Abstract. The effects of global climate change will be most dramatic in the vast Russian territory. According to the Hydrometeorological Center of Russia, there is an increase in the magnitude and frequency of extreme weather events, as well as in their damage to the ecosystems and infrastructure. To develop adaptation to climate change and mitigation of its consequences, it is necessary to promote and support activities aimed at reducing possible risks. However, there is insufficient awareness and lack of scientific background among decision-makers. Persons responsible for making decisions, stakeholders, and the general public do not often have skills and knowledge to work with climate data to elaborate an adaptation and sustainable development strategy. Some new sections of a web-system, "Climate", are tailored to provide these groups with tools, skills, and thematic information for understanding climate processes occurring in their region. A course is developed where basic concepts are explained in detail for the general public. A climate characteristics database for decision-makers is created to obtain calculated fields of the indices describing spatial distributions of extreme values of meteorological characteristics for the territory of Siberia. For work with any of the indices, it is possible to download the database in various formats for use in various desktop GIS systems.

1. Introduction

Observations and simulation results indicate that Earth's climate is changing [1], and Russia belongs to Earth's regions with the maximum observed and anticipated climate change due to the size of its territory and the fact that a large part of it is located in the high and polar latitudes. A long-term rise in the average temperature of the Earth's climate system continues, and according to authoritative climatic centers (Center Hadley; NOAA), for the whole globe 2018 was the fourth among the warmest for the entire period of instrumental observations from the second half of the 19th century. The global average annual temperature (together for land and ocean) exceeded the average for 1961-90 by 0.662 ° C according to NOAA [2]. In the territory of the Russian Federation, in general, the warming continues throughout the year and in all seasons, and its pace is much higher than the average over the globe. The average growth rate of the average annual air temperature in Russia in 1976-2018 was 0.47 ° C / 10 years. This is 2.5 times the growth rate of the global temperature over the same period: 0.17-0.18 ° C / 10 years, and more than 1.5 times the average warming rate of the surface air over the Earth's land: 0.28-0.29 ° C / 10 years. The temperature of the northern polar region grew most rapidly, especially in the last three decades: at some meteorological stations on Russia's coastline of the Arctic Ocean the growth rate of the average annual temperature in this period exceeded 1.0 ° C / 10 years.

IOP Conf. Series: Earth and Environmental Science **386** (2019) 012053 doi:10.1088/1755-1315/386/1/012053

Currently, global climate change has led to an increase in the frequency of dangerous hydrometeorological phenomena in Russia, such as floods, heavy snowfalls, and blizzards, accompanied by storm and even hurricane winds, severe long frost, ice, and late spring frost. During the warm period, torrential rains with thunderstorms, hail and wind or severe droughts are frequent. The number and strength of extreme climatic manifestations are increasing every year. For example, in many regions of Russia at the beginning of the 21st century, the frequency of catastrophic floods increased by 15% compared with the last decade of the 20th century. The report of the World Economic Forum on global risks in 2018 (http://reports.weforum.org/global-risks-2018/) includes risks associated with the climatic factor in the top ten. The increase in the magnitude and frequency of occurrence of extreme climatic phenomena has already damaged ecosystems and infrastructure. According to the latest data from Hydrometeorological Center of Russia, 2018 was the third in the last 23 years in the number of dangerous hydrometeorological phenomena that caused significant damage to natural resources, human health and livelihoods: 465 cases are reported (the maximum was in 2012 - 469) [3].

Considering all these facts, assessment of environmental, economic, political, and social impacts of global climate change on a particular region, as well as the development of measures to adapt to these changes and reduce their negative impact are needed.

2. Approach

According to the IPCC, future vulnerability depends not only on climate change but also on the development pathway. Sustainable development is believed to contribute to reducing vulnerability to climate change impacts. To succeed, adaptation must be integrated into sustainable development plans at both national and international levels. To develop an effective adaptation strategy, it is necessary, first of all, to develop educational programs, improve practical training, and increase public awareness of this problem. The system of adaptation to climate change should include various administrative levels from federal to local, taking into account regional natural and socio-economic features. This requires cooperation and coordination of actions between different administrative levels, as well as the maximum involvement of all stakeholders in defining the strategy for adaptation to climate change.

In Russia, the Climate Doctrine (http://www.meteoinfo.ru/climatedoctrine) was adopted in 2009 as a response to observed and anticipated climate change and its manifestations on different scales. It sets up the main areas of climate policy realization. Among them there is the development of a regulatory framework and the organization of state regulation; the development of economic mechanisms related to the implementation of measures to adapt and mitigate anthropogenic impacts on climate; scientific, informational, and personnel support of the development and implementation of these measures. Possible future climate changes affect the areas of responsibility of almost all federal government bodies. The doctrine defines tasks of the federal government in the development and implementation of climate policy, such as the inclusion of measures to adapt and mitigate the anthropogenic impact on climate in the medium and long term plans for the socio-economic development of the Russian Federation and the creation of mechanisms to ensure a continuous constructive dialogue between the scientific community, the public authorities responsible for decision-making, the public and the business community. According to the Climate Doctrine of the Russian Federation, a prerequisite for successful climate policy is the application of research results to assess the risks and benefits associated with the effects of climate change, as well as the ability to adapt to these effects. Implementing this policy requires the training of future specialists who will develop and apply adaptation measures. Currently, decision-makers, stakeholders, and the population, on whom the development of adequate measures to adapt and reduce the negative effects of climate change depends, do not have the skills and knowledge to work with the accumulated climate data necessary for analysis and development of an appropriate strategy. These groups need comprehensive scientifically-based information about the current and expected climate changes in their region of residence, which will prepare the public for the expected consequences and will encourage managers to find ways to adapt to them.

Currently, there is a serious lack of resources informing the stakeholders about current and expected climate changes and their consequences in Russia. Therefore, to meet the goals of the Climate Doctrine of the Russian Federation we have tried to fill this gap by developing new blocks of the previously developed web-GIS "Climate" and use it to provide climate information to assist decision-making and raise awareness of the society.

3. Method

3.1. Information-computational Web-GIS "Climate"

The information-computational Web-GIS "Climate" [4](http://climate.scert.ru/) is aimed at monitoring and analysis of ongoing and future regional climate changes. It gives online access to climate and weather models, a large set of geophysical data, and visualization tools. It is used to develop applications by distributed research teams and for research based on these applications, as well as to organize training of students and graduate students. "Climate" provides processing and analysis of sets of geo-referenced climatic and meteorological data to study climatic and environmental changes. It provides access to four categories of users: researchers, students, common public, and decision-makers (see Figure 1).

Figure 1. "Climate" start page.

A user-scientist has full access to tools for supporting the modeling and monitoring of regional climate change based on spatial data services. With the help of the graphical interface and using the GIS functionality, the user can get the results of the processing and analysis in the form of layers on a map for a selected region. The user has access to basic, pre-prepared layers, and can select the geographical area and add new layers, process geophysical data archives available in the system. To analyze the obtained results one can use zooming in and out, get the values from all layers at a point, and make additional processing of previously obtained data (for example, compare data from different layers). Along with direct studies of geophysical data, the user is allowed to perform joint research with other users, share results, and process their own data sets. "Climate" has a forum for users' communication, information exchange, and joint research. To get access to all tools for calculations and analysis, registration on the "Climate" site is required.

The web-GIS "Climate" is also used as a virtual learning laboratory providing undergraduate and graduate students with a profound understanding of changes in regional climate and environment. The educational section containing educational materials and interactive training courses allows them to receive first-hand knowledge in environmental sciences. This scientific field is experiencing a period of rapid development of the technologies of measurement and simulation, accompanied by an expansion of the conceptual and mathematical apparatus. Thus, the knowledge in the discipline grows quickly, and requires new tools and approaches to make it available for future specialists in environmental sciences. Large databases of observations and modeling results are created, and students should be trained to operate them and use computational models as tools for climate and weather forecast. The existing training programs in the disciplines of environmental sciences, as a rule, do not have time to adapt to such rapid change in the domain content. As a result, faculty graduates have a superficial knowledge of mathematical modeling of processes in the environment, do not have the required skills in modeling, data processing, and analysis of observations and modeling, and do not know how to work with big environmental data archives. The absence of this knowledge and skills necessary in modern climate science leads to the fact that the graduates are insufficiently prepared to act in the modern environmental sciences, where research requires high competence in the possession of modern information-computational tools and the ability to work in a distributed team. To reduce the effects of this lag, a special course was designed enhancing students' skills in work with modern information and computing systems [5].

An educational course on "Climate Change Monitoring and Prediction" is included in the curriculum for students of the Faculty of Geology and Geography of Tomsk State University. It covers the issues of quantitative geography of climate, a theoretical framework and tools for climate monitoring and forecast. The course contains a set of lectures devoted to basic aspects of modern climatology, including analysis of current climate change and its possible impacts, computing training with specific tasks on simulation of climate and climate change, and an information kit. The "Future Climate Analysis" training is designed to study climatic projections possible under different future scenarios, as well as the interaction of the individual components of the climate system, and statistical methods to assess the impact of global climate change on some of the parameters of the climate system. Basic skills of mathematical modeling and analysis of climate change are trained while performing the tasks. The "Climate Extremes Analysis" training describes an approach to calculating indexes of extreme values, which allow determining various characteristics of extreme climate events, such as frequency, duration, and intensity. The task is to analyze various indices and characteristics of the meteorological quantities. To carry out calculations of the indices, a series of measurements of the meteorological quantities obtained as a result of instrumental observations and as a result of global and regional climate models is used. The training also introduces a library, which contains calculated fields describing the spatial distribution of these indices for the territory of Siberia (50-65 ° N, 60-120 ° E). Calculations of climatic characteristics and statistical assessments in both parts of training are provided by the web-based GIS "Climate" tools. The course proved to be an effective way to familiarize the students with the actual situation in climate science and promote the use of modern information and communication tools.

3.2. Course for the public

Public awareness and understanding of ongoing climate change and its impacts are important steps to an effective response to this process. Implementation of a federal climate policy also means raising public awareness of the current and expected regional effects of climate change. According to a plausible scenario, changes in hydrological regime, quantity and quality of water resources, and temperature regimes will cause changes in ecological systems, agriculture, and forestry (shift of climatic zones in the northern direction and migration of wild fauna species, changes in the seasonality of growth and productivity of land in agriculture and forestry). In Siberia, the boundaries of permafrost are also shifting, which affects not only agriculture, but also the infrastructure and the fuel and energy sector. Since these changes occur gradually, it is possible to organize measures to adapt to the changing conditions and minimize their negative impacts. To reduce possible climatic risks, such measures as economical use of scarce water resources, flood control, selection of drought-resistant crops, and farming methods less vulnerable to heatwaves should be promoted among the population. Implementation of these measures requires the collaboration of scientists, decision-makers, and the public, and attempts of cooperation of these groups revealed some problems.

It became obvious that the perception by the public of climate change issues and possible risks differs from that common in climate science. Though the evidence of climate change is undeniable, a part of the public mistrusts official information. Climate change skeptics have used uncertainty in climate projections to persuade the population that climate change issues are overrated. Besides, studies show that there are several reasons for the denial or disregard of climate change. It could be due to economical reasons. In difficult times people seem to reject science [6]. Another reason for ignoring the problems associated with climate change is the fact that in some regions they are not so brightly drawn. Therefore, there is a feeling that these problems do not concern the population of this region. Often the public believes that the effects of climate change cannot occur in the near future. Last but not least, there is a lack of reliable information on climate change and its anticipated effects. The resources available in the Internet usually use scientific vocabulary, which obscures the meaning for the average user - not a specialist. In the Russian Federation's internet segment, this problem is aggravated by the fact that there are generally very few of these sites, and the existing ones are often not updated with new scientifically-reliable information confirming the ongoing climate change.

As far as the public awareness and understanding are important components of an effective response to climate change, the scientific community should translate climate change knowledge derived from research. It may be of little value unless it is put into practice. Thus, special internet resources are developed to promote this knowledge to the public. A serious lack of such resources in Russian forced us to add a special section to the "Climate" system. This section is designed to raise the public awareness of climate change and promote an understanding of the current processes. Such awareness is a necessary basis for the deployment of climate change adaptation processes on a regional scale. In general, the popularization of knowledge about climate change and the explanation of climate change trends and possible ways of adaptation to them stimulate active participation of the society in environmental protection. A basic course on climate and its ongoing processes is developed, aimed at improving the regional climate literacy of the population and stakeholders. While preparing this course, the features of the public perception of information about climate change were taken into account. It explains, in particular, the difference between weather and climate, climatic processes and factors, global climate change and its manifestations on different scales, extreme climatic events and climate risks. The course outlines and illustrates the main concepts and problems of modern climate change and its possible consequences. All materials are set out in a language accessible to non-specialists (see Figure 2). The course also includes links to popular science Internet resources on topical issues of Earth sciences. The aim is to increase the public understanding of climate risks, build resilience, and adapt to the impacts of climate change.

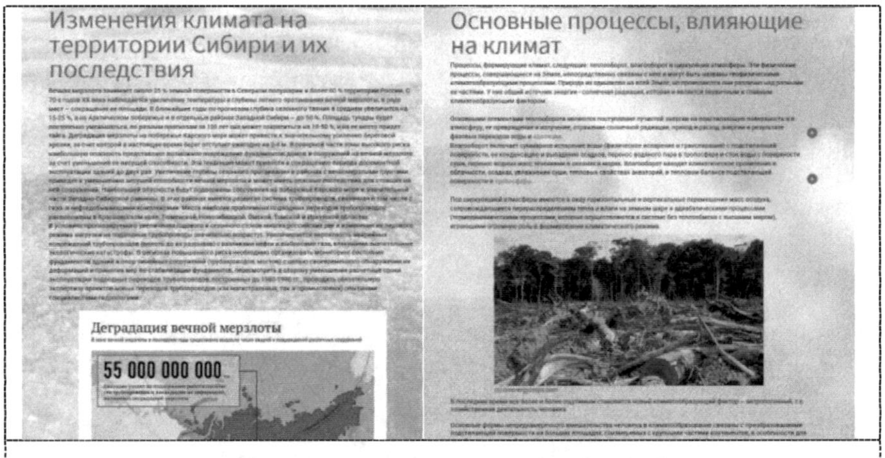

Figure 2. Pages of the course for the public.

3.3. Database for decision-makers

Climate change issues are in the spotlight now, and governments at different levels begin to develop adaptation strategies to respond effectively to current climate changes and future trends. Taking into account the ongoing and anticipated climate changes allows the authorities to reduce possible economic and social risks. At present, many adaptation projects have already been implemented over the world. For example, following the forecast of future climate change, the designers took into account the sea level rise when planning such a vital element of the infrastructure as the Confederation Bridge in Canada. It was opened on May 31, 1997, and links Prince Edward Island with the Canada mainland. The 12.9-kilometer bridge is the world's longest bridge over ice-covered water. The planners used climate scenarios to determine an appropriate height to account for a one-meter rise in the sea level and appropriate spacing between the support beams to allow for ice blocks that float down the Northumberland Strait to pass safely underneath [7]. There are also special adaptation projects on managing the coastal zone in the United States and the Netherlands. The Dutch coastal zone is extremely vulnerable to climate change and, thus, coastal management is an integral part of a national sustainable development planning [8]. Thus, adaptation decision-making requires climate knowledge and information for actions to reduce anticipated risks.

Making climate information useful to decision-makers usually requires that data and knowledge be processed and integrated through tools that range from simple graphs and maps to complex environmental models [9]. The necessity of risk identification and management revealed that decision-makers often do not have proper skills for the direct use of modern information and computing tools. Even experienced users of climate data may have difficulties choosing reanalysis or observational data sets to assess vulnerability and plan sustainable development at a regional level. Based on the experience of working with specialists from environmental organizations of the regional level, it can be concluded that only a few of them will be able to immediately use files in the netCDF format that is standard for modern quantitative climatology. Joint activities revealed that a simple way to present climate information to decision-makers is through the use of GIS.

To meet the needs of regional decision-makers and to encourage them to analyze ongoing and projected changes and develop an adaptation strategy, we propose a specially tailored information kit. It includes a glossary of climate terms, information on current climate change and future climate projections, and a database of climate characteristics including extremes, since they are supposed to cause most economic losses. Indeed, it is expected that under the influence of global warming some regions of Russia will suffer from extremely high temperatures and droughts in summer. This will affect crop yields and the state of forestry. In the north, permafrost thawing will accelerate and floods will intensify. Thus, regional decision-makers should review strategies for the development of transport infrastructure and construction projects, as well as the policy and planning of electricity supply. Accurate knowledge of the geography of extreme climatic phenomena, their frequency and intensity is needed to develop an effective adaptation strategy and measures to reduce the negative effects of extreme climatic manifestations. Since the frequency of such events is small, it is necessary to analyze such phenomena using modern approaches of the probabilistic statistical apparatus and detailed initial meteorological information accumulated during the period of instrumental observations for the territory of the region being studied. Therefore, to provide the regional decision-makers with the information necessary for the target activity, sets of relevant climatic characteristics for the Siberian region were calculated and structured as a free-access database.

The database contains calculated fields of climatic indices (http://etccdi.pacificclimate.org/index.shtml) describing the spatial distribution of extreme values of meteorological characteristics for the territory of Siberia (50-65 ° N, 60-120 ° E). These indices can be calculated both for one period (month/year) and for the whole period chosen. The indices are calculated for a period characterized by the most significant climatic changes. For different sets of source data, the start and end dates may not coincide, but the total time interval is almost the same. Also, a linear trend was calculated for each index, which shows 10-year period changes in the index over a specified time interval (see Figure 3).

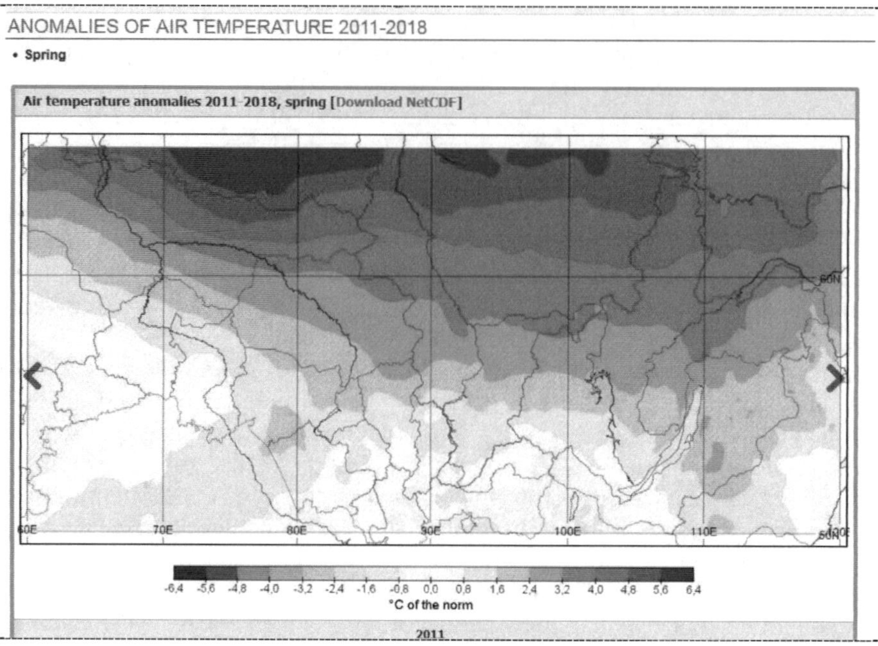

Figure 3. Database page.

If there is need for further work with any index, it is possible to download files with calculated indices in various formats (netCDF, GeoTiff, WMS, and WFS links) Special attention in the preparation of the database was given to the presentation of climate information in formats (GeoTiff files, as well as with WMS, and WFS) that can be used later in the widely used GIS (see Figure 4). The most popular and easy to work with netCDF files is Panoply developed by the NASA Goddard Institute for Space Studies. NCL (NCAR Command Language) is more specialized and has extensive tools for analyzing and visualizing netCDF files.

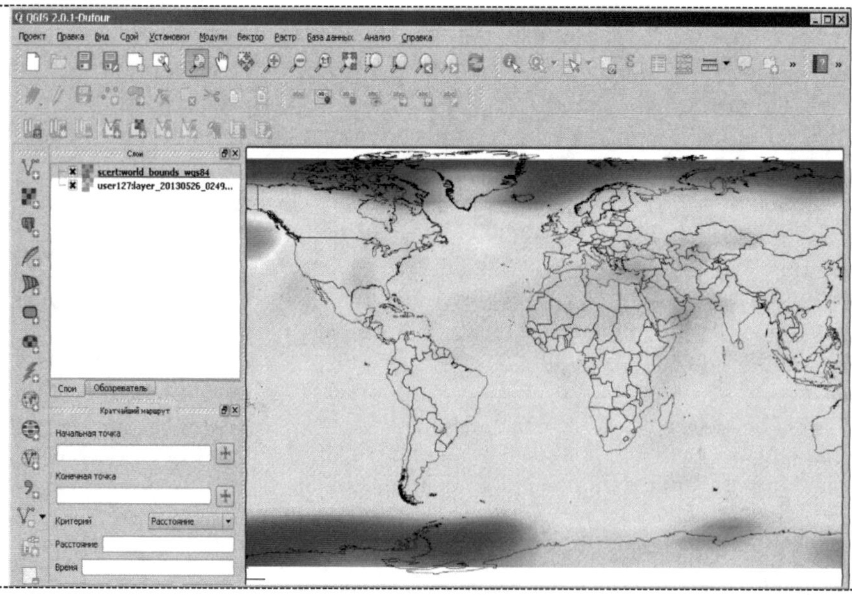

Figure 4. An example of adding the result of calculating the average air temperature as a new raster layer to QuantumGIS.

4. Conclusions

The above new sections of the system "Climate" are operating in a test mode now. They are being specially tailored to meet the needs of the general public and other regional stakeholders. Although there are a lot of Internet resources providing them with climate information, our experience shows that this information should be translated to educate and inform the population about possible climate changes and their anticipated effects. Decision-makers are also in need of scientifically-based information about the current and expected climate changes in the region of residence presented in a convenient format and with an easy tool to operate it [10]. The new sections developed for the public and decision-makers can give both targeted groups reliable information on climate issues in Russian. This is an important step to promote climate issues for the Russian-speaking community. To develop climate change awareness and mobilize these groups to find ways to adapt to the anticipated climate risks and build resilience to it, close attention is paid to the regional aspects of global climate change. This activity is in line with the Climate Doctrine of the Russian Federation. It underpins the initiatives of decision-makers and other stakeholders in different sectors of the national economy, business, and science to develop measures to adapt to climate change and mitigate its possible negative effects by implementing a unified federal climate policy.

Acknowledgments
The study was supported by basic research project AAAA-A17-117013050037-0.

References

[1] IPCC Core Writing Team, 2014 Climate Change 2014 Synthesis Report. Contribution of Working Groups I, II and III to the Fifth Assessment Report of the Intergovernmental Panel on Climate Change eds. R K Pachauri and L A Meyer (Geneva: IPCC,) p 151

[2] *NOAA National Centers for Environmental Information, State of the Climate: Global Climate Report for Annual 2018*, retrieved on June 25, 2019 from https://www.ncdc.noaa.gov/sotc/global/201813

[3] Hydrometeorological Center of Russia 2019 *A report on climate features on the territory of the Russian Federation in 2018* (Moscow: Roshydromet) p 79

[4] Okladnikov I G, Gordov E P, Ryazanova A A and Titov A G 2018 Development of a "cloud" thematic software complex for analysis of climatic and environmental changes *IOP Conf. Series: Earth and Environmental Science* **211** 012081 DOI: 10.1088/1755-1315/211/1/012081

[5] Gordova Y, Martynova Y and Shulgina T 2014 Use of the computational-informational web-GIS for the development of climatology students' skills in modeling and monitoring of climate change *Letters of Irkutsk State University* **9** 55-68

[6] Somerville R C J and Hassol S J 2011 Communicating the science of climate change *Physics Today* **64** 10 48 DOI: 10.1063/PT.3.1296

[7] Gregg R M 2010 Sea Level Rise and the Construction of the Confederation Bridge in the Gulf of Saint Lawrence [Case study on a project of Strait Crossing Bridge Limited], retrieved on July 16, 2019 from CAKE: www.cakex.org/case-studies/sea-level-rise-and-construction-confederation...

[8] VanKoningsveld M, Mulder J P M, Stive M J F, VanDerValk L and VanDerWeck A W 2008 Living with sea-level rise and climate change: a case study of the Netherlands *Journal of Coastal Research* **242** 367-379 DOI: 10.2112/07A-0010.1

[9] Grainger S, Mao F and Buytaert W 2016 Environmental data visualisation for non-scientific contexts: Literature review and design framework *Environmental Modelling and Software* **85** 299-318

[10] Liverman D. 2011 *Informing Adaptation. World Resources Repor* p 7 retrieved on July 16, 2019 from https://www.wri.org/our-work/project/world-resources-report/informing-adaptation

IOP Conf. Series: Earth and Environmental Science **386** (2019) 012054 doi:10.1088/1755-1315/386/1/012054

Implementation of artificial intelligence for analysis of long-term climate variability

E V Fedotova, V S Luferov

Moscow Power Engineering Institute, Krasnokazarmennaya 14, Moscow

ek.v.fedotova@gmail.com

Abstract. A simulation workflow is formulated to facilitate an application of modern data-driven approaches to climate problems by considering both theoretical aspects and modern tools of program implementation. A real supervised classification problem is considered as a use case. A convolutional neural network based model is developed to make the supervised classification of large-scale atmospheric circulation regimes. It has been found that a key prerequisite to ensure the success of the thus developed deep learning model is the proper matching of the data preprocessing and model architecture with the nature of the problem being considered. Particularly, applying of the seasonal anomalies transformation, tuning of the convolutional filters dimension and batch size adjustment have been found to be crucial to ensure a satisfactory accuracy of the regime recognition and a stable operation of the model. The generation of feature maps is demonstrated to be useful both as an interpretability tool and a heuristic approach to tuning of hyperparameters.

1. Introduction

The artificial intelligence (AI) methods are continuously gaining popularity during the last decades in an increasing range of applications. An attempt to use AI approaches in climate research are still quite limited. However, the modern methods of nonlinear modeling (e.g. neural networks) seem to suit perfectly well the nature of the climate problems.

First of all, the nonlinear properties of the climate system itself restrict the applicability of the linear approaches to climate modeling. Indeed, a wide range of linearization techniques may be perfectly appropriate for many practical problems. Unfortunately, the nonlinear effects not very well captured by the linear models are often of particular importance, which makes the nonlinear approaches preferable over simpler alternatives in such cases.

Second, an incredible complexity of the climate system itself determines difficulties of the conventional analysis. A classical illustration of this statement is interpretation of the eigenvalues-based climate indices. Understanding of the exact physical processes which determine the influence of the atmospheric teleconnections on regional climate is not a trivial task. The AI approaches make it possible to emulate observed interconnections and obtain some meaningful outputs even if the interpretation of these observed effects is not quite obvious.

Finally, the AI methods may alleviate one of the most typical problems of the climate research, which is only decent quality of the available data, especially on the long-term horizons. Indeed, data management is essential for any kind of models, but the adaptive features of the nonlinear AI methods determine certain toleration towards the non-perfect data quality.

IOP Conf. Series: Earth and Environmental Science **386** (2019) 012054 doi:10.1088/1755-1315/386/1/012054

The development of complex AI models is not a trivial task and requires combining an understanding of the object area with computational expertise and programming skills. Subject-specific requirements are very often neglected if the AI methods are discussed. At the same time, interference of the problem features with the requirements of the modeling infrastructure is nowadays at the very edge of the data science research.

The purpose of our work is to define a more or less general workflow to facilitate application of the AI to climate research. The next section will give a very brief overview of the available modern modeling and implementation approaches. Then we will introduce a general outline of the modelling workflow and demonstrate how to use it for a solution of a particular research problem.

2. State-of-the-art

The standard de facto of the modern data-driven modeling including artificial intelligence is an open source paradigm. The free and open source software (FOSS) movement has been progressing since the late 1970s. The creation of the Free Software Foundation in 1985 and the Open Software Foundation in 1998 greatly contributed to the wide use of open collaboration patterns and significantly changed the software development practices [1].

Rapid development of the data science software is based nowadays almost exclusively on the FOSS solutions, such as R and Python languages, PostgreSQL and Mongo databases, Kafka and Hadoop platforms. The tools relevant to the application of AI climate problems are primarily the geospatial and machine learning tools, whose main options will be briefly reviewed.

2.1. Geospatial tools

The FOSS geospatial tools began to flourish about fifteen years ago. Although the history of the oldest initiatives may span at least to the beginning of 1980, with the development of the geographical information system GRASS (Geographic Resources Analysis Support System) [2] being one of the most impressive examples. However, in 2000 extensive growth of the open geospatial community started, and the R-language geographical data community r-sig-geo [3], the OpenStreetMaps [4] project, and the Open Source Geospatial Foundation (to mention only few initiatives) were launched.

Nowadays almost every geographical computational problem can be effectively solved with the existing FOSS approaches, from assessment of the reanalysis datasets to making complex spatial-temporal interpolations and comprehensive data analysis. Indeed, the deep learning approaches naturally not belonging to the typical geospatial approaches are rarely included in the existing geospatial toolsets and usually address very specific problems.

2.2. A subsection

A wide range of the deep-learning computational methods is available via FOSS-projects. Actually, the FOSS paradigm seems to be the only approach that is flexible enough to continuously incorporate newly emerging AI methods. Wide use of the AI open source projects started in the middle of 2000 [5] and intensified a few years ago when the leading modern AI tools, such as Thensorflow [6], PyThorch [7], Caffe [8], H2O [9], and many others were created. The use of the existing AI tools has increased dramatically by their integration with the available programming approaches. The most influencing project of this kind is nowadays the Keras project [10], which incorporates a number of deep learning techniques and provides all necessary infrastructure components to run AI modeling. The Keras is written in Python, but may be imported to a number of different languages, such as R or Java, making the integration between the data processing and AI simulation as smooth as possible.

IOP Conf. Series: Earth and Environmental Science **386** (2019) 012054 doi:10.1088/1755-1315/386/1/012054

3. Problem definition

3.1. Modelling workflow
The following steps may be proposed as a more or less universal approach to the research-minded AI modeling keeping in mind the available implementation options:
[1] data management;
[2] data preprocessing;
[3] elaboration of the AI model architecture;
[4] tuning of the model hyperparameters;
[5] model learning and testing;
[6] analysis and interpretation of the modeling output;
[7] reporting and sharing of the results.

3.1.1. Data management and preprocessing. The data management, which appears quite trivial, is often the most time-consuming step. The reasons are, first, that the data amount necessary to run the data-driven models properly should be as large as possible. Another point is that the definition of the computational problem should always be fitted to the available data and the research question of interest. Preference of one algorithm over another one cannot be stated formally according to the famous "no free lunch" theorem [11]. That is why the subject expertise is a key advantage for the success of AI modeling for a particular application. Analysis of the data availability is a part of the decision making process when modeling approaches should be defined. An initial exploratory analysis and taking into consideration simpler modeling options often make sense.

Worth mentioning is the technical aspect of data organization. The two main data structures are competing options for a vast majority of the data science problems, namely, databases (relational ones, such as PostgreSQL, or document-oriented ones, such as MongoDB [11]) and dataframes which are in-built classes of R and Python. Theoretically, the databases ensure better memory management. In practice, this advantage is often leveled by the operating system memory management, which allows for a highly satisfactory performance of dataframes. The main advantage of the databases seems to be facilitation of the data reuse, for example, for collaborative data processing or when the data are intended for use by a web-application.

3.1.2. Model development and run. The AI models are quite complicated and offer many degrees of freedom. Search for a suitable model architecture requires some trials and errors. Often it makes sense to compare the results of different modelling approaches or use an ensemble of different models.

Adjustment of the proper AI model parameter values is referred to as hyperparameter tuning. A number of frameworks are available to automate this adjustment process. However, as it will be demonstrated, problem specific is essential for this purpose.

Performance may be an issue even for quite simple AI models. Parallel computing became a classical approach to overcome this performance limitation even for moderate computational resources at hand [13]. We will refer to practical implementation of parallel computations using the existing AI toolsets in the section on modeling.

3.1.3. Interpretation of the results. Proper interpretation of the output is one of the greatest challenges that nowadays restricts the use of AI results [14]. The AI models are usually too complicated to allow for a clear problem-related explanation related to all modelling details. However, some kind of proof is highly desirable to ensure reasonable behaviour of the model to use modelling results for practical purposes.

IOP Conf. Series: Earth and Environmental Science **386** (2019) 012054 doi:10.1088/1755-1315/386/1/012054

We will demonstrate a simple approach to the interpretation of particular modeling results in the next section, and show that such explanatory approaches may give some insights for improving the model structure.

3.1.4. Reporting and sharing of results. Sharing of the modelling results is perfectly consistent both with the open source paradigm and the reproducibility of the research. The modern data-science tools propose a number of sharing concepts including package maintaining [15], code and data repositories, development of web-applications [16], interactive frameworks [17], and collaborative platforms [18].

3.2. Problem being considered

The formulated workflow was implemented to automate the recognition of the global-scale atmospheric circulation regimes following, as much as possible, the code reuse principles. The main motivation of the problem was to justify the use of the classical subjective atmospheric classification as an analysis tool, which is particularly suitable for the interpretation of the output of global climate models.

One of the greatest challenges of the modern climate science is to understand the physical mechanisms behind the regional manifestation of the global climate processes. Even the most widely studied teleconnection mechanisms seem to be not quite understood yet and may be explained by different research groups in not a fully consistent way. Particularly, the long-term trends of the regional atmospheric circulation may differ dramatically when different analysis approaches are used [19].

A decisive advantage of the subjective classification schemes towards the global atmospheric circulation is a clear physical meaning of each particular regime including linking of the regional and global climate processes. It may be even more important that the global circulation is, in fact, a manifestation of the large-scale turbulence structures, which are known to be much more stable compared to the regional-scale structures usually assessed by the explanatory regional climate analysis.

4. Methods and implementation

The problem was addressed as an image recognition task. A supervised classification approach was used to emulate classification decisions normally made by an expert. This problem definition is very much like image classifications, which is quite a classical computer vision problem with a wide range of the existing development tools. Dzerdzeevskii's classification scheme was used [20] as the most well-developed subjective circulation scheme available up to now.

4.1. Data management

The sea-level atmospheric pressure was utilized as a classification base following the principles of manual classification. NOAA Twenty Century reanalysis [21] was used as the pressure input data, being the longest available homogeneous dataset on the global atmospheric circulation. The raw data netCDF archives were organized as the PostgreSQL database (Figure 1). The main motivation for this design was an option to reuse these data as a part of collaboration process.

4.2. Data preprocessing

The database was labeled according to the circulation calendar created by N.K. Kononova and N.N. Cherenkova according to the methodology proposed by B.L. Dzerdzeevskii [20]. The dataset was shuffled and split into a train part and a validation part. Four cumulative circulation classes (zonal, disturbed zonal, meridional northern, and meridional southern ones) were considered as far neglecting with finer classification types to ensure that the learning dataset is balanced. The improvement of the detailedness level will be assessed at the next iteration of the developed model. It should be noted that even the assumed quite rough cumulative classification approach may give meaningful results [22].

The specific preprocessing procedures were essential for ensuring stable operation of the model. The seasonal pressure patterns in raw data were found to overshadow the effect of the circulation regime. Thus, calculation of the seasonal anomalies was applied to the input pressure fields. The final data preparation step was standard hypercube transformation.

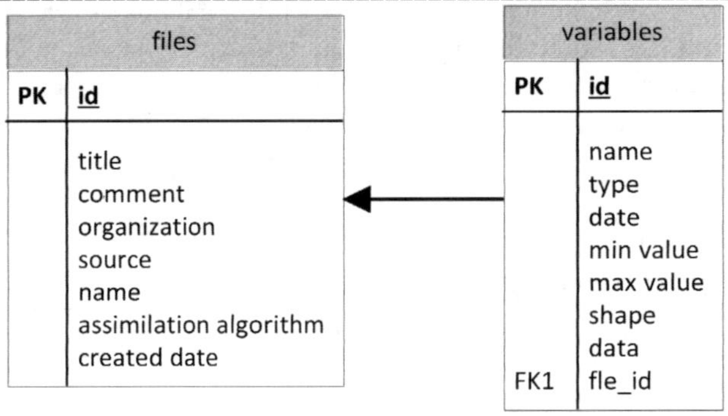

Figure 1. Database structure used to train and test the developed model.

4.3. Model architecture and implementation

The classical architecture for the image classification tasks is based on different modifications of the convolution neural network (CNN). This is one of the leading approaches of computer vision research, which allows using a wide range of available programming approaches to implement our model. We decided to use the above Keras library [10], which makes possible a seamless integration of the data preprocessing and model tuning into a Python code single block. A TensorFlow machine learning library was used as a computational backend of the model.

Parallel computation was implemented with in-built Keras parallel computing options which use the CUDA programming model via graphical processing units (GPUs). Using a GPU for paralleling allows one to accelerate the model learning up to 20-40 times as compared with the central processing unit, making it possible to run the model on a personal computer with reasonable computational speed, which is few hours on a modern high-performance video card for a single model run.

Usually CNNs are used to process pictures consisting of three layers, namely, red-, green-, and blue- color channels. Our two-dimensional pressure fields consist of a single layer corresponding to the pressure value distribution. The inlet channel of the CNN is a matrix presented as a multidimensional tensor. Each convolutional layer transforms the input tensor, and the value of channels may be different for each next layer. The number of the processing channels equals the number of feature maps extracted by the layer and should be set separately for each layer. The feature maps are the calculated response of the convolution artificial neural network to input vectors.

The convolution artificial neural network is based on the convolution operation. The convolution is a linear transformation of the input data with the following properties:

- keeping the structure of the input data due to the application of input data to each image part separately, namely, the order of elements in the one-dimensional case, mutual arrangement of pixels for a two-dimensional image;
- the property of a sparse layer, because the value of each neuron of the next layer depends only on a small fraction of the input neurons, opposite to a fully-connected network, where each neuron depends on all neurons of the previous layer;
- reuse of the same neuron weights many times, because they are applied repeatedly to different input areas.

The convolution layers are usually followed by a transformation with an activation function. The deep neural networks a non-linear rectified linear activation function (ReLU).

IOP Conf. Series: Earth and Environmental Science **386** (2019) 012054 doi:10.1088/1755-1315/386/1/012054

We have implemented the CNN model using the modern VGG16 architecture [23], which is known to ensure high classification accuracy for the image recognition problems. The model structure selected after some trials and errors is presented in Table 1. The model hyperparameter tuning could be automated with the Keras-compatible tools. However, keeping in mind that it would require to put aside a separate dataset purely for tuning purposes, we decided to tune the CNN parameters manually, which gives a chance to study the interaction between the model and the problems being considered more carefully.

The CNN model is based on combinations of the convolution and MaxPoling pairs. Two such blocks were found to be sufficient for our problem. The input layer is the ZeroPadding layer with a resolution of 91x180 equal to the resolution of the original pressure fields. The ZeroPadding layer forms a one-pixel wide empty frame across the input image to guarantee that the entire original image area will be covered by the convolution procedure.

Table 1. Architecture of the developed neural network.

Layer	*Output Shape*	*Parameter*
ZeroPadding2D	(93, 182, 1)	0
Conv2D	(93, 182, 128)	1280
MaxPooling2D	(31, 60, 128)	0
Conv2D	(29, 58, 64)	73792
MaxPooling2D	(4, 8, 64)	0
Conv2D	(2, 6, 32)	18464
Dropout	(2, 6, 32)	0
Flatten	(384)	0
Dense	(512)	197120
Dropout	(512)	0
Dense	(4)	2052

The input layer is followed by the Conv2D convolution layer, which produces 128 feature maps applying a 3x3 convolution filter followed by the ReLU activation function. Then the feature maps are transferred to a MaxPoling layer, which extracts the maximum element from each 3x3 square formed by the convolutional filters. The MaxPoling layer is not learned and serves to sub-sample the most important features extracted by the convolution layer. Besides, it serves as a compression procedure decreasing the size of the model.

The CNN model was found to work much more stably when the second convolution layer is followed by a MaxPooling layer with a 7x7 filter, instead of 3x3, allowing one to resolve cyclone-like structures in the pressure field. This idea of model design was the result of an iterative model improvement, and was proposed based on examination of the extracted feature maps.

The final convolution block is followed by a dropout regularization layer with a probability value of 0.25. The dropout layer is intended to increase the generalization abilities of the CNN model [24]. The essence of the dropout approach is including a random neuron shutdown with a given probability. In addition, the dropout procedure increases the stability of the networks to flaws contained in the input data. The concluding layer of the convolution blocks is the Flatten layer. The layer performs a conditional operation pulling the convolution results into the vector whose values are fed into the fully-coupled perceptron, that is, a dense layer with the neurons number 512. This is followed by one more dropout regularization layer with the same probability of 0.25. The output of the CNN-model is a full-connected dense layer with the number of outputs equal to the number of the classified atmospheric circulation types, which is four. The output of the model is confidence for each of the possible types calculated for a particular input pressure field (Figure 2).

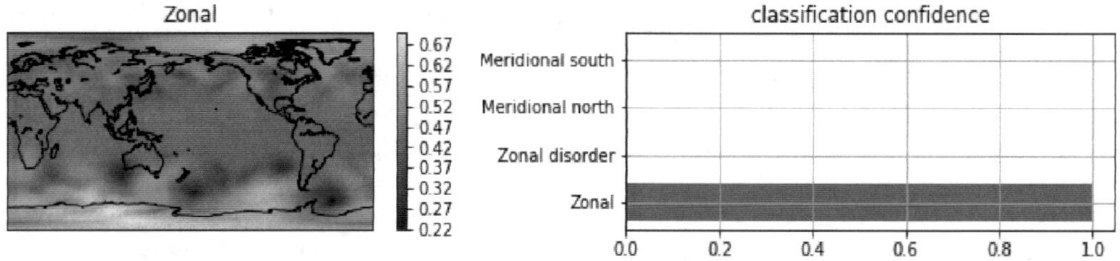

Figure 2. Example of a classification result.

4.4. Model learning and validation

The model neuron weights were set up during the training process using the algorithm of error backward propagation. The batch size parameter governed by the learning process was optimized (Figure 3).

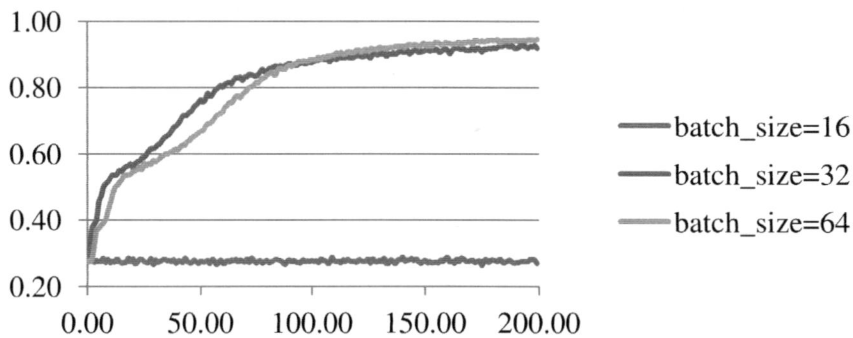

Figure 3. Effect of the batch size on the learning process as a dependence of the accuracy on the learning epoch.

This parameter determines the elements number in a training sample which is used to calculate the weights during a single iteration of the training process. Tuning of the bath size allowed us to balance the computational cost and the training error. The batch size of 64 samples was set for further calculations based on the optimization performed.

4.5. Interpretation of the results

The model was found to work on the global data better than on hemispherical ones, which is quite a surprising finding, since the used classification scheme was initially developed for the Northern Hemisphere only. Indeed, the processes in both Earth's hemispheres are tightly interconnected. However, a kind of assessment of the calculation processes inside the model was necessary to ensure that the model results are meaningful.

Extraction of the feature maps was taken as an explanatory approach. The inlet convolution layers extract the general structure of the classified images, while the deeper layers are responsible for the extraction of finer details. An example of a feature map constructed for the meridional north type is presented in Figure 4 (cyclone-like structures, as it could be expected; 7x7 filters).

5. Discussion and summary

We have demonstrated an application of a deep neural network to Dzerdzeevskii's subjective classification of atmospheric circulation regimes. The above developed convolutional neural network model was found to reproduce the classification group with a confidence higher than 90%. This means that the subjective classification is perfectly reproducible and can be used as an explanatory tool in analyzing the global climate processes, including the interpretation of the results of global climate models. The recognition of the finer circulation types should be the next step in the model development.

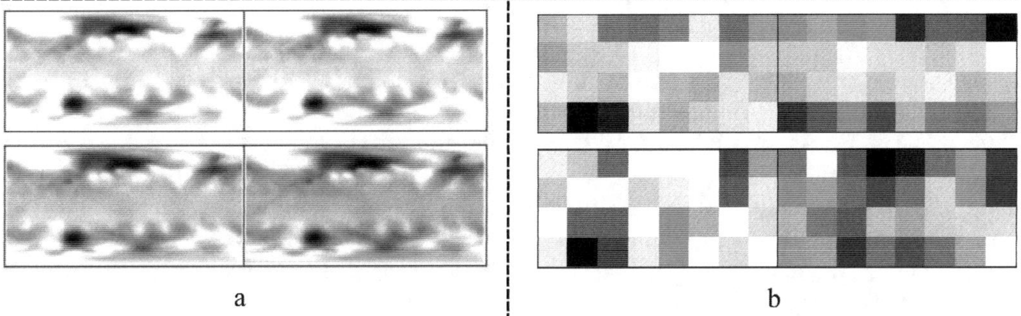

Figure 4. Example of the feature maps extracted by different layers of the model for the meridional circulation north type: a – the first convolutional layer, b – the deepest convolutional layer.

Despite the fact that the use of standard libraries has been found to work properly for the considered problem, we have also found that proper accounting for the specific problem features is essential for the successful operation of the model. This may seem somewhat contradictory to the common approach of the most AI-related works, which rarely refer to the nature of the modeled problem and usually are restricted to the use of some standard datasets to test the proposed computational methods.

The key approaches used to fit the modeling workflow to our problem are related to data preprocessing and model tuning. Particularly, we have demonstrated that the calculation of seasonal anomalies works well to separate the regime-related pressure patterns. The adjustment of the model convolution filters is another approach to ensure the proper recognition of details of the global atmospheric circulation.

The above-presented model was constructed with quite reasonable efforts thanks to the incorporation of existing computational toolsets according to the described step-by-step workflow. Further work will include the sharing of the model as a web-application. So far, the source code of the project, the results of the model tuning, and the modeling results have been made freely available under GPL v3.0 license in the repository *https://gitlab.com/luferov/ml_atm*.

Acknowledgments

The authors highly acknowledge the insightful discussions of meteorological regimes with N.K. Kononova.

This work was supported by the Russian Science Foundation (project No. 18-79-10255).

References

[1] Fogel K 2006 Producing Open Source Software How to Run a Successful Free Software Project (Boston: O'Reily)

[2] Neteler M, Bowman M H, Landa M, Metz M 2012 GRASS GIS: A multi-purpose open source GIS Environmental Modelling and Software **31** 124-30

[3] Bivand R 2014 GeoComputation and Open-Source Software: Components and Software

Component Stacks (preprint SAM 23 2011, Norwegian School of Economics)

[4] Curran K, Crumlish J, Fisher G OpenStreetMap 2012 *International Journal of Interactive Communication Systems and Technologies* **2** 1

[5] Ketkar N Introduction to Theano 35-61 in Ketkar N 2017 Deep Learning with Python (New York: Apress)

[6] Abadi M, Isard M, Murray D G 2017 A computational model for TensorFlow: an introduction Proceedings of the 1st ACM SIGPLAN International Workshop on Machine Learning and Programming Languages 1-7 (New York: ACM)

[7] Ketkar N 2017 Introduction to PyTorch 195-208 in Ketkar N 2017 Deep Learning with Python (New York: Apress)

[8] Jia Y, Shelhamer E, Donahue J, Karayev S, Long J, Girshick R, Guadarrama S, Darrell T 2014 Caffe: Convolutional Architecture for Fast Feature Embedding 2014 MM '14 Proceedings of the 22nd ACM international conference (Orlando, Florida, USA) 675-678 (New York: ACM)

[9] Laundry M 2018 Machine Learning with R and H2O (Mountain View, CA: H2O.ai, Inc)

[10] Chollet F 2018 Deep Learning with Python (Shelter Island: Manning)

[11] Wolpert D H, Macready W G 2005 Coevolutionary free lunches *IEEE Transactions on Evolutionary Computation* **9** 6

[12] Landset S, Khoshgoftaar T M, Richter A N, Hasanin T 2015 A survey of open source tools for machine learning with big data in the Hadoop ecosystem *Journal of Big Data* **2** 24

[13] Shallue C J, Lee J, Antognini J, Sohl-Dickstein J, Frostig R, Dahl G E 2019 Measuring the Effects of Data Parallelism on Neural Network Training *Journal of Machine Learning Research* **20** 1-49

[14] Ribeiro M, Singh S, Guestrin C 2016 Why Should I Trust You?: Explaining the Predictions of Any Classifier *Proceedings of the 2016 Conference of the North American Chapter of the Association for Computational Linguistics: Demonstrations* (San Diego, California: Association for Computational Linguistics)

[15] Lovelace R, Nowosad J, Muenchow J 2019 Geocomputation with R (Indianapolis: Chapman and Hall/CRC)

[16] Moraga P 2017 SpatialEpiApp: A Shiny web application for the analysis of spatial and spatio-temporal disease data *Spatial and Spatio-temporal Epidemiology* **23** 47-57

[17] Kluyver T *et al* 2016 Jupyter Notebooks—a publishing format for reproducible computational workflows *Positioning and Power in Academic Publishing: Players, Agents and Agendas. Proceedings of the 20th International Conference on Electronic Publishing* (Tokyo: IOS Press)

[18] Lokhmotov A , Fursin G 2016 Benchmarking, autotuning and crowdtuning OpenCL programs using the Collective Knowledge framework 2016 *Proceedings of the 4th International Workshop on OpenCL* (New York: ACM)

[19] Cahynova M, Huth R 2016 Atmospheric circulation influence on climatic trends in Europe: An analysis of circulation type classifications from the COST733 catalogue *International Journal of Climatology* **36** 2743–60

[20] Kononova N K 2018 Types of the global atmospheric circulation: monitoring and retrospective assessment results of 1899-2017 *Fundamental and Applied Climatology* **3** 108-23

[21] Compo G P *et al* 2011 The Twentieth Century Reanalysis Project Quarterly Journal of the Royal Meteorological Society **137** 654 1-28

[22] Kononova N K, Pimankina N V, Yeriskovskaya L A, Jing L, Weijia B, Shiyin L 2015 Effects of atmospheric circulation on summertime precipitation variability and glacier mass balance over the tuyuksu glacier in Tianshan mountains, Kazakhstan *Journal of Arid Land* **7** 5 687-95

[23] Szegedy C, Liu W, Jia Y, Sermanet P, Reed S, Anguelov D, Erhan D, Vanhoucke V, Rabinovich A 2015 Going deeper with convolutions Preprint arXiv:1602.07261

[24] Srivastava N, Hinton G, Krizhevsky A, Sutskever I, Salakhutdinov R 2014 Dropout: A simple way to prevent neural networks from overfitting *Journal of Machine Learning Research* **15** 1929-58

IOP Conf. Series: Earth and Environmental Science **386** (2019) 012055 doi:10.1088/1755-1315/386/1/012055

Analysis of aridity characteristics in Europe in the last millennium according to calculations with climatic models

N A Mirsaeva[1] **and A V Eliseev**[1,2,3]

[1] Institute of Environmental Sciences, Kazan Federal University, 18 Kremlyovskaya Street, Kazan, Russia

[2] Faculty of Physics, Lomonosov Moscow State University, 1-2 Leninskie Gory, Moscow, Russia

[3] A M Obukhov Institute of Atmospheric Physics RAS, 3 Pyzhevsky, Moscow, Russia

Abstract. Spatio-temporal distributions of aridity characteristics in Europe based on results of CMIP5 experiments are considered. According to calculations with a model, MPI-ESM-P, climate cooling in the Little Ice Age is characterized by a more humid climate in northern Europe compared with that in the pre-industrial period, although the changes are statistically insignificant. For the medieval climatic optimum, the changes in the climate humidification mode from the pre-industrial period are statistically insignificant. According to calculations with the IAP RAS CM model, cooling in the Little Ice Age is characterized by a more humid climate in the North-Eastern Atlantic compared with that in the pre-industrial period, and the changes are statistically significant.

1. Introduction

The World Economic Forum (WEF) published lists of global risks for 2017 [1]. The first position in the five most probable global risks was taken by hazardous weather events, including droughts. Drought, as a dangerous natural phenomenon, is determined primarily by meteorological factors, despite the variety of other factors contributing to its occurrence.

Recently, interest in the spatio-temporal characteristics of aridity has grown significantly. First of all, it is due to the negative impact of droughts on agricultural productivity and instability of the characteristics of aridity in a changing climate [2]. However, to attribute the observed changes to either anthropogenic influence on climate or to natural processes (both external and internal to the Earth system), it is necessary to put these changes into the context of respective changes in the past climate epochs.

2. Description and comparison of models

This paper analyzes the spatio-temporal distribution of aridity characteristics in Europe based on the results of CMIP5 experiments (Climate Models Intercomparison project, phase 5; https://cmip.llnl.gov/cmip5/) [3], which reproduce the climate over the last 1000 years. In the CMIP5 nomenclature, these simulations are referred to as 'past1000' (for 850-1849 C.E.) and 'historical' (for 1850-2005 C.E.). We used the data of monthly mean values of the surface air temperature and precipitation models MPI-ESM-P (Max-Planck-Institute for Meteorology, Hamburg) [4] and IAP RAS CM (A.M. Obukhov Institute of Atmospheric Physics, Russian Academy of Sciences) [5].

The following periods were chosen for the study:

M0: the basic, pre-industrial period (1870 - 1899 C.E.);
M1: the Medieval Climatic Optimum (1050 - 1200 C.E.);
M2: the Little Ice Age (1600 - 1750 C.E.);
M3: recent decades (1970 - 1999 C.E.).

The MPI-ESM-P is a state-of-the-art Earth system model. It implements ECHAM6 as an atmospheric general circulation model, MPIOM as an oceanic general circulation model and the module for simulating sea ice, JSBACH as a terrestrial biogeochemistry/vegetation dynamic module, and HAMOCC as an oceanic biogeochemistry module. All model compartments communicate via the OASIS coupler.

The climate model developed at the A.M. Obukhov Institute of Atmospheric Physics, Russian Academy of Sciences (IAP RAS CM) includes modules describing the state of the atmosphere, ocean, the active land layer, biogeochemical cycles, and processes associated with atmospheric electricity and atmospheric chemistry. It belongs to the class of climate models of intermediate complexity (EMICs) and participates in relevant international comparison projects. A specific feature of the model is the parameterization of synoptic variability in the atmosphere and ocean, which allows reducing the computation cost by two orders of magnitude. The model realistically reproduces climate change over a period of instrumental measurements and is used to estimate past and future climate changes on a ten-year and longer time scale [5].

Both models were forced by:
(i) changes in the parameters of Earth's orbit;
(ii) changes in total solar irradiance;
(iii) changes in the extent of agricultural land (the MPI-ESM-P also parametrizes agricultural practice);
(iv) reconstructed anthropogenic emissions of CO_2 in the atmosphere; atmospheric CO_2 concentration is calculated internally by carbon cycle routines of both models;
(v) prescribed changes of other well-mixed greenhouse gases;
(vi) tropospheric aerosol concentrations;
(vii) reconstructed optical depth of volcanic aerosols in the stratosphere.
In addition, the MPI-ESM-P takes into account solar-induced ozone variations. More detailed information on these forcings is available in [6, 7].

3. Results
The MPI-ESM-P model realistically reproduces the long-term course of surface air temperature (Figures 1 and 2).

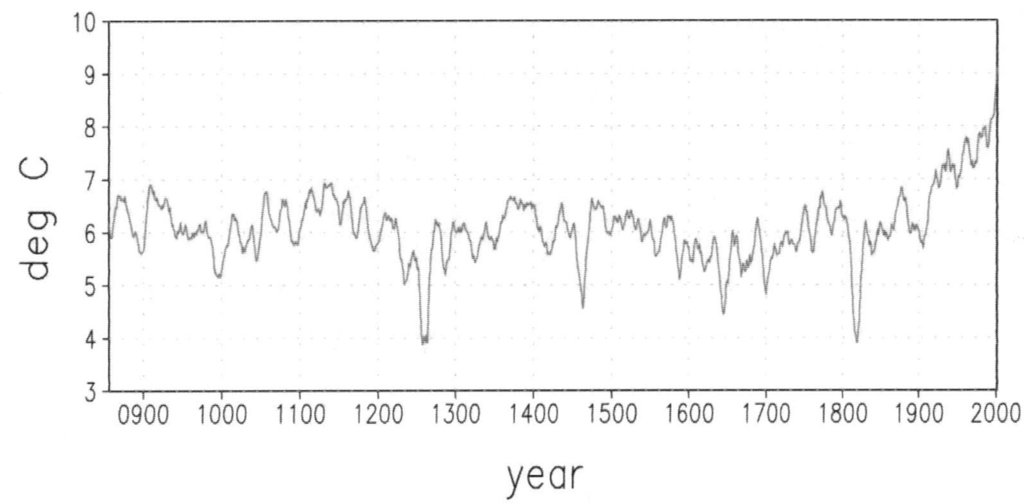

Figure 1. Annual mean surface annual temperature averaged in the region of 30-80° N, 0-90° E as simulated by MPI-ESM-P.

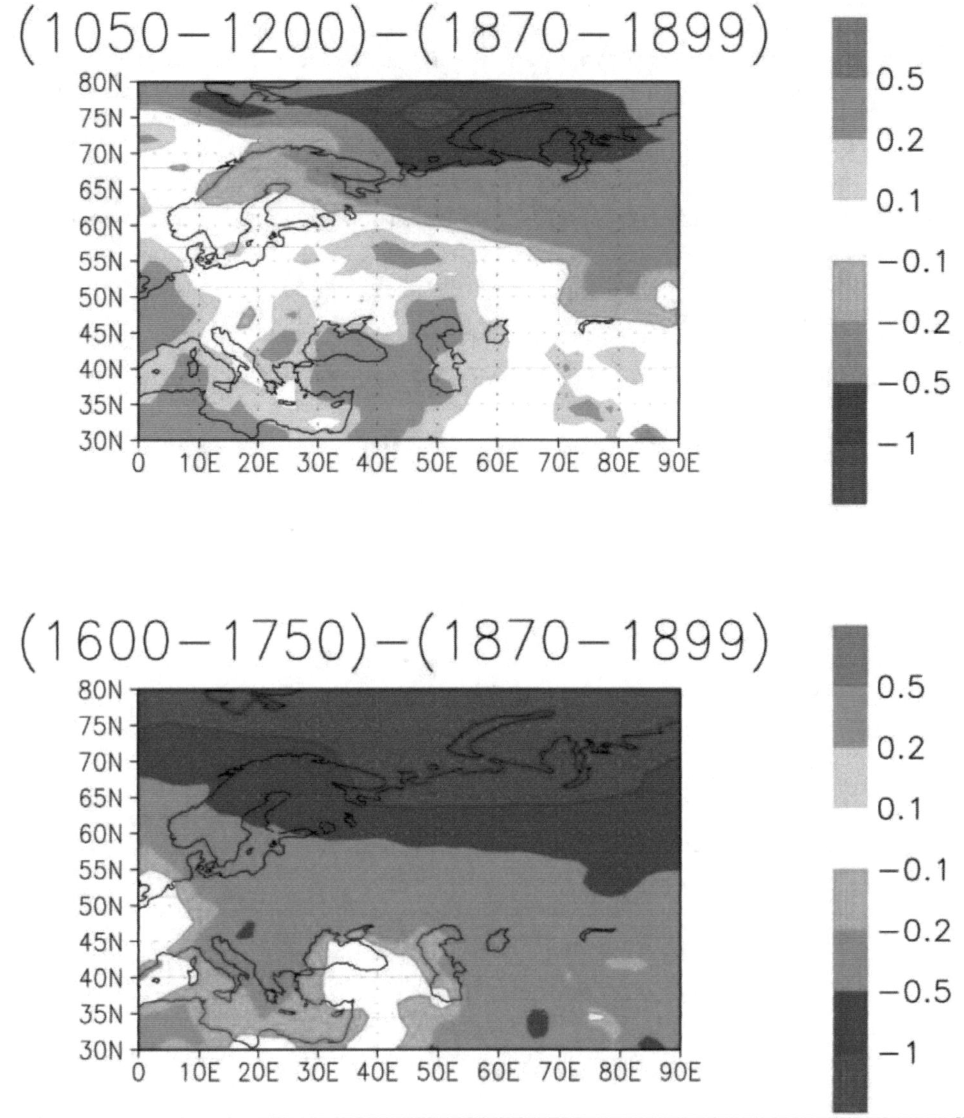

Figure 2. Annual mean surface air temperature averaged over indicated time intervals as simulated by MPI-ESM-P.

For each time interval (from M0 to M3), the Ped aridity index (PAI) [2] was calculated. We used only the data from May to August, because spring-summer droughts cause the greatest damage to agriculture and determine the functioning of natural terrestrial ecosystems. Statistical significance of the differences between time intervals M1, M2, M3 relative to M0 were estimated based on a two-sided t-criterion. The period from May to August was selected. In accordance with the typical practice of applying the PAI, the analysis was limited to Europe (which is defined in our paper as the region 30-80° N 0-90° E). We interpret positive changes of the PAI as a manifestation of a drier climate and its negative changes as a reflection of a more humid climate.

Then, for the entire summer period, PAI maps of changes between the Medieval Climatic Optimum and the pre-industrial period (M2-M0) and the Little Ice Age and the pre-industrial period (M1-M0) were constructed using the MPI-ESM-P. As an example, we present the PAI change maps for June (Figure 3*a, b*).

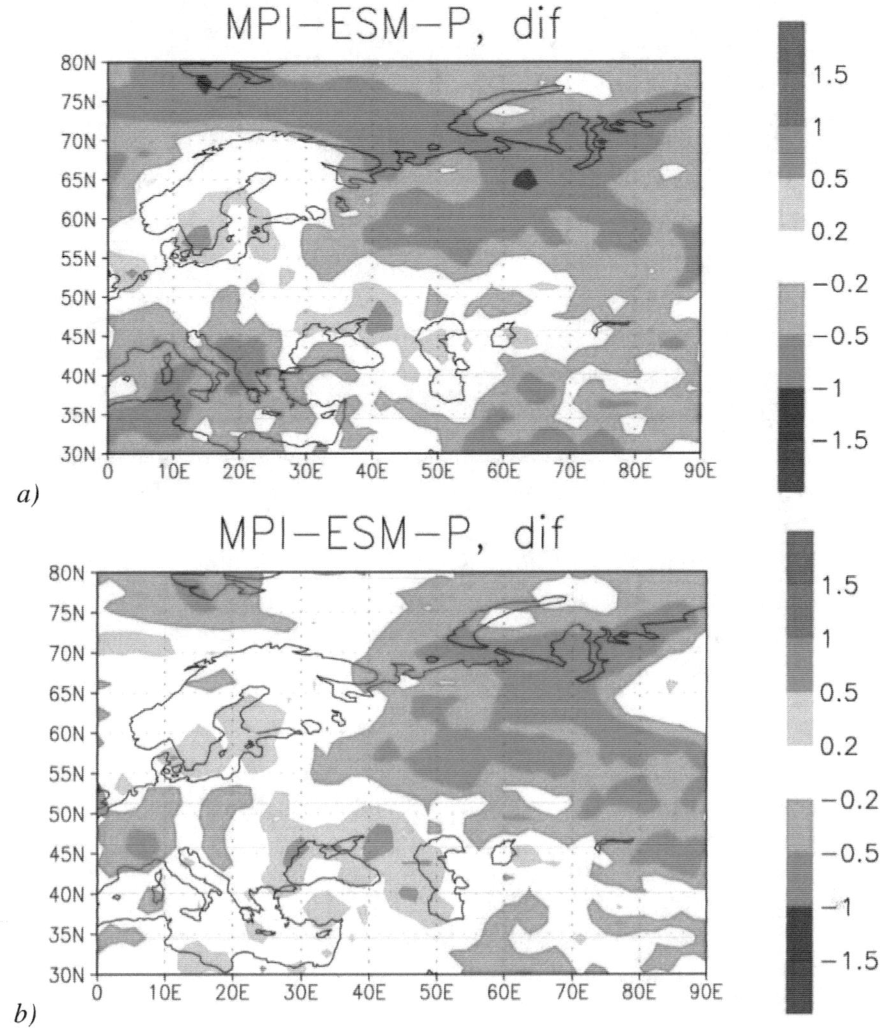

Figure 3. Changes of PAI in June between the Medieval Climatic Optimum and the pre-industrial period (M2 -M0) (*a*) and the Little Ice Age and the pre-industrial period (M1-M0) (*b*) as simulated by MPI-ESM-P.

According to calculations with the MPI-ESM-P model, the Medieval Climatic Optimum is characterized by a marginally significant humidification in the Mediterranean (Figure 3*a*). In turn, the Little Ice Age is characterized by a more humid climate in northern Europe as compared with the pre-industrial period, although the changes are not statistically significant (Figure 3*b*). Similar conclusions on the MPI-ESM-P model are typical for the whole summer period.

Similarly, PAI maps were constructed for the entire summer period of changes in the index between the Little Ice Age and the pre-industrial period (M1-M0) using the IAP RAS CM. As an example, we present the PAI change maps for June (Figure 4).

IOP Conf. Series: Earth and Environmental Science **386** (2019) 012055 doi:10.1088/1755-1315/386/1/012055

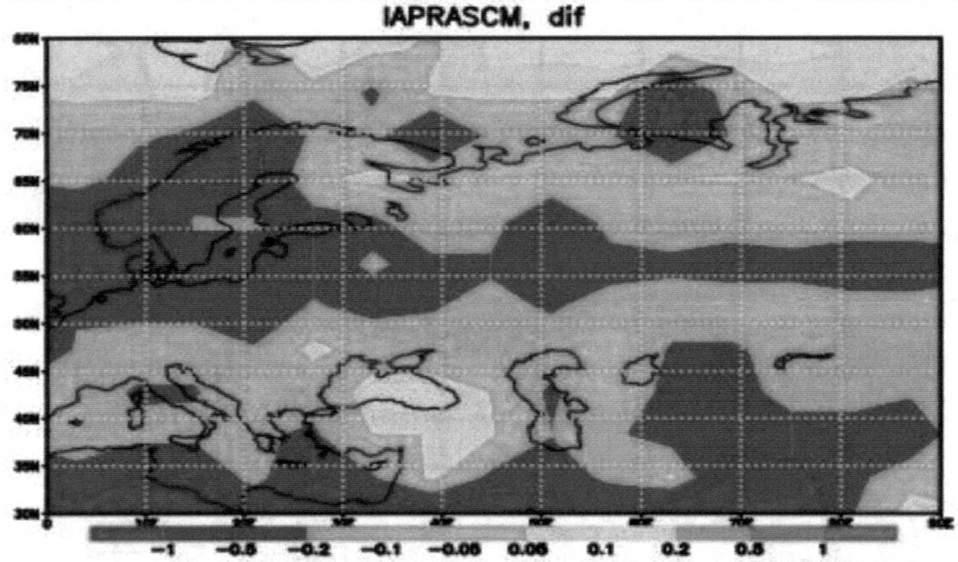

Figure 4. Changes of PAI in June, M2 -M0 as simulated by IAP RAS CM.

According to calculations with the IAP RAS CM model, cooling in the Little Ice Age is characterized by a more humid climate in the North-Eastern Atlantic compared with the pre-industrial period, the changes are statistically significant. Similar conclusions on the IAP RAS CM model are typical for the entire summer period.

Conclusions

This paper showed changes in the Ped aridity index (PAI) in Europe in the last millennium calculated with the climate model of the A.M. Obukhov Institute of Atmospheric Physics, Russian Academy of Sciences (IAP RAS CM), and described the main results obtained by using this model in comparison with corresponding results obtained by using the climate model MPI-ESM-P. It has been shown that both models realistically reproduce the pre-industrial and the Little Ice Age climates and the current state of the climate system, as evidenced by a statistically significant Student's t-test.

Both models agree on a more humid climate in Northern Europe during the Little Ice Age. However, the models differ in the longitudinal position and statistical significance of these anomalies. In particular, the more eastward location of these anomalies and their larger statistical significance in the IAP RAS CM compared to those in the MPI-ESM-P might be caused by a more simplified dynamics in the former model. An immediate possible outcome of this simplified dynamics is an underestimated moisture transport from Atlantics to Europe, which controls the longitudinal position of the humid anomalies. Another possible consequence is an underestimation of the natural variability in the model, which boosts the statistical significance of the calculated changes.

Acknowledgements
This work was supported by the Russian Foundation for Basic Research (grant 18-05-00087 A).

IOP Conf. Series: Earth and Environmental Science **386** (2019) 012055 doi:10.1088/1755-1315/386/1/012055

References

[1] *Global Risks 2017* URL: http://reports.weforum.org/global-risks-2017/

[2] Cherenkova E A 2013 Quantitative estimates of atmospheric droughts in European Russia *Izv. RAS. Geographical series* 6 76–85

[3] Braconnot P, Harrison S P, Otto-Bliesner B *et al* 2011 The Paleoclimate Modeling Intercomparison *Project contribution to CMIP5 CLIVAR Exchanges Special* **16** 2 15–9

[4] Bakker P, Stone E J, Charbit S *et al* 2013 Last interglacial temperature evolution – a model intercomparison *Climate of the Past* **9** 605 – 19

[5] Eliseev A V, Mokhov I I 2011 Uncertainty of climate response to natural and anthropogenic forcings due to different land use scenarios *Adv. Atmos. Sci.* **28** 5 1215–32

[6] Schmidt G A *et al* 2012 Climate forcing reconstructions for use in PMIP simulations of the last millennium (v1.1) *Geosci. Model Dev.* **5** 1 185–91

[7] Taylor K E, Stouffer R J, Meehl G A 2012 An overview of CMIP5 and the experimental design *Bull Am Meteorol.* **93** 485–98

AUTHOR INDEX

Institute of Physics
Dirac House, Temple Back
Bristol BS1 6BE UK

ISSN: 1755-1307
ISBN 978-1-7138-0254-9

9 781713 802549

CURRAN ASSOCIATES INC.
proceedings.com